IAVCEI Proceedings in Volcanology 1

John H. Latter (Ed.)

Volcanic Hazards
Assessment and Monitoring

With 284 Figures, Some in Colour

Springer-Verlag
Berlin Heidelberg New York
London Paris Tokyo

Dr. John H. Latter
Geophysics Division
Department of Scientific
and Industrial Research (DSIR)
P.O. Box 1320
Wellington, New Zealand

ISBN 3-540-19337-5 Springer-Verlag Berlin Heidelberg New York
ISBN 0-387-19337-5 Springer-Verlag New York Berlin Heidelberg

Library of Congress Cataloging-in-Publication Data. Volcanic hazards. (IAVCEI proceedings in volcanology ; 1) Includes index. 1. Volcanic activity prediction–Congresses. I. Latter, John H. II. Series. QE527.5.V65 1988 363.3'495 88-32779 ISBN 0-387-19337-5 (U.S.)

Printing: Druckhaus Beltz, Hemsbach/Bergstraße
Binding: J. Schäffer GmbH & Co. KG., Grünstadt
2132/3130-543210 – Printed on acid-free paper

Preface

The contributions in this book were presented, orally or as posters, at the International Volcanological Congress held in New Zealand from 1 to 9 February 1986, the centenary year of the Tarawera eruption of 10 June 1886. More than 500 people, from 29 countries, attended the Congress. Most of these works formed part of Symposium 4, "Volcanic Hazards - Prediction and Assessment", convened by J.H. Latter, R.R. Dibble, D.A. Swanson and C.G. Newhall. The collection represents over half of the published abstracts of Symposium 4, together with three papers given at the Symposium, which lacked abstracts, and two which were part of Symposium 1 on pyroclastic flow deposits. The contributions cover a good proportion of the volcanically active parts of the world, with Italy, Japan, the West Indies and the USA especially well represented. Mount Erebus, Vulcano and Rabaul are individual volcanoes which have been treated in particular detail. Unfortunately, there are no chapters in the book dealing with Africa, the Atlantic islands (except Iceland), Hawaii, Central America (except Mexico), or South America (in spite of the major disaster at Nevado del Ruiz Volcano in 1985).

The book is divided into two parts, with the first 15 chapters dealing mainly with Hazards Assessment, and the remainder mostly with Monitoring. With minor exceptions, the chapters are arranged in what I hope readers will find to be a logical order, from the general to the particular, and, where possible, works on particular regions and individual volcanoes have been grouped together. I have deliberately placed first in the book a contribution which discusses the effects of the largest and most infrequent eruptions, followed by regional studies of major long-term volcanic hazards, since the hazards and attendant risks of such eruptions are generally underestimated. Equally deliberately, the book concludes on a positive note, with a working method of prediction in use at one of the world's great volcanological observatories.

At present, less than 5% of known Quaternary volcanoes have been sufficiently well studied for estimates to be made of mean intervals between eruptions of a given volume. With few exceptions, however, these mean intervals amount to more than 500 years for eruptions of at least 1 km^3, 5000 years for those of at least 10 km^3 and 50 000 years for eruptions over 100 km^3. Such long intervals have a dual effect. On the one hand, the importance of centres which have remained dormant for a very long time is obscured to the point where the possibility of large eruptions may be regarded purely as a thing of the past. On the other hand, the danger is exaggerated at centres which last erupted on a grand scale much closer to the present time than the mean intervals between their eruptions, and this impression of potential future activity at these places tends to be confirmed by abundant signs of devastation in the recent past.

Thus at Taupo, New Zealand, where there have been two eruptions of more than 100 km^3 in the last 25 000 years, the probability of another eruption on this scale in the near future is much lower than at other, less obvious rhyolitic centres nearby, where there has been no such eruption for many tens of thousands of years. Similarly, in Kyushu, Japan, an eruption of the order of 100 km^3 is more likely at the Kakuto, Aso, or Ata Calderas than at the Kikai or Aira Calderas, where activity on this scale has occurred more recently and where its effects are vastly more conspicuous.

It is important to identify hazards and establish periodicities of eruptions for large-scale volcanic structures, e.g. Lake Taal and not just Volcano Island, Aira Caldera and not just Sakurajima, Rabaul Caldera and not just Vulcan and Tavurvur. Detailed geological mapping and tephrochronology needs to be carried out at all potentially destructive volcanoes (and it is not immediately obvious where all these are), and it should be backed up by archaeological and anthropological studies, since major volcanic eruptions ought to have left their mark on the oral tradition and collective memory of the region's inhabitants. Until this is done, large areas of the world remain threatened by what Tazieff (1967) has so aptly described as "the menace of extinct volcanoes".

Although great efforts have been made to catalogue surviving volcanoes (IAVCEI 1973; Simkin et al. 1981), it is very likely that eruptions in the near future will occur at unidentified volcanic centres. It is notable that the greatest known loss of life due to gas discharge from a volcano, that at Lake Nyos, Cameroon, in 1986, occurred in a hitherto uncatalogued volcanic district.

Once identified as a potential threat, a volcano, whether active or dormant, should be placed under surveillance. In most cases, where money is short, and particularly when the volcano shows no present signs of activity, this may simply amount to ensuring that there is a regional seismograph, installed and maintained as part of a network designed to locate tectonic earthquakes, close enough to the volcano to detect any seismic precursors to an eruption above some fixed (low) magnitude threshold. Where more resources are available, a denser seismic network, and many of the chemical, thermal and deformation measuring techniques discussed in the latter part of this book, may be appropriate. Even volcanoes remote from settled areas may prove to be a source of risk, through tephra, tsunamis or lahars affecting regions at a great distance, or through aircraft being directly endangered by high eruption plumes. Satellite surveillance, and remote infrasonic arrays, both covered by chapters in the book, are essential tools for monitoring volcanoes in uninhabited areas.

At present, there are great disparities in the extent to which effective surveillance is carried out in different parts of the world, and the majority of potentially dangerous volcanoes remain unmonitored. A great deal could be gained by more international cooperation, and the work carried out at Mount Erebus, Antarctica, which is described in two chapters in this book, could be a model for future joint studies. Much of the world's volcanological resources remain concentrated in a few well-equipped observatories, while much of the needed research is in countries which lack the necessary resources to carry it out. A sense of urgency is required to ensure that by the end of the century, a mere 12 years away, we have a better perspective of the potential for volcanic disaster on a worldwide scale.

I would like to add my personal thanks to all those who contributed to this book, and to all who refereed them. Thanks are also due to H.-U. Schmincke, R.S.J. Sparks and W. Engel for arranging publication, and in particular to R.S.J. Sparks and A. Ewart for coping with much of the editorial work. The Earthquake and War Damage Commission of the New Zealand Government, and the Insurance Council of New Zealand were very generous in their support of Symposium 4 on Volcanic Hazards, and I thank them most sincerely for this. Finally, my warm thanks to my fellow co-convenors of Symposium 4, and to J.W. Cole and fellow members of the Organizing Committee, and to all who ensured that the Congress turned out to be the success which it undoubtedly was.

References

IAVCEI (1973) Data sheets of the post-Miocene volcanoes of the world, with index maps. IAVCEI, Rome

Simkin T, Siebert L, McClelland L, Bridge D, Newhall C, Latter JH (1981) Volcanoes of the world. Hutchinson Ross, Stroudsberg, 232 pp

Tazieff H (1967) The menace of extinct volcanoes. Impact (UNESCO) 2-17:135-148

October 1988

J.H. Latter

Contents

Hazards Assessment

Monitoring

General

Thermal Infrared

Chemical

Gravity

Seismic

Infrasonic

Deformation

List of Contributors

You will find the addresses at the beginning of the respective contribution

Hazards Assessment

Volcanic Winter? Climatic Effects of the Largest Volcanic Eruptions

R. B. STOTHERS[1], M. R. RAMPINO[1,2], S. SELF[3] and J. A. WOLFF[3]

Abstract

Calculations suggest that the largest volcanic eruptions could have significant effects on global climate. We estimate the amount of sulfur volatiles that could have been released in very large eruptions by scaling up from smaller historical eruptions. The greatest well-known Late Quaternary explosive eruption, Toba (Indonesia, 75,000 years B.P.) erupted at least 1000 km^3 of magma, and may have released enough sulfur volatiles to have formed 9×10^{14} to 5×10^{15} g of H_2SO_4 stratospheric aerosols. Basaltic fissure eruptions release even greater amounts of sulfur volatiles, which can be lofted into the stratosphere in convective plumes rising above fire fountains. The Roza flow eruption (about 700 km^3 of magma) of the Miocene Columbia River Basalt Group could have produced up to 6×10^{15} g of aerosols. Distributed worldwide, these aerosol mass loadings would lead to effects ranging from a noticeable dimming of the sun to conditions similar to those described in some models of nuclear winter. Unless self-limiting mechanisms of stratospheric aerosol formation and removal are important, very large eruptions may lead to widespread darkness, cold weather, and acid precipitation. Even the minimum estimated effects of these great eruptions would represent significant perturbations of the global atmosphere.

1 Introduction

The possibility that massive amounts of soot introduced into the atmosphere by great fires in the aftermath of nuclear war might produce a climatic catastrophe or "nuclear winter" has recently attracted a great deal of attention (Turco et al. 1983, 1984; National Research Council 1985; Thompson and Schneider 1986). Some have suggested that the atmospheric aftereffects of volcanic eruptions might be used as a basis for estimating the severity of nuclear winter (Maddox 1984; Brown

[1] NASA, Goddard Space Flight Center, Institute for Space Studies, 2880 Broadway, New York, NY 10025, USA

[2] Department of Applied Science, New York University, 26 Stuyvesant St., New York, NY 10003, USA

[3] Department of Geology, University of Texas at Arlington, Arlington, Texas 76019, USA

IAVCEI Proceedings in Volcanology 1
J. H. Latter (Ed.), Volcanic Hazards

and Peczkis 1984). The relatively small amounts of aerosols produced by recent historical eruptions, however, and the considerable differences in the optical properties between volcanic aerosols (H_2SO_4 and fine ash) and the sooty smoke from urban fires make any comparisons of the atmospheric effects of historical eruptions and nuclear war highly tenuous. In other instances, namely those of the largest-scale eruptions, volcanic analogies may be relevant. Here we estimate the amounts of aerosols that might have been produced by the largest known eruptions, and the possible effects of these aerosols on the atmosphere and climate.

2 Large Explosive Eruptions

The greatest known Late Quaternary explosive eruption was the paroxysm that produced the Toba (Sumatra) ignimbrite of 75,000 years B.P., in which at least 1000 km^3 (and possibly 2000 km^3; Rose and Chesner 1986) of rhyodacitic magma (dense rock equivalent or DRE volume) erupted in less than 2 weeks time (Ninkovich et al. 1978). The widespread Toba ash layer is found in deep-sea cores over an area of more than 5×10^6 km^2; the ash and released volatiles were certainly carried high into the stratosphere. The dense ash cloud might have had a direct, but probably short-lived (weeks to a few months) effect on climate [for example, the ash cloud created by the much smaller Tambora eruption (50 km^3 DRE) in 1815 caused drastic local and regional cooling for a short period just after the eruption, see Stothers 1984a]. The total amount of the more important long-lived stratospheric H_2SO_4 aerosols created by the Toba eruption is unknown, but may be roughly estimated by using data from more recent eruptions of similar magma type (Rampino and Self 1984). Stratospheric aerosols generated by the small 1980 eruption of Mt.St. Helens, which produced about 0.35 km^3 (DRE) of dacitic magma, totaled about 3×10^{11} g. If the Toba eruption created a proportionate amount of sulfuric acid aerosols, then the stratospheric loading after that event could have been about 9×10^{14} g. If the Toba sulfur release was proportional to the output of the 1883 Krakatau eruption, which released a larger percentage of sulfur volatiles [5×10^{13} g from 10 km^3 (DRE) of dacitic magma; Devine et al. 1984; Rampino and Self 1984], then the stratospheric loading could have been as great as 5×10^{15} g.

It is important to note that these estimates are in effect "worst case" scenarios, and are made simply by extrapolating a linear increase in aerosols, assuming that the behavior of very dense aerosol clouds is not qualitatively different from that of the less dense clouds observed after historical eruptions (Rampino et al. 1985). However, differences in aerosol nucleation, saturation, and fallout in very dense clouds may affect the concentrations and atmospheric lifetimes of the volcanic aerosols. Self-limiting mechanisms could constrain the total amounts of aerosols in the stratosphere, but at present it is difficult to estimate the possible limits to stratospheric aerosol loading (Rampino and Self 1982; J.Pinto, pers. commun.). The minimum predicted effects on the atmosphere of such large eruptions, however, are still more drastic than any effects observed in the historical record.

3 Basaltic Fissure Eruptions

Basaltic volcanic eruptions may release even greater amounts of sulfur volatiles (Devine et al. 1984; Stothers et al. 1986). In the past, such "quiet" basalt eruptions were considered unlikely to produce high-altitude aerosol clouds (Lamb 1970). Recent study has shown that large, generally effusive fissure basalt eruptions, such as the 1783 Laki eruption in Iceland, have produced widespread high-altitude aerosol clouds (Sigurdsson 1982; Wood 1984). Theoretical calculations suggest that the sulfur volatiles are efficiently released and can be carried to high altitudes in convective plumes rising above vigorous fire fountains (Stothers et al. 1986).

These facts lead us to infer some interesting possible "worst case" consequences of the large flood basalt eruptions in the geologic past. A good example is the Roza flow eruption in the Columbia River Basalt Group of Miocene age. The Roza flow eruption (about 14 m.y. B.P.) produced about 700 km^3 of basaltic lava; Swanson et al. (1975) estimate that during this eruption lava issued from small intervals along a 100-km length of fissure, at a rate of about 1 km^3 per kilometer of fissure per day, for about 7 days. The estimated eruption rates of 10^4 to 10^5 $m^3 s^{-1}$ from 1- to 10-km lengths of fissure are expected to have produced fire fountains of about 1-km height (Wilson and Head 1981) and stratosphere-high (>10 km) eruption plumes (Stothers et al. 1986).

The quantity of atmospheric aerosols produced by such large basalt eruptions can be roughly estimated by scaling from the known amounts of aerosols produced by the greatest known modern fissure eruptions. The largest well-known historical fissure eruption, that of Laki (Iceland) in 1783, produced about 12 km^3 of lava (Thorarinsson 1969). Greenland ice-core acidity measurements suggest a release of about 3×10^{13} g of atmospheric sulfur from Laki (Hammer et al. 1980), in excellent agreement with the petrologically based estimate of Devine et al. (1984), equivalent to about 10^{14} g of H_2SO_4 aerosols. The sulfur emission from the Roza flow eruption can be estimated to have been $(700\ km^3/12\ km^3) \times 3 \times 10^{13}\ g = 2 \times 10^{15}$ g, equivalent to about 6×10^{15} g of H_2SO_4 aerosols.

4 Atmospheric Effects

The atmospheric and climatic effects of very dense clouds of volcanic aerosols can be estimated by comparison with the less severe aftereffects of historical eruptions. If the aerosols estimated for the large eruptions considered above were spread uniformly around the hemispheric or global stratosphere, then the incoming solar radiation would have been significantly reduced, with an associated climatic cooling. As a measure of opacity of the stratosphere, we may calculate the aerosol optical depth τ_D, which is equal to the negative natural logarithm of the attenuation of incident light: $\tau_D = -\log_e(I/I_o)$, where I_o = the initial and I = the final light intensity. For volcanic aerosols, the globally averaged optical depth is τ_D = 6.5 x

10^{-15} M_D, where M_D is the global aerosol loading in grams (Stothers 1984a,b).

For the Toba eruption, the estimated aerosol mass loading of 9 x 10^{14} to 5 x 10^{15} g is equivalent to a globally averaged peak aerosol optical depth of 6 to 33. These turbidity values may be compared with the greatest known historical perturbation of stratospheric aerosol optical depth of about 2, seen in AD 536 (Stothers 1984b). Reports of the AD 536 "dim sun" are accompanied by accounts of cold weather, failed crops, and famine in the Middle East (Stothers 1984b) and in China (Pang and Chou 1984). Clearly, an aerosol optical depth perturbation of 6 would have drastic effects on agriculture, while an aerosol optical depth of 33 would permit only a very small fraction of sunlight (less than 10^{-4}) to reach the Earth's surface. This extreme reduction in sunlight is similar to that predicted in some nuclear winter models with smoke optical depths of about 6 (Turco et al. 1984). It is important to point out that <u>aerosol</u> optical depth caused by volcanic eruptions and <u>smoke</u> optical depth from fires are not equivalent. Because of the high optical absorbance properties of sooty smoke, a smoke optical depth of about 0.2 of the aerosol optical depth has the same effect on atmospheric transparency, in the range of optical depths of interest here.

In the case of a large flood basalt eruption like the Roza flow, the estimated 6 x 10^{15} g of aerosols, if injected into the stratosphere and spread evenly around the globe, would lead to an aerosol optical depth of about 40. For the relatively high-latitude Roza eruption, the stratospheric and upper-tropospheric aerosols would most likely have been confined to the Northern Hemisphere. If this was the case, the Northern Hemisphere aerosol optical depth following the Roza eruption could have been as great as 80, allowing only 10^{-6} of incoming sunlight to reach the surface, similar to the smoke optical depths of about 15 found in some very severe nuclear winter scenarios (Turco et al. 1984).

5 Climatic Impact

What would be the climatic effects of such severe atmospheric opacity perturbations? Perhaps we may use the nuclear-winter scenarios for comparison. The original nuclear-winter studies indicated the possibility of drastic temperature decreases of up to 30°C, with subfreezing conditions for weeks to months over large portions of the Northern Hemisphere (Turco et al. 1983). More recent studies, using more sophisticated global climate models, have generally concluded that the cooling would be less severe though still significant ("nuclear autumn"), with a worst case (July) decrease of perhaps 5°C in the low latitudes (10° to 30°N) and 10° to 15°C at higher northern latitudes, lasting only a few weeks to about a month (Thompson and Schneider 1986). Some of the moderation of the cooling is related to the probable rapid removal of 75% of the smoke from the atmosphere in the course of about 30 days. In the case of volcanic aerosols in the stratosphere, however, the time needed for removal of stratospheric aerosols is several years according to historical observational data (Stothers 1984a). Therefore, significant volcanic aerosol optical depth perturbations after very large eruptions might persist for a few years, even if self-limiting effects are important.

In the geologic record it is presently difficult to detect in a direct way the predicted short-term coolings of weeks to a few years, although, in the future, analyses of annual growth bands in corals, or of other detailed records such as varved lake deposits, may be useful techniques. If positive climatic feedback mechanisms are sufficiently powerful, however, the cooling might be prolonged. It may be significant, therefore, that the Toba eruption coincides with the oxygen-isotope Stage 4/5 boundary in deep-sea sediments, a time of marked global cooling and growth of high-latitude ice sheets (Shackleton and Opdyke 1973), although any cause and effect relationship remains unproven. In the case of the Roza eruption, global temperatures indeed show a decrease in the Middle Miocene (Woodruff et al. 1981), but again no causal connection with volcanism has been established.

It is worth noting that even the minimum estimated perturbations of the atmosphere by large eruptions could cause significant problems for world agriculture (Burke and Francis 1985). The relatively small Tambora eruption in 1815 (Northern Hemisphere optical depth of $\sim$1) is associated with a decrease in Northern Hemisphere surface temperatures of about 0.5°C, and 1816 saw "the year without a summer" and "the last great subsistence crisis" in Europe (Post 1977; Stothers 1984a). Recent studies have suggested that a 1° to 2°C cooling could have marked effects on contemporary agricultural production (Parry 1985). The recurrence of a large explosive ignimbrite-forming eruption or a large flood basalt eruption of hundreds to thousands of km^3 might lead to climatic conditions that could be described as "volcanic winter", perhaps the ultimate volcanic hazard.

Acknowledgments. We wish to thank R.S.J. Sparks, H. Sigurdsson, T. Thordarson, and L. Wilson for discussions and correspondence, and the convenors of Symposium 4 for their work in putting together the interesting conference sessions.

References

Brown WH, Peczkis J (1984) Nuclear war - counting the cost. Nature (Lond) 310:455

Burke K, Francis P (1985) Climatic effects of volcanic eruptions. Nature (Lond) 314:136

Devine JD, Sigurdsson H, Davis AN, Self S (1984) Estimates of sulfur and chlorine yield to the atmosphere from volcanic eruptions and potential climatic effects. J Geophys Res 89:6309-6325

Hammer CU, Clausen HB, Dansgaard W (1980) Greenland ice sheet evidence for post-glacial volcanism and its climatic impact. Nature (Lond) 288:230-235

Lamb HH (1970) Volcanic dust in the atmosphere; with a chronology and assessment of its meteorological significance. Phil Trans R Soc Lond, Ser A266:425-533
Maddox J (1984) From Santorini to Armageddon. Nature (Lond) 307:107
National Research Council (1985) The effects on the atmosphere of a major nuclear exchange. Natl Acad, Washington DC
Ninkovich D, Sparks RSJ, Ledbetter MT (1978) The exceptional magnitude and intensity of the Toba eruption, Sumatra: an example of the use of deep-sea tephra layers as a geological tool. Bull Volcanol 41:1-13
Pang KD, Chou H (1984) A correlation between Greenland ice core climatic horizons and ancient Oriental meteorological records. Eos 65:846
Parry ML (1985) The impact of climatic variations on agricultural margins. In: Kates RW, Ausubel JH, Berberian M (eds) Climate impact assessment (SCOPE 27), studies of the interaction of climate and society, Wiley & Sons, Chichester, pp 351-369
Post JD (1977) The last great subsistence crisis in the western world. Johns Hopkins Univ Press, Baltimore
Rampino MR, Self S (1982) Historic eruptions of Tambora (1815), Krakatau (1883), and Agung (1963), their stratospheric aerosols and climatic impact. Quat Res 18:127-143
Rampino MR, Self S (1984) Sulphur-rich volcanic eruptions and stratospheric aerosols. Nature (Lond) 310:677-679
Rampino MR, Stothers RB, Self S (1985) Climatic effects of volcanic eruptions. Nature (Lond) 313: 272
Rose WI, Chesner CA (1986) Dispersal of ash in the great Toba eruption, 75,000 yrs B.P. Geol Soc Am Abs w Prog 18:733
Shackleton NJ, Opdyke ND (1973) Oxygen isotope and paleomagnetic stratigraphy of equatorial Pacific core V28-238: oxygen isotope temperatures and ice volumes on a 10^5 and 10^6 year scale. Quat Res 3:39-55
Sigurdsson H (1982) Volcanic pollution and climate: the 1783 Laki eruption. Eos 63:601-602
Stothers RB (1984a) The great Tambora eruption of 1815 and its aftermath. Science 224:1191-1198
Stothers RB (1984b) Mystery cloud of AD 536. Nature (Lond) 307:344-345
Stothers RB, Wolff JA, Self S, Rampino MR (1986) Basaltic fissure eruptions, plume heights, and atmospheric aerosols. Geophys Res Lett 13:725-728
Swanson DA, Wright TL, Helz RT (1975) Linear vent systems and estimated rates of magma production and eruption for the Yakima Basalt on the Columbia Plateau. Am J Sci 275:877-905
Thompson SL, Schneider SH (1986) Nuclear winter reappraised. Foreign Affairs (Summer 1986):981-1005
Thorarinsson S (1969) The Lakagigar eruption of 1783. Bull Volcanol 33:910-929
Turco RP, Toon OB, Ackerman TP, Pollack JB, Sagan C (1983) Nuclear winter: global consequences of multiple nuclear explosions. Science 222:1283-1292
Turco RP, Toon OB, Ackerman TP, Pollack JB, Sagan C (1984) The climatic effects of nuclear war. Sci Am 251 (Aug):33-43
Wilson L, Head JW III (1981) Ascent and eruption of basaltic magma on the earth and moon. J Geophys Res 86:2971-3001

Wood CA (1984) Amazing and portentous summer of 1783. Eos 65:410

Woodruff FS, Savin SM, Douglas RG (1981) A detailed study of the mid-Miocene isotopic record and its paleoclimatic implications. Science 212:665-668

Prediction of Volcanic Hazards on the Basis of the Study of Dynamics of Volcanic Activity, Kamchatka

I.V. MELEKESTSEV, O.A. BRAITSEVA and V.V. PONOMAREVA[1]

Abstract

The study of the history of volcanic activity over a long period of time (several 1000 years as a minimum), and the trends in the dynamics of eruptive activity and production rate, as well as the determination of the characteristic features of the present period of the life of a volcano are of decisive significance for long-term prediction of volcanic hazards. Based on these data, conclusions may be drawn on the number, frequency, type and energy of eruptions which may occur in the future.

Studies of the dynamics of volcanic activity in Kamchatka, using tephrochronological and geological-geomorphological methods and ^{14}C dating, made it possible to define groups of volcanoes regarded as belonging to the same stage of development (stage I: high and moderate levels of activity and growth of the edifice; stage II: a low level of activity and the onset of the destruction of the edifice), and to define phases of the cycle of activity, represented by the upsurge, repose or relative waning of activity. Most active volcanoes belong to stage I. The reconstruction of the cyclic regime of their activity permitted us to establish that Kliuchevskoi, Bezymianny, Karymsky and Maly Semyachik are in the period of recent cycle activity, and Avachinsky, Kizimen, Krasheninnikov, Kikhpinych and Ksudach are in the repose period of the cycle. Knowledge of the duration of such periods during previous cycles, and of the production rate of volcanoes at different stages, made it possible to determine the approximate time and parameters of future eruptions. Kliuchevskoi and Karymsky can be expected to erupt as in the present regime for a period lasting hundreds of years, Bezymianny will erupt in the style of the 1960-1980s for a period of 100 or 200 years, and a violent eruption from Maly Semyachik will probably occur in the next 50 or 100 years. The volume of erupted products of future effusive-explosive eruptions of these volcanoes will be 10^7-10^8 m^3. The repose period of Avachinsky, Krasheninnikov and Kikhpinych will last for several 100 years. The most hazardous catastrophic eruption is expected to be that of Kizimen volcano, which may occur in the next 50-100 years, and will resemble the Mount St. Helens eruption of 1980. For volcanoes which are regarded as belonging to

[1] Institute of Volcanology, Petropavlovsk-Kamchatsky, 683006, USSR

IAVCEI Proceedings in Volcanology 1
J.H. Latter (Ed.), Volcanic Hazards

stage II, both relatively minor explosive and phreatic eruptions (Mutnovsky volcano) and large, even catastrophic eruptions with a volume of erupted products of 10^8-10^9 m^3 (Khodutka, Koryaksky, Kronotsky, etc.) are possible.

1 Introduction

Prediction of the character and parameters of future eruptions of active and potentially active volcanoes, as well as of hazardous phenomena related to these eruptions, is one of the main problems of volcanic studies. The evaluation of these eruptions, and the degree of volcanic hazard, are derived on the basis of studies of the history of volcanic activity, and determination of the type, power and geological effects of eruptions. In the best case, this evaluation is made from historic records which span a few 1000 years: on the other hand, if the active volcanoes are poorly studied, this evaluation has to be made from data on the most recent eruptions, which may be as few as one or two. However, knowledge of a simple set of hazardous events which occurred in the past is needed, but this alone is insufficient for the correct evaluation of future volcanic hazards.

Of decisive significance for the prediction of future volcanic hazards is the elucidation of the trend in the dynamics of the eruptive activity and production rate of volcanoes, as well as the establishment of the characteristic features of the period of life in which the volcano is at present. Based on these data, the conclusion may be drawn on the character, frequency, type and energy of eruptions which may occur during the present period. It is clear that not all phenomena taking place during the long history of the volcano have an equal probability of occurrence in the immediate future. Accordingly, it is necessary to choose those phenomena which are characteristic of the present-day period of the life of a volcano and consider only those as potentially hazardous. Furthermore, there is a definite tendency in the region for the total energy of the most powerful eruptions to decrease with time. There is also a pronounced rhythm of volcanic occurrences, and a tendency for upper crustal extension to be significant, from at least early Holocene times (Melekestsev 1980). This leads us to the belief that no catastrophic explosive eruptions, with a mass of juvenile material of more than (4-5) x 10^9 t, will occur in the near future in Kamchatka, although in the recent past (7000-8000 years ago) eruptions with a mass of juvenile material of (10-30) x 10^9 t occurred repeatedly.

In predicting the activity of individual volcanoes, the reconstruction of their eruptive regime for a few 1000 years is of decisive significance. Unfortunately, in Kamchatka, information on historic eruptions is limited to 200-300 years, and even that is fragmentary and incomplete. We can obtain data on the activity of a volcano for a long interval of its life (ideally from its formation) only by using complex tephrochronological and geological-geomorphological methods (Braitseva et al. 1980b).

In the course of these investigations the following successive problems are considered as solved for a number of volcanoes:

(1) a detailed geological-geomorphological mapping of the volcano; (2) study of soil-pyroclastic sections at the volcano's foot, and a compilation of a summarized sequence of tephra layers, which is a geological chronicle of the history of explosive activity of the volcano; (3) dating of buried soils, wood and coals in soil-pyroclastic covers and pyroclastic flows in order to time volcanic events; (4) the determination of the position of lava flows in a summarized section of the soil and pyroclastic cover, based on comparison of tephra sections overlapping these flows with the complete reference section, by using tephra for which ^{14}C data are available; (5) the compilation of a summarized stratigraphic column, which includes all explosive and effusive products of the volcano and serves as the basis for the reconstruction of the regime of its activity. Simultaneously, lava and tephra are sampled to trace the evolution of the composition of the volcanic products. As a result, a map of the volcano is constructed where the approximate absolute age is indicated for its lava and pyroclastic flows. A series of maps and paleoreconstructions are compiled for long time spans in the history of the volcano. Using the maps and reconstructions, and isopachs of tephra, the volume of volcanics and the production rate of the volcano are estimated for separate periods of its life.

Complex tephrochronological and geological studies, made in accordance with the above procedure, allow us to: (1) determine the age of a volcano on the basis of dating of deposits underlying the pyroclastics of its first eruptions; (2) recognize and date, using the ^{14}C method, both its greatest eruptions, and the main intensified periods of its activity; (3) establish the time of flank eruptions; and (4) ultimately reconstruct the dynamics of volcanic activity with assignment to the absolute and ^{14}C scales.

2 Stages of Activity. The Stage of High and Moderate Levels of Activity, and Growth of the Edifice

Two stages can be distinguished in the development of the active volcanoes of the Kurile-Kamchatka region (Melekestsev 1973). Stage I corresponds to the time of intense volcanic activity; a positive balance of accumulation on the volcano and, as a consequence, the growth of the volume and fresh appearance of the volcano. The duration of stage I, for volcanoes of different types and dimensions, ranges from several 100 to 12,000 years. Stage II corresponds to the period of waning volcanic activity; there is characteristically a negative balance of accumulation on volcanoes in this stage. Volcanic edifices stop growing and begin to be destroyed by the processes of denudation during stage II. The duration of stage II is typically 10,000-50,000 years.

Most active volcanoes in Kamchatka are referred to stage I. Their activity is of the pronounced cyclic (rhythmic) character. Each cycle consists of two phases. On the one hand, the intensified period of activity, manifested by layers of tephra in the soil-pyroclastic cover and by pyroclastic flows and lavas, and, on the other, the repose period, or period of considerable waning of volcanic activity, when buried soils, occasionally with rare intercalations of tephra, are formed. Studies of the cyclic regime of activity over a long period of time allow one to determine the phase of the cycle in which a volcano is at present (irrespective of whether or not it is erupting at a given moment). Knowledge of the duration of these phases in the previous cycles, and of the production rate of a volcano at different stages, permit us to assess its behaviour in the immediate future, and to determine the time interval and probable parameters of future eruptions.

About two-thirds of the active and potentially active volcanoes of Kamchatka are the subject of our investigations for the purpose of prediction (Fig. 1). But because of the limitations of space we shall consider only some representative cases in this chapter.

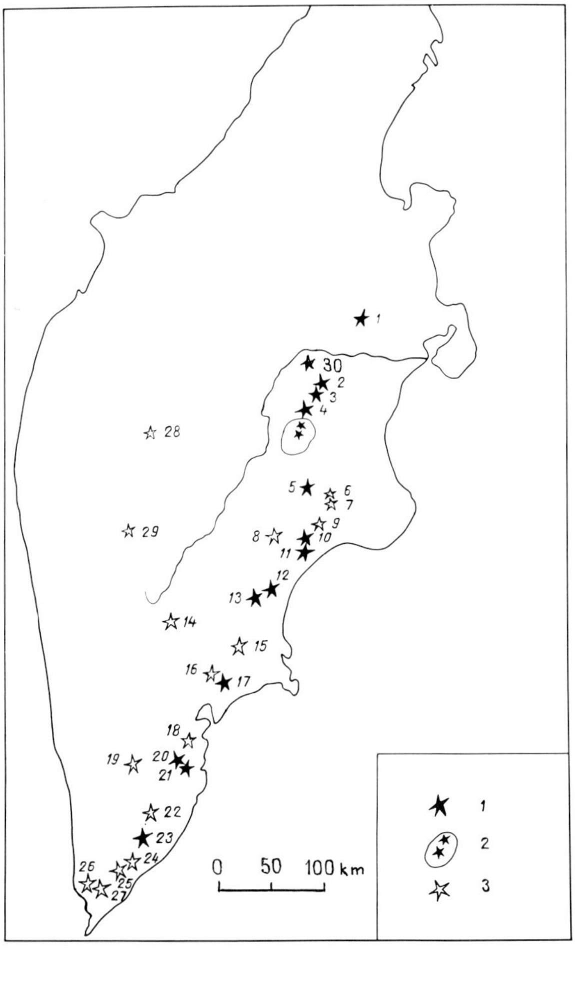

Fig. 1. Sketch map showing the distribution of active and potentially active Kamchatkan volcanoes. **1** Volcanoes for which detailed tephrochronological and geological-geomorphological studies have been carried out; **2** Tolbachik regional zone of cinder cones; **3** volcanoes for which detailed studies are planned. Volcanoes are designated by numbers: **1** Shiveluch; **2** Kliuchevskoi; **3** Bezymianny; **4** Plosky Tolbachik; **5** Kizimen; **6** Komarov; **7** Gamchen; **8** Taunshits; **9** Kronotsky; **10** Krasheninnikov; **11** Kikhpinych; **12** Maly Semyachik; **13** Karymsky; **14** Bakening; **15** Zhupanovsky; **16** Koryaksky; **17** Avachinsky; **18** Vilyuchinsky; **19** Opala; **20** Gorely; **21** Mutnovsky; **22** Khodutka; **23** Ksudach; **24** Zheltovsky; **25** Ilinsky; **26** Koshelev; **27** Kambalny; **28** Ichinsky; **29** Khangar; **30** Plosky Dalnii

2.1 The Intensified Period of the Recent Cycle of Activity (Stage I)

The following volcanoes in Kamchatka fall into this category: Kliuchevskoi, Bezymianny, Karymsky, Maly Semyachik, Shiveluch, Gorely and Plosky Tolbachik. We shall consider the first four of these, as they have been the most studied.

Kliuchevskoi is the most active and largest volcano in Kamchatka (the volume of its erupted products is 250-300 km^3, the weight is 600 x 10^9 t and its altitude measured in 1986 was more than 4800 m above sea level). The time of its origin, from tephrochronological studies, is only about 7000 years ago. Historic eruptions are known since 1698. The volcano is composed of basaltic and basaltic-andesitic rocks. About 3000 years ago the volcano apparently reached its maximum morphological volume and height, because just at that time flank eruptions began to occur on its slopes and at its foot. Since then more than 100 flank eruptions have occurred, 10 of them at intervals of 2 to 10 years between 1932 and 1983.

Judging from tephra distribution in sections of soil-pyroclastic cover at the foot of the volcano, no sharp waning or discontinuance of activity has occurred throughout its life, only periods of relatively higher and lower activity. Frequent slight and moderate eruptions (with volumes of products in the range 10^6-10^7 m^3) have predominated. Great and very great eruptions, with volumes of products of 10^8-10^9 m^3, have occurred comparatively rarely, the most recent of these being in January 1945 (Piip 1956).

Kliuchevskoi differs from other Kamchatkan volcanoes in its maximum production rate (Melekestsev 1980), which is relatively stable, both for long and for short time intervals; it fluctuates between about 5 x 10^7 and 9 x 10^7 t $year^{-1}$. Thus, throughout the whole period of its life, the average production rate amounted to 8.5 x 10^7 t $year^{-1}$, but during the period 1926-86 it was 6.3 x 10^7 t $year^{-1}$.

The high frequency of eruptions, and only minor changes in production rate indicate that Kliuchevskoi is still at stage I, and, moreover, that it is at the beginning of this stage. Therefore, in the immediate future, the character of its activity is not likely to change significantly. This was also the opinion voiced earlier by Piip (1956), who declared that moderate eruptions had occurred and would continue to occur, on the average, every few years, and that great eruptions would take place every several tens of years. Eruptions from both the central crater and from the flanks may produce a total volume of products of 0.001-0.5 km^3.

The volcanic hazards at Kliuchevskoi include lava flows, tephra falls, lahars, and volcanic and seismotectonic sector collapses. Tephra falls may be destructive at a distance of 15-20 km from the eruptive centers. Tephra falls are likely to vary in thickness from 1-2 cm near the outer boundary of the zone, to 100-500 cm in the region of the summit crater or near the foot of cinder cones produced by flank eruptions. Lava flows of terminal eruptions may be destructive at a distance of up to 15 km from the crater. The maximum probable length of lava flows produced by flank eruptions is about 10-12 km. Large volcanic-seismotectonic failures, and rockslides related to strong eruptions, will be fairly probable. One such failure occurred in January 1945, when a large funnellike structure originated near the summit of the volcano (Piip 1956). The near-summit part of the north-western sector of the volcano is the most hazardous in this respect. Lahars, affecting river valleys to distances of up to 20-30 km, depending on the power of the eruption and the position of the eruptive center, also represent a significant volcanic hazard.

Unlike Kliuchevskoi, other active volcanoes have had more complex development. Their activity is of a pronounced cyclic character, and this is the key to prediction of future volcanic activity.

Among these volcanoes, Bezymianny is of paramount interest. This volcano is well known owing to its catastrophic eruption in March 1956. Hazards produced by such eruptions highlighted the necessity of studying the history of the volcano (Braitseva and Kiryanov 1982). The active Bezymianny stratovolcano formed about 4500-5000 years ago. Estimates of age were made on the basis of the age of the Shiveluch ash marker beds which were dated by the ^{14}C method (Braitseva et al. 1984) and from calculations of the sedimentation rate of the soil-pyroclastic cover. Products from Bezymianny overlapped the more ancient Pre-Bezymianny volcano, which existed here in early Holocene times: small fragments of this ancient volcano have been preserved only in the eastern part of the massif (the "old dome" as described by Gorshkov and Bogoyavlenskaya 1965). The activity of Bezymianny between 4500 and 3000 years ago was moderately explosive, with lava flow effusions and dome extrusion; its products were basaltic-andesites and andesites.

The activity of Bezymianny has been most completely reconstructed for the last 3000 years (Figs. 2, 3). Analysis of the ash distribution, between measured sections, has demonstrated that in the upper parts of the soil-pyroclastic covers they are grouped in packets, each containing up to 12 or 15 tephra layers (see Fig. 2). These packets fix the intensified periods of Bezymianny activity, which are separated by repose intervals (marked by sandy loams, without ash from Bezymianny).

Three such intensified periods of Bezymianny activity can be distinguished (BI, BII and BIII). The first two periods, 400-700 years long, correspond to the time intervals of 2400-1700, and 1350-1000 years ago, and the third one corresponds to the period 1956 to the present time. The beginning of the intensified periods of Bezymianny activity, e.g. early BI, are characterized by moderate explosive activity, and occasionally by strong eruptions with ejections of pumiceous tephra with a volume of about 1 km^3. Lava effusions have ended all the intensified periods of Bezymianny activity. The volcanic prod-

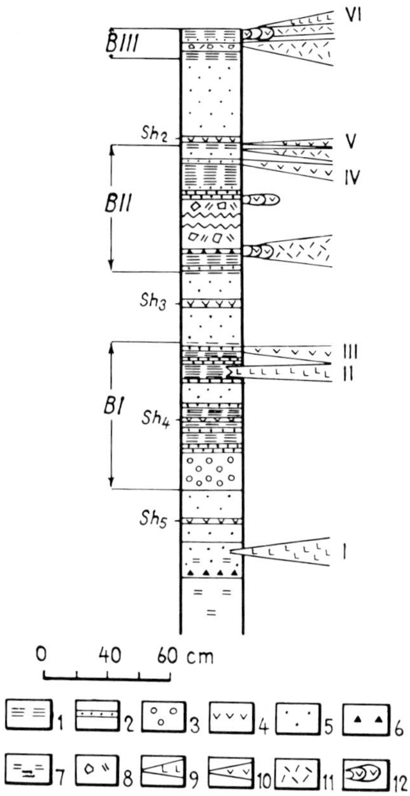

Fig. 2. Generalized section through effusive-pyroclastic deposits at the foot of Bezymianny volcano laid down during the last 3000 years. **1** Fine ash from Bezymianny; **2** coarse ash from Bezymianny; **3** pumiceous lapilli and coarse ash from Bezymianny; **4** fine ash from Shiveluch; **5** sandy loams with admixture and intercalations of ashes from Kliuchevskoi and other volcanoes of the Kliuchevskoi volcanic group; **6** lapilli of cinder and coarse ash from cinder cones on the Kliuchevskoi and Tolbachik plateaus; **7** sandy loams with admixture of ashes from Bezymianny and Kliuchevskoi; **8** landslide deposits of the Kamen volcano; **9** basaltic-andesite lava flows; **10** andesite lava flows; **11** pyroclastic flows of andesite composition; **12** andesites of extrusive domes. **I-VI** Age groups of lava flows (lava and pyroclastic flows are shown without scale). **BI-BIII** Intensified periods of Bezymianny activity. **Sh_2-Sh_5** are the Shiveluch marker ash horizons with ages: **Sh_2** ~900-1000 years ago; **Sh_3**~ 1400-1500 years ago; **Sh_4** ~2000 years ago, **Sh_5** ~2500 years ago

ucts of period BI are basaltic-andesites and andesites, whereas the volcanic products of periods BII and BIII are andesites. It is worthy of note that throughout each intensified period of Bezymianny's activity a regular change in the composition of volcanic products occurred, from more acid varieties to more basic, and again to more acid varieties. These changes occurred within the ranges of (61-54)-61% SiO_2 during the BI period, (61-56)-62% SiO_2 during BII, and from 60 to 56% SiO_2 during the present BIII period.

Superimposed on the rhythmic eruptions at Bezymianny, there were significant changes in activity about 1400 years ago (see Fig. 3). Previously, the common effusive-explosive eruptions had been characteristic, but during this last period catastrophic eruptions began. They are accompanied by the formation of pyroclastic flows and by partial destruction of the summit of the volcano. Such eruptions took place twice during the BII period. The most recent such eruption occurred in 1955-56.

At present the volcano is in the intensified period of the recent cycle of activity (see Fig. 3), which began in 1955 after approximately a 1000-year repose period. Because the

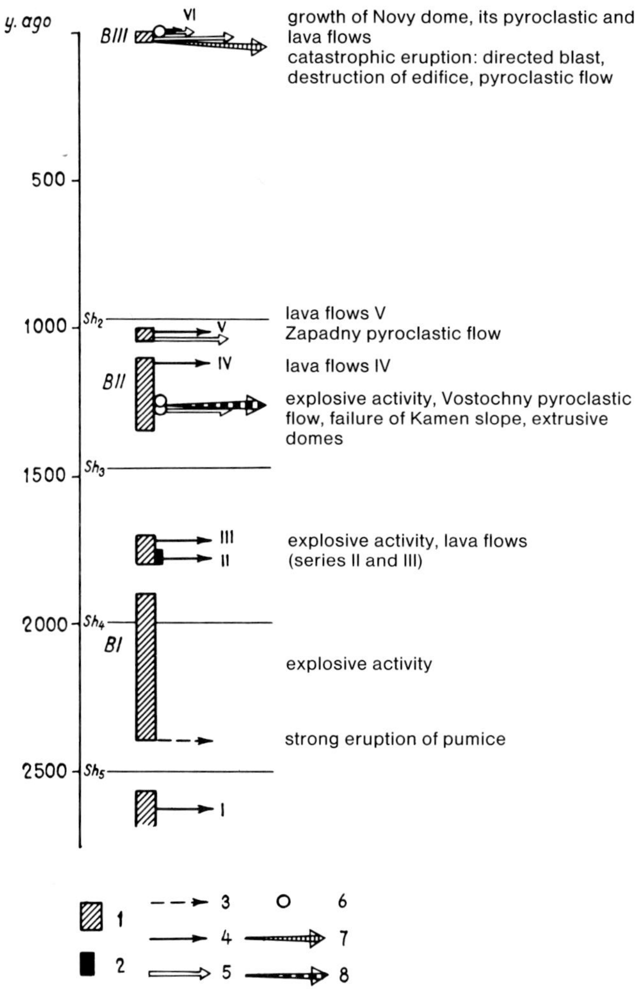

Fig. 3. Activity of Bezymianny volcano during the last 3000 years. **1** Periods of intense explosive activity; **2** period of profuse lava outflows; **3** large explosive eruption with pumice ejections; **4** lava flows; **5** pyroclastic flows; **6** extrusive domes; **7** directed blast; **8** gigantic landslide of the flank of Kamen volcano. **Sh_2-Sh_5** Shiveluch marker ash horizons. **BI-BIII** Intensified periods of Bezymianny activity. **I-VI** Age groups of lava flows

duration of the previous intensified periods of activity constituted a few 100 years, we may suggest that the duration of the present-day period will be approximately the same. A transition from the more acid andesite varieties (60% SiO_2) of 1955-56, to the more basic andesites of 1978-85 (56% SiO_2) indicates, by analogy with the previous periods, that the volcano is in the first half or in the middle of the intensified period of activity. The catastrophic eruption of the type that occurred in 1956 is unlikely to occur again in the next 100-200 years, since such eruptions take place generally only after a repose period. The activity of the volcano during the next 100-200 years will be associated with the growth of the dome in the crater which originated in 1956. Slight explosive eruptions, small and moderate explosions accompanied by partial destruction of the dome, and eruptions of pyroclastic flows with a volume of 0.001 to 0.01 km^3, together with lava flows from the dome, will be characteristic. The total volume of the material of a single eruption is unlikely to exceed 0.02-0.05 km^3, the pyroclastic flows and lava flows will extend only as far as 5-15 km and 0.2-2 km respectively, and the volume of tephra will be at most 0.01-0.03 km^3. The eastern flanks of the volcano, towards which the crater is open, will be the hazardous zone. This extends to a radius of up to 10-15 km from the summit.

Thereafter, a repose period will occur. This period will be either of short duration (a few tens of years), during the final stages of the intensified period, or it will separate the intensified periods of activity themselves, and will last for hundreds of years. In any case, after such repose periods one may expect catastrophically directed blast eruptions, accompanied by destruction of the summit, and by the eruption of pyroclastic flows with a total volume of 0.5-2 km^3.

Karymsky is the most active volcano of East Kamchatka. The history of the volcano has been reconstructed from its origin, about 6100 (5300 ^{14}C) years ago (Braitseva et al. 1980a, b; Braitseva and Litasova 1982). The volcano lies in a caldera which was formed about 8000 ^{14}C years ago, after the eruption of 8-10 km^3 of dacitic pumice. Two large intensified periods of activity can be distinguished.

Fig. 4. Generalized section through lava-pyroclastic deposits at the foot of Karymsky volcano. 1 Bombs and lapilli of pumice; 2 lapilli of pumice, coarse ash; 3 lapilli of cinder, coarse ash; 4 coarse ash; 5 stratified coarse ash; 6 stratified pyroclastics: coarse ash with lapilli, bombs and admixture of resurgent material; 7 coarse ochre ash; 8 stratified fine and coarse ashes; 9 fine ash; 10 pyroclastic flow deposits; 11 fine ashes of caldera-forming eruptions; 12 sandy loams; 13 sandy loams with coarse ash admixture; 14 buried soils; 15 underlying deposits; 16 wood and coals; 17 andesite lava flows; 18 andesite and dacite lava flows; 19 data points: a) tephra from Karymsky, b) lava from Karymsky, c) pyroclastics of caldera-forming eruptions; 20 radiocarbon dates: a) wood, b) coals, others-buried soils (in **frames** are dates determined by alkaline extractions from the same soil sample). **KRM** Deposits associated with the formation of Karymsky caldera; **KM** complex of deposits confined to the time of Karymsky's formation; **IKM, IIKM** deposits of intensified periods of Karymsky activity; **I-XI** age groups of lava flows, pm_1-pm_{12} marker layers of pumiceous tephra; **SC** cinder interlayer; 1-12 horizons of transit ashes (from other sources)

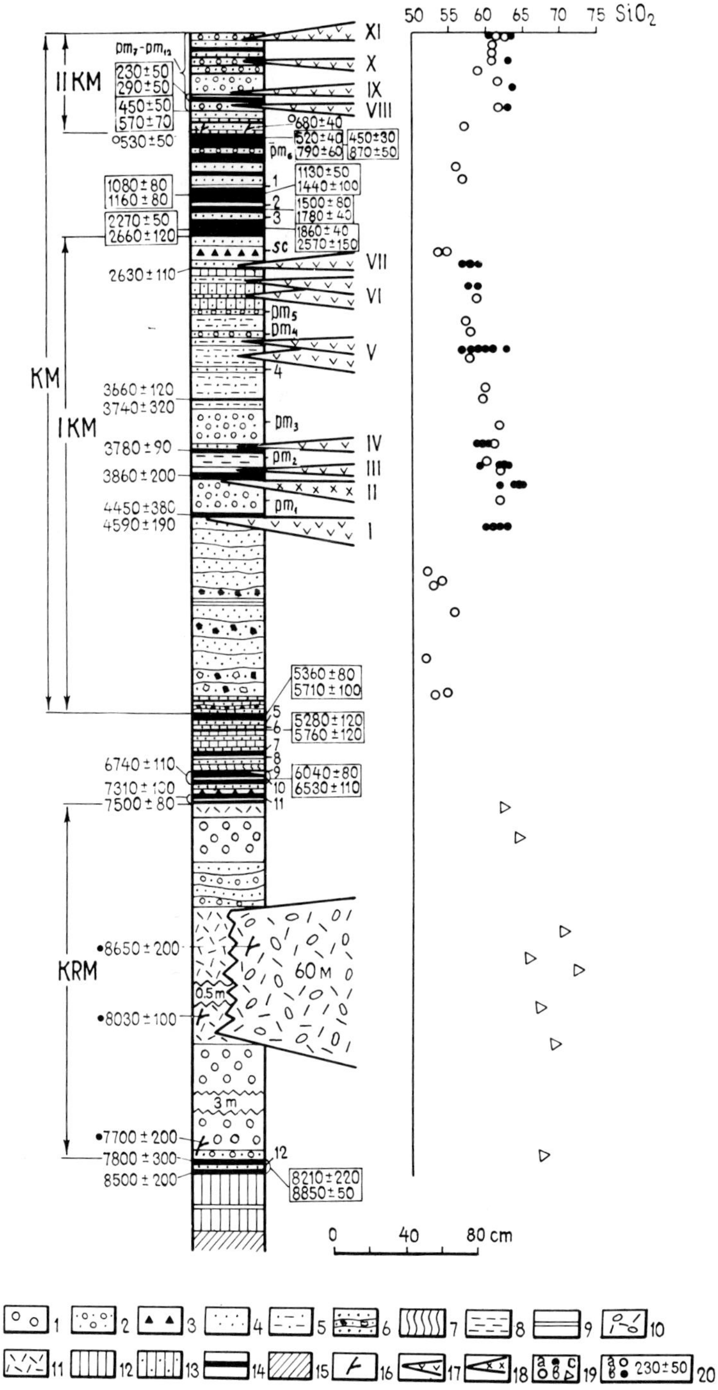

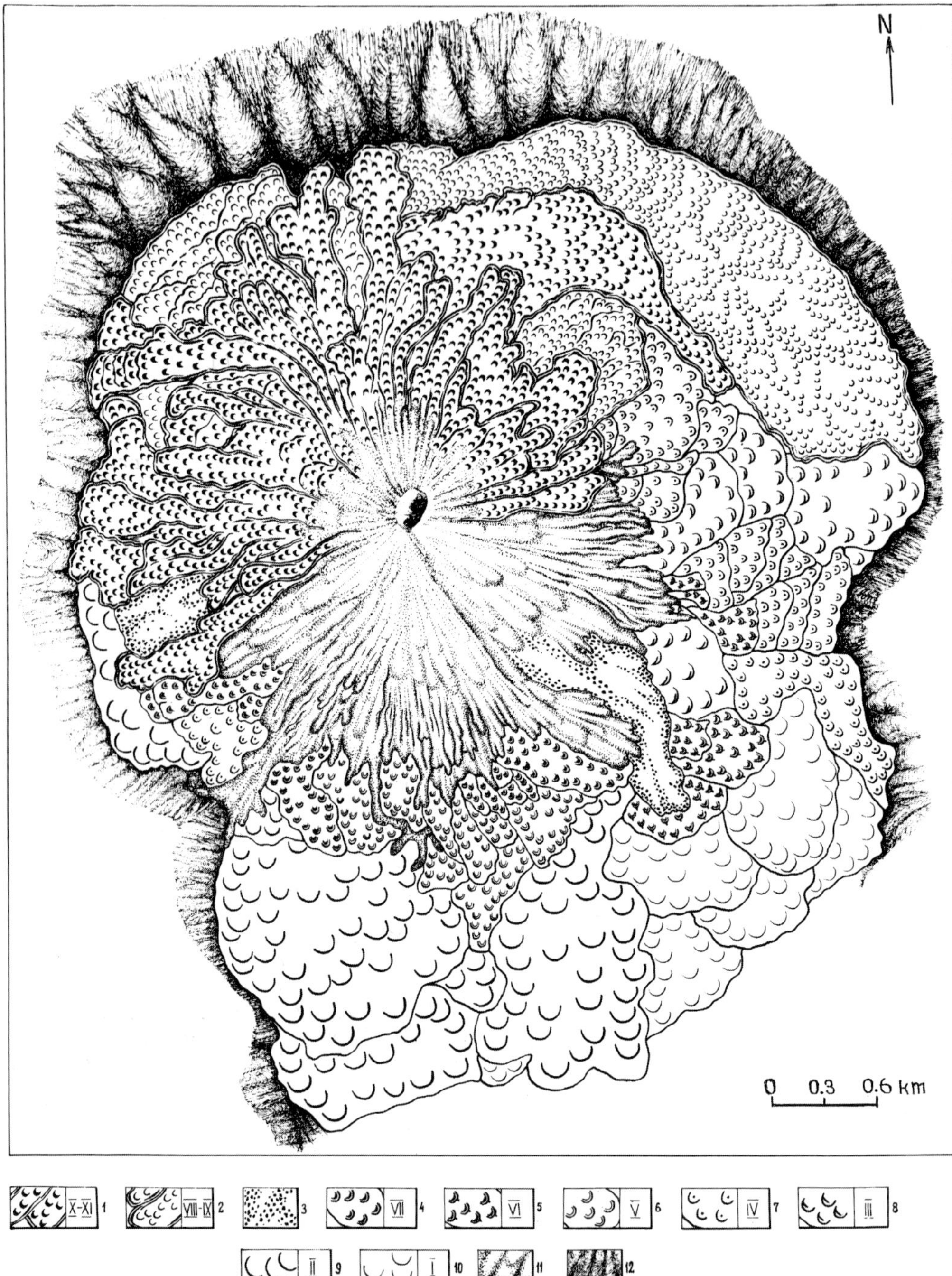
N
0 0.3 0.6 km
X-XI 1
VIII-IX 2
3
VII 4
VI 5
V 6
IV 7
III 8
II 9
I 10
11
12

The first (IKM) corresponds to the stage of cone building; it continued for 3000-3500 years (Figs. 4, 5, 6). At the beginning and at the end of this period, there was a series of frequent, slight and moderate eruptions, and effusion of lava flows of age complexes I,V-VII (see Fig. 5). The products of the initial stages of the formation of the volcano are basaltic-andesites; later products are andesites. In the middle of the intensified period of activity, several large discrete explosive-effusive eruptions occurred, accompanied by ejections of pumiceous tephra (pm_1-pm_3) and effusion of lava flows (II-IV). The intervals between the eruptions increased sharply as time passed; this was exhibited by the increasing thickness of buried soils. The volume of pumices of the greater eruptions was about 0.1 km^3 (pm_1 and pm_3). The eruption products were andesites; dacites were noted only among lavas of age complex II. The intensified period of activity ended with an explosive eruption, accompanied by ejections of basaltic-andesite scoria (s.c.).

A new intensified period of activity (IIKM) began approximately 500 years ago, after a 2300-year repose period and is still continuing. The regime of the volcano at this stage is one of pulsatory character, i.e. explosive and explosive-effusive eruptions are separated by short periods of quiescence (lasting for a few years or for a few tens of years). Ejections of pumiceous tephra (lapilli, bombs and ash of layers pm_7-pm_{12}), or of andesitic coarse ashes, take place from time to time, and lava pours out onto the northern flank and the foot of the volcano. The most recent eruptions occurred in 1970-71, and 1975-81. The volume of products of separate eruptions is of the order of 0.001-0.01 km^3.

Since the volcano is in the beginning, or in the middle of the intensified period of a new cycle of activity, the waning of its activity may scarcely be expected in the immediate future. Taking into account the duration of the periods of intense activity during the previous cycle (700-1200 years), we suggest that the activity of the volcano will continue for at least 100-200 years. The character of eruptions will remain much the same: they will be explosive-effusive eruptions, accompanied by ejections of andesitic tephra (often pumices), and by lava effusions, with the total volume of products ranging from about 0.001 to 0.03 km^3. The zone of lava flows will be limited by the rim of the caldera (radius 2.5-3 km), and the thickness of

---

Fig. 5. Map of Karymsky lava flows. 1 Lava flows of age groups XI (historic eruptions of 1963 - present time) and X (younger than 200 years); 2 lava flows IX-VIII (200-400 years ago); 3 pyroclastic avalanches ~2800 years ago; 4 lava flows VII ~2800 years ago; 5 lava flows VI ~3000 years ago; 6 lava flows V ~3300-3400 years ago; 7 lava flows IV ~4200 years ago; 8 lava flows III ~4300 years ago; 9 lava flows II ~5100 years ago; 10 lava flows I ~5400 years ago; 11 pyroclastics of the cone of the volcano and pyroclastic avalanches of historic eruptions; 12 margin of the caldera

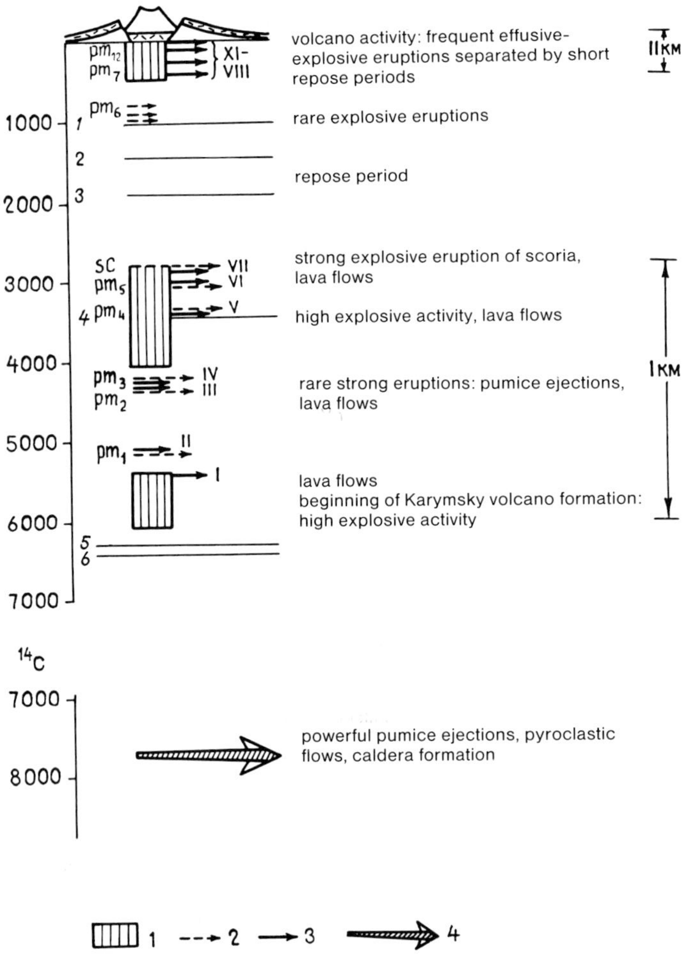

Fig. 6. Activity of Karymsky volcano. 1 Periods of explosive activity; 2 separate explosive eruptions; 3 outpourings of lava flows; 4 caldera-forming eruptions; pm_1-pm_{12}, SC pumice and cinder ejections. 1-6 transit ashes (from other sources). IKM, IIKM Intensified periods of activity

tephra of a single eruption in the caldera may range from 5 to 40 cm (up to 100-120 cm immediately at the foot of the cone in the direction of the ejection of pumice). Beyond the boundaries of the caldera, at a distance of 4-5 km from the volcano, tephra will be at most 20-30 cm thick, and at a distance of 10 km it will be about 3-5 cm thick: it will consist of coarse and thin ash.

We are led to the conclusion that the Maly Semyachik volcano may erupt in the near future. Although, since 1804, it has remained relatively calm, with only slight phreatic eruptions, the reconstruction of its history shows that it is in the period of general intensification of the recent volcanic cycle.

The volcano lies in a caldera, and consists of three coalescing stratovolcanoes, namely Paleo-, Meso- and Kaino-Semyachik. Products of the volcano are basalts and basaltic-andesites, with andesites emerging only at the final stage of Kaino-Semyachik's activity.

Four principal stages (PS, MS, IKS and IIKS; Fig. 7) can be recognized in the history of the volcano. Each is associated with significant structural-morphological transformations of the massif (Braitseva et al. 1978; Selyangin et al. 1979). The

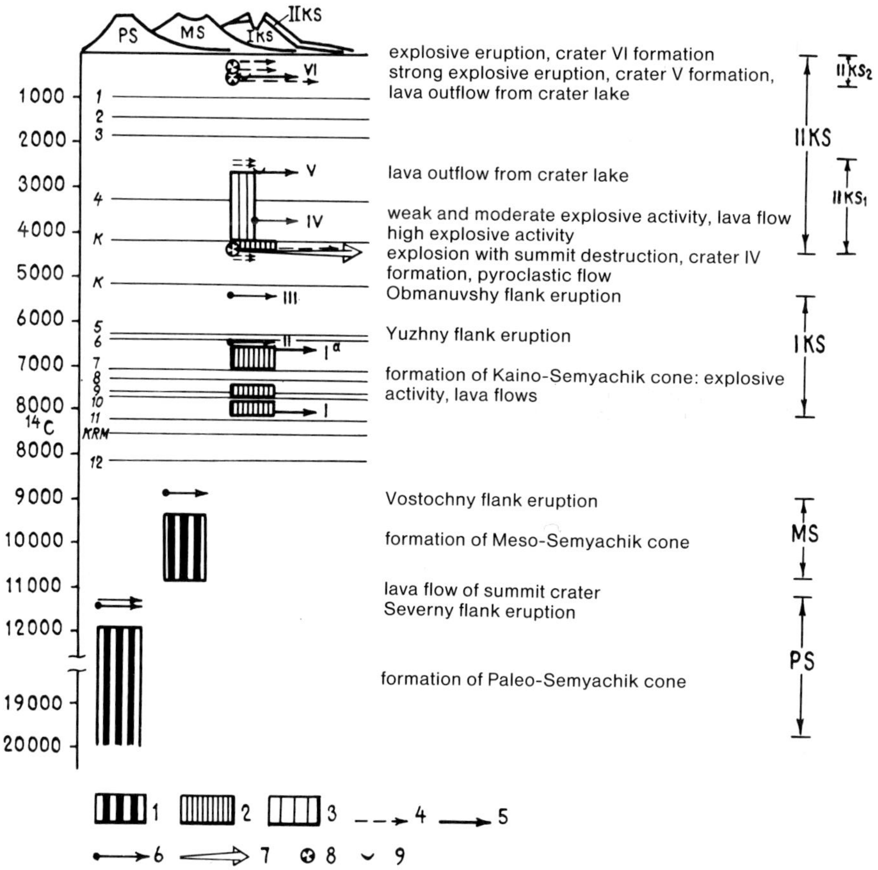

Fig. 7. Activity of Maly Semyachik volcano. 1 Periods of explosive-effusive activity; 2 periods of intense explosive activity; 3 periods of slight explosive activity; 4 separate strong explosive eruptions; 5 lava flows; 6 flank eruptions; 7 explosion, eruption of pyroclastic flow; 8 formation of explosive crater; 9 crater lava lakes. **PS,MS** Activity of Paleo- and Meso-Semyachik; **IKS,IIKS** stages of Kaino-Semyachik activity; **$IIKS_1$** and **$IIKS_2$** intensified periods of Kaino-Semyachik activity at stages IIKS. **KRM** Deposits associated with the formation of Karymsky caldera

first three stages show great similarity and correspond to the time of formation of the three stratovolcanoes, Paleo-, Meso- and Kaino-Semyachik (see Fig. 7). The intensified periods of activity at the beginning of each stage correspond to the time of building of the respective cone, and the periods of waning and cessation of activity coincide with the termination of the formation of these cones, when they reached their maximum heights and when flank eruptions began to occur. Under the conditions of free displacement of the feeding channel along a fissure, such cyclic behaviour was repeated three times without essential differences.

Had we been predicting the activity of the volcano, after stage IKS, we might have expected the formation of the next, fourth cone, near the previous one. However, continued displacement of the feeding channel ceased, and this caused an essential change in the character of volcanism. Throughout stage IIKS, the volcanic activity developed as a series of alternations of destruction of the summit of the volcano, and reconstruction of the edifice to its previous height. The eruptive channel migrated only insignificantly in the summit part of the volcano, forming a series of craters. The largest of these was the crater which formed at the beginning of stage IIKS, when an explosion occurred which destroyed the summit of the volcano; in this eruption, the height of the edifice was reduced and a pyroclastic flow was generated.

During stage IIKS, two intensified periods of activity can be distinguished, $IIKS_1$ and $IIKS_2$ (see Fig. 7). They are separated by a 2000-year repose period, during which only a few slight phreatic eruptions occurred. The last intensified period of activity, during which a few large eruptions occurred, began 500-600 years ago. The first of these eruptions produced a powerful ejection of scoria; as a result a large crater formed. The melt, which rose thereafter, formed a lava lake in the crater. Having spilled over the crater rim in different places, lava flooded the cone with a series of flows. The total volume of erupted products was 0.6 km^3. The second eruption, which occurred approximately 400 years ago, was explosive (the volume of tephra ejected was 0.1 km^3), and was accompanied by the formation of the present-day crater, in which the present thermal lake now lies. A fairly large historic eruption was observed in 1804. During the last 30 years only periodic increases in the volume and temperature of the crater lake have been observed (Gorshkov et al. 1975).

Data on the dynamics of Maly Semyachik's activity (see Fig. 7) demonstrate that, in spite of the fact that recently no eruptions of the volcano have occurred, this volcano is now in a period of general intensification of its present cycle, which, based on the duration of the past intensified periods of activity, is far from being over. The mean production rate of the volcano (0.8×10^6 t $year^{-1}$), during this $IIKS_2$ period has remained close to that of the previous period, $IIKS_1$, which averaged 1.1×10^6 t $year^{-1}$. Only the character of activity has changed: the volcanic energy has come to be released predominantly as large eruptions, separated by repose periods about

200 years long. Since the volcano at present is at the end of such a period, and since only a small amount of energy has been released since 1804, a strong eruption may be expected in the near future. Taking into account the production rate of the volcano at stage $IIKS_2$, the mass of juvenile material is expected to be about 0.5-1 x 10^8 m^3, and taking the presence of the resurgent plug into account, it may well be about 0.8-1.5 x 10^8 m^3. The eruption may be either purely explosive, or effusive-explosive. If lava flows occur, they will threaten an area within a radius of 5-6 km from the crater, and the thickness of tephra will amount to 10-20 cm at a distance of 15-20 km along the ash-fall axis. Eruption of the crater lake, which contains 0.015-0.02 km^3 of water, may generate lahars in all the river valleys.

2.2 Volcanoes in a Period of Relative Calm in Their Present Cycle

Volcanoes of this group are Avachinsky, Krasheninnikov, Kikhpinych, Kizimen and Ksudach. The prediction of their activity is different in each case, but the reconstruction of their histories permits us to suggest that the repose period of some volcanoes will continue for a long period of time (hundreds of years), while for others, on the contrary, we propose the possibility of large eruptions in the near future.

The Kizimen volcano at present is calm, showing only fumarolic activity. Kizimen is similar in morphology and parameters to Mt. St.Helens (USA) before its famous eruption of 18 May 1980. The absolute height of the volcano is 2485 m, the volume of products is about 20 km^3 and the age of the volcano is 9500-10,000 years. The main types of rocks are dacites, andesite-dacites and andesites.

A typical peculiarity of many past eruptions of this volcano is their catastrophic character, the intervals between them being as long as a few 100 or a few 1000 years. The formation of the volcano began with a gigantic explosion, followed by a profuse ejection of as much as 5 km^3 of dacitic pumiceous juvenile pyroclastics. The probable analogue of this eruption was the eruption of Santa Maria volcano in Guatemala in 1902 (Rose 1972). The eruption (or series of eruptions) ended in the formation of the extrusive dome with viscous flows 100-200 m thick in the frontal part. The next eruption, which was similar in power, occurred approximately 7000-7500 years ago. Apparently, at the end of this eruption an extrusive dome was also formed, but no lava flows of this age have been identified.

Later in its history, a few other very great eruptions took place, of which the two latest occurred 2900 and 2000 years ago. Each began with powerful explosions, followed by the ejection of juvenile pyroclastics, and ended with the formation of an extrusive dome and the effusion of lava flows of andesite composition. The total volume of erupted products of each eruption was 0.5-2 km^3. Since 2000 years ago, no great eruptions of Kizimen have been recorded. Only a few small basaltic andesite lava flows have been produced. During the 19th and 20th centuries only phreatic explosions and intermittent bursts of fumarolic activity have been observed, the last time being in 1927.

At present the volcano is a composite cone-shaped edifice with a large lava dome at the summit part of the volcano, the volume of the lava dome exceeding 0.5 km^3. A powerful jet of fumarolic gases escapes from the base of the agglomerate mantle of the dome. Altered rocks, marking the sites of previous fumarolic activity, are conspicuous near the foot of the dome and elsewhere. No recent activity has been observed in a small crater at the top of the dome.

The calculated mean production rate of the volcano is about 4 x 10^6 t $year^{-1}$ (Melekestsev 1980). Taking this production rate into account, during the last 2000 years about 8 x 10^9 t of juvenile material could be accumulating, ready for the next eruption. However, during this period, at the most, 0.05-0.10 x 10^9 t has been erupted, i.e. not more than 1% of the total available. The energy of the magma chamber has been partly released by fumaroles and thermal waters. However, our estimates indicate that the summarized effect of these processes could lead to the release of only 20-50% of the energy accumulated.

Hence, during the past 2000 years the volcano has predominantly accumulated, rather than spent, its substance and energy. Therefore, it now appears to be at a stage of preparation for a large eruption, with a volume of the order of 0.5-1 km^3. Taking into consideration the preceeding history of Kizimen's eruptive activity, its position within the zone of high seismicity, and its resemblance in geological structure and morphology to Mt. St.Helens, we may also predict to some extent the character of the future eruption. We expect the events to develop in the following way:

(1) magma ascent and its intrusion under the basement of the dome in the region of a recent fumarole with deformation of the near-summit part of the volcano; (2) a catastrophic failure of the dome in the west direction; (3) a powerfully directed blast and eruption of juvenile pyroclastics from the newly formed cirque-shaped crater. Another variant, a gigantic failure, accompanied by an eruption of similar character, triggered by a large local tectonic earthquake, is also likely.

<u>Avachinsky</u> is a large stratovolcano of the Somma-Vesuvius type, which began to form 60,000-70,000 years ago. The volume of its erupted products, taking tephra into account, is about 160 km^3. This volcano is composed of rocks ranging from basaltic andesites to dacites. For prognostic purposes, of principal interest is the history of eruptive activity of the volcano over the last 9000-10,000 years (Melekestsev and Kiryanov 1984).

Explosive eruptions in the Early Holocene (10,000-5500 years ago) were notable for great power, and for the andesite-dacite and dacite composition of their juvenile pyroclastics, a feature uncommon in its subsequent activity. The last of these eruptions, which occurred approximately 5500 years ago, was the most powerful. About 5 km^3 of pumiceous andesite-dacite pyroclastics were erupted, with pyroclastic flows up to 50 m thick, and tephra deposited over an area of a few 100,000 km^2. This strong eruption resulted in the formation of a large crater.

The next 2500-2000 years of the life of the volcano were relatively calm. The eruptive activity increased sharply only 3500-3600 years ago, when six eruptions, or short series of eruptions, occurred one after another. These were concluded 3200 years ago by an explosion and a new profuse ejection of pyroclastic flows and tephra of andesite and basaltic-andesite composition, the volume of which was 1.5-2 km^3. Thereafter, apparently, the formation of the recently active young cone of the Avachinsky volcano began, and this continues up to now. During its activity, for the first time in the Holocene, lava flows were produced. These were up to 6 km long, and 0.005-0.01 km^3 in volume.

From the formation of the young cone up to the present, no essential variations have been noted in the chemical composition and in the outer appearance of the juvenile rocks, although the volcanic activity has been manifested non-uniformly. One hundred to 300-year periods of intensified activity have alternated with 150 to 1400-year periods of relative calm. The latest intensified periods were noted 750-850 ^{14}C years ago, 400-550 ^{14}C years ago and from 1737 to 1945.

The last of them, beginning in summer 1737, coincided in time with the strongest 3-year paroxysm of earthquakes and tectonic movements known in historic time. In all, 12 or 13 predominantly explosive eruptions were recorded, the strongest of which occurred in 1737, 1779 and 1945. Only in three or four cases were small lava flows poured out. The maximum volume of pyroclastics (0.25 km^3) was erupted in 1945. During the 1737 and 1779 eruptions, 0.05-0.10 km^3 of pyroclastics was erupted, and in other cases less than 0.05 km^3 was erupted. The mean production rate of the volcano over a period of 400 years (1550-1945), which includes the duration of the preceding repose period (1550-1737), was approximately 2.5 x 10^6 t $year^{-1}$. The production rate of the previous stage (1200-1550) was about 2.1 x 10^6 t $year^{-1}$, approximately the same (2.2 x 10^6 t $year^{-1}$) as that during the early life of the young cone. It is characteristic that the final eruptions of each period were the strongest, and that their juvenile products were the most homogeneous and basic in composition. This permits us to suggest that the Avachinsky eruption of 1945, which was the largest in historic time, and whose juvenile material was represented by homogeneous basaltic-andesite scoria, was also the final eruption of the recent intensified period of activity.

Based on the reconstructed eruptive history of the Avachinsky young cone over the last 3000-3500 years, on the distinct rhythm of its volcanism and on a stable production rate, we assume that in the next several 100 years the character of its activity will not change significantly. The period of relative calm which started in 1945, is likely to continue for a long time. It will take 100 or 200 years of dormancy to prepare for the eruptions of the next intensified period of activity. The character and the power of future eruptions in this next intensified period should resemble the eruptions of the most recent similar periods of activity, i.e. explosive and effusive-explosive eruptions with a total volume of rocks from 0.001-0.01 to 0.2-0.3 km^3. Volcanic hazards will also be similar, i.e. lava flows only on the flanks and near the foot of the active cone, and tephra up to a radius of 10-15 km from the crater. An intense ash fall may be observed in a narrow (30-60°) sector, to distances of 40-50 km from the crater, provided the width of the eruptive cloud is small and the wind is strong. The probability of occurrence of lahars will be high. The most powerful lahars will pass, as before, along the valleys of the Elizovo and Khalaktyrka "dry" rivers.

However, during the inter-eruptive period, small phreatic and phreatomagmatic eruptions may take place at irregular intervals for incidental reasons (e.g. because of plugging up of the vent by debris and ice as a result of the collapse of the crater wall). Hazards produced by these eruptions will be limited to the foot of the young cone.

3 Volcanoes at the Stage of Slight Activity, and Destruction of Their Edifices

As distinct from the volcanoes of stage I, the activity of volcanoes of stage II is manifested less regularly. The duration of their repose periods is often very long and incommensurable with the duration of their intensified periods of activity, which may consist of single eruptions or a short series of eruptions.

Mutnovsky volcano is of special interest in this group because of the plan to utilize its geothermal energy potential. Mutnovsky was formed by four coalescing stratovolcanoes, built up in the second half of the Late Pleistocene age. They are composed predominantly of basalts. The maximum height of the volcano is 2324 m. Its recent activity is confined to a large double crater, shaped like a figure eight, 3 x 2 km in size, which is complicated inside by a system of smaller craters, 0.3-0.6 km in diameter.

During the last 9000-10,000 years, activity has been manifested only in the summit crater, with predominantly phreatic eruptions, no lava effusions and no ejections of large volumes of juvenile pyroclastics. This indicates that the volcano has long ago passed out of the stage of active growth, and that it is now at the stage of waning activity. The average production rate of the volcano in the Holocene, $0.1\text{-}0.2 \times 10^6$ t $year^{-1}$, has declined by a factor of 50-100 from that of the Late Pleistocene, which was $9\text{-}10 \times 10^6$ t $year^{-1}$.

During Holocene times, explosive activity characterized the volcano. Slight to moderate phreatic and phreatomagmatic eruptions, with volumes of the order of $10^6\text{-}10^7$ m^3, predominated, separated by repose periods of a few 100 years, or less frequently, a few tens of years. The juvenile material consisted of low-potassium andesites. This type of activity has continued in historic time, since the 17th century. Pyroclastics of these eruptions correspond compositionally to calc-alkaline basalts and basaltic-andesites. The general eruption frequency was one in 5 or 6 years, and moderate eruptions occurred at intervals of 50-100 years.

It is likely that the anomalously low production rate of juvenile material at Mutnovsky, compared with that at other active volcanoes, is accounted for by the fact that the main part of the energy of its magma chamber is released by steam from the crater and thermal waters. Their total capacity is about 500000 kcal s^{-1} (Vakin et al. 1976). Such speculations agree well with a wide development of hydrothermally-altered rocks in the crater and on the flanks of the volcano, and are also supported by the predominantly phreatic style of its Holocene eruptions. The recent heat capacity of the volcano exceeds, by a factor of 15-20, the average heat capacity of other active volcanoes of the Kurile-Kamchatka region which originated at about the same time, and which are at the same stage of development (Melekestsev 1980; Polyak and Melekestsev 1981). Such a discrepancy may only be accounted for by the fact that the volcano is in a state of prolonged minor eruption, which has already lasted for 25 years.

A prediction of future volcanic activity of Mutnovsky may be made, based on data available, and leads to the conclusion that three types of activity are equally probable in the next 50-100 years.

Firstly, the present style of activity at the volcano, with the energy released constantly by intensive heat liberation from steam and thermal water, may last for an uncertainly long period of time, perhaps a few tens of years.

Secondly, phreatic explosions, the energy of which will be predetermined by the strength of the roof and by the duration of the plugging up of the channel, will occur with different frequency and power. Most eruptions will not be hazardous at sites 3-5 km from the crater, because the volume of ejected material will scarcely exceed 10^7 m^3. The only exceptions may be directed phreatic explosions, which have taken place repeatedly in the past, when the material is ejected at a low angle to the horizon, and is deposited in a narrow sector, 30-90° wide. The most hazardous zone in this respect is the NE sector of the foot of the volcano, where the crater rim is lower than elsewhere.

Thirdly, we cannot entirely exclude the possibility of a moderate eruption with ejections of juvenile pyroclastics. Such an eruption would probably resemble the eruptions of the 17th-19th centuries, with the total volume of the material amounting to about 2-5 x 10^7 m^3 (Fig. 8).

Each eruption may itself be the trigger for accompanying phenomena, including lahars, sector collapses and landslides, which

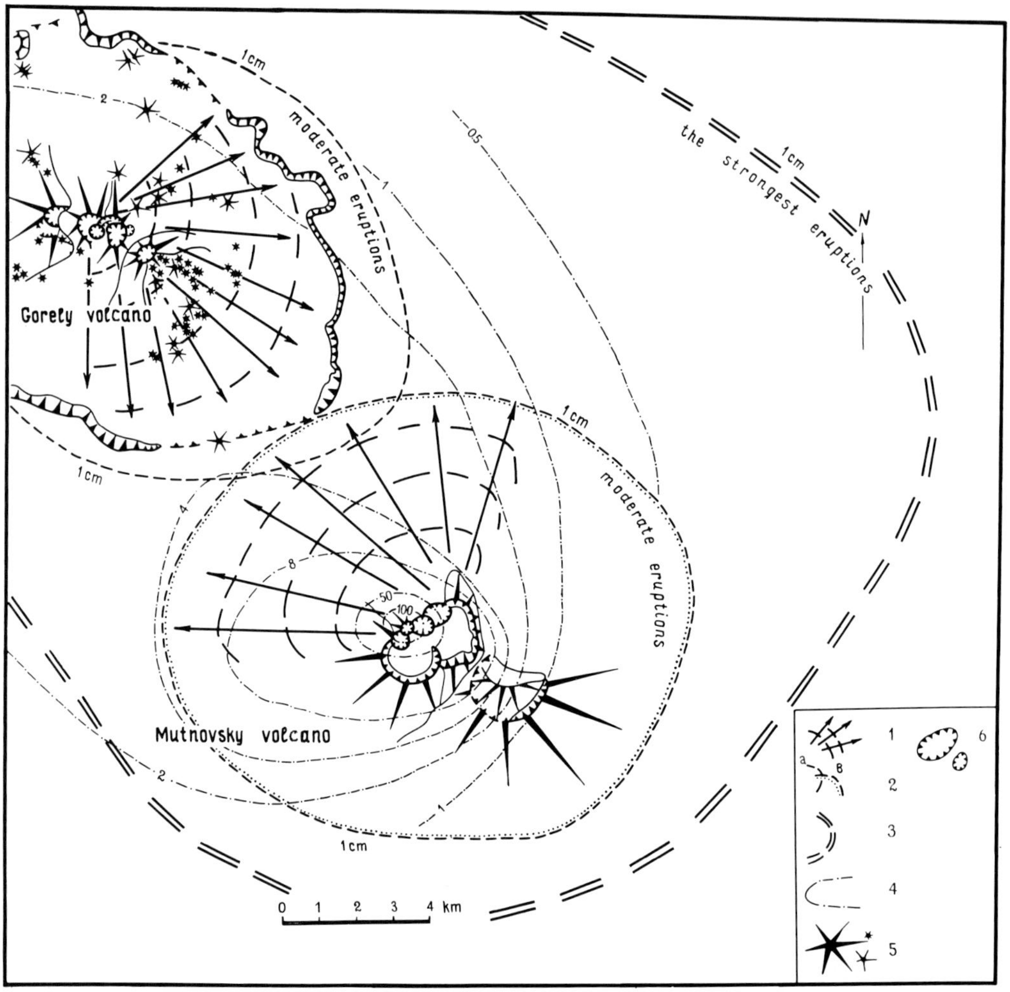

Fig. 8. Scheme showing the future hazards which may be produced by Mutnovsky and Gorely volcano eruptions. **1** Possible sectors of directed volcanic explosions; **2** 1-cm isopachs of tephra during moderate (10^7 m^3) eruptions: **a** Gorely, **b** Mutnovsky; **3** 1-cm isopachs of tephra during strong eruptions (10^8 m^3); **4** isopachs of tephra of the 1848-54 Mutnovsky eruption; **5** cones of volcanic edifices; **6** Gorely caldera and craters of both volcanoes

have taken place repeatedly during the past eruptions of the volcano (Melekestsev and Braitseva 1984). The distribution of likely future lahars and slope failures is shown in Fig. 9.

The Gorely volcano lies to the northwest of Mutnovsky. Now it is in the period of the recent cycle of activity. Its next eruptions may be hazardous for the site where the Mutnovskaya geothermal power station is under construction. This aspect is discussed further below.

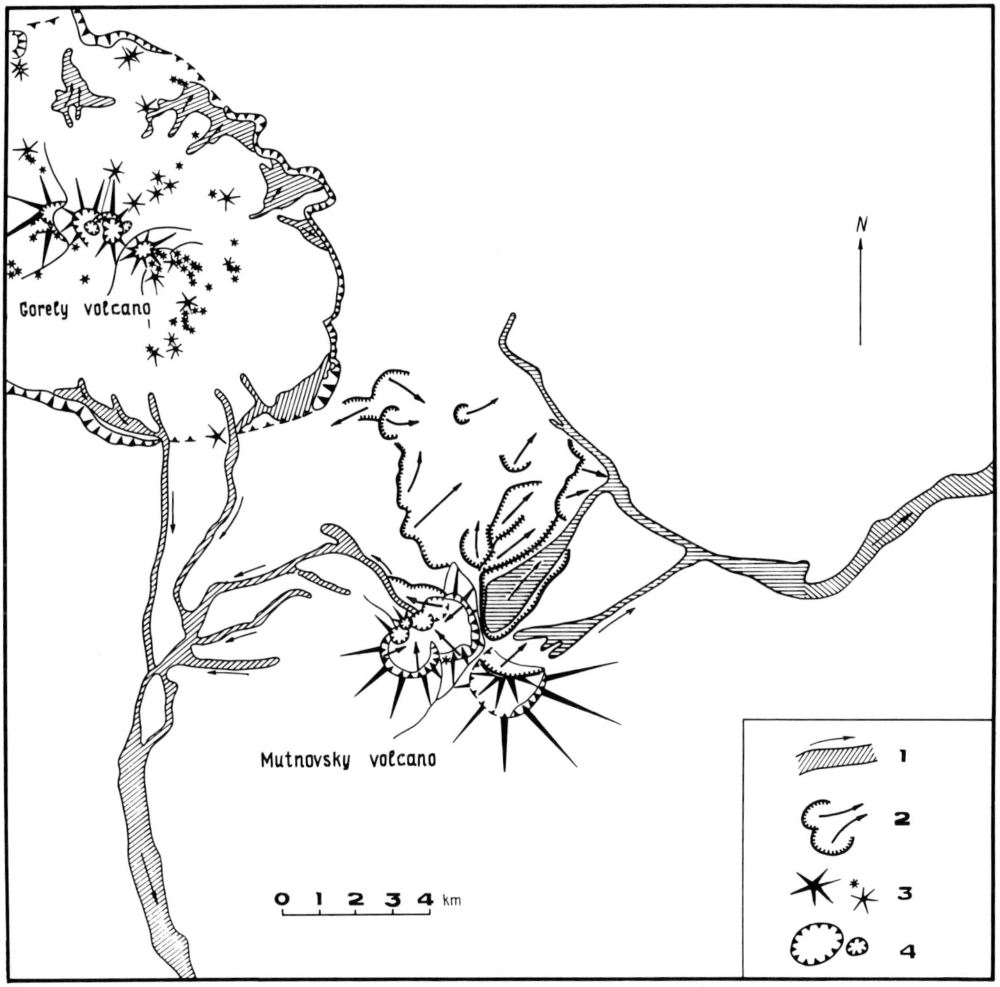

Fig. 9. Scheme showing the future hazards which may be created by lahars and volcanic-seismotectonic landslides in the region of Mutnovsky and Gorely. 1 Probable routes of lahars; 2 walls of landslide cirques, arrows indicate possible directions of movement of landslide masses; 3 volcanic edifices; 4 calderas and craters

Gorely is a volcanic ridge 7 km long WNW trending, consisting of five small coalescing stratovolcanoes. Its latest eruptions occurred in 1980-81 and 1985. The ridge is located in a large (about 100 km^2) oval caldera of the Krakatoa type, which was formed 40,000-50,000 years ago after the ejection of 100-150 km^3 of dacitic pyroclastics. The volcano itself is composed of basalts, basaltic-andesites (predominant) to dacites (insignificantly developed). The maximum height of the volcano is 1829 m, its base is at altitudes of 900-1100 m. In the near-summit part there are many craters of different sizes and ages, including the active crater (approximately in the middle of the ridge).

Study of Gorely's activity in the Holocene shows that it has long ago passed through the stage of active growth. This stage was notable for intense explosive activity, with a series of more powerful eruptions (layers of coarse scoria) and lava flow effusions. During the last 2000 years the volcano entered the stage of waning activity: effusive activity almost ceased and explosive activity was of moderate character, with the predominance of slight (10^6 m^3) and moderate (10^7 m^3) eruptions separated by a few years and 100-year intervals. During the historic stage the frequency of eruptions was one each 4-60 years but eruptions, which are marked by tephra layers and which might represent a volcanic hazard, occurred once each 50-100 years.

There are no grounds to expect significant changes in the activity of the volcano in the near future. Consequently, in the next 50-100 years a few eruptions may also take place. These eruptions will occur either rather frequently (each 5-10 years) and will be slight, of the 1985 and 1869 eruption type, and non-hazardous or they will be moderate of the 1828-32, 1929-31, 1980-81 eruption types and will occur more likely by the end of the predicted time span. The eruption energy and the amount of ejected material will apparently be limited by the duration of the repose period before the eruption. Using data on the average production rate (about 1.5×10^6 t $year^{-1}$) and the average power (70,000-170,000 kW) of the volcano in the Holocene, one can estimate approximately the mass of erupted juvenile material and the energy of the next eruptions for the repose periods lasting for 10, 50 or 100 years. They will be correspondingly 10^7 t and 1.7×10^{23} erg, 5×10^7 t and 8.5×10^{23} erg, 10^8 t and 1.7×10^{24} erg. The total amount of the erupted solid products may increase 1.5-2 times due to resurgent material, i.e. up to $1.5\text{-}2 \times 10^7$ t, $7.5\text{-}10 \times 10^7$ t and $1.5\text{-}2 \times 10^8$ t. The comparison, during the Gorely eruption in 1980-81, which occurred after a 50-year repose period, of the mass of solid products, according to our estimates, did not exceed, apparently, $2.5\text{-}3 \times 10^7$ t, their volume being about 0.02 km^3. Consequently, the mass of the products produced by this eruption was two to three times less than that calculated above and the continuing ejection of the material during the 1985 eruption is apparently related to this.

The explosive eruption of Gorely in 1980-1981 was not hazardous for the environment and properties in a radius of more than 4-5 km, where the total thickness of tephra fall did not exceed 1-2 cm. Increase in eruption energy of three to six times could increase the radius of the dangerous zone by a few kilometers (see Fig. 8). However, in this case also the thickness of tephra fall at the site of the proposed construction of the Mutnovskaya geothermal power station will unlikely be more than 3-5 cm (ash load 40-80 kg m^{-2}).

It is difficult to make any prediction of future activity for most volcanoes at stage II, especially if their eruptions over the period of the last several 1000 years have been few and irregular. An example of this kind of volcano is **Khodutka**, which, because of lack of data on its historic eruptions, was not included in the Catalogue of the Active Volcanoes of Kamchatka (Vlodavets and Piip 1959). This is a large stratovolcano in the south of the peninsula, composed of rocks of basaltic andesite-andesite composition. Its volcanic activity was manifested most intensively at the end of the Late Pleistocene and in the Early Holocene. Thereafter, during a long period of time, the volcano showed no activity. Deep barrancos began to dissect its surface, and erosion has now almost destroyed its summit crater. These features indicate that the volcano is at stage II.

However, tephrochronological and geological-geomorphological studies have established that strong explosive eruptions occurred near the foot of the volcano comparatively recently. The penultimate of these eruptions occurred approximately 2900-2800 years ago, and the last of them 1900-1800 years ago (Kirsanova and Melekestsev 1984). The volume of tephra of the most recent eruption was not less than 2-3 km^3. Eruptive centers are represented at Khodutka by chains of large explosive funnels, up to 1-1.5 km across. Discharge of thermal waters, at a temperature of up to 88°C, occurs from one of these explosive funnels.

All this suggests that the volcano should be considered potentially hazardous. The next eruption is most likely to occur within the same area at the north-western foot of the volcano, and, since it will follow a long period of quiescence, it is likely to be strong. At present it is not possible to make a more accurate prediction of the character, place and parameters of the next eruption.

In Kamchatka, there are also other potentially hazardous volcanoes which are at stage II. These are Koryaksky, Kronotsky, Opala, Koshelev, Plosky Dalnii, at all of which strong, or even catastrophic eruptions may occur. To make a prediction of these eruptions, and their associated volcanic hazards, is extremely difficult or indeed impossible.

Finally, we stress that estimations of future volcanic activity, and the type and parameters of eruptions, made on the basis of detailed reconstruction of the eruptive history of a volcano, should be the main elements in compiling all kinds of maps and schemes of volcanic hazards, as well as in making long-term predictions. However, they play a subsidiary role when precise prediction of the time of eruptions is made. This is only possible on the basis of direct volcanological, geophysical and geochemical observations of the state of volcanoes. In short, it must be based on a complex of special studies, which are united by the general term "monitoring" of volcanoes.

4 Conclusions

1. For long-term prediction of volcanic hazards, of decisive significance is the elucidation of the trends in the dynamics of eruptive activity and the production rate of a volcano. Ba-

sed on these data, conclusions may be drawn on the number, frequency, type and energy of eruptions which may occur during the present period in the life of the volcano.

2. Reconstruction of the dynamics of volcanic activity is made on the basis of tephrochronological and detailed geological-geomorphological studies, by means of ^{14}C dating. Regularities in the cyclic regime of volcanic activity are established by distinguishing the intensified periods of activity and the repose periods. A knowledge of the duration of such periods for the past cycles, and of the production rate of a volcano at different stages, makes it possible to determine the approximate time and parameters of future eruptions.

3. Groups of volcanoes at stage I (high and moderate levels of activity and growth of the edifice), and at stage II (low level of activity and onset of the destruction of the edifice) have been recognized. Most active volcanoes belong to stage I. Kliuchevskoi, Bezymianny, Karymsky and Maly Semyachik are in the intensified period of their recent cycles. Kliuchevskoi and Karymsky will erupt as at present for several 100 years, and the activity of Bezymianny, characteristic of the period of the 1960-1980s, will persist for 100-200 years. Maly Semyachik may erupt violently in the immediate future.

Avachinsky, Krasheninnikov, Kikhpinych and Kizimen are in a period of relative calm of the recent cycle. This period, for the first three volcanoes, will last for several 100 years, but for Kizimen, on the other hand, a catastrophic eruption is forecast to occur in the next 50-100 years.

4. It is difficult to make a prediction of activity for volcanoes at stage II. In some cases, Mutnovsky for example, prediction can be made with a significant degree of probability. In other cases, when the eruptions of a volcano in the Holocene are few and irregular, such prediction becomes uncertain or impossible.

References

Braitseva OA, Kiryanov VY (1982) On the past activity of Bezymianny volcano as shown by tephrochronological studies. Volcanol Seismol 6: 44-56 (in Russian)

Braitseva OA, Litasova SN (1982) On the age of Karymsky and Maly Semyachik volcanoes. Volcanol Seismol 3:92-96 (in Russian)

Braitseva OA, Egorova IA, Nesmachny IA, Selyangin OB, Sulerzhitsky LD (1978) Tephrochronological studies as the method of studying the regularities of cyclic development of volcano. Bull Volcanol Stat 54:41-53 (in Russian)

Braitseva OA, Egorova IA, Sulerzhitsky LD (1980a) Karymsky volcano. In: Masurenkov YuP (ed) Volcanic centre: structure, dynamics, substance (Karymsky structure). Nauka, Moscow, pp 243-254 (in Russian)

Braitseva OA, Sulerzhitsky LD, Egorova IA (1980b) Tephra stratigraphy and radiocarbon dating. In: Masurenkov YuP (ed) Volcanic centre: structure, dynamics, substance (Karymsky structure). Nauka, Moscow, pp 90-100 (in Russian)

Braitseva OA, Sulerzhitsky LD, Litasova SN, Grebzdy EI (1984) Radiocarbon dates of Holocene deposits of soil-pyroclastic covers of the Kliuchevskoi volcanic group. Volcanol Seismol 2:110-115 (in Russian)

Gorshkov AP, Grebzdy EI, Samoilenko BI, Slezin YB (1975) On calculation of heat and mass balance of the crater lake of Maly Semyachik volcano. Bull Volcanol Stat 51:43-58 (in Russian)

Gorshkov GS, Bogoyavlenskaya GE (1965) Bezymianny volcano and peculiarities of its eruption in 1955-1963. Nauka, Moscow, 171 pp (in Russian)

Kirsanova TP, Melekestsev IV (1984) On the origin and age of the Khodutkinskie thermal waters. Volcanol Seismol 5:49-59 (in Russian)

Melekestsev IV (1973) Types and age of active volcanoes of the Kurile-Kamchatka zone. Bull Volcanol Stat 49:17-23 (in Russian)

Melekestsev IV (1980) Volcanism and topography formation. Nauka, Moscow, 212 pp (in Russian)

Melekestsev IV, Braitseva OA (1984) Gigantic collapses at volcanoes. Volcanol Seismol 4:14-23 (in Russian)

Melekestsev IV, Kiryanov VY (1984) When will the Avachinsky volcano erupt in Kamchatka? Volcanol Seismol 6:107-111 (in Russian)

Piip BI (1956) Kliuchevskoi volcano and its eruptions in 1944-45 and in the past. Proc Labor Volcanol 11:308 (in Russian)

Polyak BG, Melekestsev IV (1981) Production rate of volcanic apparatuses. Volcanol Seismol 5:22-37 (in Russian)

Rose WI Jr (1972) Notes on the 1902 eruption of Santa Maria volcano, Guatemala. Bull Volcanol 36:29-45

Selyangin OB, Braitseva OA, Egorova IA, Sulerzhitsky LD, Nesmachny IA (1979) Geological and tephrochronological studies of recent volcano. In: Sobolev VS (ed) Problems of deep magmatism. Nauka, Moscow, pp 31-49 (in Russian)

Vakin EA, Kirsanov IT, Kirsanova TP (1976) Thermal fields and hot springs of the Mutnovsky volcanic region. In Sugrobov VM (ed) Hydrothermal systems and thermal fields of Kamchatka. Vladivostok, pp 85-114 (in Russian)

Vlodavets VI, Piip BI (1959) Catalogue of the active volcanoes of the world including solfatara fields. Part VIII Kamchatka and continental areas of Asia. International Volcanological Association, Napoli, 110 pp.

Study Models for Volcanic Hazards in Iceland

P. IMSLAND[1]

Abstract

The studies of volcanic hazards in Iceland are young and yet mostly qualitative. The volcanic zones of Iceland cover nearly 30% of its area and comprise both rift zones and non-rifting areas. These differ fundamentally in nearly all aspects of the volcanic geology. In contrast to most other volcanic regions of the dry land of the world, Icelandic volcanism is dominated by effusive basalt eruptions and consequently land use is exposed to high risk of lava flows. The reversed situation, that of explosive volcanism and a high risk from explosion materials and lahars, is the general case in most other areas, including those where volcanic hazards studies have most been practiced and its methodology established. The basalt-producing fields of Iceland do not have clearly defined centers of activity as the volcanic cones of the continents and the island arcs have. The origin of the hazards in Iceland is thus more poorly known ahead of events than in the cone cases. This means that the study models and presentation of hazards, already established in practice, cannot be immediately applied to the Iceland situation, especially not in case of its rift zones. New methods must be found, and apparently the general quantification of the volcanism and its behaviour patterns in Iceland are a premise for the evolution of a useful study and presentation model of volcanic hazards in Iceland.

1 Introduction

Iceland is a part of the worldwide oceanic rift system. It is produced by the same processes, from the same materials, and has the same gross structures as the oceanic crust. However, the productivity of Icelandic volcanism exceeds that of the surrounding ocean floor. Therefore an island has been formed, where the oceanic rift system is exposed on dry land with its volcanism, tectonics and other active geological processes. Iceland is thus an oceanic bottom in all fundamental respects and, being easily accessible to study, its generating processes can shed much light on the rest of this global feature.

---

[1] Nordic Volcanological Institute, University of Iceland, 101 Reykjavik, Iceland

IAVCEI Proceedings in Volcanology 1
J. H. Latter (Ed.), Volcanic Hazards

Other volcanic islands in the oceans are not direct parts of the rift system, not even Hawaii. Their volcanism is by nature different from that in Iceland. The volcanism on the continents of the globe differs still more from Icelandic volcanism, with the exception of the Afar region in eastern Africa, where the continental rift has already given rise to a local oceanic crustal block within the continent.

If we, in this context, exclude the unobservable and generally non-hazardous submarine volcanism, Icelandic volcanism is in many respects quite different from that of the rest of the world. As a hazardous phenomenon and a threat to the inhabitants and general land use, this poses some problems. At present the most serious one is the lack of models for its study and presentation, i.e. a methodological problem.

Before dealing with the methodology of hazards and risk studies in general and with special emphasis on the Iceland case, a summary of Icelandic volcanism is given. Volcanic hazard studies in general are then briefly commented upon and the hazardous aspects of Icelandic volcanism accounted for. This is followed by a short and simplified description of the study models or methods which have been used throughout the world. This description reveals how these models are unfit for general use in the Icelandic situation. Finally, an approach to the Icelandic case is presented, which may be expected to yield results.

2 Icelandic Volcanism

In Fig. 1 the volcanic fields of Iceland are shown. Taking notice of the distribution of earthquakes, surface tectonic features, types of rocks, volcanic style, frequency of eruptions, productivity of the volcanism and other such aspects of the active geology, as well as some general stratigraphical and structural facts, it is clear that there are fundamental differences between the various volcanic fields of Iceland. The relatively complex picture thus obtained of the active zones of Iceland calls for an explanation. The geographical arrangement of the volcanic fields alone is in opposition to a simple model of oceanic rifting through Iceland. It is, nevertheless, generally agreed among students of Icelandic geology, that this variation is best explained within the model of global tectonics.

The most recent views divide the volcanic fields into two broad categories, <u>rift zones</u> and <u>non-rifting areas</u> (see Fig. 1). Both categories exhibit some internal variation, but this is minor compared to the differences between the major categories themselves, and will not be discussed here. The main differences between the rift zones and the non-rifting areas lie in volcanotectonics, petrochemistry and volcanic productivity (to avoid misunderstanding, the major components of these terms are here accounted for as follows: volcanotectonics refers mainly to the stress pattern in the crust and the resulting earthquakes, fractures and relative movements of crustal blocks which are associated with the volcanism; petrochemistry refers to the

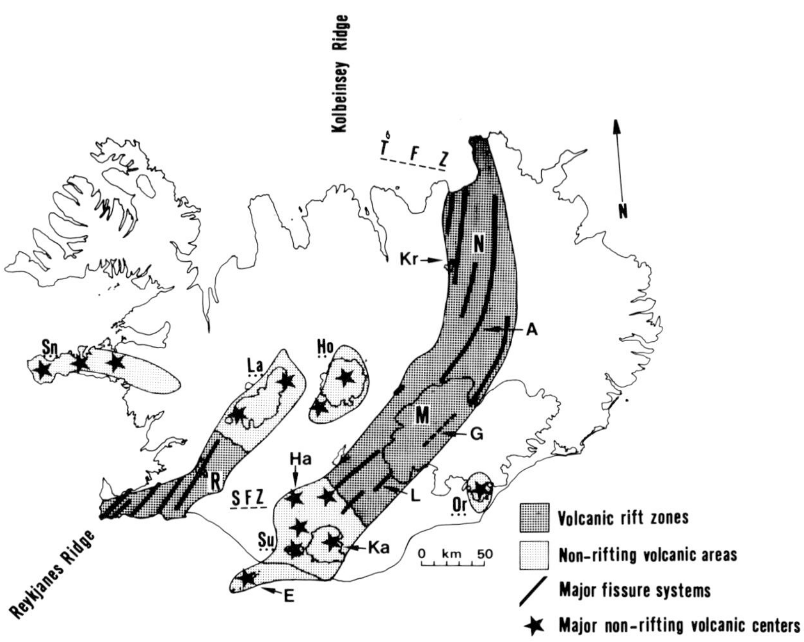

Fig. 1. The volcanic fields of Iceland and their volcanotectonic relationships. The rift zones are offset by two fracture zones: **SFZ** Sudurland Fracture Zone and **TFZ** Tjornes Fracture Zone. There are five non-rifting volcanic areas: **Sn**aefellsnes, **La**ngjokull, **Ho**fsjokull, **Su**durland and **Or**aefajokull, as indicated by the first two letters in the names. The rift zones, indicated by the first letter only, are: **N**ordurland, **M**idisland and the **R**eykjanes Peninsula. The letters: **Ha**, **E**, **L**, **G**, **A**, **Ka** and **Kr**, accompanied by an **arrow**, point towards the individual volcanoes mentioned in the text, these are: Hekla, Eldfell, Laki, Grimsvotn, Askja, Katla and Krafla respectively

general chemical characteristics, compositional variation and evolution of the magmas produced by the volcanism; volcanic productivity refers to the variation of frequency and sizes of eruptions and the total volume and volume relationships of erupted materials of volcanoes or volcanic areas, through specified periods).

The volcanotectonics of the rift zones is characterized by a long-linear parallel arrangement of structures. The eruptions typically occur on long fissures which tend to group into units, together with both the vertical faults and open fissures of the rifting movements. These units of parallel fissures are called fissure swarms. The fissure swarms tend to be arranged in an en echelon manner along the rift zones as shown by Fig. 1. For more detailed accounts of the general Icelandic volcanology and related geological features, including those described here, see Saemundsson (1978), Einarsson and Bjornsson (1979) and Imsland (1983).

Each fissure swarm appears to have an activity or life span of the order of 1 million years. During the early part of this life span the activity is characterized by an intense fissure volcanism producing so-called primitive basalts (mostly olivine-normative), which belong to the tholeiitic rock suites. As magmas these are relatively hot and fluid and the eruptions are dominantly of an effusive nature, with a weak explosive phase (or lacking such a phase). These basalt lavas are piled up on top of the preceding ones but are frequently, in spite of their high fluidity, hindered in flowing far away by the block-faulting graben tectonics associated with this early volcanic activity. Thus this fissure volcanism and its associated plutonism produce the entire oceanic crust at the rift zones of Iceland and the rest of the globe, the constructive plate boundaries. As a fissure swarm evolves, it tends to produce more evolved lavas (mostly quartz-normative basalts, but to a slight degree more silicic rock types, which also belong to the tholeiitic rock suite). This means relatively less fluid and slightly cooler lavas, with slightly higher explosivity, and more tephra production. At the same time the volcanic activity tends to concentrate on the central regions of the fissure swarms and thus to build up more pronounced topographic height, while the faulting process becomes less apparent. Thus a fissure swarm evolves towards a central volcano or a volcanic complex, with concentrated activity, increased evolution of products and decreased linear tectonic movements. As an example of such an evolved volcanic complex on the rift zone we can take Askja in central-northern Iceland (for location see Fig. 1), while the corresponding, less evolved fissure swarm stage may be exemplified by the Reykjanes Peninsula fissure swarms (location on Figs. 1 and 11). This volcano evolution is generally accompanied by a considerable decrease in the amount of volcanic material erupted in each eruption, and further it is apparently accompanied by an increase in eruption frequency. At this evolved stage a magma chamber or a magma reservoir complex is generally believed to have evolved in the crust at a relatively shallow depth below the surface of the complex. This is strongly indicated by general surface observations and geophysical measurements.

To obtain independence of the time-related concepts, fissure swarm and central volcano (or volcanic complex), both concepts are by some authors included in a new one, that of a volcanic system.

The volcanotectonics of the non-rifting areas are, by contrast, not characterized by a long-linear arrangement of features. Tectonic movements are generally less prominent than in the rift zones. The volcanism is discordantly superimposed on top of older rock formations and thus does not build up the entire crust as it does in the rift zones. The volcanism is to a large extend bound to major volcanoes or complexes, which may bear some resemblances to the great volcanic cones of the world, regarding both the pattern of activity and topographic evolution. As an example of such a volcano on the non-rifting areas, we can take Hekla (location shown on Fig. 1). The magma types of these volcanoes are more variable and the eruptive behaviour

Table 1. Icelandic Volcanism - Summary of Characteristics

	Volcanic rift zones	Non-rifting volcanic areas
Tectonics	Long-linear. Regional rifting and subsidence	Short-linear. Non-rifting
Stratigraphic relations to crust	Concordant production of entire crust	Discordant superimposition on older crust
Volcanic style	Overwhelmingly effusive eruptions	Typically mixed eruptions with highly explosive cases
Magma types	Tholeiitic magma series	Non-tholeiitic magma series. Heterogeneous suites
Areal relations of volcanic field	18% of Iceland 60% of volcanic zones	11% of Iceland 40% of volcanic zones
Productivity	High: 77.5% of total	Low: 22.5% of total

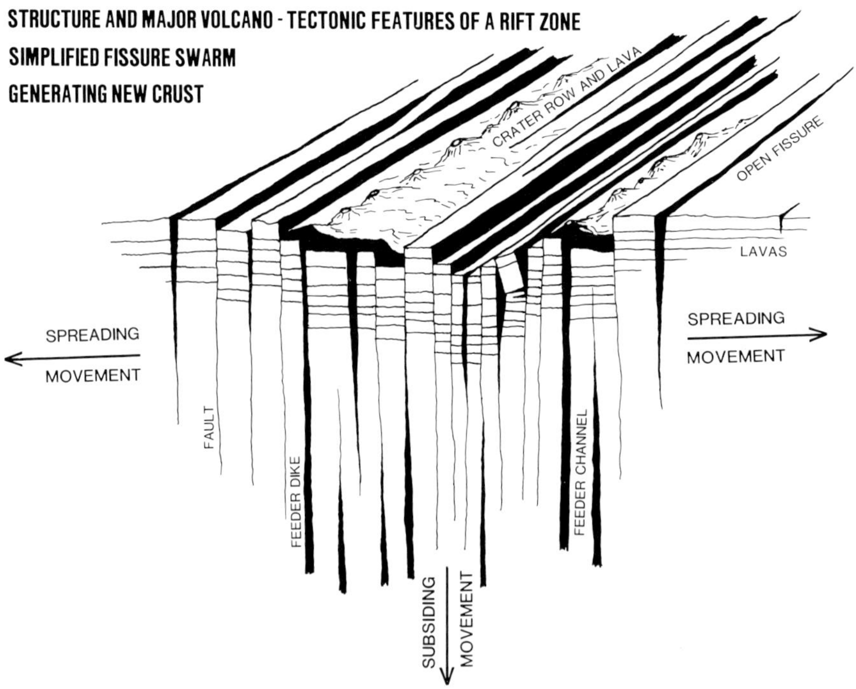

Fig. 2. The major volcanotectonic features of the surface and uppermost part of a rift zone volcanic system, while still in the early fissure swarm stage of evolution, shown schematically. The **arrows** show the movements of the crustal blocks, which result in a graben structure. The process of magma injection and extrusion, here depicted as dikes and lavas, is responsible for the generation of the entire crust, which is modified by tectonic movements and hydrothermal and metamorphic processes to obtain its final constitution

shows a complex pattern of effusive and explosive activity. Generally the magmas are relatively cold and viscous, and lavas do not therefore flow as easily as they typically do on the rift zones. On the other hand, the magmas seem to be relatively gas-rich. All this tends to increase the explosivity of the eruptions. In between the major volcanoes of the non-rifting areas, basaltic eruptions occur, but typically these take place on very short fissures or at single craters, and are of mixed eruption types. Petrochemically, the non-rifting volcanism is heterogeneous compared to the rift zone volcanism, and of non-tholeiitic compositions (transitional to alkalic, even calc-alkalic affinities). The productivity of the various non-rifting areas varies greatly, but in general it is much lower than in the case of the rift zones.

In a simplified manner the differences between the rift zones and the non-rifting volcanism are summarized in Table 1. Figure 2 shows a schematic three-dimensional view of the rift zone, indicating its concordant structure and continuous formation. For comparison, Fig. 3 shows a schematic section through a volcanic cone, depicting its discordant superimposition upon an older crust. It may be taken to represent both a non-rift Icelandic volcano and the subduction related volcanic cones.

STRUCTURE AND MAJOR FEATURES OF A VOLCANIC CONE SUPERIMPOSED ON AN OLDER CRUST

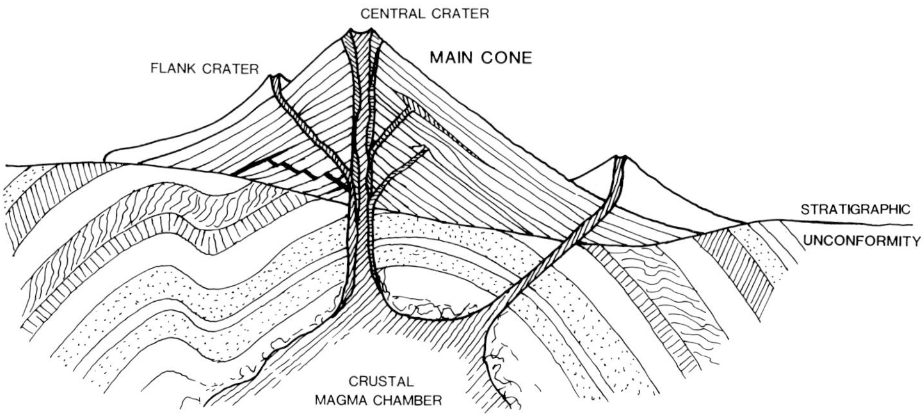

Fig. 3. Schematic presentation of a volcanic cone superimposed upon an older crust which was formed by a different process or at a different time. This is, generalized, the case with the volcanoes of the non-rifting volcanic areas of Iceland. Comparison with Fig. 2 should thus clearly indicate the great differences in the overall situation. No tectonic movements are indicated on the figure to stress that these are minimal and insignificant compared to the rift zone case

3 Volcanic Hazards and the Hazardous Aspects of Icelandic Volcanism

Volcanic hazard studies aim at gaining knowledge, as quantitative as possible, on volcanism and its interference with man's use of land. This aims at forecastings and predictions of where, when, how, and how much it will erupt in specific places, in order to allow the use of that land insofar as it is safe to do so. In such cases safe means safe for both people and property, and totally safe or safe within the risk limits chosen in each case.

The volcanic risk is more dependent on the amount and nature of land use in the volcanic region than on the volcanism itself. On the other hand, the hazard of the volcanism is in general primarily dependent on the physico-chemical status of the magmas, mainly on the temperature and viscosity, the silica content and the amount of gases they contain. How individual eruptions behave, and thus the degree of their hazards and violence, depends on these physico-chemical characteristics, on the tectonic situation at the eruption site, on the amount and rate of flow of magma and on hydrological factors.

Together with quantitative knowledge of the volcanic history of the area concerned, these parameters allow the threat of the volcanism to be estimated. Typically these hazard determining parameters tend to have a general pattern or a characteristic range of variation for individual and definable areas. Accordingly, after a preliminary study of the volcanic geology and volcanic history of an individual volcanic area, it is possible to describe in a generalized way which hazardous phenomena are most likely to occur, and to some extent to list them in order of importance.

In short, in continental and island arc volcanic areas, the physico-chemical parameters of magmas lead to explosive volcanism, relatively cold and viscous silicic magmas of relatively high gas content. In the oceanic environment they lead to effusive volcanism, relatively hot and fluid basaltic magmas of relatively low gas content. With this change in the physico-chemical nature of the magmas the hazardous aspects of the volcanism change drastically. The traditional classification of eruptions (for example see Macdonald 1972) from Icelandic and Hawaiian types through Strombolian and Vulcanian, etc. to Plinian and Peléan, primarily describes the eruption behaviour, reflecting their relationships to the physico-chemical nature of magmas. In Table 2 it is shown how such a classification of eruption behaviour can be related in a general way to major hazard factors and the relative amount of risk.

Table 2 is both a modification and extension of Whittow's (1980) scheme, which primarily accounts for the risk resulting from the explosive phenomena of the volcanism. The extension of the scheme comprises the addition of the risk resulting from lava flow hazard, which generally is reversed compared to the

Table 2. Explosive gas release and fluidity of magma and the resulting eruption behaviour related to volcanic hazard and risk

HAZARD FROM LAVA FLOW			HIGH RISK ⟷ LOW RISK		
	HAZARD FROM TEPHRA FALL, TEPHRA FLOW AND OTHER EXPLOSIVE PHENOMENA		LOW RISK ⟷ HIGH RISK		
			EXPLOSIVE GAS RELEASE		
		MAGMA FLUIDITY	LOW	MEDIUM	HIGH
HIGH RISK	LOW RISK	FLUID	PUU OOAN	BORGAN	FUJIAN
↕	↕	INTERMEDIATE	BUNGAN	STROMBOLIAN	VESUVIAN
LOW RISK	HIGH RISK	VISCOUS	SHOWA SHIN-ZANIAN	VULCANIAN	PLINIAN PELEAN

NB. Does not include hydro-magmatic or phreatic activity.

risk from the "airborne" hazards. The modification concerns the eruption classification scheme itself. As it contains some new type references, these must briefly be accounted for. Instead of using the traditional Icelandic and Hawaiian types of eruptions in the scheme, the tendency towards differences in common eruption behaviour of the basalt eruptions of these two countries (and others) is stressed in this modification, by giving new references. In Hawaii the very quiet effusion of fluid lava from a pond or a lava lake is very common after a preceding phase of fountains and curtains of fire. Such ponding is extremely scarce in Iceland, where the most common eruption type is one that takes place on a crater row. The fissure that opens up starts with fountaining and continuous or en echelon arranged curtains of fire, as in the Hawaiian case. As this initial phase ceases, the eruption continues as more or less continuous low-energy explosions on separated craters, which throw up scoria and spatters accompanying more quiet outflow of lava through an eroded pass or tunnel in the crater walls. By this mechanism a row of craters is formed along the fissure, where crater mounds of scoria and spatter (though not amounting to the Strombolian cinder cones) are built around individual vents by these degassing explosions. The basaltic fissure eruptions in Iceland are thus in many cases not as quiet as these eruptions may be in Hawaii, i.e. the gas release is more explosive. These two types of eruptions are given here, new names to stress their characteristics instead of using their traditional country reference, which stresses different aspects, and which in this context may be misleading as the eruption behaviour is not controlled by the geographical location of the volcano. Puu Ooan is used here, as a name for the quiet lava pond activity, the case of least explosive gas release of fluid magma. The Puu Oo lava pond, Hawaii, was active at the time this was written. Borgan is used here as a name for the scoria crater mound activity so typical for Iceland, the case of somewhat more explosive gas release of fluid magma. The name comes from the Icelandic word borg (pl: borgir, literally = fortress), which is very common as an ending of the local names of craters and crater rows formed by this type of activity (examples: Eldborg, Vikraborgir, Threngslaborgir). The eruption of a relatively fluid magma accompanied by highly explosive gas release is an apparently rather uncommon phenomenon. Mt. Fujisan in Japan fits this description according to Tsuja (1955). Here, this category is thus called Fujian. The likewise apparently uncommon category of intermediately viscous magma erupted with low explosive gas release is called here Bungan. The name comes from an Icelandic case, Hraunbunga (literally: hraun = lava, bunga = low dome, thus Bungan), which is a rather small lava dome of icelanditic composition, 30-50 m high. Hraunbunga is within the Krafla volcanic system in northern Iceland (for location see Fig. 1). It is postglacial but prehistoric in age, but as no tephra, which can be traced back to the eruption site, is found in its vicinity, and no signs of major explosions are visible in the formation itself, it is taken as a relevant example. The eruption of a viscous magma at low explosive gas release primarily takes place at dome extrusions. The well-known event of the Showa Shin-zan formation (Minakami et al. 1951) within the Usu volcano, Japan is a good example of such dome formation. The steam explosions of the second phase of the eruption are interpreted (e.g. by Macdonald 1972) as having been heated groundwater explosions. The latest phase of the eruption, the dome extrusion itself, was thus an eruption practically without explosive gas release. Showa Shin-zanian is thus used here as a name for such dome extrusions. The other cases of the classification scheme are well known from the literature (e.g. Macdonald 1972) and do not need any further comments in this context.

The main purpose of the scheme is to explain and underline the strong contrast between the risk taken, on the one hand, in areas, where the dominating hazardous aspects of volcanism result from basaltic effusive behaviour, and the areas of "airborne" hazards of explosive behaviour, on the other hand, i.e. the oceanic versus the continental and island arc cases.

In Iceland the hazardous phenomena of the volcanism depend, to some extent, on which volcanic field we are dealing with. In Table 3 a generalized and simplified volcanic hazards characterization for Iceland is given. Throughout Icelandic history, ~1100 years, volcanism has not been responsible for casualties to any major extent. The secondary effects of the well-known Laki eruption in 1783 (Thorarinsson 1969), the famine of the years that followed, caused the death of about 10,500 people (about 24% of the population of Iceland at that time), but no one was killed directly by the eruption. During the 1362 eruption of the ice-covered Oraefajokull volcanic cone (Thorarinsson 1958) apparently some hundreds of people were killed by an initial jokulhlaup or lahar. Otherwise casualties associated with volcanic eruptions in Iceland can be considered as individual cases, as far back as is known. During the 1973 eruption in Eldfell on Heimaey, off the south coast of Iceland, no one was killed by the activity in spite of the fact that the erup-

Table 3. Icelandic volcanism: generalization on hazards

Hazard aspect	Frequency	Size	Comments
Lava flow	Extremely common	All sizes	By far the most hazardous of all aspects of the volcanism because of its high frequency. People, live-stock and property may be threatened in individual cases.
Tephra flow	Extremely rare	Small	Not known for at least 10,000 years.
Tephra fall	Origin common in non-rifting areas, scarce in rift zones	All sizes	In cases extremely hazardous on a local scale. Vegetation and farmland primarily threatened.
Base surge	Rare	Small?	Not known in historic time except in Surtsey eruption.
Mud flow	Rare	Small	Not hazardous.
Glacial burst, Jokulhlaup	Common	Various sizes	Primarily hazardous in SE-Iceland. Live-stock, roads and bridges mainly threatened.
Gas release	Extremely common	All sizes	Generally not hazardous.
Earthquakes	Extremely common	Mostly small	May be hazardous to roads and constructions in habited areas.
Ground fissuring	Extremely common in rift zones	Various sizes	(as above)
Tsunamis	Extremely rare or absent	Small?	Apparently not hazardous.

Mud flows, jokulhlaups and earthquakes may occur without being associated with eruptions and may all in fact be more common as such. The table does not consider such non-volcanic cases.

tion broke out during the night just ~200 m from the outskirts of the town of ~5300 inhabitants. The preventive and protective measures taken during the struggle against the eruption, in order to decrease the loss as much as possible were extreme, and have nowhere been matched since (Einarsson 1974).

Hazardous phenomena related to the volcanism, other than the lava flows, which are highly stressed in this chapter due to their dominating frequency and pure volcanic origin, are several (see Table 3), but they are of varying severity. Among these phenomena are the jokulhlaups. These are restricted to the areas of the volcanic zones which are at present covered by ice caps. These areas amount to nearly 20% of the total area of the volcanic zones of Iceland. The areas may, however, deliver jokulhlaups at times when no volcanic eruptions are taking place. These jokulhlaups result from meltwater, collected in natural reservoirs, and from hydrothermal melting of the ice mass, where ice caps cover such hydrothermal fields. The distinction between "purely" volcanic jokulhlaups and the others, especially the hydrothermally caused ones, may be complex in cases, but the distinction is not of great importance seen from the viewpoint of risk. In general, jokulhlaups occur in Iceland in the order of one per year. These are mostly small and do not cause much harm. The really hazardous ones are more scarce and more difficult to predict, but among these are the enormous jokulhlaups accompanying eruptions in the volcano Katla (see location on Fig. 1). Hydrothermal fields and volcanic complexes go together in many cases. Such a complex is Grimsvotn (location on Fig. 1), which delivers jokulhlaups with ~4-year intervals. Some of these are purely the released water mass of the hydrothermal melting process while others are followed or accompanied by volcanic eruptions as well. The entire range from highly water-dominated jokulhlaups to more typical lahars or mud flows is probably present within the Icelandic spectrum of these phenomena, but their studies are incomplete.

With the modern techniques of society in general, and the continuous expansion of human activity, the variation and amount of land use within the volcanic areas of Iceland are constantly increasing and, consequently, more life and property is exposed to volcanic risk. This calls for increased concern and studies of volcanism, and its threats are part of this. Volcanic hazards studies and forecastings are becoming an increasing "must" in Iceland, but how are we to go about this? This is the main question of this chapter. The differences between Icelandic volcanism and volcanism elsewhere has been outlined to show that they are differences of nature and not only differences of degree.

The tradition of volcanic hazards studies in the world is very short. They have mostly been carried out during the last decades in countries of continental and island arc volcanism, primarily USA, Japan, the West Indies and the South West Pacific-South East Asian island arcs. In some other countries work has begun, and still more countries show interest, or have started preparation or preliminary studies. Hawaii is the main area of island volcanism, in which some effort has been put into hazards studies. Are the study models from these areas useful in the Iceland case?

4 Study Models for Volcanic Hazards in Iceland

It is in the countries mentioned above that volcanic hazard study models can best be looked for. The American model will be taken here, in a simplified version, as an example of the continental and island arc cases. This model is in use with minor local modifications in other countries of continental and island arc volcanism and is thus a sort of general hazard study model for these volcanotectonic settings.

The volcanoes in question are major cone-shaped mountains which form a chain along the Cascade Mountain Range in the western United States (Fig. 4). Each cone has an activity pattern which results in a simple hazard map. In simplified and schematic form such a map comprises several zones of different hazards, as shown by Fig. 5. A simplified risk map for Mt. Rainier is shown in Fig. 6 and a tephra fall risk map for Mt. St. Helens is shown in Fig. 7. In Iceland this kind of model can be useful with local modifications in the case of some of the major vol-

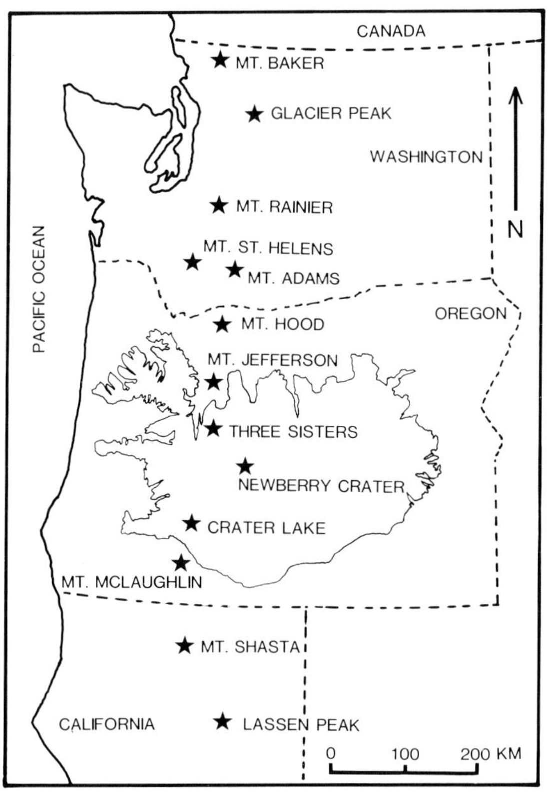

Fig. 4. Map showing the volcanic cones of the Cascade Range in USA, with Iceland inserted for size comparison. Three to five volcanic cones of the Cascade type would occupy the volcanic belt of Iceland, if it had the volcanotectonic settings of the subduction areas. Three to five hazard maps could account for volcanic hazards in Iceland

CHARACTERISTIC AND GENERALIZED HAZARD MAP FOR THE CASCADE VOLCANIC CONES

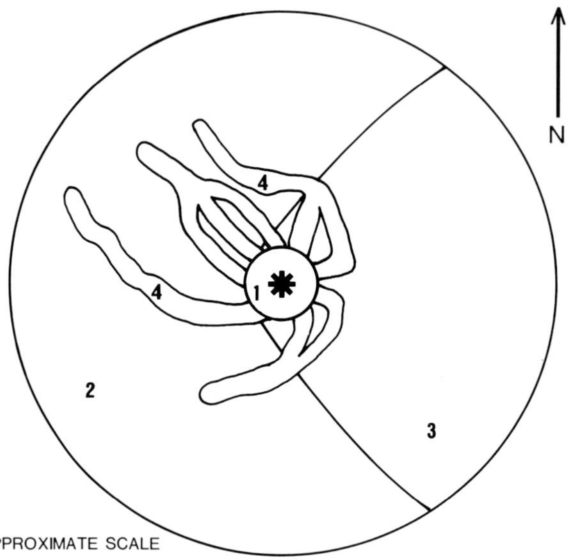

APPROXIMATE SCALE
0 20 40 KM

HAZARD ZONATION:

1. THE WHOLE RANGE OF POSSIBLE HAZARD EFFECT
2. TEPHRA FALL GREATER THAN 1/2 INCH (DEPENDING ON WIND DIRECTION)
3. GREATER TEPHRA FALL RISK CAUSED BY PREDOMINANCE OF WIND DIRECTION
4. RIVER VALLEYS AND FLOOD PLAINS SUBJECT TO MUD FLOWS AND FLOODS

Fig. 5. The shape of the volcanic cones, their known center of activity and pattern of behaviour allow the construction of a relatively simple hazard map, which is based on the eruption history of the volcano. Such a map usually comprises several concentric zones of outward decreasing hazards of "airborne" materials. A narrow sector of higher risk of such hazards is present, if prevailing winds reign in the region. The innermost zone is, as well, the area where the extrusion of lava domes and viscous flows usually takes place. The radii of comparable risk zones for different volcanoes differ with the violence of the volcanoes. In addition to the concentric zones, the map contains some irregularly shaped, long and narrow zones, located in valleys and gullies with high risk of floods and lahars and other materials, the distribution of which is controlled by the landscape (generalized on basis of Warrick 1975)

canoes of the non-rifting volcanic areas, mentioned above, e.g. Hekla. In these volcanoes the magmas are frequently rather viscous and the gas content relatively high, and thus the tephra production is relatively large and lava flows usually restricted to the very close vicinity of the mountain. The model is, however, unable to describe the volcanism in the non-rifting zones which takes place in between these major volcanoes.

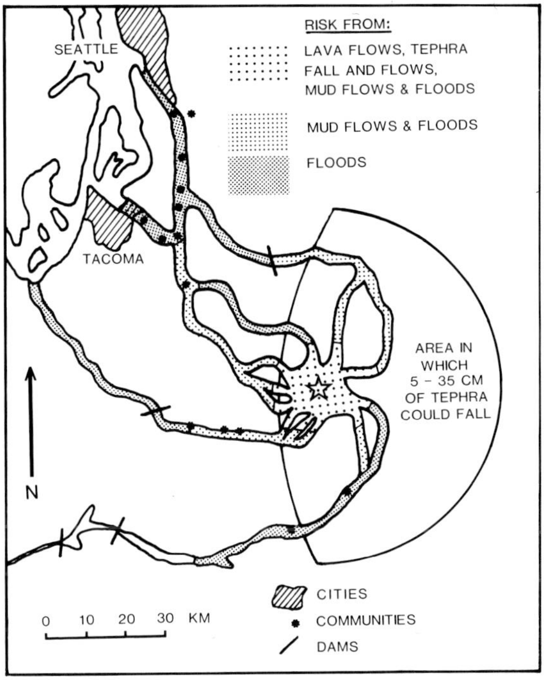

SIMPLIFIED RISK MAP OF MT. RAINIER AREA

Fig. 6. The distribution of materials, which are controlled by the landscape, differs with slope and type of material. Lava flows at such volcanoes are usually viscous and the lavas tend to be restricted to the slopes of the volcano or its close vicinity. Mud flows extend further and sediment-poor floods still further before they fade out to become harmless or enter the sea (redrawn, slightly altered, from Crandell et al., 1979)

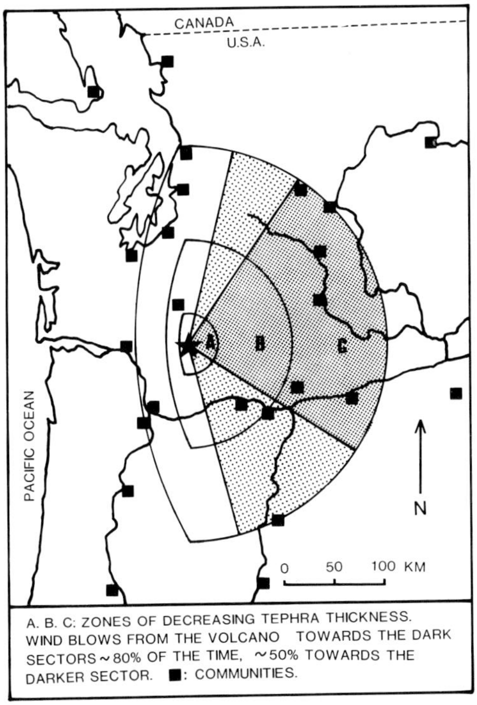

MT. ST. HELENS. TEPHRA FALL RISK MAP

Fig. 7. Where prevailing winds reign, the sectors of tephra fall can be given various weights according to the frequency of wind directions. Most of the time the wind blows within a certain wide sector and shorter times within more narrow sectors. This results in a narrow zone of high probability of tephra fall during potential eruptions, bordered on both sides by zones of lower probabilities (redrawn, slightly altered, from Crandell and Mullineaux 1978)

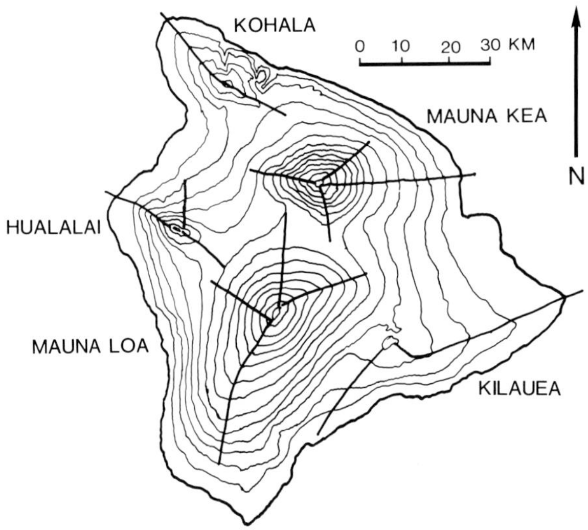

HAWAII ISLAND

SHIELD VOLCANOES AND RIFT ARMS

Fig. 8. The shield volcanoes of Hawaii have radiating arms of narrow rifts extending downslope from a summit caldera. There are usually two main rift arms, but one to two additional arms, much less active, may develop. The caldera and main rift arms are by far the most active parts of these shields. At present, Kilauea is the most active of these shields, and the world's most active volcano. In activity it is followed by Mauna Loa, which is still very active, while Hualalai has erupted only once during the 200 years of Hawaii's written history. Mauna Kea and Kohala have not erupted during this period at all. The productivity and eruption frequency thus show a clear decrease towards NW (redrawn and simplified from Macdonald and Abbott 1970)

In conclusion, this model can be used to some extent in the low productivity areas of Icelandic volcanism. These cover 40% of the total volcanic areas but are responsible for only ~22%, by volume, of the volcanic products erupted in Iceland and for ~20% of the eruptions taking place. In the rift zones, the remaining 60% of the volcanic areas of Iceland, accounting for ~78% of the volcanic production of the country and ~80% of the eruptions, the hazards cannot be forecast using this model.

The only country with rift tectonics, in which hazards and risk studies have been carried out, is Hawaii. The Hawaiian rifting does not, however, quite fit the Icelandic situation. Both the causes of Hawaiian rifting and quantities involved show that the processes are different. The Hawaiian rifting forms local transecting or radiating rift arms on the flanks of major shield volcanoes which have a clear center of activity, as shown on Fig. 8. Thus they bear some resemblance to the cone-type volcanoes. The major difference between the Hawaiian shields and the cones lies in the magma chemistry, eruptive

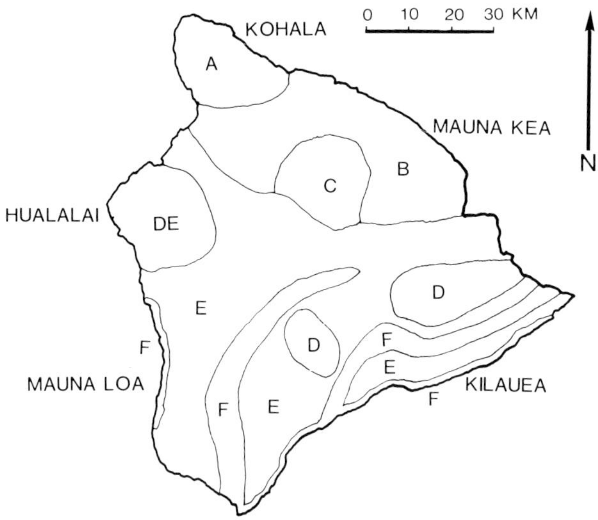

HAWAII ISLAND

ZONES OF INCREASING HAZARD FROM A THROUGH F.

Fig. 9. In accordance with the frequency pattern of the activity of the Hawaiian shields, the calderas and main rift arms of Kilauea and Mauna Loa have been allocated the highest level of risk (F) in the hazard zonation of the island. In addition to these active volcanic areas, the SW and SE coastal regions of the island are included in this high-risk zone because of a tsunami effect present (redrawn and simplified from Mullineaux and Peterson 1974)

mechanisms and the relative amount of top-crater activity. In the Hawaiian shields the volcanic activity is primarily basaltic and effusive, such that fluid lava production makes up the bulk. In the cones the explosive activity of silicic magmas plays the major role.

Volcanic hazard zonation on the island of Hawaii distinguishes seven degrees of hazard (Fig. 9). Apparently six of these are zones of relatively low hazard compared to the seventh one, which comprises the rift arms of Kilauea and two of the rift arms of Mauna Loa. Within this high-risk zone, 97.5% of historic eruptions (those within the last ~200 years) on the island have originated. About 50% of this hazard zone is covered by these historic lavas, compared to only 15% cover in the case of the next zone of lower hazard. This high-hazard zone is a long and narrow feature, less than 5 km in width, which only covers about 13% of the area of the island. For a more detailed review of the Cascade and Hawaiian volcanic hazards and risk studies see Hodge et al. (1979).

The Icelandic rift, in contrast to the Hawaiian one, is a regional feature which goes through the country from southwest to northeast as a continuation of the volcanic rift of the Atlantic ocean floor. The rift zones go through the island with a fracture zone offset in Sudurland (southern lowlands), which is of the same type and nature as the fracture zones that characterize the rift of the ocean floor in general.

The rift zones of Iceland are composed of fissure swarms where most of the tectonic and volcanic activity goes on. These are generally 40-100 km long and 2-10 km wide. They are thus much larger structures than the rift arms of the Hawaiian shields which are 10-40 km long and 1-2 km wide. In Hawaii the rift seems to be a response to the volcanism, an intraplate volcanism, i.e. an equilibrium adjustment of the enormous mass piled up on top of the oceanic crust. In Iceland the situation is reversed, i.e. a global rift transecting the oceanic crust, and the voluminous volcanism accompanying it results in an extra thick oceanic crust but without hiatus and discordance. In Fig. 10 the fissure swarms of the rift zones in Iceland and the rift arms of the Hawaiian shields are shown to the same scale. In Iceland, all the fissure swarms shown are presently active and they do not show any recognized signs of becoming extinct in the near future. In Hawaii, on the other hand, only three of the shields have been active during local historic time and in general the shields show a pattern of decreasing activity from

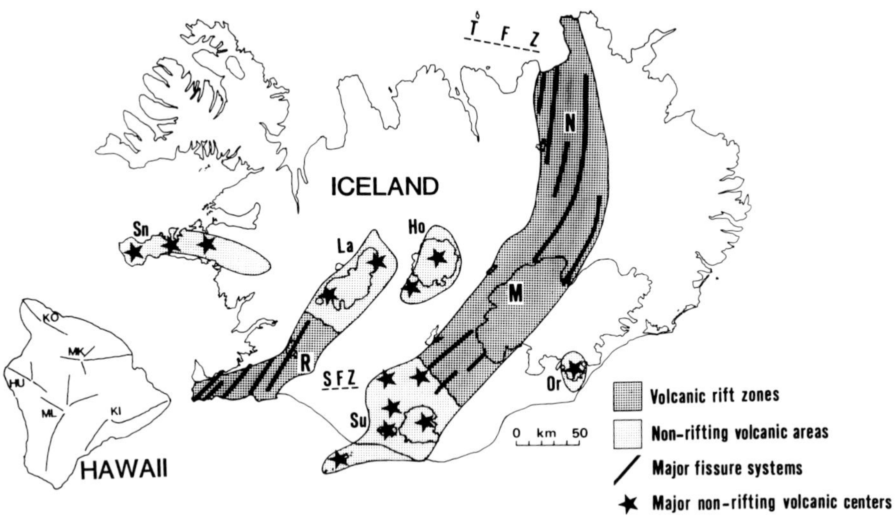

Fig. 10. The volcanic zones of Iceland (103,000 km^2) cover ~29% of the area of the country or ~30,000 km^2. In Hawaii (10,400 km^2) the F-zone covers 13% of the island's area or ~1350 km^2. If each fissure swarm of the volcanic rift zones of Iceland and the immediate vicinity of the major volcanoes of the non-rifting areas were categorically given the highest level of risk, to obtain some comparison with Hawaii, this would probably amount to at least 10,000 km^2 or ~ one third of the volcanic zones. The frequency of eruptions within these high-risk zones would, however, be much lower than it is within the F-zone on Hawaii (for abbreviations see Figs. 1 and 8)

Kilauea, through Mauna Loa to Hualalai, which only once has erupted in historic time. This pattern of decreasing activity and increasing age continues, viewed on an extended scale, northwest through the entire Hawaiian island chain. In conclusion, Hawaii cannot serve as a direct example in the case of volcanic hazard studies in Iceland.

In Fig. 11 we see the fissure swarms of the Reykjanes peninsula as these are expressed through the craters and crater rows of postglacial time (~11,000 years on the peninsula) and neglect the open fissures and faults which belong to the fissure system. Each fissure swarm is, in most geological respects, to be looked upon as a single volcanic system, but with no clearly defined center of activity and with obscure outer limits. When foreign volcanologists count the active volcanoes of the various volcanic fields of the world and arrive at a specific number, in Iceland the question "what is a single volcano?" cannot be answered. Concerning the hazards and the future eruption sites we are thus dealing with a surface of considerable dimensions where hazardous activity can take place. This is in strong contrast to the cone volcanoes of the continental and island arc hazards model previously described, which have a clear center of activity. Categorically each fissure swarm of the Icelandic rift zones could be given a hazard zone status of the highest risk as in the case of the rift arms of Mauna Loa and Kilauea in Hawaii, but this would be a qualitative ranking of hazards zones. The quantitative treatment would still be lacking.

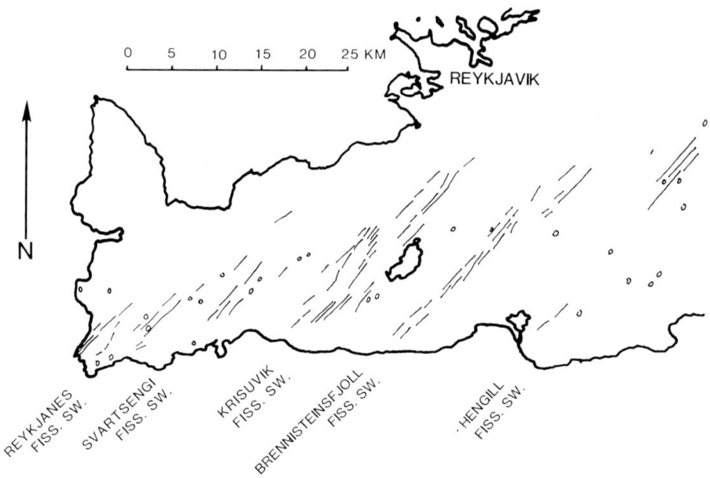

THE MAIN POSTGLACIAL ERUPTION SITES OF THE FISSURE SWARMS OF THE REYKJANES PENINSULA, SW – ICELAND.

Fig. 11. It is clear from this map how the eruptions occur mainly on fissures and how these are arranged in a parallel manner to form the fissure swarms. No eruption has taken place here for the last 700 years or so. On the Reykjanes Ridge, just off the coast towards SW, several eruptions have occurred during this period, a few miles from the coast (based on Saemundsson and Einarsson 1980)

Here we come to the major problem in the Icelandic risk and hazards studies: how do we quantify the hazards and risk and how do we account for these matters? This will apparently only be done by quantifying the volcanism itself, by measuring its productivity and analyzing the patterns in it (a study much needed for volcanism worldwide in fact). By use of the literature, some preliminary and general figures for the productivity of the Icelandic volcanic zones as a whole have been obtained. A summary of these is given in Table 4. The next step is to make some detailed, first-hand studies of the past products of volcanism to establish a better knowledge of the productivity pattern, thus testing the statements of Table 4 and eventually to increase their reliability and to narrow the quantitative knowledge down to individual fissure swarms and volcanic centers. This is an expensive and time-consuming task, which has not been planned, but which must be seriously considered.

According to preliminary studies a quantitative and thorough investigation of the morphological parameters of both craters and lavas, and the structural parameters of their products will give quantitative information on types of eruptions and their eruption mechanisms. After having established a geographical distribution of such types, the subdividing of the rift zones and individual fissure swarms into zones of defined mechanisms of eruption and types of products, is likely to be possible. Further, the establishment of the frequency of eruption types and the volume relations of the products will allow this subdivision to be given specific hazards and risk status.

Table 4. Icelandic volcanism: some figures on productivity

	Entire volcanic fields	Rift zones	Non-rifting areas
Annual production: km^3	0.01-0.02	0.008-0.016	0.002-0.004
Percent of production	100	77.5	22.5
Total area: km^2	29800	18000	11800
Percent of area	100	60	40
Average lava cover: %	29	36	17
Number of historic eruptions	225	156	69
Percentage of eruptions	100	70	30
Average time between eruptions: years	4.9	7.1	15.9

There are thus apparently good possibilities of establishing a method for the general volcanic hazards studies on the rift zones of Iceland and of making detailed hazards maps of both the rift zones and their individual fissure swarms.

At present no actual hazards map has been produced, but the first one is in preparation. It is based on the above criteria and serves as a test of the method. It accounts for the lava flow hazard only, of a single fissure swarm. The general and detailed volcanic and eruptive history of this particular swarm is better known than for other parts of the rift zone. Scrutinizing this map will show whether the method is valid or not, or to what extent it can be reconsidered or modified, if necessary.

References

Crandell DR, Mullineaux DR (1978) Potential hazard from future eruptions of Mount St. Helens volcano, Washington. US Geol Surv Bull 1383-C, 26 pp + maps

Crandell DR, Mullineaux DR, Miller CD (1979) Volcanic-hazard studies in the Cascade Range of the western United States. In: Sheets PD, Grayson DK (eds) Volcanic activity and human ecology. Academic Press, New York, pp 195-219

Einarsson P, Bjornsson S (1979) Earthquakes in Iceland. Jokull 29:37-43

Einarsson Th (1974) The Heimaey eruption in words and pictures. Heimskringla, Reykjavik, 56+32 pp

Hodge D, Sharp V, Marts M (1979) Contemporary responses to volcanism: case studies from the Cascades and Hawaii. In: Sheets PD, Grayson DK (eds) Volcanic activity and human ecology. Academic Press, New York, pp 221-248

Imsland P (1983) Iceland and the ocean floor: comparison of chemical characteristics of the magmatic rocks and some volcanic features. Contrib Mineral Petrol 83:31-37

Macdonald GA (1972) Volcanoes. Prentice-Hall, New Jersey, 510 pp

Macdonald GA, Abbott AT (1970) Volcanoes in the sea: the geology of Hawaii. University of Hawaii Press, Honolulu, 441 pp

Minakami T, Ishikawa T, Kenzo Y (1951) The 1944 eruption of Volcano Usu in Hokkaido, Japan: history and mechanism of formation of the new dome "Syowa-Sinzan". Bull Volcanol 11:45-157

Mullineaux DR, Peterson DW (1974) Volcanic hazards on the island of Hawaii. Washington, DC : US Geol Surv Open-file Report 74-239, 61 pp + map

Saemundsson K (1978) Fissure swarms and central volcanoes of the neovolcanic zones of Iceland. Geol J Spec Iss, 10:415-432

Saemundsson K, Einarsson S (1980) Geological map of Iceland, sheet 3, SW-Iceland, 2nd edn. Mus Nat Hist Icel Geodet Surv, Reykjavik

Thorarinsson S (1958) The Oraefajokull eruption of 1362. Acta Nat Isl 2 no 2, 99 pp

Thorarinsson S (1969) The Lakagigar eruption of 1783. Bull Volcanol 33:910-929

Tsuja H (1955) Geological and petrological studies of Volcano Fuji, 5: On the 1707 Eruption of Volcano Fuji. Bull Earthquake Res Inst, Tokyo Univ, Vol. 33, no. 3, pp 341-383

Warrick RA (1975) Volcano hazard in the United States: a research assessment. Boulder: Inst Behav Sci Univ of Colorado, 144 pp

Whittow J (1980) Disasters: the anatomy of environmental hazards. Penguin Books, 411 pp

Volcanic and Associated Hazards in the Lesser Antilles

M. J. Roobol and A. L. Smith[1]

Abstract

Historic records and stratigraphic studies throughout the Lesser Antilles reveal that volcanic and associated hazards may be categorized by increasing magnitude and decreasing frequency into:

Category I: Events that occur every few decades and affect only the parent volcano's immediate flanks.
Category II: Events that occur every few hundreds to thousands of years with effects extending beyond the flanks of the parent volcano, often over large parts of the island.
Category III: Events which occur every few tens to hundreds of thousands of years, and affect several islands; these may include non-volcanic activity such as sector collapse of the volcano's flanks with associated earthquakes and tsunamis.

The historic volcanic activity of the arc over the past 300 years, since widespread European settlement, can be considered in Category I with 39 events. At least six pyroclastic, fourteen phreatic/phreatomagmatic, nine submarine eruptions, four dome/lava extrusions and six volcano-seismic crises have occurred. The pyroclastic activity is represented by the historic andesite eruptions of Mt. Pelée, Martinique, and the basaltic andesite eruptions of Soufrière, St. Vincent, although other types have occurred in the past.

Category II hazards are typified by large Plinian-type eruptions which have generated widespread air falls, valley-fill pumice and ash flows (high aspect ratio ignimbrites, HARIs) and extensive ash hurricanes (low aspect ratio ignimbrites, LARIs). Examples of such activity were the eruption of Mt. Pelée around 2000 years ago, and the eruption that produced the Roseau tuff on Dominica approximately 30,000 years ago. Although this type of activity has not been witnessed by Europeans, several such eruptions occurred during the 3000-year occupation of the islands by pre-Columbian peoples. Category II hazard is also exemplified by the older geology of the Quill, St. Eustatius

---

[1] Department of Geology, University of Puerto Rico, Mayaguez, Puerto Rico 00708

IAVCEI Proceedings in Volcanology 1
J. H. Latter (Ed.), Volcanic Hazards

(Statia), where, some 30,000 years ago extrusion of rhyolitic magma onto a shallow marine shelf (about 3 km offshore from an older island) generated pyroclastic activity which became partially phreatomagmatic with increased energy release. The adjacent island was engulfed by surge deposits that eventually linked the early Quill to the older island.

Category III hazards involve major events during which, in addition to large new geological structures appearing on one island, associated seismic effects and tsunamis are likely to be experienced on adjacent islands. Examples of this type are illustrated by the geology of St. Lucia, Dominica and St. Vincent, where large arcuate depressions occur on the back-arc coasts. These structures have been interpreted as either calderas or as resulting from the gravity collapse of the steeper back-arc flanks of the islands. Whatever their origin, up to 12 km^3 of land may be consumed.

At present the reactions of the governments of the many small islands to volcanic hazards vary considerably. Most are struggling with variable success (according to their level of affluence/impoverishment) to consider and plan for the short-term or category I hazard. For Saba and St. Eustatius where, and possibly because, we have prepared detailed category I and II hazard assessment reports, these reports are "lost" in the offices of the Netherlands Antilles government in Curaçao "awaiting approval" for publication and for 6 years have been unavailable to the administration and populations of these volcanic islands.

The mitigation of volcanic disasters requires not only hazard assessment reports but a wholehearted effort by the local government including education of the population at risk. Interaction between the international scientific community and local governments in developing countries is complicated by political-social-economic factors. The whole complex provides a new subject of geohazard management, without which the hazard assessment reports are unlikely to mitigate natural disaster.

1 Introduction

The 600-km-long active arc of the Lesser Antilles (Fig. 1) is highly accessible to the western world, with excellent air services and accomodation for the tourist industry. The arc is a polyglot of British-, French- and Dutch-administered as well as recently independent islands. Some are prosperous and peaceful, others impoverished and ravaged by social unrest. Some are extensively cultivated, while others retain much of their original tropical rain forest. They provide challenging and varied problems for the earth scientist concerned with geohazard assessment. As far as volcanic hazard is concerned, such assessments are very much interplays between levels of knowledge of a particular volcanic centre, level of monitoring by resident earth scientists, the prosperity and resources of the island concerned and the contribution in funding and interest by outside organizations.

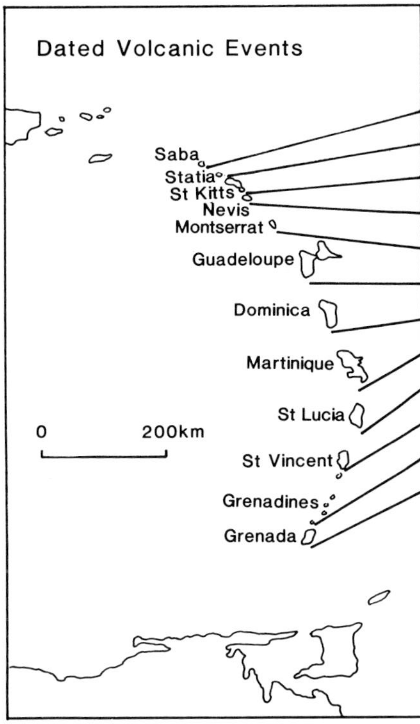

European post-1650AD A	B	C	Indian 3000–300 yearsBP	Pre-historic 50000–3000 yearsBP
			1	1
			1	3
	2		9	5
		2		
		3	1(?)	8
2[a]	8	1	9	11
	1		1	6
2	2		26	46
	1			7
7			8	9
8*				

A Magmatic eruptions
B Phreatic–Phreatomagmatic eruptions
C Volcanoseismic crises

* Submarine eruptions
a Includes 1 submarine eruption off Guadeloupe

Fig. 1. Islands of the active arc of the Lesser Antilles showing dated activity (at present state of knowledge), grouped according to European, pre-Columbian Indian, and prehistoric periods

Although the British and former British islands are monitored (seismically and volcanologically) by the Seismic Research Unit of the University of the West Indies, and the French islands by Les Observatoires Volcanologiques des Antilles of L'Institut de Physique du Globe de Paris and the Bureau de Recherches Geologique et Minieres, the islands are on the whole weak in resources, and the local populations commonly lack education in volcanic hazards. A major part of the volcanological work has been and is likely to remain, in the hands of overseas scientific organizations with independent funding. Volcanic activity during the past decade has provided an opportunity for an ad hoc group of resident and non-resident earth scientists to assess hazards and in particular to consider their interaction with today's sensation-oriented modern press (Fiske 1981). It seems likely that future activity will be monitored by IMEWS (Inernational Mobile Early Warning Systems for Volcanic Eruptions) working in conjunction with resident scientists as well as the island government through whom press releases will be coordinated.

2 Reconstruction of Volcanic History

A major requirement for hazard assessment of a particular volcano is the reconstruction of the styles and frequencies of eruptions, and the production of maps illustrating the distribution of the various deposits. Throughout the Lesser Antilles there are three approaches to these data.

1. Eyewitness Accounts. Following the European discovery of the Lesser Antilles with the arrival of Christopher Columbus (on his second voyage) at Dominica on 3 November 1493, eyewitness accounts of volcanic activity exist. However, the first 2 centuries of European settlement were very turbulent so that most of these accounts are for the past 3 centuries only. Many of these accounts are summarized in Robson and Tomblin (1966).

2. Deposits Overlying Pre-Columbian Indian Remains. For at least 3500 years before European discovery, the Lesser Antilles were occupied by Indian peoples who migrated along the arc from South America. The oldest shell midden of pre-pottery Archaic Indians, so far identified, is in southern St. Kitts, and has yielded a radiocarbon age of 4100 ±60 years B.P. (Goodwin 1978). The midden is overlain by an air-fall pumice deposit (Baker 1985). Other archaic shell remains have been dated at 2175 ±60 years B.P., also on southern St. Kitts (Goodwin 1978) and 3155 ±65 years B.P. on Saba (Roobol and Smith 1979). At the time of the arrival of Columbus, the arc was populated by pottery-making Arawak and Carib Indians. Potsherds marking the occupation sites of these later Indians are common and widespread throughout the islands. They provide a valuable stratigraphic marker for the identification of the youngest pyroclastic deposits. On Mt. Pelée, Martinique, Roobol et al. (1976) identified several occupation sites overlain by air-fall pumice deposits, and others overlain by ash hurricane or LARI deposits. The Indian sites are sufficiently common to be of use to the volcanologist for deposits as old as about 3000 years.

It is worthy of note from the point of view of putting natural hazards into perspective, that the pre-Columbian Indian population survived 3500 years of volcanic activity, but largely "disappeared" after the European occupation, often as a result of bloody fighting. For example, Bloody Point on western St. Kitts is reputedly named after a day in 1626 A.D. when over 2000 Carib Indians were slaughtered by combined French and British troops (Cotner 1975). The point made here is that volcanic and associated hazards are not necessarily the major hazards faced by the Lesser Antillean and other populations, rather political-economic and military considerations may override these.

3. Combined Stratigraphic and Geochronological Studies. The natural tropical rain forests on the islands have resulted in carbonized wood being a common ingredient of pyroclastic flow and surge deposits. Such wood can sometimes be found in air-fall deposits within the range of large ballistic bombs, which probably ignited vegetation (Roobol et al. 1985). Mud flow/lahar deposits commonly contain non-carbonized wood. Radiocarbon dating is therefore a major tool in stratigraphic reconstruction for deposits back to about 50,000 years old.

For older deposits, K-Ar geochronology has been used (Briden et al. 1979). By combining stratigraphic and geochronological studies, prehistoric events have been identified which have not been repeated in the brief interval of European settlement. For example, Carey and Sigurdsson (1980) combined on-land observations with an offshore piston-coring program to identify what is probably the largest explosive eruption in the Lesser Antilles in the past 200,000 years, when around 58 km^3 (fresh fallen volume) of felsic tephra were erupted at about 30,000 years B.P. from Dominica, apparently without the formation of a caldera. This is a somewhat improbable conclusion and perhaps further consideration should be given for a component in the subaqueous tephra of the Grenada Basin (Carey and Sigurdsson 1980; their Fig. 5) possibly derived from felsic eruptions of similar age on other islands, e.g. St. Lucia (Wright et al. 1984). Alternatively a large arcuate structure identified by Robson and Tomblin (1966) in central Dominica may mark the caldera. Roobol and Smith (1976) and Westercamp and Traineau (1983a) described the alternating styles in pyroclastic activity of Mt. Pelée. A disadvantage of the combined stratigraphic-geochronological study is that it emphasizes large volume deposits. Thin deposits, e.g. from surges and LARIs, which may also represent major hazards, may be missing from the stratigraphic record due to erosion shortly after deposition.

3 Frequency and Styles of Volcanic Activity

When the volcanic deposits of the Lesser Antilles are viewed in terms of the above three time divisions, it will be shown that the volcanic activity was not the same for each. In Fig. 1 and Tables 1 to 3, the dated volcanic eruptions of the Lesser Antilles are listed according to style of volcanic activity. The data are far from complete. Only the records for the last 300 years can be regarded as accurate. The older data are based entirely on deposits which have been dated by the radiocarbon method. The largest number of such dated deposits is found on Martinique, which reflects the amount of geological work and the number of new outcrops generated during road-building schemes. The abundance of dated deposits is in part a reflection of the affluence of an island. The poorer islands have few new road cuts and quarries, e.g. St. Lucia and Dominica. Other islands such as Nevis and Grenada appear to genuinely lack historic activity. The data for the entire arc (Fig. 1) do, however, sensibly suggest that volcanic eruptions have occurred more frequently on the larger islands in the centre of the arc, than on the smaller islands at the northern and southern ends.

The data of Tables 1 to 3 have been used to construct Figs. 2 to 4, showing the distribution of dated deposits resulting from Pelean-, St. Vincent- and Plinian-type eruptions along the arc. Carbon-bearing deposits resulting from Pelean activity are typically coarse valley-fill block and ash flow deposits and more widely dispersed surge deposits. Those from St. Vincent-type

Table 1. Dates and types of eruptions in the Lesser Antilles since widespread European settlement ca. 1650 A.D.[b]

Island / Eruptive style	Grenada	Grenadines	St. Vincent	St. Lucia	Martinique	Dominica	Guadeloupe	Montserrat	Nevis	St. Kitts	St. Eustatius (Statia)	Saba
Pelean	No historic activity				1929-32 1902-05		1798				No historic activity	No historic activity
St. Vincent			1979 1902-03 1812 1718									
Plinian												
Phreato-magmatic					1851		1809 1797					
Phreatic				1766	1792	1880	1976 1956 1903 1837 1696 1680			1843? 1692?		

Table 1 (cont.)

Island / Eruptive style	Grenada	Grenadines	St. Vincent	St. Lucia	Martinique	Dominica	Guadeloupe	Montserrat	Nevis	St. Kitts	St. Eustatius (Statia)	Saba
Volcano-seismic crises	No historic activity						1962	1966-67[c] 1933-36 1897-98	1961-62 1950		No historic activity	No historic activity
Phreatic/phreato-magmatic		1977 1974 1972 1966 1965 1953 1943 1939					1843					
Lava/dome[a]			1979 1971-72 1880? 1784									

[a] Eruptions of domes or flows lacking significant associated pyroclastic activity.

[b] Data sources are Robson and Tomblin (1966) and Simkin et al. (1981) unless otherwise indicated. All dates in years A.D.

[c] Other sources: Shepherd et al. (1971).

Table 2. Dates and types of eruptions in the Lesser Antilles during widespread pre-Columbian Indian settlement[a] (3000 to 300 years B.P.)

Eruptive style \ Island	Grenada	Grenadines	St. Vincent	St. Lucia	Martinique	Dominica	Guadeloupe	Montserrat	Nevis	St. Kitts	St. Eustatius (Statia)	Saba
Pelean	No dated activity	No dated activity		No dated activity	{310B {320G 490B {545B {590C {595B 690C 765B 1,040B 1,140C 1,230B 1,305B 1,730C 2,260B 2,490C 2,550B 2,675B 2,750C	450D	2,600E	306J	No dated activity			280K
St. Vincent			300A 320A 400A				450E 650E			2,80H 2,340H 2,420H	1,550K	

Table 2 (cont.)

Eruptive style / Island	Grenada	Grenadines	St. Vincent	St. Lucia	Martinique	Dominica	Guadeloupe	Montserrat	Nevis	St. Kitts	St. Eustatius (Statia)	Saba
St. Vincent (cont.)	No dated activity	No dated activity	467A 535A 555A 2,480A 2,700A						No dated activity	2,720H 2,860H 3,060H	1,550K	
Plinian					650C 1,140B 1,506B {1,625B {1,670C {1,800B {1,845B 1,900B {2,010C {2,015B 2,145B 2,390B {2,440C {2,447B 2,565B		510F 550G			1,750H		

Table 2 (cont.)

Eruptive style \ Island	Grenada	Grenadines	St. Vincent	St. Lucia	Martinique	Dominica	Guadeloupe	Montserrat	Nevis	St. Kitts	St. Eustatius (Statia)	Saba
Phreatic/ phreato-magmatic	No dated activity	No dated activity					1,800E		No dated activity	1,620J 1,710J		
Lava/ dome[b]							450E 1,370E 2,600E					

[a] Radiocarbon ages in years B.P. Where several dates are believed to represent one eruption they are bracketed together. For locality data and 1σ values see sources as indicated by letter suffix.
Sources: **A** Rowley (1978); **B** Smith and Roobol (in press); **C** Westercamp and Traineau (1983a), **D** Roobol et al. (1983); **E** Westercamp (1981); **F** Semet et al. (1986); **G** Robson and Tomblin (1966); **H** Roobol et al. (1981); **J** Baker (1985); **K** Roobol et al. (unpubl. a,b).

[b] Eruptions of domes or lava flows without any significant associated pyroclastic activity.

Table 3. Dates and types of eruptions in the Lesser Antilles prior to widespread human settlement[a] (>3000 years B.P.)

Eruptive style / Island	Grenada	Grenadines	St. Vincent	St. Lucia	Martinique	Dominica	Guadeloupe	Montserrat	Nevis	St. Kitts	St. Eustatius (Statia)	Saba
Pelean	No dated activity	No dated activity			{3,110C {3,130D {3,150C 3,710D {3,940C {3,980D {3,990C 4,225C 4,300D 4,410G {4,500C {4,515D {4,940C {4,990C 5,100D 5,195C 5,250C 5,650G 8,395C {11,300D {11,340G 13,500D 20,240C 25,120C 36,095C >39,000C >48,000C	25,310H	3,100K 13,500 to 14,500K 17,000 to 30,000K	17,670L 18,450L 18,600L 18,890N {19,590N {19,600N {19,630N 19,970L 20,800L 23,568M	No dated activity		7,810T	34,750T

Table 3 (cont.)

Eruptive style / Island	Grenada	Grenadines	St. Vincent	St. Lucia	Martinique	Dominica	Guadeloupe	Montserrat	Nevis	St. Kitts	St. Eustatius (Statia)	Saba
St. Vincent	No dated activity	No dated activity	{3,510A {3,590A 3,890A 3,970A {4,130A {4,165A {4,325A {4,335A 14,000A		{21,100R {21,185C {21,690R {21,750E 22,300G 22,550R 22,650R 23,540R 24,410R 25,000R 25,700G 25,980R 26,500R 26,810R 32,070F 32,300F 36,070F 37,485E 41,000E >40,000F		3,500K 4,500K 7,700K 8,500K		No dated activity	>41,730P		
Plinian				20,000B 20,980B 21,440B 22,380B 23,080B	4,060D 4,375C 4,610D 5,770C 6,630G	27,600J 28,400J 29,000J 31,000J 33,200J	30,000K >38,500K	18,390M		3,658N 4,270N	7,940T 22,240T	

Table 3 (cont.)

Eruptive style / Island	Grenada	Grenadines	St. Vincent	St. Lucia	Martinique	Dominica	Guadeloupe	Montserrat	Nevis	St. Kitts	St. Eustatius (Statia)	Saba
Plinian (cont.)	No dated activity	No dated activity		32,840B 34,200B	{7,365C 7,405C} {7,745C 7,800G} 9,175C 10,280G 24,550C				No dated activity			
Subplinian (Mansion Series St. Kitts only)										>41,140P >41,730P		
Lava/ Dome[b]							4,500K 13,500?K					

[a] Radiocarbon ages in years B.P. Where several dates are believed to represent one eruption these are bracketed together. For data and 1σ values see sources as indicated by letter suffix.
Sources: A Rowley (1978); B Wright et al. (1984); C Smith and Roobol (in press); D Westercamp and Traineau (1983a); E Bourdier et al. (1985); F Westercamp and Traineau (1983b); G Dupuy et al. (1985); H Roobol et al. (1983); J Carey and Sigurdsson (1980); K Westercamp (1981); L Smith et al. (1979); M Rea (1974); N Baker (1985); P Roobol et al. (1981); R Traineau et al. (1983); T Roobol et al. (unpub. a,b)

[b] Eruptions of domes or lava flows lacking significant associated pyroclastic activity.

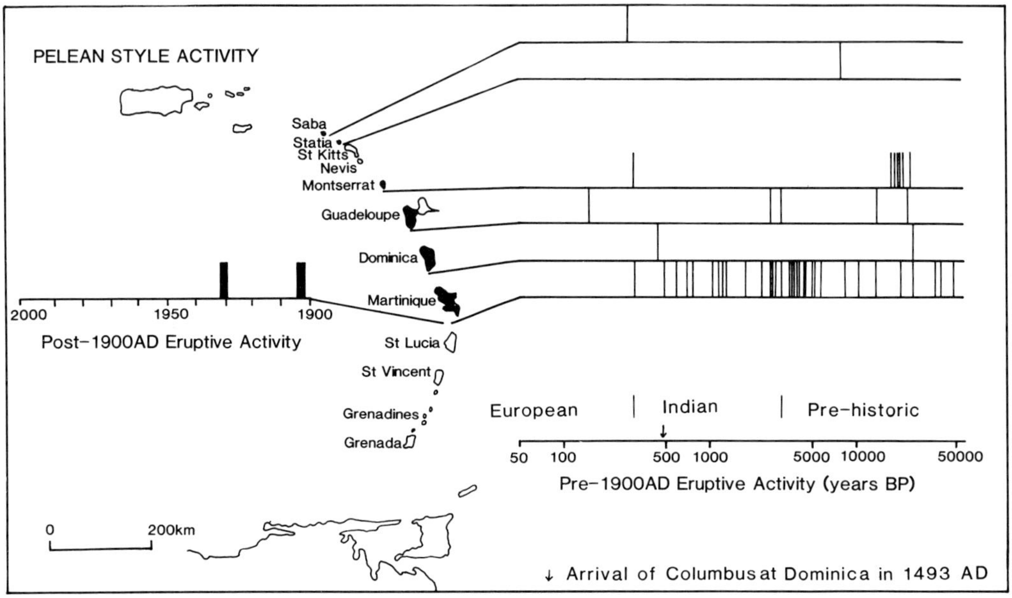

Fig. 2. Map of Lesser Antilles showing occurrences and ages of dated deposits resulting from Pelean-type activity

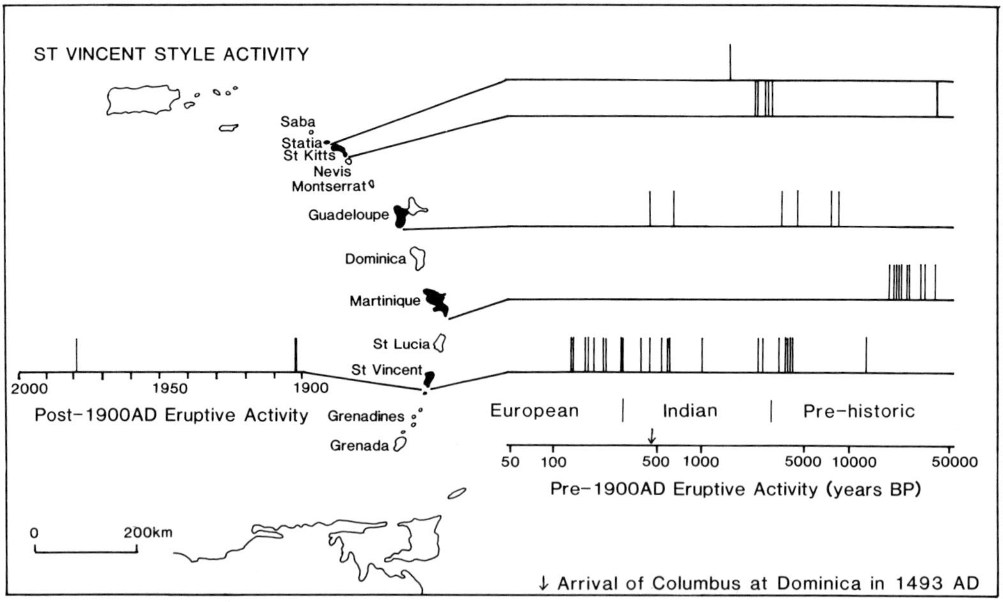

Fig. 3. Map of Lesser Antilles showing occurrences and ages of dated deposits resulting from St. Vincent-type activity

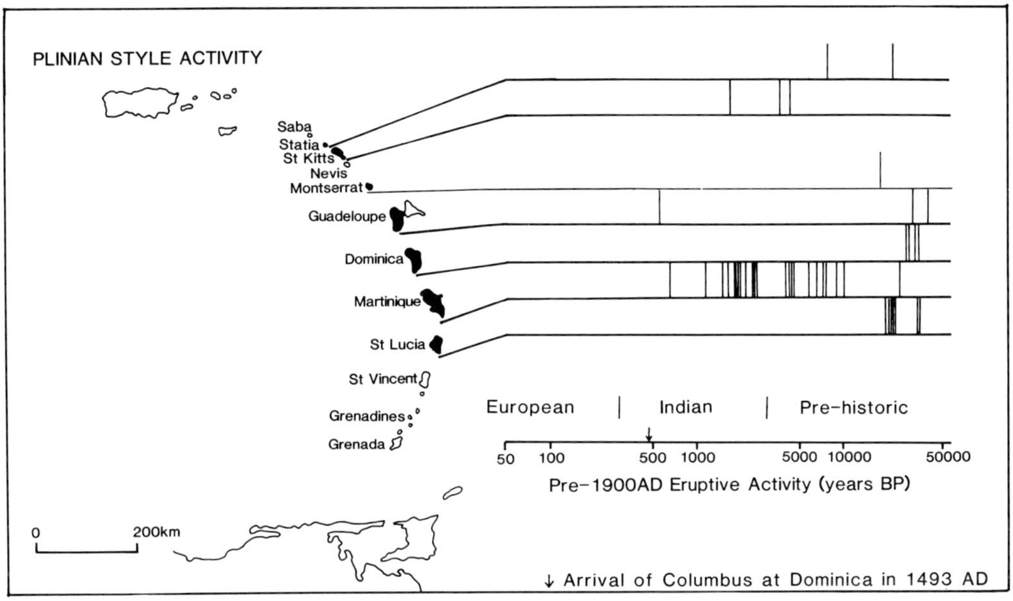

Fig. 4. Map of Lesser Antilles showing occurrences and ages of dated deposits resulting from Plinian-type activity

eruptions are scoria and ash-flow deposits and associated basaltic-andesite surge and air-fall deposits. In the case of Plinian activity, the radiocarbon-dated deposits result from ignimbrites (HARIs) and ash hurricanes (LARIs) produced by the gravity collapse of the Plinian eruption columns, together with air-fall deposits. Occurrences of dome, and phreatic/phreatomagmatic and submarine activity as well as the lithologically distinct air-fall deposits of the Mansion Series of St. Kitts (Roobol et al. 1985) are shown in Fig. 5.

Tables 1 to 3 and Figs. 2 to 4 clearly show that pumiceous Plinian-type activity has not been experienced during the period of European settlement, but was experienced by the pre-Columbian Indian peoples. This probably results from long repose intervals relative to the short period of European settlement. In addition the Lesser Antillean volcanoes appear to undergo long periods characterized by one style of activity and then alternate to another (e.g. Roobol and Smith 1976; Westercamp and Traineau 1983a).

4 Magnitude of the Volcanic and Associated Hazards

Apart from those organizations monitoring the British, former-British and French islands, the resources of most islands are so slender that neither libraries containing accounts of pyroclastic eruptions, nor residents trained in understanding them are available on the islands. For practical application to these islands, some straightforward system is needed to indicate magnitude of any impending hazard (regardless of scienti-

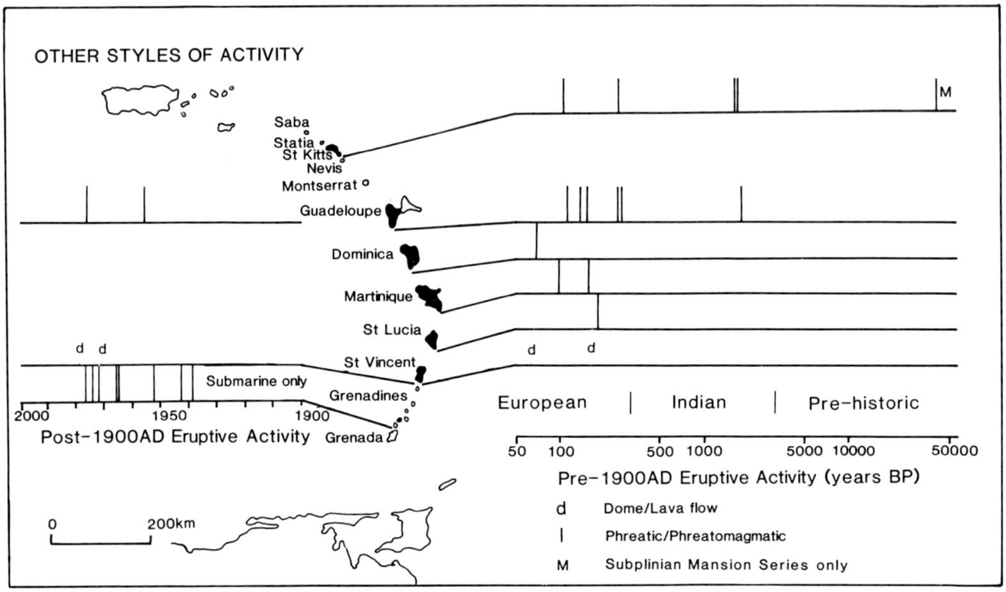

Fig. 5. Map of Lesser Antilles showing occurrences and ages of dated dome/lava, submarine eruptions, phreatic/phreatomagmatic eruptions and the sub-Plinian activity of the Mansion Series of St. Kitts. Records of submarine eruptions are from eyewitness accounts and associated seismic activity

fic details) which would be of use in identifying potential areas for evacuation. We propose here a simple three-fold system of increasing magnitude and decreasing frequency:

Category I: Events that occur every few decades and affect only the parent volcano's immediate flanks.

Category II: Events that occur every few hundreds to thousands of years with effects extending beyond the flanks of the parent volcano, often over large parts of the island.

Category III: Events which occur every few tens to hundreds of thousands of years and which affect several islands; these may include non-volcanic activity, such as sector collapse of the volcano's flanks with associated earthquakes and tsunamis.

4.1 Category I Hazard

This includes all the historic activity of the arc for the past 300 years, as recorded since widespread European settlement. At least six pyroclastic, fourteen phreatic/phreatomagmatic, nine submarine eruptions, four dome/lava extrusions and six volcano-seismic crises have occurred (Fig. 1). The pyroclastic activity is typified by the andesitic Pelean-type and the basaltic-andesite St. Vincent-type eruptions of this century. From the point of view of the risk (risk = type of hazard x number of people or properties) the category combines the smallest areas of destruction with the greatest frequency of occurrence. The volumes of material erupted in the category I hazard, as best as can be estimated from the onshore deposits, is less than 2 km^3 and usually less than 0.5 km^3. Distribution maps of typical products are shown in Fig. 6. Apart from thin layers of airfall ashes, the pyroclastic flow and surge deposits are essentially confined to sectors of the flanks of the parent stratovolcano. This is the characteristic feature of the category I hazard. Had lava flows formed, then they also would be included in category I.

However, a complication to this scheme occurred in May 1902 when Mt. Pelée, Martinique, and Soufrière, St. Vincent, erupted simultaneously. A comparison of activities is given by Roobol and Smith (1975). The 1902 activity is classified here as "multiple category I hazard" and not as category III, so that the possibility of multiple, simultaneous hazards of all categories should be considered.

Today, as the active volcanoes have mostly been identified and their styles of activity are known, it can be argued that the siting of some of the towns and villages is remarkably poor from the viewpoint of risk, considering category I hazards. There is an interesting historic reason for this, supported by the fact that the eruptions of Mt. Pelée and Soufrière, St. Vincent, earlier this century, destroyed the tropical rain forest on those sectors of the volcanoes' flanks traversed by the pyroclastic flows and surges. After the eruptions, vegetation reappeared in the form of grasslands, since it requires many decades before rain forest re-establishes itself. Plant life has been observed to re-establish itself first on the lower flanks of the volcano and then to migrate up the flanks towards the summit. During the first century after European discovery,

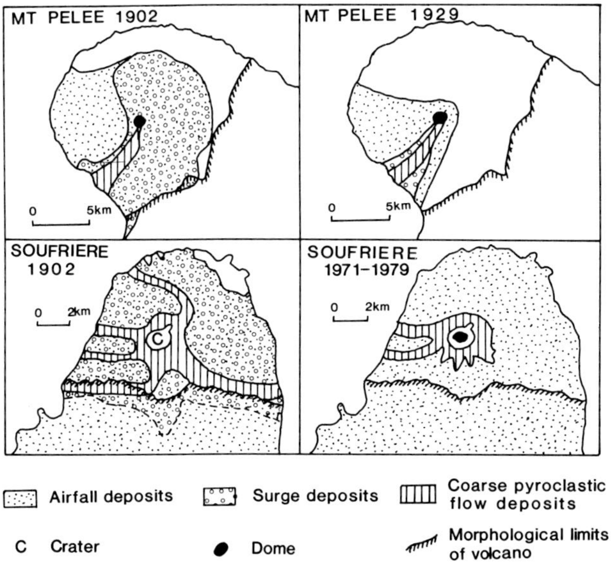

Fig. 6. Maps showing the morphological limits of Mt. Pelée, Martinique, and Soufrière, St. Vincent, with the distributions of the products of their eruptions during this century

there were difficulties in selecting sites for settlement on the islands. The different European communities fought with each other and with the resident warlike Carib Indians. It seems that in several instances settlers may have been lured onto pleasant uninhabited areas of grassland on the flanks of active volcanoes. In particular (Fig. 2) three Pelean eruptions appear to have occurred on Saba, Montserrat and at Mt. Pelée about 300 years ago (Table 2). This was after the discovery by Columbus, but before European settlement was well established. Thus LaCroix (1904) described how the early settlers and founders of St. Pierre in 1635 A.D. described the volcano as "Mt. Pelée" meaning "Bald Mountain". These early settlers had arrived shortly after a Pelean-type eruption and before vegetation had re-established itself on the top of the volcano. The deposits of this eruption have been identified and have yielded a radiocarbon age of 310 ±60 years B.P. (Smith and Roobol 1980). From the island of Saba, Hartog (1975) described how the early settlers established the village named "The Bottom" in 1636 A.D. on grassy meadows on the volcano's flanks. Andesite surge deposits containing accretionary lapilli underlying the village have been radiocarbon dated by us at 280 ±80 years B.P. (Table 2). Significantly, the andesitic surge deposit overlies debris flow deposits containing sherds of Indian pottery and is overlain by debris flow deposits containing European pottery remains. On Montserrat, the island capital of Plymouth is situated on a major block and ash-flow fan immediately below and only 3 km from a notch in the crater rim of the Soufrière Hills (map in Rea 1974). Radiocarbon ages from deposits in gullies within Plymouth show the pyroclastic flow fan to be around 20,000 years old (Table 3). The youngest radiocarbon age of 306 ±54 years B.P., and its location, is described by Baker (1985). We consider that the poor siting of Plymouth may well have a similar explanation to that of St. Pierre, Martinique, and the Bottom, Saba.

Also included in category I volcanic hazards are volcano-seismic crises and phreatic/phreatomagmatic activity which usually affect single volcanic centres and rarely leave deposits of stratigraphic importance. Ancient seismic activity is recorded in abundant Neptunean dikes revealed in road cuts on the western side of the Pitons of Carbet centre of northern Martinique. Volcano-seismic crises may also generate debris flow deposits. Phreatic/phreatomagmatic ashes are typically lithified (in otherwise unlithified sections) and crowded with accretionary lapilli and entombed gas cavities. Examples of such ashes were described from Mt. Misery, St. Kitts, by Roobol et al. (1985).

During the short period of historic activity there have been six volcano-seismic crises (Robson and Tomblin 1966; Shepherd et al. 1971). Such crises are believed to result from the movement (both up and down) of magma at high crustal levels resulting in locally felt earth tremors (Table 1). At the present time (late 1986) a seventh crisis is occurring at the Morne Patates centre on southern Dominica (J.B. Shepherd, pers. commun., 1986).

In some cases the rise of magma into high crustal levels produces not only earth tremors but also phreatic or steam explosions due to the heating of groundwater but without the eruption of a juvenile silicate component. An example was the 1975-77 activity of Soufrière, Guadeloupe (Dorel and Feuillard 1980; Heiken et al. 1980). Precursor seismic activity over 12 months was followed by the first steam explosions on 8 July 1976. Explosive activity continued for a further 8 months and seismic activity for a further 11 months.

When juvenile ash is included in the steam explosions the activity becomes phreatomagmatic. Such activity often occurs as a precursor to magmatic eruptions. What was probably entirely a phreatomagmatic event occurred on Mt. Pelée, Martinique, in 1851 (Heilprin 1903). The activity was preceded by increased fumarolic output and felt earthquakes which commenced on 10 May 1851. On 5 August explosions began and a large area of the volcano's flanks was blanketed with a thin layer of ash. During this activity the Rivière Blanche flooded and several lahars were generated. The activity continued into 1852.

The rise of magma to high crustal levels may result in the expulsion of the water table onto the ground surface leading to flooding of rivers and lahar generation. Such events occurred in 1851 on Mt. Pelée and probably recurred during the precursor activity to the 1902 eruption of Mt. Pelée, (Roobol and Smith 1975).

4.2 Category II Hazards

In this case, rather than only a sector of the parent volcano being affected, the pyroclastic deposits are more widely dispersed and extend over adjacent volcanoes, so that the areas devastated are considerably larger than in category I. On the larger islands more than a half of the total area may be affected, while on the smaller islands the entire area may be affected. The volumes of material in each of the category II hazards appears generally to be considerably larger than for the category I hazards. The only single estimate available combining onshore and submarine products from a single eruption is the 58 km^3 of tephra estimated by Carey and Sigurdsson (1980) from an eruption in Dominica approximately 30,000 years ago. Category II mainly includes non-caldera-forming Plinian activity resulting in pumiceous deposits, as experienced by the pre-Columbian Indian peoples of the arc. The Plinian activity produces pumiceous air-fall deposits dispersed by the anti-trade winds of the upper troposphere (Roobol et al. 1985), valley-fill ignimbrites (HARIs) and extensive ash hurricane deposits (LARIs), all of which affect areas greater than the parent volcano. The deposits of such an eruption of Mt. Pelée around 2000 years ago have been described by Roobol and Smith (1980) and Westercamp and Traineau (1983a, their Fig. 6). The distributions of ignimbrite deposits on Dominica (Fig. 7) and St. Lucia (Fig. 8) also demonstrate the extensive areas affected. The presence of several ash hurricane deposits (LARIs) erupted from Mt. Pelée

Fig. 7. Map of Dominica showing the distribution of the youngest pyroclastic flow deposits. (After Carey and Sigurdsson 1980 and Roobol et al. 1983)

(Roobol and Smith 1976, 1980) extend over the northern half of Martinique (Fig. 9) and also constitute a category II hazard.

The geology of the island of St. Eustatius (Statia) provides a further example of another type of category II hazard. The island comprises an older centre in the north, built of andesitic Pelean domes with aprons of lithified block and ash flow deposits. Between 32,000 and 22,000 years ago (Westermann and

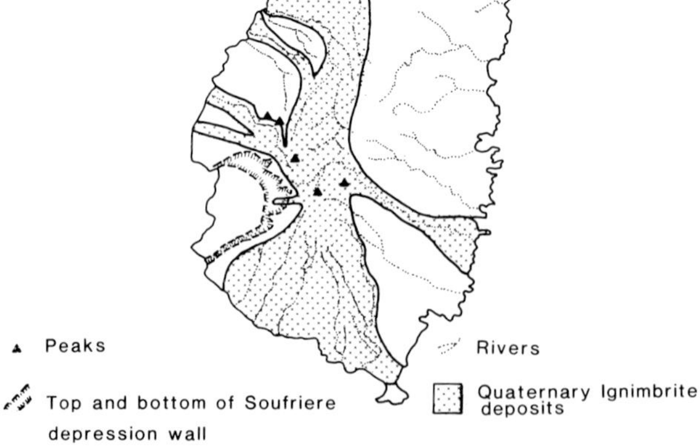

Fig. 8. Map of St. Lucia showing the distribution of the youngest pumice and ash-flow deposits. The distribution is inferred from relict patches of the deposits within radial valleys. (After Roobol et al. 1983)

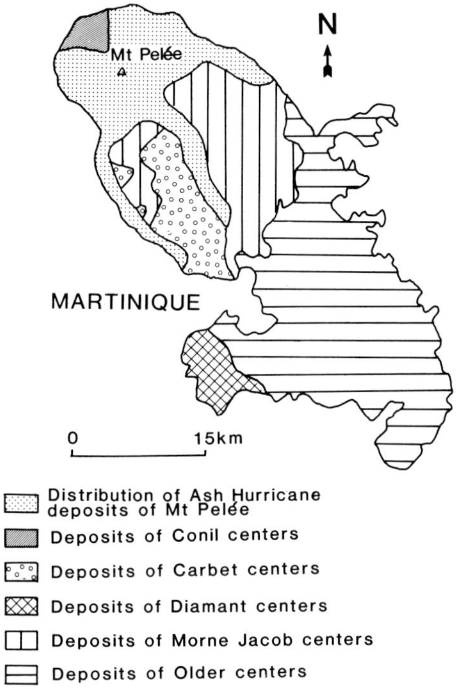

Fig. 9. Map showing the volcanic centres of Martinique (after Westercamp 1975), and the distribution of ash hurricane (LARI) deposits from Mt. Pelée

Keil 1961), a rhyolitic centre erupted 3 km offshore on a shallow-water limestone bank, and subsequently built the Quill volcano which extended the island of St. Eustatius 3 km to the southeast. The earliest pyroclastic deposits are interstratified with limestones and are exposed in the White Wall cliff at the south end of the island (Fig. 10). The limestones were uptilted at some stage during the formation of the Quill, probably by a dome. The interaction of rhyolite and seawater produced extensive base surge deposits rich in limestone clasts. These deposits linked the emerging Quill volcano with the older island to the northwest. This older island, rising to 289 m

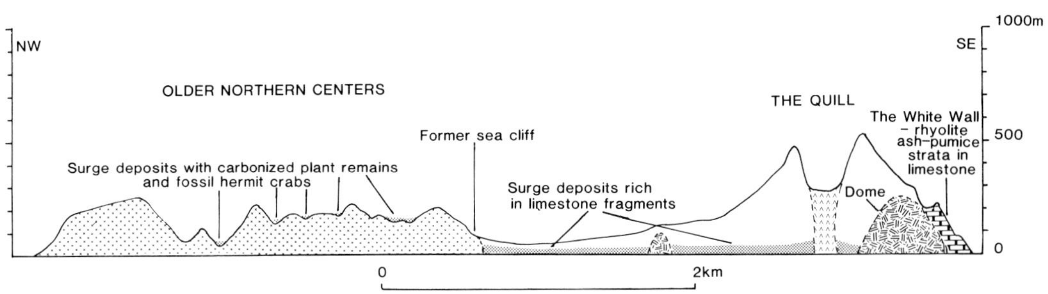

Fig. 10. Sketch section through St. Eustatius (Statia), showing how the Quill emerged 3 km offshore from the older northern centres, which were engulfed in surge deposits during the emergent phreatomagmatic stage

above sea level was engulfed by the surges. The deposits are preserved in depressions between the older hills and are rich in carbon and hermit crab remains with associated shells ("bioclasts", representing the flora and fauna of the older island). Thus a felsic shallow-water eruption near the shores of a populated island would also constitute a category II hazard.

At present, only one example of such a deposit is known in the Lesser Antilles. However, the active volcano of Kick'em Jenny is situated 8 km north of Grenada and has erupted eight times between 1939 and 1977 (Sigurdsson and Shepherd 1974; Simkin et al. 1981). The crater is situated 232 m below sea level and dredged samples reveal that it is erupting olivine basalt. Although black eruption clouds have been observed, because this centre is not situated close to populated shores and because rhyolitic activity is not expected, it does not constitute a present-day hazard on the scale of the emergence of the Quill. An unnamed submarine volcano between the islands of Guadeloupe and Marie Galante is reported to have erupted in 1843 (Robson and Tomblin 1966). As this volcano is also far from centres of population, it, like Kick'em Jenny, does not present a category II volcanic hazard. Another possible submarine centre, Hodder's volcano, situated 8 km west of Castries, St. Lucia (Robson and Tomblin 1966) was not located by a recent submarine survey and may not exist (P. Bouysse, pers. commun., 1980). A submarine centre situated 5 km offshore of west-central Martinique is reported in Case and Dengo (in press). This may be the same centre described by fishermen from the village of Carbet on the northwest coast of Martinique, who reported sightings of boiling sea from time to time offshore of the village (J.P. Viode, pers. commun., 1974). As the one or possibly two submarine centres are situated only a few kilometres off the populated shores of Martinique, detailed investigation is justified in view of the stratigraphic record of the phreatomagmatic emergence of the Quill, St. Eustatius (Statia).

4.3 Category III Hazard

This involves the formation of major arcuate structures (described as either calderas or sector collapse scars after gravity sliding) where volumes up to 12 km^3 of land may be displaced in a single event. It is assumed that associated seismic effects and tsunamis would also be experienced on adjacent islands. Only three or four such large structures are known along the arc on the subaerial parts of Dominica, St. Lucia and St. Vincent, where they have formed in the last few millions of years. They are eccentrically placed on the back-arc side of the islands and open onto the Caribbean Sea, where the island shorelines closely approach the 1000-m submarine contour (Fig. 11). It is perhaps significant that some of these structures are so large and so heavily eroded that both their number and origin are at present a matter of debate.

The best-known example is the Qualibou structure on southwestern St. Lucia which was interpreted as a caldera by Tomblin

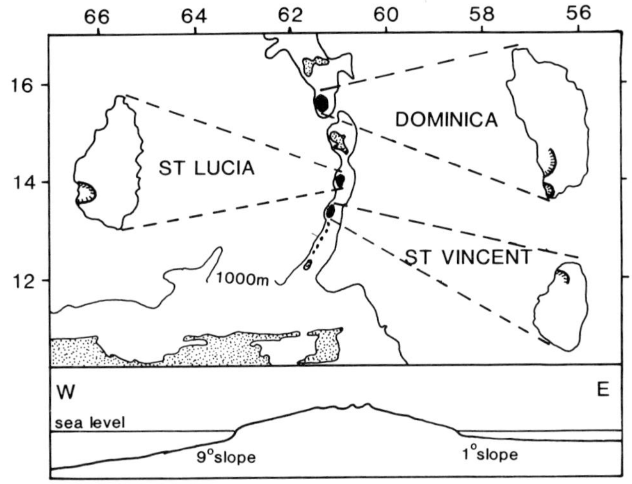

Fig. 11. Map of the Lesser Antilles arc showing how some of the larger islands, with shores near the 1000-m submarine contour, have back-arc morphological depressions, variously interpreted as calderas or gravity-slide structures

(1964). Such an interpretation for this and similar structures was questioned by Roobol et al. (1983) and an alternative of gravity-slide or sector collapse structure proposed. In the case of the Qualibou structure, this new interpretation was firmly rejected by Wohletz et al. (1986) and the caldera model re-instated.

Wohletz et al. (1986) have proved subsidence by drilling of an elliptical area around the Terre Blanche and Belfond domes. However, the elliptical area of subsidence (Wohletz et al. 1986, their Fig. 3) occupies only about half of the floor of the Qualibou morphological depression outlined by Roobol et al. (1983). Whether one is dealing with a younger subsidence within an older gravity-slide structure, or whether the whole is a caldera, lies in the identification of the vent area of the extensive pumiceous pyroclastic flow deposits. These deposits are much removed by erosion and the remnants largely hidden by dense vegetation. As mapped by Wohletz et al. (1986, their Fig. 2), the flow deposits appear to originate within the Qualibou structure. As mapped by Roobol et al. (1983) and Wright et al. (1984), they appear to originate in the central rain forest outside the Qualibou structure, in a manner similar to that demonstrated for Dominica by Carey and Sigurdsson (1980). One reason for this difference in mapping is because Roobol et al. (1983) separated the pyroclastic flow deposits into dacitic block and ash-flow deposits (resulting from Pelean activity) and dacitic pumice and ash-flow deposits (resulting from Plinian activity). This division identifies the pumiceous flow deposits as being largely absent from around the rim of the main Qualibou morphological depression and originating in the Central Highlands. However, we feel that more detailed study is required both to identify coarse near-vent breccias (Walker

1985) and to trace the submarine morphology. Whichever of the two models is considered, the subsidence of many cubic kilometres, either by gravity sliding or cauldron collapse, provides what is probably one of the major geohazards in the Lesser Antilles. Such an event would at least be accompanied by earthquakes and the generation of tsunamis within the Caribbean Sea.

5 Geohazard Management

An objective of the earth scientist in an area of natural geohazard is to collect all available evidence and produce a geohazard assessment report. This, in a volcanic area for example, will include the present status of monitoring (if any), records of historic activity and prehistoric activity as reconstructed from stratigraphic-geochronological study. Maps combining the occurrence of each type of deposit provide the basis for identifying areas for evacuation. Such a report then provides the basis for the local government to draw up evacuation plans which include such details as availability of vehicles, boats, aircraft; widths and conditions of roads, bridges and runways; alternative accomodation; population education in hazard response emphasizing evacuation sites and living procedures (shelter and food distribution); contact addresses of hazard experts and sources of financial assistance.

Many earth scientists believe that having completed the hazard assessment report, their involvement is finished. However, experience in the Lesser Antilles suggests that this is not the case, as such reports have a tendancy to be "lost" by island administrations or placed in "archival black holes" (Wadge 1985). Thus the conscientious earth scientist may be placed in a moral dilemma if the government of the country concerned fails to follow up the recommendations, and the report remains "lost", with the contents unavailable to the local administration and population. The moral dilemma arises because, during field investigations, the earth scientist will encounter, in the local population and administration, those most interested in the geohazard work: these are generally the same people who should receive training in monitoring, for example in regular temperature measurements of hot springs and fumaroles. Where wholehearted support for risk assessment is lacking, the earth scientist will inevitably be bound by secrecy clauses in contract with the local government, thereby preventing on-the-spot training of the resident population who may well be assisting the field program.

Despite much success and friendly cooperation along the arc, we report one such case history in our experience and the "loss" of a major piece of work. This involves the islands of Saba and St. Eustatius (Statia) at the northern end of the active arc of the Lesser Antilles. These small islands are administered by the Netherlands Antilles government based in Curacao, a non-volcanic island off the coast of South America. Detailed field studies were initiated in January 1976 and completed in August 1980: these were funded privately, by the National Science Foundation of the USA, and by the Netherlands Geological Survey (a contract with the latter contained the secrecy clause). A very detailed hazard assessment report was submitted to the Netherlands Geological Survey for publication on 6 February 1981 (M.J. Roobol et al., unpubl. a). The initial reaction was to divide the report into two parts, geology for publication (M.J. Roobol et al., unpubl. b), and hazard probably not for publication (M.J. Roobol et al., unpubl. c). Beyond this there has been no progress despite considerable effort involving many letters, personal visits and international telephone calls. The island populations and administrations have remained uninformed of the hazard assessment since the project commenced 11 years ago and finished 6 years ago.

The above example illustrates one case of a management problem in the practical application of modern volcanological work. Such a situation is likely to arise again in countries of the developing world as the technological and scientific gap between them and the developed countries continues to widen. Populations in some developing countries which suffer from day to day survival problems of food shortages and national unrest, are unlikely to be interested in long-term hazard reports, even if supplied freely. Further, if a natural disaster does occur after such a report has been prepared, then whatever government is in power will be open to criticism. Thus we can envisage situations where a hazard assessment report is both unwelcome and not required, even if it is provided as foreign aid.

As the number of hazard reports continues to grow, and more earth scientists become involved, more problems of this type are likely to be encountered. This complex interaction between scientists and government departments which is so critical to the successful application of hazard assessment can be termed "geohazard management". A small book, **Volcanic emergency management** (UNDRO 1985), has recently appeared dealing purely with the volcanic aspects, while the problems have been discussed at recent international volcanological symposia. We envisage geohazard management as a broader field involving all aspects of natural geohazard, including, amongst other things, earthquakes, flooding and avalanches.

6 Conclusions

Considerable progress has been made over the past 2 decades, both in monitoring the volcanoes of the Lesser Antilles, and in identifying and dating the various types of pyroclastic deposits. Hazard assessment has become a reality. The eruptions of Soufrière, St. Vincent, in 1971-72 and 1979 and that of Soufrière, Guadeloupe, in 1976, provided recent practical examples of hazard management. However, in terms of the three categories proposed here, the volumes of the deposits left by these latest eruptions were so small that they would most likely have been overlooked in a stratigraphic study. Thus it could be argued that the latest activity was so minor, insofar as it did not create extensive new deposits, as to be barely worthy of inclusion in category I. Future activity similar to the examples of category I described above, in which large sectors of the parent volcano's flanks are devastated, would require considerably more outside assistance than has been provided in recent eruptions, due to larger areas of destruction. At the present time, activity of categories II and III is not considered by the island administrations, as they have not recurred in the period of European settlement. The logistics of evacuation of major portions of the larger islands, and entire small islands, is outside the capacities of most administrations. However, interest by international organizations such as UNDRO and the availability of UN funding for projects such as the Pan Caribbean Disaster Preparedness Project (Wadge 1985), suggests that a basic structure for coordinated monitoring of future activity is now being established.

The general scientific interest in the Lesser Antilles arc remains high. For example, deposits of the 1902 eruption of Mt. Pelée, Martinique, continue to stimulate interest and details of the mechanism of the destruction of St. Pierre are still debated (Fisher and Heiken 1982; Sparks 1983; Westercamp and Traineau 1983a; Walker and McBroome 1983). This continued academic interest will enable successively more accurate hazard assessments to be made.

Today, with the post- Mt. St. Helens increase in interest in volcanology and the support of such organizations as IAVCEI (Anonymous 1986) and UNDRO, it is a relatively straightforward matter to arrange the preparation of a reasonably accurate hazard assessment report. However, it is a very different matter to ensure that such a report will be used to mitigate volcanic disaster. Here the subject of geohazard management becomes critical. Examples exist which illustrate two main areas of management failure.

In the first case upon completion of the hazard assessment report and submission by the scientists to the government organization, the report is "lost" in "archival black holes". An example of this is the loss of the detailed reports on Saba and St. Eustatius. This first-stage failure by the government of the Netherlands Antilles to approve access and publication of the report (by the Netherlands Geological Survey who have funding for publication available) results in the loss of both future planning and the establishment of a program to monitor the volcanoes and train and educate the local population (in such matters as routine temperature measurements of the hot springs).

In cases where management is sufficient to enable geohazard reports to be published, examples exist illustrating a second-stage failure, which is the failure to act on the information in the report. A recent example of this was the disaster of the Nevada del Ruiz volcano of Colombia where 25,000 people were killed on 13 November 1985. This occurred **after** the hazard had been correctly assessed by both local and visiting volcanologists and **after** a volcanic hazard map accurately delineating the areas at risk from lahars had been published (Sigurdsson and Carey 1986).

We feel that at present there is much enthusiasm for funding the preparation of volcanic hazard reports and geohazard reports in general. However, we wish to draw attention to the fact that although the scientific assessment can be made, much effort is needed to consider the complex social-political-economic factors complicating the management side. Cooperation between local governments and the scientific community appears to present the greatest weakness and requires careful consideration, otherwise much of the value of the funding and effort put into hazard assessment will be lost.

Acknowledgements. The work reported here began in 1971 and continues today. It was funded in a variety of ways, mainly by the N.S.F. (N.S.F. Grant GA-39706 and EAR 77-17064) but also by the Universities of Puerto Rico and the West Indies, the American Philosophical Society (Hays Fund Grant 106) and the Netherlands Geological Survey. Assistance was often provided by the various island governments. We thank our many friends and colleagues for their cooperation over the years. John Latter and Trevor Hatherton provided encouragement and valuable suggestions to improve the manuscript. We thank Nidia Torres for typing the manuscript.

References

Anonymous (1986) Scientists call for volcano monitoring. Geotimes 32:8-9

Baker PE (1985) Volcanic hazards on St. Kitts and Montserrat, West Indies. J Geol Soc Lond 142:279-295

Bourdier JL, Gourgaud A, Vincent PM (1985) Magma mixing in a main stage of formation of Montagne Pelée: the Saint Vincent-type scoria flow sequence (Martinique FWI). J Volcanol Geotherm Res 25:309-332

Briden JC, Rex DC, Fallar AM, Tomblin JF (1979) K-Ar geochronology and palaeomagnetism of volcanic rocks in the Lesser Antilles island arc. Phil Trans R Soc Lond SerA 291:485-528

Carey SN, Sigurdsson H (1980) The Roseau ash: deep-sea tephra deposits from a major eruption on Dominica, Lesser Antilles arc. J Volcanol Geotherm Res 7:67-86

Case J, Dengo G (eds) (in press) The geology of North America Vol. H: The Caribbean region. Geol Soc Am

Cotner J (1975) A motoring guide to St. Kitts. Heidelberg Press USA, 42 pp

Dorel J, Feuillard M (1980) Note sur la crise sismo-volcanique à la Soufrière de La Guadeloupe 1975-1977. Bull Volcanol 43:419-430

Dupuy C, Dostal J, Traineau H (1985) Geochemistry of volcanic rocks from Mt. Pelée, Martinique. J Volcanol Geotherm Res 26:147-165

Fisher RV, Heiken G (1982) Mt. Pelée, Martinique: May 8 and 20, 1902, pyroclastic flows and surges. J Volcanol Geotherm Res 13:339-371

Fiske RS (1981) Scientists and the news media: contrasting interaction during the two volcanic crises in the eastern Caribbean. Abstr IAVCEI Symp "Arc Volcanism" Tokyo, 93 pp

Goodwin RC (1978) The Lesser Antilles Archaic: new data from St. Kitts. J Virgin Islands Archaeolog Soc 5:6-16

Hartog J (1975) History of Saba. Van Guilder NV, Saba, 140 pp

Heiken G, Crowe B, McGetchin T, West F, Eichelberger J, Bartram D, Peterson R, Wohletz K (1980) Phreatic eruption clouds: the activity of La Soufrière de Guadaloupe, FWI, August-October 1976. Bull Volcanol 43:383-395

Heilprin A (1903) Mont Pelée and the tragedy of Martinique. Lippincott, Philadelphia London, 337 pp

LaCroix A (1904) La Montagne Pelée et ses eruptions. Masson, Paris, 622 pp

Rea WJ (1974) The volcanic geology and petrology of Montserrat, West Indies. J Geol Soc Lond 130:341-366

Robson GR, Tomblin JF (1966) Catalogue of active volcanoes of the world including solfatara fields. Part XX West Indies. Int Assoc Volcanol Naples, 56 pp

Roobol MJ, Smith AL (1975) A comparison of the recent eruptions of Mt. Pelée, Martinique and Soufrière, St. Vincent. Bull Volcanol 39:1-27

Roobol MJ, Smith AL (1976) Mount Pelée, Martinique: a pattern of alternating eruptive styles. Geology 4:521-524

Roobol MJ, Smith AL (1979) Archaeological implications of some radiocarbon dating on Saba and St. Kitts. Proc 8th Int Congress pre-Columbian cultures of the Lesser Antilles, pp 168-176

Roobol MJ, Smith AL (1980) Pumice eruptions of the Lesser Antilles. Bull Volcanol 43:277-286

Roobol MJ, Smith AL, Tomblin JF (unpubl a) An assessment of the volcanic hazard on the islands of Saba and St. Eustatius in the northern Lesser Antilles. Confidential Report Rijks Geologische Dienst, 205 pp

Roobol MJ, Smith AL, Tomblin JF (unpubl b) Volcanology of the islands of Saba and St. Eustatius in the northern Lesser Antilles. Part I - Geology, pyroclastic stratigraphy and petrology. Confidential Report Rijks Geologische Dienst, 165 pp

Roobol MJ, Smith AL, Tomblin JF (unpubl c) Volcanology of the islands of Saba and St. Eustatius in the northern Lesser Antilles. Part II - Volcanic hazard. Confidential Report Rijks Geologische Dienst, 62 pp

Roobol MJ, Petitjean Roget H, Smith AL (1976) Mt. Pelée and the island population of Martinique. Proc VI Int Congress pre-Columbian cultures of the Lesser Antilles, pp 46-53

Roobol MJ, Smith AL, Wright JV (1981) Ravisions in the pyroclastic stratigraphy of Mt. Misery volcano, St. Kitts, Lesser Antilles: ^{14}C ages and recognition of pyroclastic flow deposits. J Geol Soc Lond 138:713-718

Roobol MJ, Wright JV, Smith AL (1983) Calderas or gravity-slide structures in the Lesser Antilles island arc? J Volcanol Geotherm Res 19:121-134

Roobol MJ, Smith AL, Wright JV (1985) Dispersal and characteristics of pyroclastic fall deposits from Mt. Misery volcano, West Indies. Geol Rundschau 74:321-335

Rowley KC (1978) Stratigraphy and geochemistry of the Soufrière volcano, St. Vincent, West Indies. Unpubl Ph.D thesis, Univ West Indies, St. Augustine, Trinidad, 282 pp

Semet MP, Boudon G, Vincent PM (1986) Volcanological evolution of La Grande Decouverte (La Soufrière), Guadeloupe, FWI Abstr Int Volcanological Congress, New Zealand, 73 pp

Shepherd JB, Tomblin JF, Woo DA (1971) Volcano-seismic crisis in Montserrat, West Indies. Bull Volcanol 35:143-163

Sigurdsson H, Carey S (1986) Volcanic disasters in Latin America and the 13th November 1985 eruption of Nevado del Ruiz volcano in Colombia. Disasters 10:205-216

Sigurdsson H, Shepherd JB (1974) Amphibole-bearing basalts from the submarine volcano Kick'em-Jenny in the Lesser Antilles island arc. Bull Volcanol 38:891-910

Simkin T, Siebert L, McClelland L, Bridge D, Newhall C, Latter JH (1981) Volcanoes of the world. Smithsonian Institution, Hutchinson Ross, Stroudsburg; 232 pp

Smith AL, Roobol MJ (1980) Late pre-historic and historic activity of Mt. Pelée, Martinique. Abstracts IX Carib Geol Conf, pp 68-69

Smith AL, Roobol MJ (in prep.) Mt. Pelée, Martinique - a study of an active island arc volcano. Mem. Geol. Soc. America.

Smith AL, Roobol MJ, Rowley K (1979) Pyroclastic character of the active volcanoes of the northern Lesser Antilles. Trans 4th Latin Am Geol Cong, pp 24-30

Sparks RSJ (1983) Mont Pelée, Martinique: May 8 and 20, 1902, pyroclastic flows and surges-discussion. J Volcanol Geotherm Res 19:175-184

Tomblin JF (1964) The volcanic history and petrology of the Soufrière region, St. Lucia. Unpubl Ph.D thesis, Oxford Univ, 213 pp

Traineau H, Westercamp D, Coulon C (1983) Mélanges magmatiques à la Montagne Pelée (Martinique). Origine des eruptions de type Saint-Vincent. Bull Volcanol 46:243-269

UNDRO (1985) Volcanic emergency management. United Nations New York, 86 pp

Wadge G (1985) Avoiding another St. Pierre. Volcano News 22:4

Walker GPL (1985) Origin of coarse lithic breccias near ignimbrite source vents. J Volcanol Geotherm Res 25:157-171

Walker GPL, McBroome LA (1983) Mount St. Helens 1980 and Mount Pelée 1902 - flow or surge? Geology 11:571-574

Westercamp D (1975) Petrology of the volcanic rocks of Martinique, West Indies. Bull Volcanol 39:1-26

Westercamp D (1981) Assessment of volcanic hazards at Soufrière de Guadeloupe, FWI. Bull BRGM Sec IV (2):187-197

Westercamp D, Traineau H (1983a) The past 5000 years of volcanic activity at Mt. Pelée, Martinique (FWI): implications for assessment of volcanic hazards. J Volcanol Geotherm Res 17:159-185

Westercamp D, Traineau H (1983b) Carte Géologigue de la Montagne Pelée à 1/20,000: Service Geol. Natl BRGM, Orleans

Westermann JH, Keil H (1961) The geology of Saba and St. Eustatius. Uitg Natuurwet Studietering Suriname Ned Antillen 24:1-175

Wohletz K, Heiken G, Ander M, Goff F, Vuataz F-D, Wadge G (1986) The Qualibou caldera, St. Lucia, West Indies. J Volcanol Geotherm Res 27:77-115

Wright JV, Roobol MJ, Smith AL, Sparks RSJ, Brazier SA, Rose WI, Sigurdsson H (1984) Late Quaternary explosive silicic volcanism on St. Lucia, West Indies. Geol Mag 121:1-15

The Evolution of La Grande Découverte (La Soufrière) Volcano, Guadeloupe (F.W.I.)

G. Boudon[1], M.P. Semet[2] and P.M. Vincent[3]

Abstract

La Soufrière is part of a recent volcanic complex on the southern Basse-Terre island of Guadeloupe. It is bounded on the south by the Monts Caraîbes volcanoes and on the north by the Sans Toucher and Pitons de Bouillante massifs. The complex consists of two juxtaposed edifices: firstly, the composite volcano of la Grande Découverte (C.V.G.D.) of which la Soufrière, sensu stricto, represents the most recent activity; and secondly, la Madeleine eruptive ensemble, which lies south of C.V.G.D., and is of lesser extent. The construction of C.V.G.D. has been subdivided into three main stages:

1. The Grande Découverte stage (ca. 0.2 to 0.1 MA): growth, on the southern flank of Sans Toucher, of the main edifice, with dimensions of 20 x 15 km, by a sequence of effusive and pyroclastic events. This stage ended in a major Plinian eruption (several km^3), giving rise to the Pintade pumice flows and falls, and to the formation of la Grande Découverte caldera, only the northern rampart of which is still visible;

2. The Carmichaël stage (ca. 0.1 MA to ca. 3000 B.P.[4]): buildup of the Carmichaël volcano inside la Grande Découverte caldera. The end of this stage is marked by two catastrophic sector

---

Observatoires Volcanologiques, Institut de Physique du Globe de Paris, 4 Place Jussieu, 75252 Paris Cedex 05, France

1 Observatoire Volcanologique de la Montagne Pelée, Institut de Physique du Globe de Paris, Fonds Saint Denis, 97250 Saint Pierre, Martinique, France

2 Laboratoire de Géochimie Comparée et Systématique, Université Pierre et Marie Curie, 4 Place Jussieu, 75252 Paris Cedex 05, France

3 Département de Géologie et Minéralogie et L.A. 10 CNRS, Université de Clermont-Ferrand II, 5 rue Kessler, 63038 Clermont-Ferrand Cedex, France

4 ^{14}C data are reported according to usage and conventions recommended by the 12th International Radiocarbon Conference (Radiocarbon, 28, Nos. 2A and 2B, 1986). Calendar estimates (cal AD/BC dates) were obtained from Klein et al. (1982) calibration tables.

IAVCEI Proceedings in Volcanology 1
J.H. Latter (Ed.), Volcanic Hazards

collapses, dated at 11 500 and 3100 B.P., each of which was accompanied by large-scale explosions. The two events are characterized by the formation of km-size amphitheater craters (**Carmichaël** and **Amic** craters), extensive debris flows, and blast deposits (to the west and south of C.V.G.D., respectively). The older event has no magmatic component and is likened to the 1888 Bandaï-san eruption, whereas the younger event might have been similar to the 18 May 1980 Mount St. Helens eruption;

3. La Soufrière stage (ca. 3000 B.P. to the present): formation of la Soufrière volcano, sensu stricto, within the Amic crater. The present dome was emplaced at the end of the last magmatic activity, a mild pelean eruption, in cal AD 1440. Two major eruptions, in 1797 and 1976, and several minor ones have occurred in recorded history (since 1635); all were of the phreatic type.

During most of its evolution, C.V.G.D. lavas have remained of a rather constant andesitic composition. However, contrasted basaltic and dacitic products characterize single events, both in the older part of the history of the volcano, and the recent period. Trace element geochemistry indicates that C.V.G.D. had a magma supply system independent of other volcanoes elsewhere in southern Basse-Terre.

Volcanic hazard assessment has to take into account the past recurrence of collapses of sectors of the volcano. Abnormal seismicity and deformation that accompanies the development of Mt. St. Helens-type eruptions would most probably be recognized as such by the observatory monitoring networks. However, it may be much more difficult to interpret future unrest as premonitory signs of an impending Bandaï-san-type event.

1 Introduction

Guadeloupe, in the Lesser Antilles volcanic arc (Fig. 1), is made up of two main islands: Grande-Terre, to the east, is covered by Plio-Pleistocene limestones; whereas Basse-Terre, to the west, is entirely volcanic. Its position in the geodynamic evolution of the arc has recently been accounted for (e.g., Bouysse and Guennoc 1983; Bouysse et al. 1985).

La Soufrière dome, the site of the 1976-77 phreatic activity (Le Guern et al. 1980; Feuillard et al. 1983), was emplaced ca. cal AD 1440 (Vincent et al. 1979; Semet et al., in press) following mild explosive activity, and is the last recorded magmatic activity of a volcanic complex of Recent age. The complex is bounded:

1. On the south, by the Monts Caraïbes, a set of submarine volcanoes where hydromagmatic tuffs predominate. K/Ar ages of about 0.5 MA have been obtained by Blanc (1983) on local lava flows, and may perhaps be extended to the whole massif;

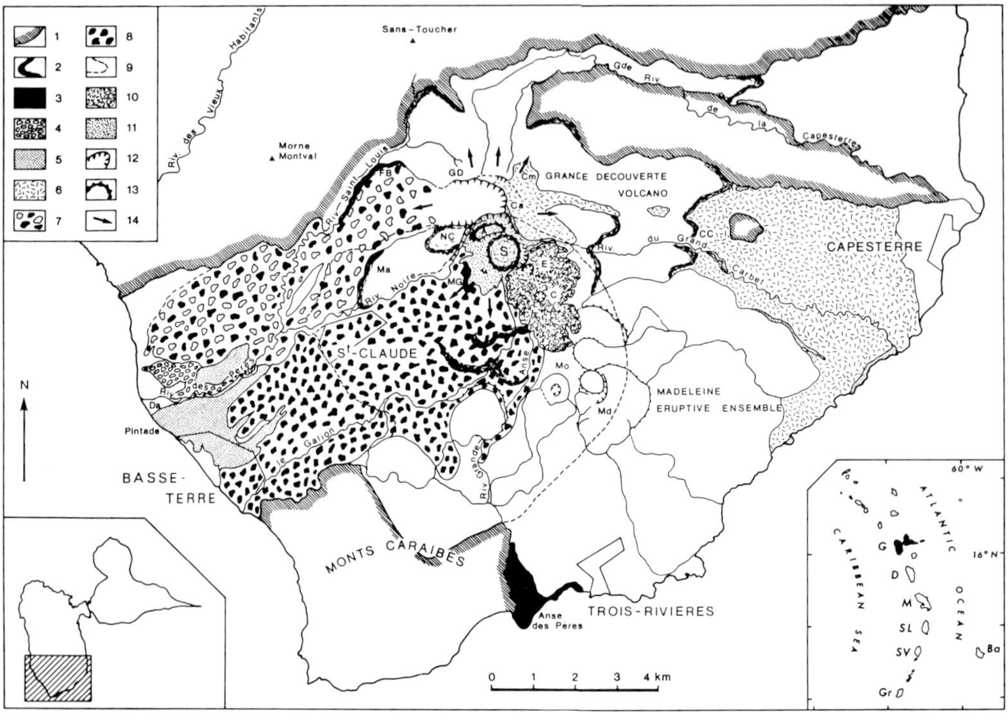

Fig. 1. Simplified geological map of the composite volcano of la Grande Découverte. The **inset** shows the situation of Guadeloupe **(G)** in the lesser Antilles island arc: **D** Dominica; **M** Martinique; **SL** Saint Lucia; **SV** Saint Vincent; **Ba** Barbados; **Gr** Grenada. Most localities cited in the text are reported on the map as abbreviations: **A** Amic dome; **C** la Citerne scoria cone; **Ca** remnants of Carmichaël volcano; **CC** third fall of the Carbet river; **Cm** Crète à Mangles lava flow; **Da** Danois quarry; **E** l'Echelle scoria cone; **FB** Fond Bernard lava flow; **GD** Grande Découverte summit; **Ma** village of Matouba; **Md** Madeleine dome; **MG Morne Goyavier lava flow; Mo** Moscou plain; **NC** Nez Cassé flow dome; **S** la Soufrière dome. Symbols on the map are as follows: **1 Limits of Recent volcanism. Grande Découverte stage: 2** flow fronts of the major lava flows, including those that underlie more recent debris flows; **3** quartz dacitic pumice flows of Anse des Pères; **4** debris flows; **5** Pintade pumice flows. Carmichaël stage: **6** undifferentiated lava and pumice flows (30 000-18 000 B.P.); **7** debris flow of the 11 500 B.P. Bandaï-san-type eruption; **8** debris flow of the 3100 B.P. Mount St. Helens-type eruption; **9** boundaries of the mapped blast deposits, and of parts of the debris flow underlying more recent deposits (3100 B.P. eruption). La Soufrière stage: **10** l'Echelle and la Citerne scoria cones; **11** undifferentiated products of the Amic and la Soufrière eruptions. **12** calderas and craters; **13** domes and flow domes; **14** directions of maximum dispersion of volcanic products of the different stages

2. On the north, by the Axial Chain, the main volcanic massif of Basse-Terre. This NNW-SSE trending chain consists of a suite of composite volcanoes. The Sans Toucher massif is the southernmost member of this suite. The construction of the chain seems to have ended about 0.6 MA ago (Blanc 1983; Gadalia et al., in prep.);

3. On the northeast, by the Bouillante Chain (Gadalia et al., in prep.), consisting of small basaltic to rhyolitic eruptive centers. The youngest deposits attributed to this chain are about 0.25 MA old (Blanc 1983).

The volcanic complex, of which la Soufrière is a part, comprises eruptive centers younger than those described above (Fig. 1). It has long been considered as a series of juxtaposed monogenetic volcanoes, aligned on a NNW-SSE trending fissure (e.g., Mervoyer 1974; Westercamp and Tazieff 1980; Dagain et al. 1981; Gadalia 1985). We propose, however, that it consists of two composite edifices:

1. The Grande Découverte-Soufrière volcano, a complex structure whose deposits cover an area of about 200 km^2. Its highest elevation, 1467 m, is la Soufrière dome which is also the highest point in the Lesser Antilles islands;

2. A volumetrically smaller eruptive ensemble, southeast of the preceding volcano. Its activity ended with the emplacement of several domes, one of which (la Madeleine dome) is a major feature of southeastern Basse-Terre.

"La Soufrière" is the toponym of the dome which was formed by the most recent activity of the Grande Découverte-Soufrière volcano. We therefore prefer to name the whole structure the composite volcano of la Grande Découverte (C.V.G.D.) or simply Grande Découverte, after the topographic name of a summit close to the center of symmetry of the edifice. This summit is also the only remnant of the primitive volcano. In addition we propose to name the Recent volcanic structures which encircle the Moscou plain (Fig. 1), SE of C.V.G.D., la Madeleine eruptive ensemble.

Ten years ago, la Soufrière entered an 18-months period of unrest, which culminated in several sizeable phreatic explosions. A polemic developed between scientists, government officials, and the press (e.g., Tazieff 1977; Barrère 1977; Bostok 1978; Sigvaldason 1978; Fiske 1979; Tomblin 1979), and it became apparent that a thorough knowledge of the volcano's past behavior was lacking. This was one of the main reasons for the difficulties in coming to a consensus of opinion on the possible development of the crisis. Since then, we and others have tried to obtain a precise knowledge of the stratigraphical and structural relationship of the volcano and of its volcanological evolution. Part of this work has involved obtaining detailed geological, petrographic, and age data for the most recent eruptions.

This chapter presents a descriptive summary of our views on the volcanological and magmatological evolution of the composite volcano of la Grande Découverte.

2 The Composite Volcano of La Grande Découverte: Stages in its Evolution

C.V.G.D. is located on the southern flank of the Sans Toucher volcanic massif (Fig. 2). The age of the start of its activity is not accurately known; it is bracketed by (1) the most recent age obtained for products of the Axial and/or Bouillante Chains, 0.244 MA (quartz thermoluminescence age on the dacite pumice of Danois quarry (Fig. 1), Blanc 1983), and (2) the oldest age obtained for products attributed to C.V.G.D., 0.143 MA (K/Ar age of an andesite flow at Fond Bernard, 1 km north of Matouba, Blanc 1983). Older lava flows and agglomerates are known but have not been dated. We assume that activity at Grande Découverte started approximately 200,000 years ago. Bouillante and/or Axial Chain volcanoes may still have been active at that time (Gadalia et al., in prep.).

Fig. 2. Oblique aerial photo of the southern Basse-Terre island of Guadeloupe viewed from the west. La Soufrière dome, partly obscured by clouds, tops la Grande Découverte volcano in the center of the photo. Two hills forming the horizon line in the right quarter are part of la Madeleine dome. The slope from the left to the center right is the Sans Toucher southern flank

The evolution of C.V.G.D may be subdivided into three main stages (Fig. 3):
1. The Grande Découverte stage (0.200-0.100 MA);
2. The Carmichaël stage (0.100 MA-ca. 3000 B.P.);
3. La Soufrière stage (ca. 3000 B.P. to the present).

2.1 The Grande Découverte Stage (0.200-0.100 MA)

Volumetrically, this is the major volcanic stage. It corresponds to the growth of the primitive edifice (Fig. 3 a). Unfortunately, it is also the least known, as most products of its activity are either hidden by later deposits or have been eroded or destroyed by catastrophic eruptions. Products of stage 1 crop out sporadically on the massif, and in the deep canyons of Rivière Noire and Rivière St. Louis (Fig. 1). A detailed stratigraphic column has not been constructed and accordingly we will only enumerate the types of products encountered.

The primitive edifice of C.V.G.D. grew through a succession of effusive and pyroclastic eruptions. Lava flows seem to be dominant; scattered outcrops, where several flows, 10 to 100 m thick, are piled up, may be found at various elevations. These flows form the skeleton of the volcano. Three K/Ar ages have been obtained by Blanc (1983): the Fond Bernard flow, 0.143 (±0.024) MA; the flow of the third Carbet river fall, 0.129 (±0.015) MA; and the Morne Goyavier flow, 0.100 (±0.035) MA. Pyroclastic deposits appear to be volumetrically less important. Limited outcrops are known over the whole surface of the volcano, mainly around its periphery. In addition to the major pyroclastic event that terminated the Grande Découverte stage (see below), other pyroclastic products are listed below.

1. Scoriaceous pyroclastic rocks interbedded with debris flows older than the Pintade pumice (G. Boudon and J. Dagain, unpublished fieldwork, March 1986). They crop out in the valleys of Rivière Noire and Rivière des Pères and may represent some of the oldest products of the primitive edifice.

2. Quartz dacitic pumice, first described by de Reynal de St. Michel (1966) at l'Anse des Pères (Atlantic coast west of Trois Rivières) where pumice flows are about 15 m thick. There, the deposits lie uncomfortably over the Monts Caraïbes volcanic formation. Thermoluminescence dating of the quartz (Blanc 1983) yields an age of 0.140 (±0.023) MA. On the basis of petrographic and geochemical similarities, Vatin-Pérignon et al. (1984), and, in more detail, Gstalter (1986), attributed some quartz dacite pumice falls of the Caribbean flank of the volcano to this episode. At the Morne Montval (Fig. 1) location, Blanc (1983) obtained a 0.108 (±0.015) MA age by the same method. Recently, new outcrops of pumice flows with the same petrographic characteristics have been found (G. Boudon et J. Dagain, unpublished fieldwork, March 1986) in the valley of Rivière des Pères, on the western flank of C.V.G.D. It therefore seems likely that the Anse des Pères quartz dacite pumice was originally deposited over a large area.

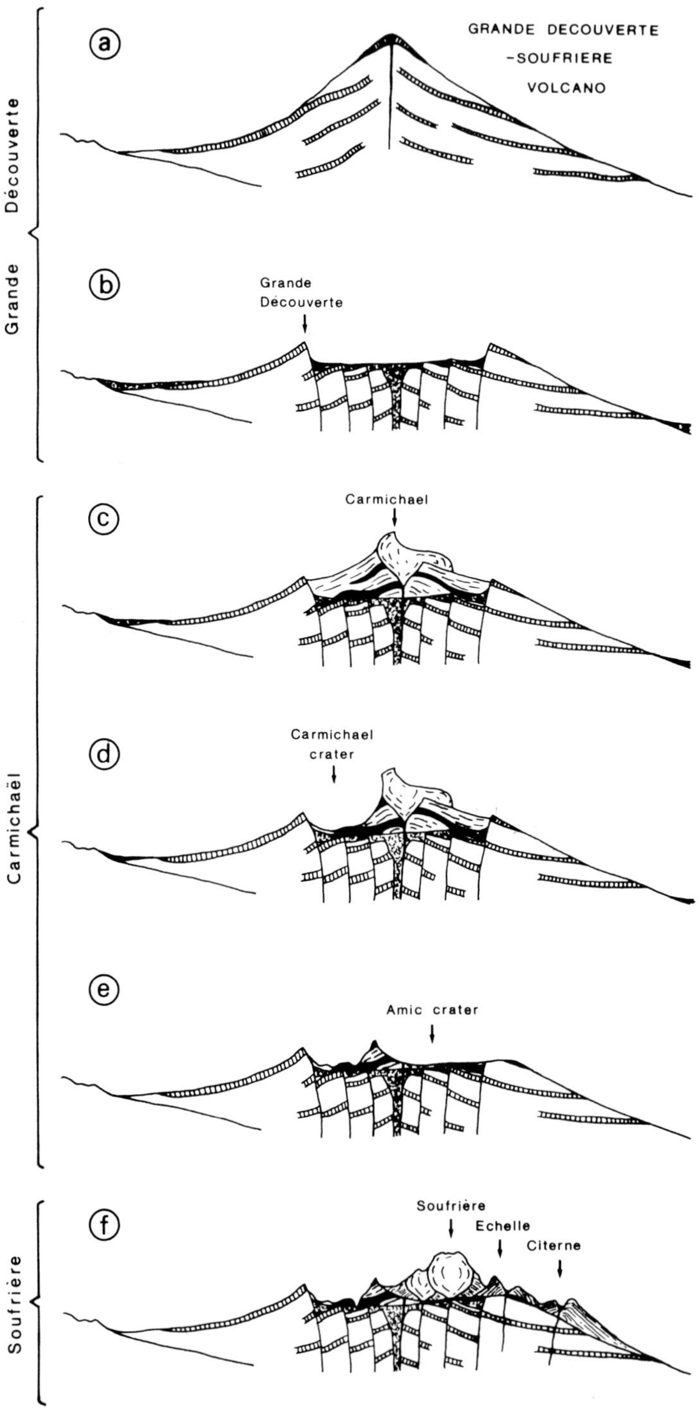
Grande Découverte
Carmichaël
Soufrière
a
GRANDE DECOUVERTE
-SOUFRIERE
VOLCANO
b
Grande
Découverte
c
Carmichael
d
Carmichael
crater
e
Amic crater
f
Soufrière
Echelle
Citerne

Major Event Ending the Grande Découverte Stage. A Plinian episode that is most probably the major pyroclastic event in the evolution of Grande Découverte occurred after the main period of growth of the primitive edifice. It is called the Pintade episode after the name of a suburb of Basse-Terre city where pumice flows are well exposed in quarries. The episode gave rise to thick flow and fall sequences.

Pumice falls show alternating meter-thick ash and pumice beds that mantle older topography. Their combined thickness is of the order of 10 m where they overlie Axial Chain deposits a few kilometers WNW of the volcano. However, fall beds are almost absent on Grande Découverte itself, and easterly trade winds have presumably contributed to their maximum dispersion to the west.

Although the outcrops of the Pintade pumice only cover a relatively small aggregated area (7-8 km^2), they are scattered over a much larger surface indicating an original extension of perhaps more than 100 km^2. Pumice flow beds, meter- to decameter-thick, form some of the Atlantic and Caribbean sea cliffs, which are over 30 m high (Fig. 4). Surge or fall deposits are rarely present, interbedded with flows. Emplacement temperatures were of the order of 400 °C (Maury et al. 1976; Zlotnicki et al. 1984) which is quite high enough to carbonize abundant wood fragments.

The total volume of the Pintade pumice can only be roughly estimated, but may be of the order of 3 km^3. A summit caldera, the Grande Découverte caldera, was formed as a consequence of the Pintade episode (Fig. 3 b). Only its northern margin remains, as it has been almost completely obliterated by subsequent volcanic activity. The caldera diameter might have been about 2.5 km.

The age of the Pintade eruption is not precisely known. It is older than the 40 000 years that can be measured by conventional ^{14}C dating (Paterne 1980; Jérémie 1980) and younger than the age of stratigraphically older lava flows (e.g., 0.100 MA) dated by the K/Ar method (Blanc 1983). Its age could be close to 0.070 MA (G. Glacon, unpubl. micropalaeontological data, 1983).

2.2 The Carmichaël Stage (0.100 MA-ca. 3000 B.P.)

The volcanic activity that we attribute to this stage took place within the Grande Découverte caldera. It corresponds to the

Fig. 3 a-f. Schematic N-S cross-sections of la Grande Découverte during successive stages of its evolution. **a** Primitive edifice of la Grande Découverte stage; **b** la Grande Découverte caldera after the Pintade eruption; **c** Carmichaël volcano at the end of its growth; **d** formation of Carmichaël crater during the 11 500 B.P. Bandaï-san-type eruption; **e** formation of Amic crater during the 3100 B.P. Mount St. Helens-type eruption; **f** present state of the volcano

Fig. 4. Pintade pumice flows in the Calbassiers quarry of Pintade (Basse-Terre city). At least five flows with no intervening fine ash are seen in the foreground cliff

growth of Carmichaël volcano (Fig. 3 c) and its almost total destruction by two catastrophic sector collapses (11 500 and 3100 B.P.). Lava flows NE and E of the Carmichaël summit (e.g., the Crète à Mangles flow to the east) and the dome flow of Nez Cassé to the south west, both of which overflowed the presumed caldera margins, are thought to be remnants of the products of this volcano. On the basis of their outcrop distribution, and from their petrographic characteristics, we also attribute pyroclastic flows and nuées ardentes deposits of the Grand Carbet valley (Fig. 1) to eruptions of the Carmichaël volcano. These were previously thought to represent activity of la Madeleine eruptive ensemble (e.g., Dagain 1981). Radiocarbon ages obtained for these deposits range from 30 000 to 18 000 B.P. (Paterne 1980; J.C. Fontes, unpubl., 1981).

A dormant period of about 7000 years apparently followed the last pyroclastic activity of Carmichaël volcano, as is suggested by the absence of volcanic products with ages in the range 18 000 to 11 500 B.P. This dormant period could correspond to protracted hydrothermal activity that preceded the first catastrophic collapse of a sector of C.V.G.D.

This first major collapse, dated at about 11 500 B.P. (Boudon et al. 1987), corresponds to the breakdown of the western sector of Carmichaël volcano. The collapse led to the emplacement of a major debris flow to the west, and was accompanied by an explosive eruption and the formation of blast deposits (Fig. 1). We believe that the horseshoe-shaped Carmichaël crater (1.5 x 1.2 km, ca. 300 m deep) which is open to the west, was formed during this event (see below). The debris flow is particularly outstanding in showing abundant mega-blocks of hydrothermally-altered rock types. It extends for about 10 km, from the mound of Carmichaël crater to the Caribbean Sea, with a difference in elevation of about 1100 m (Fig. 1). Deeply weathered blast deposits are not extensive. They are observed at a maximum distance of 6.5 km from the crater. The areas covered by the debris flow and the blast deposits are approximately 20 km^2 and 10 km^2, respectively, and their combined volume about 1 km^3. No juvenile magmatic component seems to be present. On the contrary, hydrothermally-altered rocks and ashes are abundant. Deep hydrothermal alteration is also a characteristic feature of the bottom and walls of Carmichaël crater, which does not show any signs of magmatic activity following its formation. We therefore infer that the eruption which gave rise to the 11 500-year-old debris flow was contemporaneous with the blast deposits, and that the formation of Carmichaël crater was similar to the 1888 Bandaï-san event in Japan (Sekiya and Kikuchi 1889).

A second catastrophic collapse (Boudon et al. 1984, 1987) took place some 3000 years ago. In this event, the southern sector of Grande Découverte avalanched into the Caribbean Sea (Fig. 1), 9 km away. Depressurization caused major contemporaneous explosions. A debris flow extends over 28 km^2 in the Basse Terre-St. Claude area and has a minimum volume of 0.5 km^3. Its maximum observed thickness in the deep gorge of the Galion river is about 80 m. Blast deposits have been mapped over a 60-km^2 area but may have covered some 100 km^2 at the time of the eruption. They are characterized by a complex stratigraphy with two conspicuous coarse-grained beds (average grain size of 1 to 10 cm) rich in a juvenile gray andesite. The amphitheater <u>Amic crater</u> (Figs. 1 and 3 e), formed by the collapse and explosions, is some 1.7 x 1.3 km in diameter. It cuts <u>Carmichaël crater</u> (Fig. 5) on the north, and also cuts the southeastern part of the Nez Cassé dome. It is now partially filled with eruptive products of later activity.

Radiocarbon ages of uncharred wood of the debris flow, and of carbonized wood fragments of the coarse-grained part of the blast deposits, are concentrated around 3100 B.P. (1380 cal BC). Unlike the 11 500 B.P. episode of Carmichaël volcano, this eruption had a definite magmatic component. Boudon et al. (1984) surmised that this second catastrophic eruption was similar to the 18 May 1980 Mount St. Helens event (Lipman and Mullineaux 1981), and that the juvenile andesite present in the blast and debris-flow deposits originated from a recently emplaced cryptodome.

The magmatic activity that triggered the 3100 B.P. event proceeded into the next stage.

Fig. 5. Aerial photo looking south showing Carmichaël crater **(foreground)** and remnants of Amic crater **(center)**. Carmichaël crater is open to the **right.** The northern rim of Amic crater is still well preserved. The crescent-shaped hill in **front,** and **to the right** of the fuming la Soufrière dome is the remaining part of Amic dome

2.3 La Soufrière Stage (ca. 3000 B.P. to the present)

After the 3100 B.P. cataclysmic eruption, volcanic activity took place within and on the southern margin of Amic crater (Fig. 6); this stage is attributed to activity of la Soufrière, sensu stricto. At least two eruptions took place between the 3100-year-old Mount St. Helens-type event and more recent eruptions (see below). These gave rise to the emplacement of Amic dome and to the formation of l'Echelle and la Citerne scoria cones.

Fig. 6. Carmichaël crater and la Soufrière dome viewed from the west. The 11 500 B.P. debris flow, from the top of which the photo was taken, extends to the mound of the crater. The mound is emphasized here by part of the Grande Découverte **(left skyline)** and the flow dome of Nez Cassé **(central skyline)**

2.3.1 Amic Eruption

A crescent-shaped hill (Morne Amic) partially filling Amic crater, presents a single outcrop of a massive purplish andesite. Péléan nuées ardentes of the same petrographic characteristics are observed south and west of Morne Amic within a 2-km radius. We argue that the hill is the remnant of a Péléan dome (Amic dome) which has been mostly destroyed during later eruptions of la Soufrière. The eruption that gave rise to these products has not been dated.

2.3.2 L'Echelle and la Citerne Scoria Cones

These two scoria cones were emplaced along a line oriented NNW-SSE toward the eastern wall of Amic crater. A number of fossil as well as presently active faults (Julien and Bonneton 1984) have the same general strike. This direction also corresponds to the alignment of the main volcanic centers in Basse-Terre as a whole.

L'Echelle and la Citerne cones (Fig. 7), 1397 and 1150 m high respectively, consist of interbedded scoria, ash, and phreatomagmatic deposits of alternating black and reddish color. The deposits of both cones are clearly imbricated in some of their outcrops, indicating nearly contemporaneous emplacement.

Fig. 7. Aerial view of the recent structures of la Soufrière stage. The view is looking north. From foreground to background: la Citerne and l'Echelle scoria cones, and la Soufrière dome with an active fumarole at the top

La Citerne is a regular truncated cone rising about 150 m above its base. The summit is occupied by a 50-m-deep crater some 250 m in diameter: its well-preserved geometrical appearance suggests a phreatomagmatic origin.

L'Echelle, on the other hand, is morphologically not well preserved. The 250-m-high cone is breached on its SW and N flanks, exposing short basaltic flows that are particularly rich in xenolithic blocks of an andesitic lava. In addition, the northwestern flank, cut by the crater occupied by la Soufrière dome, and the southwestern flank, show active fumarolic activity.

Neither cone has been directly dated. They are probably younger than the 3100 B.P. eruption because they are not covered by any of its deposits. Radiocarbon ages of 1800 (±100) B.P., 1750 (±?) B.P., 1700 (±75) B.P., and 1545 (±65) B.P. have been obtained (Paterne 1980; Dagain 1981; J.C. Fontes unpubl. 1981; Gadalia and Westercamp 1984) on uncharred wood fragments sampled in Ravine Chaude (2 km south of la Soufrière), Rivière Grande Anse, and the Galion river, respectively, less than 2 km from the cones, in lahars and lacustrine deposits. These may correspond to the formation of the cones.

2.3.3 Recent Activity of la Soufrière

Within the last few centuries, la Soufrière stage was characterized by a magmatic episode, namely the so-called 16th century eruption (Vincent et al. 1979; Semet et al. 1981 and in press). Historical eruptions, since 1635, have been entirely phreatic.

The most recent magmatic event at la Soufrière was a Péléan eruption that gave rise to pyroclastic deposits and to the growth of a dome. The first radiocarbon age obtained for these deposits dates back some 35 years ago (Bruet 1953). Since then, a number of new radiocarbon determinations have been made on the pyroclastic deposits (Dorel et al. 1972; Paterne 1980; Dagain 1981; J.C. Fontes, unpubl.), giving an average age of 429 (±66) B.P. Before 1982, the uncalibrated date of AD 1521 (1950-429) was used to argue that the eruption happened between Columbus' second voyage in 1493, when he landed on Guadeloupe, and European settlement in 1635. Since 1982, however, Stuiver (1982) and Klein et al. (1982) have provided a precise dendrochronological scale for radiocarbon determinations, and this leads to a revised date of cal AD 1440 (±100). The so-called 16th century eruption (see above) probably therefore took place before Columbus' visit!

The AD 1440 episode started with the explosive eruption of pumiceous dacite, followed by more basic scoriaceous ejecta. The latter products are often banded and show a mechanical mixture of scoriaceous andesite and pumiceous dacite. La Soufrière crater, partially encroaching on Amic dome and L'Echelle scoria cone most probably formed during these explosive eruptions. The episode ended with the emplacement, within the crater, of la Soufrière dome, of basic andesitic composition. ^{14}C ages of 680 (±30) B.P., corresponding to cal AD 1250 (±70) (Paterne 1980; Dagain 1981; J.C. Fontes, unpubl.), were obtained on charcoal from pyroclastic products similar to the AD 1440 deposits. This indicates that a minor explosive event might have taken place some 200 years before the latter eruption.

The pyroclastic products of these eruptions show meter-thick outcrops of fall deposits within 1.5 km of the crater. Pyroclastic flows of the same age are also found in the upper valleys of the Galion and Carbet rivers (see Westercamp and Tazieff 1980, Fig. 53, for the extension of the deposits). Their combined volume is rather small (much less than 0.1 km^3), whereas the estimated volume of the dome is about 0.1 km^3.

Semet et al. (1981, in press) have attributed the triggering and sequence of events of the AD 1440 eruption to injection of hot basic magma into a differentiated and fluid-saturated magma reservoir of dacite composition at a depth of about 6 km. The injection apparently preceded the explosive eruption by not more than a few months (perhaps only by tens of hours).

Several phreatic eruptions, which took place since 1635, characterize the historical period. Two of them (1797-98 and 1975-77) were major sequences of steam blasts that ejected dust and blocks. Accumulation of the products emitted led to the formation of small lahars in the main rivers. Minor phreatic eruptions (Lacroix 1904; Jolivet 1958; Barat 1984) also took place. Fumarolic activity was nearly continuous for the whole of the historical period.

The two major events seem to have been quite comparable (e.g., Rapport 1798; Feuillard et al. 1983; Barat 1984). Both were preceded by year-long seismic crises, and were marked by sizeable steam blasts that opened new radial fissures within la Soufrière dome or cleared existing ones. Explosive products rich in hydrothermal clays, gypsum, and pyrite (Jérémie and Pascaline 1982; H. Pascaline and J.J. Jérémie, unpubl., 1983) could all be attributed to rock types from the basement, notwithstanding the presence of a small fraction of unaltered volcanic rocks (Marinelli 1976; Brousse et al. 1977; Heiken et al. 1980). Block ejecta in the vicinity of the dome, and the highly acid dust and gases that were carried to the coast by trade winds had a severe ecological impact on mountain and tropical forest vegetation (Sastre et al. 1983).

The well-monitored recent eruption of 1975-77 (Hirn and Michel 1979; Pozzi et al. 1979; Dorel and Feuillard 1980; Le Guern et al. 1980; Heiken et al. 1980) was identified as a "still -born" magmatic eruption (Feuillard et al. 1983) on the basis of the geophysical effects and the presence of a magmatic gas phase.

3 Geochemical Evolution of C.V.G.D.

This section summarizes the petrological and geochemical characteristics of rocks from the different stages of C.V.G.D., as they bear on its evolution, and compares them with rocks from other volcanic centers of southern Guadeloupe. The data base comprises detailed studies of : (1) the AD 1440 eruption (Semet et al. 1981, in press); (2) the Pintade pyroclastic products (Blanc 1983; Vatin-Pérignon et al. 1984); (3) the quartz dacitic products of Anse des Pères (Gstalter 1986); and (4) magmatic products of the 3100 B.P. eruption (unpublished data). Additional sources of petrological and chemical data used are the general studies of Westercamp and Mervoyer (1976), Pascaline (1980), Dagain (1981), Pascaline and Jérémie (1984), Lefèvre and Cocusse (1985), and unpublished work by J.L. Joron and M.P. Semet (1980-86).

FeO^*/MgO-SiO_2, AFM, and K_2O-SiO_2 plots of the data (see Gill, 1981 for nomenclature) are presented in Fig. 8. Irrespective of age, all rocks from southern Guadeloupe apparently define single trends close to the CA-TH boundary of Gill (1981) and have relatively low K_2O concentrations. A few rock types that have TH affinities belong to the Quaternary volcanic products of the Axial and Bouillante Chains and the Monts Caraïbes, all older than C.V.G.D. The chemical similarity of C.V.G.D. rocks to others from southern Guadeloupe, and, more generally, of Island Arc series, is also reflected in their petrographic characteristics. Andesites rich in plagioclase, two pyroxenes, and oxides are the dominant rock types, whereas aluminous basalts, dacites, and rhyolites are subordinate. Olivine pheno-

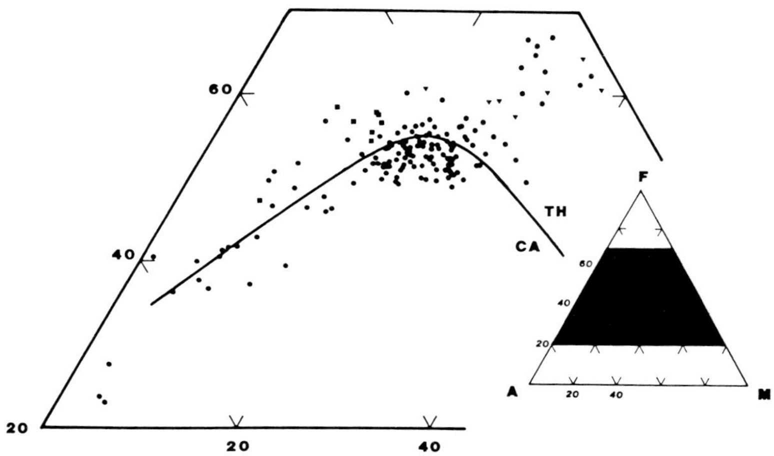

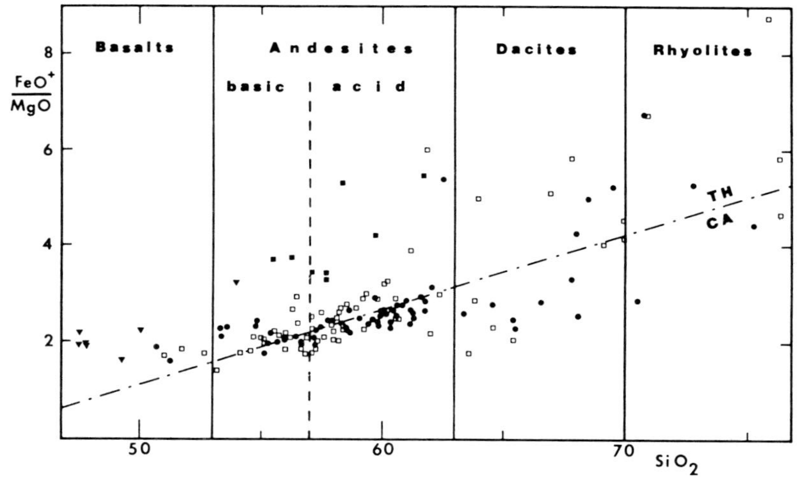

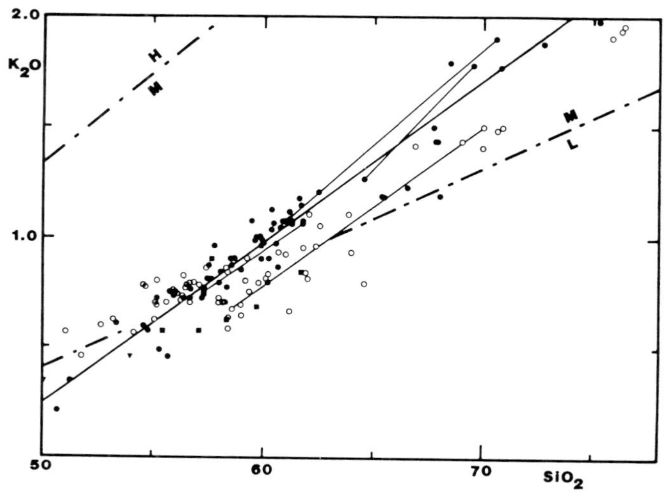

Fig. 8. Major chemical element diagrams for Quaternary volcanic rocks of southern Basse-Terre. AFM diagram: **dots**, undifferentiated rock samples from the Axial and Bouillante Chains, and from C.V.G.D.; **squares**, dated lava flows from the Axial and Bouillante Chains (0.6-1.2 MA, Blanc 1983); **triangles**, Monts Caraîbes effusive rocks. The TH-CA boundary has been drawn after Gill (1981). K_2O-SiO_2 diagram: **black dots**, Axial and Bouillante Chain rock samples; **open dots**, C.V.G.D. samples; **triangles**, Monts Caraîbes samples. **Thin straight lines** join data points for separated basic and acid parts of banded tephra. The linear regression of all data points is shown by the **heavy line**. The low (**L**), medium (**M**), and high (H) K_2O series (e.g., Gill 1981) are limited by **dash-dot lines**. FeO*/MgO-SiO_2 diagram: **black dots**, C.V.G.D. samples; **open squares**, undifferentiated Axial and Bouillante Chain samples; **black squares**, dated flows of the Axial and Bouillante Chains; **triangles**, Monts Caraîbes samples. TH-CA boundary from Gill (1981)

crysts are present in the basalts, basic andesites, and the basic parts of banded pyroclastic rocks. In some andesites, quartz showing reaction aureoles may be present with olivine microphenocrysts. Quartz phenocrysts, often accompanied by hornblende, are only present in the rare dacitic to rhyolitic pumice deposits. Quartz-bearing pumice deposits frequently contain banded lumps with a basic olivine-bearing component.

Contrary to the major chemical element data, trace element concentrations (Blanc 1983; Vatin-Pérignon et al. 1984; Gstalter 1986; J.L. Joron and M.P. Semet, unpubl.) show distinctive trends for the different southern Guadeloupe volcanic massifs. The SiO_2-Th, La-Th, Co-Th, and Co/Sc-Th plots of Fig. 9 show how these elements clearly discriminate between products of the

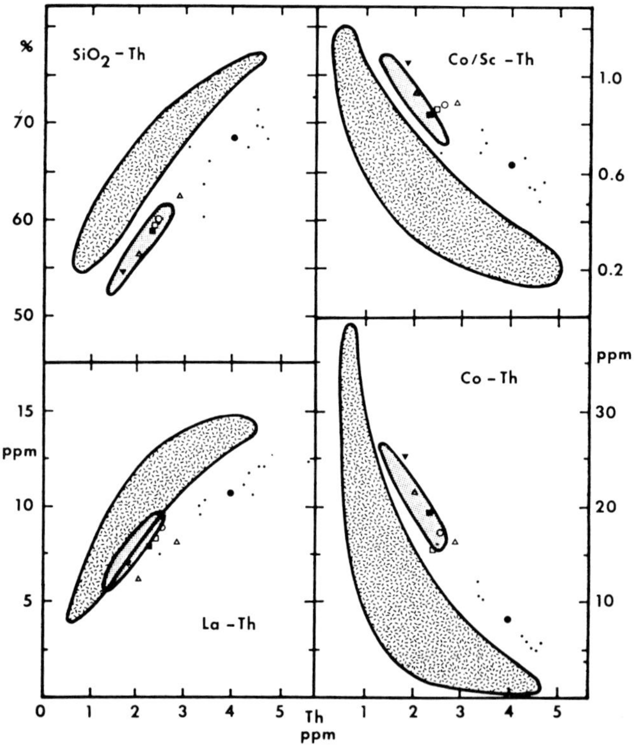

Fig. 9. Trace element data for rocks of the southern Basse-Terre volcanic series. **Unpatterned stippling,** envelope of the Axial and Bouillante Chains compositions; **patterned stippling,** la Madeleine samples; **black square,** average composition of lava flows from the primitive edifice of the Grande Découverte stage; **large dot,** average composition of whole-rock samples of the Anse des Pères pumice; **small dots,** individual pumice samples of Anse des Pères deposits; **open circle,** average Pintade pumice; **open square,** pyroclastic deposits of the Carmichaël stage; **black triangle,** average composition of l'Echelle and la Citerne scoria cones; **open triangles,** basic and acid parts of products of the AD 1440 eruption

Axial and Bouillante Chains, of the Monts Caraïbes, and those of the more recent activity. Products of la Madeleine eruptive ensemble are close to the C.V.G.D. trends, yet distinguishable. Because trace elements reflect the characteristics of the mantle magma sources and differentiation processes affecting parental magmas, better than major elements do, we surmise that the different series evolved in different magmatic plumbing systems.

During most of its volcanological evolution, C.V.G.D. emitted products whose composition did not vary much (Fig. 8). There are only minor compositional differences between lava flows of the primitive edifice (early Grande Découverte stage, ca. 0.150 MA) and the pyroclastic products that were emitted near the end of the Carmichaël stage (≤18 000 B.P.). During this whole period, effusive products have an average composition only a little less "evolved" [SiO_2 = 59.4 (±1.2)%; Th = 2.38 (±0.32) ppm] than the average Pintade pumice [SiO_2 = 61.0 (±1.0)%; Th = 2.49 (±0.30) ppm] or other pyroclastic products [SiO_2 = 60.2 (±0.5)%; Th = 2.37 (±0.25) ppm]. A notable exception is the Anse des Pères pumice (ca. 0.124 MA). This shows chemical heterogeneities at the hand-specimen (banded quartz dacite-olivine andesite pumice) and outcrop scales. The average bulk composition of the pumice is more evolved than the other products and shows a larger variability [SiO_2 = 68.7 (±3.9)%; Th = 3.91 (±0.88) ppm]. Moreover, the most basic olivine-bearing parts of banded pumice presumably are basaltic and much less evolved. However, separated fractions of the basic parts could not be obtained and analyzed.

Basic products and dacites also occur with andesites similar to those of the first two stages in la Soufrière volcanism. Basic andesites are the most abundant products of the AD 1440 eruption of la Soufrière, and basalts of l'Echelle and la Citerne scoria cones.

In conclusion, C.V.G.D. products span a compositional range as large as for all Quaternary volcanics. As in the wider case, andesites are dominant. The Grande Découverte series, however, has distinctive and restricted geochemical trends.

4 Summary

The composite volcano of la Grande Découverte evolved through a succession of episodes of growth and violent destruction. This resulted in a complex morphology from which original geometrical features are difficult to reconstruct. In addition, the heavily forested mountain area gives few exposures. We propose that C.V.G.D. is a central composite volcano, and not part of a fissural ensemble of monogenetic volcanoes aligned on a NNW-SSE trend, a hypothesis that has been advocated in the past. Our main arguments are the following:

1. C.V.G.D. is geometrically well defined as a central volcano installed on the southern flank of Sans Toucher (Fig. 2). The morphology of its northern part was mostly acquired by the

growth of the primitive edifice of the Grande Découverte stage before partial destruction by the Pintade eruption (Fig. 3);

2. The large-scale plinian Pintade event with a volume of several km^3 has resulted in the formation of the Grande Découverte caldera within which most subsequent activity has taken place;

3. Several overlapping craters (e.g., Carmichaël, Amic, and la Soufrière craters) lie within the caldera;

4. Two major recent eruptions involving the collapse of a sector of the volcano (11 500 and 3100 B.P.) are not considered possible events on minor monogenetic volcanoes. Deposits of two other debris flows older than the Pintade eruption reinforce this argument by suggesting other sector collapses;

5. The unique geochemical trends shown by products of C.V.G.D. are evidence for an unshared magma supply system.

However, if one considers the large-scale structures of the volcanic arc through the Quaternary, C.V.G.D. is indeed in alignment with the NNW-SSE trend of the main volcanic centers of southern Guadeloupe (e.g., Axial Chain) and even with the Martinique-Montserrat segment of the Lesser Antilles arc. This alignment is thought to reflect the regular structure of the Benioff zone in this segment of the Lesser Antilles arc (Bouysse and Guennoc 1983; Bouysse et al. 1985; Andreieff et al. 1987).

The growth of the primitive C.V.G.D. was followed by caldera collapse during the Pintade eruption. Volcanic activity for most of the Carmichaël and la Soufrière stages took place within the caldera. The basaltic composition of l'Echelle and la Citerne scoria cones, located on the southern margin of the presumed caldera, might have inherited their primitive chemical composition from the fact that deep supply of magma did not intersect shallow reservoirs. The evolution of C.V.G.D. is quite comparable to that of other large andesitic volcanoes (e.g., Cantagrel et al. 1981; Robin and Cantagrel 1982; Robin 1984). Sector collapses during the 11 500 and the 3100 B.P. eruptions have profoundly modified its summit morphology. More recent activity has been mostly restricted to the crater formed in the latter eruption. Historical phreatic eruptions seem to draw their energy from a hydrothermal system connected to the most recent magma conduits of la Soufrière stage.

5 Application to Hazard Assessment and Volcano Monitoring

The occurence on C.V.G.D. of at least two eruptions that involved sector collapses and an associated directed blast, one of which (3100 B.P.) involved a direct magmatic trigger, while the other (11 500 B.P.) did not, shows that this type of eruption may recur on a single volcano (see Bogoyavlenskaya et al. 1985, for another case). Volcanic hazard assessment must take such recurrence into consideration. This was not done in evaluations that were made in the past (Westercamp 1980, 1984). Furthermore, the methods of volcano surveillance might have to be reconsidered in view of the 11 500 B.P. event. In an event similar to the Mount St. Helens 18 May 1980 eruption, with which the 3100 B.P. eruption is thought to be comparable, precursory signs of an impending eruption should not escape detection by the permanent geophysical networks of la Soufrière Volcano Observatory. However, in a scenario where the mechanical instability of the volcano is not caused by shallow magma emplacement, e.g., the 11 500 B.P. outbreak, precursory signs may be much more difficult to evaluate. Premonitory signs of an impending disaster at Bandaī-san in 1888 (Sekiya and Kikuchi 1889) were restricted to a few felt seismic shocks a week before the eruption and a more severe earthquake immediately preceding it, on the morning of the 15 July 1888. Neither visible deformation nor increase in the fumarolic activity were recorded. The same premonitory signs might have preceded the 11 500 B.P. event in Guadeloupe. How is then one to evaluate the outcome of preeruptive events similar to those observed before the historical phreatic eruptions of la Soufrière, which are of the same nature and intensity as those at Bandaī-san?

Another problem may arise in forecasting the location of future volcanic events. Even though magmatic and phreatic eruptions have been localized in or on the margin of Amic crater during the last 3000 years, migration of the eruptive center within the Grande Découverte caldera is a distinct, albeit remote, possibility if one considers the past behavior of the volcano. This suggestion gains some ground from the observation that some of the seismic swarms observed since the 1975-77 crisis originated from locations several kilometers NNW of the usual site of epicenters below la Soufrière dome (M. Feuillard, pers. commun., 1985).

Acknowledgments. This chapter is the result of several years of work during which we have benefited from financial, logistic, and scientific support from many organizations and individuals. To all of these we are grateful. Heated discussions with D. Westercamp, J. Dagain, and A. Gadalia, our partners in adventure, materially helped to focus on the content of this work. We thank D. Westercamp and R. Maury for their kind reviews of a French version of this chapter. J.H. Latter and R.M. Briggs materially improved the final version. The Institut National d'Astronomie et de Géophysique (I.N.A.G.) then Institut National des Sciences de l'Univers (I.N.S.U.), and the Programme Interdisciplinaire de Recherche pour la Prévision et la Surveillance des Eruptions Volcaniques (P.I.R.P.S.E.V., C.N.R.S.) are warmly acknowledge for their continual financial support. The CORDET program of Ministère de l'Industrie et de la Recherche also provided partial support for our work. Continuing scientific and technical assistance from the personnel of Observatoire Volcanologique de la Soufrière was invaluable and cheered our hearts through pouring rain...

References

Andreieff P, Bouysse P, Westercamp D (1987) Géologie de l'arc insulaire des Petites Antilles, et évolution géodynamique de l'est-caraïbe. Thèse Doctorat d'Etat, Univ Bordeaux I, n° 921, 359 pp + Plates

Barat A (1984) Etude du rôle des eaux souterraines dans les mécanismes des éruptions phréatiques. Application à la Montagne Pelée de Martinique et à la Soufrière de Guadeloupe. Thèse 3ème cycle, Univ Bordeaux III, 232 pp + annexes

Barrère M (1977) Imbroglio scientifico-politique à la Soufrière. La Recherche 8 74:102-104

Blanc F (1983) Corrélations chronologiques et géochimiques des formations volcaniques du sud de la Basse-Terre de Guadeloupe (Petites Antilles). Début du cycle récent. Thèse 3ème cycle, Univ Sci Médic Grenoble, 171 pp + annexes

Bogoyavlenskaya GE, Braitseva OA, Melekestsev IV, Kiriyanov V Yu, Miller CD (1985) Catastrophic eruptions of the directed-blast type at Mount St. Helens, Bezymianny, and Shiveluch volcanoes. J Geodynamics 3:189-218

Bostok D (1978) A deontological code for volcanologists? Editorial. J Volcanol Geotherm Res 4, n° 1/2, 1

Boudon G, Semet MP, Vincent PM (1984) Flank-failure-directed blast eruption at Soufrière, Guadeloupe, French West Indies: a 3,000-yr-old Mt. St. Helens? Geology 12:350-353

Boudon G, Semet MP, Vincent PM (1987) Magma and hydrothermally driven sector collapses: the 3100 and 11,500 Y. B.P. eruptions of la Grande Découverte (la Soufriere) volcano, Guadeloupe, French West Indies. J Volcanol: Geotherm Res 33:317-323

Bouysse P, Guennoc P (1983) Données sur la structure de l'arc insulaire des Petites Antilles entre Ste Lucie et Anguilla. Mar Geol 53:131-166

Bouysse P, Westercamp D, Andreieff P, Baubron JC, Scolari G (1985) Le volcanisme sous-marin Néogène récent au large des côtes caraïbes des Antilles francaises:Relations avec le volcanisme à terre et évolution du front volcanique. Géologie de la France 1:101-114

Brousse R, Monneyron N, Semet MP (1977) Sur la présence de verres non altérés dans les projections de la Soufrière (Guadeloupe), durant la crise de 1976. CR Acad Sci Paris, Sér D 285:753-754

Bruet E (1953) L'âge absolu de la dernière grande éruption péléenne de la Soufrière de la Guadeloupe. Bull Volcanol, Sér II 13:105-107

Cantagrel JM, Robin C, Vincent PM (1981) Les grandes étapes d'évolution d'un volcan andésitique composite: le Nevado de Toluca (Mexique). Bull Volcanol 44:177-188

Dagain J (1981) La mise en place du massif volcanique Madeleine-Soufrière, Basse-Terre de Guadeloupe, Antilles. Thèse 3ème cycle, Univ Paris sud, Orsay, 156 pp + annexes

Dagain J, Paterne M, Westercamp D (1981) La mise en place du massif volcanique Madeleine-Soufrière, Basse-Terre de Guadeloupe, Antilles. CR Acad Sci Paris, Sér II, 292:921-926

de Reynal de Saint-Michel (1966) Carte géologique détaillée de la France. Département de la Guadeloupe. Feuilles de Basse-Terre et des Saintes. Minist Ind, Paris

Dorel J, Feuillard M (1980) Note sur la crise sismo-volcanique à la Soufrière de Guadeloupe 1975-1977. Bull Volcanol 43:419-430

Dorel J, Eschenbrenner S, Feuillard M (1972) Les volcans actifs de la Guadeloupe et de la Martinique, Petites Antilles. Bull Volcanol 36:359-381

Feuillard M, Allègre CJ, Brandeis G, Gaulon R, Le Mouel JL, Mercier JC, Pozzi JP, Semet MP (1983) The 1975-1977 crisis of la Soufrière de Guadeloupe (F.W.I.):A still-born magmatic eruption. J Volcanol Geotherm Res 16:317-334

Fiske RS (1979) A deontological code for volcanologists? A response to Derek Bostok's editorial. J Volcanol Geotherm Res 5:211-212

Gadalia A (1985) Cadre structural de l'activité hydrothermale des zones d'exploration de Bouillante et de Moscou (Guadeloupe). Rapport 85 SGN 080 GTH, 46 pp + annexes, BRGM Orléans

Gadalia A, Westercamp D (1984) Prospection géothermique de la région de Bouillante-Vieux Habitants, Guadeloupe. Rapport 84 SGN 063 GTH, 53 pp, BRGM Orléans

Gadalia A, Gstalter N, Westercamp D (in preparation) La chaine volcanique de Bouillante (Basse-Terre de Guadeloupe, Petites Antilles): identité pétrographique, volcanologique et context géodynamique.

Gill JB (1981) Orogenic andesites and plate tectonics. Springer, Berlin Heidelberg New York Tokyo, 390 pp

Gstalter N (1986) Les formations ponceuses à quartz de la région sud de Basse-Terre (Guadeloupe): pétrologie, géochimie, thermobarométrie et chimie des inclusions vitreuses. Thèse Univ Paris Sud, Orsay, n° 92, 296 pp

Heiken G, Crowe B, Mc Getchin T, West F, Eichelberger J, Bartram D, Peterson R, Wohletz K (1980) Phreatic eruption clouds: the activity of la Soufrière de Guadeloupe, FWI, August-October. Bull Volcanol 43:383-385

Hirn A, Michel B (1979) Evidence of migration of main shocks during major seismovolcanic crisis of la Soufrière (Guadeloupe, Lesser Antilles) in 1976. J Volcanol Geotherm Res 6:295-304

Jérémie JJ (1980) Age de certaines formations pyroclastiques du volcanisme quaternaire de la Basse-Terre de Guadeloupe, Petites Antilles. CR Acad Sci Paris, Sér D 290:843-846

Jérémie J J, Pascaline H (1982) Importance de la phase hydrothermale contenue dans les poussières émises lors des manifestations volcaniques explosives: cas des volcans actifs des Petites Antilles. CR Acad Sci Paris, Sér II 295:887-890

Jolivet J (1958) La crise volcanique de 1956 à la Soufrière de Guadeloupe. Ann Geophys 11:305-322

Julien P, Bonneton JR (1984) Etude néotectonique de la Soufrière de Guadeloupe. CR Acad Sci Paris, Sér II 298:907-909

Klein J, Lerman JC, Damon PE, Ralph EK (1982) Calibration of radiocarbon dates: Tables based on the consensus data of the workshop on calibrating the radiocarbon time scale. Radiocarbon 24:103-150

Lacroix A (1904) La Montagne Pelée et ses éruptions. Masson, Paris, 2 vol.

Lefèvre C, Cocusse P (1985) Etude pétrographique et minéralogique des laves du massif volcanique Madeleine-Soufrière de Guadeloupe (Petites Antilles). Implications magmatologiques, Bull Minéral 108:189-208

Le Guern F, Bernard A, Chevrier RM (1980) Soufrière of Guadeloupe, 1976-77 eruption: Mass and energy transfer and volcanic health hazards. Bull Volcanol 43:577-593

Lipman PW, Mullineaux DR (eds) (1981) The 1980 eruptions of Mount St. Helens, Washington. US Geol Survey, Prof. Paper 1250, 844 pp. U.S. Gvt. Printing Office, Washington D.C. 1981.

Marinelli G (1976) Sur la "cendre" projetée par la Soufrière (Guadeloupe) en août 1976. CR Acad Sci Paris, Sér D 283: 1731-1732

Maury RC, Mervoyer B, Westercamp D (1976) Températures de mise en place des pyroclastites fossilifères de la Montagne Pelée (Martinique) et de la Soufrière de Guadeloupe. Bull BRGM, Sér II, Sect IV, 347-352

Mervoyer B (1974) B1. Volcanisme récent et actuel de la Basse-Terre. A3. Le massif de la Soufrière. VIIème Conférence Géologique des Caraïbes, 30 juin - 12 juillet 1974, Livret-guide d'excursions dans les Antilles françaises. BRGM Orléans, 25-56

Pascaline H (1980) Géochimie des roches et des eaux de sources chaudes du massif de la Soufrière de Guadeloupe, Petites Antilles. Thèse 3ème cycle, Univ Paris Sud, Orsay, 150 pp

Pascaline H, Jérémie JJ (1983) Expérience de lessivage de poussières volcaniques récentes de la Soufrière de Guadeloupe (Petites Antilles). Bull Mus Hist Nat Paris, 4ème Sér, 5, Sect C, 191-204

Pascaline H, Jérémie JJ (1984) Proposition d'un nouveau schéma évolutif du volcanisme de la Basse-Terre de Guadeloupe. CR Acad Sci Paris, Sér II 229:31-36

Paterne M (1980) Chronologie des éruptions récentes du massif de la Soufrière (Guadeloupe - Petites Antilles). Essai de comparaison des périodes d'activité volcanique de quelques grandes régions volcaniques. Thèse 3ème cycle, Univ Bordeaux I, 165 pp + annexes

Pozzi JP, Le Mouel JL, Rossignol JC, Zlotnicki J (1979) Magnetic observations made on la Soufrière volcano (Guadeloupe) during the 1976-1977 crisis. J Volcanol Geotherm Res 5:217-237

Rapport (1798) Fait aux citoyens Victor Huges et Lebas, agents particuliers du Directoire exécutif aux Isles du Vent. Au port de la Liberté - Guadeloupe. Reprint of the 1798 edition by Soc Hist Guadeloupe, Basse-Terre, 1977, 84 pp + errata

Robin C (1984) Le volcan Popocatepetl (Mexique): structure, évolution pétrologique et risques. Bull Volcanol 47:1-23

Robin C, Cantagrel JM (1982) Le Pico de Orizaba (Mexique): structure et évolution d'un grand volcan andésitique complexe. Bull Volcanol 45:299-315

Sastre C, Baudoin R, Portecop J (1983) Evolution de la végétation de la Soufrière de Guadeloupe depuis les éruptions de 1976-77 par l'étude de la répartition d'espèces indicatrices. Bull Mus Natn Hist Nat Paris, Sér 4, 5, Sect B, Adnansonia, 1: 63-92

Sekiya S, Kikuchi Y (1889) The eruption of Bandaï-san. J Coll Sci Imp Univ Jpn, 3, pt 2, 91-172, Pl XV-XXIV

Semet M, Vatin-Pérignon N, Vincent PM, Joron JL (1981) L'éruption du XVIème siècle de la Soufrière de Guadeloupe. Mélanges de magmas et dynamismes éruptifs. Bull. PIRPSEV CNRS, Paris, n° 60, 42 pp + Tables and Figures

Semet MP, Vatin-Pérignon N, Vincent PM, Joron JL (in press) Magma mixing once more; its role in triggering the 16th century volcanic activity at la Soufrière, Guadeloupe (FWI). J Volcanol Geotherm Res

Sigvaldason GE (1978) Reply to editorial. J Volcanol Geotherm Res, 4, n° 3/4, I-III

Stuiver M (1982) A high-precision calibration of the AD radiocarbon time scale. Radiocarbon 24:1-26

Tazieff H (1977) La Soufrière, volcanology and forecasting. Nature (Lond) 269:96-97

Tomblin J (1979) Deontological code, probabilistic hazard assessment or Russian roulette? J Volcanol Geotherm Res, 5:211-212

Vatin-Pérignon N, Semet MP, Blanc F, Joron JL (1984) Petrochemistry of quaternary pumiceous pyroclastic products in southern Guadeloupe (FWI). Bull Volcanol 47-4:749-767

Vincent PM, Vatin-Pérignon N, Semet M, Cheminée JL (1979) Le dôme de la Soufrière (Guadeloupe): son âge et son mode de mise en place. CR Acad Sci Paris, Sér D 288:51-54

Westercamp D (1980) Une méthode d'évaluation et de zonation des risques volcaniques à la Soufrière de Guadeloupe, Antilles françaises. Bull Volcanol 43:431-452

Westercamp D (1984) Evaluation et zonation du risque volcanique à la Montagne Pelée, Martinique et à la Soufrière de Guadeloupe (Antilles françaises). Rapport BRGM-DSCMI, BRGM Orléans

Westercamp D, Mervoyer B (1976) Les séries volcaniques de la Martinique et de la Guadeloupe (FWI). Bull BRGM, Sér II, Sect IV, 4, 229-242

Westercamp D, Tazieff H (1980) Martinique-Guadeloupe-Saint-Martin-La Désirade. In: Guides Géologiques Régionaux, Masson, Paris, 135 pp

Zlotnicki J, Pozzi JP, Boudon G, Moreau MG (1984) A new method for the determination of the setting temperature of pyroclastic deposits (example of Guadeloupe: French West Indies). J Volcanol Geotherm Res 21:297-312

Volcan Popocatepetl: Recent Eruptive History, and Potential Hazards and Risks in Future Eruptions

C. BOUDAL and C. ROBIN[1]

Abstract

The recent history of Popocatepetl shows that violent repetitive eruptions took place during four periods, each of them lasting 1000 to 2000 years. The first one occurred before 10,000 years B.P., the second between 10,000 and 8000 years B.P., the third from 5000 to 3800 years and the fourth period started 1200 years ago. Cataclysmic events of St. Vincent-type leading to ash and scoria pyroclastic flows, and numerous air-fall deposits, alternate with a few lava flows during these periods of intense activity. In spite of the apparent quiescence of the volcano, the last period is still continuing. Prehistoric and historic eruptions repeatedly formed large volumes of hot pyroclastic flows and air falls that extended 20 km southeastwards and 10 to 15 km northeastwards. Lava flows seem to be restricted to the summit area. Pyroclastic flows, ash and pumice falls, and lava flows, are likely to be produced on the century scale or even on the decade scale. The morphology of Popocatepetl's slopes favours a wide distribution of the pyroclastic products, and many people live in sectors considered dangerous at the bottom of the volcano. In the case of pyroclastic flows similar to those of past eruptions, and directed towards the southeast, Atlixco (with 80,000 people) could be affected. If a Plinian eruption, like those produced just before the Hispanic Conquest, occurs with prevailing westerly winds, the whole region of Puebla could be affected. Risks presented by avalanches and lahars are also discussed.

1 Introduction

Popocatepetl volcano is composed of two parts, with a modern cone overlying an older composite edifice. Products of a Bezymianny-type event separate these two parts (Robin and Boudal 1987). The modern volcano is 1200 m high on its northern side, and 2000 m high on its southern side. At the summit of the cone is a 650 x 450 m wide and 250 m deep crater. A glacier, from 4300 m to 5200 m, covers about 1.5 km^2 on the northern slopes (see Fig. 8 b). There is a small dome in the bottom of the crater. Popocatepetl has been emitting gas and steam since the last eruption in 1920.

---

[1] Centre de Recherches Volcanologiques, UA CNRS n° 10 - Université Clermont II, 5 rue Kessler, 63038 Clermont-Fd, France

IAVCEI Proceedings in Volcanology 1
J.H. Latter (Ed.), Volcanic Hazards

The historic activity seems to have been limited to episodic outbursts of ash clouds. According to the Pre-Hispanic codices, one can postulate that a major eruption took place around 1350. Other episodes may have happened from 1519 to 1592 (seven eruptions), between 1642 and 1720 (five eruptions) and from 1802 to 1804 (Mooser et al. 1958). More recently, explosions occurred in the crater in 1920 (Waitz 1920b, 1921) generating large ash clouds, and an increase of fumaroles was observed lasting until 1927.

No text mentions any lava flow during the historic period. C^{14} dates obtained from soils overlain by ash and pumice falls (Lambert and Valastro 1976), and from charcoal debris (Robin 1981; Cantagrel et al. 1984) suggest important Holocene eruptions, many of them cataclysmic. According to Heine and Heide-Weise (1973), a major pumice eruption, represented by a continuous layer on the northeastern slope, is 965 ±60 years old (B.P.) agreeing with the 880 ±90 years + B.P. obtained by Robin (1981). This yellow-coloured pumice, and another pale-grey to white horizon, were together named the "superior" pumice layers (Robin 1981). Grey cinder deposits, clearly observable on the northern side, are approximately 450 years old.

2 Summary of Recent Activity

The activity of the modern volcano is characterized by an alternation of effusive and pyroclastic periods. Explosive periods culminate with cataclysmic eruptions like those observed at La Soufrière de St. Vincent and Colima volcano in 1902 and 1913 respectively (Anderson and Flett 1903; Waitz 1920a). These vertical eruptions from an open vent gave rise to collapse nuées, surge deposits and ash-flow deposits with a varying content (10 to 70% in volume) of scoriaceous bombs and minor blocks (nuées of Saint-Vincent type: NSV).

Ancient craters related to major cataclysmic events of this type have been clearly identified (Robin and Boudal 1987). They correspond to destructive stages of the two edifices composing the modern volcano, Volcan El Fraile and Summit Popocatepetl respectively (Fig. 1). These periods, characterized by great pyroclastic eruptions, have been alternating with periods of effusive activity for about the last 15,000 years.

2.1 Volcan El Fraile

Lava flows from the lower modern volcano, Volcan El Fraile, cover a surface of 125 km^2, and, if one also considers the pyroclastic flow deposits related to the Saint Vincent eruptions and Plinian air falls, the products extend over an area of about 250 km^2 (Fig. 1). After a long period of essentially effusive activity, over 20,000 to 30,000 years, three periods of intense pyroclastic activity, each marked by eruptions of the St. Vincent type, occurred during El Fraile's history (Robin and Boudal 1987). They alternate with effusive activity. During the first period, which occurred more than 10,000 years

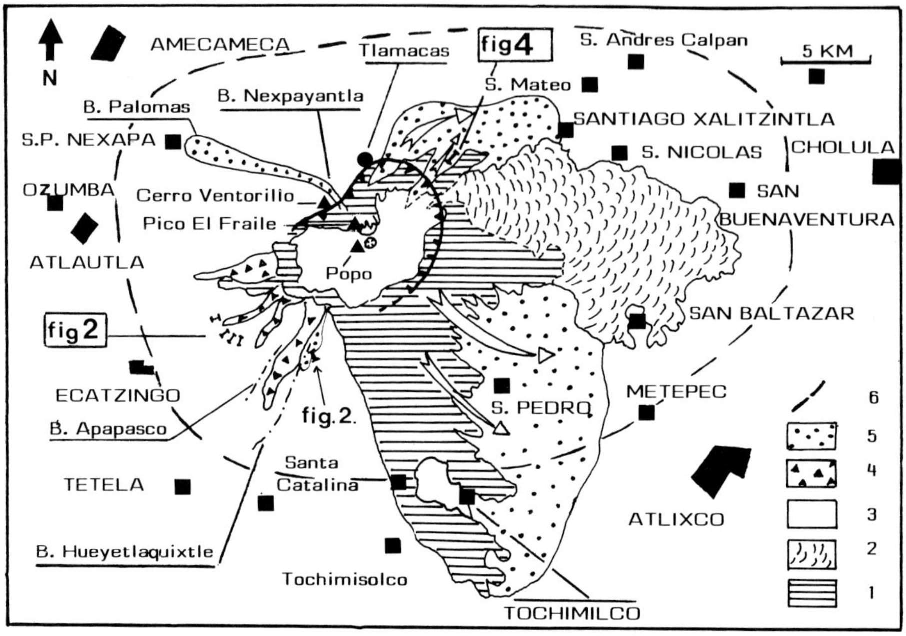

Fig. 1. Sketch map showing the extent of products related to the modern volcano. 1 Lava flows from V. El Fraile; 2 extent of the lateral Unit "II B" (lava flows); 3 lava flows from summit Popocatepetl; 4 lahars; 5 undifferentiated NSV deposits from V. El Fraile and summit Popocatepetl; 6 extent of thick air-fall deposits, especially for the "upper" yellow and grey pumice layers.
Numerous villages are built on the products of the modern volcano or in the vicinity (see also Figs. 6 and 7)

ago, the pyroclastic products were directed towards the southwest. The second period took place between ~9500 and 8000 years B.P. The third period, between ~5000 and 3800 years B.P., corresponds to the end of El Fraile's evolution. The sequences of the deposits related to this explosive activity are reported in Fig. 2 (for the first and second periods) and Fig. 3 (for the third period). Towards the southwest and west, products of the NSV were reworked by numerous lahars (see Fig. 2 c and d).

2.2 Eruptive Activity of the Summit Cone of Popocatepetl

The area covered by lava from the summit Popocatepetl amounts to only 40 km^2. From 3800 to 1200 years B.P., the activity was mainly effusive. About 1200 years ago, pyroclastic flows poured in two directions towards Santiago Xalitzintla in the northeast and towards San Pedro in the southeast. Two of these flows were dated at 1220 ±60 and 1230 ±90 years B.P. These pyroclastic flows spread widely on the Atlixco plains over about 100 km^2. Beside these flows, three pyroclastic episodes provide time markers for the history of the summit Popocatepetl:

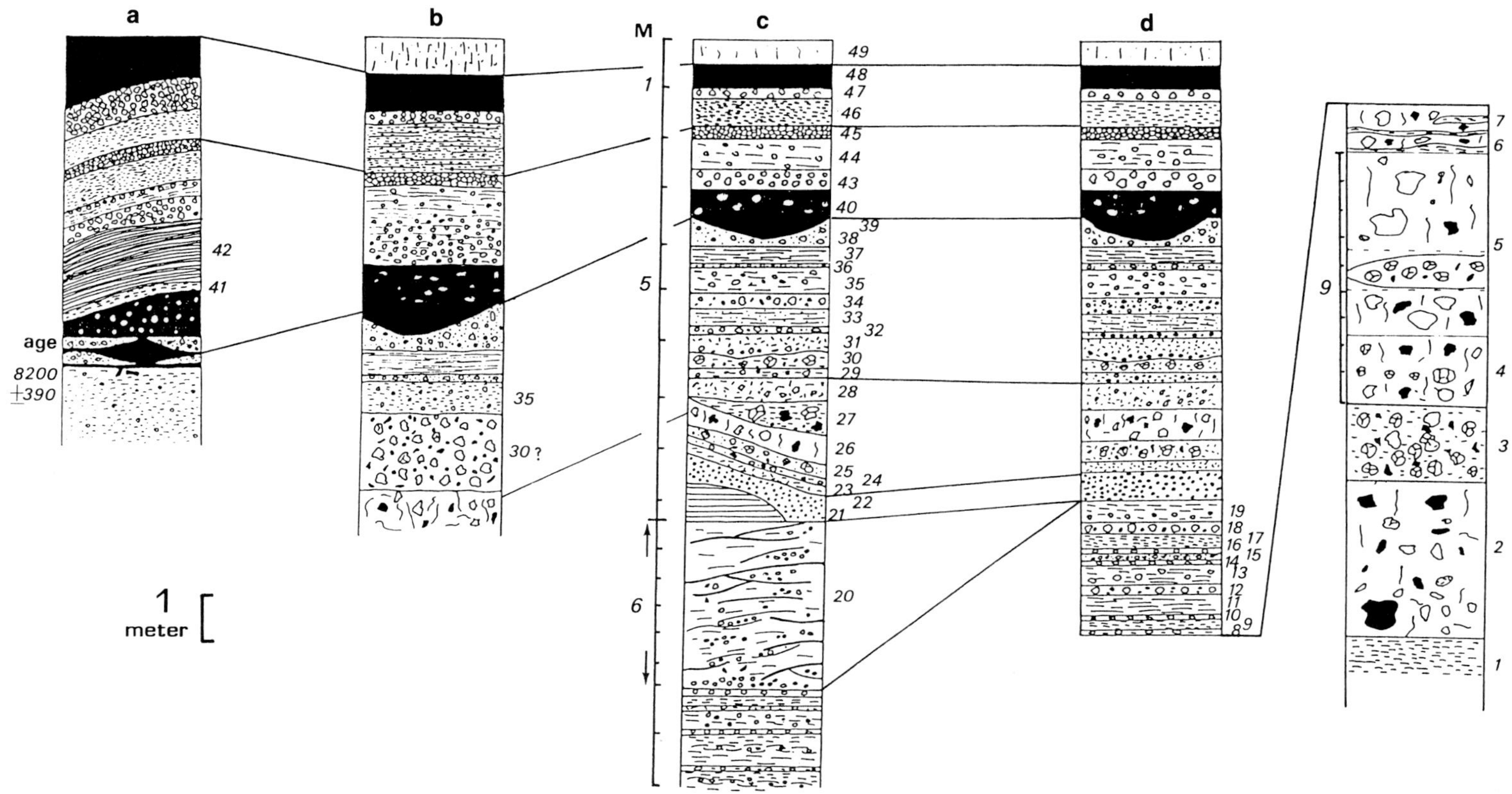

Fig. 2 a-d. Stratigraphic columns from pyroclastic deposits on the southwest slope; for location see Fig. 1. **1 to 7** Products related to the first period of intense explosive activity (more than 10,000 years B.P.: probably less than 15,000 years B.P.). **1** Ash, 60 cm; **2** lahar reworking andesite blocks and pumice, 3 m; **3** NSV deposit, pyroclastic flow deposits with abundant scoria bombs (70%), 1.5 m; **4** lahar reworking NSV products; **5** lahar with pumice-rich lenticular beds; **6** brown ash; **7** lahar deposits (like **2** and **4**). **8 to 39** Second period of pyroclastic activity (8000 to 10,000 years B.P.). **8 to 19** Ash and pumice layers, 5 cm to 40 cm thick. Total thickness: 2.4 m. Charcoal debris from correlative beds on the eastern side were dated at 9640 + 440 years B.P. **20** Thin cross-bedded surge deposits (laminites), 2 to 10 cm thick, alternating with pumice layers (5-10 cm thick), total thickness: 6 m. **21** Ash; **22** andesite lapilli (1 to 30 mm in diameter); **23** ash; **24** reworked materials; **25** ash with scoria and pumice (up to 10 cm in diameter); **26** pumice deposits with fumarolized blocks, 60 cm; **27** NSV pyroclastic flow, ash with scoria bombs and pumice. **28 to 39** Ash layers (dark, grey or brown-coloured) alternating with pumice horizons, 10 to 40 cm thick, over 3 m. **40 to 48** Horizons of grey or dark ashes (up to 1 m thick) and pumice layers related to the activity of summit Popocatepetl. **45** "Upper" yellowish pumice (965 ±60; 880 ±80 years B.P.); **47** "upper" grey pumice layer; **48** grey ash; **49** soil

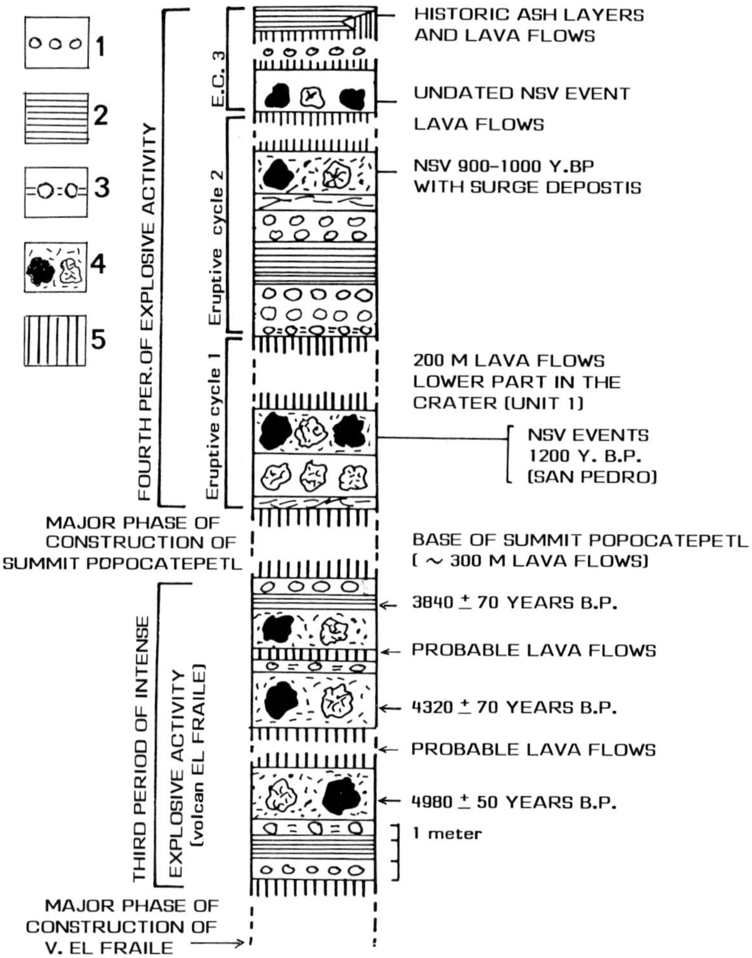

Fig. 3. Synthetic stratigraphic column of pyroclastic deposits on the northeast side, from 5000 years B.P. to Present. Detailed field sections in Robin (1981) and Boudal (1985). **1** Air-fall pumice layers; **2** air-fall ash deposits; **3** alternation of ash and pumice horizons; **4** pyroclastic flow deposits (NSV type); **5** lava flows

1. The uniform layer of yellow pumice, 30 cm to 1 m thick, dated at 965 ±60 and 880 ±80 years B.P. (see horizon 2, Fig. 4; see also upper part, Fig. 2 and its extent, Fig. 1).
2. A grey pumice horizon of about the same thickness, extent and age (horizon 5, Fig. 4).
3. The products of an NSV event, including at least four pyroclastic flows on the northern slope, above the grey pumice horizon; they were dated at 1000 ±60 years B.P.

The present crater is slightly elliptical, 650 x 450 m, with the long axis in the NE-SW direction. The edge of the crater dips from the west, where it is at an altitude of 5450 m, to

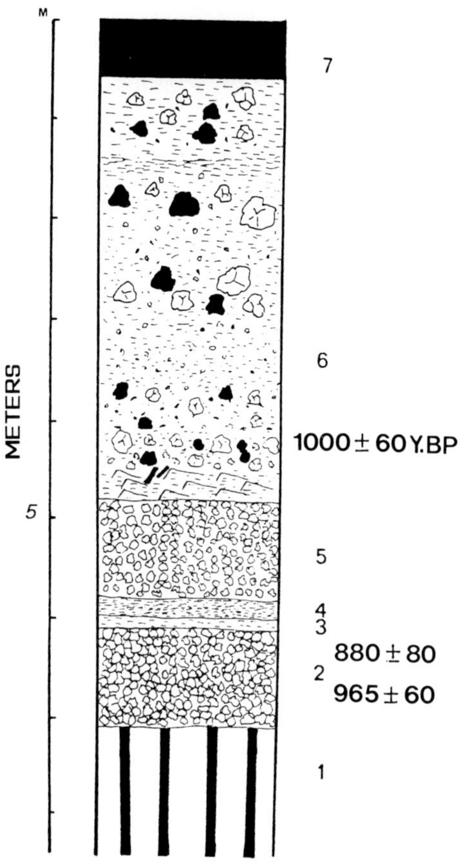

Fig. 4. Example of deposits related to an NSV event: the 1000 ±60 years B.P. eruption. Section in Barranca Cuyulula (location, Fig. 1). 1 Andesite lava flow; 2 "upper" yellow pumice layer; 3 grey ash; 4 dark ash; 5 "upper" grey pumice layer; 6 NSV pyroclastic flow deposits with surge deposits at the base: five layers of ash containing light and dark-coloured scoriaceous bombs (up to 30 cm in diameter; 10 to 70%); 7 grey ash

the east where it is 5200 m. Its walls are vertical over a distance of 250 m. The section of the western wall shows two distinct units:

1. The lower part (unit 1, Fig. 5) is made of a pile of flows and pumiceous layers which correspond to the construction of an edifice between 1200 and 1000 years B.P. The two pumiceous layers mentioned above (the yellow and the grey "superior" layers from the northeastern flank) mark the end of this period of construction.
2. Above the discordance related to the NSV dated at 1000 ±60 years B.P., two pumiceous layers, different from the yellow and grey "superior" layers, are intercalated with approximately ten lava flows. The thickness of that upper unit reaches about 100 m. One of the pumiceous layers crops out on the cone slopes. The base of that horizon contains small angular glassy blocks (10 cm maximum). Its upper part includes yellowish pumice, becoming rust-coloured upward. Deposits of this very obvious pumiceous episode cover the eastern and southwestern sides of the summit zone. Above the Las Cruces refuge, the transformation of these pumices into a chaotic flow is obvious: the more or less welded pumices contain many blocks. These deposits are interpreted as the

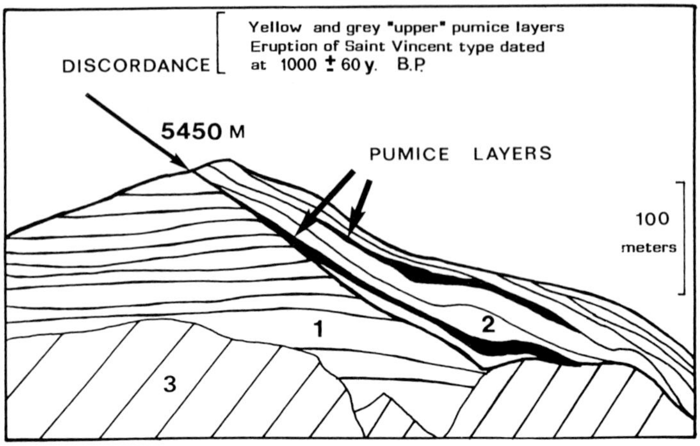

Fig. 5. Sketch map of the NW crater wall. **1** Lava flows and pumice layers of the lower unit with an age between 1200 and 900-1000 years B.P. Discordance related to the pyroclastic flow from Barranca Cuyulula. **2** The upper unit represents a pile of ten flows and pumice layers, including the rust-coloured pumice layer from the south-southeast slope; 3 scree

dense and hot base of an eruptive column which fell back near the crater. They correspond to a third, undated, St. Vincent-type event during the last period, similar to those dated at 1200 B.P. and 900 to 1000 years B.P.

From the present crater, lavas flowed into the large northern crater of the first edifice composing the modern volcano, Volcan El Fraile. Variations in the magmatic level allowed the overflow of these viscous block lavas (unit 2, Fig. 5) during the prehistoric or historic period. This suggests an intense effusive activity just before the Hispanic Conquest, or even more recently, the top of the volcano being not visible for a great part of the year.

The crater is the result of one of the last historic Plinian eruptions. It was slightly modified during the last episode in 1920-1921 and in 1927, when the bottom dome appeared. This dome is cut by a small crater whose origin is probably phreatic, which is partly filled by a small lake. In April 1984, fumarolic activity was taking place mainly all over the dome.

3 Definition of Eruptive Cycles

Only the pyroclastic products ejected far enough to fall on the lower slopes covered with vegetation can be dated by the ^{14}C method. The general lack of charcoal debris on the summit area prohibits precise determination of the length of both the eruptive and quiescent periods. Each period with strong explosive activity, lasting for about 1000 to 1500 years, is marked by at least two active cycles. Each cycle starts with cataclysmic events that release nuées of St. Vincent type. For the last period, the distinctive units of ash flows, ash and pumice falls, and the observation of the crater permit a division into three parts:

1. A first cycle starting with the emission of the San Pedro NSV (1200 years B.P.) and associated Plinian eruptions, and followed by a series of lava flows, more than 150 m thick, which were erupted from the lower part of the crater.
2. A destructive episode, responsible for the two yellow and grey pumice layers and the NSV dated at 1000 ±60 years B.P., which marks the beginning of a second eruptive cycle. The cataclysmic event corresponds to the obvious discordance in the middle part of the crater (Fig. 5).
3. A third cycle which started with the last NSV episode, probably just before the Hispanic Conquest. This cycle includes the upper lava flows in the crater.

Each cycle starts with a great explosive eruption which destroys the summit part of the cone, leading to pyroclastic flows. This activity is followed by lava flows alternating with ash and pumice eruptions, with quiet phases of a few centuries or less. At the end of the cycle, the effusive activity dominates until the next cycle starts. These cycles are a characteristic feature of the great Mexican composite volcanoes. At the Pico de Orizaba, for example, nuées of St. Vincent type occurred from 13,000 to 6000 years ago regularly every 1000 to 1500 years (Robin 1981; Robin and Cantagrel 1982), and at Colima, at least three cycles can be defined during the historic period, with cataclysmic events occurring much more frequently, every 100 to 200 years (Luhr and Carmichael, 1980; Robin et al. 1987).

4 Hazards and Risks Presented by Popocatepetl

Hazard evaluation at a weakly active, or dormant volcano, such as Popocatepetl, is essentially based on the volcanological study of its recent activity. Risk assessment, however, incorporates other parameters such as the economic development of the region and the increasing population, as well as other factors, such as the climatic conditions during an eruption, which are more difficult to define.

4.1 Nuées of Saint Vincent Type (NSV) (Fig. 6)

The peculiarity of these nuées is their ability to affect all sides of the volcano. Constricted to the valleys, the flows and surge deposits can reach populated plains surrounding the volcano, and can overspread a whole region, as happened in the case of the San Pedro nuée on the southeast side of the volcano some 1200 years ago. The high probability that these glowing avalanches will take place again, the suddenness of their emission, their speed, their temperature and their richness in gas, make this type of eruption the most dangerous for the population. Three sectors are mainly threatened by pyroclastic flows: the southwest, northeast and southeast. The path taken by products of the next eruption will also depend on the directed character of the explosion, whether from a new inclined crater, or from the present crater.

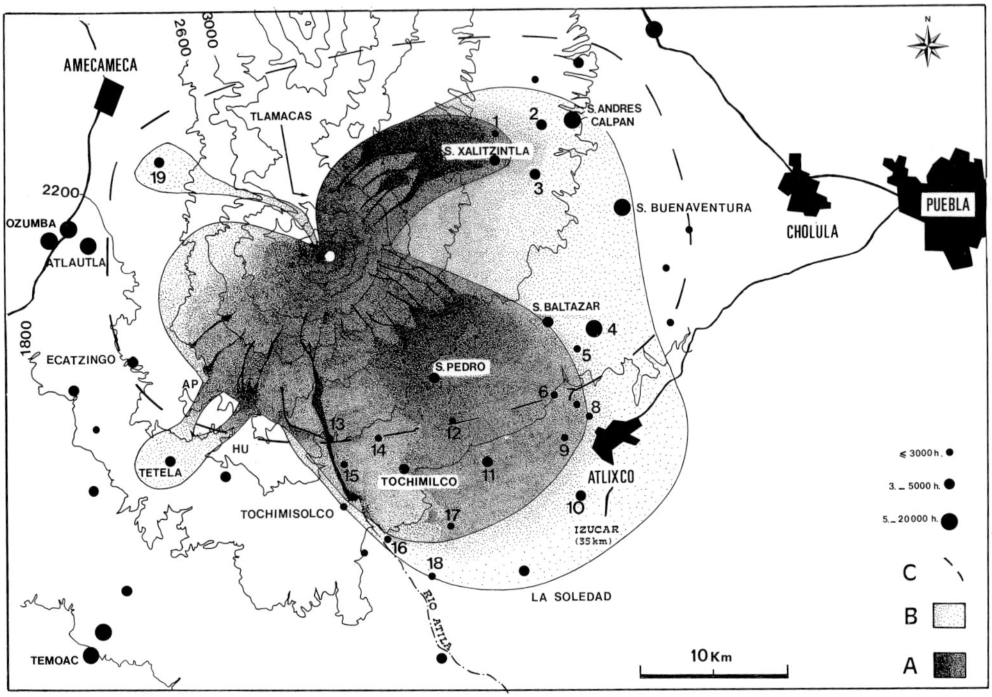

Fig. 6. Areas of potential hazards for pyroclastic flows of St. Vincent type and air falls. A High risk sector for pyroclastic flows; B minor risk sector for pyroclastic flows; C extent of air falls from the modern volcano. Locality symbol sizes are related to the size of the population (see Appendix). AP Barranca Apapasco; Hu Barranca Hueyetlaquixtle. Numbers 1 to 19 on the map refer to locality names cited in Appendix

SW Sector (Tetela-Ecatzingo). During the first two periods of explosive activity which have been recognized, nuées were directed towards the south and southwest. The most important ones reached within a short distance of Tetela, but they were stopped by the big hummocks of the debris avalanche emplaced prior to the construction of the modern volcano. They never reached zones which are inhabited today. However, owing to the morphology of this slope, two valleys could be used as channels for pyroclastic flows of greater volume, which would allow them to pass into the Tetela plains: these valleys (Barrancas) are the Apapaxco and the Hueyetlaquixtle (Figs. 1, 6)

NW Sector. The pyroclastic flow products observed in the Palomas valley (Robin and Boudal 1987) can be considered as a unique case, due to a lateral and strongly directed eruption from Volcan El Fraile, which does not seem likely to be repeated. Moreover, the very narrow Nexpayantla valley could only be used as a possible channel for an NSV resulting from a vertically directed explosion, if the collapse of the magmatic column were to take place beyond Fraile, at a distance of more than 1 km from the present crater. Thus, the Amecameca and San Pedro

Nexapa region seems to be only slightly threatened by this type of eruption.

NE Sector. This sector was used as an outflow plain for many pyroclastic flows during the third period (5000-3800 years ago). The most recent NSV also flowed in this direction, as far as 13 km from the summit. Its deposits are 1 m thick at the entrance of Santiago Xalitzintla. In the case of a vertical eruption, the northern side of the cone, which corresponds to the curved and dipping crater of Fraile, would act as a funnel and would channel the products falling in that direction. Santiago Xalitzintla and San Nicolas are villages situated in this sector with very high risk. In the case of a very broad avalanche, towns such as San Mateo, San Andreas Calpan, San Buenaventura or even Cholula, could be damaged.

The S and SE Sectors. The southeast sector constitutes a broad smooth zone with regular slopes, over which the 1200-year-old NSV spread. Several villages or small towns are built directly on its products: San Pedro (Figs. 1, 6), San Jeronimo (Fig. 8 a, b), Tochimilco, for example. The general aspect of the volcano with its regular, steep slopes on that side, and the absence of any main morphological obstacle, are principal factors pointing to the high risk in that quarter. In the case of an eruption directed towards the southeast, from an inclined crater, the town of Atlixco (80,000 people) could be reached by the pyroclastic flows.

4.2 Ash and Pumice Falls (Fig. 6)

These may be associated with the NSV pyroclastic flows, or could happen independently during a Plinian eruption. The probability of such an episode is significant (Table 1). The deposits (Figs. 2-4) are mainly composed of yellowish, grey or black cinders, light pumiceous or poorly vesiculated lapilli of moderately acid composition (andesites to dacites). They are interbedded with pyroclastic flow deposits and with lahars which reworked the products of older nuées emplaced on the southern side. For the episodes during the prehistoric period, these tephra are interstratified with lava flows on the northeast slopes of the cone, and near the crater.

Zoning of Potential Hazards. The sequences related to the activity of Volcan El Fraile are thicker and more numerous on the southwest side (Fig. 2). The thicker horizons are also the more recent ones (age <1200 years). Nevertheless, the distribution of the two upper pumice layers does not show any preferential extent. The lack of outcrop and reworked superficial materials make it difficult to identify these products beyond a distance of 20 km.

The zone which is liable to be covered by pumice falls of the same scale as those emitted 900-1000 years ago (the yellow and grey horizons) has an elliptical shape that is slightly elongated east-west. It is delimited by Ecatzingo and Atlautla in the west, Amecameca in the northwest, Tochimilco in the south

Table 1. Average occurrence of pyroclastic events since 10,000 years and rough probability of risks according to the different types of eruptions

A summary of pyroclastic events since 10000 years

	Number of listed layers corresponding to major eruptions[a]	Average periodicity over 10,000 years	Average periodicity during periods of pyroclastic activity[b]	Number of major eruptions during an eruptive cycle	The most cataclysmic
Air-fall products (ash, pumice)	40	250 years	95 years	1 to 5	An eruption like that of the upper yellow pumice layer (~950 years ago)
Nuées of St. Vincent type (and associated surges)	11 eruptions (some of which include 4 or 5 episodes)	900 years	350 years (duration of an average eruptive cycle)	1 event at the beginning of the cycle (or more events separated by short intervals)	An eruption like that of the San Pedro pyroclastic flow, 1200 years ago (extent >100 km^2)

a Layers composed of at least 20-30 cm deposits 10-15 km from the crater: some layers correspond to many successive episodes and have a total thickness of 1-2 m (for example, alternation of ash and pumice).
b Three periods marked by explosive activity: 9640 to 8200 years B.P.: 4980 to 3800 years B.P.: 1200 years B.P. to Present.

Table 1 (cont.)

B Evaluation of risks in terms of rough probability

	Almost certain	Very likely	Likely	Not very likely	Quite unlikely	Improbable
Nuées of St. Vincent type	********	***************				
Air falls	**************					
Lava flows	********	***************				
Ash flows					**************	
Domes					**************	
Pelean nuées ardentes					**************	
Mud flows		***************	***************			
Debris flow (Bezymianny-type event)			************	***********		
Cessation of the activity						*************

and San Andreas Calpan in the northeast. An ash eruption could considerably exceed this area covered by pumice, and could affect the City of Puebla. If a strong easterly wind is blowing, the whole Valle de Mexico would experience the effects of such eruption.

4.3 Lava Flows

The lavas from the summit are acid andesite and dacitic block lavas. These viscous outpourings flow very slowly (some meters an hour or a day), even along steep slopes. The lava flows from the last edifice, the Summit Popocatepetl (age <3800 years) did not spread beyond the caldera.

Zoning of Potential Hazards. During a new rise of the magma, the present form of the crater would favour outpourings towards the north, east and southeast sectors. However, if these are preceded by an explosive activity which modifies the crater's shape, they could threaten the entire circumference of the cone. Only the northwest side, protected by Pico El Fraile, seems to be totally safe from the flows. As for the risk, only the Tlamacas refuge is threatened. Thus, the damage caused by a lava emission from the crater should be limited to the destruction of the vegetation (below 3500 m) and possibly that of the refuge. The II B unit (Fig. 1) erupted from lateral vents, along a west southwest-north northeast fracture zone; the emission of flows from such lateral vents remains a possibility, but would not present any more risks for the human population. Moreover, if the formation of new fracture during a strong seismic crisis, such as those at the origin of the II B lateral unit (Robin and Boudal 1987) occurs, the quantity of available magma remains an unknown parameter. Should such an eruption occur, numerous villages would be threatened. Yet the risks incurred by these villages would mainly come from the seismic activity before the eruption and not from the eruption itself, a major lava outpouring being quite unlikely.

4.4 Domes and Associated Pyroclastic Flows

Apart from the small dome at the bottom of the crater, no domes were found anywhere on the modern volcano. The absence of domes and of Pelean nuée ardente deposits is a feature of Popocatepetl, in contrast to the other great Mexican volcanoes. The risks presented by this type of eruption seem to be very low (Table 1).

One must not, however, disregard the possibility that the intracrater dome might grow and overtop the crater rim. Nevertheless, this would occur so slowly that there would be time for volcanologists to assess which inhabited zones were threatened. The formation of a large-scale Pelean nuée ardente would be an exceptional event in the history of the volcano.

4.5 Lahars (Figs. 7 and 8)

Southern Sector. The distribution of lahars indicates that they have not reached zones which are densely populated today. However, a lahar of larger volume than any so far identified could easily follow the Barranca Hueyetlaquixtle and overflow into Tetela. Towards the south, the Rio Atila (Figs. 6, 7) could channel such lahars, especially in the event of very strong rain after a pyroclastic eruption. The southeast slopes are not protected from lahars and a village such as San Pedro would be at risk in the event of a Plinian eruption during the rainy period, from May to November.

Northeastern Sector. Laharic formations extend widely over the northeast plain at the bottom of the volcano in the San Buenaventura-San Andres Calpan-San Mateo area. They are 10 m or more thick and consist of reworked ash and pumice. They are related to Volcan Iztaccihuatl (Fig. 8 a). Lahars reworking prehistoric NSV deposits from Popocatepetl are also known in the northeast as far as Santiago Xalitzintla (Robin 1981). During a new eruption, especially an explosive event, lahars would be generated by the melting of the glacier located on the northeast slope of

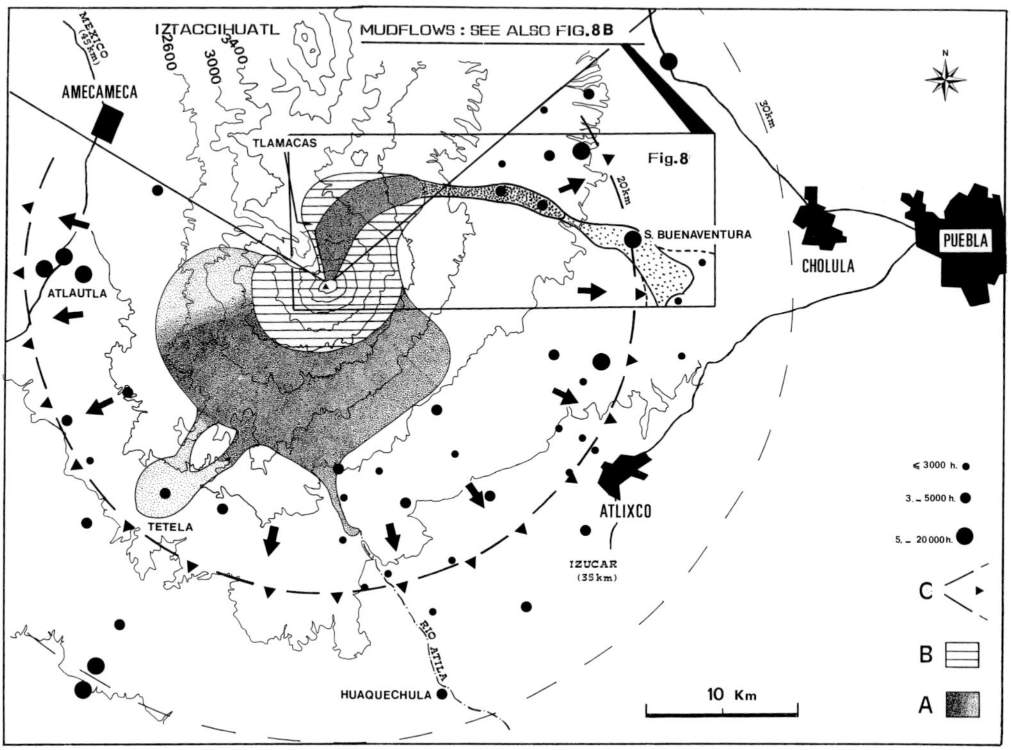

Fig. 7. Areas of potential hazards for lahars, lava flows and debris avalanches. A High risk sector for mud-flow deposits (see also Fig. 8); B high risk sector for lava flows; C delimitation of the area for a possible debris avalanche. The insert shown is the area covered by Fig. 8

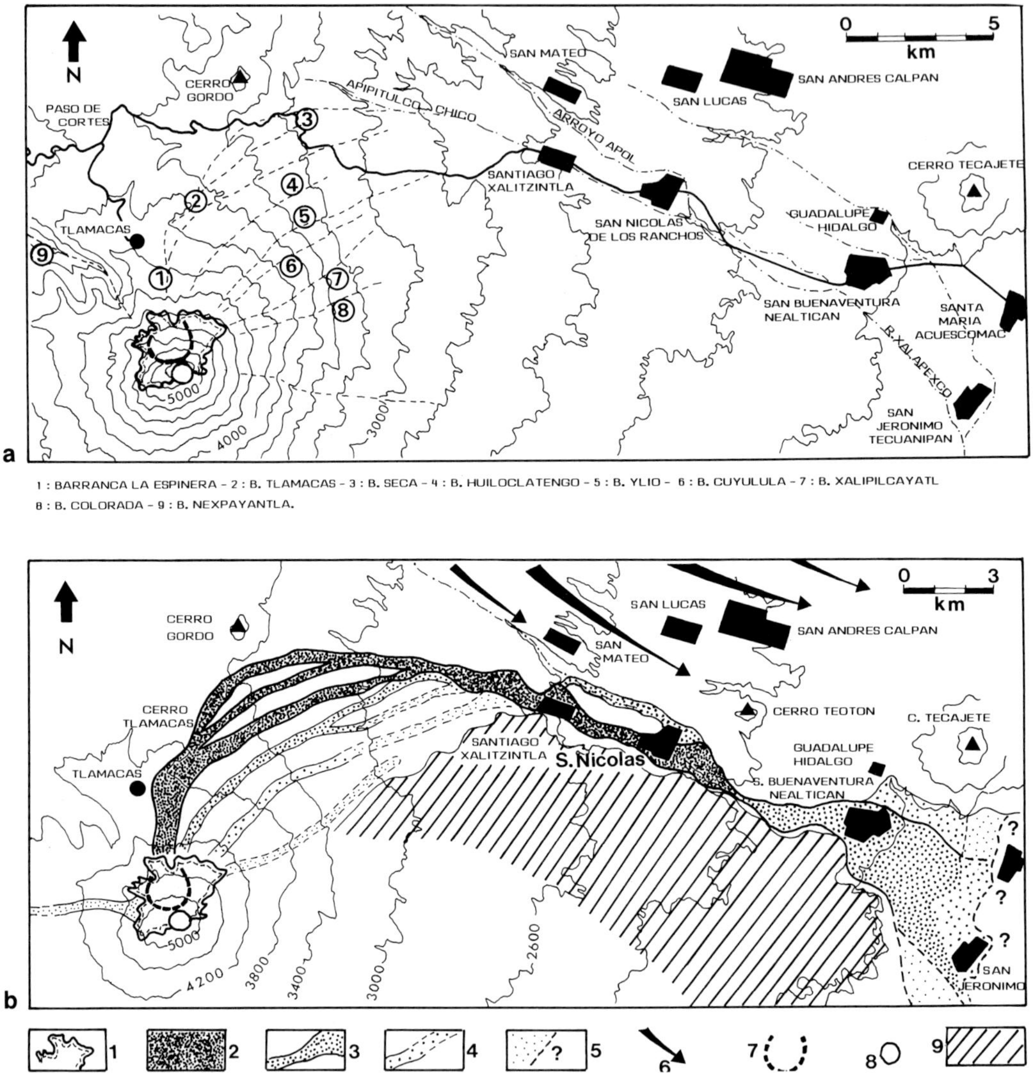

Fig. 8 a, b. Areas of potential hazards for lahars related to the melting of the glacier, northern side of Popocatepetl. a The main topographic features and name localities. b 1 Limit of permanent snow and ice; 2 high risk sector; 3 high risk sector in case of a major eruption (note that if a barrier of rocks, trees and mud accumulates west of Santiago Xalitzintla, part of the lahars may be diverted to the north to follow the Apol river). Moderate (4) and low (5) risk sectors and their possible extension in case of a major eruption. 6 Lahar deposits from V. Iztaccihuatl; 7 ancient large crater of V. El Fraile containing the glacier; 8 present crater; 9 lava flows of unit II B

the volcano (Fig. 8). The evaluation of risk presented by the melting of the glacier on the northern side of Popocatepetl requires more data, especially a precise knowledge of its shape and thickness. The distance and the area overflowed by lahars would depend upon numerous factors, principally the volume of water produced. However, most, if not all, of the lahars would be directed towards the northeast plains by numerous barrancas (B. Tlamacas, B. Huiloclatengo, B. Ylio, etc.). The ash and pumice on this slope would be easily remobilized. In addition, the steep drop and regular dip of the slope from the edge of the glacier to Santiago Xalitzintla, would also ensure that lahars would be large in this area. Thus, despite the lack of data about the glacier, we can postulate that Santiago Xalitzintla, San Nicolas and San Buenaventura are at risk from lahars (Fig. 8).

4.6 Voluminous Ash Flows

The only known large ash flows are associated with the Bezymianny-type event (Robin and Boudal 1987). Considering the evolutionary stage reached by Popocatepetl (Boudal 1985), the probability of such a phenomenon, which generally occurs just before the "caldera stage" in other Mexican volcanoes (Robin 1981) will take place again is practically nil. In such a case, the Plinian activity that occurs with this type of eruption would be stronger than the one that characterizes the episodic emptying of the upper part of the shallow reservoir (normal Plinian eruptions).

4.7 Eruption of Bezymianny Type (Fig. 7)

It is difficult to assess the likelihood of the repetition of this type of event during the evolution of a single volcano. The only possibility for evaluating the risks of a debris avalanche is to list all geological, morphological and structural factors which could favour such an eruption. The terminal cone of Popo has steep slopes (30-35°) and is partially convex on the west side, whereas the slopes on the east and south sides are straighter. The cone fills the caldera, and its summit is higher than the summit of the primitive volcano. The intrusion into it of a cryptodome could lead to a quick imbalance of its upper parts. Moreover, two other factors must be considered:

1. The existence of the small intracrater dome supposes that the central vent is obstructed, and that the formation of a cryptodome is therefore likely.
2. The fracture zone at the origin of the II B unit (Fig. 1) may favour the emplacement of such an intrusion on the flank of the volcano.

S and SE Sectors. Robin (1984) has shown that volcanism has migrated from north to south. This, and the morphology of the volcano, suggests that the southeast side is the most threatened by the spreading of a debris flow. The absence of obstacles also suggests that a debris flow, even of comparatively small volume (2-3 km^3), might reach a maximum distance of more than 30 km, thus threatening the city of Atlixco.

W Sector. As mentioned above, the convex shape of the terminal cone is a critical factor in the imbalance of this slope. As the slopes of the ancient volcano are very steep, a new debris avalanche would extend over the Ozumba plains, after passing the caldera edge. Around the northwest, the remains of the primitive volcano would protect the Amecameca region. Nothing prevents the spreading of a debris avalanche towards the southwest (Ecatzingo region).

E Sector. In both the east and south sectors no obstable would stop the avalanche. The great drop (3500 m), and the regular dip of the Valle towards Puebla down to low altitude (1500 m), are important factors in spreading the products of a landslide in that area (Figs. 6, 7). All the towns situated at less than 25 km from the volcano (for example Huejotzingo and Atlixco) would be directly involved, but cities such as Cholula and Puebla are not totally safe from other risks related to an eruption of this type such as a blast.

N Sector. The formation of a cryptodome on this side is not very likely. Should one occur a debris avalanche might develop, either to the northeast or to the northwest.

5 Appendix. Volcanic Risk: Probability and Number of People Likely to be Affected

As Westercamp (1980) suggests for the Soufrière de Guadeloupe, one can evaluate the risks in terms of rough probabilities (Table 1). The latter points out the danger represented by the pyroclastic flows of Saint Vincent type (NSV) and Plinian activity. In the case of a more violent eruption, such as an event of Bezymianny type, the hazard is greater, but the probability is much lower. An estimation of the populations which would be affected by Popocatepetl's activity has been carried out. It is subdivided according to the type of eruptions and distance from the volcano. This estimation shows that there is a large population at risk and that an eruption of only moderate size can cause much material damage and endanger the lives of many people. Damage to agriculture would extend over a radius of about 20 km.

Pyroclastic Flows of St. Vincent Type (ash and scoria pyroclastic flows with bombs and associated surges and air falls). They represent the major danger for the population taking into account the probability of another eruption of this type and the extent of products of former eruptions.
The towns listed below are located in areas at risk. They are divided into two groups according to the level of risk; towns marked with an asterisk are at risk even in the case of minor eruptions. Numbers in parentheses preceding the name of the towns refer to their location in Fig. 6.
Santiago Xalitzintla (4000 inhabitants)*; (3) San Nicolas (5000)*; San Andres Calpan (15 000); (2) San Lucas (5000); (1) San Mateo (3000); San Buenaventura (8000); San Baltazar (4000)*; (6) Metepec (4000)*; (7) El Leon (3000)*; (8) Cabrera (3000)*; (9) Axocopan (5000)*; San Pedro (4000)*; (11) San Jeronimo (5000)*; Tochimilco (5000)*; Atlixco (80 000); (17) Tejupa

(3000)*; (10) La Trinidad (5000); La Soledad (2000); (13) Santa Catalina (3000)*; (5) San Martin Tlapala (2000). San Juan. (4) Tianguismanalco (6000). (12) Agricola + Guadalupe (3000)*; (14) Magdalena (2000)*; Tochimisolco (2000); (18) San Miguel Aguacomulican (1000); Tetela (5000); (19) San Pedro Nexapa (5000); (16) Huilango (1000); (15) San Martin Zacatempa (5000)*.
Total: 198 000 people, of whom 60,000 live in a particularly hazardous zone.

Air Falls. The direct effects of major pumice and ash falls would affect the whole zone represented by the broken line in Fig. 6. Twenty or so localities are included in this sector, a population of about 100 000 people. Ash-fall deposits of minor thickness could extend over the Valles of Puebla and Mexico.

Lava Flows. Only the Tlamacas refuge could be affected.

Domes. The risks for the population are very low and restricted to the summit zone if small Pelean nuées ardentes occur.

Lahars. Two inhabited sectors would be affected: Santiago Xalitzintla (4000 inhabitants) and San Nicolas (5000) in the northeast and the Rio Atila zone with Santa Catalina (3000) and San Martin (2000) in the south. The whole circumference of the cone to a distance of about 10 km from the summit could be affected if an eruption were to occur during the rainy months (from May to October).

Voluminous Ash-Flows. Very low risk of affecting populated regions.

Eruption of Bezymianny Type (debris avalanche). The extent of damage and the towns threatened would depend on the direction taken by the avalanche.
More than 100 000 people would be affected (including Atlixco) in the case of a sliding of the upper part of the volcano towards the southeast. If the sliding occurred towards the west, about 50 000 people would be affected (including towns like San Pedro Nexapa, Ozumba, Atlautla and Ecatzingo). In the case of sliding towards the east two scenarios are proposed:
1. In the case of a "minor" eruption (flank failure restricted to the upper part of the edifice) about 200 000 people would be affected.
2. Should a major event destroy the whole east flank of the cone; about 1 million people would be at risk, taking into account the location of two big cities (Cholula and Puebla).

Acknowledgements. Financial support was provided by the P.I.R. P.S.E.V. (Programme Interdisciplinaire de Recherche pour la Prévision et la Surveillance des Eruptions volcaniques) C.N.R.S. I.N.S.U. We thank Dr. J. Guerrero, Director of the Institute of Geology, University of Mexico, for his assistance during fieldwork. We also thank Mrs. Delibrias and J. Labeyrie, Director of the Centre des Faibles Radioactivités de Gif-sur-Yvette (C.N.R.S.-C.E.A.), for the C_{14} data. We are grateful to Prof. Mac Birney and to P.M. Vincent for the helpful comments during the preparation of this work.

References

Anderson TH, Flett JS (1903) Report on the eruption of the Soufrière in Saint Vincent, 1902, and on a visit to Montagne Pelée in Martinique. Phil Trans R Soc Lond, Ser A 22:353-553

Boudal C (1985) Pétrologie d'un grand volcan andésitique mexicain: le Popocatepetl. Thèse 3ème cycle, Univ Clermont II, 225 pp

Cantagrel JM, Gourgaud A, Robin C (1984) Repetitive mixing events and Holocene pyroclastic acitvity at Pico de Orizaba and Popocatepetl (Mexico). Bull Volcanol Spec Iss magma mixing 47-4 (1):735-748

Heine K, Heide-Weise M (1973) Junguartäre fürderfolgen des Malinche Vulkans und des Popocatepetl (Sierra-Nevada) Mexico. Münster forsch. Geol Paleont V 31/32:303-322

Lambert PW, Valastro S (1976) Stratigraphy and age of upper Quaternary tephra on the northwest side of Popocatepetl volcano, Mexico. Am Quatern Assoc, 4th Biennal meet Abstr, p 143

Luhr JF, Carmichael ISE (1980) The Colima volcanic complex, Mexico. Part I: Postcaldera andesites from volcan Colima. Contrib Mineral Petrol 71-4:343-372

Mooser F, Meyer-Abich H, Macbirney AR (1958) Catalogue of the active volcanoes of the world. Int Volc Assoc Part VI:26-30

Robin C (1981) Relations volcanologie-magmatologie-géodynamique: application au passage entre volcanismes alcalin et andésitique dans le Sud Mexicain. Thèse Doctorat d'Etat Univ Clermont-Fd Ann Univ Clermont II, Ser 2, Vol 31, 503 pp

Robin C (1984) Le volcan Popocatepetl (Mexique). Bull Volcanol 47-1:1-23

Robin C, Boudal C (1987) A gigantic Bezymianny-type event at the beginning of modern volcan Popocatepetl. J Volcanol Geotherm Res 31:145-130

Robin C, Cantagrel JM (1982) Le Pico de Orizaba (Mexique): structure et évolution d'un grand volcan andésitique complexe. Bull Volcanol 45-4:299-316

Robin C, Camus G, Cantagrel JM, Gourgaud A, Mossand Ph, Vincent PM (1987) Eruptive history of the Colima volcanic complex. J Volcanol Geotherm Res 31:99-113

Sean Bulletin (1986) Popocatepetl (Mexico): increased fumarolic activity in summit crater. Vol 11-1:10-11

Siebert L (1984) Large debris avalanches: characteristics of source area, deposits, and associated eruptions. J Volcanol Geotherm Res 22:163-197

Waitz P (1920a) "Nubes ardientes" observadas en las erupciones del Jorullo (1859), Ceboruco (1870) y del volcan de Colima (1913). Mem Soc Ant Alzate, Tomo 37, Mexico: 267-277

Waitz P (1920b) La nueva actividad y el estado actual del volcan Popocatepetl. Mem Soc Ant Alzate, Tomo 37, Mexico: 295-313

Waitz P (1921) Popocatepetl again in activity. Am J Sci, 5th Ser, V 1:81-85

Westercamp D (1980) Une méthode d'évaluation et de zonation des risques volcaniques à la Soufrière de Guadeloupe, Antilles françaises. Bull Volcanol 43-2:431-452

Use of Computer-Generated Movies to Present Volcanic Hazards

B. A. Blackerby[1], H. L. Lindblom[2] and G. R. Roquemore[2]

Abstract

The approximately 3 MA rhyodacitic Haiwee Tuff appears to document the largest explosive eruptions in the Coso Volcanic Field of southeastern California. The tephra was produced by Plinian activity which erupted an estimated minimum dense rock equivalent of 1.18 km^3 as pyroclastic fall and 0.43 km^3 as pyroclastic flows and surges. A study of these deposits has permitted interpretation of the sequence of events during this activity.

A computer-generated movie has been produced recreating parts of this eruptive sequence. The eruption site chosen for this movie was one of the areas currently considered a possibility for future eruptions, thus allowing the computer model to interact with present-day topography. In the movie three events are displayed, the maximum calculated: (1) pyroclastic fall; (2) ballistic activity; and (3) pyroclastic flow/surge. A time scale has been used for the fall and flow/surge events. Velocities of these two events vary and a visual indicator shows the elapsed time for them. Such a film is a potentially useful dynamic means of presentation for both scientists and nonscientists.

1 Introduction

The Coso Volcanic Field (CVF) in southeastern California (Fig. 1) has been intermittently active for the last 4 million years. Activity has been dominantly bimodal, with basaltic and rhyolitic eruptions cumulatively producing approximately 35 km^3 of lava in the past 4 MA (Duffield et al. 1980). It has not been possible to date the last activity in the field directly, but present estimates, based largely on morphology and similar evidence, suggest the latest eruptions may have occurred less than 30,000 years ago. Hot springs, fumaroles, high heat flow, and on-going though usually low-level seismic activity indicate that the field may be dormant. In addition, recent studies (Sanders and Kanamori 1984) show a possible magma body only 3 km below the surface of nearby Indian Wells Valley.

---

[1] Department of Geology, California State University, Fresno, California 93740, USA

[2] Naval Weapons Center, China Lake, California 93555, USA

IAVCEI Proceedings in Volcanology 1
J. H. Latter (Ed.), Volcanic Hazards

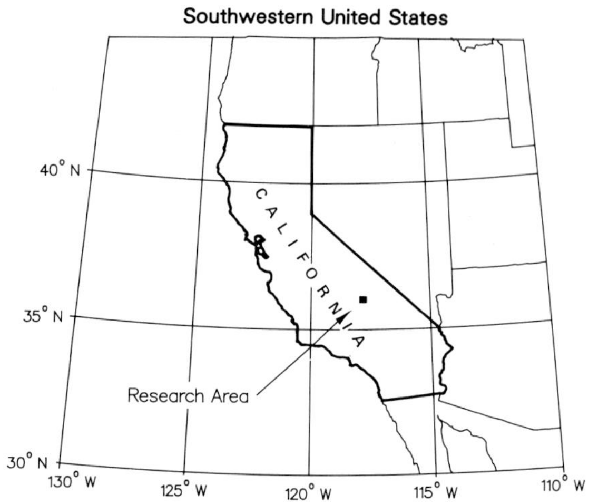

Fig. 1. Index map of the southwestern United States showing location of the research area

With this background, the present investigation was undertaken as part of a Volcanic Hazards Study Program sponsored by the Naval Weapons Center (NWC), at China Lake, California. The NWC encompasses almost the entire CVF, as well as the potential magma body.

2 Concept of the Program

The present investigation had two parts:

1. To determine and characterize a possible "worst case" eruption scenario for the field, based on past geologic history and setting.
2. As far as possible, to devise a means, or models, which would allow a computer to recreate all or parts of this "worst case" eruption wherever desired using present-day topography.

The worst case scenario was chosen as a geologically probable extreme approach to hazard evaluation. A smaller case can always be modeled if the largest reasonable case can. The 3 ma rhyodacitic Haiwee Tuff appears to document the largest explosive eruptions in the 4 ma history of the CVF and was therefore taken as a worst case.

Computer graphics were selected for hazard prediction and presentation because of the speed with which many different locations and combinations of eruption parameters could be examined. NWC has excellent computer facilities and a digitized topographic base is available for the region.

3 Worst Case Characterization

The Haiwee Tuff was examined in the field and laboratory in order to characterize the eruptive activity and suggest possible eruption parameters. These investigations showed that at least nine eruptive events could be recognized, including Plinian pyroclastic falls, pyroclastic flows and surges, and intervals of strongly ballistic activity. An estimated minimum dense rock equivalent of 1.18 km^3 (fresh fallen volume = 5.89 km^3) was erupted as pyroclastic fall and 0.43 km^3 (fresh deposited volume = 2.15 km^3) as pyroclastic flows and surges (Blackerby et al. 1984; Roquemore and Blackerby; 1985).

The largest events of each of the following types were chosen to be recreated: (1) pyroclastic fall; (2) ballistic activity; and (3) pyroclastic flow/surge (these two were not distinguished because they are both modeled the same way by the computer). Parameters estimated and used for these events are given in Table 1.

4 Computer Recreation

The computer can be used not only to predict hazard limits but also to prepare a movie of an event in progress. This is done by allowing the computer to generate a series of many intermediate frames between a beginning point (the start of an eruptive event) and the final limits, recording these frames photographically, then projecting them at a rate of 24 frames s^{-1} to produce an animated sequence.

The actual methods and algorithms used to generate the plots are described in a forthcoming paper (Lindblom, in prep.). Briefly, the methods used for the three hazards are summarized below.

Table 1. Parameters used in the computer recreation

A Pyroclastic fall	
Column height	25 km
Wind velocity	20 m s^{-1}
Wind direction	S 85 E
Semi-major axis	30 min
Fall velocity	45-6.25 m s^{-1}
Event time	75 min
Scene time (in recreation on film)	60 s
B Ballistic apron	
Radius around vent	6.2 km
Scene time (arbitrarily chosen)	60 s
C Pyroclastic flow/surge	
Height of collapsed column	3.819 km
Energy cone angle (vertical angle from most distal edge of flow to top of collapsed column)	7.8°
Velocity of flow/surge	145-0 m s^{-1} (av 129 m s^{-1})
Event time	4 min
Scene time	90 s

4.1 Pyroclastic Fall

This mimics existing empirical data, specifically the deposits of material as shown on a map as a result of field studies. The method used is to create an ellipse, modified by various parameters to achieve the desired final shape about the postulated vent (Fig. 2). Although the process may seem (and is) inelegant, it works very well in simulating the events on film. Note that the parameters of wind direction and velocity assumed here are those suggested by the original fieldwork, but that these can be changed in the computer recreation.

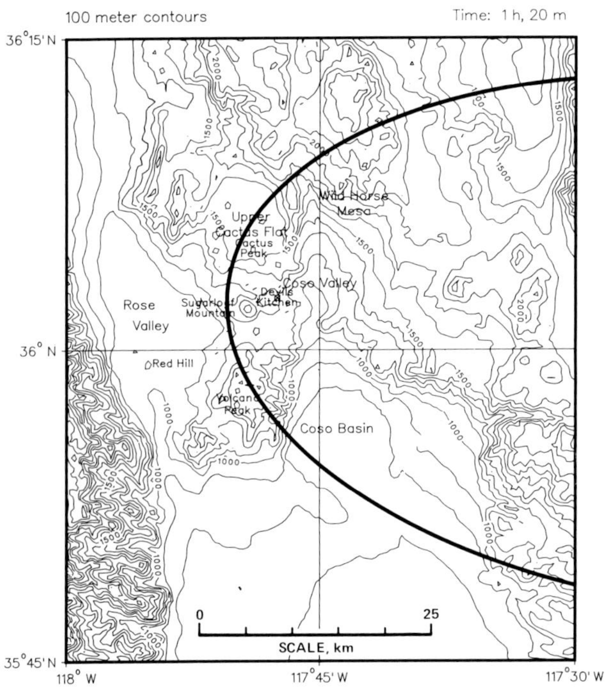

Fig. 2. An example of the computer-generated ellipse for pyroclastic fall at 1 h and 20 min after the start of the event

4.2 Ballistic Apron

The distribution of recognizable ballistic fragments up to 6.2 km from the vent within a single pyroclastic fall unit of the 3 MA Haiwee Tuff caused us to consider the ballistic fall as a special type of hazard. Here, again, the computer program attempts only to animate the event and recreate the empirically determined distribution (Fig. 3). Two pseudo-random numbers were generated for each fragment represented. One of these gave the ejection bearing and the other the distance from the vent. This second number also determined the size of the dot on the screen representing the fragment, as fragment diameter is seen in the field to be inversely proportional to distance ejected.

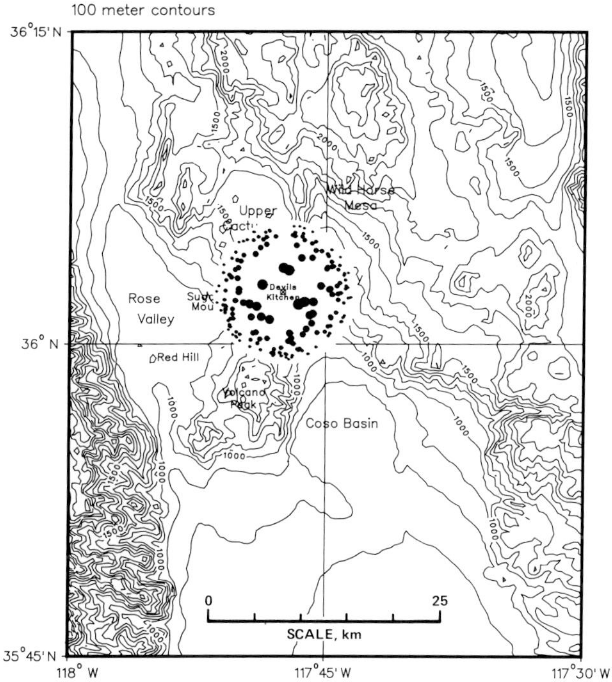

Fig. 3. An example of the computer-generated ballistic apron with increasingly smaller fragments farther away from the vent

4.3 Pyroclastic Flow/Surge

Both pyroclastic flow and surge are computer-modeled in the same way, so they were not distinguished in the recreation. The basic model is one originally devised and used by Malin and Sheridan (1982). This model assumes column collapse as the generating mechanism, with consequent conversion of potential energy to kinetic energy, and an "energy cone" which delineates the limits of flow/surge. This is the most complicated of the computer programs used but it also is in many respects the most realistic (Fig. 4 a, b, c).

For two of the three events (fall, flow/surge), velocities were selected which were empirically reasonable. In addition, the velocities were varied during the event, as could be the case in an actual eruption. Thus, the flow/surge sequence begins with an initial velocity near the vent of 145 m s^{-1} and decreases to 0 m s^{-1} at the outer limit. Selection of velocities allowed choosing a time scale for the events and adding a visual timer to the sequences, showing elapsed time from the beginning of the event. The timer for fall reads in minutes; the timer for flow/surge reads in seconds. No time scale was used for the ballistic event. Besides showing possible speeds with which these events can happen, the timer graphically illustrates how much, or how little, reaction time there may be for the events.

For the movie the eruption site was moved to Devil's Kitchen, one of the areas currently considered a possibility for future eruptions. This allowed the computer model to interact realistically with present-day topography.

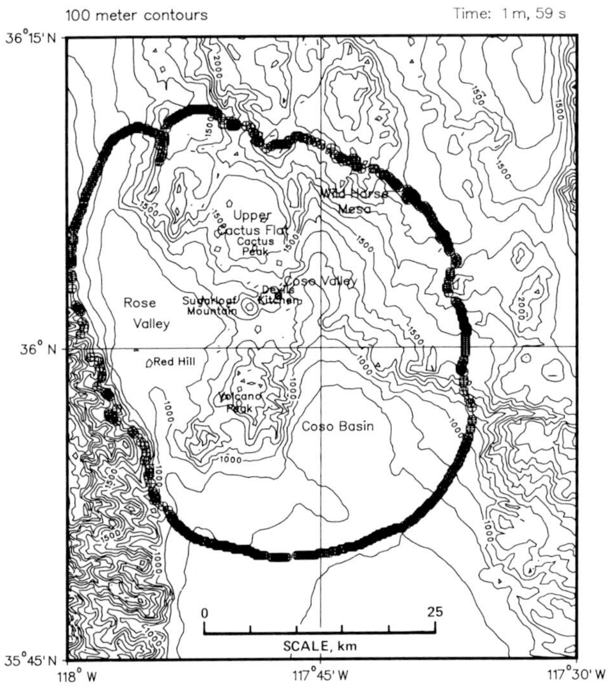
100 meter contours
Time: 1 m, 59 s
36°15'N
36° N
35°45'N
118° W
117°45'W
117°30'W
Upper
Cactus Flat
Cactus
Peak
Wild Horse
Mesa
Coso Valley
Devils
Kitchen
Rose
Valley
Sugarloaf
Mountain
Red Hill
Volcano
Peak
Coso Basin
0
25
SCALE, km

Fig. 4 a

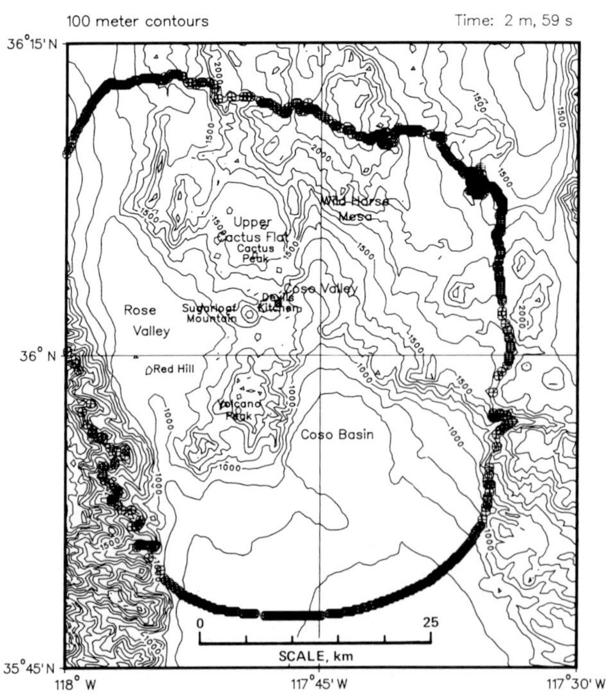
100 meter contours
Time: 2 m, 59 s
36°15'N
36° N
35°45'N
118° W
117°45'W
117°30'W
Upper
Cactus Flat
Cactus
Peak
Wild Horse
Mesa
Coso Valley
Devils
Kitchen
Rose
Valley
Sugarloaf
Mountain
Red Hill
Volcano
Peak
Coso Basin
0
25
SCALE, km

Fig. 4 b

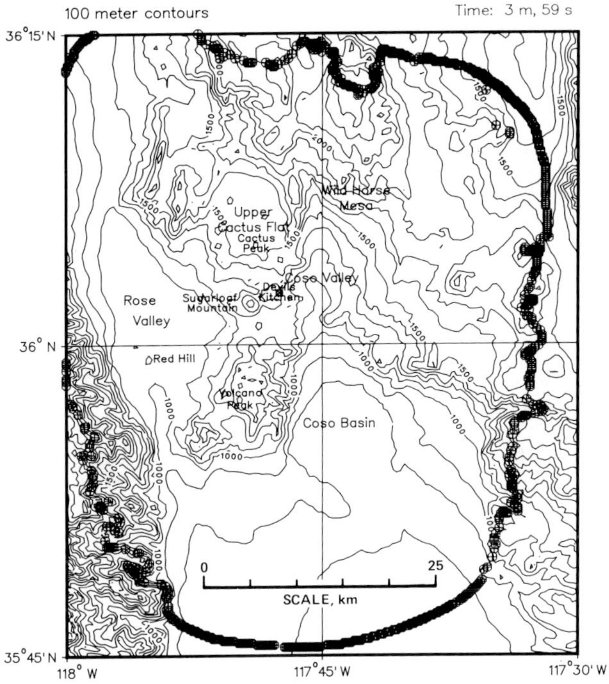

Fig. 4 c

Fig. 4 a-c. a The pyroclastic flow/surge front at 1 min and 59 s after the start of the event. b The pyroclastic flow/surge front at 2 min and 59 s. c The pyroclastic flow/surge front at 3 min and 59 s

An explanatory text at the beginning of the film and between each event was added to the computer programming as well as an index map and references. The final "film" was run on the computer and then processed by the photo lab at NWC. The entire film takes approximately 7 min running time.

5 Adaptability to Other Areas

In its present form, the program requires moderately sophisticated computer facilities. The software was developed on a Digital Equipment Corporation (DEC) VAX 11/780 cluster; the final computer runs were made on a larger Sperry 1100/83 mainframe computer. An Information International Incorporated COMP 80 microfilm recorder was used to generate 35-mm color film which was then reduced to 16 mm in the photographic laboratory. The authors feel that the program can be modified to run on a VAX-size computer and perhaps even on an advanced microcomputer. Such modification may be pursued to make the technique available to regions where facilities are more limited. Meanwhile, it is possible that films such as the present one could be produced for other selected areas, by special arrangement, using the NWC facilities. Availability (or creation) of a digitized topographic base is essential.

6 Conclusion

Although the final product of this investigative phase has been a computer-generated movie, it should not be forgotten that all the calculated and assumed parameters which were used in producing the movie were based on conventional field and laboratory studies of a volcanic unit. The method of estimation of hazard limits is slightly unconventional, the method of presentation perhaps more so. But, as noted in the film, we believe that this is a dynamic means of presentation for both scientists and nonscientists. This "movie of an eruption before it happens" has the potential for strong visual impact on the public in general and in particular on political officials, planners, and disaster coordinators, who might be concerned with an eruption. Such a movie also impressed us as an interesting way to present a geologic story.

A copy of the 16-mm Haiwee Tuff Movie and two videotapes of it (one each in beta and VHS formats) have been prepared for loan to interested individuals. They are available for a slight fee to cover costs of mailing. Inquiries should be addressed to Dr. Glenn R. Roquemore.

References

Blackerby BA, Roquemore GR, Zellmer JT (1984) The Haiwee Tuff: Coso volcanic field, southeastern California. Abstracts with Programs, Cordilleran Section, Geol Soc Am, Cordilleran 16, 5, p 270

Duffield WA, Bacon CR, Dalrymple GB (1980) Late Cenozoic volcanism, geochronology, and structure of the Coso Range, Inyo County, California. J Geophys Res 85:2381-2404

Malin MC, Sheridan MF, (1982) Computer-assisted mapping of pyroclastic surges. Science 217:637-639

Roquemore GR, Blackerby BA (1985) Ballistically transported fragments within pyroclastic fall deposits - potential uses and assumptions. Abstracts with Programs, Cordilleran Section, Geol Soc Am, Vol. 17, No. 6, p 405

Sanders C, Kanamori H (1984) Anomalous shear wave attenuation in the shallow crust beneath the Indian Wells Valley-Coso Region, California. EOS 65, 65, 45, p 1009

Mt. Etna (Sicily): Volcanic Hazard Assessment

G. Forgione[1], G. Luongo[1,2] and R. Romano[2]

Abstract

In recent times in Italy many research programs on active volcanoes have been devoted to surveillance and volcanic hazard, because the density of inhabitants around the main active volcanoes produces a high risk. At Mt. Etna the principal volcanic hazards, on the basis of the historical activity, are lava flows. A preliminary hazard map for effusive activity is proposed. This is obtained starting from well-determined volcanological data (a geological map at a scale 1:50,000 is available) and utilizing geophysical results on the structure and dynamics of the area. The pattern of the main feeder structures, the probability of lava-flow invasion and of the opening of eruptive vents (fissures and/or cones) for the different zones of the volcano are shown.

1 Introduction

Mount Etna is Europe's largest volcano and sits in a complex and incompletely understood tectonic setting: the Calabrian Arc, which is thought to be an arc-trench system which developed from the subduction of the Ionian plate in a NW-SE direction beneath the southern Tyrrhenian Sea. However, the subducted plate might be considered as a passive element of the dynamics of the area, rather than as an active one because the European plate (Tyrrhenian basin) overthrusts the African plate (Ionian Sea). The stress-strain field for southern Italy, inferred from geological and geophysical data, is consistent with that of a bending plate fixed at one end (Fig. 1: Gaudiosi et al. 1984). This stress field is produced by the continuing collision between Africa and Europe which started when the Tyrrhenian Sea began to open 10 million years ago (Boccaletti and Manetti 1978; Calcagnile et al. 1981; Scandone 1979). The acting stress field is dominantly compressive on the inner side of the arc (Aeolian volcanism), and extensive on the outer side of the arc (basaltic volcanism of Mt. Etna) (Fig. 2). The bending of the plate produces the clockwise rotation of Sicily, and the opening of a chasm between the eastern coast of Sicily and

1 Osservatorio Vesuviano, Ercolano, Napoli, Italy
2 Istituto Internazionale di Vulcanologia, Catania, Italy

IAVCEI Proceedings in Volcanology 1
J.H. Latter (Ed.), Volcanic Hazards

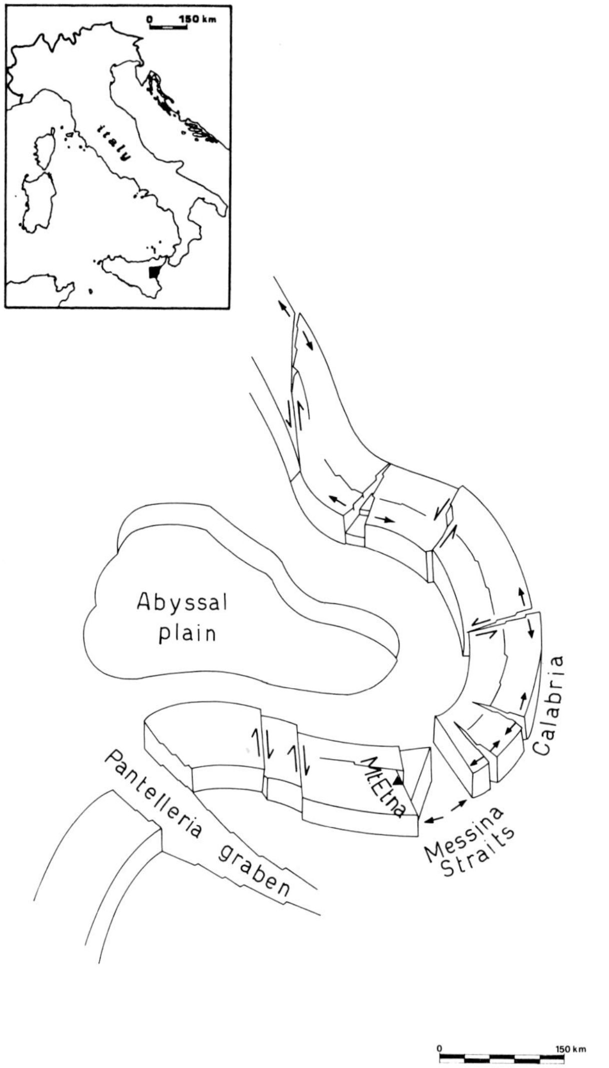

Fig. 1. The bending model for southern Italy. (After Gaudiosi et al. 1984)

the southern Calabria peninsula. These extensional processes are thought to cause volcanism at Etna and also give rise to the general migration of the main feeding conduits of the Etnean volcanic area from SE to NW (Ancient Alkalic Eruptive Centres, Trifoglietto Unit, Mongibello Unit; Romano 1982). Because of the uprising of a large volume of magma in the Etnean area, the direction of the principal stresses in the upper part of the crust has changed: the maximum principal stress direction has become vertical, and the strain reaches a maximum value along belts 120° apart. Eruptive vents (fissures and/or cones) are prevalently located along these belts (Fig. 3).

The initial products of Mt. Etna, about 500,000 years old, are subalkaline lavas of both submarine and subaerial character.

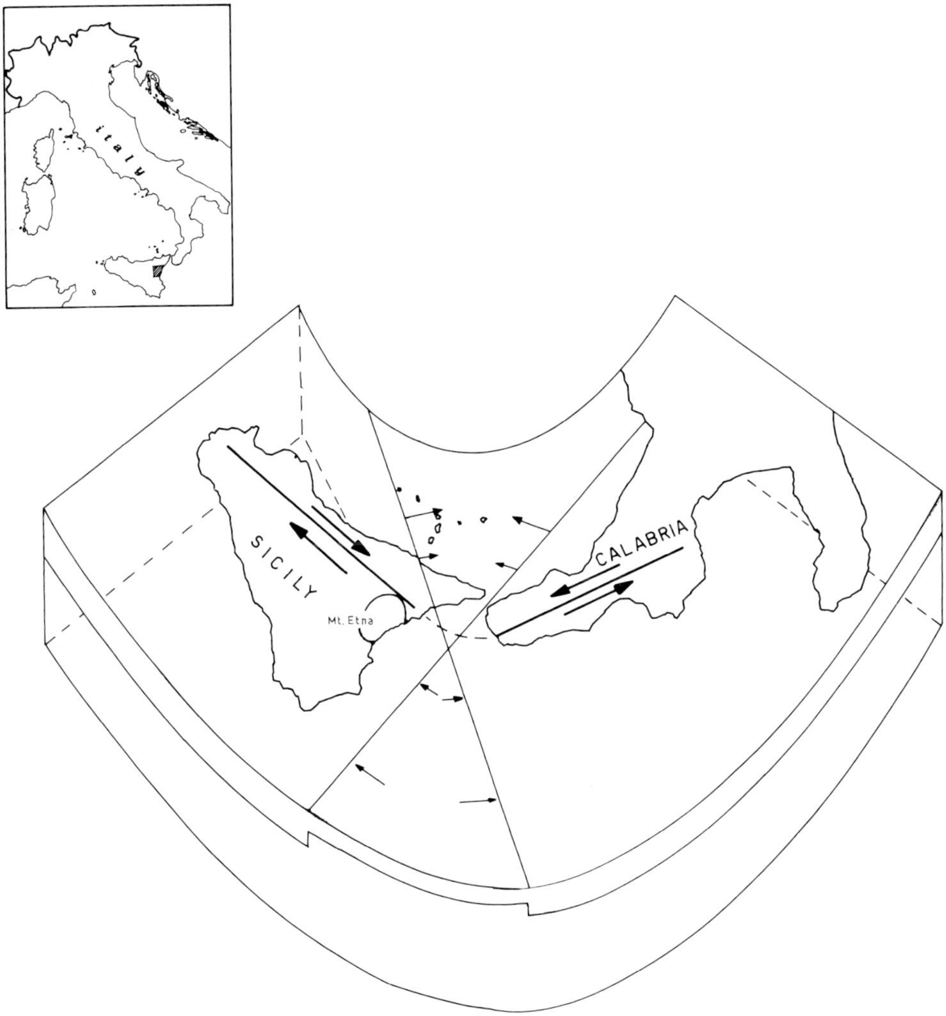

Fig. 2. The bending of the plate produces compressive stresses on the inner side of the arc, tensile stresses on the outer side of the arc and the clockwise rotation of Sicily. (After Gaudiosi et al. 1984)

Between about 200,000 and 100,000 years ago, alkali-basaltic volcanics were erupted, with the formation of small isolated centres (Ancient Alkalic Eruptive Centres). After that time, there were two periods when more extensive differentiation of the magma occurred and the explosivity increased notably. This activity produced two large stratovolcanoes (the Trifoglietto Unit and Ancient Mongibello), formed of numerous secondary eruptive centres. This stage lasted until 8,000 years ago, when prevalently effusive activity started (Recent Mongibello) with a lesser degree of evolution in the alkali-basalt products (Fig. 4: Romano et al. 1979; Romano 1982).

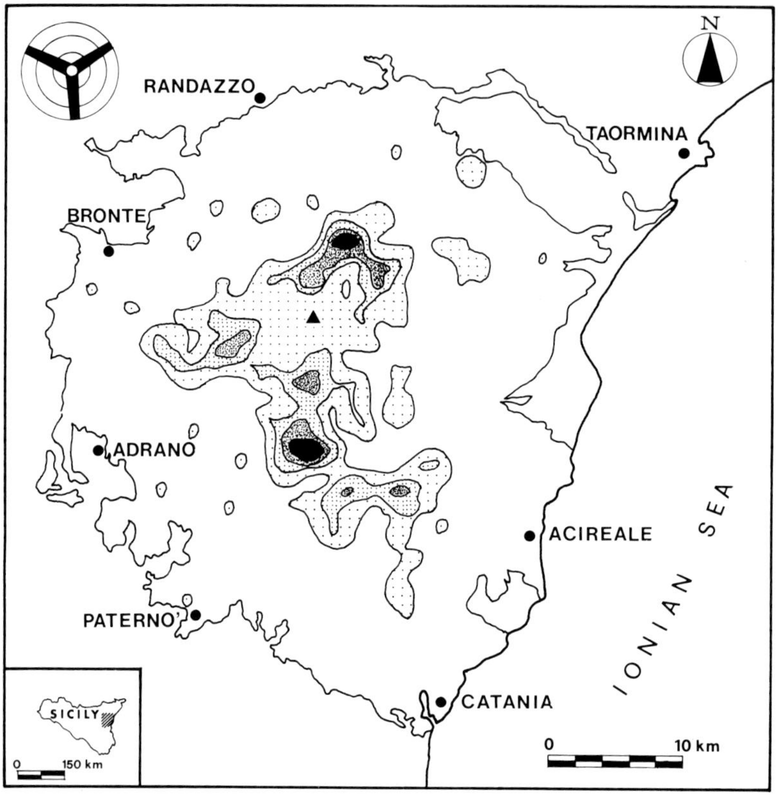

Fig. 3. Density distribution of eruptive vents on Mt. Etna. The **dark areas** of greatest density contain more than two centres km^{-2}. The other areas (**dotted less and less closely**) indicate a lower density (after Rittmann 1973). The **circle** represents the principal stress direction with belts 120° apart. The Central Crater is indicated by a **black triangle**

2 Discussion

During its evolution, Etnean volcanism has varied in its degree of hazard in accordance with changes in its prevalent eruptive activity. While the initial eruptive events were mainly effusive, with low volcanic hazard, the later eruptive activities of the Trifoglietto Unit and Ancient Mongibello were explosive with high volcanic hazard (Fig. 4). During the activity of Ancient Mongibello there were also eruptive episodes with the formation of lahars and pyroclastic flows (Romano 1982).

The present-day hazard of Mt. Etna (Recent Mongibello), on the basis of the historical activity, is due essentially to lava-flow invasions from flank eruptions (Frazzetta and Romano 1978; Guest and Murray 1979; Chester et al. 1985). The eruptions on Mt. Etna have been classified (Rittmann 1973) according to the site where they occurred and the way in which the activity evolved, as <u>terminal</u>, <u>subterminal</u>, <u>flank</u> and <u>eccentric</u>. The

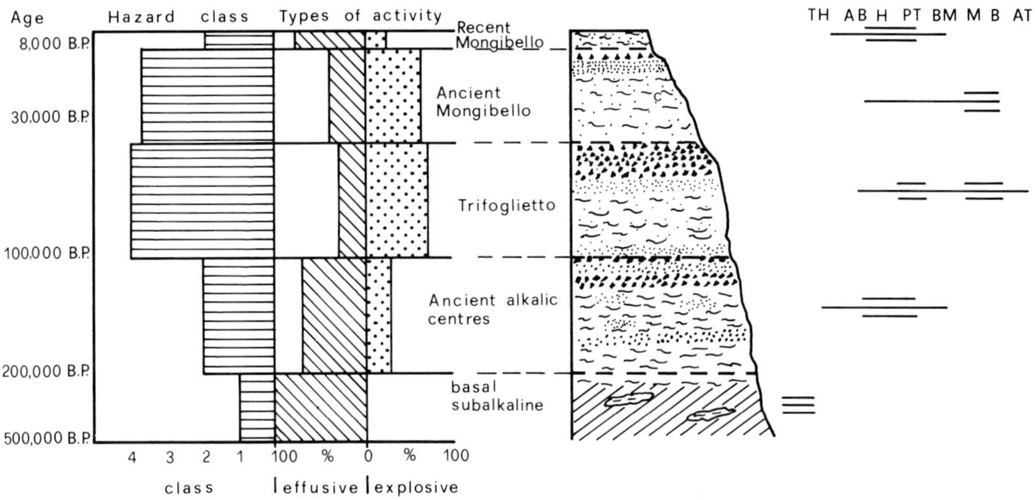

Fig. 4. Right: An idealized stratigraphic cross-section of the volcanic units found in the Etnean area. At **top right** the various types of volcanics are indicated: **TH** tholeiitic basalts; **AB** alkali-basalts; **H** hawaiites; **PT** phonolitic tephrites; **BM** basic mugearites; **M** mugearites; **B** benmoreites; **AT** alkali-trachytes. The **triple lines** indicate a greater frequency of those types of volcanics in that particular period. **Left:** The variations of volcanic hazard in the various periods of the evolution of the Etnean area volcanism. In the stratigraphic sketch section the volcanic products are stylized with: **wavy lines** (lava flows); **small triangle and dots of various size** (pyroclastics); **diagonal lines** (clays). Class of explosivity: **1** 0%; **2** 0% to 30%; **3** 30 to 60%; **4** >60%

study of recent eruptions, in comparison with volcanological data of the previous activity, has shown that the Etnean eruptions can be broadly divided into two overall categories: quiet eruptions, characterized by slow and continous lava effusion, with effusion rates $<10\ m^3\ s^{-1}$, which usually last for several months, or even years; and paroxysmal eruptions, characterized by an intense explosive-effusive activity for a short period of time, usually days to a few weeks, with effusion rates from 10 to 100 $m^3\ s^{-1}$ (Romano and Sturiale 1982).

Table 1 shows the most recent historical eruptions, from 1978 to 1985. Data for the greatest historical eruption (1669) are given for comparison.

Figure 5 shows the frequency of all eruptions which occurred from 812 A.D. to 1985, as a function of their energy. The energy output has been defined by utilizing the relation:

$$E = M\ (\Delta T \cdot C + H),$$

where M is the mass, ΔT the temperature difference between lava and environment (1000 °C), C is the specific heat (0.25 cal/°C),

Table 1. The recent eruptions from 1978 to 1985(2)

Eruption (year)	Duration (days)	Flank	Altitude vents (m a.s.l.)	Maximum length of flows (km)	Altitude reached by flows (m a.s.l.)	Area covered by flows (km^2)	Volume lava emitted (10^6 m^3)	Rate, (m^3 s^{-1})	Energy (10^{17} J)
1978(1)	37 29/IV-5/VI	Eastern	3000 2875- 2575	4	1625	1.8	27.5	8.5	0.88
1978(2)	6 24/VII-30/VII	Eastern	2725 2700- 2520 2300	3.5	1650	0.75	4	7.7	0.13
1978(3)	12 18/XI-30/XI	Eastern	2600 1675	5	1100	2.75	11	10.6	0.36
1979	6 3/VIII-9/VIII	Eastern; north-eastern	3000 1850 1700- 2150	8	870	2.5	7.5	14.5	0.22
1981	6 17/III-23/III	North-western	2550 1400- 1300 1150	7.5	600	6	30	58	0.97

Table 1. (cont.)

Eruption (year)	Duration (days)	Flank	Altitude vents (m a.s.l.)	Maximum length of flows (km)	Altitude reached by flows (m a.s.l.)	Area covered by flows (km^2)	Volume lava emitted (10^6 m^3)	Rate, (m^3 s^{-1})	Energy (10^{17} J)
1983	131 28/III-6/VIII	Southern	2300	7.5	1080	6	100	8.8	3.2
1984	172 28/IV-16/X	Eastern	3000	3	2000	1.5	10	0.7	0.32
1985(1)	3,123 8/III-10/III 12/III-3/VII	Southern	3000 2500	3	1830	2.2	30	2.8	0.97
1985(2)	1,4 25/XII 28/XII-31/XII	Eastern	2600	3.5	1675	0.7	0.7	1.6	0.022
The greatest historical eruption									
1669	122 2/III-11/VIII	Southern	850- 800	15	0	37.5	937.5	14	30

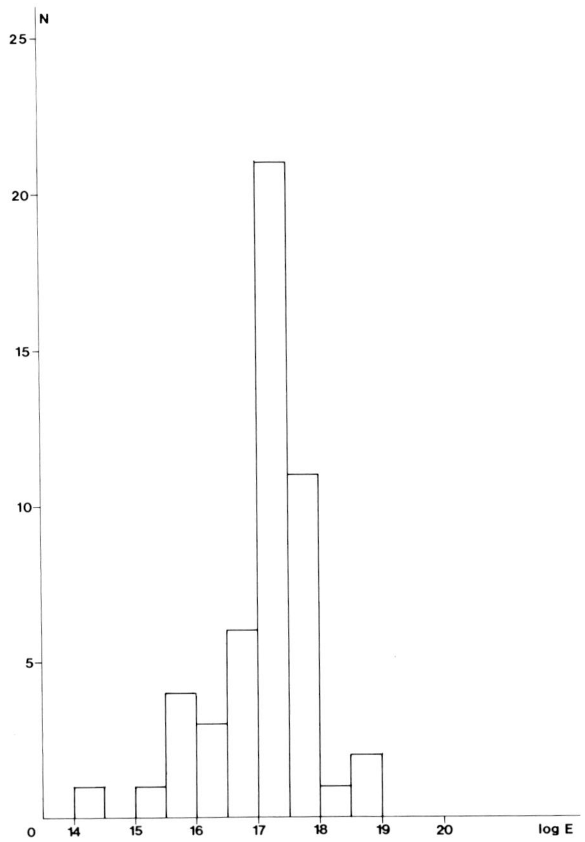

Fig. 5. Frequency of volcanic events as a function of their energy. **N** Number of volcanic events; **log E** logarithm of energy in Joules

H the latent heat (50 cal/°C). M has been calculated with a density of 2.6 g cm^{-3}; volumes of lava are those reported by Romano and Sturiale (1982). Then, the distribution of the number of eruptions has been calculated according to their energy. The greatest number of events (21) is found in the energy range 0.8×10^{17}-2.8×10^{17} J.

From the analysis of the spatial distribution of the eruptions considered, it is seen that the greatest number of them, within the log energy range 17-17.5, are located along all the magmatic feeder structures without there being one preferential structure. The same is seen in the case of the eruptions with the highest and lowest energy. Moreover, there is no apparent relationship between the type of eruption (paroxysmal or quiet), on the one hand, and the energy output on the magmatic feeder structures, on the other hand. The two types of eruption may be related to the particular chemical and physical magmatic conditions, or to the stress acting in the area at the time of the eruption.

The cumulative amount of thermal energy released (Fig. 6) shows three trends with different rates separated by a period with low volcanic activity. The first one has very low volcanic ac-

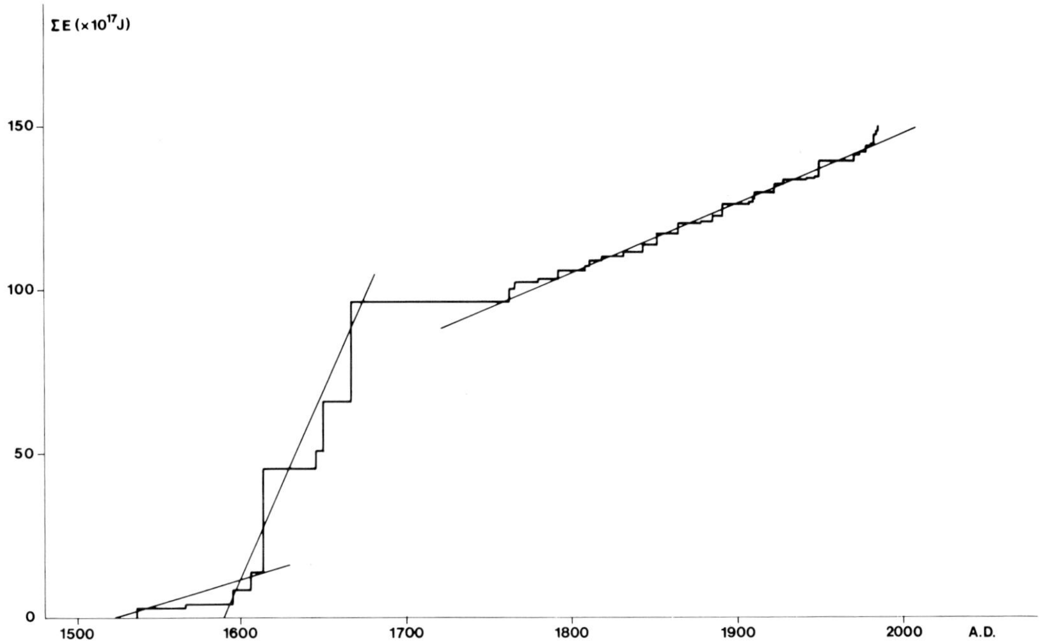

Fig. 6. Cumulative amount of thermal energy released at Mt. Etna volcano from 812 A.D. to 1985

tivity and the last moderate activity, while in the second trend the largest eruptions occur. In general, the same trends of eruptive energy released were obtained by Wadge et al. (1975), but the level of energy they report is different to ours because they utilize different estimates for the lava volume emitted.

The difference between the first and second trends in Fig. 6 is considered as related to the dynamics of eastern Sicily. During the 17th century, a large amount of magma evidently migrated to the surface (see the 1669 eruption) and there was a pause in this phase, probably due to the change in the dynamics of the region as the large earthquake of 1693 (M = 7.1), which violently shook eastern Sicily, can show (Progetto Finalizzato Geodinamica-CNR 1985). After that time, the dynamics of the area quietened, as the low seismicity shows. Only during the 20th century did the regional seismicity increase again (Fig. 7; Barbano et al. 1981).

The time-space distribution of the Etnean eruptions allows zones with different levels of hazard to be classified. The volcanic hazard can be considered variable in one and the same area depending very much on the local morphology and on the distance from potential eruptive vents. On the basis of the location and frequency of past eruptions, a map has been drawn of the probability of eruptive fissures and/or cones opening up and of lava-flow invasion; such a map shows the zones with in-

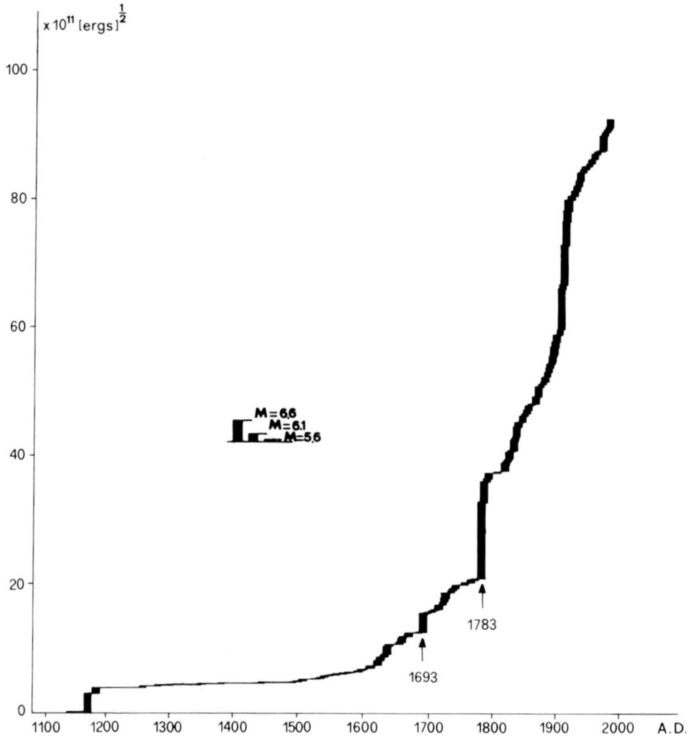

Fig. 7. Benioff diagram. Earthquakes localized in Sicily and Calabria from 1000 A.D. to 1978. (After Barbano et al. 1983)

creasing degrees of volcanic hazard (Fig. 8). In Fig. 8 four zones of greater or lesser probability of the opening up of eruptive vents are drawn:

Zone 1: no opening of eruptive vents has occurred in recent time;
Zone 2: at least one eruptive vent has opened up per km;
Zone 3: 2.5 eruptive vents have opened up per km;
Zone 4: 3 or more eruptive vents have opened up per km.

In the same figure (Fig. 8) three areas of increasing probability of lava invasion are drawn:

Area A: little probability;
Area B: medium probability;
Area C: great probability.

These three areas are delimited by a lower boundary, representing the maximum extension of the lava cover from the very first activity of the volcano; an intermediate boundary, related to flank and eccentric eruptions; and an upper boundary, referring essentially to terminal and sub-terminal eruptions.

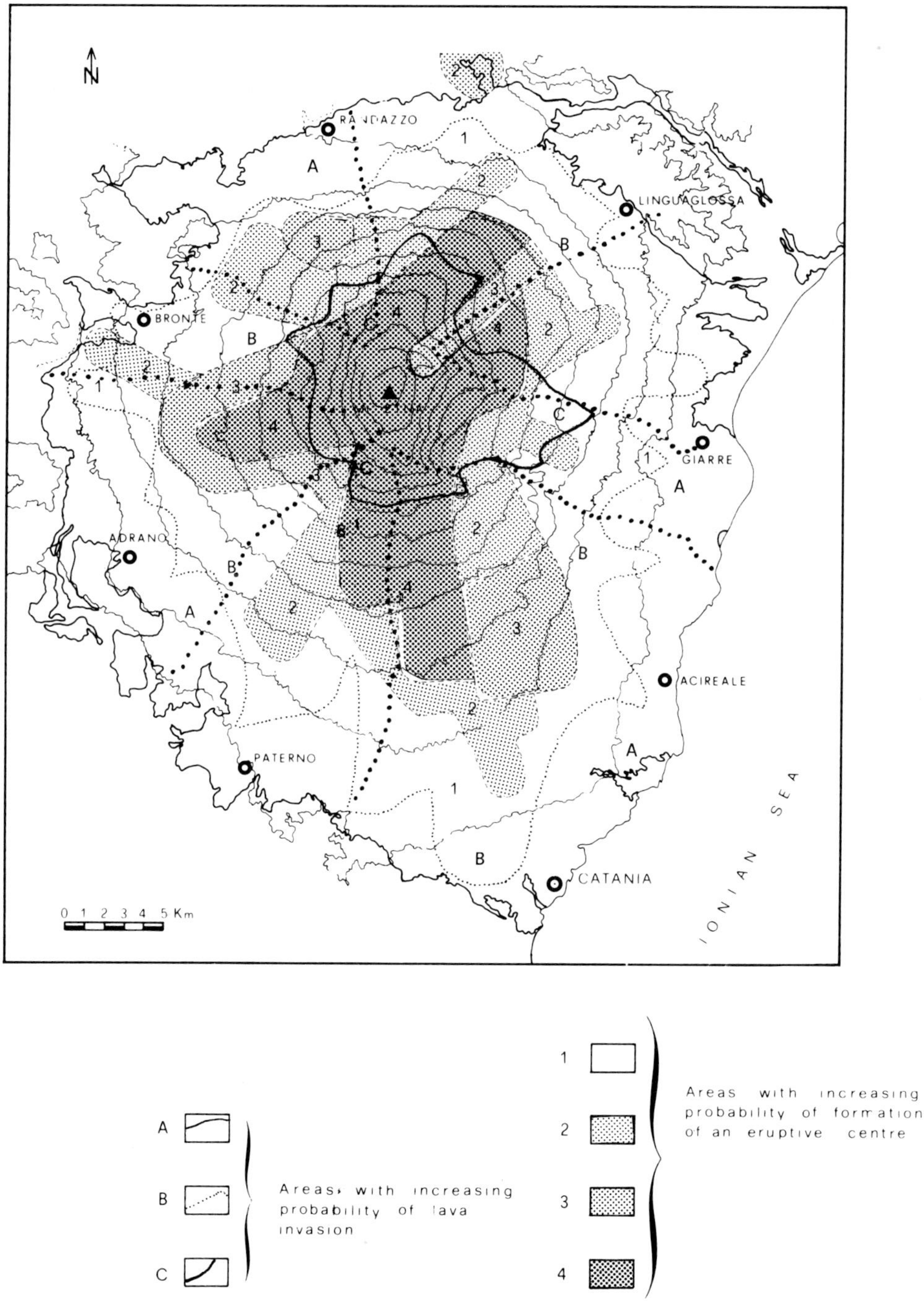

Fig. 8. Map of potential volcanic hazard. Dotted lines delimit radial sectors with different volcanic hazard; the boundaries are traced on a morphological basis

The second and third of these boundaries have been obtained from consideration of the ease with which lava can flow in a given area (potential flow paths) due to the surface morphology present, and of the frequency in historic times of the same area being covered by lavas.

The seismic surveillance of Mt. Etna shows that the level of activity before and during paroxysmal eruptions is higher than that accompanying quiet eruptions (Cosentino et al. 1984; Cosentino and Lombardo 1984; Glot et al. 1984; Guerra et al. 1976; Patané et al. 1984). Notwithstanding the moderate energy of the earthquakes which precede and accompany the eruptive activity, the seismic hazard may be high due to the shallowness of the hypocentres. In recent times (Christmas 1985) a 4.5-magnitude earthquake localized in the eastern flank of the volcano produced the collapse of a small hotel and the death of one person. On the basis of the historical and present-day seismicity and the local and regional tectonics, a preliminary seismogenetic map is drawn (Fig. 9), which shows that the zones with high seismic hazard are located in the eastern part of the volcano. The complex picture is due to the contribution of a tensile stress which is produced by the opening of the Messinian chasm and by a local tensile stress due to the uprising of large magmatic masses. The first stress field acts prevalently along the Ionian coast while the second acts in three belts in the SE, SW and NE parts of the volcano. However, this seismogenetic map is only an oversimplification of the real strain field, because other tectonic trends are not well understood as, for example, the E-W faults observed in the NE sector of the volcano. The complexity of the active superficial tectonics is due also to the variable geometry of the sedimentary basement and to the changes of the thickness of the volcanics in the different parts of the volcanic area.

3 Conclusions

The bending model gives a sufficiently real picture of the tectonics of the Mt. Etna volcano. The bending of the plate produces a tensile stress field that creates the conditions for the formation of this volcanic area. This process gives rise, in general, to the overall migration of the main feeding conduits in the volcanic area from SE to NW. A local stress field is added to the regional one. Magmatic pressure has superimposed a local volcanic stress field on Etna. Three extensional rift zones, 120° apart, have been created and magma is preferentially erupted along these zones. High seismic hazard areas occur along the coastal line and the rupture zones induced by the local stress field.

A detailed analysis of the eruptive history of the volcano in the last centuries evidences the high probability of new fissures and eruptive cones opening up, and thus of there being lava invasion, in the southeastern flank. There is a corresponding relative safety on the southwestern flank. Unfortunately, recent population censuses show the southeastern and eastern flanks, with their apparently high seismic hazard, to be the

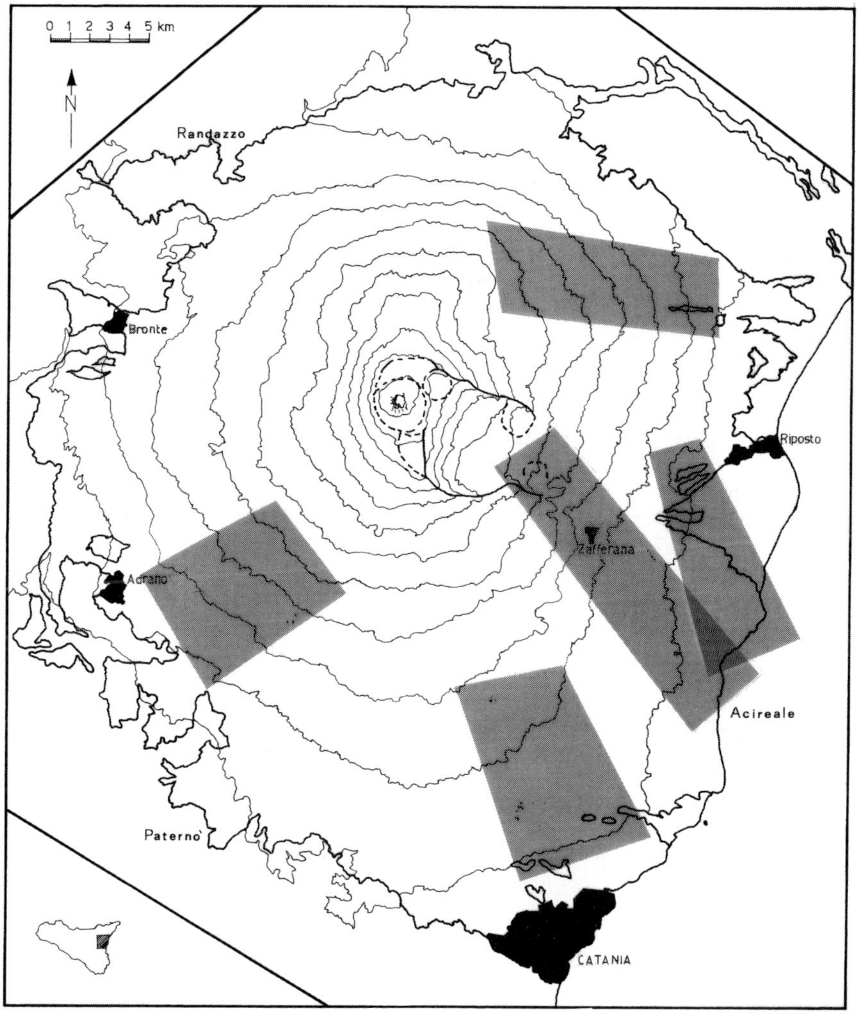

Fig. 9. A tentative seismogenetic map showing the structures with high hazard. These are prevalently located in the eastern part of the volcano, where a regional stress is acting. The **sheaves** indicate the zones of major concentration of faults and fractures together with the epicentres of major earthquakes

most densely populated and to contain the most industrial plants and public structures of social interest.

The other sectors with a lesser volcanic hazard, the northeastern (Linguaglossa) and northwestern flanks (Bronte and Randazzo) are less densely populated (ca. 50,000 inhabitants), although they contain agricultural crops of a certain prestige (pistachio nuts and vines).

References

Barbano MS, Carrozzo MT, Chirenti A, Cosentino M, Lombardo G, Riuscetti M (1981). Sismicità di Calabria e Sicilia. Rend Soc Geol It 4:529-535

Boccaletti M, Manetti P (1978). The Tyrrhenian Sea and adjoining regions. In: Nairn AEM, Kanes WH, Stehli FG (eds) The ocean basins and margins, vol 4B. Plenum Publishing Corporation, New York, pp 149-191

Calcagnile G, Fabbri A, Farsi F, Gallignani P, Gasparini C, Iannaccone G, Mantovani E, Panza GF, Sartori R, Scandone P, Scarpa R (1981). Structure and evolution of the Tyrrhenian basin. In: Proceedings of the Commission Internationale pour l'Exploration Scientifique de la Mer Méditerranée. 27,8:197-208

Chester DK, Duncan AM, Guest JE, Kilburn CRJ (1985) Mt. Etna: the anatomy of a volcano. Chap 8:296-336, Chapman Hall, London

Cosentino M, Lombardo G (1984) A model for eruptive mechanisms of Mt. Etna from the study of seismicity from 1978 to 1983. Bull Volcanol 47:977-993

Cosentino M, Gresta G, Lombardo G, Patane' G, Riuscetti M, Schick R, Viglianisi A (1984) Features of volcanic tremors on Mt. Etna (Sicily) during the March-August 1983 eruption. Bull Volcanol 47:929-939

Frazzetta G, Romano R (1978) Approccio di studio per la stesura di una carta del rischio vulcanico (Etna-Sicilia). Mem Soc Geol It 19:691-697

Gaudiosi G, Luongo G, Ricciardi GP (1984) A bending model for the Calabrian Arc. In: Cassinis R (ed) Problems and methods for lithospheric exploration. Plenum Publishing Corporation, New York, pp 191-199

Glot JP, Gresta S, Patane' G, Poupinet G (1984) Earthquake activity during the 1983 eruption. Bull Volcanol 47:953-963

Guerra I, Lo Bascio A, Luongo G, Scarpa R (1976). Seismic activity accompanying the 1974 eruption of Mt. Etna. J Volcanol Geotherm Res 1:347-362

Guest JE, Murray JB (1979). An analysis of hazard from Mount Etna volcano. J Geol Soc Lond 136:347-354

Patané G, Gresta S, Imposa S (1984). Seismic activity preceding the 1983 eruption of Mt. Etna. Bull Volcanol 47:941-952

Progetto Finalizzato Geodinamica - CNR (1985) Catalogo dei terremoti italiani dall'anno 1000 al 1980 Quaderni de la Ricerca Scientifica Vol. 114, 2B, 239 pp. Postpischl D. (ed), Bologna

Rittmann A (1973). Structure and evolution of Mt. Etna. Phil Trans R Soc Lond, Ser A 274:5-16

Romano R (1982) Succession of the volcanic activity in the Etnean area. Mem Soc Geol It 23:27-48

Romano R and Sturiale C (1982) The historical eruptions of Mount Etna (volcanological data). Mem Soc Geol It 23:75-97

Romano R, Sturiale C, Lentini et al. (1979). Carta geologica del Monte Etna, scala 1:50,000. CNR-Istituto Internazionale di Vulcanologia, Catania; Progetto Finalizzato Geodinamica

Scandone P (1979) Origin of the Tyrrhenian Sea and Calabrian Arc. Bull Soc Geol It 98:24-34

Wadge G, Walker GPL, Guest JE (1975) The output of the Etna volcano. Nature (Lond) 255:385-387

The 1982–83 Eruption of Galunggung (Indonesia): A Case Study of Volcanic Hazards with Particular Relevance to Air Navigation

A. GOURGAUD[1], G. CAMUS[1], M.-C. GERBE[1], J.-M. MOREL[1], A. SUDRADJAT[2] and P.M. VINCENT[1]

Abstract

The 1982-83 eruption of Mt. Galunggung (Java) was characterized by three successive phases of different eruptive styles, which each generated distinctive types of deposits. The first are channelled ash and scoria flow deposits. The second are phreatomagmatic explosion breccias, air fall and base surge deposits, correlated with the opening of Walirang maar crater. The third are Strombolian deposits, associated with a small lava flow.

The explosive eruptions increased in violence, while paradoxically the magma composition evolved from andesite to primitive basalt. This change is related to the superficial water. Increasing interaction of groundwater with magma may be correlated with periods of quiescence between explosive phases. Quiescent periods varied from a few days to a week during phase 1, and 1 to 3 weeks together with a lowering of the level of magma disruption during phase 2. Such evolution can occur very late in the course of an eruption, after long periods of quiescence during which increased water supply is possible and when the observers' alertness may well have slackened. Consequences for aircraft navigation are considered in the light of the two jet plane incidents related to the hydromagmatic phase.

1 Introduction

Galunggung stratovolcano is located in west Java, about 100 km southeast of Bandung. It has a horseshoe-shaped caldera (Fig. 1) open to the southeast, probably generated by a major collapse of the southeastern flank a few 1000 years ago. Debris-avalanche deposits cover about 175 km^2 (Escher 1925), with a hummocky topography ("Ten Thousands Hills" near the town of Tasikmalaya). Before 1982, Galunggung had erupted in 1822, 1894 and 1918 (Kusumadinata 1979). The historic eruption record of the volcano is characterized by quiescent intervals of several

---

[1] Centre de Recherches Volcanologiques, Observatoire de Physique du Globe, U.A. 10 C.N.R.S. et Université Blaise Pascal. 5, rue Kessler, 63 038 Clermont-Ferrand Cedex, France

[2] Volcanological Survey of Indonesia (V.S.I.), Jl Diponegoro 57, Bandung, Indonesia

IAVCEI Proceedings in Volcanology 1
J.H. Latter (Ed.), Volcanic Hazards

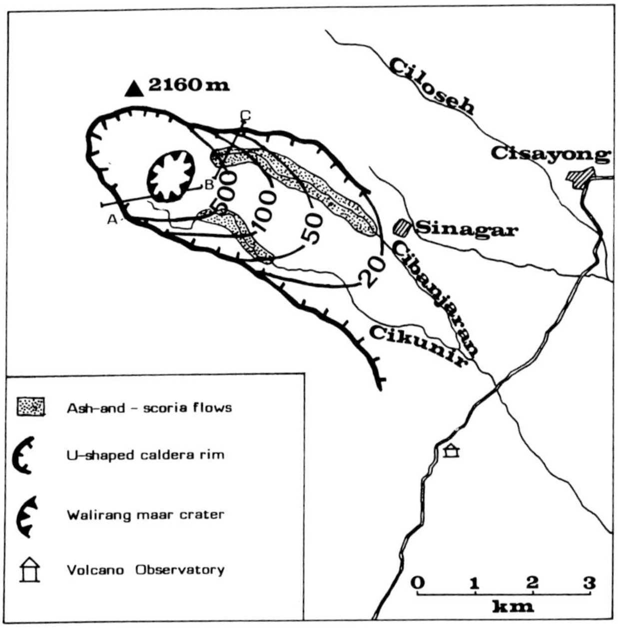

Fig. 1. Tephra isopach map of air-fall and surge deposits (cm) produced by the 1982 hydromagmatic eruption (phase 2) of Galunggung, inside the caldera. Ash and scoria flows from phase 1 are also reported. The ABC section is shown in Fig. 3

decades, interrupted by brief eruptions lasting only a few hours or days. The last eruption of 1982, 9 months long, seems to be exceptional. Short-lived eruptions in 1822 generated nuées ardentes and lahars (with 4000 victims) followed by the extrusion of a dome a few months later. In 1894, the eruption destroyed the 1822 dome but no nuées ardentes were reported. In 1918, a new lava dome (Gunung Jadi) emerged above the crater lake surface, and this episode can be considered as the continuation of the 1894 eruption.

Two jet plane incidents were reported in June and July 1982 during the last eruption. Since 1973, 24 aircraft incidents have been reported worldwide and a working group has been created by the International Organization of Civil Aviation.

2 Chronology of the 1982-83 Eruption

The 9-month-long eruption of Galunggung was monitored by V.S.I. Seismic, ground deformation and geochemical data were recorded. Three phases of activity were recognized (Katili and Sudradjat 1984; Sudradjat and Tilling 1984; this work).

Phase 1 (5 April to mid-May). The first explosion began at dawn on 5 April. Phase 1 generated ash plumes up to 6 km high (Fig. 2) and nuées ardentes that travelled up to 5 km down the Cibanjaran and Cikunir rivers (Fig. 1). Lahars developed during, or shortly after, some eruptions. At first, the quiescent intervals between major eruptions were several days long (typically about 3 days). Later, however, the repose interval became longer (typically about 1 week). During this first period, the eruptions removed about 20% of the total volume of the 1918 dome (Fig. 3).

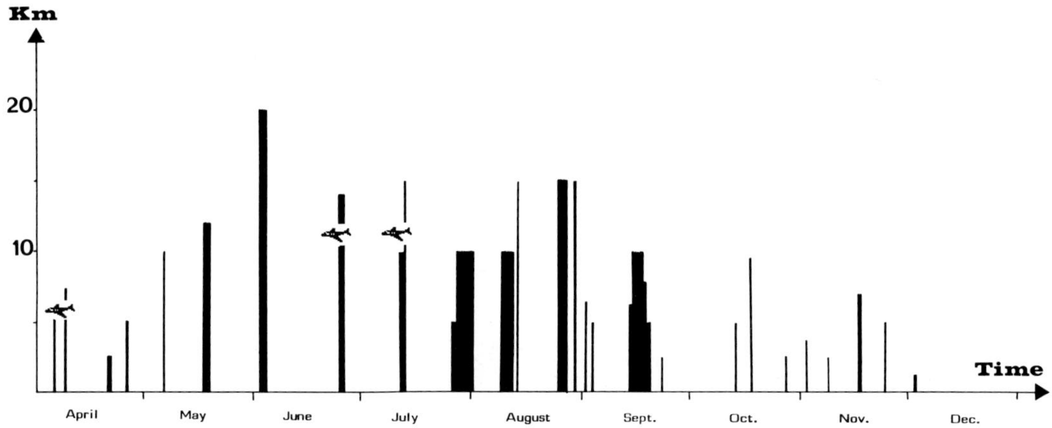

Fig. 2. Diagram showing the height of the ash column versus time. Mild explosions are omitted

Phase 2 (mid-May to October). The most violent explosions took place during this period and ash columns reached a height of some 20 km (Fig. 2). The interval between eruptions increased from 1 to 3 weeks. As much as 80% of the 1918 dome was destroyed and the explosions hollowed out a crater 800 m wide and

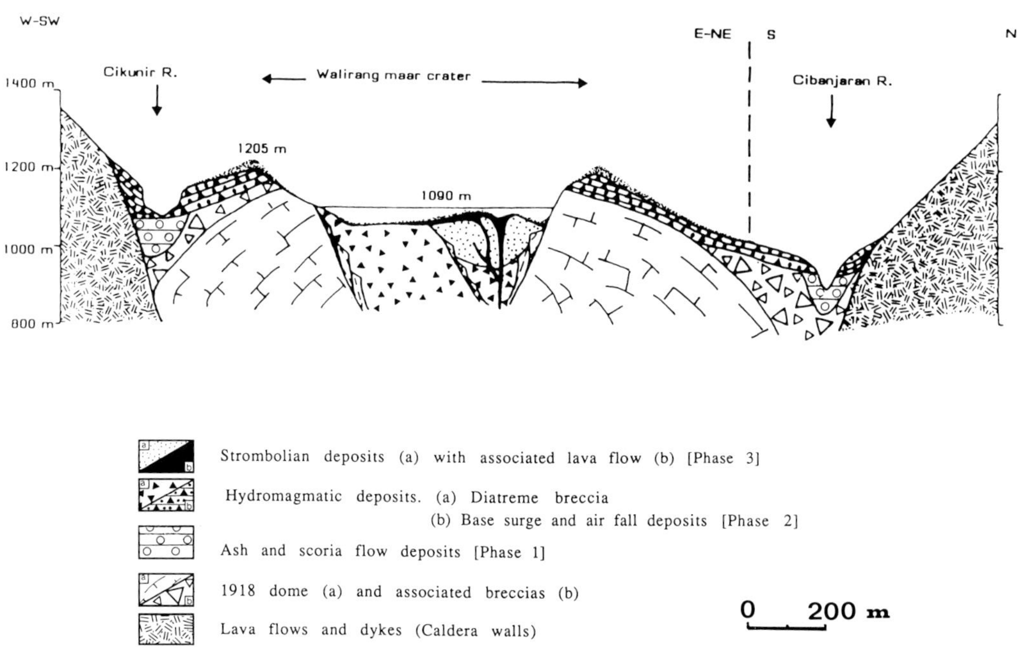

Fig. 3. Section through the Galunggung eruptive system in 1986

150 m deep. This is called Walirang crater and is now interpreted as a maar crater (Fig. 3). We emphasize the phreatomagmatic character of the second phase, previously described as Vulcanian (Katili and Sudradjat 1984): white-coloured plumes, base surges at the lower part of ash-laden eruptive columns, cock's tail explosions, lightning in a water-saturated atmosphere. By September, the vigour of the eruption began to decline, as evidenced by smaller plumes (reaching only 6-km height), and a greatly reduced volume of ash falls. Base surge clouds became infrequent and travelled no further than 2 km. From September to November, gradually, the eruptive activity became Strombolian.

Phase 3 (November to 8 January). This phase was characterized by an abundance of incandescent lava fragments ejected from the crater which showered the entire slope of Walirang crater. Ash eruptions decreased and plumes reduced to heights less than 6 km (Fig. 2). A small cinder cone (60 m high) grew inside the maar crater and was built by Strombolian activity (Fig. 3). A minor effusive activity occurred from December to 8 January inside the Walirang crater. Since January 1983, Galunggung has been inactive except for two small phreatic explosions in January 1984. A water lake has progressively filled the maar crater. In December 1984, only 5 m of the cinder cone were above the water line and in November 1986 the whole cone was submerged (Fig. 3).

3 Tephra Studies: Deposits of the 1982-83 Eruption

Pyroclastic flows, lahars, surge and air-fall deposits of the eruption probably exceed 0.37 km^3 (Katili and Sudradjat 1984). Only 0.014 km^3 are preserved within the caldera. The eruptive material consists of deposits laid down by nuées ardentes, phreatomagmatic and Strombolian activity, related respectively to phases 1, 2 and 3. Field relationships between these deposits are shown in Fig. 3. Grain-size determinations, percentages of magmatic and xenoclastic components, SEM images and density measurements of the magmatic component were made on the deposits related to each phase (Table 1, Figs. 4 and 5).

3.1 Ash and Scoria Flow Deposits (Nuées Ardentes)

The pyroclastic flows were channelled in the Cibanjaran and Cikunir valleys (Fig. 1), inside the U-shaped caldera, and extended relatively short distances (6 km maximum). The volume of these flows is estimated to be about 12 x 10^6 m^3, including the related fall deposits. Some of the deposits were immediately reworked, and there is a progressive transition to mud lahars. Several flow units, separated by thin ash beds or muddy beds, can be distinguished. These flow units consist of subrounded scoria and bombs in a slighty compacted matrix which is black to grey in colour. They lack internal stratification, sorting or grading. Thickness and grain size increase rapidly towards the crater, where some units reach 15 m in thickness, with bombs 2 to 3 m in diameter.

Table 1. Characteristics of the three eruptive phases. M = million

Phase	Volume of the deposit (M m^3) including distal falls	Magmatic component (%)	Volume of the magmatic component (M m^3)	Density of the magmatic component	Mass of the magmatic component (M t)	Duration of the phase (days)	Apparent daily magma supply (t day^{-1})	Duration of the eruptive phases (h)	Intensity (M m^3 h^{-1})
1	12	97.5	11.7	2.13	25	38	0.65	12	0.99[a] 0.97[b]
2	356	77	274	1.93	529	200	2.63	312	1.02[a] 0.87[b]
3	3	97	2.9	1.68	5	45	0.1	19	0.16[a] 0.15[b]

[a] including xenolithic phase;
[b] magmatic component only

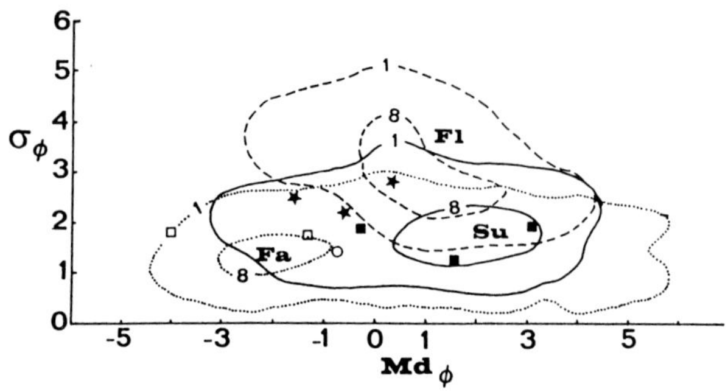

Fig. 4. Plot of the 1982-83 pyroclastic deposits of Galunggung in a ϕ versus Md_ϕ diagram. Contours after Walker (1971) for the air falls (**Fa**) and flows (**Fl**), after Sparks and Walker (1973) for the surges (**Su**).
*: Ash and scoria deposits (matrix <-5 ϕ), phase 1; ■ and □: respectively base surge and air-fall deposits, phase 2; o: Strombolian deposits, phase 3

Grain-size data, for matrix (<-ϕ) of three different flow units have been reported in the $\sigma\phi$ versus $Md\phi$ diagram (Fig. 4); the representative points are close to the pyroclastic flow field (after Walker 1971). SEM images of pyroclasts from these ash and scoria flows show the high vesicularity of the material (Fig. 5).

3.2 Phreatomagmatic Deposits (Phase 2)

The phreatomagmatic products are related to the opening of the Walirang maar crater which truncates the pre-existing substratum. This crater was mainly opened between 17 May and 27 June (Katili and Sudradjat 1984). A second maar-type crater formed inside the first, after the 28-31 July eruptive period (Fig. 3). The deposits are well exposed on the inner wall of the Walirang crater and between the two valleys of Cibanjaran and Cikunir. The thickness decreases from about 50 m in the tuff ring to 4 m, 500 m east of the crater (Fig. 1).

The proximal tuff-ring facies consists of an explosion breccia and air-fall deposits with outward dipping beds. This overlies the 1918 dome and the associated collapse breccia. The explosion breccia consists of coarse, angular, lithic fragments (derived from the 1918 dome), 10 to 30 cm in diameter, mixed with sandy material. The breccia beds are 50 cm to 1 m thick and are commonly interbedded with ash-fall deposits.

Distal deposits are thinly bedded and mainly exposed in the cliffs of the Cibanjaran and Cikunir rivers. They are dominantly composed of pyroclastic surge deposits and ash-fall beds. The surge deposits show a wide spectrum of bed forms, antidunes, and massive and planar sand waves (Wohletz and Sheridan, 1979). Numerous bed-form transitions per vertical meter indicate rapid-

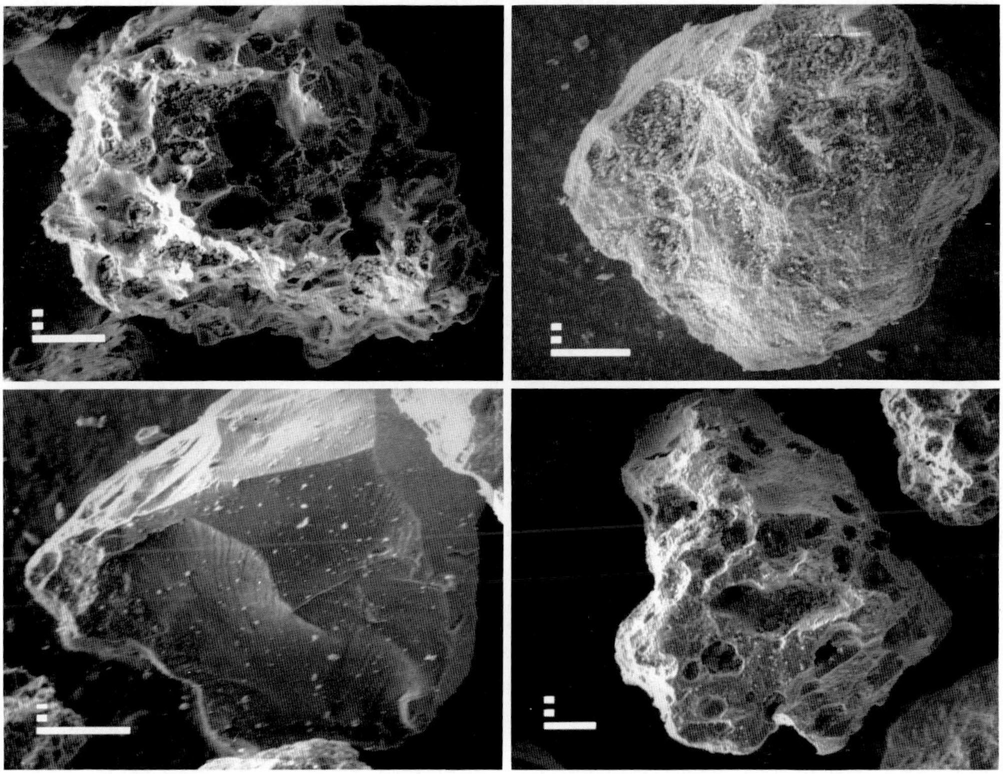

a b c d

Fig. 5 a-d. SEM images of Galunggung juvenile particles. Scale: 0.1 mm. **a** Ash and scoria flow, phase 1. Partly rounded pyroclast with high vesicularity (0.01-0.1 mm diameter vesicles). **b** Hydromagmatic phase 2, surge deposit. Block-shaped grain with adhering dust particles. The planar surfaces have been rounded and the vesicularity is low. **c** Strombolian phase 3, angular sideromelane fragment with no vesicles. Such blocky morphology is typical of phreatomagmatic ash grains: phreatomagmatic explosions were reported during the Strombolian phase. Note the absence of adhering dust particles. **d** Strombolian phase 3. Typical sideromelane lapillus with smooth, glassy surfaces. The sample is highly vesicular with 0.02-0.2 mm diameter vesicles

ly changing depositional characteristics. U-shaped channels (Fischer 1977) are common and cauliflower bombs (associated with bomb sags) have been observed also.

The yellow-grey to ochre colour of the phreatomagmatic deposits of Galunggung is due to the abundance of xenolithic material: fumarolized and oxidized fragmented blocks (Table 1). Phase 2 is characterized by a high xenolithic content (about 25%), which contrasts with xenolith contents lower than 5% in phases 1 and 3.

Some fine ash beds have entrapped vapour on emplacement (vesiculated tuffs); they were nearly saturated with water at the time of the deposition (Lorenz 1974). The presence, within the deposit, of beds of accretionary lapilli, and the persistence of a muddy deposit coating the inner walls of the avalanche caldera for 2 years after the end of the eruption, also indicate a wet environment. The temperature of the deposit, inferred from infrared spectrometry on charcoal samples, is about 175°C. This is in agreement with the field measurements (150°C, 1 week after deposition; Katili and Sudradjat 1984). The vegetation on the southern slopes of the pre-existing dome was not charred, also indicating a low emplacement temperature for the deposit. Trees were severed at half their heights and debarked on the blast side. All these data are consistent with a predominantly base-surge emplacement.

Grain-size analyses have been made on base surge and air-fall deposits from hydromagmatic distal deposits; data are plotted in Fig. 4. SEM images from these deposits show that blocky grains, rounded with adhering particles (Fig. 5), and vesicular grains are common.

3.3 Strombolian Deposits (Phase 3)

Black Strombolian lapilli and bombs are found over a limited area around the Walirang crater, at distances of up to 2-3 km. Whereas the transition between products of phases 1 and 2 was sudden, that between phases 2 and 3 was gradual. The deposits reflect the progressive change in style of the activity through the eruption, from phreatomagmatic to Strombolian. Juvenile lapilli and bombs form an increasing proportion of the deposits upward through the observed outcrops. Generally, the lower units are well-bedded blocks and ash-fall deposits with a high component of non-juvenile rock debris (oxidized fragments of red to yellow colour from the 1918 dome). Around Walirang maar crater, the upper units are poorly bedded Strombolian deposits. Black scoria are mixed with ribbon bombs, spindle bombs and also some cauliflower bombs.

Grain-size analyses are characteristic of fall deposits (Fig. 4). Mixing of Strombolian and hydromagmatic products during phase 3 has been established by SEM observations (Fig. 5).

4 Role of Groundwater in the 1982 Eruption Progress

Explosive historic activity of Galunggung destroyed previous extrusions and hollowed out new craters which were subsequently filled by lakes. Such activity can be related essentially to phreatomagmatism, which seems to be a constant factor in the recent history of the volcano.

The most striking event during the Galunggung eruption was the increase, during the early part of May, in the violence of the explosions (Fig. 2), even though the bulk chemistry of magma evolved from andesite (with about 58% SiO_2) to primitive basalt (with 47% SiO_2). Thus, the violence of phase 2 must be related to the influence of groundwater. If this had not been the case, explosive activity would have decreased with the observed magmatic evolution. We attempted to make a rough estimate of the volumes emitted during each phase. The bulk volume is well established (0.37 km^3). The volume of the phase 3 deposit can be determined with good precision, but that of the phase 1 deposit is more approximate. The results are summarized in Table 1, together with the apparent corresponding magma supply rate and the real discharge rate (intensity) during the eruptive phases. In these computations we believe that the major part of the fine distal deposits belongs to phase 2. In spite of the rough estimates of these values, it is clear that magma supply was significantly higher during phase 2.

Nevertheless, a transition to a higher water/magma ratio is clear. The period from May to November corresponds to the dry season in Java. At this time, therefore, weather change cannot be invoked to explain the apparent increase of the water/magma ratio. The main difference between phases 1 and 2 lies in the fact that the quiescent intervals increased from several days to about 3 weeks duration. We believe that at the end of phase 1 the level of the magma in the pipe had fallen, and that the long periods of quiescence allowed more water to come into contact with the magma body, even though the water supply was nearly constant. In this way the effective water/magma ratio could have been higher during phase 2 than during phase 1. Seismic data are in agreement with our interpretation. Numerous B-type earthquakes during the first 2 weeks of the eruption may mark the position of a shallow magma front, less than 2 km below the surface. When the conduit was opened wide, the magma front retreated to a depth about 6-8 km.

5 Jet Plane Incidents

At least 24 eruptions have resulted in air safety incidents with jet planes. Glassy (ceramic) deposits on turbine blades and vane cooling ports were observed only three times (Tracy 1983; Smith 1984), on 20 June 1980 during the Mount St. Helens eruption and on 24 June and 13 July 1982, during the second phase of the Galunggung eruption. In these last cases, all four engines of the Boeing 747 jets failed. Fortunately, the pilots were able to restart them, after escaping from the plumes, and were able to land safely at nearby airports. Thereafter, international aircraft routes over Indonesia were changed.

The investigations made after the incidents (Pratt and Whitney Aircraft United Technologies 1982) concluded that the events in which there was loss of power to several engines were the result of stalled compressors. They were direct results of blockage of compressor operating lines, due to the accumulation of "ceramic" on the first-stage turbines. These deposits have been attributed to melting of the ash as it passed through the engine combustion chamber followed by deposition on the cooler vanes. The technical reports do not mention the chemical or mineralogical composition of these ashes and ceramics. The only indication (Dlida 1982) suggests "heavy accumulation of deposits, predominantly quartz..." in the filters.

Crude calculations (Smith 1984) confirm that particles <0.05-0.07 mm, with a low melting point can undergo partial or total melting at the turbine inlet with a residence time of 1/1000 s. We emphasize that such deposition on engine surfaces has not been observed in most cases of encounters with volcanic clouds. For example, on 5 April in 1982 at Galunggung, at the beginning of the first phase characterized by a differentiated magma (with about 55-58% SiO_2), an aircraft encountered a plume but suffered no such deposition in the engine. The June and July incidents occurred while the magma was more basic (<51% silica content) and were therefore characterized by a higher melting point. The major differences between the April and June-July products lie in the abundant, chilled, juvenile component, the high xenolithic content, the abundant ashy fraction, and the wet environment in the cloud.

X-ray studies of a base-surge bed deposit of phase 2 shows, as crystalline components, mainly calcic plagioclase and olivine: pyroxenes and Fe-Ti oxides are not detected. The abundance of largely crystallized xenolithic material is an unfavourable factor for melting of the ashes. So we must conclude that three modifications due to water-magma interaction can be considered as favourable for the melting, and can then result in jet plane incidents:

1. The presence of water;
2. The presence of chilled glass as juvenile component;
3. The extreme pulverization and consequent fine grain size of the materials.

6 Conclusions

Our main conclusion concerns the role of hydromagmatism in the increase of the violence of the explosions during the course of an eruption and considerations of jet plane incidents. The problem of the gradual filling of the Walirang maar crater by a lake is a major hazard for the next eruption.

Hazards reported from hydromagmatic eruptions have been emphasized recently (Fischer and Schmincke 1984; Moore et al. 1966; Kienle et al. 1980; Self et al. 1980). The Galunggung case study shows that phreatomagmatic activity may occur very late in the course of an eruption. This drastic change of activity, with explosive renewal, occurs after a long period of rest, during which renewed groundwater supply to the vicinity of the magma is possible, and perhaps when the alertness of the observers may have slackened.

The most striking consequences of the Galunggung eruption were the two similar jet plane incidents, fortunately without casualties. Airborne radar is unable to detect dilute volcanic plumes, because of the low transmitted power, the moderate sensitivity and the low refractivities of the clouds. These two incidents, in late June and mid-July 1982, occurred during the hydromagmatic phase 2, involving aircraft flying through the ash and vapour laden plume. In the case of the Galunggung activity, the plume reached an abnormal height for a mafic magma.

Intervention of superficial water should therefore be taken into account in all volcanological forecasting, including that of magmatic eruptions, and perhaps more especially in the following environments:

1. Tropical and equatorial areas where volcanoes are particularly saturated with water;
2. Island environments, because of the ease with which groundwater may fill depressions;
3. Depressed areas, such as calderas and avalanche craters, filled with sedimentary or pyroclastic material.

Acknowledgements. This work was supported by the Programme Interdisciplinaire de Recherche sur la Prévision et la Surveillance des Eruptions Volcaniques, C.N.R.S., France, the Centre de Recherches Volcanologiques (Université Blaise Pascal, Clermont-Ferrand, France) and by the Volcanological Survey of Indonesia. We thank Dr. J. Latter, Dr. R.S.J. Sparks and an anonymous reviewer for helpful comments and constructive suggestions, Y. Blanc (Clermont-Fd) for X-ray data and J. Clemens for English improvement.

References

Dlida RP (1982) United technologies Pratt and Whitney aircraft, Ref JT9 - 72-00/ PSE:TSB:2-12-8-1, 4 p

Escher BG (1925) L'éboulement préhistorique de Tasikmalaja et le volcan Galunggung. Liedsche Geologische Mededeclingen 1:8-21

Fischer RV (1977) Erosion by volcanic base surge density currents: U-shaped channels. Geol Soc Am Bull 88:1287-1297

Fischer RV, Schmincke HU (1984) Pyroclastic rocks. Springer Berlin, Heidelberg, New York, Tokyo, 472 p

Katili JA, Sudradjat A (1984) Galunggung - the 1982-83 eruption. VSI publ, 102 p

Kienle J, Kyle PR, Self S, Motyka J, Lorenz V (1980) Ukinrek maars. Alaska. 1. April 1977 eruption sequence, petrology and tectonic setting. J Volcanol Geotherm Res 7:11-37

Kusumadinata K (1979) Galunggung, data basar gunungapi Indonesia. Catalogue of reference on Indonesian volcanoes with eruptions in historical time. VSI spec publ, pp 178-189

Lorenz V (1974) Vesiculated tuffs and associated features. Sedimentology 21:273-291

Moore JG, Nakamura K, Alcaraz A (1966) The 1965 eruption of Taal volcano. Science 151:955-960

Self S, Kienle J, Huot JP (1980) Ukinrek maars. Alaska. II. Deposits and formation of the 1977 craters. J Volcanol Geotherm Res 7:39-65

Smith W (1984) Effects of volcanic ash on air operations. 15 p: attachment F to VAW memo 2. Volcanic Ash Warning Study group from I.C.A.O. Quebec, Canada

Sparks RSJ, Walker GPL (1973) The ground-surge deposit: a third type of pyroclastic rock. Nature (Lond), 241:62-64

Sudradjat A, Tilling R (1984) Volcanic hazards in Indonesia: the 1982-83 eruption of Galunggung. Episode. Vol. 7. 2:13-19

Tracy RF (1983) Flight operations engineering bull., Pratt Whitney. 83-1, 3 p

Walker GPL (1971) Grain-size characteristics of pyroclastic deposits. J Geol 79:696-714

Wohletz KH, Sheridan MF (1979) A model for pyroclastic surge. Geol Soc Am Spec Paper 180:177-194

Some Notes on False Alarms of Volcanic Activity and Mud Flows

A. P. ALCARAZ[1]

Abstract

Reports on the observation of certain physical phenomena believed to be related to volcanic activity are often made, especially when an eruption of some magnitude has just occurred. Many reports are unfounded and born of peoples' apprehension over volcanic eruptions. This chapter comments on the substance of such reports, and offers explanations for their misconceptions based on the writer's Philippine experience. Mud flows or lahars are briefly discussed as these can be very destructive, and are constant sources of danger to those living in the vicinity of active volcanoes, e.g. the Mayon volcano in the Philippines.

1 Introduction

Quite often after a volcanic eruption has taken place, and has been given wide coverage by the media, people become volcano conscious and as a result reports are made alleging that volcanic activity of some sort has been observed. These reports, when investigated by volcanologists, almost always prove to be unfounded. Some of these "false alarms", commonly reported in the Philippines are now described.

2 False Volcanic Flow

Hunters often burn patches of underbrush to attract wild game to the salty ashes left by the fire. If this happens on the higher slopes or near the summit of a volcano, people on the lowland, noticing the glare of the fire, may become alarmed and send in a report that the volcano is showing signs of activity.

This can easily be resolved by a pair of good binoculars. There is no real "fire" emitted by volcanoes. The term "volcanic fire" is figurative and refers to incandescent materials ejected during eruptions, although momentary flames may be noted as these red-hot ejections accompanied by small amounts of combu-

1 Former Chief Volcanologist, Philippine Commission on Volcanology (Now the Philippine Institute of Volcanology and Seismology: PHIVOLCS), Philippines

IAVCEI Proceedings in Volcanology 1
J. H. Latter (Ed.), Volcanic Hazards

stible gases come into first contact with air at the crater. However, fires may be started in the vicinity or downslope by hot scoria fall and lava flows coming into contact with trees and bushes.

If lava (incandescent rock material) has nearly reached the crater or summit of a volcano, then "crater glow" may be observed at night. The glow will be broader and persist longer than the glow from a camp fire. Crater glow, incidentally, is a positive sign that an eruption is imminent if the volcano in question has been quiet for some years. A good example of this is Mayon, an 8000-ft-high, active Philippine volcano. Six years after it last erupted in 1978, crater glow was observed a day before its 1984 eruption began.

3 Drying and Dying of Vegetation

Some reports may allege that vegetation on the slopes or on a certain sector of a mountain is drying up, and that volcanic activity is feared to be the cause. Such reports are not necessarily confined to volcanoes. While resurgent volcanic activity has been preceded in some instances by drying of vegetation in the area where volcanic activity eventually broke through, there are, nevertheless, other possible causes of this phenomenon. Some non-volcanic causes are: (1) less soil moisture becomes available to plants because of prolonged, rainless days, and as a result shallow-rooted plants are affected; (2) lowering of the local water table due to deep excavations, or excessive pumping of water wells; (3) deforestation; and (4) changes in the drainage system due to man-made activities, or as a result of a strong earthquake.

If the drying of vegetation is due to heating up of the ground by sub-surface, upward movement of magma or highly heated volatiles, there should be other accompanying indications, such as slight tremors and rumbling sounds. This is especially so if the volcano has not erupted for many years. Mount Hibok-Hibok in southern Philippines manifested perceptible tremors and some rumbling noises several days before the start of its Pelean activity in 1948 after a dormancy of 51 years.

Another occasion of vegetation drying up, or even dying due to certain manifestations related to volcanism is that attributable to passive gas emissions and seepages. This occurrence has been observed at some inactive volcanoes. On the northwestern slopes of Cuernos de Negros, an inactive volcano in the vicinity of the southern Negros geothermal field, several areas with standing tree stumps and dead animals were observed, in which strong sulphurous odors were noticed. The local inhabitants call such an area "kaipohan", meaning in the Philippine Cebuano dialect a place of dead rats. It is believed that recently formed or re-opened fractures (for example, as a result of an earthquake) allowed noxious volcanic gases to escape or seep to the surface, thus causing the vegetation and small wild life to die in the area.

4 Pseudo-Ash/Steam Emissons

An added fear for people living near an erupting volcano is the formation of new craters or an opening from which steam and ashes could be emitted. Such a possibility makes them apprehensive, for if this happens, it could mean that the area of eruptive activity has broadened and has thus placed additional areas in danger. Therefore, they are alert to report any observations that might support this fear.

Some reports of ashes being ejected from alleged new vents have turned out to be unfounded for the short, small, dust columns observed were dust devils or "ipo-ipos" in the Philippine vernacular, meaning swirling winds. In areas that have received appreciable ash deposition from an on-going or very recent eruption, swirling winds may pick up the volcanic ash and raise it vertically upward giving the impression from afar of a rising ash column. This becomes more convincing if observed several times as originating from the same area. In the summer months following the destructive September 1965 phreatomagmatic eruption of Taal volcano in the Philippines, several reports were made by persons on the western shore of Lake Taal (the lake surrounding the volcano) that ash clouds were observed rising from the Caua area of Volcano Island. Investigation revealed that these rising ash columns were only dust devils. Topographic conditions cause winds to swirl in the Caua area, thus producing the phenomenon.

Reports of new steam vents, on the other hand, turned out to be steam generated by hot lava flows reaching a body of water. This was the case when the lava of the 1968 Strombolian Taal eruption flowed into the water-filled collapse pits formed during the 1966 phreatomagmatic eruption, and when the lava flow reached Lake Taal itself. In an eruption of Mayon, another active Philippine volcano, similar reports are sometimes received from persons who have mistaken steam emissions at the front of lava flows as new steam vents, especially when lava has come into contact with water in large gullies, or after heavy rain. In some instances, minor steam explosions have been observed to occur in very fresh lava flows, e.g. when the rapidly advancing flow may have entrapped a small, but deep pond of water, thus bringing about conditions conducive to hydrothermal eruptions.

5 Electrical Storms

Thick or dense eruption clouds of the Vulcanian type are often accompanied by electrical displays. Thus, people may associate electrical storms with eruptions if the volcano has in fact erupted recently. However, thunderstorms naturally quite often pass or occur over a volcano even if it is not erupting. Fear is nevertheless felt by the people in the vicinity of an active volcano when this happens, for the thunderstorm may hide the volcano from direct observation. All they can see and hear are the lightning flashes and the claps of thunder.

Thunderstorms over a volcano in an eruptive state naturally seem to be more intense and terrifying than a normal electrical storm. Towards the end and just after the Hibok-Hibok (a volcano in southern Philippines) Pelean eruptions of December 1951, severe thunderstorms were quite frequent over the newly formed dome, and these kept the people edgy.

Thunderstorms can easily be distinguished, however, from electrical displays inside an eruption cloud, even if the volcano is obscured from view by thick clouds or heavy rain. The peals of thunder during a thunderstorm are likely to be staccato and reverberating in character, when compared with the muffled sounds of electrical discharges associated with Vulcanian eruption clouds, or the booming noise of volcanic explosions.

If the thunderstorm should occur while the volcano is not erupting or ejecting ash, then the raindrops will be ash-free just like ordinary rain. However, rain may contain ash particles when the thunder clouds happen to be seeded with ash from the erupting volcano.

Ashy rains unrelated to rain cloud formations were observed to fall on Volcano Island and its vicinity during the 1966 phreatomagmatic eruption of Taal. Apparently, the steam mixed with the ash ejections condensed upon reaching cooler air levels above the volcano and the precipitation that ensued was ash-charged. The ashy rain was cold and caused no harmful effects to human or animal life.

Pelting by large ash particles and by small scoria or pumice fragments may occur during an eruption. Large globules of ash are called pisolites or accretionary lapilli. They consist of concentric layers of ash particles formed due to turbulence inside an ash column sustained by rapid succession of ash ejections. Their formation and structure are similar to that of hailstones formed inside strong convecting air masses which give rise to the thunder heads often seen in summer. Danger from the fall of these volcanic tephra, usually observed on the leeward side of the erupting volcano is minimal, unless the eruption is of catastrophic magnitude.

6 Mud Flows or Lahars

This phenomenon should not really be included in this discussion of false alarms of volcanic activity since mud flows or lahars, as the Indonesians call them, are real hazards and there are no pseudo or false equivalents. Their inclusion, however, is justified as authorities or people living in the vicinity of active volcanoes may become complacent about this danger once the eruption is over.

Mud flows are torrents of mud and volcanic debris that rush and churn their way rapidly downslope of the volcano during every heavy downpour. These flows are especially dangerous during and after a protracted eruption characterized by voluminous ash ejections, and the deposition of larger pyroclastic clasts from the crater. These ejecta pile up in large quantities on the upper slopes of a volcano. During heavy rains the ash is easily turned into mud, which then starts to flow downhill, gathering momentum as it descends. Because of its speed and density the mud flow is able to carry along with it large boulders, trees and other debris that it may encounter. Some, descending from high volcanoes, have reached places tens of kilometers away.

The destruction that mud flows wrought can be very extensive, not only by individual large flows, but in the combined effects of a whole series of mud flows. Geologically, the effects of mud flows on the landscape are both destructive and constructive. At the steeper slopes of the volcano, because of their speed and high density, mud flows possess tremendous erosive power and thus are able to cut deep gullies and easily cross low obstructions along the way. At this portion of their descent, the effect on the terrain is geologically destructive. However, on reaching the volcano's apron there is a slackening of speed, and consequently a decrease in their energy. Onward the flows are then able to follow existing channels, depositing their load of debris along the way until their energy is totally dissipated. Because of this deposition, former channels can become choked with debris and thus force succeeding mud flows to follow other or new channels of flow. This leads to braiding of mud-flow deposits.

While the deposition by mud flows of volcanic debris at the base of the volcano is, geologically speaking, a constructive process, it is destructive in the humanitarian sense as it causes devastation of what might be valuable arable lands. Braiding of mud-flow deposits tends to enlarge the affected areas, and renders difficult any engineering scheme to contain deposition of the debris.

In the Philippines, mud flows are the gravest danger posed by eruptions of Mayon. The symmetrical form of this volcano with its lower slopes tangentially blending with its apron, attest to the major role mud flows played in bringing this about. Deep gullies or barrancos debouch into coalescing alluvial fans built up by innumerable mud flows of past eruptions.

The 1984 Vulcanian eruption of Mayon, while similar in pattern to those of 1928, 1938, 1947, 1968 and 1978, was different in that much more ash was ejected, almost continously for several days. This caused large amounts of ash and larger sizes of ejecta to accumulate on the slopes, which later gave rise to extensive mud flows with every heavy rain that occurred over the area. The flows on the southeastern sector of the volcano were particularly voluminous, and repeatedly cut at several points the Legaspi-Tabaco provincial highway skirting the volcano's base.

It should be borne in mind that steam ejected with the ash can also, upon condensation, cause rain to fall over the volcano and precipitate mud flows. Thus on several occasions, mud flows were observed to occur during an eruption of Mayon in spite of good weather conditions prevailing at the time.

In retrospect, it was good that Mayon erupted in 1984, only 6 years after its 1978 eruption. This should erase from peoples' minds the erroneous belief that Mayon erupts only every 10 years. There is no such 10-year cycle evident from the intervals of Mayon eruptions since 1616. With this false notion of regularity in the frequency of eruptions of a volcano disproven, the authorities and people concerned cannot then afford to be complacent and lax in the surveillance of an active volcano, especially if it has had destructive eruptions in the past.

Explosive Volcanic Activity Generated from Within Advancing Silicic Lava Flows

J. H. FINK[1] and C. R. MANLEY[1,2]

Abstract

After initial explosive activity, eruptions of silicic magma often decrease in violence and associated hazards due to a state of gentle degassing during effusion. Recent field mapping, microscopic, and geochemical analyses of drill core samples from the Inyo and Valles Scientific Drilling Programs indicate that during the advance of some silicic lava flows the concentration of volatiles may increase substantially, due to the migration and concentration of water vapor released by crystal growth and microfracturing. As a result, the explosive hazards related to emplacement of silicic lava flows may become greater rather than less as flow length increases, even though the overall volatile content of the extrusion is constant or decreasing. This idea is supported by the occurrence of explosion craters on the distal surfaces of large rhyolite flows, and by observations of nuées ardentes generated through the collapse of dacite lava flow fronts several kilometers from their vents. Volatile migration may also be responsible for the rise of diapirs of dark pumiceous lava to the flow surface. Hence, monitoring the surface appearance of an active silicic lava flow may provide indications of the potential for endogenic explosive activity.

1 Introduction

The hazards associated with erupting silicic magmas depend in large part on their dissolved water contents. Silicic eruptions typically begin with explosive ejection of volatile-rich magma, followed by the quieter effusion of coherent degassed lava flows. Studies of the juvenile water contents of rhyolite lava flows and the tephras that immediately preceded them (Eichelberger and Westrich 1981), and of the isotopic character of the water (Taylor et al. 1983) suggest that after initial explosive activity the magma degasses to an extremely low volatile content in the conduit, and reaches the surface as a coherent lava flow.

[1] Geology Department, Arizona State University, Tempe, Arizona 85287, USA
[2] Present address: Geology Department, Stanford University, Stanford, California 94305, USA

IAVCEI Proceedings in Volcanology 1
J. H. Latter (Ed.), Volcanic Hazards

However, some of these "degassed" silicic lava flows themselves appear in turn to have generated explosive activity, sometimes several kilometers away from the primary magmatic vent. Such pyroclastic flows are sometimes referred to as the Merapi type (Macdonald 1972; Fischer and Schmincke 1984; Zen 1985), after the Indonesian andesitic volcano where they are most common. Since the volatile content of lava generally decreases as a flow advances, the generation of explosive activity several kilometers from a vent seems enigmatic. Why should the lava's volatile content become sufficiently great to generate an explosion after it has already degassed during surface flow for several weeks, months, or even years?

In this chapter we present textural evidence from the surfaces and interiors of several silicic flows that supports a model of volatile migration within active lavas. We suggest that if the redistributed volatiles become sufficiently concentrated near a flow front, pyroclastic flows may be generated. Concentration away from the front may result in disruption of the flow surface by explosion craters or by rising diapirs of distinctively dark pumiceous lava. The appearance on the surface of an active silicic flow of either of these latter features thus may provide a warning of the increased potential for Merapi-type pyroclastic flows.

2 Secondary Explosive Activity from Silicic Flows and Domes

In September 1973, collapse of the over-steepened front of the El Brujo dacite flow (Santiaguito Volcano, Guatemala) 2 km from the vent resulted in the rise of an 800-m-high ash column and the generation of a destructive nuée ardente that traveled an additional 3 km downslope (Fig. 1) (Rose et al. 1976). Study of the deposits showed that the nuée ardente consisted of a small avalanche restricted to a stream valley, and a pyroclastic flow or surge that denuded and devasted a much more extensive area.

An earlier pyroclastic flow appeared to have issued from the nearby Caliente vent of Santiaguito in April 1973 (Rose 1973). At the time, Caliente was filled by a dacite dome that was not significantly affected by the eruption. Later mapping of the April deposits indicated that they may have come from the distal end of the Caliente lava flow rather than from the Caliente vent (Rose et al. 1976). This conclusion is supported by the fact that the dacite dome remained largely intact.

Scanning electron microscopic (SEM) examination of shards from the El Brujo nuée deposit showed them to be highly vesicular (Rose et al. 1976); since the lava flow itself was observed to be generally nonvesicular, this suggests vesiculation in the pyroclastic flow during emplacement. Nuées of comparable volume, generated by column collapse at primary vents at nearby Fuego and Arenal volcanoes, covered considerably less area than that from the front of the El Brujo flow, and contained ash particles with far fewer ruptured surface vesicles (Rose et al. 1976, p. 11). Rose et al. (1976) interpreted these vesicles as indicating that "active vesiculation may have been less impor-

Fig. 1. Aerial photograph of the El Brujo lava flow, Santiaguito Volcano, Guatemala, taken 3 October 1973. The light-colored avalanche scar at the flow front is the source of the 15 September nuée ardente. Field of view is 3.6 km across (Rose et al. 1976)

tant in the [Fuego and Arenal] eruptions". These observations suggest that the destructiveness of the El Brujo pyroclastic flow was due to a relatively high volatile content in the disrupted lava. Since no prior cooling in an eruptive column took place, the higher temperature of the El Brujo material may also have contributed to the magnitude of devastation.

Several other pyroclastic flows generated from the collapse of lava flows and domes have been documented or suspected. Perret (1937) suggested that Mount Pelée's catastrophic nuée of 1902, as well as several during the activity of 1929, formed by the collapse of a rhyolite dome at the summit. Mount Merapi in Indonesia has produced numerous pyroclastic flows by the collapse of andesitic lava domes (Zen 1985). Throughout late 1986, renewed activity at Merapi Volcano included generation of many pyroclastic flows (T. Casadevall, pers. commun., 1987) On-going monitoring by the Indonesian and U.S. Geological Surveys in 1987 may be able to help establish the mechanisms of formation of these flows.

In October 1986, a new pyroclastic flow was generated from the Caliente flow at Santiaguito (SEAN Report, December, 1986). Early reports (W. Rose, pers. commun., 1987) indicate that this explosive eruption also originated from near the front of the advancing lava flow. Fieldwork at Caliente is planned for the summer of 1987. In 1986, observers at Mount Augustine, Alaska, witnessed dozens of small nuées ardentes associated with the collapse of the margin of the growing dacite dome (J. Kienle, pers. commun., 1986). Another nuée observed in 1976 at Mount Augustine also may have formed by the collapse of the dacite dome that was growing at the time (Stith et al. 1977). Unwitnessed collapse of the front of the Mount St. Helens dacite dome in May 1985 and May 1986 produced block and ash deposits that may have been emplaced as pyroclastic flows. (D.A. Swanson, pers. commun., 1986).

A similar pyroclastic flow deposit of Pleistocene age has been identified (S. Self, pers. commun. 1987) associated with the 140 ka Banco Bonito obsidian flow in Valles Caldera. At a location where the lava flow directly overlies previous Plinian fall and surge deposits from the primary vent, a thin layer of ash occurs immediately beneath the base of the lava in a channeled depression in the older deposits. This ash layer, which contains subrounded lava and pumice clasts in its upper portion, appears to be the product of a small pyroclastic flow that had its source in the advancing toe of the lava flow (S. Self, pers. commun., 1987).

3 Mechanisms of Pyroclastic Flow Formation

Pyroclastic flows may be roughly divided into two types (e.g., Smith and Roobol 1982; Fischer and Schmincke 1984): those formed by the collapse of eruptive columns, and those formed by the gravitational or explosive disruption of lava flows or domes. Pyroclastic flows formed by column collapse are generally more voluminous, more energetic, more widely dispersed, more easily identified in the geologic record, and pose hazards to larger areas around volcanoes than those formed by dome collapse. Consequently, column-collapse pyroclastic flows have been the subject of many more studies, both theoretical and field-based, than those related to dome collapse.

Pyroclastic flows resulting from dome collapse may be highly dangerous because their formation may be totally unanticipated: weeks, months, or even years after an extrusion began, a quietly advancing lava flow or dome may collapse to produce a destructive pyroclastic flow. These pyroclastic flows may have very high temperatures and, along with their associated mud flows, may travel distances up to tens of kilometers away from a volcano. They are particularly hazardous for volcanologists studying quietly growing domes that have been assumed to have lost their explosive potential.

Owing to the lack of direct observations, fewer models of pyroclastic flow formation by dome collapse are available than for column collapse. Two explanations for the generation of dome-related pyroclastic flows have been suggested: (1) the sudden release of pressure, such as by gravitational collapse of a flow front, allows lava from the interior of a dome or flow to vesiculate explosively and disaggregate (e.g., Lacroix 1904; Perret 1937; Taylor 1958); or (2) the collapsing avalanche of hot lava blocks falling off an over-steepened dome incorporates and heats cold air without substantial vesiculation (e.g., McTaggart 1960; Moore and Melson 1969). Over-steepening of domes is known to have triggered numerous pyroclastic flows at Merapi Volcano in Indonesia, but the source of explosivity has not been determined (Zen 1985). Mobility of possible pyroclastic flows generated from the Mount St. Helens dome is also debatable. As pointed out by Rose et al. (1976), the extent of vesiculation in a nuée ardente appears proportional to the silica content of the lava, but this vesiculation may not significantly enhance its mobility. Rather they suggest that vesiculating nuées ardentes create more extensive blast-devastated zones. Rhyolite lavas and domes should therefore have the greatest tendencies to generate destructive pyroclastic flows.

4 Internal Textures of Rhyolite Flows

Examination of two drill cores from the 550-year-old Inyo Obsidian Dome rhyolite flow in California, one core from the 140 ka Banco Bonito rhyolite flow in New Mexico, and several gullies through two dissected 8-million-year-old rhyolite domes in Arizona, has revealed a complex, yet consistent, zonation of pumiceous, glassy, and devitrified textural layers in all these extrusions (Fig. 2) (Fink and Manley 1987; Manley and Fink 1987). Most of these zones can be related to processes of effervescence at the flow surface, crystal growth in the flow interior, and quenching near the upper and lower flow margins. However, a distinctive zone of coarsely vesicular pumice (CVP) that was not clearly recognized prior to the examination of the drill cores appears to play a crucial role in the generation of secondary explosive activity like that observed at Santiaguito.

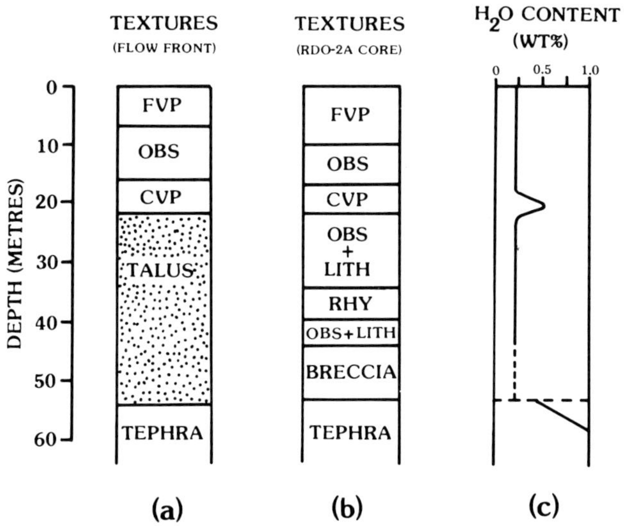

Fig. 2 a–c. Typical zonation in a rhyolite flow, based on RDO-2A drill core through Obsidian Dome. **FVP** finely vesicular pumice; **OBS** obsidian; **CVP** coarsely vesicular pumice; **LITH** lithophysae; **RHY** devitrified rhyolite. Water contents shown are of fresh glass samples from the same core (Fink and Manley 1987)

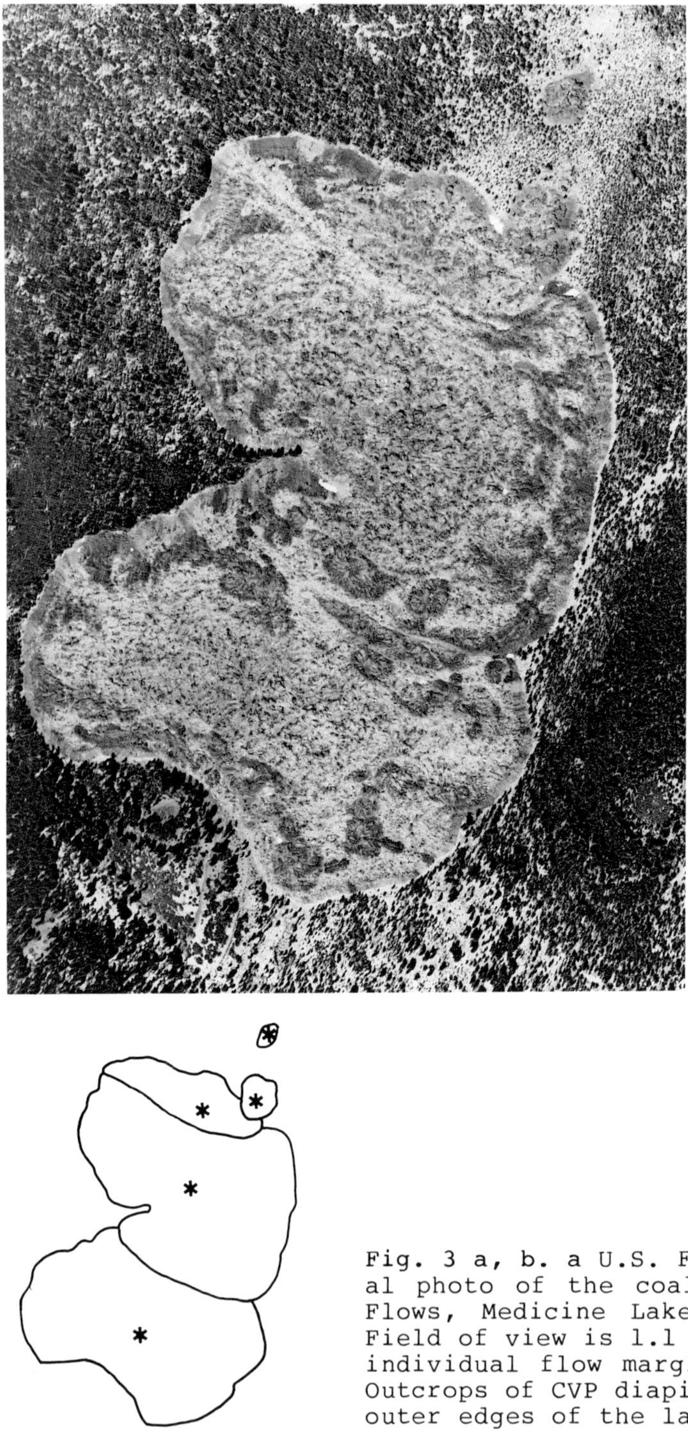

Fig. 3 a, b. a U.S. Forest Service aerial photo of the coalesced Crater Glass Flows, Medicine Lake Highland Volcano. Field of view is 1.1 km. b Sketch of the individual flow margins and vent areas. Outcrops of CVP diapirs occur around the outer edges of the largest three domes

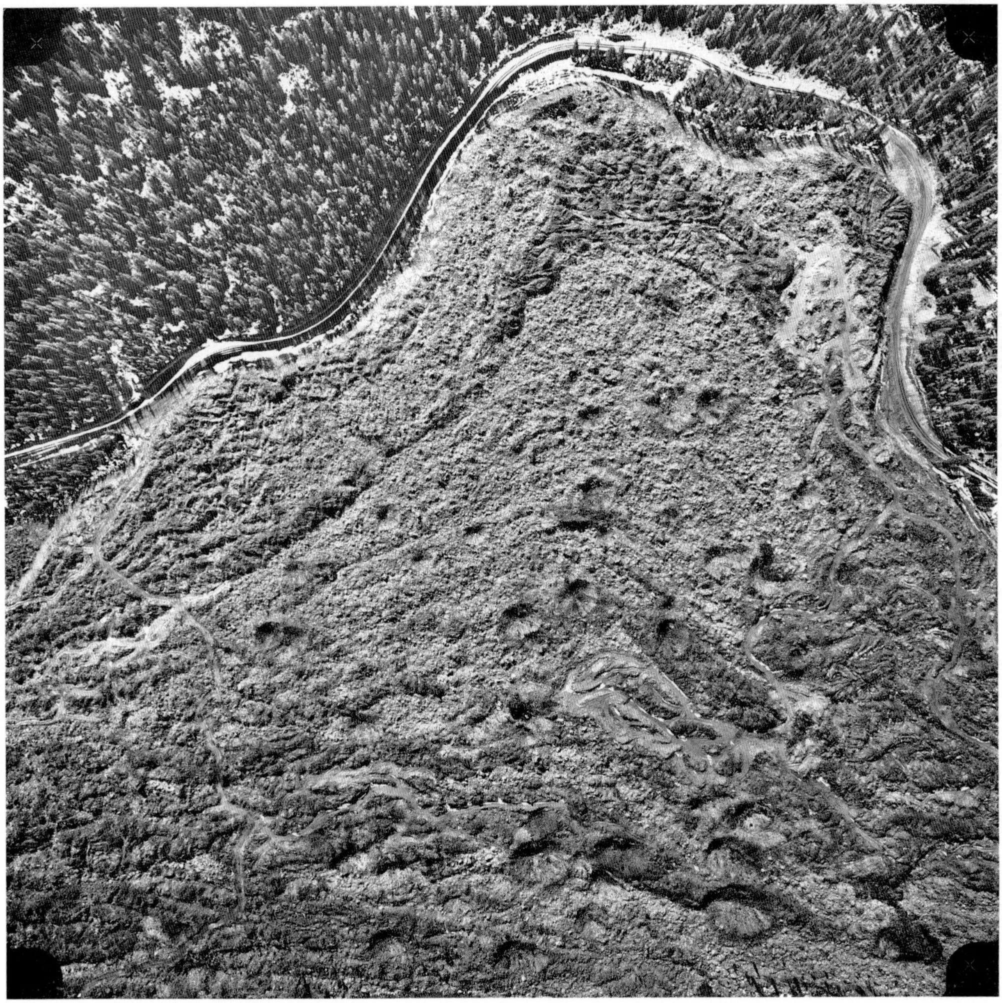

Fig. 4. U.S. Forest Service aerial photo of the southwestern portion of Glass Mountain, Medicine Lake Highland Volcano, northern California, showing explosion pits common there. The craters range from 7 to 11 m deep. Field of view is 700 m

The textural zonation revealed by the drill cores through the Obsidian Dome and Banco Bonito rhyolite flows consists of a finely vesicular pumice carapace (FVP), underlain successively by obsidian (OBS), coarsely vesicular pumice (CVP), obsidian with spherulites, devitrified rhyolite (RHY), more obsidian with spherulites, and basal breccia. The obsidian layers form where rapid cooling inhibits diffusion and prevents crystal growth. The transition from FVP to OBS is controlled by the depth at which overburden pressure suppresses vesiculation. Formation of the FVP in the surficial lava begins as soon as it is extruded from the vent and decompressed to atmospheric pressure. This

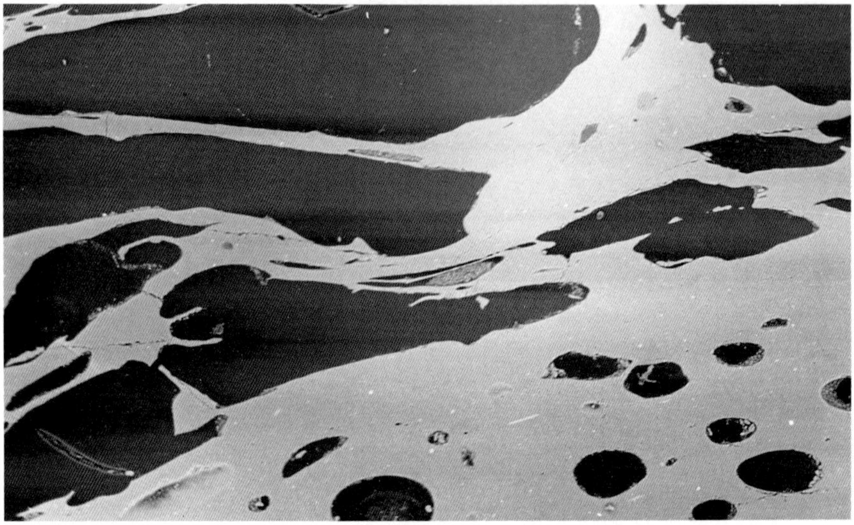

a

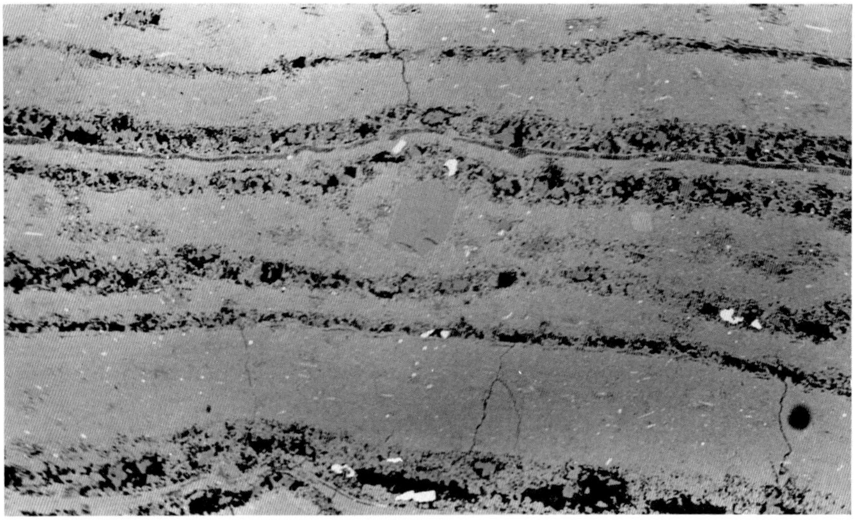

b

Fig. 5. SEM pictures showing a cracks and bubbles in CVP layer, and b partially crystalline interior of the flow (RHY). Both samples from the RDO-2A drill core of the Obsidian Dome: a from 17.1 m depth in flow; field of view = 2.25 mm; b from 36.9 m; field of view = 5.5 mm

process gently inflates the carapace of the flow and in no case appears to occur explosively. The thickness of the rigid crust, composed of the FVP and upper-OBS layers, is determined by the rapid decrease with depth of the lava's temperature-dependent yield strength.

The CVP layer ranges from 3 to 5 m in thickness, and is located beneath the rigid crust at depths of from 17 to 40 m in the different flows. Observations of the collapsed front of the El Brujo flow revealed a similar scoriaceous carapace and interior glassy zone, each between 5 and 10 m thick (W. Rose, pers. commun., 1985). Measurements of samples from the Inyo drill cores show that the CVP layer has up to three times greater juvenile water content and up to three orders of magnitude higher porosity than the rest of the flow (Eichelberger et al. 1986). Diapirs of CVP are present on the distal surfaces of many moderate to large volume rhyolite flows (Fig. 3) (Fink 1983), and have also been observed on dacite flows at Medicine Lake Highland Volcano, northern California. Endogenic craters seen on rhyolite flows (Fig. 4) appear to extend inward as far as the CVP zone and are concentrated in areas containing CVP diapirs.

SEM examination of the Inyo and Valles drill cores (Fig. 5), and mapping of numerous flow surfaces, led Fink and Manley (1987) to propose that the CVP zone develops its high porosity and water content as a result of the upward migration of volatiles through transient microcracks from the crystallizing interior of the flow to the base of its rigid surface crust. This crystallization takes the form of spherulite growth, which commences while the flow is still active and continues through post emplacement cooling (Manley and Fink 1987). In the absence of crystal growth, volatiles might still be able to diffuse into microcracks from the glassy lava. According to this model, both the thickness and buoyancy of the CVP layer should increase with flow length. Hence, the growing CVP layer may eventually produce diapirs that rise to the flow surface or may generate over-pressures capable of forming explosive craters in the upper surface or triggering pyroclastic flows at the flow front (Fig. 6). Even if the CVP layer in a given lava flow does not overpressurize significantly, its high porosity and low strength would contribute to efficient pulverization of the lava during collapse of the flow front. This could greatly enhance heat exchange and vesiculation in the nuée ardente and thus increase its destructiveness regardless of triggering mechanism.

5 Discussion

We have suggested that the formation and destructiveness of pyroclastic flows may be promoted by the concentration of volatiles within advancing silicic lavas, and that the endogenic explosive potential of these lavas may actually increase as they move further from their source vents. Volatile concentration requires that gases released in the flow interior be prevented from escaping to the atmosphere. The surfaces of rhyolite lava flows are characteristically brecciated due to tensile fracturing, but these cracks do not generally penetrate beneath the glassy zone at the base of the flow's crust. Conditions that cause tensile fractures to deepen, such as advance over a steep slope or enhanced cooling due to precipitation, may allow the premature escape of volatiles from beneath the cooling crust, analogous to the degassing through pipes in welded tuffs (e.g., Sheridan 1970). The crystallization of hydrous minerals may also reduce the endogenic explosive potential of a lava flow by removing available volatiles, but this process probably does not operate in lavas more silicic than andesite.

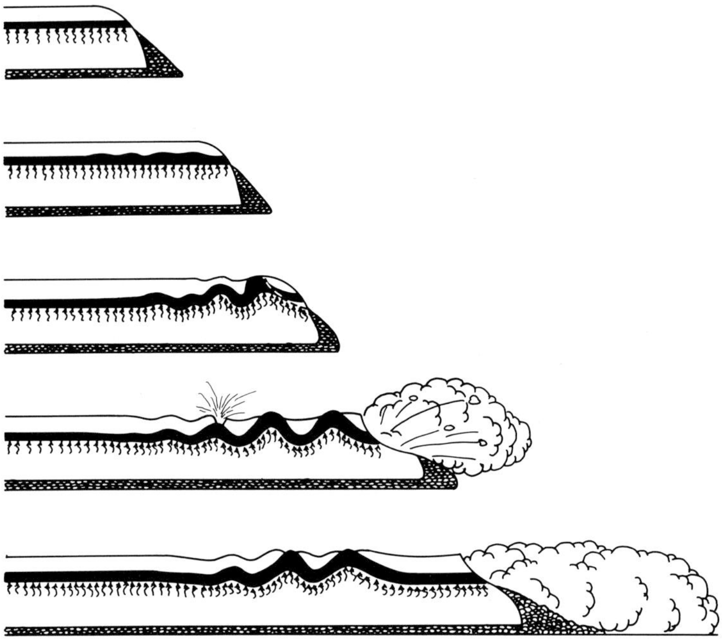

Fig. 6. Schematic diagram showing increasing potential for endogenic pyroclastic activity in an advancing silicic lava flow due to formation and growth of the CVP layer (**black**). **Arrows** represent volatiles rising into the CVP layer. As the CVP layer thickens, it forms diapirs that rise to the flow surface, and generates explosion pits. Flow-front collapse exposing the CVP layer may lead to generation of pyroclastic flows

The concentration of volatiles within an advancing lava flow may occur without any surface manifestations. However, when the internal CVP layer becomes sufficiently vapor-enriched it may rise diapirically to the flow surface. The appearance of these diapirs in a moving flow may thus provide a warning of the increased potential for explosive activity. Further study of the relationships among lava texture, microstructure, and volatile contents may allow these endogenic hazards to be better understood.

Acknowledgements. The authors thank Ian Nairn and John Latter for helpful comments on the manuscript; Bill Rose, Don Swanson, Tom Casadevall, Robin Holcomb, Juergen Kienle, and Steven Self for discussion about pyroclastic flows generated by dome collapse; and David Krinsley and Pat Trusty for assistance with the SEM analyses. Research was supported by DOE grant DE-FG02-85ER13320, NASA grant NAGW-529, and NSF grant EAR-8309500.

References

Eichelberger JC, Westrich HR (1981) Magmatic volatiles in explosive rhyolitic eruptions. Geophys Res Lett 8:757-760

Eichelberger JC, Carrigan HR, Westrich HR, Price RH (1986) Nonexplosive silicic volcanism. Nature 323:598-602

Fink JH (1983) Structure and emplacement of a rhyolitic obsidian flow: Little Glass Mountain, Medicine Lake Highland, northern California. Geol Soc Am Bull 94:362-380

Fink JH, Manley CR (1987) Pumiceous and glassy textures in rhyolite flows and implications for eruption and emplacement. In: Fink JH (ed) The emplacement of silicic domes and lava flows. Geol Soc Am Spec Paper 212:77-88

Fischer RV, Schmincke H-U (1984) Pyroclastic rocks. Springer, Berlin Heidelberg New York Tokyo, 472 pp

Lacroix A (1904) La Montagne Pelée et ses eruptions. Masson, Paris 662 pp

Macdonald GA (1972) Volcanoes. Prentice-Hall, Englewood Cliffs, NJ, pp 1-510

Manley CR, Fink JH (1987) Internal textures of rhyolite flows as revealed by research drilling. Geology 15:549-552

McTaggart KC (1960) The mobility of nuée ardentes. Am J Sci 258:369-382

Moore JG, Melson WG (1969) Nuées ardentes of the 1968 eruption of Mayon Volcano, Philippines. Bull Volcanol 33:600-620

Perret FA (1937) The eruption of Mt. Pelée 1929-1932. Carnegie Inst Wash Publ 458:1-126

Rose WI Jr (1973) Nuée ardente from Santiaguito Volcano, April 1973. Bull Volcanol 37:365-370

Rose WI Jr, Pearson T, Bonis S (1976) Nuée ardente eruption from the foot of a dacite lava flow, Santiaguito Volcano, Guatemala. Bull Volcanol 40:53-70

Self S, Wolff JA, Kircher DA, Ivanovich M, Kyle PR (1986) Age, emplacement mechanisms and petrology of the youngest eruption from the Valles Caldera, New Mexico. EOS (Am Geophys Union Trans) 67:1249-1250

Sheridan MF (1970) Fumarolic mounds and ridges of the Bishop Tuff, California. Geol Soc Am Bull 81:851-868

Smith AL, Roobol MJ (1982) Andesitic pyroclastic flows. In: Thorpe RS (ed) Andesites. Wiley and Sons, New York, pp 415-433

Stith JL, Hobbs PV, Radke LF (1977) Observations of a nuée ardente from St. Augustine volcano. Geophys Res Lett 4:259-262

Taylor BE, Eichelberger JC, Westrich HR (1983) Hydrogen isotopic evidence of rhyolitic magma degassing during shallow intrusion and eruption. Nature (London) 306:541-545

Taylor GA (1985) The 1951 eruption of Mount Lamington, Papua, Austr Bur Min Resour Geol Geophys Bull 38:1-117

Zen MT (1985) Destruction, formation, and growth of the Merapi domes. Bull Jurusan Geol 14:7-13

The May 1915 Eruptions of Lassen Peak, California, I: Characteristics of Events Occurring on 19 May

D. B. EPPLER[1] and M. C. MALIN[2]

Abstract

Eyewitness accounts, conventional and computer-assisted analysis of photographs, and sedimentological studies indicate the likely occurrence of a volcanic blast associated with the eruption of a dacite lava flow and the emplacement of a major lahar at Lassen Peak, California, on the night of 19 May 1915. Although early investigators arrived at this conclusion, it was subsequently abandoned in favor of an interpretation that stressed the generation of the lahar by heat exchange between a dacite lava flow and the residual snow pack. The existence of a deposit with many of the characteristics of those formed by volcanic blasts, including charred organic material, and the planimetric distribution of the effects of the eruption are the strongest evidence for a volcanic blast on the night of 19 May. The cause of the blast is, however, problematic. The preferred candidate mechanism involves the physical disruption and avalanching of the dacite lava down the steep eastern slope of Lassen Peak, with intimate mixing and heat exchange between snow and hot fragments of dacite causing an explosive production of steam and leading to a volcanic blast and a subsequent lahar.

1 Introduction

The Lassen Peak eruptions of 19 and 22 May 1915 were the most violent in a series of eruptions that began a year earlier and continued until late 1917. During May 1915, eruptive activity included volcanic blasts, formation of major lahars, and the emplacement of a dacite lava flow (Table 1). The eruption on the afternoon of 22 May 1915 was witnessed by many people in the region. However, the activity on 19 May took place at night, and consequently was not directly observed.

Of principal concern in this report is the nature of the eruption that occurred on the night of 19 May. Diller (1916) investigated the effects of the May 1915 eruptions during the fol-

1 Geology Group, MS D462, Los Alamos National Laboratory, Los Alamos, NM 87545, USA
Present address: Site Integration Branch, Science Applications International Corporation, 101 Convention Center Drive, Las Vegas, Nevada, 89120, USA
2 Department of Geology, Arizona State University, Tempe, AZ 85287, USA

IAVCEI Proceedings in Volcanology 1
J.H. Latter (Ed.), Volcanic Hazards

Table 1. Times of known events during the May 1915 eruptions

Date	Time	Description
14 May	All day	"New crater. Large eruption. Smoke-east ...Fire seen for first time by 8 persons...lasted all night...seen just before daybreak [observations made from Manton]" (Day and Allen 1925)
15 May	2:30 A.M.	"Light seen at top of Lassen at 2:30 A.M." (Day and Allen 1925)
17 May	Evening	Persons on all sides of the volcano reported a glow on clouds above the summit crater (Day and Allen 1925)
19 May	Evening	"...after dark a steady glow of light was seen shining on cloud of smoke for several hours...Fire lava seen on top at 9:00 P.M...." (Day and Allen 1925) Personnel at power station in Manton report seeing incandescent boulders rolling down west side of Lassen around sunset (Loomis 1926) These observations indicate that the dacite lava flow that now covers a portion of the summit was erupted between 14 and 19 May
20 May	≈1:00 A.M.	Ranchers in Hat Creek Valley forced to move to high ground because of extensive flooding of the creek (Loomis 1926) The extensive flooding was apparently caused by the emplacement of a major lahar into Hat Creek
22 May	4:45 P.M.	"Terrific eruption. Incomparable with any former eruption...column of steam reached 30,000 feet" (Day and Allen 1925) This eruption caused the most damage of any previous eruptions in the series; details in Diller (1916), Day and Allen (1925), Eppler (1987)

lowing summer. He concluded that a steam blast, associated with the eruption of a dacite lava flow, had melted the residual snow pack on the northeast flank of the Peak and had triggered the large lahar that occurred during the early morning of 20 May. Day and Allen (1925) drew similar conclusions, although they felt that the snow was melted by the deposition of tephra, rather than by a hot steam cloud. Finch (1929, 1930) disputed the claims of Diller (1916) and Day and Allen (1925), primarily on the basis of the absence of tree blow down in the photographs taken by B.F. Loomis on the afternoon of 22 May 1915. Finch instead proposed that the lahar had been triggered by snow melted during the extrusion of the dacite lava. Finch (1930) specifically argued against the occurrence of a volcanic blast on the night of 19 May 1915. Williams (1932) adopted Finch's conclusions, and the understanding of these eruptions has remained essentially unchanged since his work (E. Richards, pers. commun., 1982). The characteristics of the eruption that occurred late on the afternoon of 22 May 1915 are discussed in Eppler (1987), and a more expanded treatment of the eruptions of May 1915 and a discussion of the lahars that they generated may be found in Eppler (1984).

Gorshkov (1963) introduced the term volcanic blast to describe destructive, laterally-directed explosive events that are distinct from vertical eruption columns. Two distinctive characteristics are important in the conceptual model of directed blast events: (1) horizontal or near-horizontal ejection angles, and (2) lack of a strong channelization by topography (Gorshkov 1963). For this work, a volcanic blast is defined as the effect of a violent eruption that causes significant destruction to the area surrounding the volcanic edifice, with an inferred significant radial component to the forces associated with the event. Any sediment, either juvenile or accidental in origin, that is emplaced during a volcanic blast will be termed "blast deposit."

The re-evaluation presented here is prompted by continuing concerns about future volcanic hazards in the Cascade Range, and as a test of recent concepts and techniques developed to examine processes in physical volcanology through field and laboratory analyses. Several factors complicate this evaluation: the larger eruption of 22 May covered essentially the same area as was affected on 19 May; the 19 May pyroclastic deposits were relatively thin and were immediately covered by a volumetrically larger lahar; and all the deposits have been subjected to 70 years of erosion and modification.

Three sources of information were examined in this study: (1) limited eyewitness observations of Lassen Peak made prior to and on the night of 19 May, which permit a rough estimation of the setting and chronology of events; (2) photographs taken by B.F. Loomis immediately after the 19 May eruption, which provide some spatial constraints on the type of eruptive activity; and (3) deposits emplaced during the May 1915 eruptions that can constrain the sediment transport characteristics of the events that occurred on 19 May.

2 Eyewitness Observations of Lassen Peak, 14-20 May 1915

Most observations of activity at the summit of Lassen Peak during the 1914-1917 eruptions were made from Mineral, California, about 20 km south of the summit, and from Manton, California, about 25 km to the west (see Fig. 1). Neither of these localities permits direct observation of the northeast flank of the volcano, which was most affected by the 19 May eruption. Additional observations were occasionally made by individuals who climbed to the summit of the peak during periods of relative quiescence, or who traveled to the northeast flank of the volcano along the Emigrant Trail. A complete list of observed eruptions was compiled by Day and Allen (1925) and reproduced in Eppler (1984).

Eyewitness accounts are notoriously difficult to decipher, as the descriptions are usually biased toward the observer's previous experience and degree of scientific literacy. The qualitative nature of these observations (e.g., "medium-sized eruption" or "one of the largest eruptions seen") make them difficult to use for constraining eruption phenomena and processes. However, they can help establish the timing of events and document changes in the style of eruption, especially when such changes are seen by many observers.

One such change occurred during the third week of May, 1915. For the first time in the year since the start of the eruption cycle, "fire" was reported at the summit of Lassen Peak during

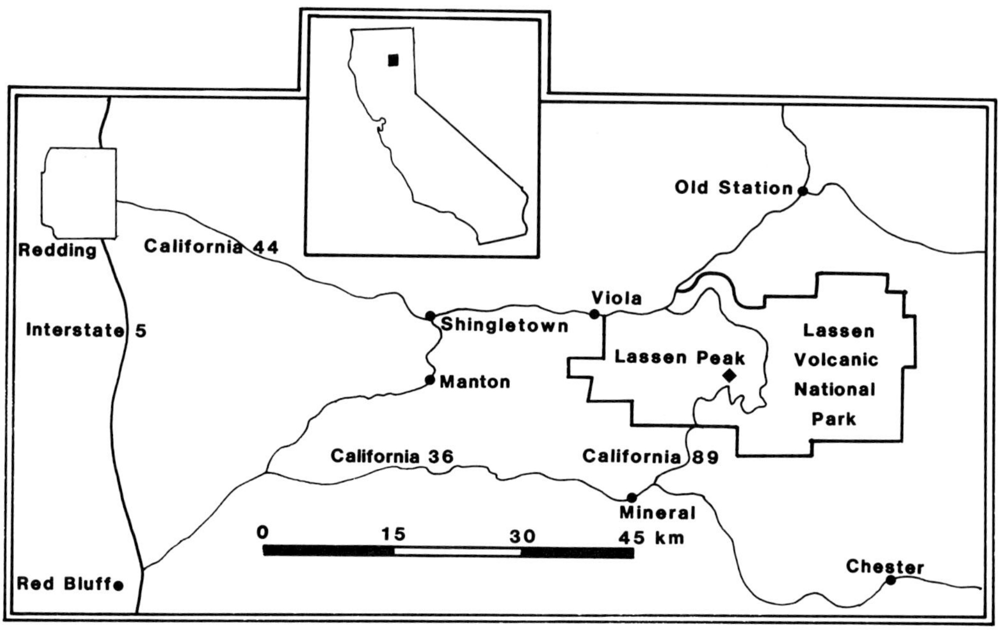

Fig. 1. Map of locations around Lassen Peak from which the 1914-1917 eruptive activity was observed

the early afternoon of 14 May. This was noted initially by a group of eight people in Manton, including the postmistress, who recorded the event. It was a sufficiently anomalous event for her to telephone as many people as she could, and over 100 people saw the eruption during the night and morning hours of 14-15 May (Day and Allen 1925).

No report was made of events on the 16th, presumably because of poor visibility. A "red glow" was observed on the clouds above the summit on the evening of 17 May (Day and Allen 1925). On 19 May, at about 9:00 P.M., observers in Manton reported seeing incandescent material at the summit, and people in Viola observed incandescent boulders rolling down the west flanks of the volcano, presumably derived from the dacite lava flow that erupted at the summit and was emplaced partly down the west and northeast flanks of the volcano (Day and Allen 1925; Loomis 1926). These are the only reports of incandescence for the Lassen Peak eruptions, and they become more important when considered in light of subsequent events. No observations subsequent to those of 19 May mention "fire" or summit "glows" other than those attributed to a glow on the clouds above the summit at sunset or sunrise.

At around 1:00 A.M. on 20 May, about 4 h after the evening observations of incandescent boulders described above, ranchers about 30 km from Lassen Peak in the lower reaches of Hat Creek were forced to move to high ground due to a rapid increase in the discharge of the creek (Loomis 1926). This increase in discharge was presumably caused by the emplacement of a major lahar in the upper reaches of the creek. No other real-time eyewitness observations were made of the 19 May eruptions.

3 Photographic Evidence of Activity on 19 May

The 1914-1917 eruptions of Lassen Peak were photographed by many local amateur and professional photographers. The photographs of B.F. Loomis, in particular, are of a quality that has allowed them to serve as a primary data source for deciphering the events of these eruptions (Diller 1916; Day and Allen

Fig. 2 a, b. a The northeast flank of Lassen Peak photographed on 22 May 1915 by B.F. Loomis from the position marked on Fig. 5, several hours prior to the 22 May eruptions. Peaks labeled **A** and **B** can be used as registration points when comparing Fig. 2 a and Fig. 3. The label **"hot rock"** was written on the original glass plate by Loomis over the image of a piece of 19 May dacite lava and alludes to the fact that the rock was still too hot to touch ≈48 h after it was erupted. **b** Sketch drawn from Fig. 2 a, showing areas discussed in text: **1** steam cloud rising from the summit area; **2** lahar deposit emplaced on the night of 19 May; **3** area inundated by the 19 May lahar; **4** area believed to have been deforested on the night of 19 May; **5** area unaffected by the 19 May eruption; **6** area apparently blanketed by tephra and partially cleared of snow during the 19 May eruption

a

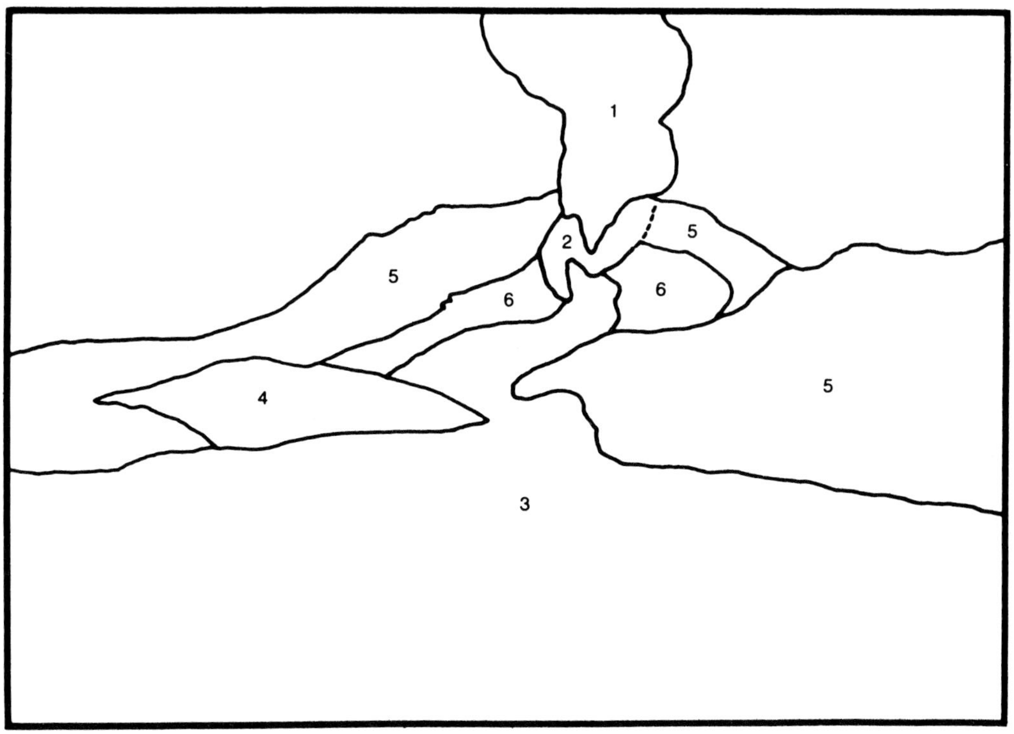

b

1925). Loomis' negatives are 6.5" x 8.5" glass plates, exposed through a 6" focal length lens. The field of view, about 70°, is approximately equivalent to that of a 25-mm lens used on a 35-mm format camera. Many of Loomis' photographs are reproduced in Loomis (1926), reprinted in 1979 by the Loomis Museum Association.

On 22 May, Loomis traveled to Emigrant Pass, about 5 km northeast of the summit of the volcano. At this location, he photographed and recorded his observations of the effects of the 19 May eruptions, just prior to the eruption of 22 May. Figure 2 a is a photograph of the northeast flank of Lassen Peak taken from Emigrant Pass by Loomis on 22 May, and Fig. 2 b is a sketch drawn from the photograph, delineating the areas discussed below. The time of day is estimated to be between 12:00 and 3:00 P.M., on the basis of shadow orientations. Prior to 19 May, the Peak was largely covered by snow, with some admixed ash from eruptions that occurred during late winter and early spring. Contemporary comparison of snow pack at Chester, California ($\approx$30 km southeast of Lassen Peak; see Fig. 1), and at Mineral ($\approx$20 km south; see Fig. 1) with that observed at Lassen Peak suggest that in May 1915 there was probably $\approx$15 m snow on the upper slopes of the peak (Day and Allen 1925) and as much as 5 m in the vicinity of Emigrant Pass (E. Richard, pers. commun., 1983). Accordingly, the surface above the tree line that was unaffected by the 19 May eruption should appear white except around rock outcrops, as it does in Fig. 2.

In Fig. 2, the summit is obscured by vapor cloud (Area 1), but several important features can be seen on the flank of the volcano. Area 2 extends downslope from the summit into gullies that form the head of Lost Creek. This area has sharp boundaries that appear to follow, and be confined by, Lost Creek. In the photograph, this area has a complex variation of gray tones, suggesting a rugged surface morphology. This area is interpreted as coarse-grained, bouldery lahar deposits that were emplaced in the upper reaches of Lost Creek on the night of 19 May.

In the foreground of Fig. 2 is a meadow, locally called Jessen Meadow, that has been inundated by the lahar deposit that was triggered by the 19 May eruption. The lahar appears to have been confined generally within the channel of Lost Creek until reaching the break in slope where Lost Creek enters the meadow. At this point, a portion of the lahar left the channel, spread out into the meadow and continued onto the divide between Lost and Hat Creeks.

Area 4 is a ridge to the southeast of Lost Creek that is clear of tree cover, in sharp contrast to slopes to the north of Lost Creek. It is not clear whether this ridge was deforested as a result of the 19 May eruption. A Loomis photograph taken in 1910 of Lassen Peak from Jessen Meadow (Fig. 3) shows that the ridges surrounding the volcano were heavily forested at that time. Due to the position of the camera, Fig. 3 does not show most of the area designated as Area 4 in Fig. 2 b, but the density of the forest throughout the photograph suggests that this

Fig. 3. The area affected by the 19 May 1915 eruptions photographed by B.F. Loomis in 1910. Lost Creek debouches into the meadow at the break in the tree line (see **arrow**) and flows from left to right along the base of the ridge beyond the fence line. Peaks labeled **A** and B are the same peaks labeled in Fig. 2 a

ridge supported a normal population of trees in mid-1910. There is no record of a violent eruption occurring prior to 19 May, nor was the area inhabited or subject to logging operations. Accordingly, it seems probable that the ridge was deforested by the 19 May eruption.

Extending outward from the summit is a light gray area (Area 6) that appears to be shaped like a fan with its apex at the summit, extending northeast toward the camera. This gray area has well-defined though feathery edges that do not appear to be confined by topography, and appears to be overlain in the summit area by the lahar deposits of Area 2. The margin of Area 6 is generally darker than the interior. The interior is also more variable in tone than the margin. This area is interpreted as snow covered by a blanket of tephra at the edges, with bare ground exposed in the interior. This interpretation is based on the apparent overall shape of the area, its lack of topographic confinement, and the feathery nature of its lateral boundaries. The variation of gray tones may be the result of progressively greater clearing of snow toward the axis of the fan, which allowed more of the surface texture previously covered by snow to show through.

4 Computer Processing of Loomis' Photographs

Loomis' photographs are of sufficient quality and interest to warrant digital computer processing to aid the understanding of spatial relationships between geographic features and the effects of the 19 May eruption. The specific goal of this processing was to provide a map-referenced guide to field investigations of the 19 May eruptions by geometrically rectifying Loomis' 22 May photograph to an overhead map perspective. In particular, it permitted the actual shape of Area 6 (described above) and its relationship to the physiography of the north flank of the volcano to be viewed. Relatively simple image processing techniques were employed, as outlined below.

First, a print of Loomis' 22 May photograph was digitized using a video camera attached to a Gould IP8400 image array processor operating on a VAX 11/750 minicomputer. This turned the print into a digital image that could be mathematically processed by the computer to change the viewing geometry of the original image, and to enhance contrast and improve the detail of the image. The resulting 512 x 512 x 8 bit image was processed to enhance detail and scene contrast. The results are shown in Fig. 4 a.

Next, a digital version of the 1:63,360 scale topographic map of the Lassen Peak area was created by tracing the contour lines on a GTCO X-Y digitizing table. Algorithms were applied to create an artificial three-dimensional surface from these contours, and to artificially illuminate this surface to produce a synthetic shaded relief image seen from a position approximately equal to the one from which Loomis photographed Lassen Peak on 22 May 1915. This image is shown in Fig. 4 b.

Finally, identical positions were located on both the shaded relief image and the digitized version of Loomis' photograph. This allowed the digital photograph to be meshed with the ter-

---

Fig. 4 a-d. Computer processed images of Lassen Peak: **a** digitized Loomis photograph taken 22 May 1915 from the same location as Fig. 2; **b** digitized terrain model of the Lassen Peak area rotated down to the viewing perspective of **a**; **c** digitized Loomis photograph after projection onto the terrain model of the Lassen Peak area and rotation of both data sets to an overhead perspective. Superimposed on the image are topographic contours with 200-ft contour interval. The strong lineation seen on the processed image is the result of foreshortened boundaries in the unprocessed image being stretched as it is rotated to an overhead perspective; **d** boundary of the area affected by the 19 May 1915 eruption superimposed on the digitized terrain model

ROCK
LASSEN PEAK, CALIFORNIA: PHOTOGRAPH TAKEN BY B. F. LOOMIS ON
22 MAY 1915, THREE DAYS AFTER THE 19 MAY ERUPTION-LAHAR AND
ONLY A FEW HOURS BEFORE THE BLAST AND LAHARS OF 22 MAY.
PHOTOGRAPH DIGITIZED FROM THE BOOKLET "ERUPTIONS OF LASSEN PEAK"
AND COMPUTER ENHANCED AT ARIZONA STATE UNIVERSITY.
14-DEC-85 17:56:41 ASU/IPF

a

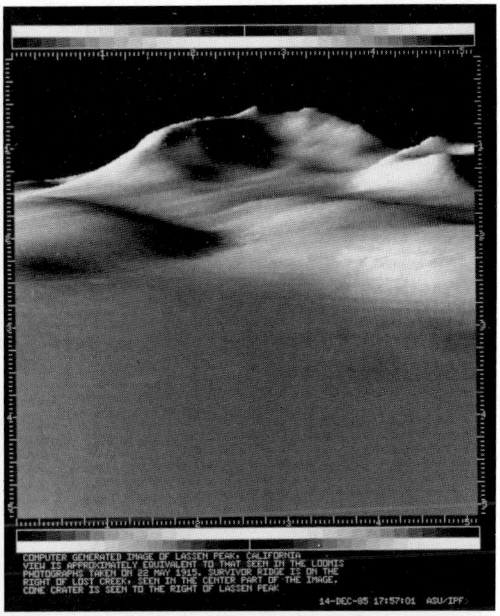
COMPUTER GENERATED IMAGE OF LASSEN PEAK, CALIFORNIA
VIEW IS APPROXIMATELY EQUIVALENT TO THAT SEEN IN THE LOOMIS
PHOTOGRAPHS TAKEN ON 22 MAY 1915. SURVIVOR RIDGE IS ON THE
RIGHT OF LOST CREEK, SEEN IN THE CENTER PART OF THE IMAGE.
CONE CRATER IS SEEN TO THE RIGHT OF LASSEN PEAK
14-DEC-85 17:57:01 ASU/IPF

b

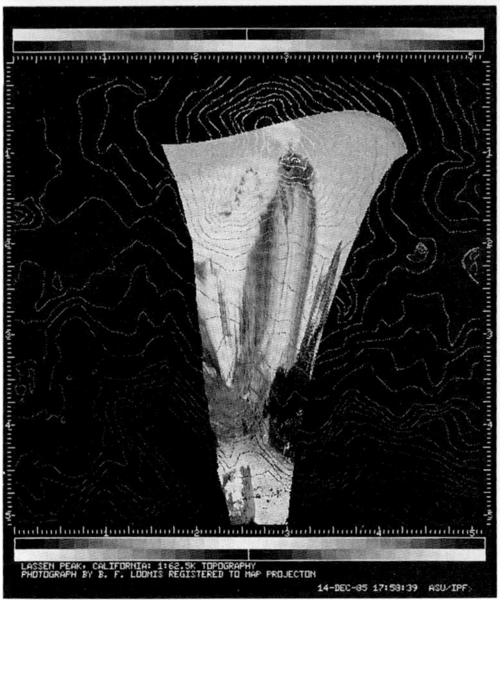
LASSEN PEAK, CALIFORNIA: 1:62.5K TOPOGRAPHY
PHOTOGRAPH BY B. F. LOOMIS REGISTERED TO MAP PROJECTON
14-DEC-85 17:58:39 ASU/IPF

c

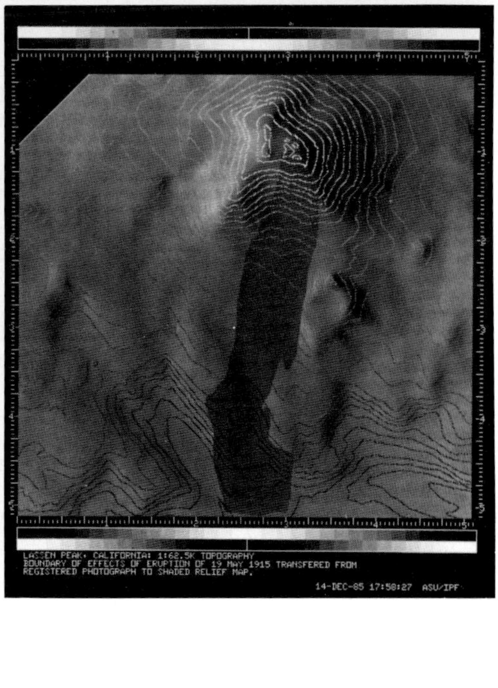
LASSEN PEAK, CALIFORNIA: 1:62.5K TOPOGRAPHY
BOUNDARY OF EFFECTS OF ERUPTION OF 19 MAY 1915 TRANSFERED FROM
REGISTERED PHOTOGRAPH TO SHADED RELIEF MAP.
14-DEC-85 17:58:27 ASU/IPF

d

rain model, and allowed the photograph to be geometrically manipulated along with the digital terrain model. Both data sets were geometrically reprojected to produce an image of the photograph seen from an overhead, orthographic perspective. This, in effect, allowed viewing of Loomis' photograph as if it were taken from a position above the volcano, rather than looking obliquely across the terrain. Although this causes stretching and distortion of the photograph, it permits viewing of the geometrical relationships between the areas described in the previous section. The reprojected photograph is shown in Fig. 4 c, with the digital topography overlain to relate the features seen on the photograph to the topography. There is a sufficiently good spatial correlation between known topographic features seen on the photograph and the contour expression of these features to give confidence in geometric relationships seen on the reprojected photograph. The exterior boundary of Area 6 was outlined and the digital photographic image was then removed, in order to show only its shape and relationship to the topography (Figs. 4 d and 5).

Of particular interest is the strong linear shape of Area 6. The relatively large length to width ratio ($\geq$4:1) is unlike other volcanic blasts, such as that which occurred at Mount St. Helens (e.g., see Kieffer 1981, p 392, and Crandell and Hoblitt 1986). Area 6, which in oblique views appears consistent with the fan-shaped blast zones seen around other volcanoes, actually resembles more the planimetric form of a region affected by more topographically confined flows, such as lahars. The path also resembles that taken by many snow avalanches that begin along the eastern ridge crest at the summit of Lassen Peak. The area lies entirely within the zone delineated by tree blow down associated with the 22 May eruption, reinforcing field experience that there remain few localities where 19 May tephra deposits may be preserved without having experienced the effects of the 22 May volcanic blast. In addition, the distal portion of Area 6 lies within the area that was inundated by the lahar that was triggered by the 19 May activity and was emplaced at $\approx$1.00 A.M. on 20 May (see Table 1). The lateral extent of the affected area in the lower meadow (i.e., close to the camera) is uncertain, and is marked by question marks in Fig. 5. This queried boundary represents the lateral limit of data available in Loomis' photograph. At present, we have been unable to find either photographic or written records that might indicate how far outside the queried boundary the area affected by the 19 May eruption extended.

5 Tephra Associated with the 19 May Eruption

Primary fragmental deposits of the 19 May eruption are rare within the area affected by the 22 May eruption. In most localities, no distinctive pyroclastic deposit can be found beneath the 22 May blast deposit, which is itself typically less than 10 cm thick (Eppler 1987). Two arguments can be offered to account for the absence of a 19 May blast deposit: (1) it never formed, or (2) it was scoured away and incorporated into the subsequent 19 May lahar deposit, or the 22 May blast deposit.

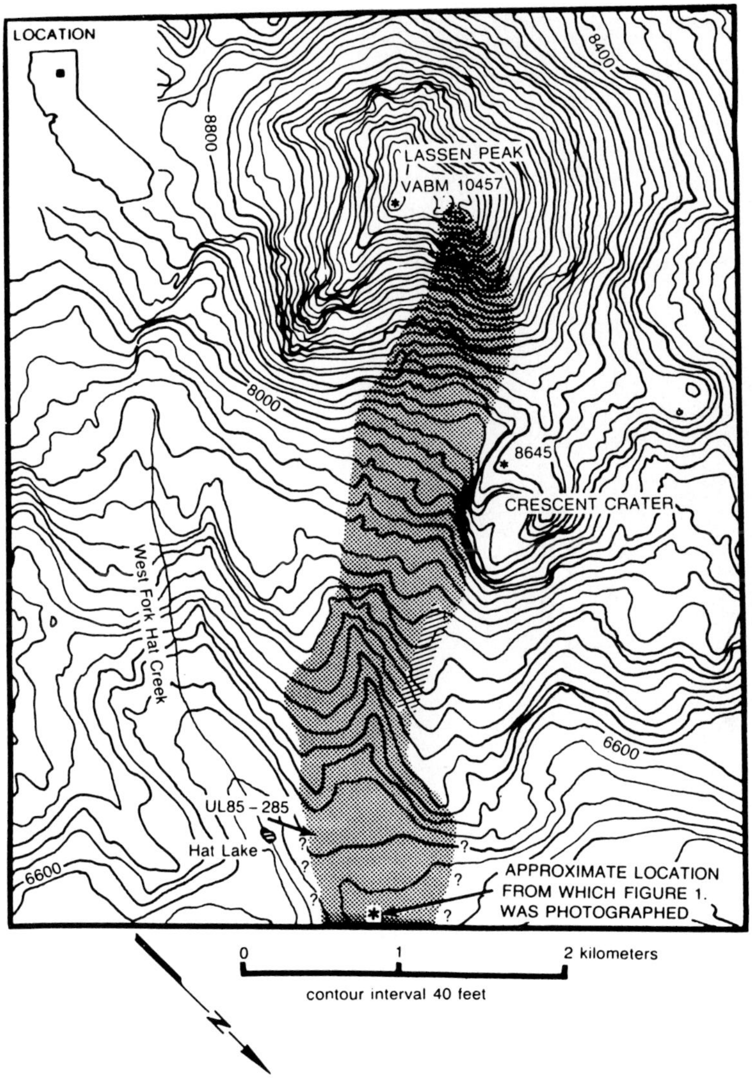

Fig. 5. Sketch map of the area affected by the 19 May eruption prepared using Fig. 3 d. The **question marks** lie along the boundary defined by the edge of Loomis' photograph, and represent the limit of data available from the photograph. It is uncertain how far laterally this boundary extended to the northeast and southwest. Note the location of **UL85-285** northwest of Hat Lake, and the approximate location from which B.F. Loomis photographed the area

Although it is possible that rain fell after the 19 May eruption, it seems unlikely that sufficient precipitation occurred between the early morning hours of 20 May and the morning of 22 May to allow complete erosion of the 19 May deposits.

Fortunately, a location about 250 m north of Hat Lake (Fig. 5, labeled UL85-285) preserves a sequence of deposits associated

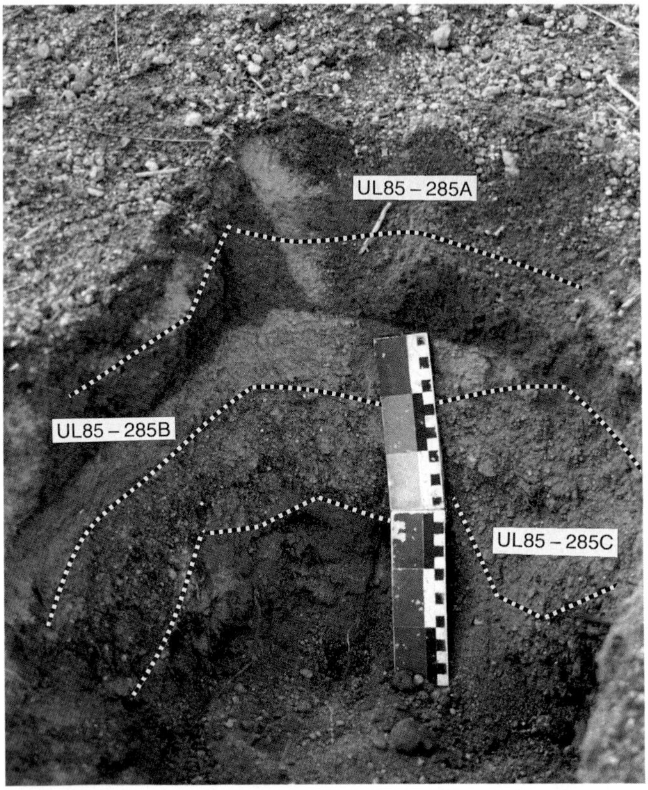

Fig. 6. Photograph of the pit excavated at UL85-285, showing the relationship between May 1915 eruption products. The uppermost deposit is the 22 May volcanic blast deposit (UL85-285A); the intermediate deposit is the marginal deposit from the 19 May lahar (UL85-285B); the lower deposit is interpreted as the volcanic blast deposit from the 19 May eruption (UL85-285C). The bottom of the pit is in pre-May 1915 soil

with the May 1915 eruptions. Included is a tephra deposit whose stratigraphic position, described below, suggests that it was emplaced on 19 May. This location was protected by a small ridge from experiencing the full force of erosion by lahars within the Lost Creek drainage basin, and received only thin marginal overflow from the 19 May lahar. The specific locality is within an inlier surrounded but not covered by a lahar deposit emplaced late in the 22 May eruption sequence. These lahar deposits appear to have subsequently protected the site and have preserved a record of events that occurred on 19 May. Excavation of the inlier to a depth of 50 cm reveals four distinct deposits (Fig. 6). The lowermost is a soil formed from a mixture of fluvially- and glacially-deposited detritus derived from Lassen Peak, and pyroclasts erupted from the Chaos Crags ≈1080 years B.P. (D. Trimble, pers. commun. 1982). Organic debris associated with the pre-1915 forest cover is no longer visible as a discrete layer, and was probably incorporated into the tephra deposit from 19 May.

Above the old soil are three May 1915 deposits. The uppermost (UL85-285A in Fig. 6) has a flat upper surface an irregular lower contact. The deposit ranges from 10-15 cm in thickness. No sedimentary structures were visible in the limited exposure of the pit. Median grain size (Md_{Φ}) is 0.5Φ and the sorting coefficient (σ_{Φ}) is 2.1Φ, indicating poor sorting. Binocular and scanning electron microscope examination of grains indicate that the pyroclasts are similar in composition and morphology to those found in many locations affected by the 22 May volcanic blast (Eppler 1987). This uppermost deposit was most likely emplaced by the 22 May blast cloud (Eppler 1987).

Immediately below UL85-285A is another layer (UL85-285B in Fig. 6) that has irregular upper and lower contacts and is also 10-15 cm thick. The deposit has poorly developed normal grading. Md_{Φ} for an aggregate sieved sample is 1.1Φ, and it is poorly sorted with a σ_{Φ} of 2.4Φ. The deposit includes black, glassy grains with large plagioclase phenocrysts that appear to have been derived from the 19 May dacite lava flow; altered grains of dacite erupted from the Chaos Crags and Lassen Peak prior to the 1914-1917 eruptive cycle; and tephra derived from the underlying deposit. The 19 May dacite lava grains are distinguished from the older Lassen dacite by fresh, glassy surfaces and lack of hydrothermal alteration. Scanning electron microscope examination of the grains from UL85-285B show the deposit to be composed mostly of lithic and mineral fragments that have been partially rounded by grain to grain collisions during transport (Fig. 7 a). Rare pumices present in the deposit are also partially rounded and have a sizeable population of smaller particles, most likely clay minerals or clay-sized particles, within the vesicles (Fig. 7 a). The rounding of the grains, and the presence of fine-grained sediment in void spaces, suggests transport within a wet slurry composed of abundant clay- and silt-sized particles. The presence of grains derived from the 19 May dacite lava flow, the condition of constituent grains in the deposit, and the similarity of this deposit to the 19 May lahar deposits seen at many localities along Lost and Hat Creeks (Eppler 1984) make it reasonable to conclude that this unit is either material emplaced by splash from the 19 May lahar, or by runoff associated with the dewatering of the same lahar deposit after emplacement.

The lowermost deposit of interest at this locality (UL85-285C in Fig. 6) unconformably overlies the soil horizon. It is coarser grained than the upper deposits, and has a speckled appearance caused by the presence of large plagioclase phenocrysts in clasts >0.5 cm. The deposit is generally ungraded, although it occasionally appears to have poorly developed reverse grading. Additionally, in some portions of the pit it is matrix supported, while in others it appears to be clast supported. The deposit ranges in thickness from 4-15 cm. Grain size analyses yield an Md_{Φ} of 0.4Φ and σ_{Φ} of 1.7Φ, indicating poor sorting.

Fig. 7 a-f. Scanning electron photomicrographs of samples from UL85-285: a pumice and lithic grains that have been partially rounded during transport. Also note the fine-grained sediment filling void spaces and adhering to individual grains. b Plagioclase grain coated with vesicular glass. Note the stretched vesicles and general lack of chipping and edge modification. c Pumice grain with small, primarily spherical vesicles. d Pumice grain with spherical vesicles substantially larger than in c. e Glass fragment that has undergone some chipping and edge modification during transport. f Plagioclase grain that has been partially rounded during transport

Macroscopically, UL85-285C consists of clasts of the 19 May dacite lava; fine-grained, dark gray pumice; and intimately intermixed pieces of charred and uncharred organic material (Table 2). Binocular and scanning electron microscope examination of the deposit indicates that a portion of the fine-grained sediment is fresh, glassy pumice with a wide range of vesicle sizes and proportions (Fig. 7 b-d). The vesicles in the pumice are often flattened and stretched, and in thin section this deformation gives the appearance of flow banding. There do not appear to be any juvenile pumices that strongly resemble the juvenile pumices erupted on 22 May; neither does there appear to be a continuum between the vesicular grains and more massive lithic grains. Mineral grains as well as pumices show some indication of grain to grain collision during transport (Fig. 7 e,f), but less than seen on similar clasts within the 22 May deposit (Eppler 1987). Either collisions were less vigorous on 19 May than on 22 May (Eppler 1987) or there may have been fewer collisions. There is no evidence of transport by fluvial processes, and there are few adhering silt- or clay-sized particles within the pumice vesicles.

The organic material in UL85-285C consists of small red fir needles, pieces of bark, and segments of twigs. The majority of fir needles retain their shape but have been completely charred. The pieces of bark are only partially charred, and in most cases, the area of charring is greatest at the surface, decreasing toward the interior. Several pieces of bark are completely charred but retain their internal structure. Such effects indicate temperatures in excess of 250°C applied for periods of at least several minutes (Winner and Casadevall 1981) under oxygen-poor conditions.

The stratigraphic position, composition, and the presence of fresh, glassy pyroclasts and carbonized organic material suggest that the lowermost unit is a deposit emplaced by eruptive activity on 19 May 1915. The mechanism of emplacement of the deposit is unclear. Waitt (1981) described a basal gravel facies emplaced by the 18 May 1980 blast deposit (layer A1) at Mount St. Helens that resembles UL85-285C. Waitt's layer A1 is massive to weakly stratified and in general consists of a sandy to gravelly layer of juvenile dacite, accidental lithic fragments, and wood fragments. This deposit ranges from ungraded to normally graded in some locations to reversely graded in others. Waitt (1981) attributed this layer to deposition by a highly turbulent pyroclastic density flow. We suggest that the deposit at UL85-285C may have been deposited by a similar type of flow. However, inability to document areal variations due to the lack of additional exposures makes determination of such a specific mechanism of emplacement impossible. For the purposes of the following discussion, we prefer to use the term "blast deposit" to describe UL85-285C. In this case, the term is used to describe a deposit for which a specific fluid dynamic model of emplacement cannot be determined, but which was emplaced as part of a violent, probably explosive eruption.

Table 2. UL85-285C grain count data

Component	Number of grains	Percent	Description
19 May dacite lava	199	40.0	Subhedral plagioclase phenocrysts, zoned, rounded and embayed; smaller broken plagioclase phenocrysts, rounded biotite, with opaques at rim, also very rare biotite pseudomorphic after olivine; rounded and embayed clinopyroxene some with rims of amphibole; rare subhedral hornblende; all set in vuggy/vesicular groundmass with devitrified glass, feldspar needles and opaque oxides; vugs appear to be undeformed
Fresh pumice	111	22.3	Phenocrysts of euhedral-subhedral plagioclase; euhedral amphibole; set in vesicular-vuggy, glassy groundmass with minor needles of feldspar, opaque oxides; voids have been flattened and deformed, giving the pumice a flow-banded appearance in plane light
Pre-1915 dacite	104	20.9	Similar mineralogy to 19 May dacite lava; distinguished by the absence of glass and extensive hydrothermal alteration causing Fe-staining of groundmass
Mineral grains	84	16.8	
Plagioclase	56	11.2	
Quartz	26	5.2	
Opaque oxides	2	0.4	
Total	498	100.0	

6 Discussion

Table 3 summarizes some of the evidence that has been developed concerning the eruption of 19 May. The eruption appears to have been preceded by about 1 week of anomalous activity, which culminated in the extrusion of a dacite lava flow onto the upper slopes of Lassen Peak. The small number of eyewitness observations and times recorded for the events indicate that perhaps 4 h elapsed between the extrusion of the dacite lava flow onto the upper slopes of the volcano and the emplacement of the la-

Table 3. Summary of characteristics of the 19 May eruption

Characteristic	Implication
Two juvenile components are present in sample UL85-285C: 19 May dacite lava and fresh, glassy pumice	The eruption involved production of juvenile pumice as well as disruption of the 19 May dacite lava flow
Mineralogy of the juvenile components appears to be extremely similar	Both components may have been derived from the same source
Vesicles in the pumices were deformed	Pyroclasts were plastic enough to deform when subjected to stress during the eruption
Organic material intimately mixed with pyroclasts in the deposit	Organic material was mixed into the deposit during transport/ deposition by the emplacement process
Needles pyrolized	Deposit was >250°C immediately after deposition (Winner and Casadevall 1981) Needles charred in an oxygen-poor environment after burial
Large pieces of bark not completely charred	There was sufficient heat in the deposit to char small, slender needles, but insufficient to completely char large, equant pieces of bark
Pyroclasts not strongly abraded	Grain to grain collisions were either rare or of a low kinetic energy
Boundary of the disturbed area is dissimilar to disturbed areas from point-source explosions	The eruption did not occur as a point-source type of explosion, but has the characteristics of a mass-movement event

har that caused the rapid increase in discharge on Hat and Lost Creeks. Photographs suggest that the hills below the northeast flank of Lassen Peak supported a normal growth of trees that was removed sometime between the summer of 1910 and the morning of 22 May 1915. In the photographs, the snowpack on the upper northeast flank of the volcano appears to have been blanketed by a deposit of light gray material, and to have had some areas stripped of their winter snow accumulation. The shape of this zone of deposition and stripping is highly constrained by the relief of the area, resembling the path of a rapidly emplaced mass movement. A highly fragmented layer of debris, including organic material that has been partially to completely carbonized after deposition, and lithic fragments that appear to have been been emplaced while still sufficiently plastic to deform, was found in a location consistent with emplacement on 19 May 1915. The stratigraphic and compositional relationships of this layer, and the thermal conditions inferred from the alteration of the organic material, are consistent with its emplacement by a high temperature volcanic blast prior to 22 May. These observations suggest that the 19 May eruption included a violent, probably in part explosive eruption, as originally suggested by Diller (1916) and Day and Allen (1925). This event occurred in addition to the known extrusion of a dacite lava flow and the generation of a lahar.

The exact mechanism by which the 19 May volcanic blast was generated is unknown. The shape of the affected zone is not like that of other volcanic blasts attributed to directed events (e.g., Mt. Pelée, 1902, in Fisher and Heiken 1982), or where hydrodynamic pressures have dominated the distribution of effects (e.g., Mount St. Helens, 1980, in Kieffer 1981). Rather, the resemblance of the affected zone to a mass movement suggests that the blast behaved as a "ground hugging" flow. However, no geomorphic evidence of the source of this flow remained on 22 May, suggesting that its generation did not result from the collapse and subsequent avalanching of a major part of the volcanic edifice, nor from the excavation of a crater on the east face of Lassen Peak.

It seems plausible that the blast was in some way associated with the emplacement of the dacite lava flow. Two possibilities can be readily suggested. First, the flow may have collapsed, producing a landslide which traveled down the steep eastern slope of Lassen Peak, similar to, but with a significantly greater volume than the incandescent boulder falls that were observed moving down the gentler western slopes. Under such conditions, the lava would have broken up and become intimately mixed with the more than 15-m-deep snowpack, generating an explosive production of steam in a fuel-coolant reaction similar to that described by Colgate and Sigurgeirsson (1973), and produced under experimental conditions by Wohletz and McQueen (1984). The large volume change brought about by the water to steam reaction would have fragmented the lava further, enhancing the heat transfer and steam production. The shape of many of the vesicles indicates that some vesiculation and deformation took place in the pyroclasts after their formation. It is not clear, however, what contribution the continued growth of the vesicles made to the magnitude of the blast. A similar set of circumstances (i.e., one involving the breakup of the hot dacite cryptodome and its intimate mixing with the water in Spirit Lake) has been proposed by Moore and Rice (1984) to explain the more energetic pulse of the 18 May 1980 eruption of Mount St. Helens that occured nearly 8 km north of the volcano and almost 2 min after the initial summit explosion.

The second possibility, suggested by the presence of pumice, the degree of vesiculation of that pumice, and the range in shapes and sizes of the vesicles, is that the dacite lava flow produced a pyroclastic flow. Rose et al. (1976) reported on the eruption of a pyroclastic flow from the toe of an active dacite lava flow at Santiaguito Volcano in Guatemala. The pyroclastic flow, which was triggered when the lava flow collapsed during extrusion (Rose et al. 1976), produced many features resembling those described above, particularly strong similarities in the shape of the area devastated, the proportions of various grain types in the flow deposit, and similarity in vesicle shape and degree of vesicularity in the two deposits. It is clear from the Santiaguito eruption that silicic lava flows can produce violent events that emplace volumetrically significant pyroclastic deposits without the involvement of an underground magma chamber. The additional factor at Lassen Peak of mixing of hot dacite pyroclasts with snow probably increased the violence of the event.

The temperature inferred by the presence of pyrolized organic material in UL85-285C, the fragmentation characteristics of materials within the deposit, and evidence of transport with some energetic collisions between particles, is consistent with either of these hypotheses. A third possibility, that the deposit could have been generated by a phreatic explosion at the summit, seems much less likely, given the shape of the affected area as seen in Figs. 4 and 5, the presence of juvenile pumice, and the lack of a crater that would have formed in such an explosion.

The data on the 19 May eruption that are presently available do not allow the determination of the exact genetic mechanism for the eruption of 19-20 May, and for the subsequent emplacement of the pyroclastic deposit at UL85-285. However, they strongly suggested that a violent, high temperature event took place on the night of 19-20 May, as was suggested originally by Diller (1916) and Day and Allen (1925).

Acknowledgements. This research* was supported by the National Science Foundation Environmental Geoscience Grant EAR8313091, and a grant by the Associated Students of Arizona State University. We are extremely grateful of the National Park Service at Lassen Volcanic National Park, in particular to Chief Naturalist Richard Vance and Assistant Chief Naturalist Ellis Richard for their assistance in the conduct of this project, and to Diana Michna for her assistance in sample preparation. We also thank Bob Christiansen, Mike Clynne, Steve Self, and Grant Heiken for helpful discussion. This work benefited substantially from critical reviews by Jamie Gardner, Mike Clynne,

* Originally submitted 16.6.86; Resubmitted 3.2.1987

Pete Lipman, John Latter, and an unnamed reviewer, and from continued encouragement from Jamie Gardner, Steve Mattson, Dave Jorgenson, Peg Youngman, and Donna Marron.

References

Colgate SA, Sigurgeirsson T (1973) Dynamic mixing of water and lava. Nature (Lond) 244:552-555

Crandell DR, Hoblitt RP (1986) Lateral blasts at Mount St. Helens and hazard zonation. Bull Volcanol 48:27-38

Day AL, Allen ET (1925) The volcanic activity and hot springs of Lassen Peak. Carnegie Inst Washington Publ 360:1901

Diller JS (1916) The volcanic history of Lassen Peak. Science 43:727-733

Eppler DB (1984) Characteristics of volcanic blasts, mudflows and rock-fall avalanches in Lassen Volcanic National Park, California. Ph. D. Thesis, Arizona State University, Tempe, 1-261

Eppler DB (1987) The May 1915 eruptions of Lassen Peak, II: May 22 volcanic blast effects, sedimentology and stratigraphy of blast and lahar deposits, and characteristics of the blast cloud. J Volcanol Geotherm Res 31:65-85

Finch RH (1929) The origin of Lassen mudflows. Volcano Ltr (224):1

Finch RH (1930) Mudflow eruption of Lassen mudflows. Volcano Ltr (266):1-3

Fisher RV, Heiken GH (1982) Mt. Pelée, Martinique: May 8 and 20, 1902 pyroclastic flows and surges. J Volcanol Geotherm Res 13:339-371

Gorshkov, G (1963) Directed volcanic blasts. Bull Volcanol 20:83-88

Kieffer SW (1981) Fluid dynamics of the May 18 blast at Mount St. Helens. In: Lipman PW, Mullineaux DR (eds) The 1980 Eruptions of Mount St. Helens, Washington. US Geol Surv Prof Paper 1250:379-400

Loomis BF (1926) Pictorial history of the Lassen volcano. Loomis Museum Assoc 1-96

Moore JC, Rice CJ (1984) Chronology and character of the May 18, 1980 explosive eruptions of Mount St. Helens. In: Explosive volcanism: inception, evolution and hazards. Natl Acad Sci 133-142

Rose WI Jr, Pearson T, Bonis S (1976) Nuée ardente eruption from the foot of a dacite lava flow, Santiaguito Volcano, Guatemala. Bull Volcanol 40-1:1-16

Waitt, RB Jr (1981) Devastating pyroclastic density flow and attendant air fall of May 18 - stratigraphy and sedimentology of deposits. In: Lipman PW, Mullineaux DR (eds) The 1980 eruptions of Mount St. Helens, Washington. US Geol Surv Prof Paper 1250:439-460

Williams H (1932) Geology of Lassen Volcanic National Park, California. Dept Geol Sci, Calif Univ Bull 17:195-385

Winner WE, Casadevall TJ (1981) Fir leaves as thermometers during the May 18 eruption. In: Lipman PW, Mullineaux DR (eds) The 1980 eruptions of Mount St. Helens, Washington. US Geol Surv Prof Paper 1250:315-322

Wohletz KH, McQueen RG (1984) Experimental studies in hydrovolcanism. In: Explosive volcanism: inception, evolution and hazards. Natl Acad Sci 158-169

Discrimination Between Debris Avalanches and Other Volcaniclastic Deposits

T. Ui[1]

Abstract

Criteria are presented on how to discriminate a debris avalanche deposit from other volcaniclastic formations. A debris avalanche deposit is morphologically characterized by hummocky topography and may be steep-sided cliffs at the distal and lateral edges. The source area of the debris avalanche is characterized by a horseshoe-shaped crater. Additional evidence to identify a deposit as a debris avalanche is the presence of block facies and matrix facies. Block facies consists of coherent, unconsolidated or poorly consolidated fragments of a volcanic body which commonly preserve their internal stratifications and/or intrusive contacts. Conjugate fractures develop within block facies that originates from massive and compact igneous bodies. A matrix facies is made of a mixture of various kinds of volcanic clasts derived from various parts of a source volcano. Patches of soil, stream gravels and pieces of wood may be picked up during transportation.

Lahar (mud flow and debris flow) deposits have textures and internal structures similar to the matrix facies of a debris avalanche deposit. Lahar deposits do not contain debris avalanche blocks. Large boulders within a lahar deposit are generally surrounded by finer-grained material. Boulders concentrate towards the upper surface of the deposit, forming reverse grading. No steep cliff forms at the distal and lateral edges of a lahar.

The surface of a nuée ardente deposit undulates more regularly than that of a debris avalanche deposit. The slope angle of undulation is not so steep as in the case of a debris avalanche deposit. Shrinkage cracks are developed on the surface of juvenile fragments.

Glacial moraines show occasionally hummocky topography similar to that of debris avalanche deposits. However, no fractured blocks are contained within moraines. Striations and grooves are common on the surface of the larger clasts within a moraine.

---

1 Department of Earth Sciences, Faculty of Science, Kobe University, Nada, Kobe 657, Japan

IAVCEI Proceedings in Volcanology 1
J.H. Latter (Ed.), Volcanic Hazards

1 Introduction

Debris avalanches represent the rapid flow of debris that generally form from landslides (Varnes 1978). The landslides may result from volcanic activity, an earthquake, heavy rainfall or the rapid melting of snow. The type of volcanic activity which may cause a debris avalanche is deformation of a volcanic edifice due to dome intrusion (Voight et al. 1983), phreatic explosion (Nakamura 1978) and slumping of a caldera wall that produces megabreccia and mesobreccia (Lipman 1976). Before the catastrophic eruption of Mount St. Helens in 1980, the difference between debris avalanche and lahar deposits was not clearly defined, and consequently many debris avalanche deposits have been included within a variety of lahar deposits (Grange 1931; Mason and Foster 1956; Neall 1976; Siebert 1984). Some debris avalanche deposits have been interpreted as pyroclastic flow deposits (Kofu Basin Quaternary Research Group 1969; Yamagishi 1977), as lava flows (Bossard 1928) or as moraines (Williams 1949), because of similarities in their texture or internal structure. Since the eruption of Mount St. Helens in 1980, however, some volcaniclastic formations have been reinterpreted to be of debris avalanche origin (Crandell et al. 1984; Francis et al. 1985).

The aim of this chapter is to clarify the difference between a debris avalanche deposit and other deposits that have similar texture, structure and morphology. These deposits include lahar (volcanic mud flow and debris flow) deposits, small-scale pyroclastic flow (nuée ardente) deposits and moraines. There are commonly many different kinds of volcaniclastic formations on the outer flanks of volcanic edifices. Knowledge of the mode of origin of these volcaniclastic formations is important for the assessment of volcanic hazards.

2 Geomorphic Evidence

The surface morphology of a volcaniclastic deposit is important for judging its mode of origin. A debris avalanche deposit is characterized by hummocky topography (Fig. 1 A). The shape of individual hummocks is variable and irregular, and is probably controlled by the shape of individual debris avalanche blocks beneath the surface (Ui et al. 1986b). No overall trend of alignment is found, although local parallelism of the long axes of hummocks has been described and discussed (Tsujimura and Kiuchi 1936; Mizuno 1958; Crandell et al. 1984). The volume and height of hummocks is larger in the proximal part of the deposit, and decreases towards the distal end (Crandell et al. 1984; Ui et al. 1986a; Fig. 11 of Glicken 1986). Natural levees formed on both sides of the 1980 Mount St. Helens deposit (Voight et al. 1983) (Fig. 1 D). Francis et al. (1985) described the natural levee that rises to 40 m above the debris avalanche deposit at Socompa volcano, in northern Chile. Steep slopes commonly form at the distal margin of an debris avalanche deposit. The most excellent example of a distal slope is found at the Blackhawk landslide, California (Fig. 1 of Shreve 1968), although this example is a debris avalanche formed in a non-

Fig. 1 A-F. A Hummocky surface of debris avalanche deposit, north of Mount Shasta, California. B Terminal moraine of Tasman Glacier, near Mount Cook, South Island, New Zealand. C Flat surface of 1985 Nevado del Ruiz lahar deposit at Armero, Colombia. The direction of flow was towards the photographer. D Natural levee of 1980 Mount St. Helens debris avalanche deposit, near Elk Rock, Washington. Source of the avalanche is towards the **right**. E Block facies surrounded by matrix facies within a debris avalanche deposit, north of Mount Shasta, California, 5 km northwest of Weed, California. F Fractured pattern in a block facies develops only in the massive part of an andesitic lava section. Fractures are scarce at the clinkered part. An abandoned quarry, 2 km north of Konoura, Kisakata debris avalanche, Chokai Volcano, northern Japan

volcanic area. In contrast to debris avalanches, the surface of a large-scale lahar deposit is relatively flat (Fig. 1 C). No hummocks are visible. Discrete large boulders may remain on the surface of a lahar deposit (Crandell 1971). The edge of the deposit has a lower relief than that of a debris avalanche deposit of similar scale.

The surface of a nuée ardente deposit is generally flat. However, dunelike structures are present on the surface of some nuée ardente deposits (Ui 1975). Their gentle and regularly undulating pattern differs from that of hummocks of debris avalanche deposits (Fig. 1 A). The pattern of natural levees described from small-scale pyroclastic flows, for example at Komagatake in 1929 (Kuno 1941) and at Mount St. Helens in 1980 (Rowley et al. 1981), shows channels and tonguelike shapes. Such a channelled pattern is different from that seen in the levees of debris avalanches.

The shape of some terminal moraines associated with mountain glaciers is very similar to the hummocks of debris avalanche deposits. A typical example is shown in Fig. 1 B, east of Mount Cook, South Island, New Zealand (Fitzharris et al. 1982), although it should be noted that this deposit has formed in a non-volcanic area. Accordingly, it is difficult to discriminate between a debris avalanche deposit and a moraine from the surface morphology of a deposit alone.

3 Internal Structures of Deposits

The internal structure which characterizes a debris avalanche deposit was first described by Mimura et al. (1971). They described two lithologically different portions of a debris avalanche deposit, which they called megablock and main facies. Later, Crandell et al. (1984) identified the same parts of a debris avalanche deposit from ancestral Mount Shasta and called these lithologies, respectively block facies and matrix facies (Fig. 1 E). Block facies (Fig. 1 F) consists of relatively coherent, unconsolidated or poorly consolidated fragments (debris avalanche blocks) derived from the source volcano. The fragments preserve their original internal structure, although they may be deformed, faulted, fractured or thoroughly shattered (Mimura et al. 1982; Crandell et al. 1984; Fig. 50 of Glicken 1986; Fig. 2 of Ui and Glicken 1986). Matrix facies consist of an unsorted and unstratified mixture of various volcanic clasts. Some additional clasts may be picked up from ground surface during movement of the material (Mimura et al. 1982; Crandell et al. 1984). The proportion of individual volcanic clasts within the matrix facies is also variable. Smaller debris avalanche blocks and isolated clasts, which are petrographically the same as nearby larger debris avalanche blocks, may concentrate close to the boundary of the block facies and matrix facies (Mimura et al. 1982). Stretched patches of soil, picked up from the ground surface during movement, or soft tuffaceous materials (remnants of debris avalanche blocks), may be included within a matrix facies (Ui 1983). Fragments of splintered tree trunks may also be included in matrix facies (Voight et al. 1983). These wood fragments are generally not carbonized (Mimura et al. 1982; Voight et al. 1983).

A lahar deposit consists of various types of clasts, and its texture and internal structure are similar to the matrix facies of a debris avalanche deposit. No internal structures similar to those of the block facies of a debris avalanche are present. No fracture or faults are visible within a lahar deposit (Fig. 2 B). Some deposits show reverse grading (Fig. 2 A).

Nuée ardente deposits are small-scale pyroclastic flow deposits structurally characterized by dense juvenile fragments (Aramaki and Yamasaki 1963). No internal structures similar to those of the block facies of a debris avalanche are present. The deposit is generally massive, without any grading or laminated structure.

Moraines are made of a mixture of clasts picked up from the source area of a glacier. Some parts of moraines are similar in texture and internal structure to the matrix facies of debris avalanche deposits. However, no internal structures similar to the block facies of the debris avalanches are present.

4 Fracture Pattern

Difference in fracture pattern within an exposure is useful for discriminating the origin of volcaniclastic formations. A block facies within a debris avalanche deposit is generally fractured; a type of the resultant fractures is referred to as jigsaw cracks (Ui 1983). The fracture pattern is different from the pattern of cooling joints of lava flows, chilled margins of dikes or juvenile fragments of pyroclastic flow deposits. A fracture within the block facies of a debris avalanche deposit is not so smooth or flat as the cooling joint of a massive igneous body (Fig. 2 D) (Ui 1983). The interval between cracks (the density of cracks) is variable (Fig. 2 D). Jointed planes usually remain closed, but occasionally open wide, probably as a result of deformation during movement of the debris avalanche (Fig. 2 D). Conjugated joint patterns have been observed in the massive part of block facies within several debris avalanche deposits (Ui et al. 1986a; Fig. 2 B of Ui and Glicken 1986), suggesting the existence of local compressional stress.

The direction of a cooling joint at the surface of a juvenile fragment within a nuée ardente deposit is normal to the surface of the juvenile clast. The population of cracks is greater at the quenched surface (Fig. 2 E).

The fracture pattern due to a meteorite impact is similar to the block facies of a debris avalanche deposit made of massive formations (Fig. 2 C), although such examples within volcanic formations have not been reported.

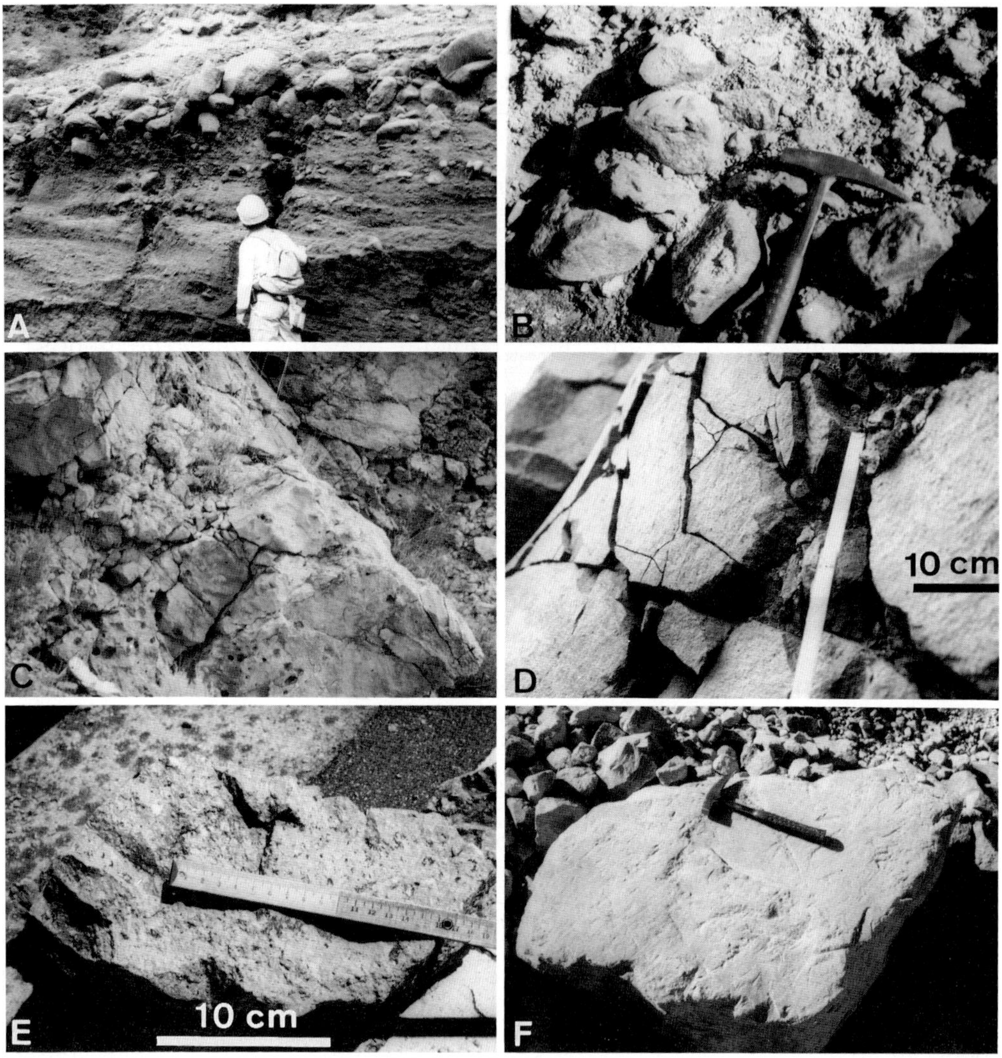

Fig. 2 A–F. A A section of a small-scale lahar deposit showing reverse grading. Near Gongen-yama, Sakurajima volcano, western Japan. B A section of lahar deposit at Mangatoetoenui River Gorge, east of Mount Ruapehu, New Zealand. C Fracture pattern at the rim of the Meteor Crater, Arizona. D Fractures within block facies formed normal to the cooling joint of a lava flow; same exposures as Fig. 1 **F.** E Cooling crack on the surface of a juvenile fragment dug out of the Nawa nuée ardente deposit, Daisen volcano, western Japan. F Streaks and grooves observed on the surface of a large clast within a moraine. Bunny Flat, south of Mount Shasta, California

5 Morphology of Individual Clasts

Roundness of clasts, and the presence of percussion marks or streaks on the surface of clasts, are keys to discriminate the origin of the deposit. Clasts within massive and fractured debris avalanche blocks are angular to subangular. The degree of roundness is variable in case of clasts within block facies made of volcaniclastic material. Clasts within matrix facies of debris avalanches may be slightly abraded. Roundness has been shown to increase towards the distal area in some debris avalanche deposits (Ui et al. 1986a). Well-rounded clasts may be abundant within matrix facies. Such clasts were probably picked up from the valley floor during the transportation process (Ui and Glicken 1986).

The degree of roundness of juvenile clasts within a nuée ardente deposit depends on the vesicularity of the clasts. Some deposits include angular to subangular and faceted juvenile clasts. Part of such clasts shows shrinkage cracks on its surface. Some other deposits include subrounded to rounded and internally vesiculated juvenile clasts (Ui 1975).

Percussion marks are commonly found only at the junction of two cracks within the block facies of a debris avalanche deposit (Fig. 2 D) (Ui et al. 1986a). Smaller fragments, formed by percussion, are usually enclosed among larger clasts. Percussion marks are also common on the surface of larger clasts within lahar deposits.

Scratches are also associated with percussion marks in the case of a lahar deposit. Scratches form due to collision between individual larger clasts during flowage. Such a structure is visible only in a newly formed deposit. Some clasts within moraines show faceted and polished surfaces with grooves (Fig. 2 F).

6 Summary

It is possible to discriminate debris avalanche deposits from other volcaniclastic formations by means of their irregular surface topography and the existence of block facies and matrix facies. Jigsaw cracks are common within the block facies of some deposits. Percussion marks are common, but are limited only to the junction between two crack planes. The mixing process in matrix facies can be incomplete. Lahar deposits consist of a blended mixture of various clasts. The amount of accidental clasts picked up during transportation is larger for a lahar than for a debris avalanche. The surface of a lahar deposit is rather flat. Nuée ardente deposits are characterized by juvenile clasts with shrinkage cracks. Clasts within moraines are commonly faceted and grooved.

Acknowledgments. The author would like to thank P.W. Lipman for his valuable comments, when this paper was read at the IAVCEI meeting. Many of the author's colleagues have assisted in fieldwork for this study. Among them, S. Kawachi and H. Glicken are particularly thanked for helpful and stimulating discussions. Reviews by H. Glicken, J.H. Latter and P.C. Froggatt helped to improve and clarify the manuscript. Expenses for this study in New Zealand, United States and Colombia were supported by a grant from the Ministry of Education, Science and Culture, Japan (No. 58-KO-186) and by Tokyo Broadcasting System Inc. Expenses of this research in Japan were partially supported by a Scientific Research Grant, Ministry of Education, Science and Culture, Japan (No. 58020012, Principal Investigator, S. Aramaki, University of Tokyo).

References

Aramaki S, Yamasaki M (1963) Pyroclastic flows in Japan. Bull Volcanol 26:89-99

Bossard L (1928) Origin of conical hills in the neighbourhood of Mount Egmont. New Zealand J Sci Tech 10:119-124

Crandell DR (1971) Postglacial lahars from Mount Rainier Volcano, Washington. US Geol Surv Prof Paper 677:1-73

Crandell DR, Miller CD, Glicken HX, Christiansen RL, Newhall CG (1984) Catastrophic debris avalanche from ancestral Mount Shasta volcano, California. Geology 12:143-146

Fitzharris BB, Mansergh GD, Soons JM (1982) Basins and lowlands of the South Island. In: Soons JM, Selby MJ (eds) Landforms of New Zealand. Longman Paul, Auckland, pp 343-366

Francis PW, Gardeweg M, Ramirez CF, Rotherry DA (1985) Catastrophic debris avalanche deposit of Socompa volcano, northern Chile. Geology 13:600-603

Glicken H (1986) Rockslide-debris avalanche of May 18, 1980, Mount St. Helens Volcano, Washington. Ph. D. Dissertation, University of California, Santa Barbara, pp 1- 303; also in press as US Geol Surv Prof Paper

Grange LI (1931) Conical hills on Egmont and Ruapehu volcanoes. New Zealand J Sci Tech 13:376-384

Kofu Basin Quaternary Research Group (1969) On the proposal of the Nirasaki type pyroclastic flow and the genesis of conical hills. J Geol Soc Jpn 75:485-489 (in Japanese)

Kuno H (1941) Characteristics of deposits formed by pumice flows and those by ejected pumice. Bull Earthq Res Inst 19:144-149

Lipman PW (1976) Caldera-collapse breccia in the western San Juan Mountains, Colorado. Geol Soc Am Bull 87:1397-1410

Mason AC, Foster HL (1956) Extruded mudflow hills of Nirasaki, Japan. J Geol 64:75-83

Mimura K, Fujimoto U, Hyuga T, Ichikawa S, Kawachi S, Taneichi M (1971) Natural remanent magnetism of Nirasaki pyroclastic flow. Bull Volcanol Soc Jpn 15:146-147 (in Japanese)

Mimura K, Kawachi S, Fujimoto U, Taneichi M, Hyuga T, Ichikawa S, Koizumi M (1982) Debris avalanche hills and their natural remanent magnetization - Nirasaki debris avalanche, central Japan. J Geol Soc Jpn 88:653-663 (in Japanese)

Mizuno Y (1958) Geomorphology of Okinajima mudflow area - special reference to mudflow hill. Tohoku Geography 11(1):22-24 (in Japanese)

Nakamura Y (1978) Geology and petrology of Bandai and Nekoma volcanoes. Sci Rep, Tohoku Univ, Ser 3, 14:68-119

Neall VE (1976) Lahars as major geological hazards, Bull Int Assoc Eng Geol 14:233-240

Rowley PD, Kuntz MA, Macleod NS (1981) Pyroclastic flow deposit. US Geol Surv Prof Paper 1250:489-512

Shreve RL (1968) The Blackhawk landslide. Geol Soc Am Spec Paper 108:47

Siebert L (1984) Large volcanic debris avalanches: characteristics of source areas, deposits, and associated eruptions. J Volcanol Geotherm Res 22:163-197

Tsujimura T, Kiuchi S (1936) Geomorphology of volcanic mud flow deposit. Kagaku 6:288-290 (in Japanese)

Ui T (1975) Origin of so-called mud flow deposit at the northwestern foot of Gassan Volcano. In: Dewa-Sanzan Hayama, 15-20 (in Japanese)

Ui T (1983) Volcanic dry avalanche deposits - Identification and comparison with nonvolcanic debris stream deposits. J Volcanol Geotherm Res 18:135-150

Ui T, Glicken H (1986) Internal structural variations in a debris-avalanche deposit from ancestral Mount Shasta, California, USA. Bull Volcanol 48:189-194

Ui T, Kawachi S, Neall VE (1986a) Fragmentation of debris avalanche material during flowage - evidence from the Pungarehu Formation, Mount Egmont, New Zealand. J Volcanol Geotherm Res 27:255-264

Ui T, Yamamoto H, Suzuki-Kamata K (1986b) Characterization of debris avalanche deposits in Japan. J Volcanol Geotherm Res 29:231-243

Varnes DJ(1978) Slope movement types and processes. In: Schuster RL, Krizek RJ (eds) Landslides analysis and control. Natl Acad Sci, pp 11-33

Voight B, Janda RJ, Glicken H, Douglass PM (1983) Nature and mechanics of the Mount St. Helens rockslide-avalanche of 18 May 1980. Geotecnique 33:243-273

Williams H (1949) Geology of Macdoel quadrangle. California Div Mines Bull 151:7-60

Yamagishi H (1977) Pyroclastic flow deposit from the Shikaribetsu Volcanoes, Hokkaido, Japan. Rep Geol Surv Hokkaido 49:37-48 (in Japanese)

The 1984 Collapse and Debris Avalanche Deposits of Ontake Volcano, Central Japan

K. ENDO, M. SUMITA, M. MACHIDA and M. FURUICHI[1]

Abstract

The Western Nagano Prefecture Earthquake (M = 6.8) triggered the collapse of the southern ridge of Ontake Volcano, central Japan, on 14 September 1984. The collapsed rock fragments, with a volume estimated as 3.4×10^7 m^3, became a debris avalanche which travelled a distance of 13 km, descending a total height of 1600 m. By means of geological observations and photo interpretations, three stratigraphic units of the debris avalanche/debris flow deposits were distinguished: Flows I, II, and III. These consist of lithofacies A (breccia and its powder), B (less fluid mixture of breccia and mud), and C (very fluid mixture of breccia and mud). Flow I is the main debris avalanche, which left hummocky hills in the deposition area and longitudinal and transverse ridges on the lava plateaus that were overflowed. Lithofacies, preserved within Flow I on the plateaus overflowed, are characterized by facies A in the proximal parts, and by alternations of facies A and B in the more distal parts. The transformation in lithofacies of Flow I is due to incorporation of vegetation and water-bearing surface materials on the way to the deposition area. Flow II consists of the secondary flow deposits, composed of facies B, which overlie Flow I in the valley floor, and are attached cohesively to the steep valley slopes. Flow III, composed of facies C, consists of fluid debris flow deposits, found only in the valley floor. It eroded Flows I and II, creating flat surfaces. Wet and mobile materials, incorporated into the avalanche and debris flows during their passage to the deposition area, occupy 40% of the total volume of those deposits, which was estimated at 5.6×10^7 m^3.

Introduction

On 14 September 1984 a number of debris avalanches and debris flows were triggered around the southern flank of Ontake Volcano (35° 53'N, 137° 29'E, 3063 m) in Nagano Prefecture, central Japan (Fig. 1) by the Western Nagano Prefecture Earthquake (Naganoken-seibu Earthquake:35° 49'N, 137° 34'E, 2 km in depth). The earthquake had a magnitude of 6.8, and took place

[1] Department of Applied Earth Sciences, Nihon University, Sakura-jousui, 3-25-40, Setagaya-ku, Tokyo 156, Japan

IAVCEI Proceedings in Volcanology 1
J.H. Latter (Ed.), Volcanic Hazards

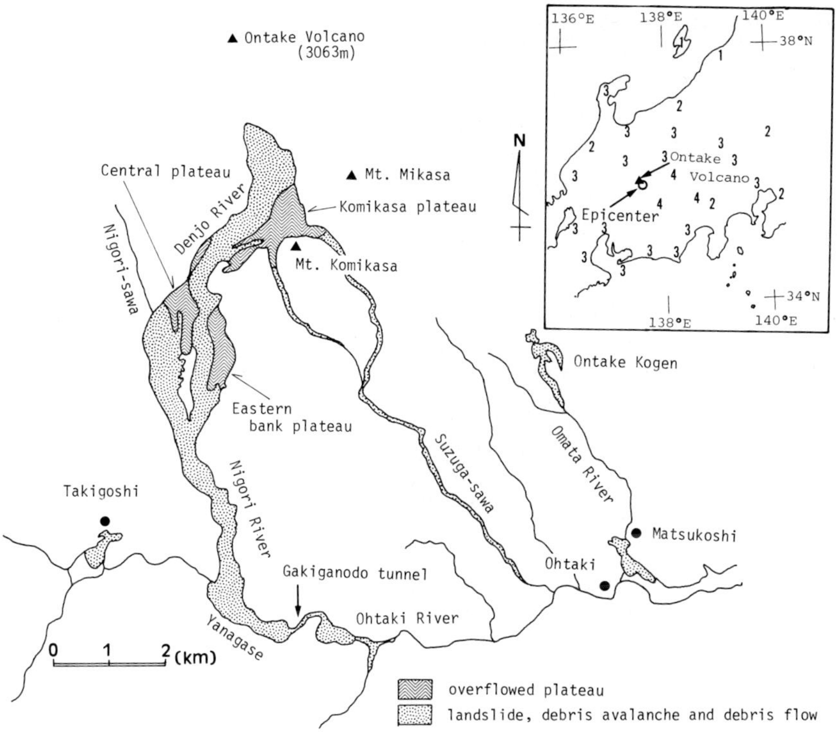

Fig. 1. Locality map showing the source area and distribution of debris avalanches and debris flows triggered by the Western Nagano Prefecture Earthquake. The **inset** shows the locality of the epicenter and seismic intensity of the earthquake

at 8:48.9 A.M. on 14 September 1984, 9 or 10 km south-east of the volcano. The seismic intensity of the main shock (according to the Meteorological Bureau of Japan) is shown in Fig. 1. Masaki et al. (1985) inferred that the intensity around the epicenter was 6, based on seismic records of ground motion, damage to houses and ground, overturns of gravestones, and a questionnaire survey. Aftershocks occurred very frequently for a week after the main shock. About 4700 earthquakes were recorded by Nagoya University in the period up to 31 October, from the date of the earthquake (Iida 1985).

The collapse of one of the southern ridges of Ontake Volcano gave rise to the largest of these debris avalanche and debris flow deposits, and is called the "Ontake Collapse" (or "Ontake Landslide"). The top of the break-away rim is about 2550 m in altitude, situated on a ridge adjacent to the head of the Denjo River. The ridge is composed of andesitic lava, and welded and non-welded scoriaceous pyroclasts, belonging to the Upper Group of the Younger Ontake Tephra Formation (Kobayashi 1974, 1985). The collapsed area is oval in shape, measuring 1300 x 450 m, and covers an area of about 4×10^5 m^2 (Fig. 2). The total volume of the collapse was estimated as 3.4×10^7 m^3 (Nagaoka et al. 1985).

Fig. 2. Vertical view of the source area of the debris avalanche, from an aerial photo by Asahi Koyo Co. (locality shown in Fig. 3)

The Ontake Collapse gave rise to a debris avalanche which flowed down the valley of the Denjo River for about 5 km, down the valley of the Nigori River for about another 4 km, and buried the valley of the Ohtaki River for a distance of about 3 km (Fig. 1). The maximum distance which the avalanche travelled was about 13 km, and the maximum height through which it fell was about 1600 m (Fig. 3). Ten people were killed by the avalanche.

Many Japanese researchers surveyed various aspects of the avalanche, and discussed it on the basis of the field evidence, eyewitness accounts, and analytical data (e.g. Okuda et al. 1985; Suwa et al. 1985; Hirano et al. 1985; Moriya 1985; Kisodani Subgroup 1985; Nagaoka et al. 1985; Kobayashi 1985; Matsuda and Ariyama 1985; Momikura et al. 1985; Tanaka 1985; Tanaka et al. 1985).

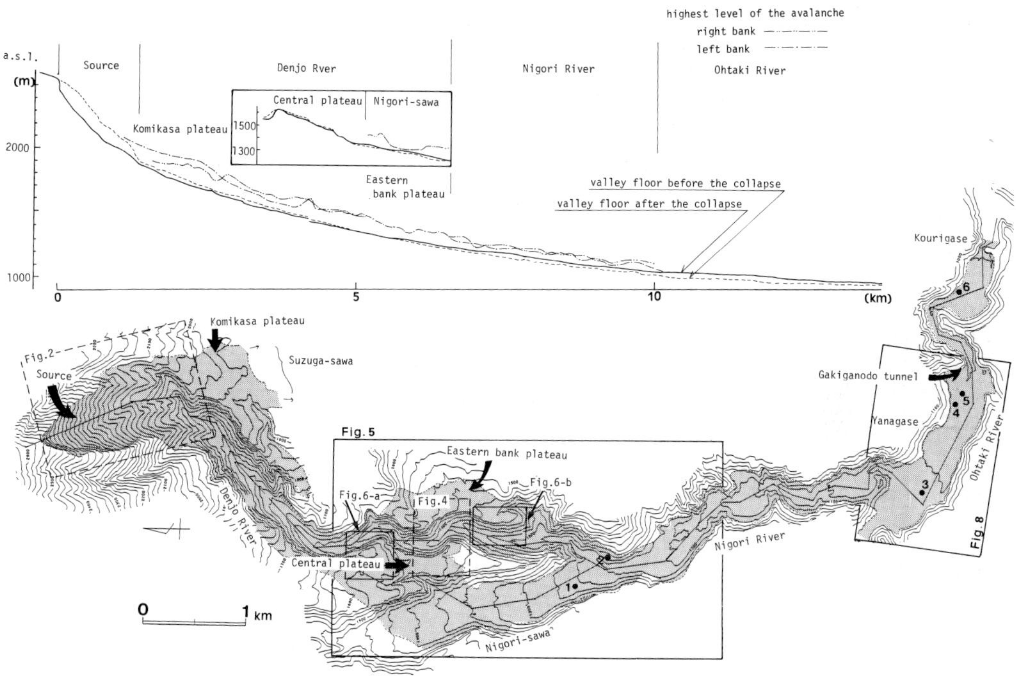

Fig. 3. Longitudinal profiles and distribution map of the Ontake Collapse and the debris avalanche. Boxes in the map give locations of Figs. 2, 4, 5, 6 a, 6 b, and 8. 1 - 6 Sample localities for rock fragment composition analysis (Fig. 13)

In this chapter, the lithology, microtopography, grain size, and flow units of the debris avalanche deposits are described and interpreted. A preliminary version of this work is Endo et al. (1986).

2 Eyewitness Accounts of the Avalanche Deposits

2.1 Initiation of the Debris Avalanche

The debris avalanche occurred immediately after the earthquake, and terminated before it reached the village of Ohtaki. Therefore, only a very few people actually saw it emplaced.

The initiation of the collapse was not observed directly, as it was foggy. One person, however, about 1 km east of the source, heard a great sound like thunder, at 8:49 A.M. on 14 September (Okuda et al. 1985). This suggests that the initiation of the collapse was almost simultaneous with the earthquake.

2.2 Occurrence Time and Velocity of the Debris Avalanche

At Gakiganodo tunnel, Yanagase (Fig. 3), about 10 km below the source, two people saw the avalanche flowing down and burying the valley of the Ohtaki River at 8:56 A.M., and between 9:00 and 9:05 A.M. it struck a forestry workshop. From this, Awata et al. (1984), Okuda et al. (1985), and the Public Works Research Institute (1985) considered that the avalanche took about 7 min to flow a distance of about 10 km after the collapse took place. Thus, the average velocity was about 70 to 90 km h^{-1}.

At Gakiganodo tunnel, one of the same two people observed that the deposition level of the avalanche was about 15 m lower than the level of the water-carrier bridge which was broken by an aftershock at 8:57 A.M. The bridge was about 38 m above the riverbed before the avalanche.

2.3 Second and Third Flows

At about 10:30 A.M., one of two eyewitnesses saw a debris flow running up and over the small ridge (marked with a small peak of 1113.9 m in Fig. 8) situated at the junction of the Nigori and the Ohtaki Rivers (Public Works Research Institute 1985). According to a video taken about 11:30 A.M., the deposition level had risen to 2-3 m below the water-carrier bridge, which was the final level that it reached.

A mud flow running down the valley of the Ohtaki River was also recorded by the video at this time (Public Works Research Institute 1985).

Judging from the eyewitness accounts mentioned above, the following can be summarized.

1. The main flow of the debris avalanche (Flow I), triggered directly by the earthquake, came to rest about 10 min after it was generated. Its average velocity was about 80 km h^{-1}.
2. A secondary flow (Flow II) passed down the Nigori River into the Ohtaki River at about 10:30 A.M. The deposition surface rose to about 10 m around the Gakiganodo tunnel.
3. One of the fluid debris flows ran down the Ohtaki River at about 11:30 A.M. Flow III may consist of a number of units of fluid, small-scale debris flows.

3 Lithologic Features of the Avalanche Deposits and the Associated Flows

3.1 Stratigraphic and Lithologic Classification

The deposits of the avalanche and associated flows can be divided stratigraphically into three units: Flow I, II, and III (Table 1). Their lithofacies are classified into three types (A, B, and C in Table 2). Lithofacies A corresponds to the avalanche deposit of Mount Shasta. Lithofacies B and C are un-

Table 1. Flow units and their lithofacies

	Near the source (Denjo River)	Around confluence of Denjo and Nigori Rivers	Yanagase Kouri-gase (Ohtaki River)
Flow I	Facies A	Facies A + B	Facies A + B
Flow II	Facies B	Facies B	Facies B
Flow III	Facies C	Facies C	Facies C

sorted mixtures of subrounded breccia and mud, and are equivalent to the matrix facies (Crandell et al. 1984). Lithofacies B, however, has rugged surfaces, whereas lithofacies C shows smooth surfaces.

Flow I, the main debris avalanche, is characterized by megablocks, hummocky surfaces in the deposition area, levees at lateral margins, and stripes on the overflowed plateau. The first three of these are major criteria for the identification of volcanic dry avalanche deposits (Ui 1983). Lithologic features of Flow I belong mainly to lithofacies A, and partly to lithofacies B. The Flow I deposit is well preserved as elongated stripes or ridges of blocks, on the lava plateau which it overflowed. Hummocky hills formed only on the valley floor in the deposition area.

Flow II, mainly characterized by lithofacies B, overlies the Flow I deposit. It is composed of a mixture of breccia and mud.

Flow III, composed of lithofacies C, cuts into Flows I and II to make a lower terrace.

Table 2. Classification of lithofacies

Facies A
Shows block structures of angular fragments of lava and welded pyroclasts which compose the source area: scarce in silt and clay

Facies B
Dark brown, humic, and mixed deposits, rich in mud, various kinds of weakly subrounded breccia, and crushed plant fragments: has rigid surfaces with many protrusions of timber

Facies C
Grayish-brown, mixed deposits, composed of poorly sorted sand and mud with fine breccia: more fluid than the other two facies: makes flat surfaces.

3.2 Along the Upper Parts of the Streams (Denjo River)

Komikasa Lava Plateau. The upper part of the collapsed body flowed over the Komikasa Plateau and left stripes (sometimes forming low ridges) composed of variously colored breccia and scoriaceous pyroclasts. Part of the debris which overflowed reached into the valley heads of Suzugasawa and produced small-volume debris flows.

Flow I: The stripes (or ridges) are generally elongated parallel to flow directions, or show reversed U-shapes. Whitish and purplish-gray stripes consist of angular fragments of andesitic lava, mainly with a diameter of 10 to 30 cm. Reddish stripes consist of scoriaceous pyroclasts which are partly welded. Each deposit is very thin; a thickness of only 15 to 60 cm was measured in two test pits around the central part of the plateau. Brown stripes include a mixture of various kinds of breccia, and blocks of weathered tephras and soils. Compared with the other areas described, blocks of weathered tephras and soils are well preserved. In the marginal area of the overflowed plateau, weathered tephras and soils are dominant, as blocks and pieces of decomposed matrix which form a marginal facies of A.

Flow II: A dark brown muddy breccia is found crossing the southern part of the plateau in narrow bands from north to south. Since it covers deposits of Flow I, it seems that part of a secondary debris flow in the Denjo valley splashed onto the plateau. The deposits contain many large tree trunks. The distribution and the flow direction suggest that a secondary flow was generated from the source, possibly triggered by aftershocks.

Valley of the Denjo River. The main body of the collapsed blocks ran down into the valley, travelling at an estimated speed of 90 to 95 km h^{-1}. The flow initiated deepening and widening of the valley. As a result, bared and scored surfaces of the older lava flows are exposed along the whole length of the steep slopes of the valley (Fig. 4).

Northern Central Lava Plateau. When the most rapidly moving part of the avalanche turned south at the first sharp bend of the Denjo River, part of it overflowed onto the northern Central Lava Plateau, where it left longitudinal stripes (ridges), and flowed into the valley of the Nigori-sawa (Figs. 1, 4, 5, and 6).

Flow I: The plateau was entirely covered by thin deposits of breccia which show longitudinal stripes (mostly forming low ridges) parallel to the flow direction (Fig. 6): each are about 5 to 20 m wide and 250 to 400 m long. The width increases towards the lower parts of the streams. Individual stripes are pale bluish-gray, gray, reddish, purplish-gray, or black in color, these being the colors of the breccia composing each stripe. Brown-colored stripes, which occur between the stripes mentioned above, are composed of a mixture of blackish and reddish or purplish-gray breccia. Along the northern margin of the

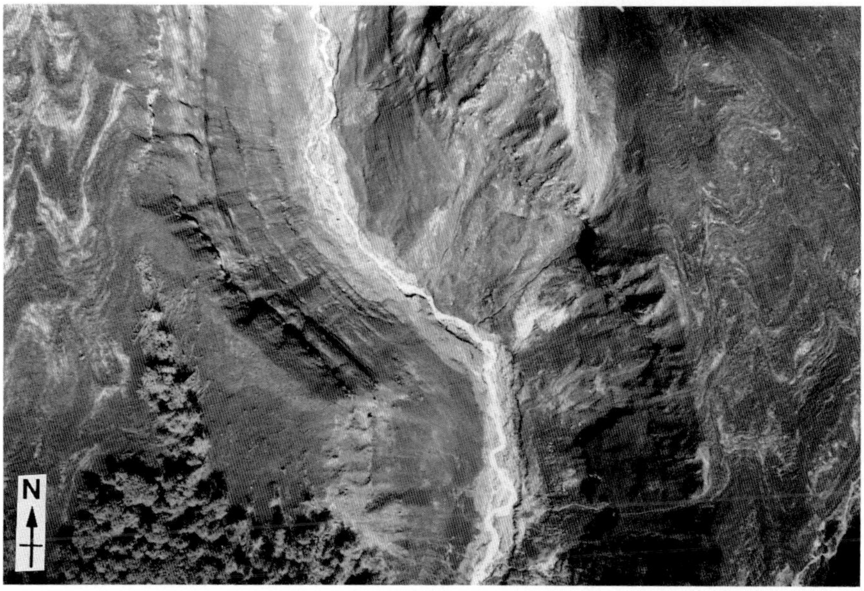

Fig. 4. Vertical view of scored valley slopes of the Denjo River (**center**) and overflowed lava plateaus (**left:** Central Plateau, **right:** Eastern Bank Plateau), from an aerial photo by Asahi Koyo Co. (locality shown in Fig. 3)

Flow I deposits, surface materials such as soil and weathered tephra blocks are abundant. The occurrence of the longitudinal stripes suggests that blocks in the high velocity material that overflowed were dragged and stretched along the flow direction to form elongated stripes (or ridges), and that the avalanche was not wet enough for all the various blocks to become thoroughly mixed.

Eastern Bank Lava Plateau. At the second sharp bend, the avalanche and the secondary flow climbed over a small ridge and the Eastern Bank Plateau. The avalanche (Flow I) plucked out forests and surface soils, and left thin deposits on the plateau to form wavy ridges with a number of wrinkles (Fig. 5). Flow II crossed the center of the plateau (Fig. 5).

Flow I: The avalanche deposits formed wavy ridges with wrinkles which consist of alternations of whitish-gray breccia (facies A) and a brown mixture (facies B), as shown in Figs. 4 and 6. Ridges and wrinkles are perpendicular to the flow direction. On the whole, facies B is dominant, is rich in timber, and comprises small blocks of surface materials. Towards the margin, the matrix of the deposits becomes rich in brown weathered tephra (marginal facies of A): a levee with a height of 2 to 3 m is developed along the margin.

Flow II: Thin dark brown deposits (facies B), composed of breccia with a uniform muddy matrix, were deposited by Flow II in a

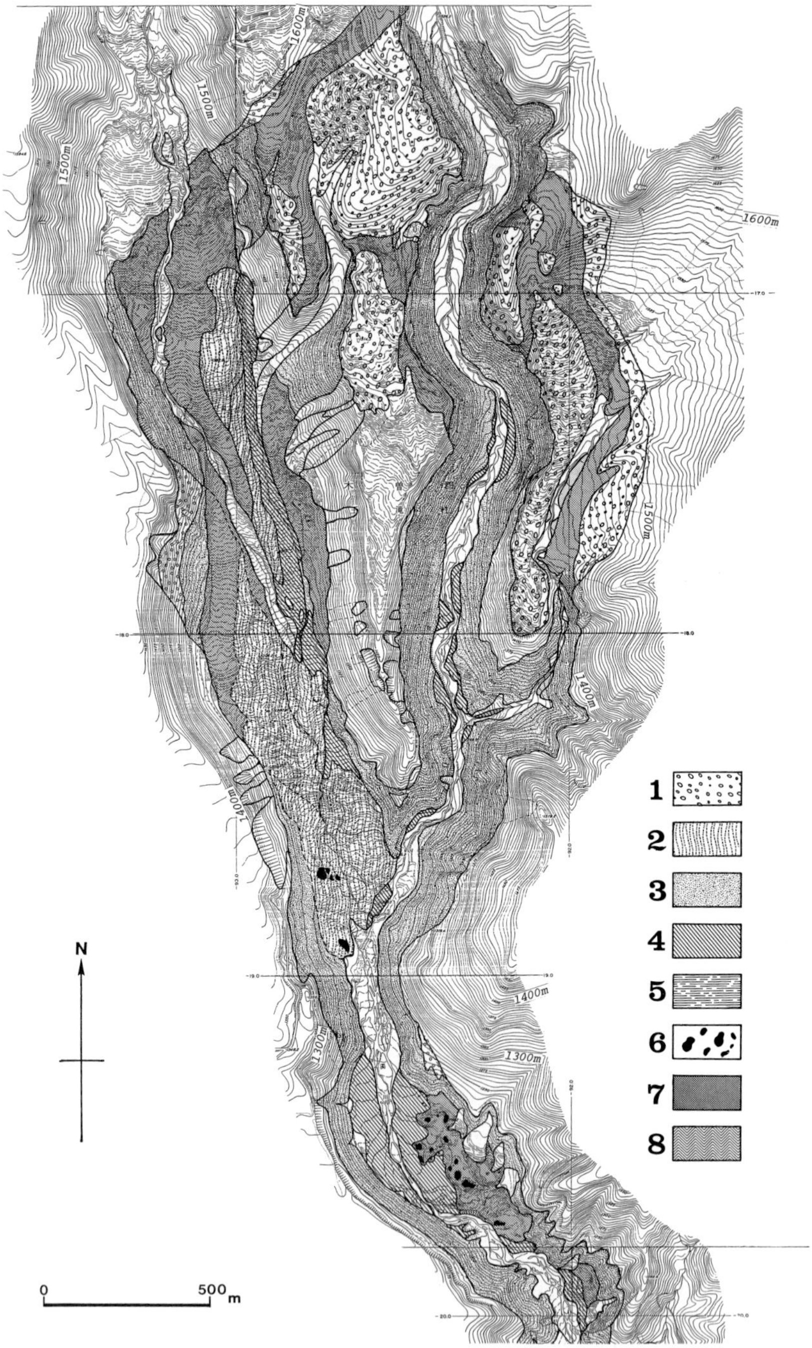
1600m
1500m
1500m
1600m
1500m
1400m
1400m
1400m
1300m
1300m
N
1
2
3
4
5
6
7
8
0
500 m

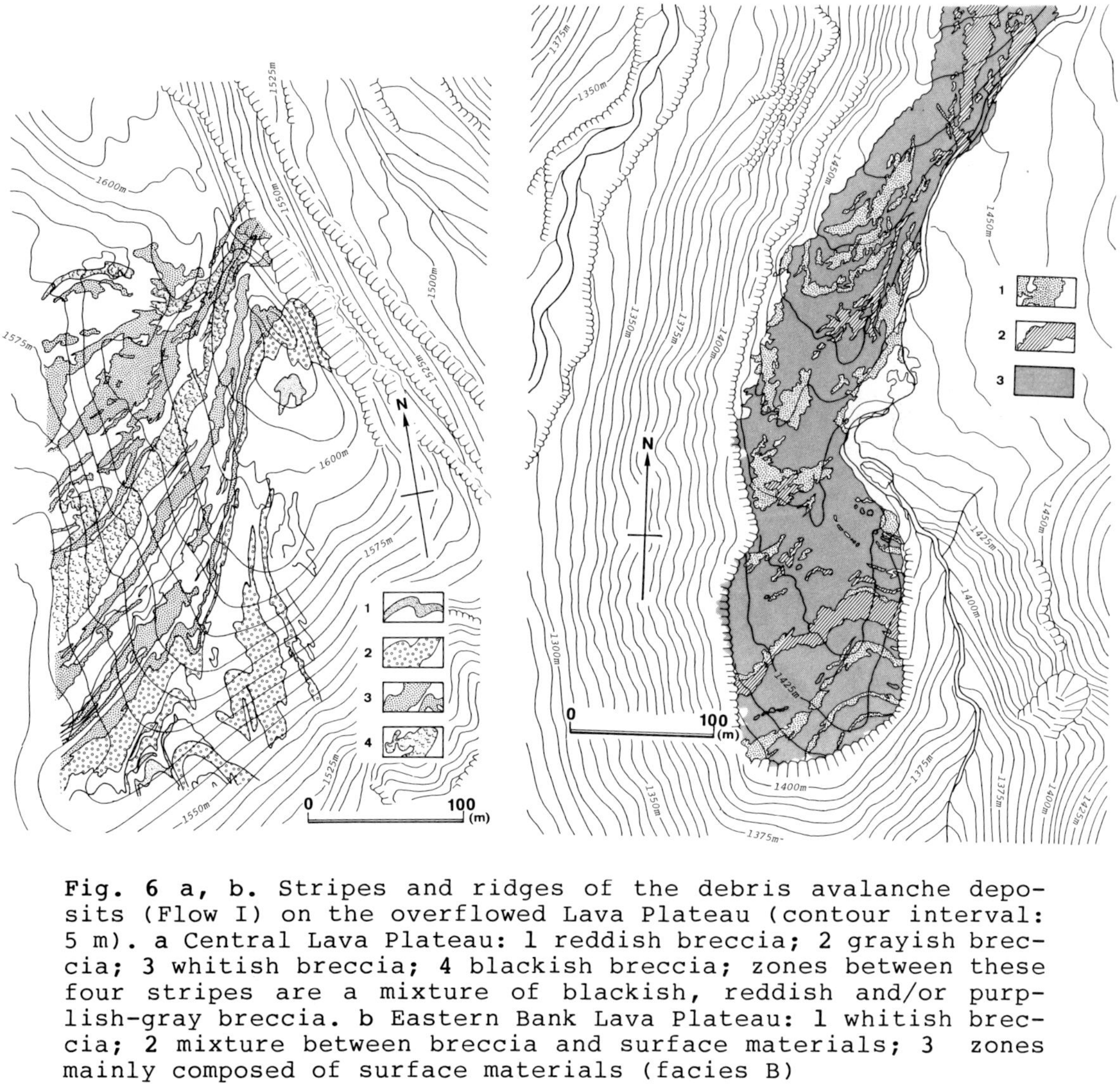

b

Fig. 6 a, b. Stripes and ridges of the debris avalanche deposits (Flow I) on the overflowed Lava Plateau (contour interval: 5 m). a Central Lava Plateau: 1 reddish breccia; 2 grayish breccia; 3 whitish breccia; 4 blackish breccia; zones between these four stripes are a mixture of blackish, reddish and/or purplish-gray breccia. b Eastern Bank Lava Plateau: 1 whitish breccia; 2 mixture between breccia and surface materials; 3 zones mainly composed of surface materials (facies B)

Fig. 5. Distribution of the debris avalanche and associated deposits around the junction of the Denjo and the Nigori Rivers (topographical map by the Geographical Survey Institute): contour interval: 5 m. Legend (for Figs. 5 and 8): 1 Flow I (on lava plateau and valley slope); 2 Flow I mixed with Flow II (facies B); 3 Flow II (on valley slope); 4 Flow III; 5 pond; 6 hummocky hills of Flow I; 7 Flow II (facies B) overlying Flow I in deposition area; 8 levee of Flow I covered with Flow II

narrow band across the plateau. Similar deposits are attached to the valley slope surrounding the southern Eastern Bank Plateau, where the highest level of Flow II is clearly marked (Fig. 5).

Southern Central Lava Plateau. When it reached the third bend, the avalanche overflowed again over the southern Central Plateau, leaving wavy ridges with a number of wrinkles (Fig. 4).

Flow I: The wavy ridges, perpendicular to the flow direction, consist of alternations between variously colored breccia (facies A) and brown material consisting of a mixture of grayish breccia and surface materials (facies B). Breccia of facies A is dominantly whitish-gray, but bluish-gray, reddish, black, and purplish-gray breccia also make up the ridges. Many wrinkles (pressure ridges) are apparent, especially in the brown zones. These features, as well as those of the Eastern Bank Lava Plateau, indicate that the debris which overflowed plucked out surface materials, including timber, and moved to the south comparatively slowly, almost coming to a stop to form wavy pressure ridges and wrinkles.

Flow II: Brownish massive mixtures partly overlie the deposits of Flow I. This suggests stratigraphically that a secondary flow overflowed onto the plateau after the deposition of Flow I.

3.3 Along the Middle Courses of the Streams (Nigori-sawa and Nigori Rivers)

Part of the debris avalanche which passed across the Central Plateau came to rest mostly in the valley of the Nigori-sawa. Deposition here was helped by the fact that the main flow from the Denjo River had dammed the gorge of the Nigori River, at a point a little lower than the confluence of the Denjo River and the Nigori-sawa, the main flow extending back a short way into the Nigori-sawa (Fig. 1). As a result, a number of wrinkles (pressure ridges) formed on the deposits of Flow I in the Nigori-sawa, and hummocky hills were formed near the confluence. Along the Nigori River, below this area, both erosion and deposition occurred. Most deposition surfaces, including hummocky hills, were covered with brown muddy deposits of Flow II. Distribution of Flow II deposits, attached to the valley slopes at sharp bends, indicates that the surface of Flow II was remarkably tilted, so that the highest traces of the debris flow (Flow II) marked on the slopes intersect those of Flow I.

Flow I: Deposits in the Nigori-sawa are characterized by a weak mixing of breccia separated from blocks with muddy materials. Near the confluence, deposits of hummocky hills are composed of breccia and/or pyroclasts, and in some places thin layers or lenses of humic materials are intercalated.

Flow II: Deposits of Flow II consist of an unsorted mixture of a muddy matrix and breccia, covering hummocky hills and attaching to valley slopes. Compared with facies B deposits of Flow

I, facies B deposits of Flow II are well mixed, and richer in breccia and twigs.

Nigori River. The debris avalanche (Flow I) that flowed down the Nigori River is estimated to have travelled at a velocity of 50 to 70 km h^{-1}, as determined by calculations based on the tilted surface heights at sharp bends of the avalanche. Its headward portion probably eroded the riverbeds and valley slopes, and the middle portion and/or tails buried the valley, mainly behind the bends or the gorges. The maximum thickness of the deposits is about 40 m. The upper parts of the deposits were truncated by Flow III to form flat surfaces, except for protruding hummocky hills capped with facies B deposits of Flow II. In one of the deposition areas, deposits of facies A, more than 10 m thick, were exposed through deep undercutting by streams, which took place probably within a few days after the avalanche.

Figure 7 shows deposits of Flows I and II on the valley slope of the Nigori River about 400 m lower than the confluence with the Denjo River (locality number 2 in Fig. 3), and about 50 m above the riverbed. In this outcrop the deposits of facies A consist of an alternation of thin layers of breccia of pyroclasts and muddy surface materials of facies B. These deposits of Flow I are clearly covered by the mixed deposits of breccia and muddy matrix of Flow II.

Both Flows I and II overtopped the ridge, marked by the 1113.9 m peak, near the confluence of the Nigori and Ohtaki Rivers (Fig. 8), and flowed into the wide valley of the Ohtaki River.

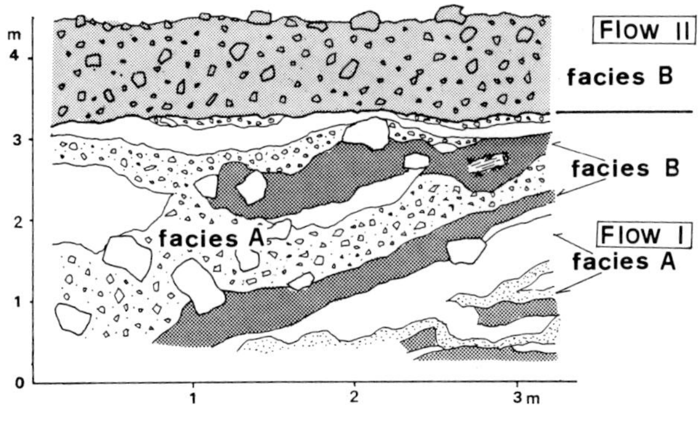

Fig. 7. Outcrop of the debris avalanche deposit attached to the valley slope of the Nigori River (locality No. 2 in Fig. 3)

3.4 Along the Ohtaki River

The valley of the Ohtaki River was buried by deposits of Flows I and II, up to a thickness of 20 to 40 m, as checked by bore-hole data (Public Work Research Institute 1985). The deposition surfaces are covered mainly by deposits of facies B and C (Figs. 8 and 9). Hummocky hills are occasionally found protruding from the surfaces, and levees and wrinkles parallel to the flow direction are distributed mainly along the margin (Fig. 8).

In Fig. 10, the deposits composing a hummocky hill in the eastern part of Yanagase (between locality numbers 3 and 4 in Fig. 3) are illustrated. The layered deposits of facies A generally consist of separated breccia and rock powder, although blocks bearing jigsaw structures are partly preserved. These deposits of Flow I are covered by facies B deposits of Flow II in which lenses of pyroclasts of facies A are incorporated.

In the western part of Yanagase, facies B deposits more than 5 m in thickness include a number of lenses of facies A pyroclasts. Although they are possibly deposits of Flow II, it is rather difficult in this area to discriminate facies B deposits of Flow II clearly from those making up the matrix of Flow I deposits.

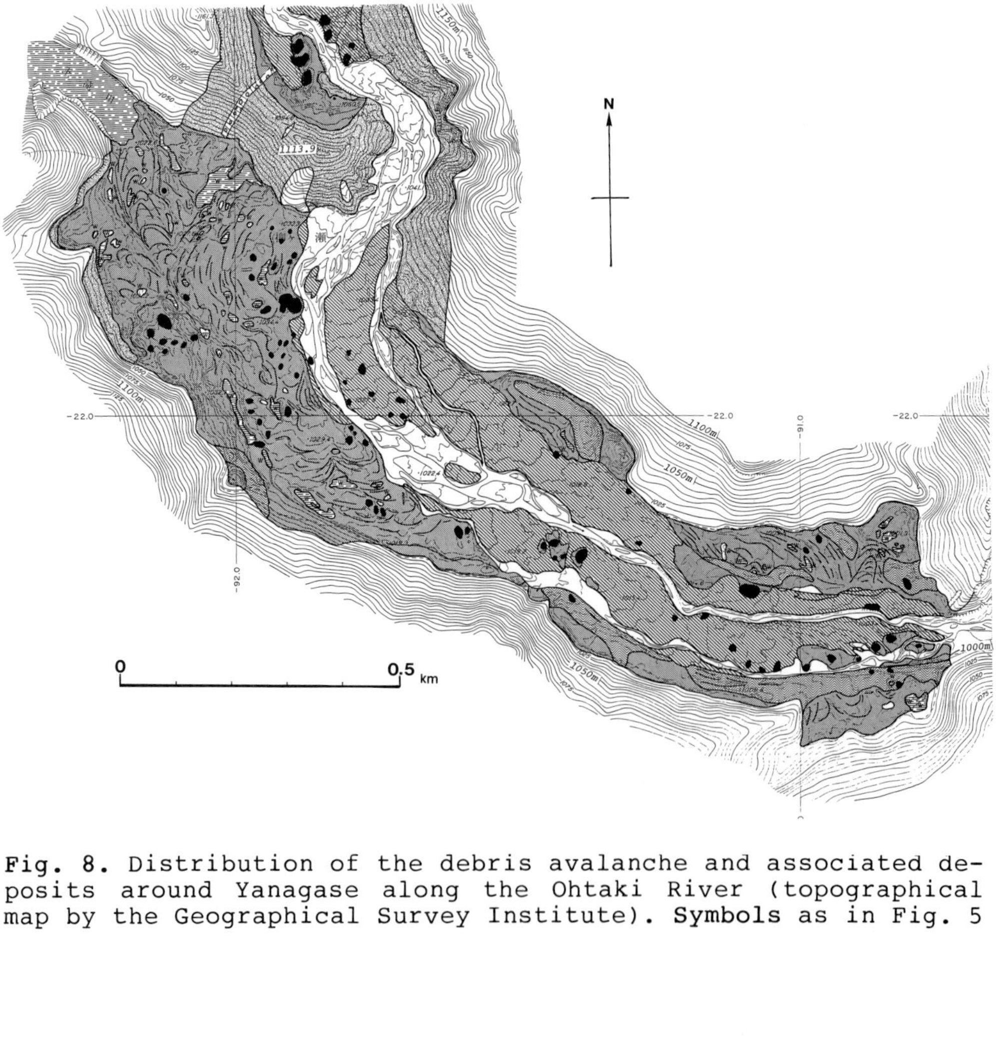

Fig. 8. Distribution of the debris avalanche and associated deposits around Yanagase along the Ohtaki River (topographical map by the Geographical Survey Institute). Symbols as in Fig. 5

Fig. 9. A hummocky hill of Flow I and rugged surface of Flow II in front. Yanagase (about 0.5 km ESE of locality No. 3 in Fig. 3)

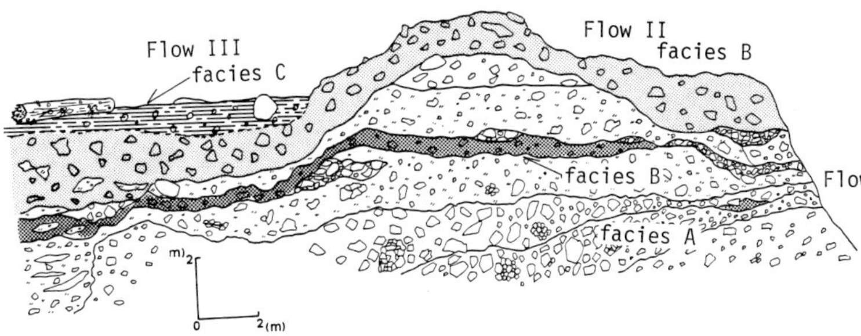

Fig. 10. Outcrop of a hummocky hill of Flow I, Yanagase (locality No. 5 in Fig. 3)

4 Sedimentological Characteristics of the Three Flow Units

4.1 Flow I

In the upper parts of the streams, Flow I deposits consist mainly of breccia or pyroclasts derived from the source area. From the middle to the lower parts of the streams, the lithologies of Flow I were transformed into a mixture of facies A and B. In the deposition area, in Yanagase, borehole data obtained by the Public Work Research Institute (1985) show the dominance of facies B or the matrix facies.

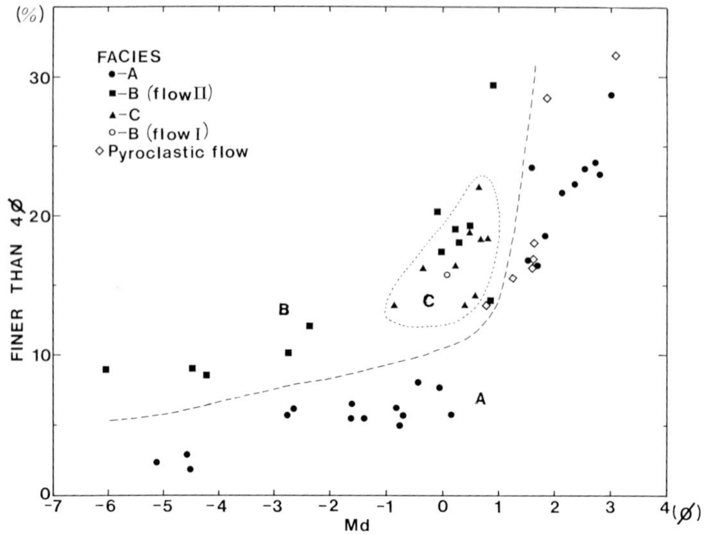

Fig. 11. Fraction finer than 4 vs Md for the samples of each lithofacies. Pyroclastic flow deposits from Kutcharo and Shikotsu Calderas are also plotted for comparison

As shown in Fig. 11, the grain size of facies A is characterized by the following two types: (1) the lava block type, scarce in fine-grained material: (2) the pyroclastic type, which was originally rich in fine-grained material. The grain size of the latter resembles that of pyroclastic flow deposits (Fig. 11). There are no systematic changes of grain size with distance from the source, except for the facies A breccia in the overflowed plateaus, where there is a remarkable scarcity of fine materials. Therefore, the grain size of facies A tends to reflect the types and conditions of rocks at the source.

In contrast to the facies B deposits in Flow II, the grain size of facies B deposits in Flow I shows great local differences in mud content.

4.2 Flow II

Lithofacies B from Flow II is comparatively uniform; it is dark brown in color, with a muddy matrix, rich in fragments of crushed twigs and coarse-grained breccia. Figure 12 shows that crushed twigs are included only in deposits of facies B and C. It should be noted that the deposits of Flow II make rugged surfaces and wrinkles, with occasional occurrences of large protruding tree trunks. The clasts have a tendency to be slightly rounded and are spherically shaped, in contrast to the angular breccia of Flow I. Moreover, various kinds of rock clasts are included throughout the deposit, including gray, black, purplish-gray and red andesitic rocks, scoria, and basement rocks (Fig. 13). In the distal deposition area, Yanagase

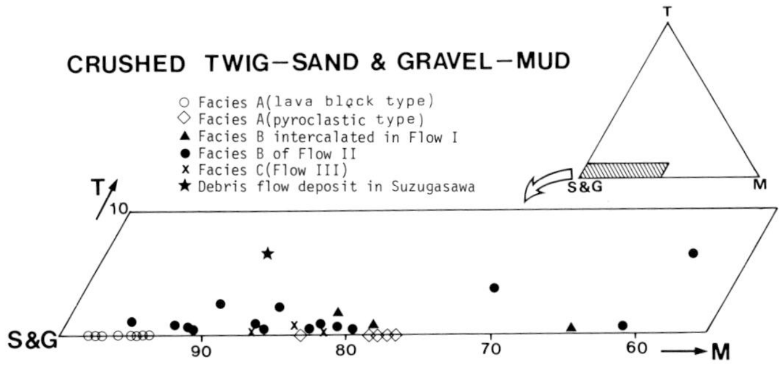

Fig. 12. Ternary diagram showing the composition of crushed twig (T), sand plus gravel (S & G), and mud fractions (M) for the samples of each lithofacies

and Kourigase, the deposit exhibits an increase in the content of coarse-grained breccia.

In the upper and middle parts of the streams, Flow II had a variety of courses (Fig. 14), gathering secondary debris flows in various places. These may have contributed to the growth of the flow.

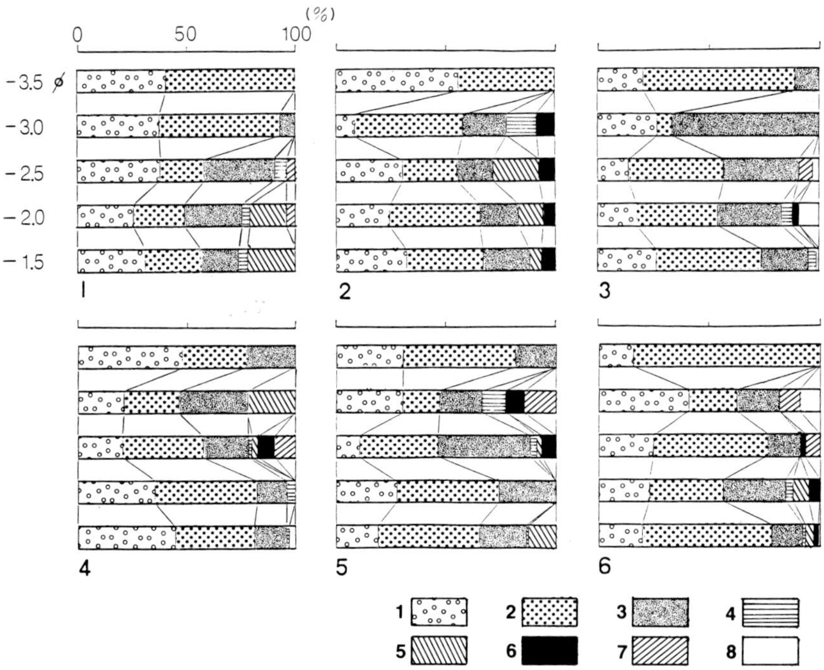

Fig. 13. Composition of rock fragments for Flow II samples (sampling localities 1-6 are shown in Fig. 3). 1 reddish breccia (including reddish scoria); 2 black breccia (including black scoria); 3 gray lava; 4 grayish-white lava; 5 black glassy rock; 6 pumice; 7 basement rock; 8 others

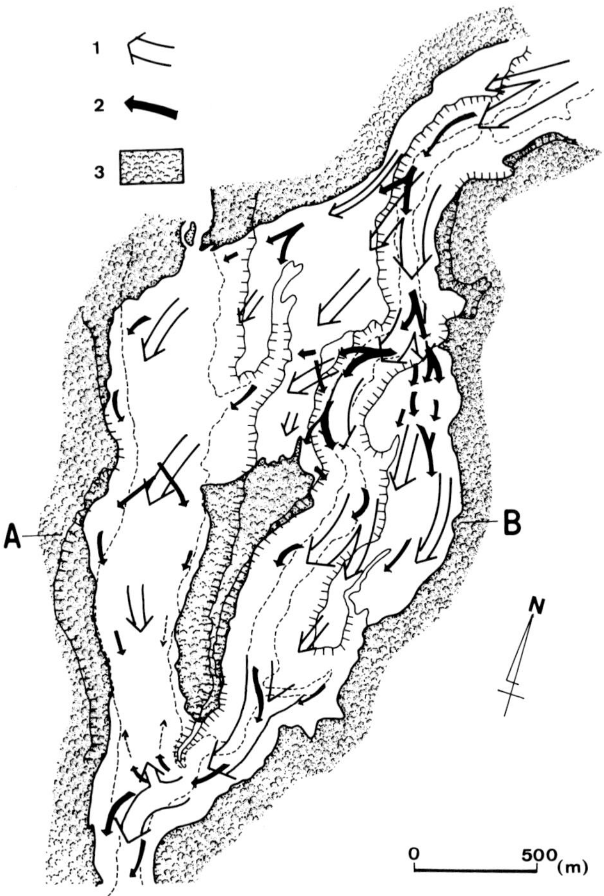

Fig. 14. Reconstructed flow directions of Flows I and II around the Denjo-Nigori junction. 1 Flow I; 2 Flow II; 3 vegetation cover remaining. Location same as Fig. 5

4.3 Flow III

Lithofacies C from Flow III is also uniform throughout the course of the deposit: it is dark gray in color, scarce in coarse-grained breccia, and dominated by a muddy matrix, and is less than 1 m in thickness, thus making flat surfaces. A large number of tree trunks lie on the surface parallel to the flow direction. These facts indicate that the deposits of Flow III were very fluid and were saturated with water.

5 Discussion

The debris avalanche of Flow I was initiated by the collapse of a ridge of Ontake Volcano. It started as a rockslide avalanche composed of blocks of andesitic lava fragments and scoriaceous pyroclasts, with a volume estimated as 3.4×10^7 m^3 at the

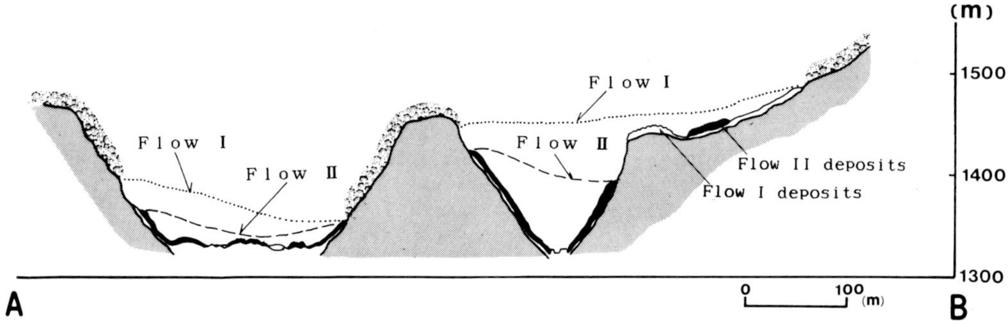

Fig. 15. Reconstructed profiles of Flows I and II along profile A-B shown in Fig. 14

source. During the emplacement, however, it was transformed into a mixture of drier blocks and wet matrix, by incorporation of vegetation and water-bearing surface materials. Cross-section profiles of Flows I and II were reconstructed in their middle courses (Fig. 15).

By comparing the earlier topographical maps with those made by the Geographical Survey Institute in 1985 after the earthquake, we can discuss the process of transformation quantitatively. This involves the degree to which the avalanche eroded and incorporated various kinds of surface materials, including river-bed materials, soils, weathered volcanic ashes and remains of vegetation on the valley slopes and plateaus, mainly along the Denjo and Nigori Rivers. The volume of surface materials plucked out from both the valley slopes and bottoms, and the plateau surfaces, was estimated as 2.31×10^7 m^3: this total volume is subdivided as follows (in 10^7 m^3): Denjo River 1.22, Komikasa Plateau 0.26, Central Plateau 0.17, Eastern Bank Plateau 0.15, and Nigori River 0.51. The total volume of the deposits in the valleys and on the plateaus is about 5.57×10^7 m^3: this is subdivided as follows (in 10^7 m^3): Nigori River 2.69, Ohtaki River 2.03, Nigori-sawa 0.62, and overflowed lava plateaus 0.23.

This estimate shows that wet and mobile materials, incorporated into the avalanche during its passage down the valleys, amounted to about 40% of the total volume. In the final deposition area, wet and muddy materials seem to have become dominant in the matrix, as shown by the borehole data.

Acknowledgements. We are deeply indebted to Dr. J.H. Latter and Dr. V.E. Neall for constructive review of the manuscript. We thank Prof. T. Kobayashi of Toyama University and Mr. E. Hayakawa of Phoenix Geoengineering Co., for their help with the fieldwork. We are grateful to the Geographical Survey Institute (Japan) for providing the topographical maps before and after the collapse. Financial support was provided by a grant from Nihon University to Endo.

References

Awata Y, Harayama S, and Endo H (1984) Preliminary report on the 1984 Nagano-ken Seibu Earthquake[1]. Chishitsu News 364:20-31 (in Japanese)

Crandell DR, Miller CD, Glicken H, Christiansen RL, Newhall CG (1984) Catastrophic debris avalanche from ancestral Mount Shasta volcano, California, Geology 12:143-146

Endo K, Sumita M, Machida M, Furuichi M (1986) Debris avalanche deposits by the 1984 collapse of Ontake Volcano, central Japan. Volc Prod and Quatern Geol 1:27-49

Hirano M, Ishii T, Fujita T, Okuda S (1985) Geomorphological and geological characteristics of 1984 landslide hazard in Ohtaki Village, Nagano Prefecture, Japan. Disaster Prevention Res Inst. Kyoto Univ 28 B-1:519-532 (in Japanese with English abstract)

Iida K (1985) Summary of the investigation of earthquake and damage caused by the Western Nagano Prefecture Earthquake, 1984. Synthetic Invest Earthq Damage Western Nagano Prefecture Earthq 1984: 1-10 (in Japanese)

Kisodani Subgroup, Matsumoto Basin Collaborative Research Group (1985) The land-slide and mud-flow disasters caused by the Naganoken-seibu Earthquake and the Quaternary geology of the southern slope of Mt. Ontake. Earth Sci 39:89-104 (in Japanese with English abstract)

Kobayashi T (1974) The petrochemical characteristics of Ontake Volcano. J Coll Lib Arts, Toyama Univ 7:71-85

Kobayashi T (1985) Geohistory of Ontake Volcano and Geology of Collapsed area by Naganoken-seibu Earthquake[1]. Kazantai-nokaitai oyobi soreni-tomonau Dosha-idou (Jpn Geomorphological Union): 48-58 (in Japanese)

Masaki K, Taniguchi H, Ikeda K, Muramatsu I, Taga N, Togashi Y (1985) Intensity distribution based on seismic records of ground motion, damage to houses and ground, overturns of gravestones, and questionnaire survey. Synthetic Invest Earthq Damage Western Nagano Prefecture Earthq 1984: 83-88 (in Japanese)

Matsuda T, Ariyama T (1985) Debris avalanche deposits of Ontake Volcano, caused by the 1984 Western Nagano Prefecture Earthquake. Bull Earthq Res Inst, Univ Tokyo 60:281-316 (in Japanese with English abstract)

Momikura Y, Yasuda S, Sakaki Y (1985) A prediction method for landslides by an earthquake and its appraisal based on the Naganoken-seibu Earthquake. Tsuchi-to-kiso 33:41-46 (in Japanese with English abstract)

Moriya I (1985) 1984 Large-scale rockslide of southern slope of Ontake Volcano and the debris avalanche deposit[1]. Chikyu (Monthly) 7:369-373 (in Japanese)

Nagaoka M, Hoshino Y, Takei Y, Ichikawa K, Saito F (1985) Geomorphological characteristics and factors of the large-scale landslide and detritus flow on the south slope of Ontake Volcano caused by the Naganoken-seibu Earthquake, 1984. UJNR 17th, Tsukuba (in Japanese)

---

[1] English title translated by K. Endo

Okuda S, Okunishi K, Suwa H, Yokoyama K, Yoshioka R (1985) Restoration of motion of debris avalanche at Mt. Ontake in 1984 and some discussions on its moving state. Disaster Prevention Res Inst Kyoto Univ 28 B-1:186-199 (in Japanese with English abstract)

Public Work Research Institute (1985) Behaviour of debris associated with the Ontake Collapse[1]. Kazan-tai-no-kaitai oyobi soreni-tomonau Dosha-idou (Jpn Geomorphological Union): 88-101 (in Japanese)

Suwa H, Okunishi K, Okuda S, Takahashi H, Hasegawa H, Takada M, Takaya S (1985) Depositional characteristics of the debris avalanche at Mt. Ontake in 1984. Disaster Prevention Res Inst Kyoto Univ 28 B-1:108-121 (in Japanese with English abstract)

Tanaka K (1985) Features of slope failures induced by the Naganoken-seibu Earthquake, 1984. Tsuchi-to-kiso 33, 5-11 (in Japanese with English abstract)

Tanaka K, Ohyagi N, Inokuchi T, Uehara S, Takahashi H (1985) Rock avalanche and slope failures induced by the Naganoken-seibu Earthquake, 1984. UJNR, 17th, Tsukuba (in Japanese)

Ui T (1983) Volcanic dry avalanche deposits - identification and comparison with nonvolcanic debris stream deposits. J Volcanol Geotherm Res 18:135-150

---

[1] English title translated by K. Endo

Monitoring

The Detection Capability of Explosive Eruptions Using GMS Imagery, and the Behaviour of Dispersing Eruption Clouds

Y. SAWADA[1]

Abstract

The imagery taken by the Japanese Geostationary Meteorological Satellite (GMS) can detect eruption clouds generated by relatively large eruptions, and can trace the time variations of the dispersing eruption clouds. There are some problems to be solved with the further accumulation of observational data from GMS imagery, from the ground and from the air; these concern the accurate determination of the maximum altitude of an eruption cloud from the GMS infrared data and of the shape of an eruption cloud, and its recognition and identification from surrounding clouds.

The observations and the analyses of the eruption cloud data by GMS imageries should, however, be one of the most effective methods to realize the global monitoring of volcanic eruptions, to evaluate explosion intensity, to estimate quickly the total mass of ejecta in the atmosphere and on the earth's surface, and to investigate the dynamics of ascending and spreading eruption clouds in the atmosphere and in the stratosphere.

1 Introduction

The occurrence of an explosive volcanic eruption can be detected by observation of its eruption cloud, and the sequence of the activity can be monitored by the variation in time of the spreading cloud. The detection and observation of eruption clouds by satellite image is one of the methods for global monitoring of volcanic explosions. However, an explosion cloud is liable to spread, change its extent, and disappear in the atmosphere within a relatively short time, and it is therefore difficult to monitor its dispersal by means of imagery from polar-orbiting satellites, such as LANDSAT and NOAA, which have long time intervals between images, in spite of their high ground resolutions or their small instantaneous fields of view (IFOV). The Japanese Geostationary Meteorological Satellite (GMS) routinely returns its image 14 times a day, and is thus effective for observation of a spreading eruption cloud, in spite of its rather large IFOV.

---

[1] Seismological and Volcanological Department, Japan Meteorological Agency 1-3-4 Ote-machi, Chiyoda-ku, Tokyo 100, Japan

IAVCEI Proceedings in Volcanology 1
J.H. Latter (Ed.), Volcanic Hazards

GMS images, taken between late 1977 and 1985, are presented here as examples which have been used to observe explosive activity, to evaluate its intensity, and to study some characteristics of spreading eruption clouds. The use of GMS images for observation of eruption clouds can be an effective means of global monitoring of volcanic explosions, can enable forecasts to be made of where ejecta will fall, and can provide early warning of the presence of airborne tephra for safe navigation of aircraft. The eruption cloud data are also useful for estimating the mass of ejecta, and for studying the mechanism of spreading, dispersal, and ascent of eruption clouds.

2 Eruption Clouds Detected by GMS Imagery

GMS is located at an altitude of about 35,800 km over the Equator at longitude 140° E. The first GMS was launched in 1977, and the present GMS is the third in the series. The IFOV at nadir is 5 km for infrared (IR) and 1.25 km for visible (VIS) image data. GMS imagery is obtained 14 times a day at intervals ranging from 30 min to 4 h. Images of both photographic and digital data are regularly obtained.

Examples of eruption clouds detected by GMS images are shown in Figs. 1-3. Successive stages in the expansion and dispersal of eruption clouds by the prevailing winds can be recognized in these images. By the observation of the extent of eruption clouds, it is also possible to judge whether or not successive volcanic eruptions have taken place. Dispersed eruption clouds drift to distances of as much as several hundred km from their source volcanoes, and it is very difficult to observe the whole extent of such a widespread eruption cloud from the ground or from aircraft.

The detection of volcanic eruptions by means of GMS images was very difficult, because of the satellite's limited spatial resolution, and because small eruption clouds at low altitude, at the initial stage of a volcanic eruption, were often missed on account of the long intervals between images. Furthermore, it is difficult to identify eruption clouds among surrounding water clouds, although there is a tendency for the albedo of the surface of an eruption cloud to be lower than that of a water cloud (Sawada 1983a, b, 1985).

3 The Detection Capability of Eruption Clouds Using GMS Imagery

According to the SEAN Bulletin and the Bulletin of Volcanic Eruptions, 75 volcanoes erupted in the field of view of GMS between late 1977 and 1985, and the total number of volcanic eruptions over this 9-year period was 227. The total number of eruptions for which eruption clouds were detected on the GMS images was 31, at 23 volcanoes. Therefore, the detection rate of eruption cloud by GMS image against the total number of eruptions was about 14%: this is shown in Table 1. Good eruption cloud data were obtained in only about 7% of the cases. This low detection rate was mainly due to the masking effect of

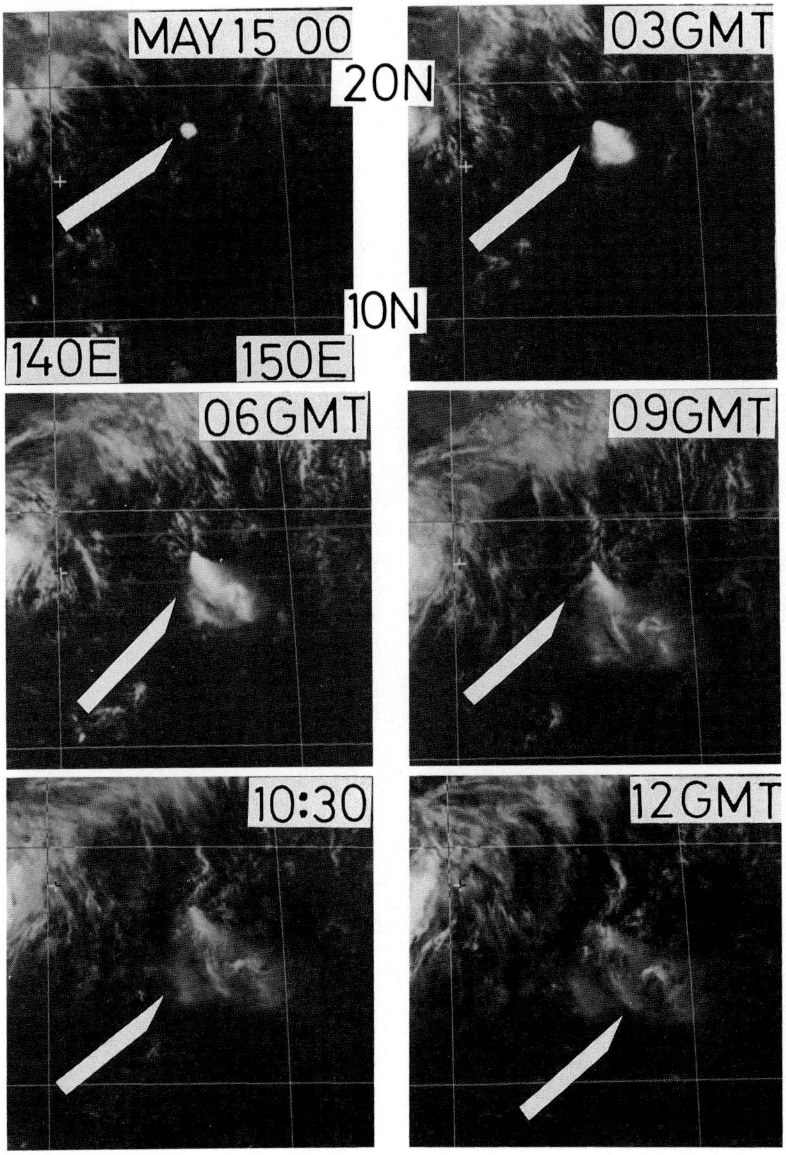

Fig. 1. Eruption cloud of Pagan volcano detected in GMS IR images between 00 h GMT and 12 h GMT on 15 May 1981. The almost circular extent of the eruption cloud generated shortly after the occurrence of the eruption is seen in the image taken at 00 h GMT, and its spreading and dispersal is well monitored through 12 h GMT

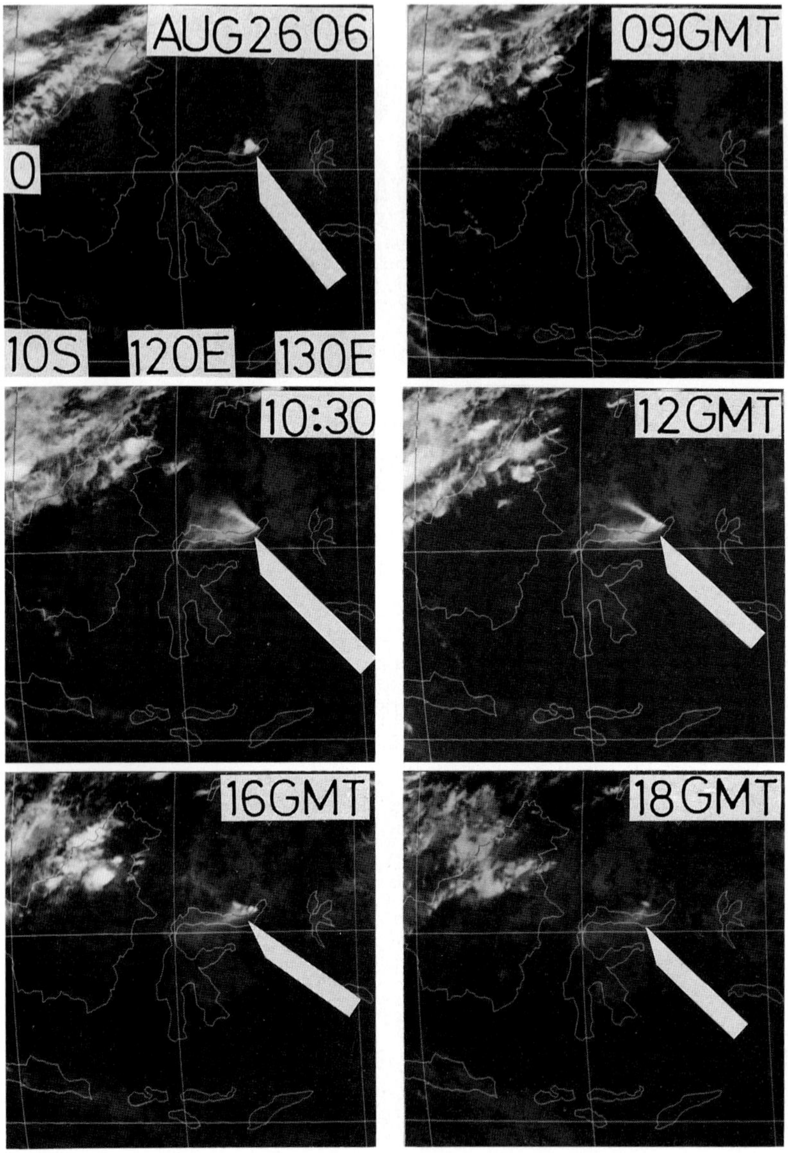

Fig. 2. Eruption cloud of Soputan volcano on GMS IR images between 06 h GMT and 18 h GMT on 26 August 1982. A bright-toned and triangle-shaped extent of the eruption cloud was detected at 06 h GMT. The cloud spread in two directions, NW and W of the volcano, between 10 h 30 m GMT and 12 h GMT, owing to different wind directions at different altitudes, and then dispersed

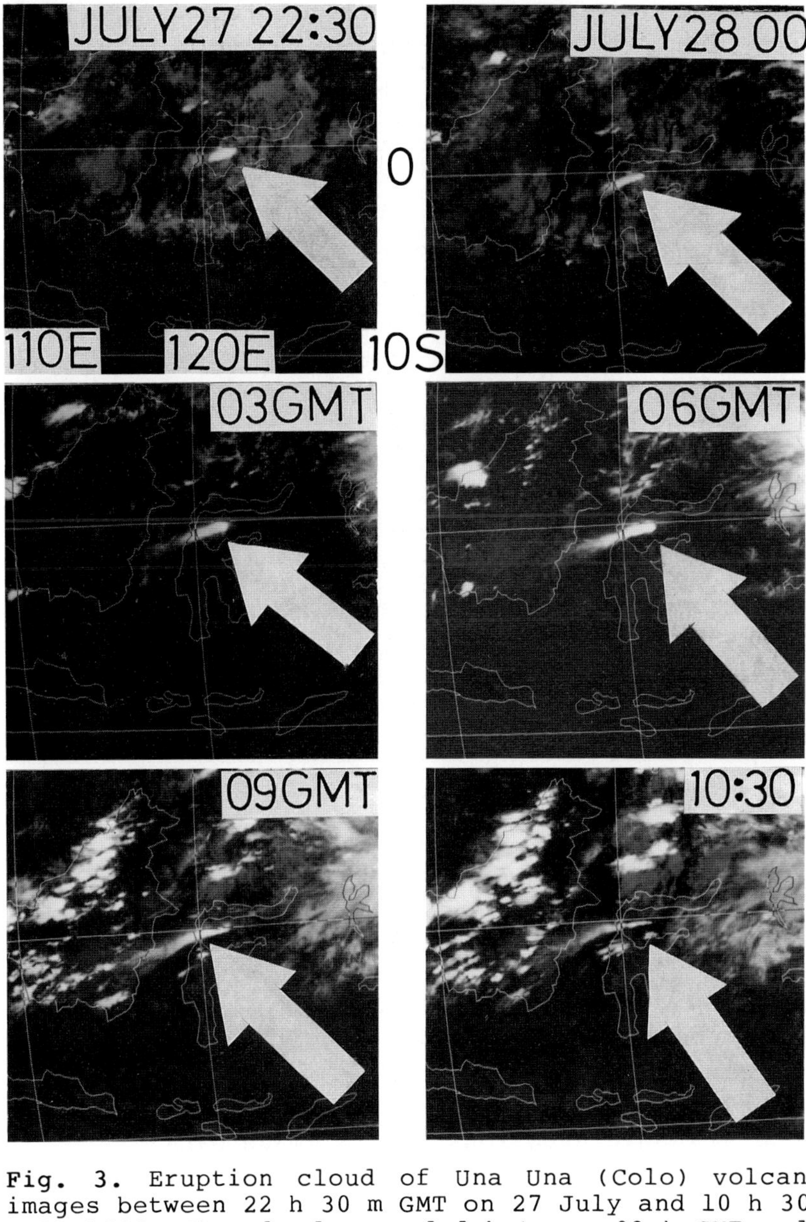

Fig. 3. Eruption cloud of Una Una (Colo) volcano on GMS IR images between 22 h 30 m GMT on 27 July and 10 h 30 m GMT on 28 July 1983. The cloud expanded between 03 h GMT and 06 h GMT on 28 July 1983, but kept its bright-toned appearance for more than 12 h, indicating the occurrence of strong successive volcanic eruptions

Table 1. Annual number of volcanoes which erupted between late 1977 and 1985 in the field of view of GMS, and the annual number and proportion of eruptions for which eruption clouds were detected by GMS images. The numbers in brackets denote the eruptions in which eruption clouds reached heights greater than 4 km. GOOD means well-detected eruption cloud data, unobscured by surrounding atmospheric clouds

Year	Eruption	Detected Eruption Cloud			
		ALL	GOOD	ALL	GOOD
1977[a]	18 (0)	0	0	0.0%	0.0%
1978	32 (3)	1	0	3.1	0.0
1979	29 (2)	1	0	3.4	0.0
1980	36 (8)	6	2	16.7	5.6
1981	23 (5)	3	2	13.0	8.7
1982	21 (8)	6	5	28.6	23.8
1983	20 (8)	4	1	20.0	5.0
1984	23 (12)	6	3	26.1	13.0
1985	25 (5)	4	2	16.0	8.0
Mean	25.2 (5.7)	3.4	1.7	13.7	6.6
Total	227 (51)	31	15	-	-

[a] 1977 November-December

other clouds and the limitation of GMS resolution, but all the big eruption clouds which reached altitudes higher than 10 km, generated by strong volcanic eruptions, could readily be detected on GMS images. Under good weather conditions, it is possible to detect an eruption cloud with a diameter of 20 km and an altitude of 5 km. The locations of the volcanoes for which eruption clouds were detected by GMS image data are shown in Fig. 4.

4 Observation of Volcanic Eruptions by GMS Image Data

Time variations of the maximum horizontal length of an eruption cloud measured on the GMS images, and the highest altitude of an eruption cloud estimated on the basis of temperature data are useful data for the study of the sequence of a volcanic eruption.

Usually, the surface temperature of a water cloud is obtained as the blackbody temperature calculated on the basis of the infrared brightness measured by the IR detector (Meteorological Satellite Center, 1980). When the cloud is dense, there is no radiance effect underlying the surface, and we can put the emissivity at 1.0. Therefore, assuming the surface temperature

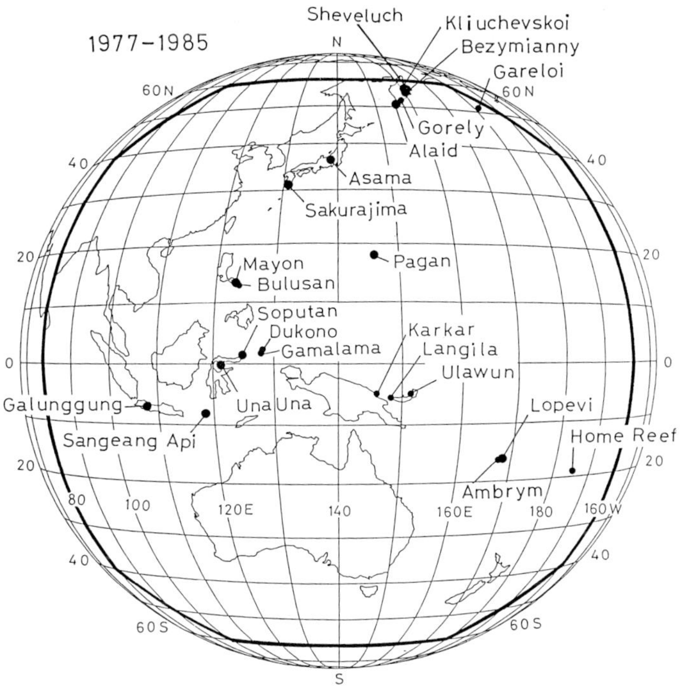

Fig. 4. Location of volcanoes at which eruption clouds were detected by GMS imagery between late 1977 and 1985. **Larger solid circles** indicate good eruption cloud data, unobstructed by atmospheric cloud

of the cloud to be that of the surrounding atmosphere, the blackbody temperature of the cloud surface represents the altitude of the cloud. However, a cloud is not always dense. In the case of the actual determination of the blackbody temperature of the surface of a water cloud, the emissivity ranging from 0.3 to 1.0 is used for the individual water clouds, on the basis of the empirical results for various kinds of water clouds, and the top altitude of the cloud can be exactly determined (Meteorological Satellite Center 1980).

It is considered that the blackbody temperature obtained from GMS IR data using 1.0 for the emissivity is accurate in the case of a dense eruption cloud. However, in the case of a thin eruption cloud, the blackbody temperature obtained is affected by the radiance of the underlying surface, and it certainly shows a temperature warmer than that at the actual altitude. Unfortunately, in cases of dispersing eruption clouds, we have no available observational data on the density and the empirical results of the emissivity. For the determination of the distribution of the blackbody temperature on the surface of an eruption cloud, the emissivity value 1.0 is tentatively used in this paper.

Assuming the surface temperature of an eruption cloud to be that of the surrounding atmosphere, the same as in the case of a water cloud, the altitude of the eruption cloud is estimated from an air-temperature profile obtained by radiosonde near the volcano.

When the eruption cloud does not reach the altitude of the tropopause, the coldest surface temperature of the eruption cloud represents the highest altitude, because the air temperature decreases upwards under the tropopause.

Analyzed results of eruption clouds during the 1983 Una Una (Colo) eruption are shown in Fig. 5. As seen in the lower portion of Fig. 5, we can judge comparatively well the end of a volcanic eruption by observation of the detachment of the eruption cloud from the location of the volcano; this is, however, not an accurate method, because an eruption cloud stays over a volcano during calm wind conditions, or under complicated wind conditions when there are different wind directions at higher and lower altitudes.

Some examples are found that show a decrease in the highest altitudes of eruption clouds towards the declining stages of eruptions (compare the relationships in time between the cross marks and the rectangles, especially in the cases of the erup-

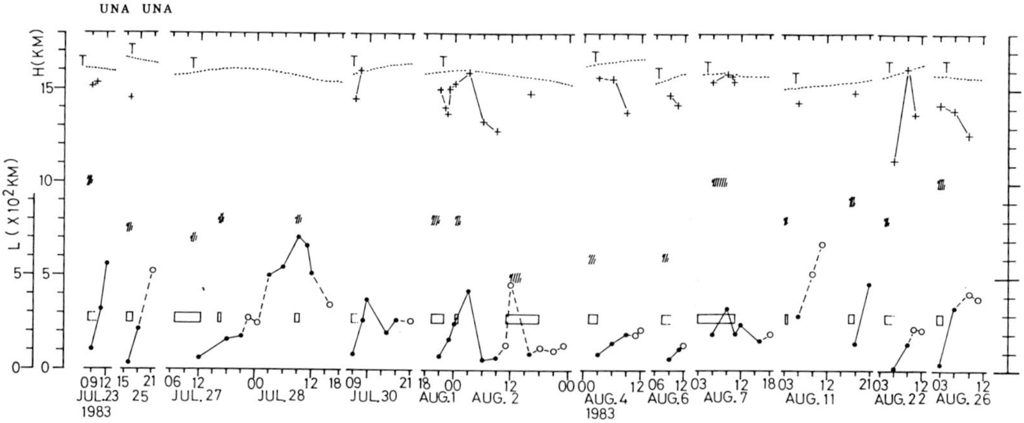

Fig. 5. Time variation of the altitude estimated from GMS infrared data (**crosses** refer to the H axis, upper case), and maximum horizontal length (**solid** and **open circles** refer to the L axis), of eruption clouds at Una Una (Colo), based on analyses of GMS images obtained during the 1983 eruption (lower case). **Dotted line** with **T** at the top of this figure means altitude of tropopause. **Solid** and **open circles** indicate that eruption clouds were respectively connected to location of the volcano, and detached from it. **Hatched areas** and **rectangles** respectively indicate the highest altitude of the eruption cloud and the duration time of individual volcanic eruptions, based on ground observations

tions in August, 1983, except during the period from August 1st to 2nd, in Fig. 5). A decrease in the height of an eruption cloud may precede the occurrence of pyroclastic flows or of ignimbrites, but at least, in the case of a normal explosion, observation of the highest altitude of an eruption cloud will be effective for making judgements as to the end or decay of eruptive activity because of its decrease.

The highest altitude of an eruption cloud, estimated from GMS and radiosonde data, did not coincide with the results of ground observation, being generally larger (see the difference between the cross mark and the hatched portion in Fig. 5). There may be a possibility that the cloud temperature from GMS IR data or the data from ground observation is being underestimated, but we must accumulate more observations before the reason for this can be decided.

5 Some Characteristics of the Behaviour of Spreading Eruption Clouds

The rising velocity of eruption clouds could not be determined by GMS image data because the images generally missed the short stage of rapid ascent soon after the occurrence of an explosion. The horizontal eddy diffusivity, estimated using Gifford's method (1959), showed large values, amounting to 10^9-10^{11} cm^2/s. The horizontal spreading velocity of the major portion of eruption clouds has values higher by several m/s than those of ambient wind velocity, obtained from the radiosonde data near the source volcano (Sawada 1983a). These measurements were obtained by measurement of the amount of movement of the edge of the eruption cloud, taken in a series of GMS images at Pagan volcano. This is an interesting phenomenon when we consider the spreading process of an eruption cloud in the atmosphere, but there still remains the problem of measurement error concerning the determination of altitude and the edge of eruption clouds in GMS image data.

The characteristic time variation of the surface temperature profile along the long axis of Una Una's eruption cloud obtained during the period from August 1st to 2nd is shown in Fig. 6. The eruption cloud spread in SW direction from this volcano, and this profile almost represents the altitude profile of the cloud along its long axis. The beginning and the end of the eruption were observed from the ground (see rectangles in Fig. 5). The eruption cloud did not show steady spreading or dispersal, but an increase in its highest altitude (E in Fig. 6) even after the volcanic eruption had ended. The time variation of the isothermal domain of the Una Una eruption cloud is shown at -10°C intervals in Fig. 7.

The eruption cloud continued to expand and the domain of highest altitude (coldest temperature) maintained its size for several hours after the end of the volcanic eruption (see rectangle in Fig. 7). This is an example of the case where we cannot simply judge the cessation time of an eruption activity only from eruption cloud data on GMS imagery. According to the

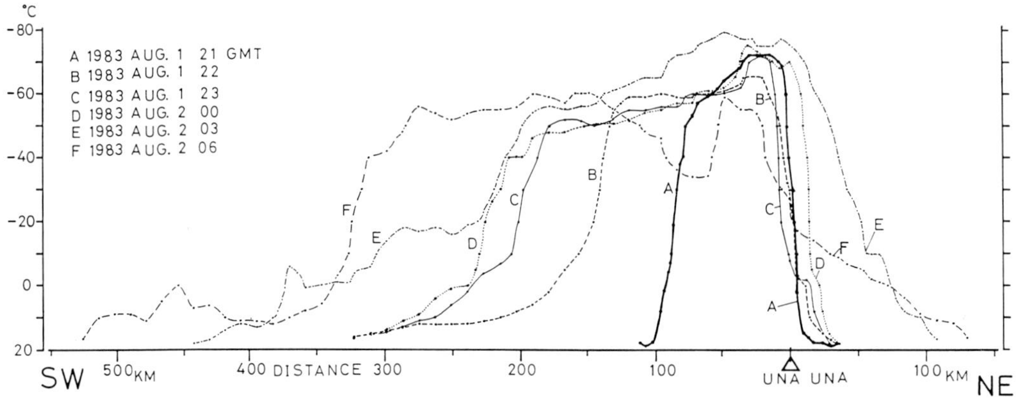

Fig. 6. Time variation of the profile of surface-temperatures of Una Una (Colo) eruption clouds along their long axes extending from NE to SW from the volcano, on GMS images taken between 21 h GMT on 1 August and 06 h GMT on 2 August 1983. **Triangle** marks the location of the volcano

ground observation, the eruption which occurred between 00 GMT and 01 GMT on August 2nd was a short-duration event, and another eruption had occurred at 19 GMT and ended at 22 GMT on August 1st, about 2 h before the former eruption (see rectangles in Figs. 5 and 7). We may consider that the eruption

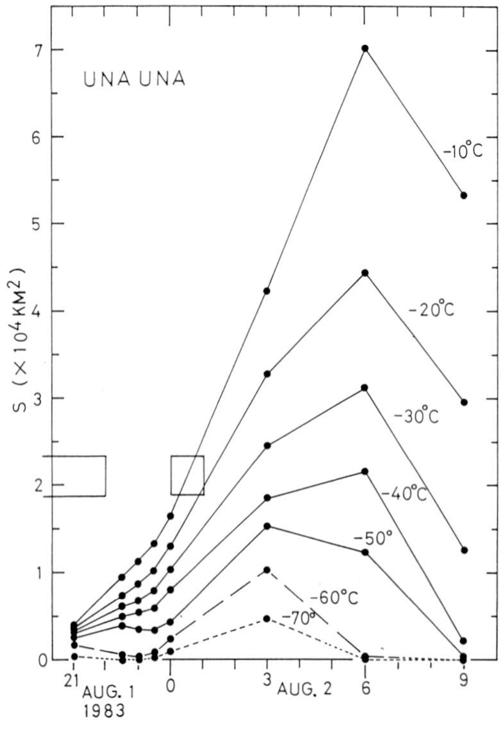

Fig. 7. The time variation of the isothermal domain for -10°C intervals, in the Una Una (Colo) eruption cloud, as recorded by GMS images between 21 h GMT on 1 August and 09 h GMT on 2 August 1983. **Rectangles** denote duration times of individual eruptions, determined by the ground observations. (After the SEAN Bulletin and the Bulletin of Volcanic Eruptions)

cloud which had remained over the volcano from the former eruption joined with the eruption cloud from the latter event, and the joined cloud gained in expanding force and could increase its height and size after the end of the latter eruption.

These phenomena may suggest that a spreading eruption cloud bears some heat source within itself, and can maintain its maximum altitude, expand in the atmosphere and drift with a slightly faster velocity than the surrounding wind speed, even after the end of a volcanic eruption. Of course, this consideration has to be examined by the further accumulation of more accurate observational data.

6 Thermal Energy Release Estimated by Eruption Cloud Data

In order to evaluate the intensity of a volcanic explosion, the thermal energy release, based on eruption cloud data, was estimated using Briggs's method (1969), which assumes that the eruption cloud in the atmosphere behaves as a rising plume. The results are compiled in Table 2. The thermal energy of juvenile ejecta was also calculated, using Yokoyama's method (1957),

Table 2. Estimated values of thermal energy release by eruption cloud (T.E. by cloud), total mass of ejected material (Total mass), and thermal energy release by juvenile material (T.E. by ejecta). Eruption period and total mass in this table are respectively based on ground observations (the SEAN Bulletin and the Bulletin of Volcanic Eruptions). T.E. by cloud and T.E. by ejecta are calculated using Briggs's (1969) and Yokoyama's (1957) methods and Nakamura's proposal (1965), respectively

Volcano	Eruption period	T.E. by cloud	Total mass	T.E. by ejecta
		erg	m^3	erg
Alaid	1981 IV...V	7×10^{22}	3×10^{8}	3×10^{24}
Pagan	1981 V	4×10^{22}	9×10^{7}	9×10^{23}
Asama	1982 IV	4×10^{19}	6×10^{4}	6×10^{20}
Soputan	1982 VIII,IX,XI	2×10^{22}	4×10^{6}	1×10^{23}
Galunggung	1982 IV...X	7×10^{23}	2×10^{8}	2×10^{24}
Lopevi	1982 X	1×10^{21}	1×10^{6}	2×10^{22}
Una Una	1983 VII...VIII	2×10^{23}	2×10^{7}	3×10^{23}
Soputan	1984 V,VIII	1×10^{22}	1×10^{7}	1×10^{23}
Mayon	1984 IX...X	8×10^{22}	ca. 10^{8}	ca. 10^{24}

with the abbreviated factor values proposed by Nakamura (1965). An approximate proportional relationship is obtained between both thermal energy releases and the thermal energy release based on eruption cloud data, which is almost one-tenth of that estimated from the ejecta. This result indicates that analyses of eruption cloud data by GMS imagery are useful for evaluation of eruption size, and can enable an estimate to be made of the total mass of the ejected materials.

7 Conclusions

The results presented here show that observation of eruption clouds by GMS imagery can provide an effective method for global monitoring of volcanic explosions, evaluation of explosion size, and approximate estimation of total mass of ejected materials. More detailed analyses of eruption cloud data have been successfully conducted using polar-orbiting satellite imageries such as NOAA (e.g., Malingreau and Kaswanda 1986), but GMS imagery can well observe time variations of eruption clouds which spread widely over a large region because of its short intervals of image taking. GMS imagery can also provide useful data for investigations on the spreading mechanism of these clouds.

Acknowledgements. The author expresses his gratitude to Dr. John H. Latter of DSIR and the other referee who made the thoughtful revisions in this manuscript; he also expresses his cordial thanks to Dr. Chris Newhall of USGS for his helpful advice on the presentation of this paper.

References

Briggs GA (1969) Plume rise, AEC Critical Review Series, TID - 25075, US Dept Commerce, Natl Tech Inf Serv Springfield, VA. 1-81

Gifford FA (1959) Smoke plumes as quantitative air pollution indices. Int J Air Poll 2:42-50

Malingreau JP, Kaswanda (1986) Monitoring volcanic eruptions in Indonesia using weather satellite data: the Colo eruption on July 28, 1983. J Volcan Geotherm Res 27:179-194

Meteorological Satellite Center (1980) The GMS user's guide. Issue 1:11-15 and 53-59

Nakamura K (1965) Energies dissipated with volcanic activities - classification and evaluation. Bull Volcan Soc. Japan II. 10th Anniversary Volume 81-90 (in Japanese with English abstract)

Sawada Y (1983a) Analysis of eruption clouds by the 1981 eruptions of Alaid and Pagan volcanoes with GMS images. Pap Meteorol Geophys 34:307-324 (in Japanese with English abstract)

Sawada Y (1983b) Attempt at surveillance of volcanic activity by eruption cloud image from artificial satellite. Bull Volcan Soc Japan II: 28, 357-373 (in Japanese with English abstract)

Sawada Y (1985) Detection capability of eruption clouds and observation of eruption sequences by GMS images. J Rem Sens Soc Japan 5:207-224 (in Japanese with English abstract)
Scientific Event Alert Network, Smithsonian Institution (1977 - 1985) SEAN Bulletin Vol. 2-10, Nos. 1-12
Volcanological Society of Japan, IAVCEI, IUGG (1979-1986) Bulletin of volcanic eruptions Nos. 17-23
Yokoyama I (1957) Energetics in active volcanoes. 2nd Paper, Bull Earthquake Res Inst 35:75-97

Volcanic Activity in Japan Monitored by the Japan Meteorological Agency Using Detected Precursory Phenomena

Y. Sawada[1], Y. Tanaka[1] and M. Seino[2]

Abstract

There are 76 potentially active volcanoes in Japan. The systematic volcanological work by the Japan Meteorological Agency (JMA) has been carried out since early 1960. The activity at 17 active volcanoes has been routinely monitored, and other volcanoes have been temporarily observed by mobile observation teams. During the period 1961-1985, 28 volcanoes erupted, and 17 volcanoes showed volcanic unrest such as earthquake swarms, fuming increase, etc. During these 25 years, therefore, there were an average of 5.5 eruptions a year, and 7.7 cases of volcanic unrest a year. According to the JMA observations up to now, the most clearly defined precursory event is the increase of seismicity before a volcanic eruption, but this pattern does not always occur. The next clearest precursor is the increase of intensity of fuming that has been observed at several volcanoes. It is difficult to predict smaller eruptions, as the clear precursor activity is lacking. In the cases of volcanoes which have been dormant for long periods, prior accumulation of relevant observational data is insufficient.

1 Introduction

There are 66 potentially active terrestrial volcanoes and 10 active submarine volcanoes in Japan. The activity at 17 of them has been routinely monitored, and other volcanoes have been temporarily observed by mobile observation teams of the Japan Meteorological Agency (JMA). Systematic instrumental volcanological work by JMA has been carried out since early 1960, and many volcanic eruptions have been experienced since then. This chapter summarizes volcanic activity during the period 1961-1985 and highlights several features of precursory events monitored or detected by the JMA.

1 Seismological and Volcanological Department, Japan Meteorological Agency, 1-3-4 Ote-machi, Chiyoda-ku, Tokyo 100 Japan

2 Seismology and Volcanology Division, Meteorological Research Institute, 1-1 Nagamine, Tsukuba, 305 Japan

IAVCEI Proceedings in Volcanology 1

J.H. Latter (Ed.), Volcanic Hazards

2 Volcanological Work of JMA

A summary of the volcanological work carried out by JMA is given in Table 1. Routine monitoring at 17 volcanoes involve seismological, visual and field observations. Seismic events, and surface-volcanic phenomena such as eruption sequences, rising clouds, fume temperatures, etc. are monitored: tilt measurements are made at four volcanoes. Mobile observation teams with similar equipment have also conducted emergency or temporary observations at other volcanoes. Information on volcanic eruptions or volcanic phenomena such as an increase in seismic activity or fuming is disseminated to the public. The observational data obtained are published quarterly in JMA's Volcanological Bulletin. Fundamental research on volcanic activity, and prediction of volcanic eruptions, is undertaken at the Meteorological Research Institute.

There are other national research institutes, government agencies and universities which undertake volcanological research and make observations in Japan. The data obtained are discussed by the Coordinating Committee for Prediction of Volcanic Eruptions, which reaches a consensus on the current state of the volcanoes and issues information to the public.

3 Volcanic Activity in Japan During the Period 1961 to 1985

In all, 45 volcanoes, including submarine ones, were active during the 1961-1985 period, representing volcanic eruptions at 28 volcanoes, and periods of unrest, such as earthquake swarms, increases in gas emission, etc. at the remaining 17. A summary of the annual volcanic activity over the 1961-1985 period is given in Fig. 1, and locations of all 76 of the volcanoes are shown in Fig. 2. The main magma type of Japanese volcanoes is andesite, so that most eruptions involve explosive activity. There are several volcanoes where dacite magma is erupted in growing lava domes, and others where basaltic lava flows are produced.

Though a volcano could have several eruptions or periods of unrest per year, we herewith count the annual activity as one eruption volcano/year, when the volcano erupted and also showed unrest in one year, and as one unrest volcano/ year, when the volcano showed only unrest in one year. Of the total matrix formed by the 76 active volcanoes, times the 25 years in the interval considered, there were 331 volcano/years of activity. This total is made up of 138 eruption years and 193 unrest years. During these 25 years, therefore, an average of 13.2 volcanoes per year were active in Japan, with 5.5 volcanic eruptions and 7.7 periods of unrest per year.

4 Precursory Events Monitored by JMA

According to JMA's volcanological observations up to now, the most clearly defined precursory event before volcanic eruptions has been an increase of seismic activity. For example, there

Table 1. Volcanological work by Japan Meteorological Agency (JMA)

Routine observation at 17 active volcanoes	
	Asama, O-sima, Aso, Sakura-zima, Me-akan, To-kati, Tarumai, Usu, Hokkaido-koma-ga-take, Azuma, Adatara, Bandai, Nasu, Kusatu-sirane, Miyake-zima, Unzen, Kirisima
Seismic events	Seismological observation by wired or wireless - telemetered electromagnetic seismograph
Volcanic cloud and eruption phenomena	Visual observation from observatory. Color TV system at 1 volcano (Kusatu-sirane)
Surface state at active site	Field inspection with thermometers, pH meter, gas detector tubes and visual observation
Ground tilt	Tilt observation with telemetered biaxial or pendulum tilt meter at 4 volcanoes (Sakura-zima, Aso, Asama, O-sima)
Temporary observation at non-instrumented volcanoes, and emergency observation during active or eruptive state by mobile volcanic observation teams	
Issue of volcano information to public on current volcanic activity	
The Coordinating Committee for Prediction of Volcanic Eruptions meets to discuss:	
observational data and techniques concerning volcanic activity and prediction of volcanic eruption, with universities, national research institutes and government agencies, and issues agreed conclusions on current volcanic activities	
Research work for fundamental investigation on volcanic activity and prediction of volcanic eruptions	
The Volcanological Bulletin of JMA is published	
quarterly, giving the observational results of JMA's routine, temporary, and emergency observations	

VOLCANO \ YEAR	61	62	63	64	65	66	67	68	69	70	71	72	73	74	75	76	77	78	79	80	81	82	83	84	85
8. 2 RYUKYU ISLANDS AND KYUSYU																									
2 OKINAWA-TORISIMA							◎	◎																	
3 SUWANOSE-ZIMA	●	●	●	●	●	●	●	●	●	●	●	●	●	●	●	●	●	●	●	●	●	●	●	●	●
5 KUTINO-ERABU-ZIMA						◎	▽	◎	◎			▽	◎	◎		◎				◎		▽			
7 KAIMON-DAKE							□	□																	
8 SAKURA-ZIMA	●	●	●	●	●	●	●	●	●	●	●	●	●	●	●	●	●	●	●	●	●	●	●	●	●
9 KIRISIMA	□	▽			□	□		□	□		◎		□	□			□	□	▽		▽□	▽	□		□
10 UNZEN		□		□		□		□	□	□	□	□	□	□	□▽	□	□	□		□	□	□		□	□
11 ASO	◎	◎	◎	◎	●	◎	◎	◎	◎	▽	◎	◎	◎	◎	◎	◎	◎	◎	●	●	▽	▽	▽□	▽□	◎
13 TURUMI														▽	▽										
8. 3 HONSYU																									
2 HAKONE						□	▽	□																	
4 ON-TAKE																			◎	▽					
7 YAKE-DAKE		◎	◎	▽				□	□																□
9 NIIGATA-YAKE-YAMA		◎	◎											◎	▽								◎		
11 ASAMA	●	▽		▽□	●		▽						●	▽□			□	□		□	□	●	●		
12 KUSATU-SIRANE																◎	□					◎	◎	▽□	
15 NASU	□	▽	◎											□		□	□				□				□
17 ADATARA																									▽□
18 AZUMA						▽		▽				□					◎	▽							
19 ZAO		□				▽	▽																		
21 KURIKOMA			▽	▽																					
22 TYOKAI														◎											
23 AKITA-KOMAGA-TAKE		□								●	●	▽			▽	▽									
24 IWATE			▽			▽						▽							▽						
25 HATIMANTAI													□												
27 IWAKI									▽	▽		□	□				□	▽							□
29 OSORE-YAMA								□																	
8. 4 IZU-MARIANA ISLANDS																									
1 O-SIMA	●	●	●	●	●	●	●	●	●	●	▽	▽□	▽	●	□	□	□	□		□			□	□	□
2 NII-ZIMA		□	□			□	□	□															□		□
3 KOZU-SIMA					□	□	□	□																	
4 MIYAKE-ZIMA		●	□▽											□		□							●	▽	
7 NEAR BEYONNAISE ROCKS										▲	△								△	△					
9 TORI-SIMA	□	□	□	□	□																				
a NISINO-SIMA													●	●	△			△							△
b KAITOKU SEA MOUNT																								▲	
12 IWO-ZIMA							◎	▽	◎						▽	▽		◎		◎		◎			
c HUKUTOKU-OKA-NO-BA		△					△	△				△	▲	▲	△	△	△	△	△	△	△	△	△	△	△
d HIYOSI SEA MOUNT															▲		△	△							
e NIKKO SEA MOUNT																			△						
f HUKUZIN SEA MOUNT								▲					▲	▲		△	△	△	△	△	△	△			
8. 5 HOKKAIDO																									
3 USU																	●	◎	□○	□○	□○	□○			
4 TARUMAI	▽	▽										□	□	□	□	□	□	◎	◎	□	□▽		▽		
5 TOKATI	◎	●	□					□▽	□					▽□										▽	▽
6 DAISETU																	▽								
7 ME-AKAN		◎		◎	◎	◎	▽						□		□				□	□		□			
8 ATOSANUPURI					□																				

VOLCANIC ERUPTION ● MAGMATIC ◎ PHREATIC ▲ SUBMARINE

VOLCANIC UNREST □ SEISMIC ▽ FUMING ○ GROUND DEFORMATION △ SEA WATER DISCOLORATION

Fig. 1. Summary of volcanic activity monitored by JMA during the period 1961-1985. Eruptions are divided into three types: magmatic, phreatic and submarine; and periods of volcanic unrest into four types: seismic, fuming, ground deformation and seawater discoloration events. **Numbers** assigned to the volcanoes are those in the List of the World Active Volcanoes (Volcanological Society of Japan, IAVCEI 1971): **small alphabetical letters** denote that those volcanoes have not been catalogued in the list

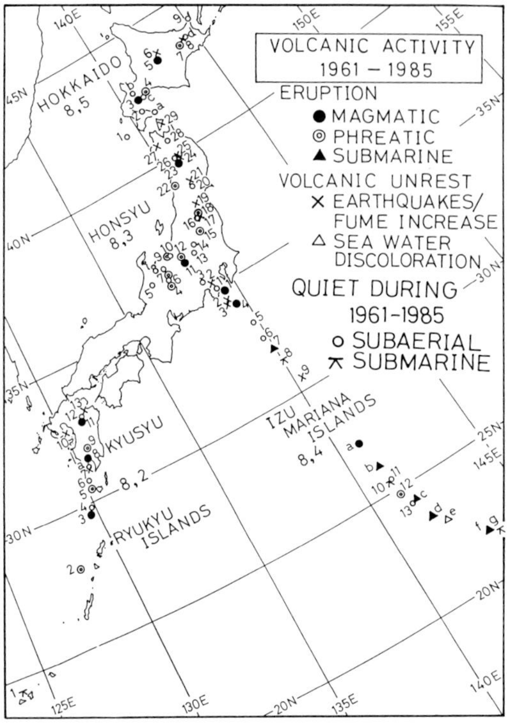

Fig. 2. Distribution of volcanoes which erupted or experienced volcanic unrest during the period 1961-1985, and the remaining potentially active volcanoes which did not show their volcanic activity (expressed as **QUIET** in this figure)

was a swarm of volcanic earthquakes about 30 h before the 1977 Usu eruption, and there were increases of seismicity before and during the Miyake-Zima flank fissure eruptions of basalt in 1940, 1962 and 1983 (Tanaka et al. 1984). In the case of the 1983 Miyake-Zima eruption, the precursory seismic activity commenced about 1 h before the outbreak. There have been increases or swarms of shallow volcanic earthquakes which preceded the increases of frequencies of individual explosion at Sakura-Zima volcano, which has been active, with intermittent explosions, for more than 30 years. It is considered that the earthquake swarms are caused by extrusions of lava mass on the crater bottom, which have capped the vent, and the lava mass extruded at the vent in the summit crater has been observed from the air after the occurrence of earthquake swarms several times since 1967. However, these events do not always occur before all of the individual explosions.

Increases in the number of small earthquakes occurred before weak phreatic explosions in the cases of the 1978 Tarumai eruptions, and the 1982 and 1983 Kusatu-Sirane eruptions. There were also instances of volcanic tremor shortly before those eruptions at Tarumai and Kusatu-Sirane volcanoes, but its relationship to eruptions was not as well defined as the swarms of earthquakes.

The next clearest precursory phenomenon was the increase of fuming, as observed in the 1926 and 1962 Tokati eruptions and the appearance of new fuming, as detected before the 1970-1971 Akita-koma-ga-take eruption. A gradual increase in the size of the plume was also noticed several months before the 1982 and 1983 Asama eruptions.

Seismological data were used successfully to forecast the course of the long period of activity of Usu volcano between 1977 and 1982, during which there were frequent earthquakes and periods of ground upheaval after explosions, and the end of the activity was successfully predicted (Yokoyama and Seino 1979; Yokoyama et al. 1981; Seino 1983). There were also the cases of the 1926 and 1962 Tokati eruptions, which were apparently triggered by tectonic earthquakes, and the decline of the 1970-1971 Akita-koma-ga-take eruption, which similarly followed a tectonic earthquake. However, no remarkable precursory phenomena were detected in the cases of the small phreatic explosions of the 1974 Tyokai and the 1979 On-Take eruptions, which occurred after long dormant intervals: however, neither of these volcanoes was routinely monitored.

Table 2 summarizes and classifies the precursory phenomena detected, and Table 3 summarizes precursory events in relation to differences in the magma type of the eruptions. In the cases of dacite eruptions, clear precursory seismic events were experienced. Magmatic or phreatic explosions at several andesitic volcanoes also showed clear precursory increases in seismicity before eruptions. Increase in gas discharge or the size of volcanic plumes were also effective precursors of several volcanic eruptions.

Table 2 . Classification of precursory events detected by JMA's work during the period 1961-1985. The volcanoes with the eruption years (in brackets) denote that their precursory phenomena were weak and their relationships to eruptions were not well defined

Item	Volcanic Eruption
Increase of volcanic earthquake	Usu (1977-1978), Miyake-zima (1983), Tokati (1962), [Tarumai (1978)], [Kusatu-sirane (1982, 1983)]
Occurrence of volcanic tremor	[Aso (1979)], [Tarumai (1978)]
Cloud increase or new fuming	Tokati (1962), Akita-koma-ga-take (1970), [Asama (1982, 1983)]
Not observed	Tyokai (1974) - Erupted after 153 years quiescence, On-take (1979) - First eruption in historical record
Tectonic earthquake	Tokati (1926, 1962), Akita-koma-ga-take (1970)

Table 3. Summary of precursory events in Japan based on JMA data. The eruption characteristics and the intensity are respectively expressed with the symbols and categories used in the Bulletin of Volcanic Eruptions

Magma	Volcano	Date	Remarks
Dacite	Usu	1977 VIII	o ↑ ⩍ (g) Swarm of volcanic earthquakes about 30 h before eruption. The course of activity was successfully estimated from seismological data.
Andesite	Tokati	1962 VI	o ↑ (g) Increase of fume temperature since mid-1957 and occurrence of ground deformations about 6 months before eruption. M = 7.0 tectonic earthquake took place about 2 months before eruption.
	Akita-koma-ga-take	1970 IX...X	o ↑ ⇝ (m) New fuming activity about 20 days before eruption. Explosion activity sharply decayed after M = 6.2 tectonic earthquake.
	Asama	1982 IV and 1983 IV	o ↑ (m) Gradual increase of volcanic plumes about 3 months before eruption.
	Tarumai	1978 V	o ↑ (l) Increase in the number of volcanic earthquakes about 2 months before eruption.
	Aso	1979 VI...IX	o ↑... ↑ (m) Amplitude variation of volcanic tremor before strong explosion.

Table 3 (cont.)

Magma	Volcano	Date	Remarks
Andesite	Kusatusirane	1982 X, XII and 1983 VII,XI,XII	o ↑ ⊍ (l) Increase of small volcanic earthquakes 5-9 days before eruption.
Basalt	Miyakezima	1962 VIII	o=↑≈> (g) Swarm of volcanic earthquakes 1-2 h before eruption.
		1983 X	o=↑&↑≈> (g) Swarm of volcanic earthquakes about 1 h before eruption

5 Some Examples of Characteristic Precursors

5.1 1926 and 1962 Tokati Eruptions

A comparison between the sequences of these two eruptions at Tokati volcano is shown in Fig. 3. There were differences of duration times, but increases of fuming, with melting and burning of sulfur sublimates, due either to the fume increase or to rise in ground temperature, were observed in both cases. Small phreatic explosions took place sporadically, followed by big magmatic eruptions. A sharp increase of volcanic earthquakes was also observed, especially in the case of the 1962 eruption. It was interesting that tectonic earthquakes occurred during both the eruptions. During the 1926 activity, a magnitude M = 6.8 earthquake occurred about 200 km southeast of the volcano, about 3.5 months after the major eruption on 24 May, and about 3 days before the next big eruption on 8 September. During the 1962 activity, a magnitude M = 7.0 earthquake took place about 170 km southeast of the volcano, about 66 days before the big eruption on 29 June.

By contrast, increased fuming was observed from 1974 to 1975 and from 1984 to the present, but on neither occasion has this been followed by eruptions. Seismicity during these periods was relatively low. A sharp increase of minor volcanic earthquakes occurred in May 1968, shortly after the occurrence of a magnitude M = 7.9 earthquake about 310 km south-southeast of the volcano, and a high level of volcanic seismicity continued through to the end of 1969, with an increase in fume emission, but no eruption followed.

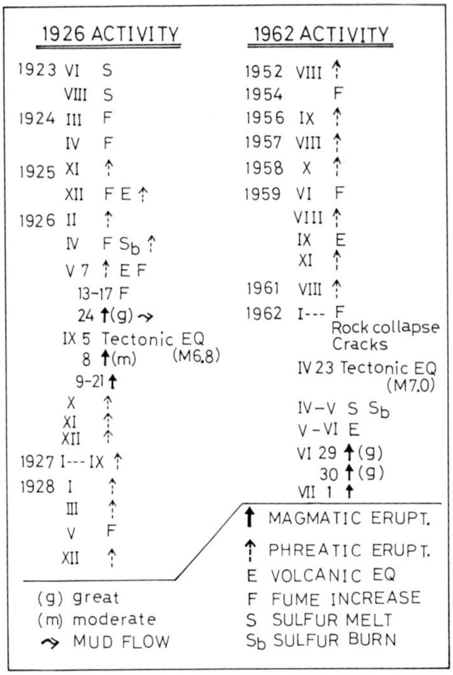

Fig. 3. Brief comparison of sequences of precursory events before the 1926 and 1962 Tokati eruptions. We can see that fuming has intensified its activity and tectonic earthquakes took place during and before both eruptions. Date of the occurrence or duration period of the individual activity is partly shown, but, in the case of the intermittent or occasional occurrence of the activity, only the month is shown. Intensity of the eruption: **g** (great) and **m** (moderate), according to the category in the Bulletin of Volcanic Eruptions

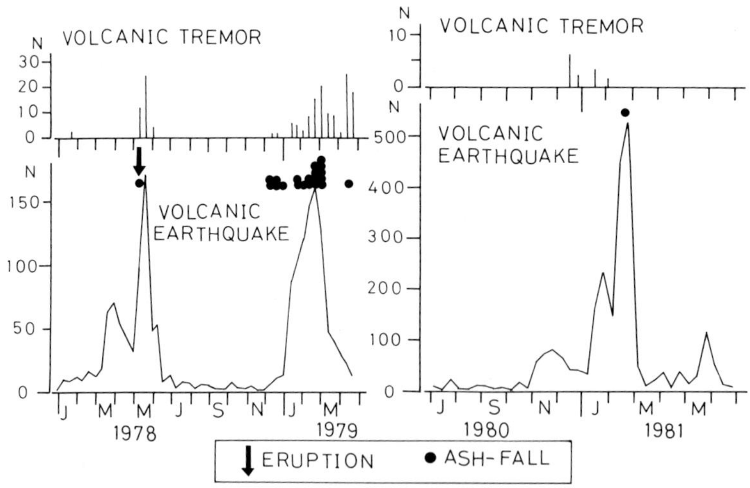

Fig. 4. Increase of number of volcanic earthquakes preceding a small phreatic explosion in 1978, and a minor ash ejection in 1981, at Tarumai volcano. The volcanic earthquakes are small and shallow events, and the volcanic tremors are the isolated events in which individual duration time is mostly under 1 h. **N** for both events means the number, every 10 days in each month, counted on the seismogram. Volcanic earthquakes frequently occurred during the phreatic explosion in 1978, and the minor ash ejections in the period 1978-1979 and 1981: volcanic tremors also occurred during the period of the high seismicity

5.2 1978, 1979 and 1981 Tarumai Activities

Increases of minor volcanic earthquakes took place, as shown in Fig. 4, before a small phreatic explosion and minor ash ejections in 1978 and 1981 at Tarumai volcano. After the eruption in 1978 and ash ejections in 1979, there was a continuing high level of seismic activity, signalling the possibility of a bigger volcanic eruption. However, only one minor ash ejection took place in 1981. Seismicity has increased during the ash ejections in 1979, but there was no clear precursory increase of volcanic earthquakes. Isolated volcanic tremor took place closely before the eruption in 1978, during the ash ejections in 1979 and before the ash ejection in 1981, as shown in Fig. 4, but their accurate relationships to the activities are not well defined.

5.3 1910, 1943-1945 and 1977-1982 Usu Eruptions

A schematic comparison between the sequences of these three eruptions is shown in Fig. 5. Almost identical patterns were repeated during the three periods of activity. Precursory earthquakes occurred first, followed by explosions, accompanied by remarkable ground upheavals, due to intrusion of dacite lava domes.

The explosions and the ground upheavals in 1910 and 1944-1945 took place at the outer slopes of the summit crater of the volcano, but the 1977-1982 activity has continued inside the summit crater. The explosions in 1910 were only phreatic events, but the 1944 and 1977 eruptions have been magmatic. These three eruptions of the volcano in 1910, 1944 and 1977 have been followed by earthquakes, including felt shocks. However, the time intervals prior to the eruptions, and the duration periods of the seismic activities, varied. In the 1910 eruption, the earthquakes occurred several days before the first explosion, but the frequency decreased after the occurrence of the explosions and during the ground upheaval due to the intrusion of a cryptodome. The explosions in 1944 were followed by earthquakes several months before, and the ground upheaval due to the intrusion of a cryptodome commenced at least 4 months before the explosions. The seismicity decreased to a low level during the ground upheaval due to the intrusion of a cryptodome and the explosions, but again increased the frequency during the growth of the dacite lava dome from late 1944 to late 1945. However, in the case of the 1977 eruption, the earthquake swarm took place about 30 h before the first explosion. Then, occasional explosions took place until late 1978, but the seismicity decreased almost steadily with the decrease of the upheaval rate of the ground.

The accurate determination of the occurrence time of the ground upheavals in these three eruptions is difficult, but it has been confirmed that the ground deformation in the 1910 eruption, such as cracks, occurred at least about a half day before the eruption, and the ground upheaval, in the 1944 eruption, occurred about 4 months before the occurrence of the first explosion.

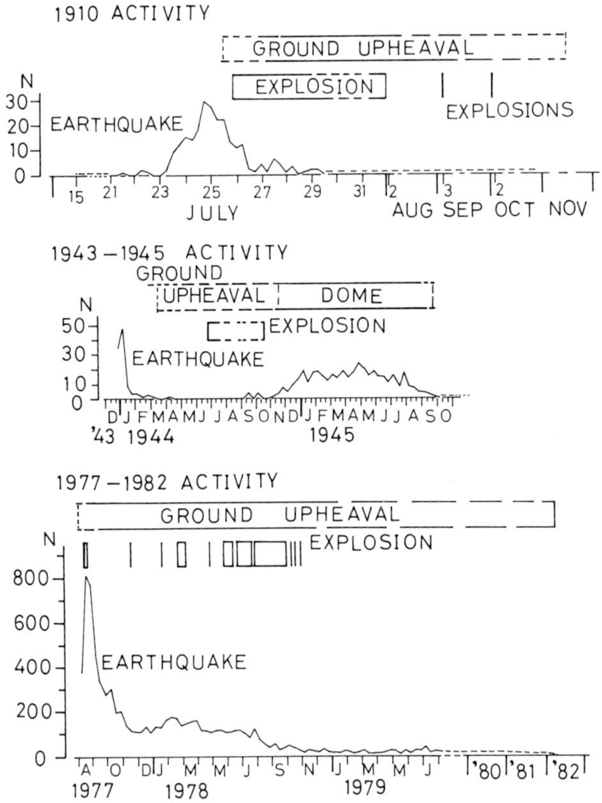

Fig. 5. Schematic comparison of the precursory sequences, showing seismic activity, volcanic explosions and ground upheaval due to growth of cryptodomes or a dacite dome, in the 1910, 1943-1945 and 1977-1982 Usu eruptions. Earthquakes have preceded the explosions and ground upheaval in these three activities. The time interval for the 1977-1982 activity (lower part) is expressed monthly during the period from August 1977 to July 1979, thereafter being simplified. The ground upheaval sequences in 1910, 1944-1945 and 1977-1982 have been respectively confirmed and measured by visual observations, levellings and theodolite surveys. The rectangle and vertical bar for the explosion denote respectively that the intermittent eruptions and a single event have taken place. Seismological data for the 1910 eruption, every 6 h, are from Omori (1911), and those for the 1943-1945 and 1977-1982 activities, every 10 days, are after Seino (1983) and Yokoyama and Seino (1979)

Ground deformation, such as cracks on the ground surface prior to the explosion in August 1977, was observed on the same day, at least several hours before the explosion.

We conclude that a sharp increase in earthquakes, including felt shocks, is the commonest and most effective precursor, and the ground deformations such as cracks followed by ground upheaval are also an effective precursor at this volcano.

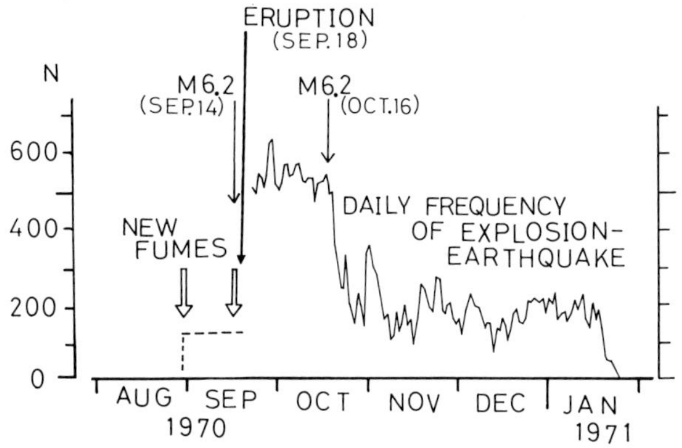

Fig. 6. Time variation of daily number of explosion earthquakes accompanying Strombolian explosions, and the occurrence of new fuming at the eruption site after about 20 days, and tectonic earthquakes, during the 1970-1971 Akita-koma-ga-take eruption. After the occurrence of the second tectonic earthquake, the frequency of the Strombolian explosions suddenly decreased

5.4 1970-1971 Akita-Koma-Ga-Take Eruption

Strombolian eruptions with lava flows took place in 1970-1971 at Akita-koma-ga-take, after a pause of about 38 years. The Strombolian eruptions have been well monitored as explosion earthquakes by seismographs that have been temporarily operated around this volcano since 20 September 1970, 2 days after the occurrence of the eruption. During the observation, the frequency of volcanic earthquakes at this volcano has been at a very low level, and we could not determine a distinct precursory activity.

The clearest precursor of this activity was the appearance of new fuming at the eruption site about 20 days before the eruption began. Another interesting event was the occurrence of two tectonic earthquakes before and during the eruption. The first earthquake was a magnitude M=6.2 event which took place about 200 km southeast of the volcano about 4 days before the beginning of the eruption on 18 September 1970 (see Fig. 6). The second earthquake was an event of the same magnitude about 60 km south of the volcano on 16 October 1970. The Strombolian explosions declined suddenly, shortly after the second earthquake, and ceased altogether about 3 months later.

5.5 1979-1980 Aso Eruption

During eruptions at Aso volcano, the amplitude of continuous volcanic tremor sometimes increases. There were Strombolian eruptions and successive ash ejections in mid- to late 1979, and four strong phreatic explosions during 1979-1980 (see Fig. 7). Periods during which the amplitude of volcanic tremor decreased sharply, lasting from several hours to about 10 days before the occurrence of each of the three phreatic explosions,

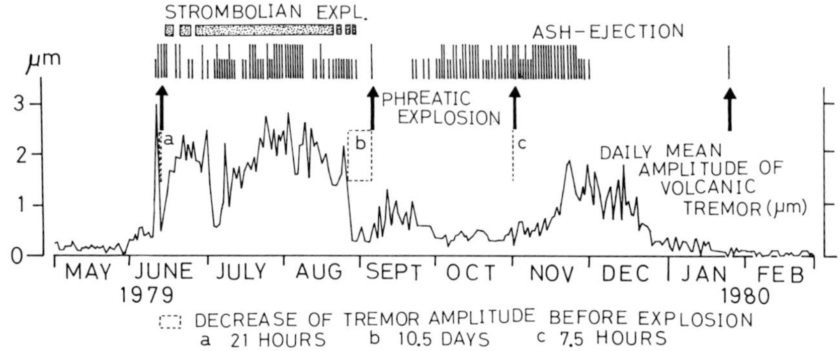

Fig. 7. Time variation of the daily mean ground amplitude in μm of continuous volcanic tremor during the Strombolian explosions and frequent ash ejections between mid-1979 and early 1980 at Aso volcano. Sharp decreases in the amplitude of volcanic tremor were observed three times (**a**, **b** and **c**) before strong phreatic explosions (**black arrows**). But, in the case of the phreatic explosion in 1980, this was not observed. The **rectangle** (at the **top**) denotes the period of the occurrence of Strombolian explosions. The intensity of the individual ash ejections belongs to category 1 (m in total), in the Bulletin of Volcanic Eruptions, but the scale of the intensity is divided into two grades, based on the approximate area and quantity of ash fall around the crater: a **long bar** for a strong ash ejection, and a **short bar** for a weaker event

were observed (a, b and c in Fig. 7), but in the fourth explosion, this was not apparently observed: the phreatic explosion occurred during a long period of declining tremor.

6 Conclusion

Volcanological observations have been systematically carried out by JMA since early 1960, during which time many eruptions have been observed. Some eruptions showed almost the same patterns or sequences of precursory events and types of activity as those experienced in the past: others did not repeat the same precursory phenomena or their patterns, and there were some in which there were differences in the time intervals between precursory events and eruptions. Some volcanoes changed their pattern of eruption, and their precursory phenomena also varied, compared to those observed earlier.

Precursory phenomena were well observed at volcanoes which erupted frequently. In recent years, no very large eruptions have occurred in Japan. Smaller eruptions are difficult to predict as clear precursors are often lacking, especially in cases of eruptions of volcanoes which have been dormant for long periods, where there has been insufficient accumulation of observational data.

Acknowledgements. The authors express their cordial thanks to Dr. Rodney Grapes, Victoria University, and Dr. John Latter, D.S.I.R., for their kind advice and for their thoughtful reviews on the authors' manuscript.

References

Volcanic activities in Japan during 1961-1985 compiled in this chapter are based on the following reports:
Japan Meteorological Agency (1961 - 1986) Volcanological Bulletin, Vols. 1 - 24 (Nos. 1 - 4), Vol. 25 (Nos. 1 - 2).
Japan Meteorological Agency (1985) Observation Report at Observatories during July - December, 1985 (in Japanese), unpublished.

Omori F (1911) The Usu-san eruption and earthquake and elevation phenomena, Bull Imp Earthq Inv Com 5, 1:1-38, 3:101-107

Seino M (1983) Seismic activity accompanying the 1977-1978 eruption of Usu Volcano, Japan. Papers Meteorol Geophys 34:105-141

Tanaka Y, Nishide N, Takahasi M (1984) Relation between the seismic activities around Miyakejima Volcano and the eruption of the volcano, Bulletin of the Volcanological Society of Japan-The 1983 Eruption of Miyakejima. Vol 29. Second Ser, Spec Iss, pp 45-54 (in Japanese with English abstract).

Yokoyama I, Seino M (1979) Prediciton of development in the 1977-78 activities of Usu Volcano with consideration for energy. discharge. J Fac Sci, Hokkaido Univ, Ser VII (Geophysics), Vol 6, 1:187-200

Yokoyama I, Yamashita H, Watanabe H, Okada Hm (1981) Geophysical characteristics of dacite volcanism - the 1977-1978 eruption of Usu Volcano. J Volcanol Geotherm Res 9:335-358

A Summary of Precursors to Volcanic Eruptions in Papua New Guinea

J. MORI, C. MCKEE, B. TALAI and I. ITIKARAI[1]

Abstract

A review was made of the 50 eruptions since 1930 at 16 volcanoes in Papua New Guinea to summarize the available information on precursory activity. The data show that a variety of seismic, ground deformation, thermal and chemical precursors have occurred at many of the volcanoes. Seismic precursors were the most commonly reported (11 cases), and for volcanoes that were instrumented with seismometers, seismic precursors were identified for 8 of 14 eruptions. The occurrence of felt earthquakes before eruptions appeared more often at the volcanoes that had not been active for more than 50 years and had higher silica compositions.

1 Introduction

There has been a wide variety of volcanic eruptions in Papua New Guinea, ranging from continuous, relatively passive, lava effusion to violent explosions. Since the time of European settlement in the 1870's there have been about 80 eruptions reported from 17 volcanoes. We have reviewed 50 significant eruptions that have occurred at 16 volcanoes from 1930 through 1985 to summarize the available information on precursory activity. In addition there have been at least four occasions when unusual activity was noted at a volcano but was not followed by an eruption. Since the 1930's there have been professional volcanologists working in the country, so that from this period there is much more reliable information than prior to 1930. Information about precursors to volcanic eruptions in Papua New Guinea has also been previously compiled in Cooke and Johnson (1978) and Lowenstein (1982b).

Continuous instrumental surveillance of volcanoes in the country was generally begun in the mid-1970's, except for Rabaul and Manam which were instrumented in the 1940's and 1950's respectively. Therefore, much of the information compiled comes only from reports of local inhabitants and subsequent visits by volcanologists. For the early time period when few volcanoes were instrumented, the information on precursors is certainly incomplete. However, there is still evidence for clear precur-

[1] Rabaul Volcanological Observatory, Rabaul, Papua New Guinea

IAVCEI Proceedings in Volcanology 1
J. H. Latter (Ed.), Volcanic Hazards

sory activity for several of the eruptions. This historical information is important for recognizing precursors at specific volcanoes, since individual volcanoes may repeatedly undergo similar patterns of pre-eruptive activity.

2 Data

Much of the information summarized in this chapter is contained in reports of the Geological Survey of Papua New Guinea and the Australian Bureau of Mineral Resources and monthly reports of the Rabaul Volcanological Observatory. Also included are some new compilations of seismic data from Ulawun volcano and seismic and tilt data from Manam volcano.

Table 1 summarizes the information on all the volcanoes for which precursors were observed. The volcanoes are grouped by region (Fig. 1), which reflect different tectonic settings and chemical compositions (Cooke and Johnson 1978). The precursors have been devided into four groups: seismic, ground deformation, thermal and chemical. A blank in Table 1 indicates that there was no observed precursor. An entry of "None" indicates there were instrumental observations but no precursor was recognized. Table 2 lists activity at dormant volcanoes which

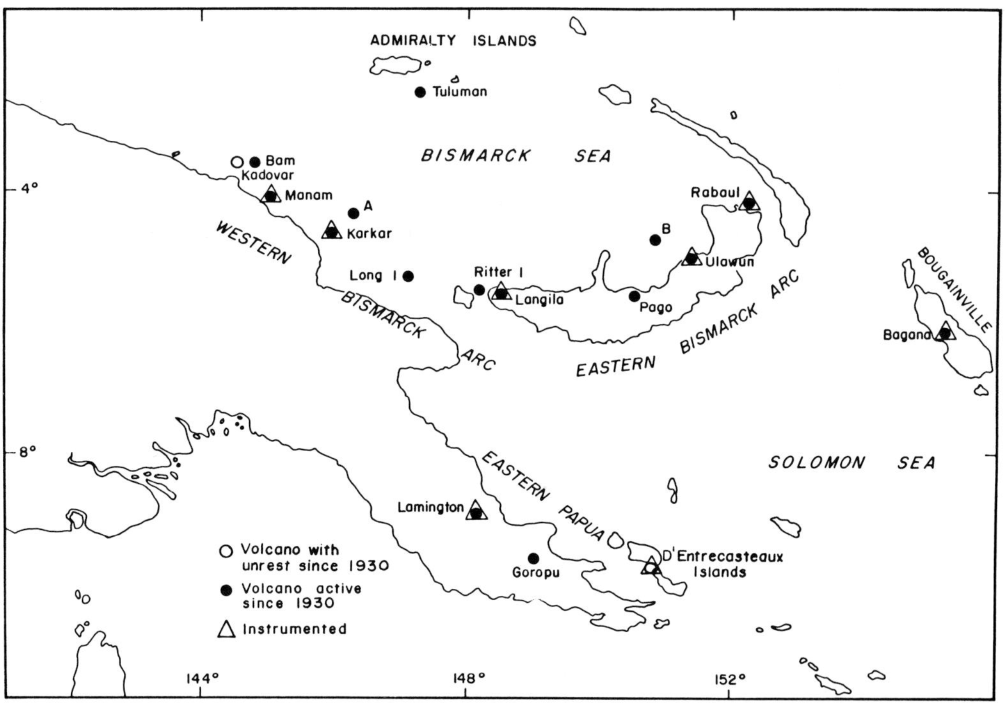

Fig. 1. Locations and groupings of the volcanoes in Papua New Guinea, which have been active or have shown signs of unrest since 1930. A and B are unnamed submarine centers

Table 1. Summary of volcanoes active in Papua New Guinea during 1930-1985 which have had observed precursors[a]

	Seismic precursors	Ground deformation precursors	Thermal precursors	Chemical precursors	Eruptions
Western Bismarck Arc					
Manam (Basalt to andesite strato-volcano, height 3000 m from seafloor)[1,2,3,4,5,6,7,8]					
1936-39	----	----	----	----	Intermittent weak to moderate Strombolian activity Ash columns Lava ejections (several 100 m) Pyroclastic flows
1946-47	----	----	----	----	Moderate Strombolian activity Ash columns Lava ejections Lava flow (2×10^7 m^3)
1956-66	(There was intermittent eruptive activity throughout this 10-year interval, with stronger activity during the following four periods)				
1956-58	----	Inflationary tilt 4 mo. prior to 1957 eruption Inflationary tilt 5 mo. prior to 1958 eruption	----	----	Intermittent weak to moderate Strombolian activity Ash columns to 7000 m Lava ejections to 300 m Lava flows (2×10^7 m^3) Pyroclastic flows (10^7 m^3)
1960	Increase in felt events 3 mo. prior to March eruption	Inflationary tilt 1 mo. prior to March eruption	----	----	Moderate Strombolian eruption Ash columns Lava ejections to 300 m Lava flow (4×10^6 m^3) Pyroclastic flows

1964	None	Uncertain	----	----	Moderate Strombolian eruption Ash columns Lava ejections to 700 m Lava flow (1×10^6 m^3) Pyroclastic flows
1965-66	None	----	----	----	Moderate Strombolian eruption Ash columns Lava ejections to 700 m Lava flow (2×10^6 m^3) Pyroclastic flows
1974-75	None	10-20 µrad inflationary tilt in 10 yr. prior to eruption 6 µrad tilt in 3 mo. prior to eruption	----	----	Moderate Strombolian eruption Ash columns to 1500 m Lava ejections to 600 m Lava flows (3×10^6 m^3) Pyroclastic flows (3×10^6 m^3)
1982	None	4 µrad tilt in 3 mo. prior to eruption	----	----	Moderate Strombolian eruption Ash columns to 7000 m Lava ejections Pyroclastic flows (10^4-10^5 m^3)
1984	Increase and abrupt decrease in seismicity 6 mo. prior to eruption	None	----	----	Moderate Strombolian eruption Ash columns to 2000 m Lava ejections to 700 m Lava flow (5×10^5 m^3) Pyroclastic flows (5×10^6 m^3)

[a] References:
(1) Taylor (1958a); (2) Taylor (1963); (3) Branch (1967); (4) Palfreyman and Cooke (1976); (5) Cooke and Johnson (1978); (6) Scott and McKee (1984); (7) Cooke (1976); (8) McKee (1984); (9) Johnson et al. (1972a); (10) McKee et al. (1976); (11) McKee and Wallace (1981); (12) McKee et al. (1981a); (13) Ball and Johnson (1976); (14) Cooke et al. (1976); (15) Taylor et al. (1957); (16) Johnson et al. (1971); (17) Palfreyman et al. (1981); (18) Cooke (1981c); (19) McKee et al. (1981b); (20) Cooke (1975); (21) Talai (1980); (22) McKee (1982); (23) McKee (1987); (24) Fisher (1939); (25) Kizawa (1951); (26) Fisher (1976); (27) McKee et al. (1984); (28) McKee et al. (Chap. 24); (29) Mori et al. (Chap. 25); (30) Taylor (1958b); (31) Baker (1946).

Table 1 (cont.)[a]

	Seismic precursors	Ground deformation precursors	Thermal precursors	Chemical precursors	Eruptions
Karkar	Andesite strato-volcano, height 3000 m from seafloor, summit calderas 3.5 and 5.5 km diameters. Recent eruptions from Bagiai cone, height 300 m from caldera floor)[9,10,11,12]				
1974-75	----	----	Change in position of fumaroles and small temp. increase during 10 yr. prior to eruption	----	Moderate Strombolian eruption Ash columns to 1500 m Lava ejections 1500 m Lava flows (7 x 10^7 m^3)
1979	Seismicity increase 2.5 yr. prior to eruption Periodic tremor appeared 6 mo. prior to eruption Increase in No. of low-freq. events 2 weeks prior, and slight increase of tremor amplitude 2 days prior to explosion of Mar. 8.	----	Fumarolic acitvity increase 6 mo. prior to eruption	----	Phreatic explosions Eruption column to 3000 m Volume of ejecta 10^8 m^3
Long Island (Basalt to andesite volcanic complex, with water-filled caldera 12.5 x 10 km. Recent eruptions from Motmot Island, a basaltic cone, height 300 m from caldera floor)[5,13,14]					
1940-43	----	----	----	----	Uncertain eruption

1953-55	----	----	----	----	Intermittent Surtseyan activity
1968-69	----	----	----	----	Intermittent Surtseyan activity 300 x 180 m island formed
1973-75	----	----	Temperature increase 2 yrs. prior to eruption	----	Intermittent Surtseyan and Strombolian activity Lava ejections to 150 m Lava flow (5 x 10^3 m^3)
Langila (Andesite composite cone, height 1330 m a.s.l.)[15,16,17]					
1954	----	----	Increase of fumarolic activity 10 yr. prior to eruption	Increase of acid gases 10 yr. prior to eruption	Vulcanian explosions Ash columns to 5000 m
1960-61	----	----	----	----	Vulcanian explosions New crater formed Lava flow (2 x 10^4 m^3)
1967-68	----	----	----	----	Vulcanian explosions Lava flow (5 x 10^4 m^3)
1973-75	----	----	----	----	Moderate Vulcanian and Strombolian activity Lava flows (10^7 m^3)

[a] References:
(1) Taylor (1958a); (2) Taylor (1963); (3) Branch (1967); (4) Palfreyman and Cooke (1976); (5) Cooke and Johnson (1978); (6) Scott and McKee (1984); (7) Cooke (1976); (8) McKee (1984); (9) Johnson et al. (1972a); (10) McKee et al. (1976); (11) McKee and Wallace (1981); (12) McKee et al. (1981a); (13) Ball and Johnson (1976); (14) Cooke et al. (1976); (15) Taylor et al. (1957); (16) Johnson et al. (1971); (17) Palfreyman et al. (1981); (18) Cooke (1981c); (19) McKee et al. (1981b); (20) Cooke (1975); (21) Talai (1980); (22) McKee (1982); (23) McKee (1987); (24) Fisher (1939); (25) Kizawa (1951); (26) Fisher (1976); (27) McKee et al. (1984); (28) McKee et al. (Chap. 24); (29) Mori et al. (Chap. 25); (30) Taylor (1958b); (31) Baker (1946).

Table 1 (cont.)[a]

	Seismic precursors	Ground deformation precursors	Thermal precursors	Chemical precursors	Eruptions
Langila (cont.)					
1977-1985	None	----	----	----	Moderate Vulcanian and Strombolian activity Lava flows (3×10^6 m^3)
Eastern Bismarck Arc					
Ulawun (Basalt to andesite strato-volcano, height 2300 m a.s.l.)[9,14,18,19,20,21,22,23]					
1941	----	----	----	----	Small explosion
1958	----	----	----	----	Uncertain ash emissions
1960	----	----	----	----	Weak ash emissions
1963	----	----	----	----	Weak ash emissions
1967	----	----	----	----	Moderate Strombolian eruption Ash columns Lava ejections
1970	----	----	----	----	Moderate Strombolian eruption Ash column to 5000 m Lava ejections to 450 m Lava flow (8×10^6 m^3) Pyroclastic flows (5×10^6 m^3)
1973	----	----	----	----	Moderate Strombolian eruption Ash column to 2000 m

1973 (cont.)					Lava ejections to 600 m Lava flow (8-10 x 10^6 m^3) Pyroclastic flows (3-5 x 10^6 m^3)
1978	Swarms appeared 17 mo. and 9 mo. prior to eruption Small events appeared 1 week prior to eruption	----	----	----	Moderate Strombolian eruption Ash columns to 2000 m Lava flow (7-9 x 10^6 m^3) Pyroclastic flows (8-10 x 10^6 m^3)
1980	Seismicity increase several days prior to eruption	----	----	----	Strong Vulcanian explosions Ash column to 20,000 m Ash volume 10^7-10^8 m^3
1984	Strong seismicity increase at least 21 mo. prior, then decrease 9 mo. prior to eruption Seismicity increase 10 days prior to eruption	----	----	----	Moderate Strombolian eruption Ash column to 2000 m
1984-85	Seismicity increase 5 days prior to eruption	----	----	----	Weak Strombolian eruption Lava flows (10^5 m^3)

[a] References:
(1) Taylor (1958a); (2) Taylor (1963); (3) Branch (1967); (4) Palfreyman and Cooke (1976); (5) Cooke and Johnson (1978); (6) Scott and McKee (1984); (7) Cooke (1976); (8) McKee (1984); (9) Johnson et al. (1972a); (10) McKee et al. (1976); (11) McKee and Wallace (1981); (12) McKee et al. (1981a); (13) Ball and Johnson (1976); (14) Cooke et al. (1976); (15) Taylor et al. (1957); (16) Johnson et al. (1971); (17) Palfreyman et al. (1981); (18) Cooke (1981c); (19) McKee et al. (1981b); (20) Cooke (1975); (21) Talai (1980); (22) McKee (1982); (23) McKee (1987); (24) Fisher (1939); (25) Kizawa (1951); (26) Fisher (1976); (27) McKee et al. (1984); (28) McKee et al. (Chap.24); (29) Mori et al. (Chap. 25); (30) Taylor (1958b); (31) Baker (1946).

Table 1 (cont.)

	Seismic precursors	Ground deformation precursors	Thermal precursors	Chemical precursors	Eruptions
Ulawun (cont.)					
1985	Seismicity increase 5 days prior to eruption	----	----	----	Moderate Strombolian eruption Ash columns to 8000 m Lava ejections to 500 m Lava flows (2 x 10^6 m^3) Pyroclastic flows (5-10 x 10^6 m^3)
Rabaul (Basalt to rhyolite volcanic complex with 9 x 14 km sea-filled caldera. Recent eruptions from Tavurvur, a 220 m (a.s.l.) andesite-dacite cone, and Vulcan, a 226 m (a.s.l.) dacite cone)[24,25,26,27,28,29]					
1937 (Vulcan and Tavurvur)	Felt earthquakes several days (to weeks?) prior to eruption	Reefs raised 2 m several hours prior to eruption Uplift possibly started several weeks prior to eruption	----	----	Vulcan: Plinian eruption Ash column to 10,000 m Formed pumice and ash cone Pyroclastic flows Volume of ejecta 3 x 10^8 m^3 Tavurvur: Moderate explosions Ash column to 1000 m Volume of ejecta 3-4 x 10^6 m^3
1941-43 (Tavurvur)	None prior to 1941, felt event 1 mo. before explosion in 1943	7.5 cm subsidence from 2 yr. prior to eruption	Increased fumarole activity from 10 mo. prior to eruption	Increased acid content of gases from 6 mo. prior to eruption	Intermittent explosions

1941-43 (cont.)			300° temp. increase from 6 mo. prior to eruption		
1971-85	Strong increase in felt and instrumental seismicity	Large inflationary ground deformation	Change in fumarolic activity	----	No eruption
Eastern Papua					
Mt. Lamington (Andesite strato-volcano, height 1680 m a.s.l., breached summit crater 1.3 km diameter)[30]					
1951	Felt earthquakes from 5 days before paroxsymal eruption	----	Increased vapor emissions 6 days before paroxsymal eruption	----	Peleaen eruption Ash columns to 17,000 m Pyroclastic flows devastated 175 km^2 Lava dome formed (10^8 m^3)
Goropu (Trachy-andesite cinder cone, height 650 m a.s.l.)[31]					
1943-44	Felt earthquakes several days prior to eruption	----	----	----	Moderate explosive eruption Formed 150 m cinder cone Pyroclastic flows devastated 30 km^2

[a] References:
(1) Taylor (1958a); (2) Taylor (1963); (3) Branch (1967); (4) Palfreyman and Cooke (1976); (5) Cooke and Johnson (1978); (6) Scott and McKee (1984); (7) Cooke (1976); (8) McKee (1984); (9) Johnson et al. (1972a); (10) McKee et al. (1976); (11) McKee and Wallace (1981); (12) McKee et al. (1981a); (13) Ball and Johnson (1976); (14) Cooke et al. (1976); (15) Taylor et al. (1957); (16) Johnson et al. (1971); (17) Palfreyman et al. (1981); (18) Cooke (1981c); (19) McKee et al. (1981b); (20) Cooke (1975); (21) Talai (1980); (22) McKee (1982); (23) McKee (1987); (24) Fisher (1939); (25) Kizawa (1951); (26) Fisher (1976); (27) McKee et al. (1984); (28) McKee et al. (Chap. 24); (29) Mori et al. (Chap. 25); (30) Taylor (1958b); (31) Baker (1946).

Table 2. Summary of dormant volcanoes in Papua New Guinea which have shown signs of unrest during 1930-1985[a]

	Seismic precursors	Ground deformation precursors	Thermal precursors	Chemical precursors	Eruptions
Western Bismarck Arc					
Kadovar (Andesite strato-volcano, height 1000 m from seafloor)[1]					
1976-78	----	----	New thermal area with temp. to 100°C	Increase of acid gases	No eruption
Eastern Papua					
D'Entrecasteaux Islands (Complex of volcanic cones and thermal areas on Fergusson, Dobu, Normanby and Neumara Islands)[2,3,4,5]					
1953-57	Numerous felt earthquakes but possibly not related to volcanic activity	----	----	----	No eruption
1982	None	----	Increase in temp. and volume of hot spring	----	No eruption
1984	None	----	New hot springs formed and probable discharge of hot water into the sea	----	No eruption

[a] References:
(1) Wallace et al. (1981); (2) Taylor (1955b); (3) Reynolds; (1957a); (4) Lowenstein (1982a); (5) Johnson (1984).

Table 3. Summary of volcanoes active in Papua New Guinea during 1930-1985 which have not had observed precursors[a]

Western Bismarck Arc

Bam (Andesite strato-volcano, height 2200 m from seafloor)[1,2,3,4,5,6]

1947	Small explosions
1954-60	Small explosions Ash columns to 3300 m

Submarine Center A[7]

1944	Uncertain eruption
1951	Uncertain eruption

Ritter Island (Submarine caldera, only small remnant island above sea level)[8]

1972	Moderate submarine eruption
1974	Moderate submarine eruption

Eastern Bismarck Arc

Pago (Dacite ash and scoria cone with recent lava flows, height 742 m a.s.l., within Witori Caldera)[9]

1933	(Considered to be part of eruption that began in 1911) Moderate explosions Ash columns

Submarine Center B[7,10]

1951	Uncertain submarine eruption
1970	Uncertain submarine eruption

---

[a] References:
(1) Best (1954); (2) Taylor (1955a); (3) Reynolds (1957b); (4) Davies (1972); (5) Cooke and Johnson (1981); (6) Johnson et al. (1972b); (7) Johnson et al. (1981); (8) Cooke (1981a); (9) Cooke (1981b); (10) Cooke and Johnson (1978).

Table 3 (cont.)

Admiralty Islands	
Tuluman Islands (Rhyodacite submarine volcano with several small cones projecting above sea level)[11,12]	
1953-57	Moderate submarine eruption formed small islands Lava flows (>10^6 m^3)
Bougainville	
Bagana (Andesite strato-volcano, height 1760 m a.s.l.)[10,13,14]	
1937-38	Moderate explosion Ash columns
1943-53	Lava effusion with occasional explosions Ash columns to 10,000 m Pyroclastic flows
1960-67	Lava effusion with occasional explosions Ash columns to 3000 m Pyroclastic flows
1970-85	Lava effusion (10^5-10^6 m^3/yr) with occasional explosions Ash columns to 5500 m Pyroclastic flows

[a] References:
(10) Cooke and Johnson (1978); (11) Reynolds and Best (1976); (12) Reynolds et al. (1980); (13) Bultitude (1976); (14) Bultitude (1979).

have shown signs of unrest but have not had eruptions historically. For completeness, Table 3 lists information on all the other volcanoes that had eruptive activity during 1930 to 1985, but for which no precursors were recognized.

3 Specific Examples of Precursory Activity

The following are summaries of activity at some of the volcanoes which have shown clear examples of precursors, with emphasis on the volcanoes which have been instrumented with seismographs and/or tiltmeters prior to eruptions. Also included are three cases of activity that appeared to be precursory but did not lead to eruptions. More detailed descriptions for most of the entries in Tables 1, 2 and 3 are given in Mori et al. (1986).

Manam

Manam is a strato-volcano of basaltic to andesitic composition. Its summit is about 3000 m above its seafloor base which has a diameter of about 25 km (McKee 1981). Throughout historical time there has been nearly continuous eruptive activity. Since 1930 the more intense periods of activity, which have been counted as separate eruptions, have been 1936-39, 1946-47, 1956-58, 1960, 1964, 1965-66, 1974-75, 1982 and 1984. The time period from 1956-1966 is regarded as a single eruption by Cooke and Johnson (1978), since there was some magmatic activity reported in every year. However, we have counted the four more intense episodes during this time as separate eruptions, because there were recognizable precursors to some of these periods of activity.

1956-58 and 1960 Eruptions

There is a fairly long history of tilt measurements at Manam from 1957 (Scott and McKee 1984). The early tiltmeter was located at Waris, 4 km east of the summit. Taylor (1958a) reported a slow inflationary tilt which led to a more rapid inflation and a major eruption in August 1957. He also reported a slow inflation of more than 150 μrad during the 5 months preceding the climactic eruption of 1958 (Taylor 1963). In addition, a large inflationary tilt of about 200 μrad from 1960 to 1966 was reported, during which time eruptions occurred in each year (Taylor 1963; Branch 1967; Palfreyman and Cooke 1976).

Along with the tilt changes, Taylor (1963) reported that large fluctuating seismic amplitudes and a large number of felt events gave warning of an increase in eruptive activity during March 1960.

1974-75 Eruption

Since 1966, tilt has been measured with 5-m-long water-tube tiltmeters located at Tabele Observatory, 4 km southwest of the summit. Figure 2 shows the radial and tangential components of tilt from 1968 through 1985. The large changes due to the 1970 Madang earthquake (M_S 7.0, 150 km away) and 1972 Josephstaal earthquake (M_S 6.5, 90 km away) have been removed. One can see a long-term inflationary trend of a few microradians per year starting in the mid-1960's and continuing until the paroxysmal eruptions in 1974.

A short-term tilt increase of 6 μrad in the 7 months before the 1974 eruptions can also be clearly seen on the radial component of tilt of Fig. 2. There is a discrepancy between the large tilt changes of a few hundred microradians recorded in the 1950's to 1960's and the recent measurements which show much smaller changes. The recent measurements have been made at about the same distance from the summit, although the different instrument site and difference in size of the eruptions could account for the difference in measurements.

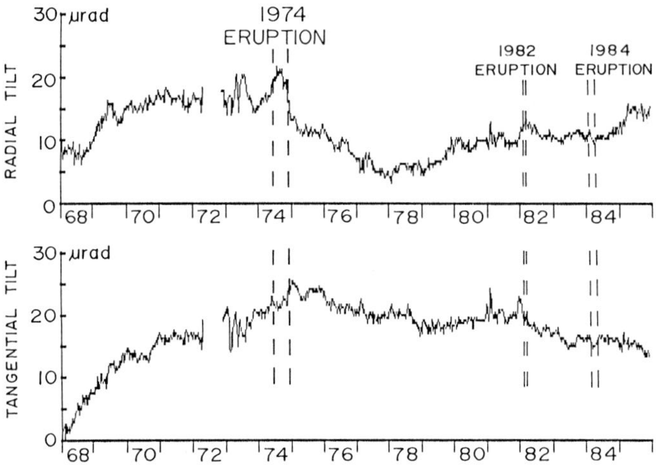

Fig. 2. Radial and tangential tilt recorded on water-tube tiltmeters 4 km from summit of Manam volcano

1982 and 1984 Eruptions

Prior to the 1982 eruption there was a short-term tilt increase of 4 μrad in 3 months, similar to the tilt change before the beginning of the 1974-75 eruption. However, there was no significant tilting prior to the 1984 eruption, which was stronger than the activity in 1982, producing lava flows (5 x 10^5 m^3) and pyroclastic flows (5 x 10^6 m^3). In 1982, there were no lava flows and the volume of the pyroclastic flows was 10^4 to 10^5 m^3.

Following the 1974 eruption the long-term increase seen on the radial component of tilt re-started in 1978 and continued through 1985. This is similar to the long-term inflation seen before the 1974-75 eruption (Fig. 2). Since there was no significant deflation following the 1982 and 1984 eruptions, this long-term inflation may be precursory to a larger eruption similar in size to the 1974-75 event (Mori et al. 1987).

Since the seismicity at Manam consists of nearly continuous low-frequency (1-3 Hz) earthquakes, it is difficult to make meaningful event counts. Instead, the level of seismic activity is quantified by amplitude measurements. Amplitude scalings are obtained by averaging peak to peak amplitudes of the five largest events in each hour of the record. In Fig. 3, bars showing the range of the smallest to the largest values of hourly averages are plotted for each day from 1980 through 1985. There were no obvious precursory signs before the 1982 eruption.

The increase in amplitude which started in March 1983 and then abruptly stopped in July 1983 could possibly be interpreted as a precursor to the 1984 activity, although it would be difficult to separate such features unambiguously from those that occurred in 1980 and 1981 and were not followed by increase in eruptive activity.

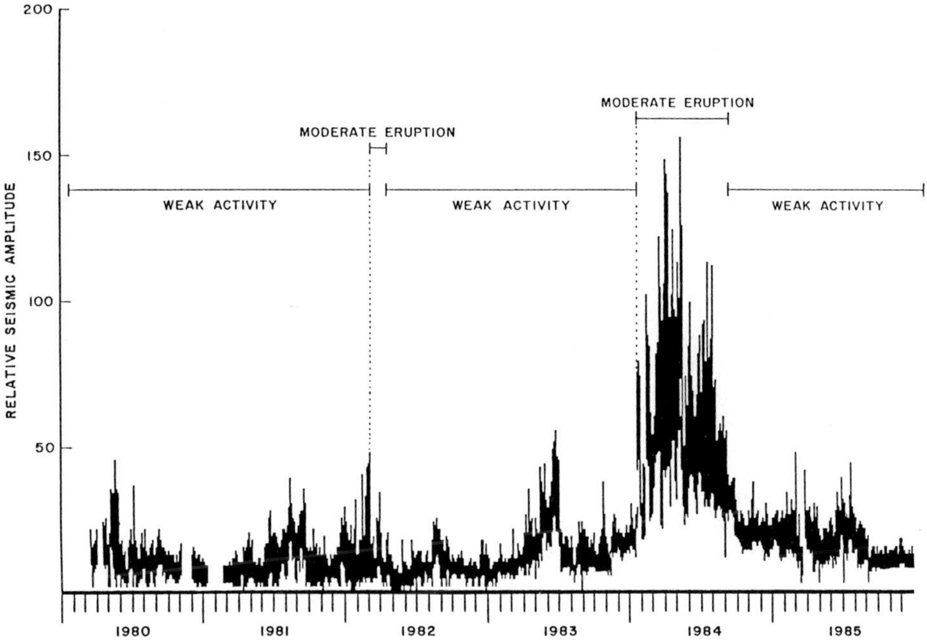

Fig. 3. Seismic amplitudes recorded 4 km from the summit of Manam volcano. The average amplitude for the five largest events in each hour are calculated. The bar shown in this graph for each day represents the range of these hourly measurements

Karkar

Karkar is an andesitic volcano whose summit rises about 3000 m above its seafloor base, which has a diameter of about 40 km. There is a summit caldera 3.2 km in diameter nested in an older caldera 5.5 km in diameter. Recent eruptions occurred in 1974-75 (McKee et al. 1976) and 1979 (McKee et al. 1981a) from Bagiai cone and a new crater, both located near the middle of the inner caldera.

1974-75 Eruption

There were possible thermal precursors to the 1974-75 eruption. Taylor noted that between 1965 and 1970 the strongest fumarolic activity had shifted from the upper to the lower rim of the summit crater (Johnson et al. 1972b). Also, temperatures measured in 1971 and 1972 were slightly higher compared to readings in 1963 (McKee et al. 1976). The subsequent eruptive phases in 1974-75 opened up new vents in Bagiai cone and produced an estimated 7 x 10^7 m^3 of lava (McKee and Wallace 1981).

1979 Eruption

The precursory signs before the 1979 eruption were much more apparent (McKee et al. 1981a). After a seismograph was installed in 1975, several types of seismic events were observed leading up to the 1979 eruption. In mid-1976 it was noted that local seismicity consisting of low-frequency (~3 Hz) earthquakes began to increase (see Fig. 4). At the beginning of July 1978 continuous tremor began to appear and counting of individual events became difficult. From July 1978 the increasing level of seismicity is shown by the tremor amplitude in Fig. 4. Accompanying the increase of seismicity , stronger fumarolic activity was also noticed. In late August 1978 increased emissions were noted from Bagiai's summit and vapor plumes were noted from the southeastern flank of Bagiai. By late September there was a widespread fumarolic area (0.8 km^2) at the southeastern foot of Bagiai, near the center of which there was a zone of incandescence emitting blue vapors.

The seismic tremor amplitude continued to increase through most of October. At this time "banded tremor", periodic intervals of strong and weak tremor which averaged about 90 min, became particularly striking (see Fig. 3, McKee et al. 1981a). Temperatures measured in October at the incandescent zone were

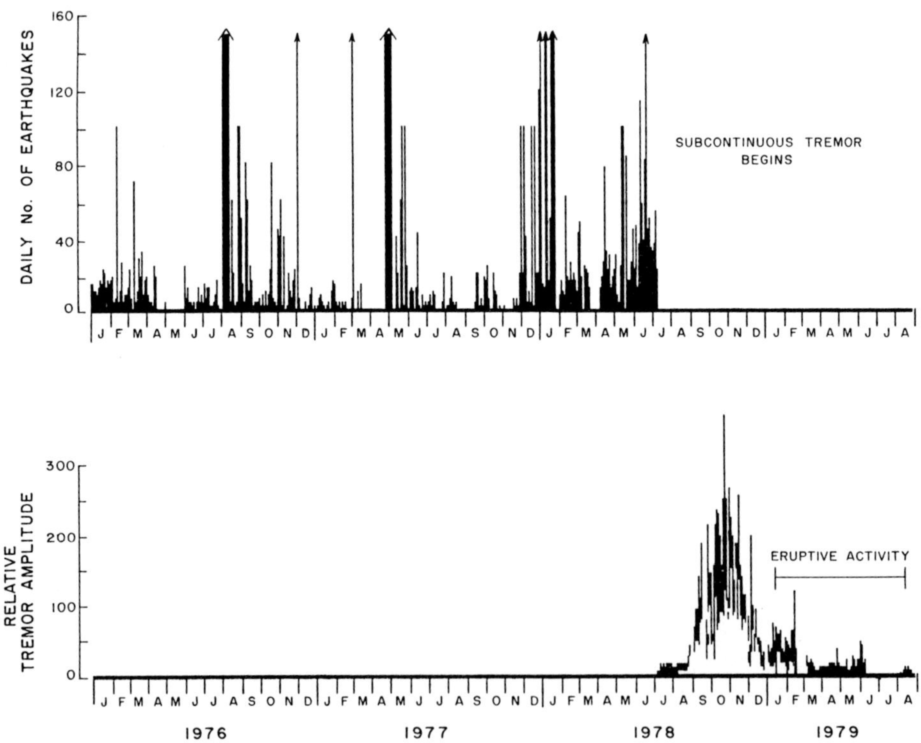

Fig. 4. Daily number of earthquakes and tremor amplitude recorded 7 km from the vent of Karkar volcano. The smallest event recorded is about M 1.5

710°-900°C. The eventual eruption occurred from this fumarolic area.

Towards the end of October the seismic amplitudes began to decrease, and continued to decline through January 1979 when phreatic explosions began. The largest explosion was on 8 March and was preceded by 2 weeks of increased numbers of low-frequency events (15-40 per day). On 6 March, 2 days before the explosion, the tremor amplitude increased slightly, although it was still much smaller than the tremor recorded during the previous few months. Following a heavy rain, there were strong white vapor emissions forming a column to a height of about 2000 m which accompanied the increase in tremor.

The eruption lasted for 7 months producing about 10^8 m^3 of tephra. There was no eruptive activity during the times of high amplitude tremor, although the incandescent zone near the foot of Bagiai was first noted in September 1978. It is suggested that the tremor was associated with the transport of high pressure steam and water in a hydrothermal system heated by a rising (but non-eruptive) magma body (McKee et al. 1981a). This is consistent with the observations that the 1979 activity consisted of phreatic explosions.

Ulawun

Ulawun is a 2300-m-high strato-volcano of basaltic to andesitic composition with a base diameter of about 20 km, located on the north coast of central New Britain. There was a minor eruption in 1941 and since 1958 eruptive periods lasting a few days to a few weeks have occurred at intervals of 0.3 to 5 years (Johnson et al. 1972a; Cooke et al. 1976; Cooke 1981c; McKee et al. (1981b).

A seismograph was installed at Ulamona, 10 km northwest of the summit in December 1976 and seismic precursors have been recognized 3-10 days before all five eruptions since that time, and in addition, there were more extended periods of swarms 1-2 years before eruptions in 1978 and 1984-85.

1978 Eruption

Prior to the eruption of May 1978, swarms of low-frequency (~3 Hz) earthquakes were recorded in December 1976 and August 1977. A few small events per day began to appear in late April 1978, and larger events resembling short bursts of tremor were recorded on 2-3 May and a few hours before visible eruptive activity began on 7 May. An intensification of tremor on 10 May may have been associated with lava effusion from a fissure on the lower eastern flank. About 20 high-frequency events were recorded at Ulamona and 140 km away in Rabaul on 9-10 May. It is suggested that these were associated with the opening of a conduit for the lava effusion (McKee et al. 1981b).

1980 Eruption

In 1980 small earthquakes were recorded on 3, 5 and 6 October and on the night of the 6th, low-frequency events became more numerous as a faint glow was noted from the summit. On the morning of the 7th, strong explosions produced pyroclastic avalanches and a large eruption column 20 km in height (McKee 1982).

1984, 1984-85 and 1985 Eruptions

To illustrate the seismic activity prior to the eruptions in 1984 and 1985, Figs. 5 to 7 show the seismic amplitudes and the number of volcanic earthquakes from 1983 to 1985. Seismic amplitudes were calculated in the same manner as for Manam. The seismometer was moved to a new location 3 km from the summit in late 1981. From the time of installation, large numbers of low-frequency earthquakes were recorded, so it is difficult to determine when this relatively high level of seismicity began. By the beginning of 1984 the seismicity had decreased to considerably lower levels. During 1983, the average size of the

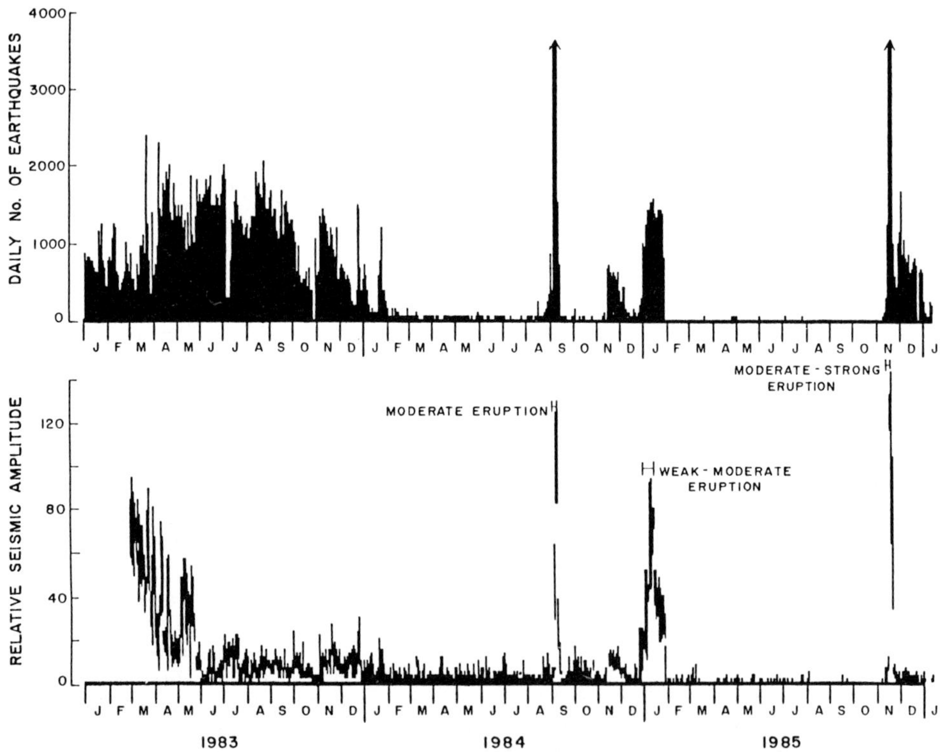

Fig. 5. Daily number and amplitudes of earthquakes recorded 3 km from the summit of Ulawun. The smallest event recorded is about M 1.5

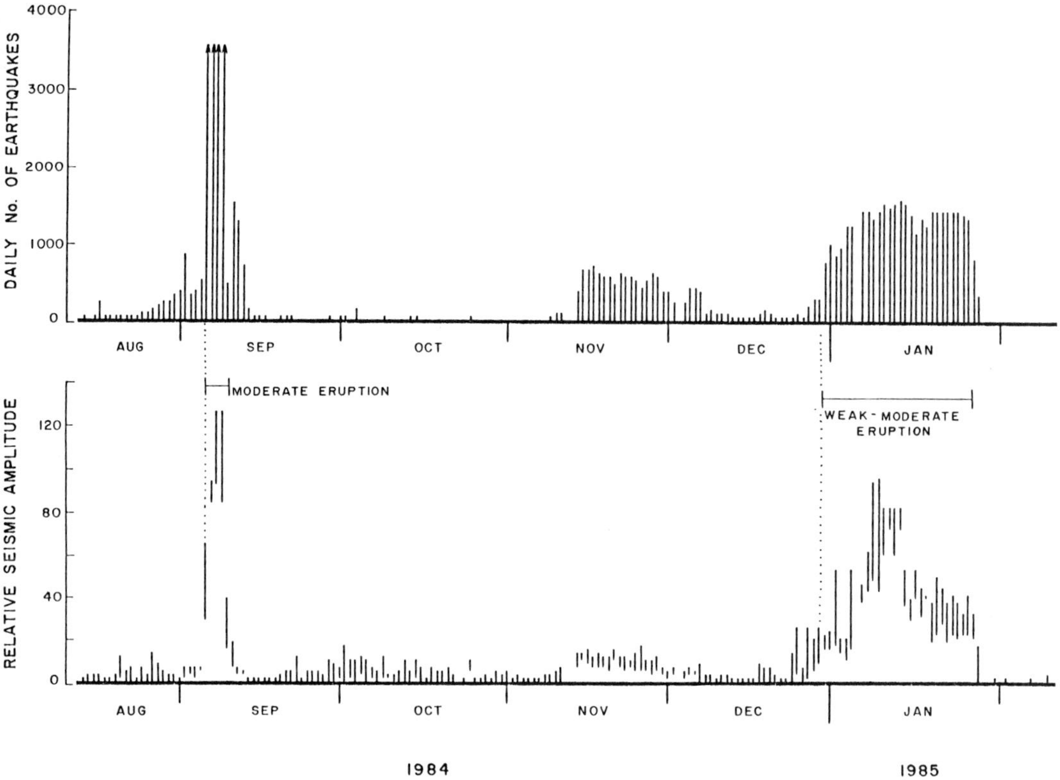

Fig. 6. Detailed version of Fig. 5 which shows the time periods of the 1984 and 1984-85 eruptions at Ulawun

earthquakes decreased, as seen by the drop in amplitudes, but the numbers of events remained high until the beginning of 1984. This extended period of strong seismicity during 1981-83 may be considered a possible precursor to the eruptive activity of 1984 and 1985, but with the short data base this cannot be confirmed.

In August 1984 the numbers of earthquakes began to increase again (see Fig. 6 which is an enlarged version of Fig. 5). In 10 days the increase in seismicity led to a moderate Strombolian eruption during which the summit crater was partially filled with molten lava. In mid-November the seismicity rose again but died away by mid-December with no eruptive activity. In late December the amplitudes and numbers of earthquakes increased once again, and 7 days later a 1-month-long period of weak eruptive activity began (Fig. 6).

The seismicity level decreased after January and remained low until mid-November 1985. Starting on 12 November the seismicity began to increase again and steadily built up for 5 days (Fig. 7), leading to an eruption which produced moderate explosive activity, moderate pyroclastic flows and small lava flows (10^6 m^3).

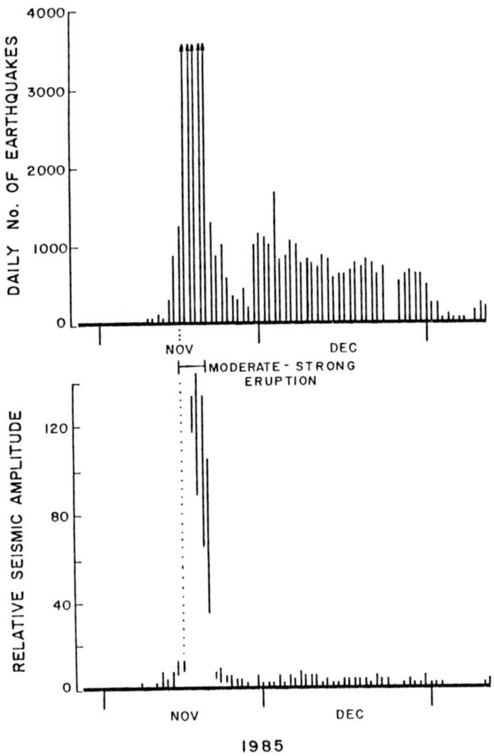

Fig. 7. Detailed version of Fig. 5 which shows the time period of the 1985 eruption at Ulawun

Rabaul

Situated on the northeastern end of New Britain Island, the 9 x 14 km Rabaul Caldera contains four centers that have had historical eruptions. The erupted products of the caldera contain a wide range of compositions from basalt to rhyolite, although historical eruptions have been of high-silica andesite or dacite. The most recent eruptions were in 1937 (Vulcan Cone formed and simultaneously Tavurvur erupted) and 1941-43 (Tavurvur produced intermittent explosions).

1937 Eruption

Widely felt earthquakes were reported before the eruption in 1937. On 28 May, a day before the eruption began, a moderate earthquake (probably about magnitude 5 to 6) was strongly felt throughout the Rabaul area, although smaller tremors had reportedly been felt for a few weeks in the Vulcan area. On the morning of the 29th, continuous earthquakes were felt throughout the Rabaul area, producing small landslides and minor damage to buildings. Also, by the morning of the 29th the seafloor near Vulcan was uplifted, and fringing reefs were exposed about 2 m above sea level. The timing of this event is not clear, as some reports say the uplift occurred on the morning before the eruption while other reports indicate that it may have started as much as a few weeks before the eruption.

The explosive eruption of Vulcan began on the afternoon of the 29th, killing 500 people during the first 30 h of strong activity. The eruption of Tavurvur, opposite the caldera, began on 30 May. The total volume of the eruptive products was estimated to be 4 x 10^8 m^3 (Fisher 1939). Prior to a similar eruption of Vulcan (Island) and Tavurvur in 1878, felt earthquakes were also reported (Fisher 1939).

1941-43 Eruptions

Prior to the 1941-43 eruption, new solfataric areas were noticed in the Tavurvur crater with temperatures of around 100°C in August 1940. In December 1940 the temperature was 224°C and continued to rise, reaching 392°C 3 days before Tavurvur began to erupt in June 1941 (see Fig. 8 in Fisher 1976). During the time of the temperature increase, the composition of the gas became increasingly more acidic, as increasing amounts of sulphur dioxide and hydrogen chloride were emitted (Fisher 1976).

A study of tide gauge records at Rabaul and sea level measurements near Tavurvur indicated that the area near the volcano subsided 7.5 cm during the 2 years prior to the eruption (Fisher 1976). However, seismographs, operating at a magnification of 150, and tiltmeters, sensitive to 5 µrad of arc, installed together at locations 1 and 8 km from Tavurvur, did not show any precursory activity.

Kizawa (1951) reported some precursory seismicity, including a felt earthquake, a month before a moderate explosion at Tavurvur in 1943.

1971-1985 Unrest

Beginning in late 1971 earthquakes began to occur more frequently within the caldera. Levelling measurements begun in 1973 showed that parts of the caldera were being uplifted at a rate of about 9-10 cm per year. In late 1983 the seismic activity (including felt events) and ground deformation rates greatly increased. Also new fumarolic activity starting in 1983 about 2 km northwest of Tavurvur killed some vegetation. The large number of earthquakes (often hundreds per day) along with large ground deformations (4 to 8 cm/month of uplift and 50 to 80 microradians/month of tilt) suggested the strong possibility of an imminent eruption (McKee et al. 1984). However, no eruption occurred during the peak in seismic activity and by mid-1985 the seismicity and ground deformation rates had dropped back to pre-1983 levels. Although no eruption occurred, the period of unrest at Rabaul was associated with magma intrusion(s) at shallow depths within the caldera (Mori et al. Chap. 25, this Vol.; McKee et al. Chap. 24, this Vol.). This points out the difficulty in distinguishing between precursors which may lead only to intrusive events from precursors that lead to eruptions.

Langila

Langila is a group of four overlapping composite cones up to 1330 m in height on the eastern flank of the extinct volcano Talawe which is located at the western end of New Britain (Palfreyman et al. 1981). Recent eruption products were of andesitic composition. There has been continous intermittent activity since 1954, following a relatively inactive period which probably lasted about 30 years (Palfreyman et al. 1981).

1954 Eruption

The resumption of activity in 1954 (Taylor et al. 1957) was preceded by increased fumarolic activity, which was first noted in 1944, together with signs of a possible small eruption between 1938 and 1944. Increased vapor emissions and considerable destruction of vegetation were noticed in 1952. Acidic gases were being emitted from vents in the crater at temperatures around 200°C. On 1 May 1954 yellowish-white vapor was noticed from crater 2 and increased in volume until 15 May. The explosive activity began on 18 May 1954.

Since 1976 a seismograph has been operated near Langila and low-frequency tremor is often recorded during periods of strong activity. However, no obvious seismic or other precursors have been noted before either periods of explosive activity or before effusive phases.

Kadovar

Kadovar is a small volcano with a height about 1000 m above its seafloor base which has a diameter of about 9 km. It has probably not been active since 1700 or earlier (Wallace et al. 1981). A strong increase of fumarolic activity in 1976-77 was thought to be an eruption precursor but it did not lead to an eruption (Wallace et al. 1981).

1976-77 Unrest

In September 1976, a moderately strong fumarolic area developed on the southern side of the island with vents producing sulphur dioxide and hydrogen chloride. By April 1977 trees and vegetation had been killed in a 15,000 m^2 area. Re-growth of vegetation had begun by mid-1978, although locally, temperatures up to 100°C were still measured until September 1978.

Before the fumarolic area appeared, there had apparently been a thermal disturbance off the southern shore of the island in May or June 1976. Villagers reported feeling weak tremors (probably regional earthquakes) and that the sea had been discolored. A yellow discoloration of the seawater close to shore persisted until late 1978.

D'Entrecasteaux Islands

This region, which includes Dobu, Neumara, the southeast portion of Fergusson and the northwest portion of Normanby Islands contains several volcanic cones and thermal areas. The submarine area between the islands is possibly a caldera (Lowenstein 1982a). There have been no historical eruptions within this area although there have been numerous earthquakes and changes in the hot spring activity.

1953-57 Unrest

During the 1950's there were frequent earthquakes that were strongly felt in the area. It is unclear whether the earthquakes were directly related to volcanic activity. However, volcanologists were concerned enough to build a permanent observatory at Esa'ala on Normanby Island.

4 Discussion

4.1 Seismic Precursors

The most commonly reported precursors in our study are seismic (Table 1). This is partly because earthquakes can be reported by the local inhabitants and the detection of larger events does not depend on instrumentation. In addition, many of the volcanoes have been instrumented with a seismograph in recent years. Table 4 summarizes the occurrences of seismic precursors for each volcano. Also included are the distances to the nearest population centers which may report felt events and information about the seismographs used for monitoring the volcano. Of the 50 eruptions studied, there were 11 cases of precursory seismicity. For the volcanoes that have been instrumented with seismometers there have been 8 recognized cases of seismic precursors out of 14 eruptions, although most of these were recognized in hindsight. The high rate of occurrence of seismic precursors justifies emphasis on seismic surveillance of the volcanoes. However, the patterns and types of the seismic activity differ considerably at the various volcanoes. For example, the recent seismicity at Rabaul consists almost entirely of high-frequency tectonic-like events, the precursory seismicity at Ulawun volcano consists of mostly low-frequency (~3 Hz) earthquakes, and at Manam the seismicity is a nearly continuous tremor (1-3 Hz), with few discrete events. Further study is needed to clarify the characteristics of the seismicity at each volcano and to understand their significance to the eruptive mechanisms.

4.2 Open and Closed Conduit Systems

One of the factors which may control the precursory activity is the openness of the conduit to the surface. Volcanoes which have been completely inactive between eruptions, and therefore probably have closed vents, tend to show stronger precursory

Table 4. Detection of seismic precursors[a]

Volcano	Nearest Villages Dist. (km)	Pop.	Eruptions with precursory felt events	Seismograph date gain dist.	Eruptions with instrumental precursors
Kadovar	1	300	0 of 0	None	----
Bam	2	400	0 of 2	None	----
Manam	4-5	5000	1 of 9	1957-65 10x 4 km	1 of 2
				1965-present 1000-3000x 4 km	1 of 4
Karkar	10-15	23,000	0 of 2	1975-present 2500x 7 km	1 of 1
Submarine center A	35	12,000	0 of 2	None	----
Long Island	10-18	1000	0 of 3	None	----
Ritter Island	10-12	500	0 of 2	None	----
Langila	8-10	300	0 of 5	1976-79 4000x 10 km	0 of 1
				1980-present 4000x 9 km	c
Pago	10-15	>1000	b	None	c
Ulawun	10	1000	0 of 12	1977-81 1000x 11 km	2 of 2
				1981-present 1000x 3 km	3 of 3

[a] Distances of villages and seismographs are measured from the volcano summit or vent. The periods of seismograph operation are listed only if the instrument was operating for an extended period. There have also been numerous occasions of shorter recording periods, particularly during eruptive activity.

[b] There is no entry under felt events for Pago, because the eruption began in 1911, well before the beginning of the study period.

[c] Eruption in progress when seismograph was installed.

Table 4 (cont.)

Volcano	Nearest Villages Dist. (km)	Pop.	Eruptions with precursory felt events	Seismograph date gain dist.	Eruptions with instrumental precursors
Submarine center B	40	500	0 of 2	None	----
Rabaul	1-10	>10,000	1 of 2	1940-1942 150x 1 and 7 km	0 of 1
				1942-1943 100x 2.5 km	c
				1943-45 100x 9 km	----
				1953-present 1-11 stations 1000-6000x 1-10 km	----
Lamington	7-20	>10,000	1 of 1	1965-1972 1000x 15 km	----
				1971-present 3000x 3 km	----
Goropu	18	Several 100	1 of 1	None	----
Tuluman	4-7	Several 100	0 of 1	None	----
Bagana	11-13	100	0 of 4[d]	1984-present 2000x 1.5 km	c

[c] Eruption in progress when seismograph was installed.
[d] Bultitude (1976) reported numerous felt earthquakes before the 1960-67 eruption, but these were probably the magnitude 6 tectonic earthquake of 10 September 1959 and its aftershocks.

signs (e.g. Rabaul, 1937; Lamington, 1951) than those at which the vents remain open. This category would also include dormant volcanoes which have not had historical eruptions but are still considered to have potential for eruptions (e.g. Kadovar, D'Entrecasteaux Islands).

We have used the reports of felt earthquakes to compare precursory activity at volcanoes with open and closed conduits. The occurrence of felt earthquakes is a precursor that can be evaluated for most of the eruptions. The information in Table 4 shows that most of the volcanoes have people living close to the summit who would be able to report larger earthquakes. There were 40 eruptions from 10 volcanoes for which there were population centers 10 km or less from the vent. Using the criterion that a volcano is considered to have a closed conduit if it has not erupted in the preceding 50 years, there were six eruptions at these volcanoes from vents that were previously closed (Karkar 1974-75, Ritter Island 1972, Langila 1954, Rabaul 1937, Lamington 1951 and Tuluman 1953-57). There was also one eruption (Goropu 1943-44) from a closed vent from which the nearest village was 18 km away but there were still reports of felt events. Of these seven eruptions from closed conduits, three (43%) had precursory felt earthquakes. In contrast, there were 33 eruptions from open conduits, of which there was only one case (3%) of precursory felt events.

Volcanoes with closed conduits tend to contain the pressures of expanding magma chambers, so that increasing stresses within the volcanic structure give rise to earthquakes and significant ground deformations. Also, since the rising magma must open new pathways to the surface, there may be a considerable amount of fracture formation and associated precursory seismicity (Minakami 1974). Another related factor is that volcanoes with closed conduits that are able to contain the increasing stresses will also tend to produce more infrequent larger eruptions. Since the eruptions are larger the precursory seismicity and ground deformation would also be more apparent.

Volcanoes with open conduits that have a low level of continuous activity [e.g. Manam, Bagana and Langila (post-1954)] show less obvious precursory activity before times of strong eruptive activity. This may be partly because it is difficult to distinguish possible precursors from the constant background level of activity (as in the case of the Manam seismicity), and partly because, with an open system, there is less buildup of pressure due to the rising magma. However, for an open-conduit volcano recognition of fumarolic and thermal activity might be a more reliable guide to future activity, since gases originating from the magma body can more easily escape to the surface. This might have been the case with the fumarolic changes observed before the 1941 Tavurvur eruption in Rabaul, in which case the conduit system may have still been relatively open following the 1937 eruption.

4.3 Chemical Composition

It is interesting to note that the volcanoes that produced precursory felt earthquakes also tend to be composed of rocks with higher silica content. Using the same 40 eruptions described above, for which there are villages within 10 km of the vent, Fig. 8 shows the percentage of the eruptions that had precursory felt earthquakes for three ranges of silica content. Al-

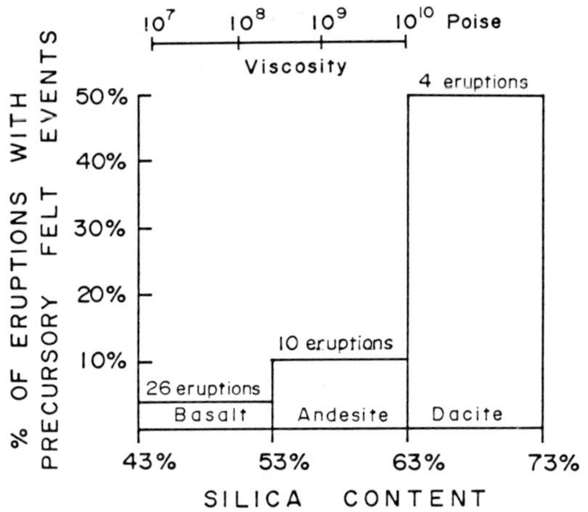

Fig. 8. Percentage of eruptions with precursory felt earthquakes plotted for three ranges of silica composition. The corresponding viscosities, estimated using the method of Bottinga and Weill (1972) are shown in the upper portion of the figure

though our data set includes only four cases of precursory felt events, there appears to be a correlation between precursory seismicity and increasing silica content. A principal physical attribute of magmas of high silica content is their high viscosity. Minakami (1974) stated that more viscous magmas are likely to produce more noticeable precursors because of the larger pressures the magma will sustain before eruption. The range of viscosities for these magmas was calculated from the chemical compositions of the erupted products using the method of Bottinga and Weill (1972). Assuming a temperature of 1200°C, the estimated viscosities are shown in the upper portion of Fig. 8.

In Fig. 8 we do not take into account the size of the eruptions. This may be misleading, since in our data set the larger eruptions, which are the ones more likely to have noticeable precursors, also tend to be the eruptions with more acidic compositions. However, we found it significant that one of the larger eruptions, the 1974-75 Karkar event which produced 7 x 10^7 m^3 of lava, was of low-silica andesite composition and did not have any precursory felt events although there were 23,000 people living 10 to 15 km from the vent.

The chemical composition of the magma and the degree to which the conduit of the volcano is closed between eruptions are not totally independent factors. The volcanoes with the higher silica contents also tend to have larger and more infrequent eruptions. So there are related physical reasons for both the silica content and the closed state of the conduit to be correlated with precursory seismicity.

4.4 Comparison of Eruptions

At present, we do not have a good understanding of the mechanics of the volcanoes of Papua New Guinea, so the prediction of eruptions must be based largely on applying empirical relationships derived from observations of previous eruptions. This practice assumes that the eruption sequence, including any precursory activity, will be similar to the eruptions of the past. The data base in Papua New Guinea does not cover a long enough period to make reliable judgements about the repeatability of the eruptions, but a few comments can be made on the volcanoes that have produced two or more eruptions in the study period. From the brief descriptions of the eruptions in Tables 1 and 3, the volcanoes can be divided into two groups: those that have produced generally similar eruptions (Bam, Manam, Ritter, Bagana) and those that have displayed significant variations in eruptive style and magnitude (Karkar, Long Island, Langila, Ulawun, Rabaul).

The four volcanoes in the first group may be good candidates for the application of empirical relationships for predicting eruptions. For the remaining five volcanoes predictions based on observations of past eruptions is more difficult.

5 Conclusions

Of the 50 eruptions studied there were 11 cases of precursory seismicity ranging from about 18 months to 1 day before an eruption. Eight of these cases were recorded instrumentally (Manam 1960, 1984, Karkar 1979, Ulawun 1978, 1980, 1984, 1984-85, 1985) and four cases had felt reports (Manam 1960, Rabaul 1937, Goropu 1940, Lamington 1951). There was at least one case of increased seismicity (both felt and instrumentally recorded) which did not lead to a subsequent eruption (Rabaul 1983-85).

There were six cases of increased thermal activity at new or known fumaroles, ranging from 10 years to a few days before eruptions (Karkar 1974, 1979, Long Island 1973, Langila 1954, Rabaul 1941, Lamington 1951). The chemical composition of the fumarolic gases emitted before the Langila 1954 and Rabaul 1941 eruptions also became progressively more acidic. There was at least one case of strong fumarolic increases without a subsequent eruption (Kadovar 1976).

Ground deformation was noted before eruptions at two volcanoes. Reefs appeared above sea level at least 8 h before the 1937 Rabaul eruption. Significant tilt changes were observed instrumentally several years to days before eruptions at Manam in 1957, 1960, 1974 and 1982.

Closed conduit systems appear more likely to exhibit clear precursory activity because the volcano is able to build up higher stresses than in an open conduit. For related reasons, volcanoes with magmas of higher silica compositions, and thus higher viscosity, would also sustain higher levels of stress and tend to have stronger seismic and ground deformational precursors.

Acknowledgements. Many helpful comments were provided by J. Cassidy, J. Latter, R.W. Johnson and P. Lowenstein. The authors publish with the permission of the Secretary of the Department of Minerals and Energy.

References

Baker G (1946) Preliminary note on volcanic eruptions in the Goropu Mountains, southeastern Papua, during the period December 1943 to August 1944. J Geol 54:19-31

Ball EE, Johnson RW (1976) Volcanic history of Long Island, Papua New Guinea. In: Johnson RW (ed) Volcanism in Australasia. Elsevier, Amsterdam, pp 133-148

Best JG (1954) Preliminary report, Bam Island, Madang District, New Guinea. Aust Bur Min Res, Record 1954/59, pp 10

Bottinga Y, Weill DF (1972) The viscosity of magmatic silicate liquids: a model for calculation. Am J Sci 272:428-475

Branch CD (1967) April 1964 eruption of Manam Volcano. Aust Bur Min Res Rep 107, pp 35

Bultitude RJ (1976) Eruptive history of Bagana Volcano, Papua New Guinea, between 1882 and 1975. In: Johnson RW (ed) Volcanism in Australasia. Elsevier, Amsterdam, pp 317-336

Bultitude RJ (1979) Bagana Volcano, Bougainville Island geology, petrology, and summary of eruptive history between 1875 and 1975. Geol Surv Papua New Guinea Mem 6, pp 35

Cooke RJS (1975) Ulawun. Bull Volcanol Eruptions 13:22-23

Cooke RJS (1976) Manam. Bull Volcanol Eruptions 14:18-19

Cooke RJS (1981a) Eruptive history of the volcano at Ritter Island. In: Johnson RW (ed) Cooke-Ravian volume of volcanological papers. Geol Surv Papua New Guinea Mem 10:115-124

Cooke RJS (1981b) Eruptions at Pago volcano, 1911-1933. In: Johnson RW (ed) Cooke-Ravian volume of volcanological papers. Geol Surv Papua New Guinea Mem 10:135-146

Cooke RJS (1981c) Notes on the activity at Ulawun volcano, 1700-1958: results of a literature search. In: Johnson RW (ed) Cooke-Ravian volume of volcanological papers. Geol Surv Papua New Guinea Mem 10:147-152

Cooke RJS, Johnson RW (1978) Volcanoes and volcanology in Papua New Guinea. Geol Surv Papua New Guinea Rep 78/2, pp 47

Cooke RJS, Johnson RW (1981) Bam Volcano: morphology, geology, and reported eruptive history. In: Johnson RW (ed) Cooke-Ravian volume of volcanological papers. Geol Surv Papua New Guinea Mem 10:13-22

Cooke RJS, McKee CO, Dent VF, Wallace DA (1976) Striking sequence of volcanic eruptions in the Bismarck volcanic arc, Papua New Guinea, in 1972-75. In: Johnson RW (ed) Volcanism in Australasia. Elsevier, Amsterdam, pp 149-172

Davies RA (1972) Eruptive history of Bam Island Papua New Guinea, with notes on instrumental observations. Geol Surv Papua New Guinea Note on Investigation 72-024, pp 18

Fisher NH (1939) Geology and vulcanology of Blanche Bay, and the surrounding area, New Britain. Territory of New Guinea, Geol Bull 1, pp 68

Fisher NH (1976) 1941-42 eruption of Tavurvur volcano, Rabaul, Papua New Guinea. In: Johnson RW (ed) Volcanism in Australasia. Elsevier, Amsterdam, pp 201-210

Johnson RW, (1984) Volcanological Inspections in Papua New Guinea, February 1984. Geol Surv Papua New Guinea Rep 84/4, pp 41

Johnson RW, Davies RA, Palfreyman WD (1971) Cape Gloucester area, New Britain: volcanic geology, petrology, and eruptive history of Langila craters up to 1970. Aust Bur Min Res, Record 1971/14, pp 34

Johnson RW, Davies RA, White AJR (1972a) Ulawun volcano New Britain. Aust Bur Min Res, Bull 142, pp 42
Johnson RW, Taylor GAM Davies RA (1972b) Geology and petrology of Quaternary volcanic islands off the north coast of New Guinea. Aust Bur Min Res, Record 1972/21, pp 127
Johnson RW, Everingham IB, Cooke RJS (1981) Submarine volcanic eruptions in Papua New Guinea: 1878 activity of Vulcan (Rabaul) and other examples. In: Johnson RW (ed) Cooke-Ravian volume of volcanological papers. Geol Surv Papua New Guinea Mem 10:167-180
Kizawa T (1951) Volcanic tremor and tilting of the ground (first paper). Kenshin Jiho 15:18-34 (in Japanese)
Lowenstein PL (1982a) Report on the occurrence of a new thermal spring at Nuemara Island in the D'Entrecasteaux Group, Milne Bay Province. Geol Surv Papua New Guinea Technical Note, TN 17-82, pp 6
Lowenstein PL (1982b) Problems of volcanic hazards in Papua New Guinea. Geol Surv Papua New Guinea Rep 82/7, pp 62
McKee CO (1981) Geomorphology, geology, and petrology of Manam volcano. In: Johnson RW (ed) Cooke-Ravian volume of volcanological papers. Geol Surv Papua New Guinea Mem 10:23-38
McKee CO (1982) Ulawun. Bull Volcan Eruptions 20:26
McKee CO (1984) Manam. Bull Volcan Eruptions 22:23-25
McKee CO (1987) Ulawun. Bull Volcan Eruptions 24:19-20
McKee CO, Wallace DA (1981) Lava fields in the inner caldera of Karkar volcano. In: Johnson RW (ed) Cooke-Ravian volume of volcanological papers. Geol Surv Papua New Guinea Mem 10:49-62
McKee CO, Cooke RJS, Wallace DA (1976) 1974-75 eruptions of Karkar volcano, Papua New Guinea. In: Johnson RW (ed) Volcanism in Australasia. Elsevier, Amsterdam, pp 173-190
McKee CO, Wallace DA, Almond RA, Talai B (1981a) Fatal hydroeruption of Karkar volcano in 1979: development of a maar-like crater. In: Johnson RW (ed) Cooke-Ravian volume of volcanological papers. Geol Surv Papua New Guinea Mem 10:63-84
McKee CO, Almond RA, Cooke RJS, Talai B (1981b) Basaltic pyroclastic avalanches and flank effusion from Ulawun volcano in 1978. In: Johnson RW (ed) Cooke-Ravian volume of volcanological papers. Geol Surv Papua New Guinea Mem 10:153-166
McKee CO, Lowenstein PL, de Saint Ours P, Talai B, Itikarai I, Mori JJ (1984) Seismic and ground deformation at Rabaul caldera: prelude to an eruption? Bull Volcanol 47:397-411
McKee CO, Johnson RW, Lowenstein PL, Riley SJ, Blong RJ, de Saint Ours P, Talai B (1985) Rabaul Caldera, Papua New Guinea: volcanic hazards, surveillance, and eruption contingency planning. J Volcanol Geotherm Res 23:195-237
Minakim T (1974) Prediction of volcanic eruptions. In: Civetta L, Gasparini P, Luongo G, Rapolla A (eds) Physical volcanology. Elsevier, Amsterdam, pp 313-333
Mori J, McKee CO, Talai B, Itikarai I (1986) Precursors to volcanic eruptions in Papua New Guinea. Geol Surv Papua New Guinea Rep 86/27, pp 39
Mori J, McKee CO, Itikarai I, Lowenstein P, Talai B, Patia H (1987) Recent inflationary tilt at Manam Volcano, Papua New Guinea as recorded on a water-tube tiltmeter. J Volcanol Geotherm Res 33:361-362

Palfreyman WD, Cooke RJS (1976) Eruptive history of Manam volcano, Papua New Guinea. In: Johnson RW (ed) Volcanism in Australasia. Elsevier, Amsterdam, pp 117-131
Palfreyman WD, Wallace DA, Cooke RJS (1981) Langila volcano: summary of reported activity, eruptive history, and eruptive periodicity from 1961 to 1972. In: Johnson RW (ed) Cooke-Ravian volume of volcanological papers. Geol Surv Papua New Guinea Mem 10:125-134
Reynolds MA (1957a) Volcano-seismic phenomena in eastern Papua since 1939. Aust Bur Min Res, Record 1957/14, pp 14
Reynolds MA (1957b) Investigation of conditions on Bam Island, January 1957. Aust Bur Min Res, Record 1957/42, pp 11
Reynolds MA, Best JG (1976) Summary of the 1953-57 eruption of Tuluman volcano, Papua New Guinea. In: Johnson RW (ed) Volcanism in Australasia. Elsevier, Amsterdam, pp 287-296
Reynolds MA, Best JG, Johnson (1980) 1953-57 eruption of Tuluman Volcano: rhyolitic volcanic activity in the northern Bismarck Sea. Geol Surv Papua New Guinea Mem 7, pp 44
Scott BJ, McKee CO (1984) Deformation, earth tidal influences, and eruptive activity at Manam volcano, Papua New Guinea, 1957-1982. Geol Surv Papua New Guinea Rep 84/3, pp 56
Talai B (1980) Ulawun. Bull Volcan Eruptions 18:21-22
Taylor GAM (1955a) Report on Bam Island volcano and an inspection of Kadovar and Blup Blup. Aust Bur Min Res, Record 1955/73, pp 9
Taylor GAM (1955b) Notes on volcanic activity and thermal areas in the D'Entrecasteaux Islands. Aust Bur Min Res, Record 1955/75, pp 11
Taylor GAM (1958a) The eruptive trend of Manam Volcano. Aust Bur Min Res, Record 1958/73, pp 11
Taylor GAM (1958b) The 1951 eruption of Mount Lamington, Papua. Aust Bur Min Res, Bull 38, 117 pp
Taylor GAM (1960) An experiment in volcano prediction. Aust Bur Min Res, Record 1960/74, pp 17
Taylor GAM (1963) Seismic and tilt phenomena preceding a Pelean eruption from a basaltic volcano. Bull Volcanol 26:5-11
Taylor GAM, Best JG, Reynolds MA (1957) Eruptive activity and associated phenomena, Langila volcano, New Britain. Aust Bur Min Res Rep 26, pp 53
Wallace DA, Cooke RJS, Dent VF, Norris DJ, Johnson RW (1981) Kadovar volcano and investigations of an outbreak of thermal activity in 1976. In: Johnson RW (ed) Cooke-Ravian volume of volcanological papers. Geol Surv Papua New Guinea Mem 10:1-13

Eruptions, Eruption Precursors and Related Phenomena in the Lesser Antilles

J.B. SHEPHERD[1]

Abstract

Brief revised descriptions are given of all volcanoes in the Lesser Antilles which are believed to have a potential for future eruptions. Accounts of all historic eruptions have been examined with a particular view to identifying phenomena precursory to eruptions. All eruptions which have occurred in historic time and which are sufficiently well described for conclusions to be drawn have been preceded by precursory phenomena lasting for periods ranging from a few days to 14 years. The main precursory phenomena have been swarms of local earthquakes, increased fumarolic activity and ground deformation. Other possible precursory phenomena such as changes in fumarolic gas chemistry, and gravity and magnetic changes have not been observed, perhaps because measurements in sufficient detail have rarely been made. The usefulness of these precursory phenomena in predicting eruption onsets is reduced by the fact that false alarms outnumber genuine precursory sequences by a factor of two or three to one.

Regional tectonic earthquakes of magnitude greater than about 6.0 which occur at shallow depth close to the volcanoes of the northern Lesser Antilles generally have a significant effect on volcanic earthquake swarms, but none of the swarms so affected has ended in an eruption. There is no evidence for eruption periodicity either at individual volcanoes or in the arc as a whole.

1 Introduction

There are at least 7 and possibly as many as 25 active volcanoes in the Lesser Antilles island arc. The precise number depends on the definition of "active" which is used and on the criteria used to decide how many of a number of closely spaced vents should be regarded as separate volcanoes. The smaller number refers to volcanoes which are known or are strongly suspected to have erupted within the historic period, while the larger number refers to volcanoes which are believed to have

[1] Seismic Research Unit, The University of the West Indies, St. Augustine, Trinidad

IAVCEI Proceedings in Volcanology 1
J.H. Latter (Ed.), Volcanic Hazards

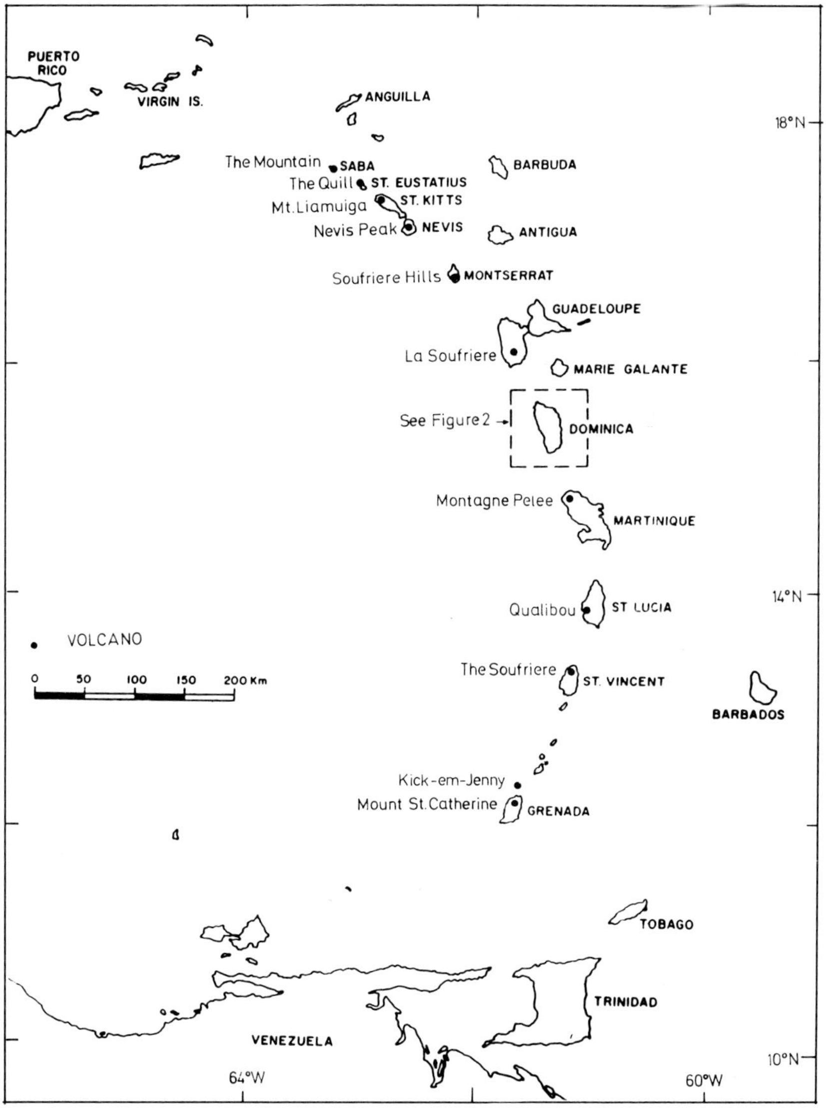

Fig. 1. Outline map of the Lesser Antilles showing locations of the main volcanic centers

erupted within the past few tens of thousands of years and which are believed to have the potential for future eruptions. The Appendix gives brief descriptions of volcanoes which fall into the larger category and uses a slightly modified form of the numbering system used by Simkin et al. (1981). Geographical locations of the volcanoes are shown in Figs. 1 and 2. The volcanoes which have no known historic eruptions are judged to be active on one or more of the following criteria:

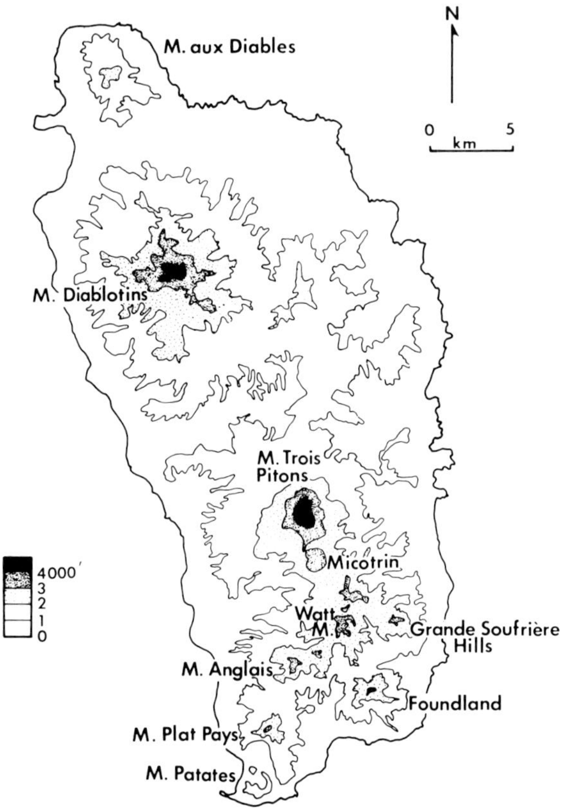

Fig. 2. Topographic map of Dominica showing the locations of ten post-Pleistocene volcanoes

1. Contemporary fumarolic activity;
2. The occurrence of local earthquakes of probable volcanic origin;
3. A youthful morphology similar to that of other nearby volcanoes known to be active on other grounds.

In comparison with the previous catalogues of Robson and Tomblin (1966) and Simkin et al. (1981) there are several differences. Two submarine volcanoes (1600-07 and 1600-15) have been omitted because careful investigations have shown that they do not exist (Bouysse and Sigurdsson 1982). The list of volcanoes in the island of Dominica has been extended because recent fieldwork by Sigurdsson (unpublished field notes 1981), Roobol et al. (1983) and Wadge (1984, 1985) has shown that each of the volcanoes described by Robson and Tomblin as the Valley of Desolation (1600-10) and Morne Patates (1600-11) consists of a group of closely spaced but probably independent volcanoes. The numbering system has been modified slightly for Grenada and the southern Grenadine islands. In this region there are a number of explosion craters which extend from the Ile de Caille in the southern Grenadines to St. George's Harbour in the south (Arculus 1973). No eruptions have been observed but many of the craters have an extremely youthful morphology and hot-spring activity occurs throughout the region. Simkin et al. (1981) include the Ile de Caille as a separate volcano (1600-16B) but until all of these craters have been studied more carefully it is probably more logical to describe them as a single volcanic field.

2 West Indian Volcanic Eruptions

There is some uncertainty about the number of eruptions which have taken place in the Lesser Antilles even in the historic period. Most of the islands were visited by Columbus in the late 15th century but at that time they were inhabited by Carib Indians who did not encourage European settlement and did not keep complete records of volcanic activity. Settlement proceeded at different rates in different islands so that some islands such as Guadeloupe and St. Kitts have comprehensive written histories dating back to the 16th century, whereas others such as St. Vincent and Dominica have hardly any detailed written history before the end of the 18th century. For this reason, fairly minor eruptions in some of the islands are sometimes better described than major eruptions in others, and minor eruptions in islands such as St. Vincent may have escaped attention entirely. In addition, careful reading of the original accounts from the 16th to 19th centuries shows that writers of this period often regarded the three major natural hazards of the region, earthquakes, volcanic eruptions and hurricanes, as equivalent, or at least closely interrelated, phenomena. Hurricanes are described as earthquakes, earthquakes as volcanic eruptions and all other possible combinations. Occasionally, these confusions have found their way into the scientific literature. For example both Perret (1939) and Mc Gregor (1949) mention a submarine volcanic eruption in 1831 between Barbados and St. Vincent without citing original sources for the information. The location is, geologically, a most unlikely one for a volcanic eruption and no other references to this eruption are to be found in contemporary newspapers or other sources. However, a major earthquake occurred in this region on 4 December 1831 (Robson 1964), and newspapers in Barbados erroneously reported it as a "volcanic disturbance". It is highly likely that further distortions converted the earthquake into a submarine eruption close to Barbados.

Also, many of the volcanoes are in a state of continuous fumarolic activity which varies considerably in intensity with time, and it is not always clear from the historic accounts when the activity reached such a peak that it should properly be described as a phreatic eruption. In the Appendix brief descriptions are given of all periods of increased activity which are known, together with descriptions of any observed premonitory phenomena and other associated activity. Long periods of related but intermittent eruptive activity, such as that of Montagne Pelee from 1902 to 1907, are here regarded as single eruptions.

It is often assumed that the typical West Indian volcanic eruption involves large volumes of volatile-rich magma with associated explosions and glowing avalanches. This is hardly surprising since the almost simultaneous eruptions of Montagne Pelee and the Soufrière of St. Vincent which began in 1902 are classical examples of these types of eruptions, and the supposed association has caused extreme anxiety during all the volcanic crises which have occurred in the Lesser Antilles during recent decades. All the volcanic islands are small with high densities of population close to the volcanoes, and public consciousness of the tragedies which occurred in Martinique and St. Vincent in 1902, when almost 40,000 lives were lost, is high. For these reasons, the maintenance of an effective volcanic monitoring system is given a high priority.

In fact the Appendix shows that, amongst the known eruptions on land, fairly mild phreatic eruptions (10 or possibly 12 in number) have been about twice as common as explosive magmatic ones (six) and that at least one eruption (that of Soufrière of St. Vincent in 1971-72) was entirely non-explosive. Phreatic eruptions have occurred in at least four, possibly five, volcanoes, whereas magmatic eruptions have been confined to Pelee and the Soufrière of St. Vincent.

3 Eruption Precursors

All the eruptions about which sufficient information exists are now known to have been preceded by premonitory phenomena of one sort or another, although these phenomena were often recognized only in retrospect. The premonitory phenomena which have been observed have been mainly of three types:

1. Premonitory earthquakes. These occurred before 11 of the 13 eruptions which have occurred on land since 1797.
2. Increased fumarolic activity. Observed before 4 of the 13 eruptions.
3. Ground deformation. Observed before one eruption.

From these observations the prospects for volcanic prediction in the future may appear to be good, and it is certainly true that if the pattern is continued, then no future eruption will occur entirely without warning. However, there are certain features which reduce the usefulness of these observations. The principal feature is that one or more of the symptoms (1), (2) and (3) have been observed on numerous occasions without being followed by an eruption. For example over 3000 earthquakes were felt close to the Soufrière Hills volcano between 1932 and 1938, and fumarolic activity increased enormously (Perret 1939; Powell 1938). This volcano-seismic crisis was far more severe than any which has preceded an eruption in the Lesser Antilles, and yet no eruption followed. Earthquake swarms have occurred close to many volcanoes of the Lesser Antilles since 1797 without being followed by eruptions, and at least 15 of these have been comparable in intensity with any that have preceded eruptions. Severe increases in fumarolic activity have been observed on five occasions, each occasion coinciding with a local earthquake swarm, and severe inflation of the volcano was observed on one of these occasions (Sheperd et al. 1971). Thus, although premonitory symptoms appear to be a necessary feature of impending West Indian eruptions they are not sufficient indicators.

Related to this is the fact that there is no clear relationship between the intensity of the premonitory symptoms and the type or intensity of the ensuing eruption (if any). The volcano-seismic crisis in Guadeloupe in 1975-77 (Feuillard et al. 1983) was much more intense than the periods of premonitory symptoms which preceded the violent magmatic eruptions of Montagne Pelee and the Soufrière of St. Vincent (Anderson and Flett 1903; Lacroix 1904), but was followed by a mild series of phreatic explosions only.

Finally it is clear that, even within a small region such as the Lesser Antilles, different volcanoes behave in quite different ways. This means both that symptoms premonitory to an eruption at one volcano may be quite different from those at a similar nearby volcano, and also that conditions which are quite normal at one volcano may be grossly abnormal at another. No general rules can be formulated which are applicable to the whole region, and each period of crisis at each volcano must be treated on its own merits. Superficially the Soufrière of St. Vincent and Montagne Pelee are similar volcanoes which behave in similar ways, but the symptoms premonitory to eruptions of the two volcanoes have been quite different. At the Soufrière each of the eruptions of 1812, 1902 and 1979 was preceded by a period of significant local earthquakes, but not by any significant increase in fumarolic activity. At Montagne Pelee, the opposite has generally been the case. None of the modern eruptions has been preceded by significant earthquake activity, but each has been preceded by the formation of new fumaroles close to the vent.

For both volcanoes there is some indication that the length of time over which the premonitory symptoms develop, and their severity, give some indication of the severity of the oncoming eruption. At the Soufrière in 1812 and 1902, felt earthquakes began over a year before the first explosion, and the eruptions which followed were roughly comparable in intensity (Anderson and Flett 1903). The smaller explosive eruption of 1979 was preceded by 9 months of local earthquakes but none was felt until a few hours before the first explosions (Shepherd et al. 1979). The seismic activity associated with the minor eruption of 1971-72 was almost negligible (Aspinall et al. 1973). At Pelee new fumaroles began to form 14 years before the first explosions of 1902 (Lacroix 1904), whereas the period premonitory to the smaller eruption of 1929 lasted only a few months (Perret 1937). However, there has been at least one important exception. The 1945-46 earthquake series at the Soufrière was comparable both in intensity and length with that of 1901-02 (Senn 1946) and was much more severe than those of 1971 or 1979, yet no eruption followed.

The dangers of applying experience gained at one volcano to a situation at a different volcano were brought out clearly during the sequence of events in Guadeloupe in 1977. The earthquake series which began in 1975 was more severe than that of 1901-02 in St. Vincent so that, with St. Vincent as a model, the probability of dangerous explosive activity was very high. On the other hand, the earthquake series and associated fumarolic activity were less severe than in Montserrat in 1932-38 so that, with Montserrat as a model, the probability of dangerous explosive activity was very low. In the event, neither model was appropriate. The crisis culminated in a series of minor phreatic explosions which were much less severe than the 1902 St. Vincent eruption, but much more severe than the culmination of the Montserrat crisis.

4 Other Related Phenomena

In the historic period there have been several instances of apparent correlations between volcanic and/or seismic activity in different parts of the Lesser Antilles. The best known of these is the almost simultaneous eruption of the Soufrière of St. Vincent and Montagne Pelee in May 1902. Until recently, this correlation appeared to be even stronger than it really is since Robson and Tomblin (1966) mention two other eruptions of nearby volcanoes at about the same time. Oceanographic investigations at the site of one of these supposed eruptions (that of Hodder's volcano 1600-07) have shown that there is no sign of any recent volcanic activity or any volcanic feature close to the site of the supposed eruption (Bouysse and Sigurdsson 1982) and the supposed eruption of La Soufrière de la Guadeloupe in 1902 is spurious. Lacroix who visited Guadeloupe at the time specifically points out that widespread <u>but unjustified</u> rumors of an eruption in Guadeloupe circulated during 1902 and 1903 (Lacroix 1904 p 61). Similar rumors circulated in the West Indies during the St. Vincent eruptions of 1971 and 1979 when local newspapers confidently reported that Montagne Pelee was erupting also, so that reports of simultaneous eruptions in the past have to be treated with great caution.

Nevertheless, the coincidence between the eruptions of the Soufrière and Montagne Pelee remains real, although not necessarily significant. Other coincidences which have been noted are:

1. Simultaneous eruptions of La Soufrière de la Guadeloupe (1809-12) and the Soufrière of St. Vincent [1812-14 (?)].
2. A phreatic eruption close to Watt Mountain volcano in Dominica in 1880 and a simultaneous increase in fumarolic activity and possible phreatic eruption at the Soufrière of St. Vincent.
3. The eruption of Pelee in 1929-33 immediately followed by earthquake series and increase in fumarolic activity in Montserrat (1932-38) and Dominica (1937-39), Kelsick (1939) and an eruption of Kick 'em Jenny volcano (1939).

Simple probability theory suggests that these coincidences are not as striking as they appear to be at first sight. If we define as a volcanically active year a year in which, somewhere within the region, an eruption is taking place or has ended within the past year, then there have been 34 volcanically active years since 1797 (the statistics are almost certainly incomplete before that date). The number of years during the same period in which an eruption has begun on land is 12, so that if eruptions begin in random years, regardless of whether the year is active or not, then the probability that an eruption will begin in any random year is approximately:

12/188 = 0.0638.

The expected number of volcanically active years in which eruption begins should therefore be:

0.0638 x 34 = 2.2.

In fact, there have been three volcanically active years, 1880 (possibly), 1812 and 1902, in which a second volcanic eruption has begun, and this is not significantly different from the chance expectation.

Similarly if we define a seismically active year as a year in which an earthquake swarm is taking place somewhere within the region, or has ended within the past year, then there have been at least 42 seismically active years since 1797. The expected number of seismically active years in which an eruption should begin (excluding eruptions immediately following a swarm in the same island) is:

0.0638 x 42 = 2.7.

There have in fact been 3 years (1836, 1902 and 1907) in which an eruption has begun following an earthquake swarm in a different island, which again is not significantly different from the chance expectation. All of the coincidences between eruptions and/or volcano-seismic disturbance in different islands can therefore be explained by chance without looking for underlying causative mechanisms. This does not mean that underlying causative mechanisms do not exist, but Ockam's Razor (Jeffreys 1948 p 315) shows that we should assume that the coincidences are random until it is proved otherwise.

5 Tectonic Earthquakes and Volcanic Earthquake Swarms

The relationship between tectonic earthquakes and earthquake swarms which may or may not end in eruptions is much stronger. A possible early example of a major earthquake followed by an eruption occurred in 1690 when a major earthquake in the Leeward Islands was followed by possible eruptions of Mt. Liamuiga and La Soufrière de la Guadeloupe but the surviving accounts of both the earthquake and the supposed eruptions are so fragmentary that no definite conclusions can be drawn. The only account of the eruption of Mt. Liamuiga is by Sir Hans Sloane (1694) who states that "St. Christopher, one of the Caribee islands was heretofore much troubled with earthquakes, which upon their eruption of a great mountain of combustible matter which still continues, wholly ceased and have never been felt there since". St. Kitts (St. Christopher) was occupied by both English and French settlers at the time and several Royal Navy ships, whose logs survive, were in the vicinity but no further accounts of this eruption have been found. The accounts of the Guadeloupe eruption are to be found in Rapport (1798) and other French sources quoted by Lacroix (1904 pp 63-65) and could equally well describe the fumarolic activity which has carried on continuously throughout historic time. A similar incident may have occurred in 1843, when a major earthquake in the Leeward Islands was followed by a possible eruption of Mt. Liamuiga. The case here is a little stronger since Capadose (1845) reported a spiral column of white smoke rising from the crater and Baker (1963 p 94) discovered traces of deposits which may have been erupted at this time. A supposed eruption of a submarine volcano (Unnamed 1600-15) close to Guadeloupe at this time is now known to be spurious. Bouysse and Sigurdsson (1982) state that there are no volcanic rocks at the location of the supposed eruption. The phenomena described by de Blainville (1843) were almost certainly the effects of the tsunami which was generated by the earthquake.

Early writers seem to have assumed as a matter of course that earthquakes and volcanic eruptions over wide areas were related, through unspecified processes of crustal adjustment. Thus Anderson and Flett (1903) related the 1902 St. Vincent and Montagne Pelee eruptions to a major earthquake in Guatemala on 19 April 1902 and the 1812 eruption to the New Madrid earthquakes in the Mississippi valley. If the net is cast as wide as this, then it is inevitable that correlations will be found, since the probability that there will be a major earthquake somewhere in the western hemisphere in any given year is close to unity. However, more direct and plausible relationships have been observed during the past few decades when seismograph coverage of the world in general, and the Lesser Antilles in particular, has improved considerably. These effects have been observed in three islands of the northern Lesser Antilles , Nevis, Montserrat and Dominica.

The first example occurred during the 1933-38 earthquake series in Montserrat and was discussed by Powell (1938) and Perret (1939). The first earthquakes were reported felt in March 1933, and gradually increased in number and intensity until October 1935 when 275 earthquakes were felt at intensities up to VIII on the modified Mercalli scale. Steam emission from the fumaroles surrounding the Soufrière Hills volcano increased simultaneously. On 10 November 1935 a tectonic earthquake of magnitude 6.25 occurred about 40 km from Montserrat. The earthquake was the biggest to occur close to Montserrat in the present century, and it had an immediate and dramatic effect on the earthquake swarm. The number of earthquakes felt during November rose to 480, but then fell to 67 in December, and 18 in January 1936. Many of the earthquakes felt during November were probably true aftershocks of the 10 November earthquake, and we can reasonably conclude that the numbers of volcanic earthquakes were significantly reduced by the major event.

Since 1935 similar effects have occurred after almost every shallow earthquake of magnitude greater than about 6.0 which has occurred within 50 km of either the Soufrière Hills volcano or Nevis Peak volcano, 40 km to the northwest. In November 1951, an earthquake of magnitude 6.1, with its epicenter 40 km from Nevis Peak was followed almost immediately by a swarm of earthquakes clustered at shallow depth beneath the volcano, some of which were felt at intensities up to MMVII (Willmore 1952). An almost exactly similar sequence occurred after an earthquake of magnitude 6.4 on 2 November 1961 (Robson et al. 1962). In January 1966 a volcanic earthquake swarm, accompanied by increase in fumarolic activity, began in Montserrat (Shepherd et al. 1971) following an earthquake of magnitude 5.6 on 14 January. A further sequence began on 16 March 1985 when an earthquake of magnitude 6.2 at a distance of 30 km from the Soufrière Hills was followed by a swarm of more than 100 volcanic earthquakes immediately below the volcano on 6 and 7 May.

The only major earthquake close to these volcanoes during the past 80 years which did not have a noticeable effect on them occurred on 8 October 1974. This earthquake was of magnitude 6.4 (m_b) with an epicenter at 17.37°N, 61.99°W which is at a distance of about 70 km from both volcanoes. At the time, a minor earthquake swarm was in progress close to the group of volcanoes in southern Dominica (1600-10 A to D) 200 km to the south. Before the earthquake, volcanic events had been occurring at a rate of one or two per day but on 10-11 October 1974 over 200 volcanic earthquakes occurred within a period of about 8 h. There is also a possible, but because of the time lag more tenuous, connection between this earthquake and a major swarm which began close to La Soufrière de la Guadeloupe in mid-1975 and culminated in a series of phreatic explosions in 1976-77 (Feuillard et al. 1983). These occurrences can less easily be dismissed as sheer chance since only one earthquake of magnitude greater than 6.0 in the northern Lesser Antilles has failed either to be followed by a volcanic earthquake swarm at either Nevis Peak or the Soufrière Hills or to modify a swarm already in progress, and the exception had an apparent effect on volcanoes further to the south.

The relationship between strong tectonic earthquakes and volcanic earthquake swarms in the northern Lesser Antilles appears to be a real one, but there is no particularly obvious physical reason for it. One possible explanation is that suggested by Robson et al. (1962) and elaborated by Robson and Barr (1963), i.e. that this type of volcanic earthquake is caused by changes in the shape of a magma body in response to changes in the regional stress system. A full test of the suggestions of Robson and Barr would require detailed study of the focal mechanism of the volcanic earthquake which has not been possible to date. However, the general idea is supported by the distribution of earthquakes within the region. The distribution of earthquakes in the northern Lesser Antilles, from Saba to the northern part of Guadeloupe is quite different from that in the southern Lesser Antilles where volcanic earthquake swarms are less common generally, and no correlation with tectonic earthquakes has been observed. In the northern Lesser Antilles, there is a Benioff zone dipping from northeast to southwest, on which earthquakes occur at depths of 140-160 km directly beneath the volcanic islands (Wadge and Shepherd 1984). However, earthquakes associated with the Benioff zone account for less than one-quarter of the very recent seismicity in this region. More commonly earthquakes occur at very shallow depths within the upper part of the overriding Caribbean plate and have strike-slip or normal mechanisms (McCann and Sykes 1984). All of the earthquakes which have affected the northern volcanoes fall into this category. These authors suggest that such earthquakes are caused by stresses induced in the overriding plate as aseismic ridges to the east of the Lesser Antilles enter the subduction zone. These earthquakes indicate that the stress differences in the part of the crust which contains the Leeward Islands volcanoes are normally close to the values required for shear fracture during tectonic earthquakes and, by the mechanism of Robson and Barr, are therefore close to the values required to cause faulting in the vicinity of a magma body. Although this explanation is rather qualitative, it provides the basis for an explanation of the difference between the northern and southern parts of the arc. Shallow intraplate earthquakes have not been observed to date in the southern Lesser Antilles.

Another possible explanation is that bodies of magma beneath the volcanoes are affected directly by shaking caused by strong earthquakes. This seems to be a less likely explanation since there have been delays of days to weeks between the causative tectonic earthquakes and the resultant volcanic events. However, the direct effect of earthquake shaking of a magma body has been observed on one occasion. This was on 23 March 1979 when an earthquake swarm, which eventually led to the eruption of the Soufrière volcano, was already in progress in St. Vincent, and an earthquake of magnitude 6.0 occurred close to the Dominican Republic 1100 km north of the Soufrière. The earthquake generated an exceptionally strong T-phase with a predominant frequency of about 1 Hz. Seismographs on other volcanoes to the north and south of St. Vincent recorded the T-phase normally for about 5 min, but at the Soufrière it triggered a 3-h burst of volcanic tremor and a short series of strong local earthquakes (Shepherd and Aspinall 1982). There was then a break in local earthquake activity until 9 April, when the series restarted and escalated steadily until explosions began on 13 April.

Apart from the possible examples of 1690, 1843 and 1976, no instance has yet been observed in the Lesser Antilles in which a tectonic earthquake has triggered a volcanic eruption. In 1979, the pre-eruptive earthquakes had already begun and other premonitory phenomena had already been observed. For this reason the tectonic earthquakes cannot be described as true eruption precursors. However, the period of detailed observation so far has been short in comparison with both the mean inter-eruption period of these volcanoes, and with the return period for truly great ($M \geq 8.0$) earthquakes. It is possible that the effect on the volcanoes of the next great earthquake may be much more dramatic.

6 Eruption Periodicity

Perret (1939) and MacGregor (1938, 1949) both suggested that there is a periodicity in the occurrence of volcanic disturbances in the Lesser Antilles, with a fundamental period close to 30 years. MacGregor suggested also that the disturbances oscillated regularly from one end of the arc to the other, so that the most likely location of the next disturbance could be predicted from the previous two. There is no evidence either for periodicity or for oscillation in the data presented in the Appendix. Both theories depend heavily on very doubtful eruptive phenomena, such as the supposed submarine eruption close to Barbados in 1839. For the arc as a whole (as distinct from any individual volcano), the frequent and irregular eruptions of Kick'em Jenny submarine volcano (which was unknown to both Perret and MacGregor) eliminate any trace of either periodicity or oscillation.

7 Conclusions

All volcanic eruptions on land in the Lesser Antilles, about which sufficient information is available for definite statements to be made, have been preceded by precursory phenomena over periods of time ranging from a few months to 14 years. The main precursory symptoms have been local earthquakes and increases in the overall rate of fumarolic activity. No significant eruption-related variations in fumarolic gas chemistry were detected on the one occasion on which the necessary measurements were made (La Soufrière de la Guadeloupe 1976, Feuillard et al. 1983). Ground deformation was observed before one of the two eruptions when measurements were made (Soufrière of St. Vincent 1979, Fiske and Shepherd 1982) but not before the other (Guadeloupe 1976). For future volcanic monitoring of Lesser Antillean volcanoes gas chemistry and ground deformation must be regarded as promising, but unproven, techniques. At individual volcanoes, the intensity of the precursory symptoms and the length of time for which they last give some indication of the type and intensity of the ensuing eruption, but do not provide an infallible guide. The prospects for prediction of the onset of future West Indian eruptions are extremely good, provided that it is accepted that a fairly high false-alarm rate is inevitable.

Tectonic earthquakes of magnitude greater than about 6, which occur close to the volcanoes of the northern Lesser Antilles almost always have an effect on volcanic seismicity at one or other of them. The effect has not been observed in the southern Lesser Antilles and its cause is unknown, although it is probably related to different stress regimes in the different parts of the arc. There is no evidence to suggest that there is any regular periodicity in Lesser Antillean eruptions, or that activity oscillates regularly from one end of the island arc to the other.

Appendix Active Volcanoes of the Lesser Antilles

1600-01
Name: The Mountain
Synonyms: None known
Location: Island of Saba, Netherlands Antilles
17.63° N, 62.23° W. Elevation 887 m
Eruptions and precursors: None known
Other activity: Hot springs
Monitoring system: None

1600-02
Name: The Quill
Synonyms: None known
Location: Island of St. Eustatius (Statia), Netherlands Antilles
17.48° N, 62.95° W. Elevation 1157 m
Eruptions and precursors: None known
Other activity: None known
Monitoring system: None

1600-03
Name: Mt. Liamuiga
Synonym : Mt. Misery
Location: Island of St. Christopher (St. Kitts),
Federation of St. Christopher and Nevis
17.37° N, 62.80° W. Elevation 1157 m
Eruptions and precursors:
1. (Doubtful) Between 1690 and 1692; followed major tectonic earthquake
2. (Doubtful) February 1843; followed major tectonic earthquake
Other activity: Hot springs. Rare local earthquakes
Monitoring system: Seismograph 9 km from summit

1600-04
Name: Nevis Peak
Synonyms: None known
Location: Island of Nevis,
Federation of St. Christopher and Nevis
17.15° N, 62.58° W. Elevation 985 m
Eruptions and precursors: None known
Other activity: Severe earthquake swarms in 1831-35, 1926, 1947-48, 1950-51, 1961-62. Swarms of 1950-51 and 1961-62 may have been triggered by tectonic earthquakes. New hot spring formed 1953. Other hot springs active throughout historic period
Monitoring system: Seismograph 2 km from summit. Dry tilt

1600-05
Name: Soufrière Hills
Synonym : Chance's Mountain
Location: Island of Montserrat,
B.W.I. 16.72° N, 62.18° W.
Eruptions and precursors: 1. (Doubtful). Early 17th century. Date based on a single carbon date and ambiguous local records

Other activity: Severe earthquake swarms in 1897-1900, 1933-38 and 1966-67, each accompanied by increases in fumarolic activity. Inflation of volcano observed in 1966-67. Continous fumarolic activity and occasional minor earthquake swarms
Monitoring system: Seismograph 3 km from summit. Dry tilt

1600-06
Name: La Soufrière de la Guadeloupe
Synonyms: None known
Location: Island of Basse-Terre (Guadeloupe proprement-dite) French West Indies
16.05° N, 61.67° W. Elevation 1467 m
Eruptions and precursors: A major magmatic eruption which resulted in the formation of the present summit dome occurred in 1580 ±50 A.D. Since then all eruptions have been phreatic
1. 1690 (Feuillard et al. 1983) or 1680 (Robson and Tomblin 1966). Phreatic. If 1690, then this eruption followed a major tectonic earthquake
2. September 1797. Phreatic. Preceded by over 1 year of felt earthquakes
3. June 1809 to 1812. Phreatic. Earthquakes and increased fumarolic activity preceded first explosions
4. 3 December 1836 to February 1837. Phreatic. Preceded by felt earthquakes
5. 20 October 1956 to December 1956. Phreatic. Preceded by several days of felt earthquakes
6. 8 July 1976 to July 1977. Phreatic. Preceded by over 1 year of felt earthquakes and strong increase in fumarolic activity. Ground deformation and changes in gas chemistry sought but not detected
Other activity: Permanent seismic activity characterized by periods of a few days to a few weeks in which a number of small earthquakes are recorded by the seismographs, separated by quiet periods of a few weeks to a few months. Continuous fumarolic acitvity
Monitoring system: Extensive seismograph network, continous tilt recording and dry tilt stations. Chemical monitoring. Magnetometry

1600-08
Name: Morne au Diable
Synonyms: None known
Location: Commonwealth of Dominica.
15.62° N, 61.45° W. Elevation 861 m
Eruptions and precursors: None known
Other activity: Severe earthquake swarms October 1841 and 1893 were associated with either this volcano or 1600-09. Hot springs
Monitoring system: Seismograph 7 km from summit

1600-09
Name: Morne Diablotins
Synonyms: None known
Location: Commonwealth of Dominica.
15.50° N, 61.42° W. Elevation 1421 m

Eruptions and precursors: None known
Other activity: Earthquake swarms of 1841 and 1893 may have been associated with this volcano. Hot springs
Monitoring system: Seismograph 4 km from summit

1600-10A (See note 2)
Name: Morne Trois Pitons
Synonyms: None known
Location: Commonwealth of Dominica.
15.37° N, 61.34° W. Elevation 1387 m
Eruptions and precursors: None known
Other activity: Frequent shallow earthquakes
Monitoring system: Seismograph 7 km from summit

1600-10B
Name: Morne Macaque
Synonym : Micotrin
Location: Commonwealth of Dominica.
15.33° N, 61.33° W. Elevation 1380 m
Eruptions and precursors: None known
Other activity: Frequent shallow earthquakes. Hot springs
Monitoring system: Seismograph 8 km from summit

1600-10C
Name: Grande Soufrière Hills
Synonyms: None known
Location: Commonwealth of Dominica.
15.31° N, 61.29° W. Elevation 1040 m
Eruptions and precursors: None known (but see 1600-10D)
Other activity: Shallow earthquakes. Hot springs
Monitoring system: Seismograph 10 km from summit

1600-10D
Name: Watt Mountain
Synonym : Morne Watt
Location: Commonwealth of Dominica.
15.30° N, 61.32° W. Elevation 1224 m
Eruptions and precursors: 1. A phreatic explosion occurred on 4 January 1880 in the Valley of Desolation solfatara field less than 1 km from the summit of this volcano. The explosion was preceded by a short period of felt earthquakes
Other activity: Shallow earthquakes. Hot springs
Monitoring system: Seismograph 5 km from summit

1600-10E
Name: Morne Anglais
Synonym : Ouliaboune
Location: Commonwealth of Dominica.
15.28° N, 61.34° W. Elevation 1122 m
Eruptions and precursors: None known
Other activity: None. Judged active on grounds of morphology
Monitoring system: Seismograph 5 km from summit

1600-10F
Name: Foundland
Synonyms: None known

Location: Commonwealth of Dominica. 15.27° N, 61.30° W. Elevation 970 m
Eruptions and precursors: None known
Other activity: None. Fairly advanced erosion suggests that this volcano may not have the potential for future eruptions
Monitoring system: Seismograph 9 km from summit

1600-11A
Name: Morne Plat Pays
Synonyms: None known
Location: Commonwealth of Dominica.
15.26° N, 61.30° W. Elevation 960 m
Eruptions and precursors: None known
Other activity: Shallow earthquakes. Severe swarms in February to April 1849, and October 1937 to April 1938 may have been associated with this volcano. Swarms in June to December 1974, and March 1986 included earthquakes directly beneath summit
Monitoring system: Seismograph 9 km from summit

1600-11B
Name: Morne Patate
Synonyms: Morne Patates, Potato Mountain
Location: Commonwealth of Dominica.
15.22° N, 61.37° W. Elevation 520 m
Eruptions and precursors: None known, but carbon-dated pyroclast flow suggests eruption in early 16th century.
Other activity: Hot springs. Seismic activity as for 1600-11A
Monitoring system: Seismograph 10 km from summit

1600-12
Name: Montagne Pelee
Synonym : Mt. Pele
Location: Island of Martinique,
French West Indies
14.82° N, 61.17° W. Elevation 1397 m
Eruptions and precursors:
1. 22 January 1792. Phreatic. No information about precursors.
2. 5 August 1851 to 1852. Phreatic. Preceded by increased fumarolic acitvity but no felt earthquakes reported
3. 24 April 1902 to 1907(?). Explosive magmatic eruption followed by interspersed dome-building, explosions and pyroclast flows.
New fumaroles began to form near vent 14 years before first explosions. Minor premonitory earthquakes a few weeks before onset of eruption
4. 16 September 1929 to 1932. Similar eruption to that of 1902 but on a smaller scale.
Inceased fumarolic activity a few weeks before onset and minor earthquakes a few days before onset

1600-14
Name: Sulphur Springs
Synonym : Qualibou (used in scientific literature but does not appear on maps and is unknown in Saint Lucia)

Location: Island of Saint Lucia (formerly St. Lucia)
13.83° N, 61.05° W. Elevation 77 m
Eruptions and precursors: 1. Phreatic explosion 1766. No information about precursors.
Other activity: Hot springs in extensive depression (caldera?). Occasional single shallow earthquakes but no severe swarms known
Monitoring system: Seismograph on wall of depression

1600-15
Name: The Soufrière of St. Vincent
Synonym : Morne Garou (early literature only)
Location: Island of St. Vincent, state of St. Vincent and the Grenadines.
13.33° N, 61.18° W. Elevation 1178 m
Eruptions and precursors:
1. 26 to 29 March 1718. Major magmatic eruption with vertical explosions and pyroclast flows.
No information on precursors
2. 27 April 1812 to 1814(?). Major magmatic eruption with vertical explosions and pyroclast flows. Activity did not cease finally until 1814.
Preceded by over 1 year of felt earthquakes but no significant change in fumarolic activity
3. 6 May 1902 to 1903. Major magmatic eruption with vertical explosions and pyroclastic flows. Preceded by over 1 year of strong felt earthquakes but no significant fumarolic activity until onset
4. (?)October 1971 to March 1972. Lava extrusion built dome in summit crater lake. No explosions. No premonitory symptoms observed but back extrapolation of the rate of minor earthquake activity during the eruption suggests that minor earthquakes began several months before lava extrusion
5. 13 April to October 1979. Phreatomagmatic explosions initially with later dome extrusion.
Minor premonitory earthquakes observed 9 months before onset. Inflation of volcano observed. Increase in crater lake temperature
Other activity: Fumarolic activity continuous throuout historic time. Strong increases in about 1780 and 1880 may have been accompanied by minor lava extrusion or phreatic eruptions. Strong earthquake series 1945-46 but background seismicity generally negligible
Monitoring system: Extensive seismograph network. Dry tilt stations. Thermal and occasional chemical monitoring

1600-16
Name: Kick'em Jenny
Synonyms: None (nearby island marked on maps as Kick'em Jenny is not this volcano)
Location: Submarine, close to southern Grenadine islands, state of Grenada.
12.30° N, 61.63° W. Elevation -160 m (1983)
Eruptions and precursors:
1. 24 July 1939. Vertical explosions broke sea surface. Water-borne acoustic signals detected

2. 5 to 6 October 1943. From acoustic signals only
3. 30 October 1953. From acoustic signals only
4. 25 October 1965 to 6 August 1966. From acoustic signals only
5. 5 July 1972. From acoustic signals only
6. 5 to 6 September 1974. Sea surface disturbance but no eruption column. Acoustic signals detected
7. 4 November 1977. Acoustic signals only.

Premonitory symptoms not detectable at this volcano

Monitoring system: Seismograph 29 km from volcano. Acoustic signals detectable throughout Lesser Antilles

1600-17
Name: Mt. St. Catherine
Synonyms: None known
Location: Grenada
12.15° N, 61.67° W. Elevation 840 m
Eruptions and precursors: None known
Other activity: Hot springs
Monitoring system: Seismograph station at summit. Thermal monitoring

1600-17A
Name: Grenada explosion craters
Synonyms: Include the Ile de Caille, Levera pond, Lake Antoine, Grand Etang, St. George's Harbour, Queen's Park and many others
Location: Grenada and the southern Granadines
Eruptions and precursors: None known
Monitoring system: Seismograph at Mt. St. Catherine. Thermal monitoring

Notes:

1. The word "Soufrière" is of French creole origin and is used in the West Indies to describe individual hot springs as well as volcanoes. Since many volcanoes include the word in their names, these names are always given in full.
2. Volcanoes 1600-10A to 10F form a closely spaced complex of domes and strato-volcanoes. These were described as a single volcano by Robson and Tomblin (1966) and called the Valley of Desolation.

References

Anderson T, Flett JS (1903) Report on the eruptions of the Soufrière in St. Vincent, in 1902, and on a visit to Montagne Pelee, in Martinique. Phil Trans R Soc Lond 2000:353-553

Arculus RJ (1973) The alkali basalt andesite association of Grenada, Lesser Antilles. Ph.D. thesis, University of Durham

Aspinall WP, Sigurdsson H, Shepherd JB (1973) Eruption of Soufrière volcano on St. Vincent island 1971-72. Science 181: 117-224

Baker PE (1963) The geology of Mt. Misery volcano, St. Kitts. Ph. D. thesis Oxford

Bouysse P, Sigurdsson H (1982) The "Hoddrr Phenomenon" of 1902: no active volcano off St. Lucia (Lesser Antilles). Mar Geol 50:M29-M35

Capadose H (1845) Sixteen years in the West Indies. Newby, London

DeBlainville C (1843) Sur un volcan qui a fait eruption entre la Guadeloupe et Marie Galante. CR Acad Sci Paris 16:1083-84

Feuillard M, Allegre CJ, Brandeis G, Gaulon R, LeMouel JL, Mercier JC, Pozzi JP, Semet MP (1983) The 1975-77 crisis of a Soufrière de la Guadeloupe, FWI: A still-born magmatic eruption. J Volcanol Geotherm Res 16:317-334

Fiske RS, Shepherd JB (1982) Ground deformation studies of Soufrière volcano, St. Vincent 1977-80. Science 216: 1125-1126

Jeffreys H (1948) Theory of probability. Oxford University Press, Oxford, pp 1-411

Kelsick JO (1939) Report on recent earthquake activity in Montserrat and Dominica by the clerk in charge of the instruments. Dep Agric, Plymouth, Montserrat

Lacroix A (1904) La Montagne Pelée et ses éruptions. Acad Sci Paris, pp 1-662

MacGregor AG (1938) The Royal Society Expedition to Montserrat, BWI. The volcanic history of Monserrat with observations on Mt. Pele in Martinique. Phil Trans R Soc Lond 229:1-87

Mac Gregor AG (1949) Prediction in relation to seismo-volcanic phenomena in the Caribbean volcanic arc. Bull Volcanol 2-8:69-86

McCann WR, Sykes LR (1984) Subduction of aseismic ridges beneath the Caribbean plate: implications for the tectonics and seismic potential of the northeastern Caribbean. J Geophys Res 89:4493-4519

Perret FA (1937) The eruption of Mt. Pelée 1929-32. Carnegie Inst Washington, pp 1-125

Perret FA (1939) The volcano-seismic crisis at Montserrat 1933-37. Carnegie Inst Washington

Powell CF (1938) The Royal Society expedition to Montserrat, BWI - Final Report. Phil Trans R Soc Lond 237:1-34

Rapport F (1798) Fait aux citoyens Victor Hugues et Lebas, agents particuliers du Directoire executif aux Iles du Vent. Au port de La Liberté. Reprint of the 1798 edition by Soc His Guadeloupe, Basse-Terre, 1977

Robson GR (1964) An earthquake catalogue for the eastern Caribbean 1530-1960. Bull Seismol Soc Am 54:785-832

Robson GR, Barr KG (1963) The effect of stress on faulting and minor intrusions in the vicinity of a magma body. Bull Volcanol 27:1-16

Robson GR, Tomblin JF (1966) Catalogue of the active volcanoes of the world: Part XX West Indies. Int Assoc Volcanol Rome, pp 1-56

Robson GR, Barr KG, Smith GW (1962) Earthquake series in St. Kitts/Nevis 1961-62. Nature (Lond) 195:972-974

Roobol MJ, Wright JV, Smith AL (1983) Calderas of gravity-slide structures in the Lesser Antilles island arc? J Volcanol Geotherm Res 19:121-134

Senn A (1946) A geological investigation at Soufrière volcano. British Union Oil Company Bridgetown, Barbados

Shepherd JB, Aspinall WP (1982) Seismological studies of Soufrière volcano, St. Vincent 1953-79: implications for volcanic surveillance in the Lesser Antilles. J Volcanol Geotherm Res 12:37-55
Shepherd JB, Tomblin JF, Woo DA (1971) Volcano-seismic crisis in Montserrat, West Indies, 1966-67. Bull Volcanol 35-1:143-163
Shepherd JB, Aspinall WP, Rowley KC, Pereira JA, Sigurdsson H, Fiske RS, Tomblin JF (1979): The eruption of Soufrière volcano, St. Vincent, April-June 1979. Nature (Lond) 282:24-28
Simkin T, Siebert L, McLelland L, Bridge D, Latter JH (1981) Volcanoes of the world. Hutchinson Ross, New York
Sloane H (1694) A letter from Hans Sloane M.D. and F.R.S. with several accounts of the earthquakes in Peru, October the 20th 1687 and at Jamaica, February 19th 1688 and June the 7th 1692. Phil Trans R Soc Lond 18:78-100
Wadge G (1984) A preliminary volcanic hazards study of Dominica, West Indies. Special publication 1984/5 Seismic Research Unit, UWI, St. Augustine, Trinidad
Wadge G (1985) Morne Patates volcano, southern Dominica, Lesser Antilles. Geol Mag 3:253-260
Wadge G, Shepherd JB (1984) Segmentation of the Lesser Antilles subduction zone. Earth Planet Sci Lett 71:297-304
Willmore PL (1952) The earthquake series in St. Kitts/Nevis in 1951-52. Nature (Lond) 169:770-772

Patterns of Volcanic Activity of Piton de la Fournaise (Réunion Island, Indian Ocean). A Synthesis Based on Monitoring Data Between 1980 and July 1985, and on Historic Records Since 1930

J.-F. LÉNAT[1]

Abstract

Piton de la Fournaise (Réunion Island, Indian Ocean) is an intraplate oceanic basaltic volcano. A survey of historic records since 1930 reveals a mean eruption frequency of one eruption every 14 months and a mean magma output rate of 0.34 $m^3\ s^{-1}$. Most of the eruptions occur in the central zone of the volcano.

A volcanological observatory was established in 1980. Three eruptions or eruptive cycles have occurred since that date: in 1981 (three eruptive phases), in 1983-84 (two eruptive phases) and then in 1985-86 (ten eruptive phases and two intrusions to November 1986, at the time of writing). For the period considered in this chapter (1980-July 1985), the physical behaviour of the volcano can be summarized as follows:

1. Between the eruptions the background seismicity remained negligible. The first phases of the eruptions were preceded by 2 to 4 weeks of moderate shallow seismic activity beneath the summit, with less than 20 events per day, at depths of 1 to 3 km. Immediately before the eruptions there was 1 to 2.5 h of intense seismic activity with 50 to 100 events per hour. Subsequent eruptive phases were preceded only by 1 to 5.5 h of intense seismic activity.

2. A deformation network has been monitored on a regular basis since June 1982. No long- or short-term inflation occurred before the initial 1983 outbreak, but the first eruptive phase of the 1985 crisis was preceded by several months of summit inflation. In both eruptive cycles, the subsequent phases were preceded by summit inflation.

3. In July 1985, following a summit intrusion which ended before reaching the surface, a strong seismic crisis occurred at depths of 1 to 6 km below sea level beneath the east flank.

---

[1] Centre de Recherches Volcanologiques, OPGC and U.A. 10, Université Clermont II, 5 rue Kessler, 63038 Clermont Ferrand Cédex, France
(Formerly at Observatoire Volcanologique du Piton de la Fournaise, IPG Paris)

IAVCEI Proceedings in Volcanology 1
J.H. Latter (Ed.), Volcanic Hazards

The pattern of seismicity and deformation, as well as structural data, suggest the presence of a shallow magma reservoir. The top of this reservoir is believed to lie in the same depth range as the pre-eruptive seismic crises (1-3 km). The lack of significant seismicity, whether deep, intermediate, or shallow between eruptive cycles, and the lack of systematic renewed inflation after the eruptions, seems to indicate that the transfer of magma to this shallow reservoir takes place intermittently. Petrological data suggest that it has not been resupplied since 1977.

The asymmetric setting of Piton de la Fournaise in which the west part is buttressed by Piton des Neiges volcano and the sea-facing flank is unbuttressed, explains the tendency of the east flank to move seawards. This is inferred from tectonic surface features. The dilatation of the central area, through successive emplacements of intrusions, is the active mechanism that causes strain to accumulate on the east flank, which is relieved by seaward movement. Following this model, the strong seismic crisis beneath the east flank in July 1985 is tentatively interpreted as an episode of strain release associated with an eastward displacement of the east flank.

1 Introduction

La Réunion is a volcanic island located in the southern part of the Mascarene Basin, 800 km east of Madagascar. It is composed of two subaerial volcanoes, and is the emergent part of a broad, elliptical topographic high which rises more than 7 km from the seafloor, and is 240 x 200 km in diameter at its base. Piton des Neiges volcano, which occupies the northwestern two-thirds of the island, is a dormant volcano. Its initial activity was basaltic, with this stage lasting from more than 2 million years ago to 0.43 m.y. ago. This activity was followed by eruptions of alkaline differentiated lavas between 0.350 and 0.022 m.y. ago (MacDougall 1971; Gillot and Nativel 1982; Chevallier and Vatin-Pérignon 1982). The other volcano in the southeastern part of the island is the Piton de la Fournaise, which began to erupt more than 3.5 m.y. ago (MacDougall 1971), and is still one of the most active volcanoes in the world.

A volcanological observatory was opened at Piton de la Fournaise in 1980. The purpose of this chapter is to describe the seismicity and the deformation associated with the volcanic crises during the first 6 years of geophysical monitoring of the volcano. Statistical data of historic activity since 1930 (the data before this date, since the first settlement in the 1640s, cannot be considered as sufficient and reliable enough for our purpose) are used in conjunction with the monitoring data collected since 1980 to describe the processes of magma transfer and storage in order to identify criteria of prevision of future activity at the volcano.

2 Geological Setting

Chevallier and Bachèlery (1981) recognized four main stages in the evolution of Piton de la Fournaise. Each of the three earlier stages ended with the formation of a caldera. Most of the recent volcanic activity has occurred within the youngest caldera called the Enclos (Fig. 1), which probably formed less than 5000 years ago (Bachèlery 1981).

The greatest density of eruptive fissures is found in the central part of the Enclos caldera, near the central cone, which is 400 m high and 3 km in diameter. Around the summit area, fissures are either concentric or radial to the main summit crater. On the floor of the caldera the fissures cluster in two rift zones trending N 10° (E) and N 170°. The Grand Brûlé (Fig. 1), an extension of the caldera depression to the coast and beyond, is interpreted as the result of a gravity sliding of the seaward flank (Vincent and Kieffer 1978; Bachèlery 1981; Lénat and Aubert 1982; Duffield et al. 1982; Chevallier and Bachèlery 1981). The north and south (N 10° and N 170°) rift zones continue outside the caldera, curving to the northeast and the southeast respectively. Eruptions outside the caldera along these structures are not frequent, averaging only two per century during historic times. The last two of these erup-

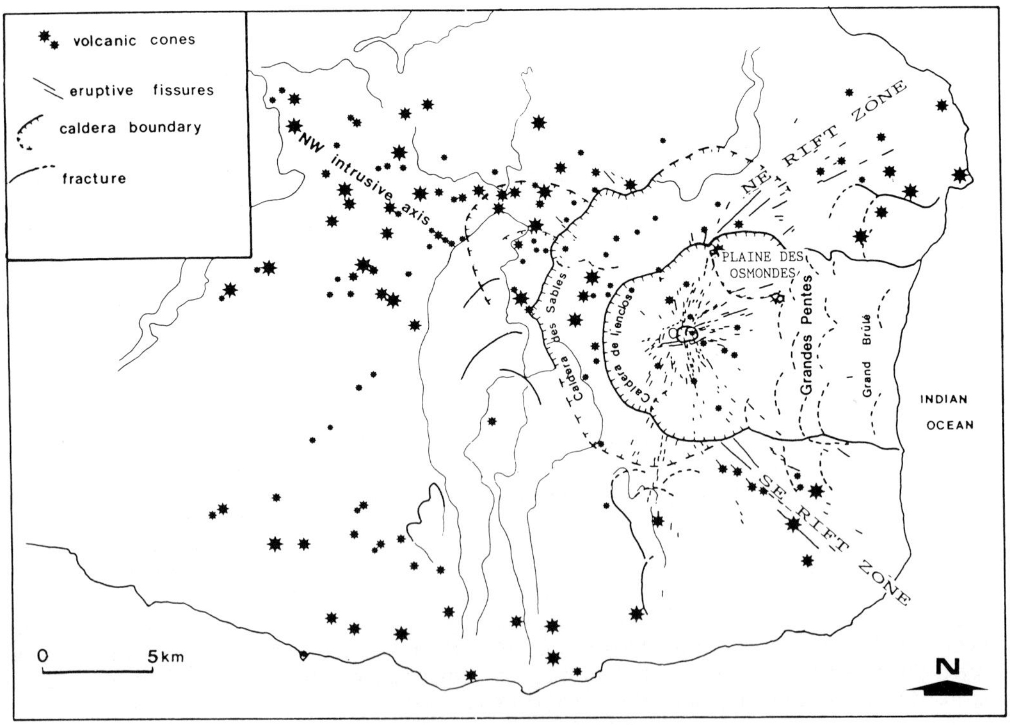

Fig. 1. Map of Piton de la Fournaise showing main structural and volcanic features (courtesy of P. Bachèlery)

tions occurred in 1977 on the NE rift zone (Kieffer et al. 1977) and in 1986 on the SE rift zone (Delorme et al. 1986).

Another well-developed fissure zone, striking N 75°, extends between the summit area and the southern edge of the Plaine des Osmondes (Fig. 1). The latter is a circular depression, 2.5 km in diameter, which is interpreted as an actively subsiding part of the Enclos caldera. The N 75° axis continues to the southwest of the summit where it merges with the south rift zone.

The present summit crater (Dolomieu crater) is about 1 km in diameter, and was formed by a collapse of the summit area, between 1927 and 1934, and not by explosive activity. The combination of a collapsed summit, a system of arcuate open fissures around the crater and a system of radial fissures, suggest the presence of a shallow magma reservoir. This is supported by geological, petrological and geophysical data. For Bachèlery and Lénat (in Lénat 1987), the effect of a magma reservoir, with its top lying at depths ranging from 0.7 to 1.3 km, would account for the surface tectonic pattern. The mineral paragenesis of most of the lavas erupted within the Enclos caldera is that of a very low pressure equilibrium, in the range of 1 kb or less (Kornprobst et al. 1984): these petrological data, therefore, fit the presence of a reservoir at depths of about 3 km or less from the surface, as also aeromagnetic anomalies in which attenuation observed above the summit area can be interpreted by the presence of rocks at a temperature above the Curie point at shallow depth (Lénat and Aubert 1982).

3 Eruption Frequency and Magma Production

More than 97% of the historic eruptions have occurred within the Enclos caldera. The few eruptions outside the caldera were located along the SE and NE rift zones. Figure 2 shows the observed vent density per km^2 within the Enclos caldera. Although the density may be slightly underestimated in the zones frequently flooded by lavas, the map clearly shows that the highest probability of eruption is restricted to the central cone and the adjacent portions of the N 10° and 170° rift zones.

Historic records of Piton de la Fournaise activity began in 1708 with the first dated eruption, after the island had been inhabited for about half a century. About 200 eruptions have been reported since that time, but because of the poor accessibility, the relative remoteness of the volcano from settlements on the island and the poor weather that often prevails on the windward side, many eruptions may have passed unnoticed. If we consider the last 50 years or so, the period during which the record of the eruptions can be considered as nearly complete owing to frequent observations by air and land, the mean eruptive frequency is about one eruption every 14 months. The true rate may be slightly higher, because it is probable that even recently (before the building of the observatory) some short eruptions were not observed (if, for example, they occurred during bad weather periods).

Fig. 2. Map showing the vent density per km^2 within the Enclos caldera (from P. Bachèlery and J.-F. Lénat, unpubl. report from Piton de la Fournaise volcano observatory). The density is slightly underestimated in the areas frequently flooded by lava flows (summit crater; NE, E, SE and S parts of the caldera). The map shows the concentration of the vents in a restricted area around the central area. This pattern is interpreted as reflecting the presence of a shallow center of magmatic pressure (a shallow magma reservoir). The N-S elongation of the contours coincides with paths of the rift zones within the Enclos; the N 75° fissure zone is also obvious

Figure 3 shows the cumulative number of eruptions since 1930 for Piton de la Fournaise, and for two very active, well-known basaltic volcanoes, Kilauea (Hawaii) and Mount Etna (Sicily). For Piton de la Fournaise, it is difficult in some cases to determine whether two or more consecutive outbreaks should be considered as different phases of the same eruption, or as separate eruptions. For this reason two curves have been drawn. Curve 1 includes all the outbreaks, while curve 2 only those that we consider as truly distinct eruptions. We observe that the eruption frequency of Piton de la Fournaise has been similar to that of Kilauea and Mount Etna during the period considered. At Kilauea a large number of intrusions, which in many respects are equivalent to eruptions, have been monitored in the rift zones since 1970. In the case of Piton de la Fournaise such data are lacking before 1980. For Etna, no record of intrusions is available to our knowledge.

However, a pronounced difference appears between these volcanoes when the activity is measured by the amount of erupted magma. Figure 4, which shows the cumulative volume erupted by Piton de la Fournaise, has been constructed from the updated estimates of volume given by Bachèlery (1981), who carefully mapped the different flows and estimated their mean thickness. In the case of two flows, his estimated volumes have been compared to volumes measured by photogrammetry, and the discrepancy between the two methods was found to be only 10 to 20% (Lénat, Bachèlery, Blum and Egels, in Lénat 1987). The cumulative volume curve can therefore be regarded as a relatively good estimate of the amount of erupted magma. The total magma budget of Piton de la Fournaise before 1980 cannot be determined owing to the lack of data on the volume of magma emplaced in the volcano as intrusions. The estimated magma output rate of Piton de la Fournaise is half that of Etna (Wadge and Guest 1981) and about one thenth to two tenths that of Kilauea

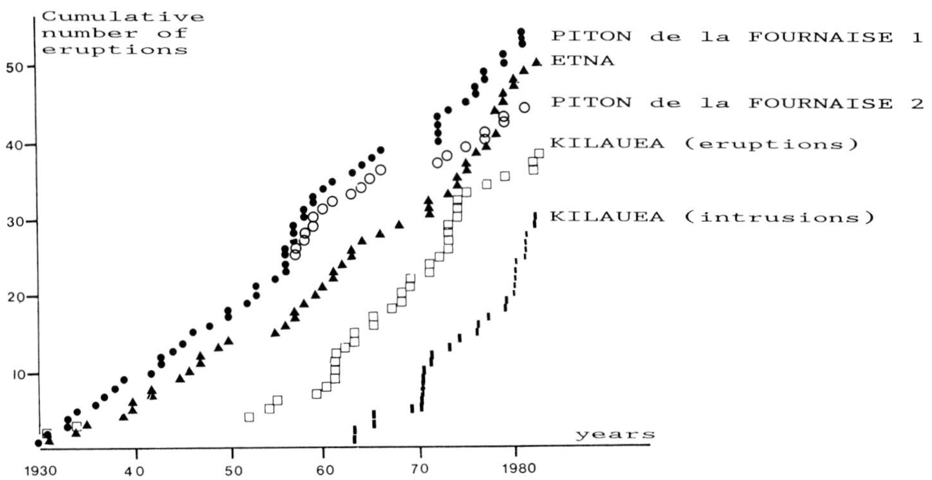

Fig. 3. Cumulative number versus time for all the eruptions at Piton de la Fournaise, Kilauea and Mount Etna since 1930 (Lénat and Bachèlery 1985). The intrusions at Kilauea are also represented. The curve **Piton de la Fournaise 1** takes into account all the eruptive phases, and that marked **Piton de la Fournaise 2** only the eruptive cycles (which can have several eruptive phases, see text). The data come from compilations by Bachèlery (1981), Tanguy (1980), Klein (1982) and Simkin et al. (1981) (Lénat and Bachèlery 1988)

(Swanson 1972; Dzurisin et al. 1984), if the volume of the intrusions is excluded.

Examined in detail, Fig. 4 shows significant variations of the magma output rate during the period considered. Lénat and Bachèlery (1985) have distinguished three main types of activity periods: (1) the large effusions of oceanite of 1931, 1961 and 1977 which appear as the major steps in the curve; (2) the periods with a high mean rate of magma output, for example during the 1961-66 period and also during the 1937-39, 1945-48 periods and in 1953, 1973 and 1977; and (3) the intervening periods with a significantly lower magma output rate.

The different types reflect differences in the rate and processes of magma supply and storage within the volcano. Ludden (1977) first recognized this possibility. He also noticed the periodic recurrence of large oceanite eruptions, but his data concerning the erupted volumes were less complete than the present set because much new geological and geophysical work has since been done.

The interpretation of the cumulative volume curve has been recently discussed by Lénat and Bachèlery (1985); their main conclusions will be repeated below, after the examination of the physical behaviour of the volcano since 1980.

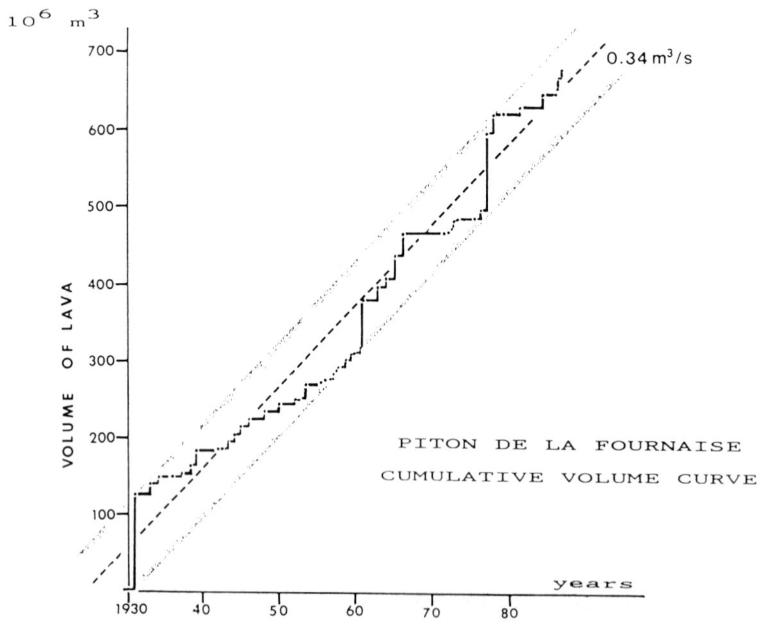

Fig. 4. Cumulative curve of the volume of lavas erupted since 1930 at Piton de la Fournaise (Lénat and Bachèlery 1985). The major steps of the curve are the large oceanite eruptions of 1931, 1961 and April 1977

4 Seismicity and Deformation Associated with Volcanic Activity Between 1980 and Mid-1985

Hereafter we will use the terms eruption or eruptive crisis to designate a complete eruption cycle (i.e. multiphase eruption) and the term phase for the individual outbreaks (or for the intrusions not followed by eruption).

Three eruptive crises occurred between the establishment of the volcanological observatory in 1980 and mid-1985. They were separated from preceding crises by long intervals without any eruptive activity (18, 30 and 16 months respectively). All were multiphase crises. The last crisis, which began in June 1985, is still continuing with its tenth eruptive phase at the time of writing (November 1986), but only the first eruptive phase in June 1985 and the subsequent intrusion on 9 July 1985 will be considered here.

Figures 5, 7 and 8 summarize the data related to the different phases. All the seismic signals so far observed at Piton de la Fournaise are high frequency earthquakes and harmonic tremors which are associated with the effusion of lava at the surface. Low frequency events such as those which are often described on other active volcanoes have not been observed or identified.

4.1 Behaviour Between the Eruptive Cycles

One of the most intriguing observations during this 5-year period has been that the seismic activity between eruptive crises remained virtually negligible. The first phase of each eruption was preceded by 2 to 4 weeks of moderate, shallow seismic activity beneath the summit (the pre-eruptive crises shown in Fig. 7), and by 2.5 to less than 1 h of intense seismic activity (the intrusive crises shown in Fig. 8). Virtually no deep, intermediate or shallow seismicity occurred during the rest of the time, between the end of an eruption and the start of the next pre-eruptive crisis.

The pattern of deformation was also found to be distinct from the classic "inflation-deflation" cycles observed on volcanoes in Hawaii and on Krafla volcano in Iceland. A deformation network has been monitored on a regular basis only since June 1982. No long- or short-term inflation occurred before the initial outbreak of the 1983-84 eruption. Following the 1983-84 eruption, progressive deflation occurred in the area which had been previously uplifted during the two intrusive phases of the crisis (Lénat et al. in press -a-). This slow deformation was similar to that observed between June 1982 and early December 1983 in the summit area, and is thought to be related to mechanical relaxation, or to thermal concentration of the zone of the intrusions (see for example Fig. 11 a).

The lack of renewed inflation and seismicity between the eruptive cycles seems to indicate that the transfer of magma to the intravolcanic reservoir is a discontinuous process at Piton de la Fournaise.

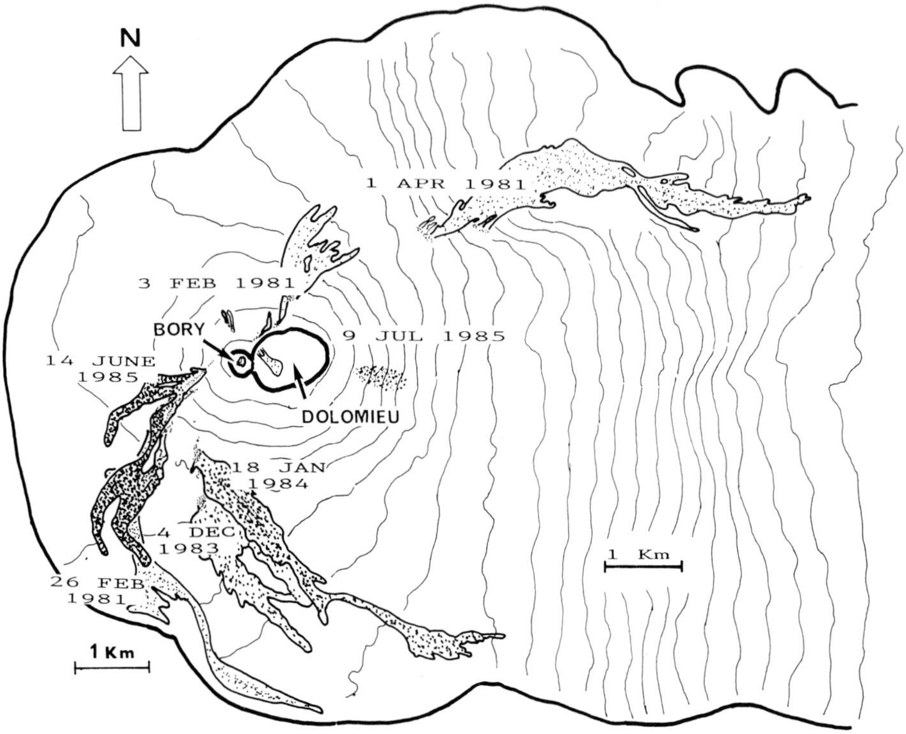

Fig. 5. Map of the eruptive fissures and lava flows between 1980 and July 1985. The location of the 9 July 1985 intrusion, which ended close to the surface east of the summit, is also shown. Elevation contour interval is 100 m

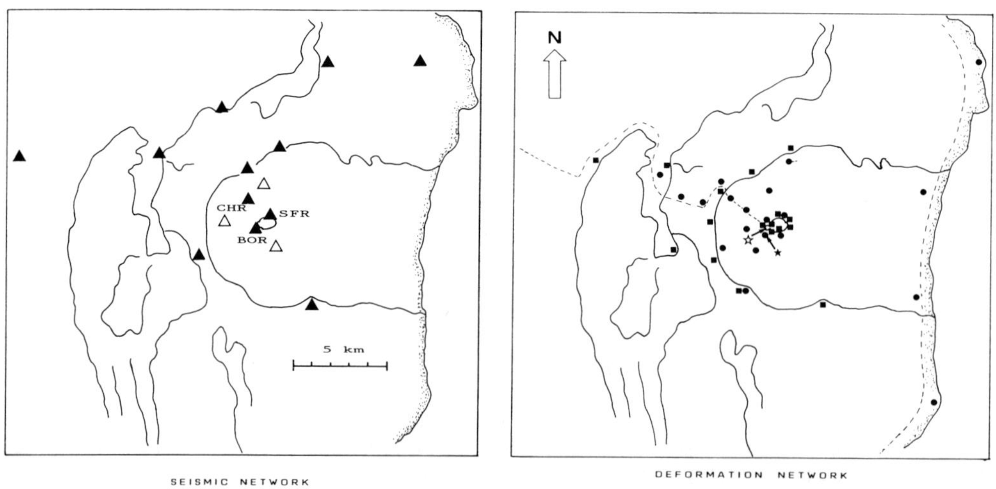

Fig. 6. Maps showing the seismic and the deformation networks (**black triangles:** permanent telemetered seismic stations; **open triangles:** mobile seismic stations; **circles:** spirit level tilt stations; **squares:** benchmarks of the EDM network; **black stars:** tiltmeter; **open star:** extensometer; **dashed line:** leveling lines)

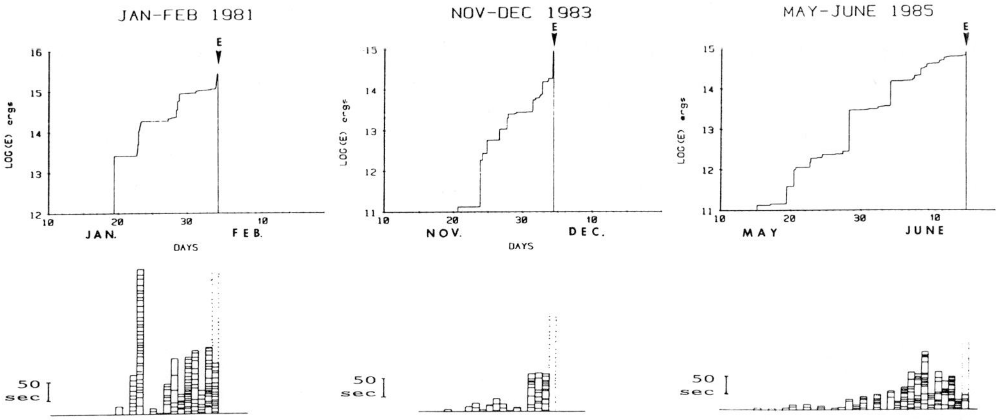

Fig. 7. Comparison of the three pre-eruptive seismic crises of 1981, 1983 and 1985. The upper diagrams represent the logarithm of the cumulative seismic energy, derived from earthquake magnitudes during these crises, and lower diagrams show the daily number and duration of earthquakes (each box represents one event and the box height represents the signal duration of each event at station BOR for the events above M = 0)

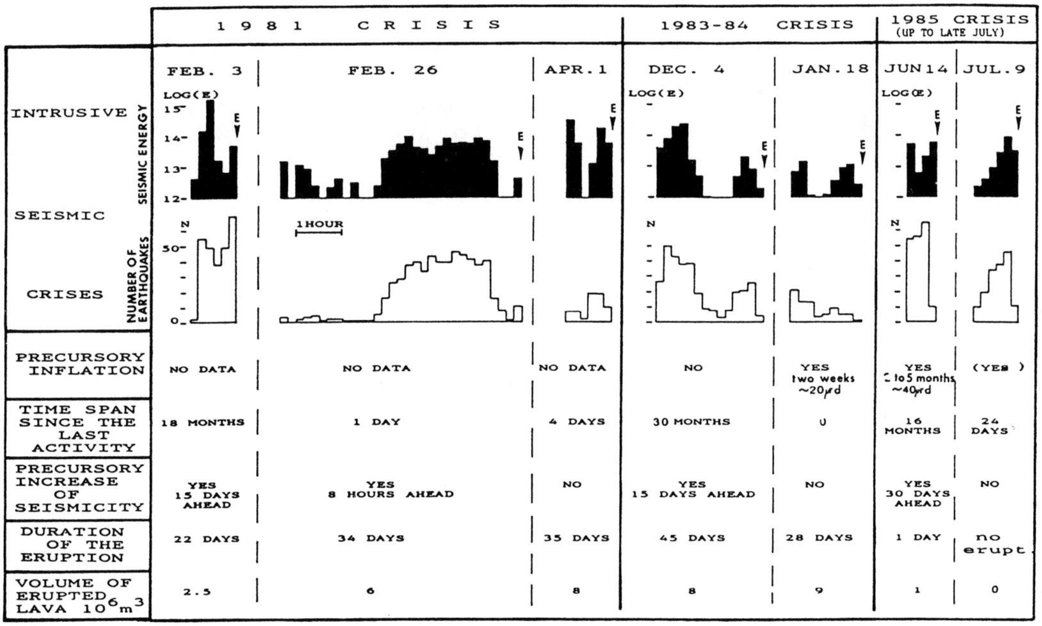

Fig. 8. Summary of seismicity immediately preceding the six eruptions between 1980 and July 1985: the 9 July 1985 intrusion is also shown. These short-term precursors are referred to as "intrusive crises" (see text)

4.2 Pre-Eruptive Seismic Crises

Although there are some differences between the three observed pre-eruptive seismic crises (Fig. 7), it is important to note their overall similarity. The three crises were located beneath the summit area at depths ranging mostly from 1 km to sea level (Fig. 9 shows the example of the 1985 pre-eruptive crisis; see also Fig. 16). This observation suggests that the focus of magmatic pressure remained in a relatively restricted area during this period. If this inference is correct, then it provides an important constraint on the horizontal extent of the shallow reservoir.

The durations of the crises are also similar (Fig. 7), if allowance is made for the lower threshold of detection of small seismic events in 1985 relative to the two former pre-eruptive crises. The addition of two stations near the summit made it possible to identify small seismic events which could not have been distinguished from small rockfalls or other seismic noise with only the first summital seismic station; therefore most of the low energy events recorded during the first 2 weeks of the 1985 crisis would not have been detected during the previous crises.

Note that the cumulative energy released is of the same order for the three crises [$15 > \log(E) > 14.5$]. This suggests that the rate of pressure buildup was similar for the different crises. Such behaviour would be expected if the same process (whether volatile or magmatic pressure increase) were to occur at the same rate, under the same lithostatic pressure (the crises were all at similar depths), in the same medium. This inference must be taken into account if we are to understand how the crises are triggered.

Fig. 9 a, b. Diagram showing the located earthquakes of the pre-eruptive and intrusive seismic crises before 14 June 1985 eruption. The **black symbols** represent events of the pre-eruptive crisis, between 15 May and 14 June, and the open stars events of the intrusive crises between 11 h 27 min and 12 h 01 min on 14 June. **The different black symbols** refer to different periods of the pre-eruptive seismic crisis which are not discussed further here. **CHR, SFR** and **BOR** are the seismic stations close to the summit (**BOR** is also a tiltmeter station). Note that the seismicity is mostly concentrated in the same area as the inflation (Fig. 11 d). The eruptive fissures are represented by **heavy straight lines**. The **stippled area** on the map shows the zone of maximum deformation during the intrusive crisis, as suggested by the data from the EDM and dry-tilt networks after the event. On the two cross-sections, the **stippled pattern** shows the inferred path of the intrusion. Elevation contour interval is 50 m. The hypocentral determinations represented by **triangles** and **stars** have been provided by Hirn (pers. commun.)

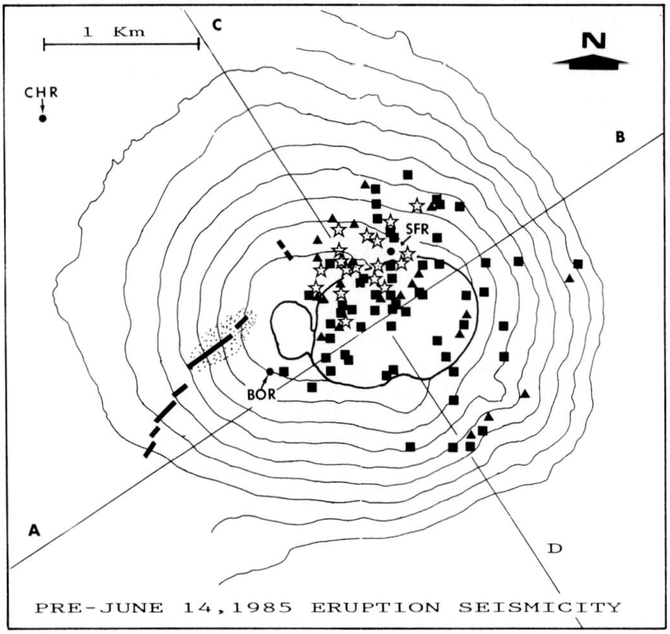
1 Km
C
N
CHR
B
SFR
BOR
A
D
PRE-JUNE 14,1985 ERUPTION SEISMICITY

a

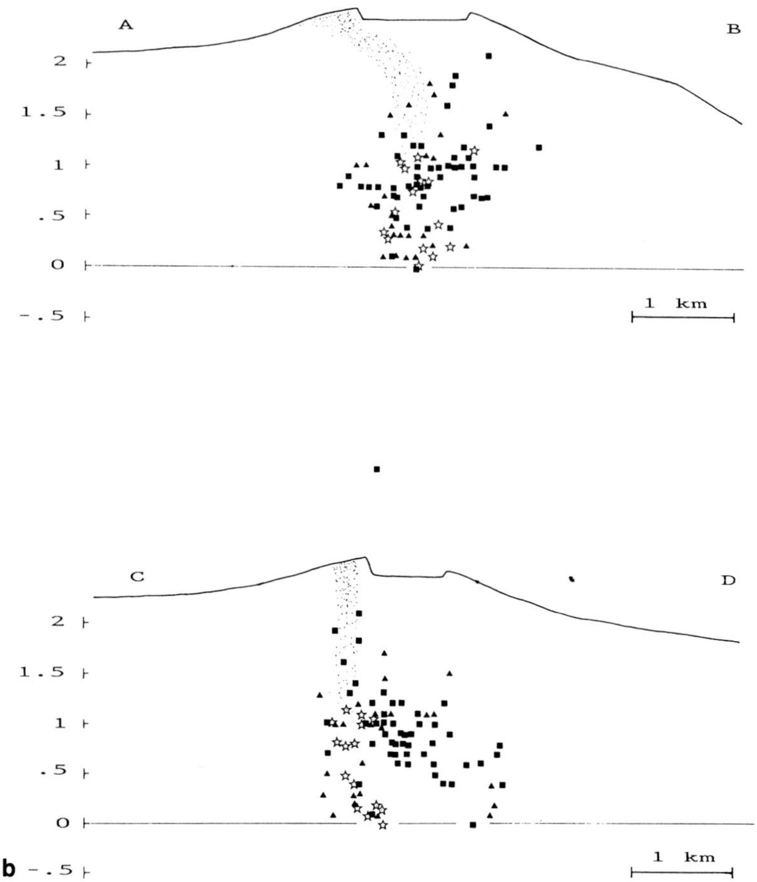
A
B
2
1.5
1
.5
0
-.5
1 km
C
D
2
1.5
1
.5
0
-.5
1 km

b

4.3 Pre-Eruptive Deformation

A crucial part of the deformation network was remeasured a few hours before the 3 December 1983 outbreak, and confirmed, as deformation data from a few days previously had done, that the pre-eruptive seismic crisis was not associated with any measurable deformation (Lénat et al. in press -a-). This pattern is unusual for basaltic volcanism. By contrast, the second phase of this eruption, which occurred 1.5 months later, was preceded by approximately 2 weeks of summit inflation unaccompanied by seismic activity.

For the other documented pre-eruptive period, the period before the 14 June 1985 outbreak, the pattern of deformation was quite different, and much more typical of this kind of volcanism. It is summarized in Fig. 11. Until mid-December 1984, slow subsidence of the summit was observed (see above). In early February 1985, this pattern of deformation began to change and in late April, it became clear that the northwest part of the summit was inflating. This inflation was not associated with any seismic activity. A pre-eruptive seismic crisis began on 15 May but until 28 May the seismicity was very low (zero to two events of magnitude <1 per day). Hence nearly aseismic inflation had occured, lasting for several months in early 1985.

After 28 May the seismic activity increased significantly and the rate of deformation also increased rapidly. Examined in detail, the deformation between 28 May and 13 June reveals a complex pattern which suggests migration of magma over small distances (Lénat et al. in press -b-). The resulting center of inflation (Fig. 11 d) was consistent with the location of the earthquakes before and during the 14 June 1985 intrusion (Fig. 9 a).

4.4 Seismicity and Deformation During Intrusive Phases

During the 1980-July 1985 period, all the outbreaks were preceded by short episodes of intense seismicity (Fig. 8). The durations of these episodes ranged between less than 1 h (14 June 1985) to 2.5 h (3 December 1983) except for the 26 February 1981 episode which was significantly longer (5.5 h). For six later outbreaks, between August 1985 and November 1986, in the central zone, similar seismic episodes have been reported (Delorme et al. 1985, 1986). Three of these depart from the average pattern; the 2 and 28 December 1985 crises were extremely short (about 15 min), and the 13 July 1986 crisis was preceded by only a few events of very low energy.

These episodes of sustained seismicity are accompanied by rapid ground deformation (Lénat et al. in press -a and b-; see Fig. 10 for example) indicating magma movement at shallow depths; these seismic episodes are therefore reported as "intrusive crises".

One of their characteristics was the suddenness of their onset; they started unexpectedly, without any precursory variation in the pattern of seismicity. An exception is the 26 February 1981 intrusive crisis where, after the harmonic tremor of the preceding eruptive phase had stopped, seismic activity (a few events per hour) resumed 8 h before the onset of the intrusive sequence.

On the other hand, it should be noted that before some of the phases that occurred later in 1985 and in 1986 (Delorme et al. 1985, 1986), when the outbreaks occurred several weeks or months after the end of the previous eruptive phase, a moderate seismicity sometimes resumed several days before the intrusive seismic crises, these latter examples may be regarded as intermediate between a true new eruption and a new phase of an eruption. Therefore the distinction between eruptions and eruptive phases may appear as artificial in some cases.

A detailed analysis of the 14 June 1985 seismic intrusive crisis by Hirn (pers. commun.) has revealed some interesting features of the mechanism of summit intrusions. Most of the events of this crisis were located at the same restricted range of depth as the events of the pre-eruptive crisis. As for the previously monitored intrusions, and in spite of a very dense seismic network (Fig. 6), no clear upward migration was observed. The hypocenters were distributed along a N 30°E trending and SE dipping zone in the northern part of the summit area (Fig. 9), and the focal mechanisms indicated that normal faulting with a large left lateral shear component dominated the

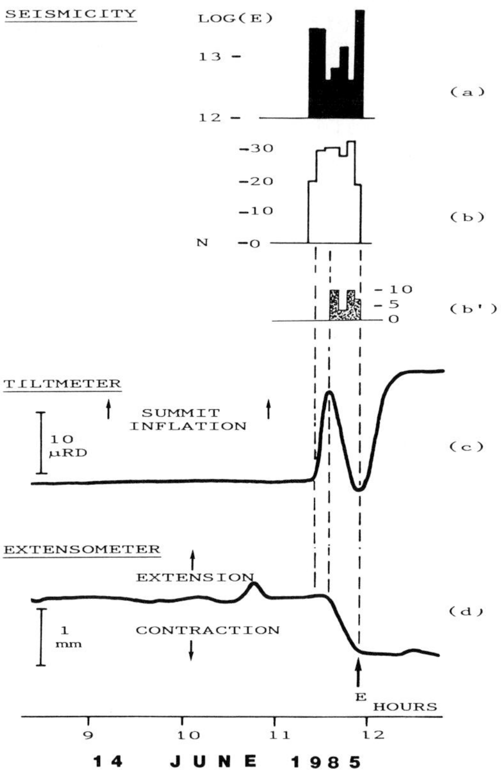

Fig. 10. Chronological diagrams of seismicity and deformation during the 14 June 1985 intrusion: **a** logarithm of seismic energy per 5-min intervals; **b** number of earthquakes per 5-min intervals (the threshold of detection of the shallow earthquakes beneath the summit is about M=0); **b'** number of small shocks recorded only at BOR seismic station per 5-min intervals; **c** tilt record of BOR tiltmeter. The component of tilt monitored is radial to the summit crater; **d** record of the extensometer in Bory crater (the monitored fissure is roughly parallel to the wall of the Dolomieu crater); E marks the time at which the eruption began

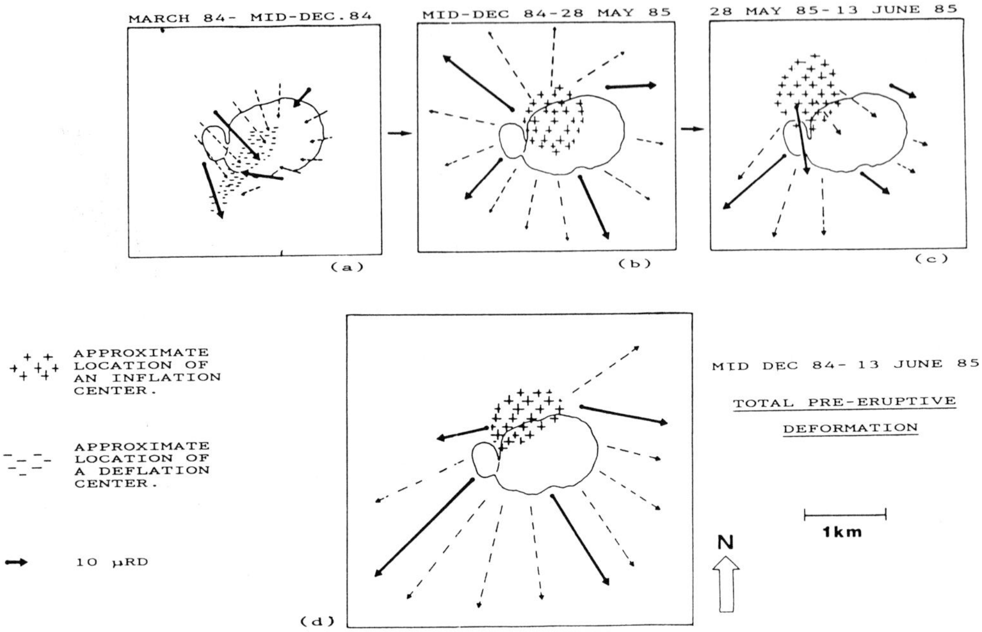

Fig. 11 a-d. Observed tilt variations at four optical level tilt stations ("dry-tilt" stations) in the summit area. **Solid arrows** represent the vectors at the tilt stations; arrows point away from the center of inflation. To make the tilt patterns more legible, hand-interpolated vectors (**dashed vectors**) have been drawn

seismicity. Therefore this seismicity could not correspond to a simple re-opening of existing cracks or the progressive opening of magma-filled fissures (A. Hirn, pers. commun.). However, such a pattern of normal faulting is consistent with the pattern of cracks and fissures at the surface (Bachèlery et al. 1983). These observations appear to conflict with the fact that the movement of magma towards the surface throughout the intrusing crisis is shown by the synchronous deformation recorded by the BOR tiltmeter (Fig. 10). It is also surprising that the eruptive fissures did not reach the surface in the epicentral zone but nearly 1 km westward of it.

At the beginning of the intrusive sequence, the BOR tiltmeter showed that the summit was rapidly inflating, probably as magma was intruded in the hypocentral zone. The pattern of deformation later changed when the tilt reversed and the extensometer (monitoring a main fissure in the Bory crater) started to show a contraction perpendicular to the fissure. Simultaneously, weak shocks began to be recorded only at the BOR seismic station which is near the area where the eruptive fissures opened later. This deformation pattern lasted until the onset of the effusion of lava. During the first tens of minutes of the eruption, while the seismic records were saturated with harmonic tremor, the tilt reversed again.

We believe that the main hypocentral volume lies approximately at the top of the magma reservoir complex, and that the seismicity is triggered by magmatic and volatile overpressure and mainly reflects the conditions of the local stress field. The tiltmeter record clearly shows that the onset of the intrusion is associated with magma migration. The next stage of the intrusion is marked by the reversal of the tilting and by two phenomena consistent with the arrival of magma at shallow depth: the closing of surface fissures and weak seismic shocks recorded only at the station within 200 or 300 m of the location of the future eruptive fissures. Therefore the reversal of the tilt is better explained by deformation related to a shallow intrusion west of the station than by a deflation of the summit area. This latter explanation cannot, however, be ruled out, since the reversal was recorded on only one tiltmeter: similarly, the reversal of tilting after the onset of the eruption can be explained by a decrease in magmatic pressure below the fissures or alternatively by continuing inflation beneath the summit.

The fact that the migration of magma in the last kilometer or so generates only weak earthquakes suggests that as the magma approaches the surface, it reaches a low stress field. For the 14 June 1985 intrusion, the elongated epicentral zone and the set of eruptive fissures coincide with a tensional tectonic system that was recognized by Bachèlery et al. (1983). This system has apparently also controlled the development of the eruptive fissures of recent eruptions (1957, July 1979, February 1981). Therefore it is probable that the 14 June intrusion was somehow "funnelled" or "trapped" in this tectonic system. This model of "capture" of upward migrating magma by a tectonic system may probably be applied to most of the summit intrusion: it is also, for example, suggested by the 9 July 1985 crisis (see below). This would explain why the eruptions do not necessarily occur above the root of the intrusion (i.e. the epicentral zone of the earthquakes) but at some unpredictable location along a tectonic system (unless an extremely dense seismic network allows one to follow the migration of the very low energy earthquakes associated with the intrusion at shallow depth). Alternatively, the magma erupted on 14 June 1985 could be distinct from the magma of the intrusion in the hypocentral zone, which would not then have reached the surface (A. Hirn, pers. commun.). Although this hypothesis cannot be eliminated with the available data, it is difficult to believe that two disconnected batches of magma were involved in the intrusive sequence since the end of the seismic crisis almost coincided with the release of magma at the surface (Fig. 10).

The case of the deformation associated with the 14 June 1985 intrusive sequence represents the most complete set of data acquired at Piton de la Fournaise for this type of event. It is probably typical of most of the intrusions in the summit area of Piton de la Fournaise.

The remeasurement of the EDM and dry-tilt networks after the intrusive events shows that the summit remains permanently deformed after the crisis. This permanent deformation is attributed (Lénat et al., in press -b-) to the emplacement of intrusive bodies at shallow depths (the feeder dikes of the eruptions and possibly secondary intrusions). The main bodies of the shallow intrusions are commonly not located below the epicentral areas (Figs. 9 a and 14 a).

4.5 The Case of an Intrusion not Followed by an Eruption

Examples of intrusive events not followed by eruptions have been frequently reported at Kilauea volcano (Klein et al. 1987) and at Krafla volcano (Einarsson and Brandsdottir 1980; Tryggvason 1980). The first recognized event of this type at Piton de la Fournaise took place during the 9-10 July 1985 crisis.

Figure 12 shows the chronology of the seismicity and the deformation during this event which was similar to the intrusive

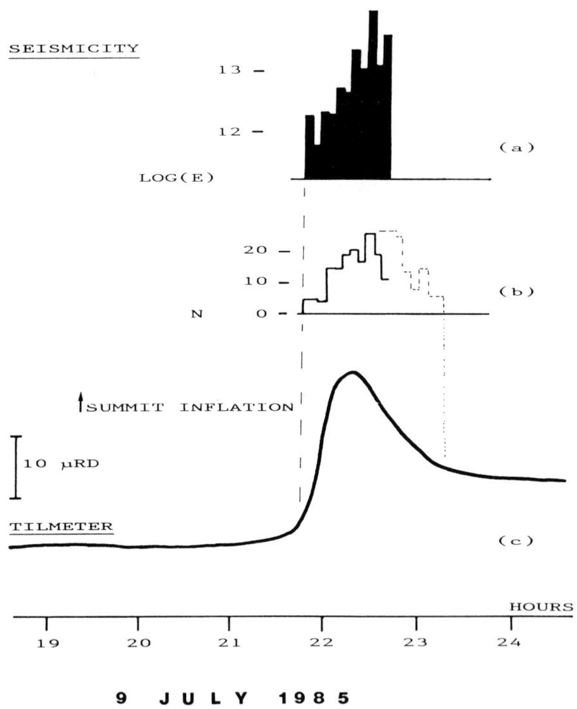

Fig. 12 a-c. Variation with time of seismicity and deformation during the 9 July 1985 summit intrusion: **a** logarithm of seismic energy per 5-min intervals at Bory station; **b** number of earthquakes (above M=0) per 5-min intervals. The dashed line shows the number of earthquakes at a mobile station closer than the Bory station to the location of the shallow intrusion (see Figs. 6 and 14), and which has thus recorded more low magnitude events during the last period of the intrusive event. **c** Tilt record of BOR tiltmeter. Monitored component of tilt is radial to the summit crater; the start and the end of the deformation sequence are correlated with the seismic activity

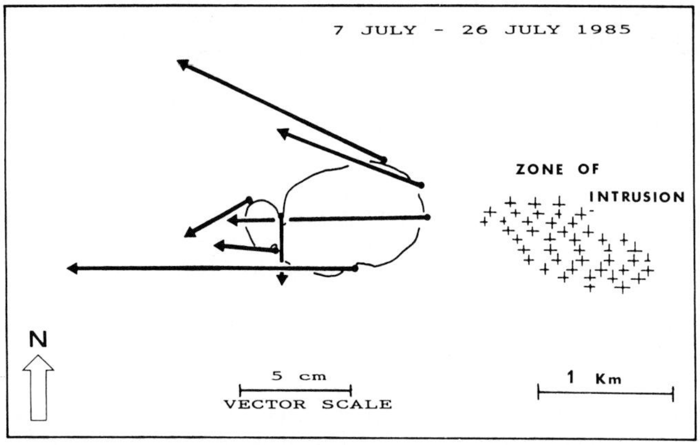

Fig. 13. Displacement vectors in the summit area between 7 and 26 July 1985. This pattern is similar to the tilt pattern observed at the four summit optical level tilt stations on 11 July 1985. The observed deformation is mostly related to the 9 July 1985 summit intrusive sequence (see Figs. 12 and 14)

sequences that had been observed previously. The comparison of the levelling and EDM data before and after this event (Fig. 13) shows that an intrusive body was emplaced at very shallow depths about 1 km east of the summit crater. The earthquakes, on the other hand, clustered on the south edge of the summit crater (Fig. 14). This situation resembles the 14 June intrusive sequence described above, where the hypocentral zone was also distinct from the zone of the shallowest part of the intrusion. In the case of the 9 July intrusion, the hypocentral zone and the shallow intrusion zone may also be associated with one of the tensional tectonic systems recognized by Bachélery et al. (1983).

4.6 An East Flank Seismic Crisis

At the end of the 9 July 1985 summit intrusive sequence, a new crisis began on the east flank of the volcano (Figs. 14 and 15). This new crisis was characterized by a type of activity that had not previously been observed since the seismic network began operating in May 1980. It surpassed previous crises by two orders of magnitude, both in the number of events, and in the energy released (peak magnitude = 3). Initially most hypocenters were concentrated at a distance of about 3 km east-northeast of the summit and at a depth of 2 to 5 km below sea level (Fig. 14); later, after the climax of the crisis, the hypocentral zone spread over the entire Grandes Pentes area.

The interpretation of this crisis is not certain. Lateral movement of magma into the area of the flank crisis may explain why the summit intrusion ended before reaching the surface (A. Hirn, pers. commun.). Alternatively, Lénat et al. (in press -b-) favour a volcano-tectonic hypothesis inspired by the analogy with the well-documented seismicity of the free flank of Kilauea, where the south flank of the volcano is moving seawards, while the north remains buttressed by Mauna Loa (Swanson et al. 1976).

At Kilauea volcano, the intrusion in the rift zones are systematically accompanied and followed by south flank seismicity principally at depths ranging from 4 to 10 km (Koyanagi et al. 1972). The study of the deformation associated with these crises has shown that the flank seismicity corresponds to the release of compressive stresses induced in the structure by the forceful intrusion of magma in the rift zones, and to the corresponding displacement of the unbuttressed south flank (Swanson et al. 1976; Lipman et al. 1985). Similar behaviour has been inferred for the east flank of Piton de la Fournaise based on pronounced volcano-tectonic similarities with Kilauea such as the presence of a free and a buttressed flank, the frequency of intrusion in the central area and the presence of the Grandes Pentes-Grand Brulé slumped structure (Vincent and Kieffer 1978; Chevallier and Bachèlery 1981; Duffield et al. 1982; Lénat and Aubert 1982). The 10 July seismic crisis could therefore correspond to the release of stresses accumulated in the edifice by a succession of intrusion over several years. The shallow intrusion at the east of the summit would thus have had a trigger effect. A similar sequence comprising a shallow intrusion east of the summit followed by a flank seismic crisis occurred in March 1986 (Delorme et al. 1986). On this occasion the sequence was followed, or accompanied by an intrusion, and by an eruption downslope along the SE rift zone, with further seismic activity on the east flank. Such a sequence seems analogous to the behaviour of Kilauea. Unfortunately, the deformation network at Piton de la Fournaise is not yet well enough designed to monitor the deformation of the east flank (because of the difficulty of access); thus crucial deformation data are lacking to constrain the interpretation of the east flank crises.

5 Discussion

The data described above, concerning historic activity and the pattern of the seismicity and the deformation during the first 5 years of surveillance, lead us to infer a preliminary model

---

Fig. 14 a, b. Representation of the located earthquakes of the 9 July 1985 summit intrusion (**square symbols**), 10 July (**triangles**) and those on the following days (**lozenges**) during the east flank seismic crisis. Elevation contour interval is 100 m. The hypocentral determinations represented by **black squares** and **black triangles** have been provided by Hirn (pers. comm.) and computed using the data of both permanent and mobile stations, whereas the other earthquakes have been computed at the observatory with the data of the permanent network only (see Fig. 9). The **stippled area** represents the intrusion, as determined from deformation data

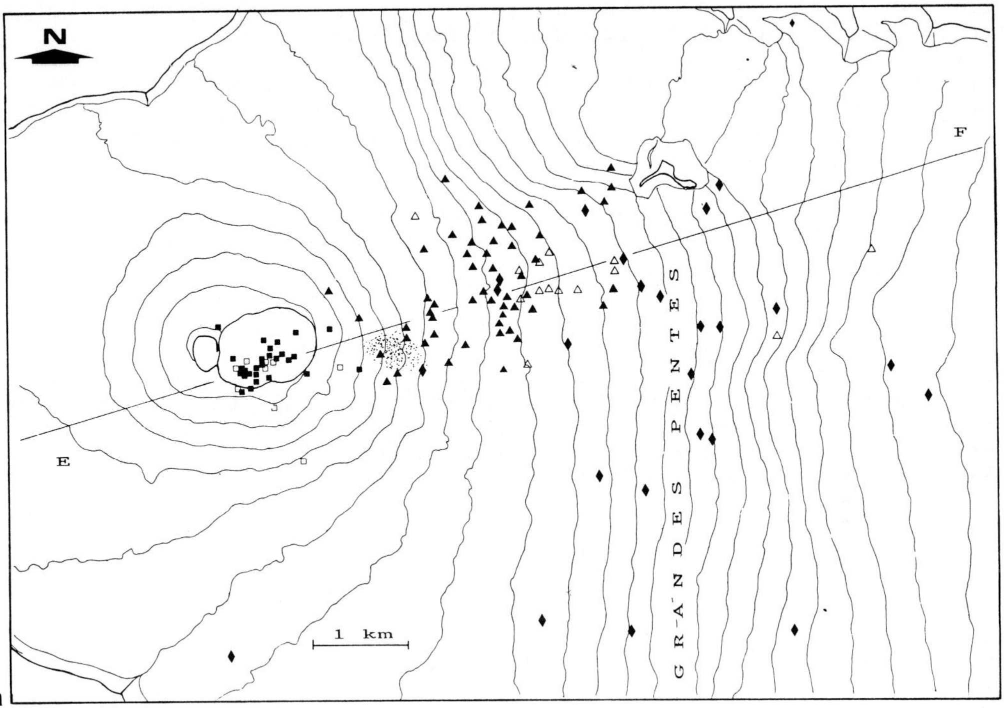

a

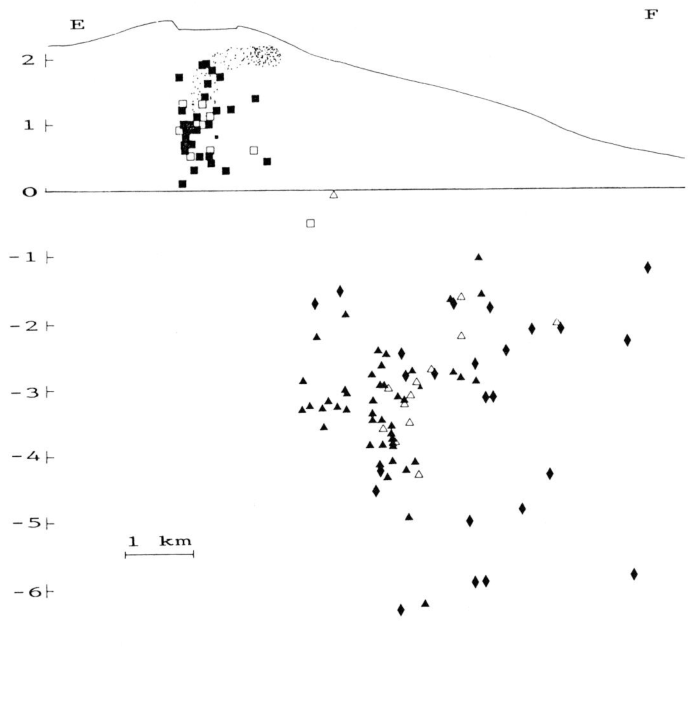

b

for the volcanism of Piton de la Fournaise, from which criteria to predict future activity can be deduced.

One key element of the model is the shallow magma reservoir complex beneath the summit, and moreover the inferences that can be made on its location, its storage capacity and the dynamics of magma supply from deeper levels.

All the earthquakes located within the massif of piton de la Fournaise between 1980 and mid-June 1985 are shown in Fig. 16. Except for a few earthquakes located outside the Enclos caldera, almost all the earthquakes were shallow events that occurred beneath the summit during the pre-eruptive and intrusive crisis reported above. In spite of the fact that the accuracy of hypocentral determination improved as the seismic network was expanded and upgraded between 1980 and 1985, and that therefore the size and the shape of the cloud of hypocenters is not tightly controlled, it is evident that all the crises occurred in a relatively small volume located between 1 and 3 km beneath the summit, with a main concentration at around 1.5 km depth. This range of depth is consistent with previous estimates of depth for a shallow magma reservoir the presence of which is inferred on the basis of structural, petrological and aeromagnetic data (Lénat 1987; Kornprobst et al. 1984; Lénat and Aubert 1982). The volume where the seismicity occurs is probably situated around the upper part of the storage complex, in a zone that undergoes fracturing in response to magmatic and fluid overpressure. The complex pattern of deformation observed before 14 June 1985 (Lénat et al. in press -b-) shows that the top of the reservoir may be composed of a plexus of interconnected storage units. Owing to the relatively small horizontal extension of the epicentral area (about 1 km in diameter), it can be assumed that the reservoir is not a large structure. It is, however, difficult at present to assess its vertical extension: the virtually aseismic zone beneath the cloud of shallow earthquakes (Fig. 16) can be interpreted as the bottom of the reservoir zone or as a molten or plastic zone of the reservoir (only two earthquakes were located deeper, in the range from -5 to -9 km, beneath the summit area during the period considered).

The virtual lack of deep, intermediate or shallow seismicity between eruptive cycles, and the lack of systematic renewed inflation after the eruption cycles, seems to indicate that the transfer of magma to the shallow reservoir is a discontinuous process at Piton de la Fournaise. No evidence of refilling of the reservoir could be found in the seismicity or in the deformation before the 1983-84 eruption. No further evidence of magma transfer was observed after this eruption, until the early 1985 summit aseismic inflation, which showed that overpressure was developing in the northeast part of the roof of the reservoir. The problem is clearly to determine whether or not magma has been transferred from deep levels to the shallow reservoir during the 1980-85 period. Lénat and Bachèlery (1985) have proposed:

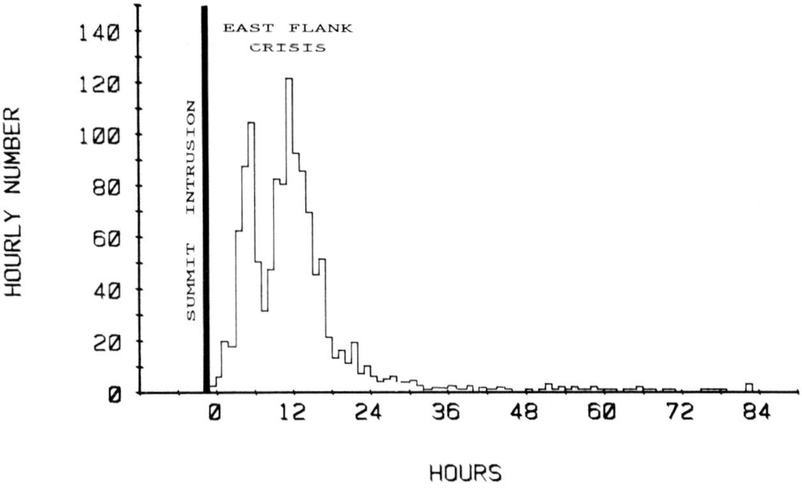

Fig. 15. Hourly number of earthquakes (above about M = 0.5) during the 10 July 1985 east flank crisis following the 9 July summit intrusion

1. That for the last 50 years or so, the very large oceanite eruptions of 1931, 1961 and 1977 (Fig. 4) reflect major pulses of internal magma transfer. These account for about half of the erupted volume of lava of the 50 or so eruptions since 1930.
2. That less considerable eruptions of oceanite or olivine basalts (1937-39, 1945-48, 1961-66, 1973) represent smaller pulses of resupply to the chamber.
3. That during the intervening periods, in which the magma output rate is low, the summit reservoir is hardly resupplied at all. They believe that the summit reservoir has not been refilled since the 1977 oceanite eruption.

This hypothesis is consistent with the geophysical data given above, and also with the petrological data which support the idea that the lavas erupted since 1977 may be derived from the same batch of magma evolving at low pressure (Boivin et al. 1984; P. Boivin, pers. commun.). On the other hand, it should be noted that a preliminary interpretation of the gravity variations between 1981 and 1985 (J.N. Delatre and M.F. Esnoult, pers. commun.) assumes the emplacement of a large amount of magma at shallow depth (1 to 3 km) beneath the summit area.

Another important factor that has to be taken into account in order to understand how Piton de la Fournaise works is the interaction between the gravitational stress field of the volcano and the strains generated by its volcanic activity.

The west flank of the volcano is buttressed by the rest of the island, whereas its sea-facing flank is unbuttressed. Accordingly the strains due to the emplacement of the intrusion should be accomodated by seaward displacement of the east

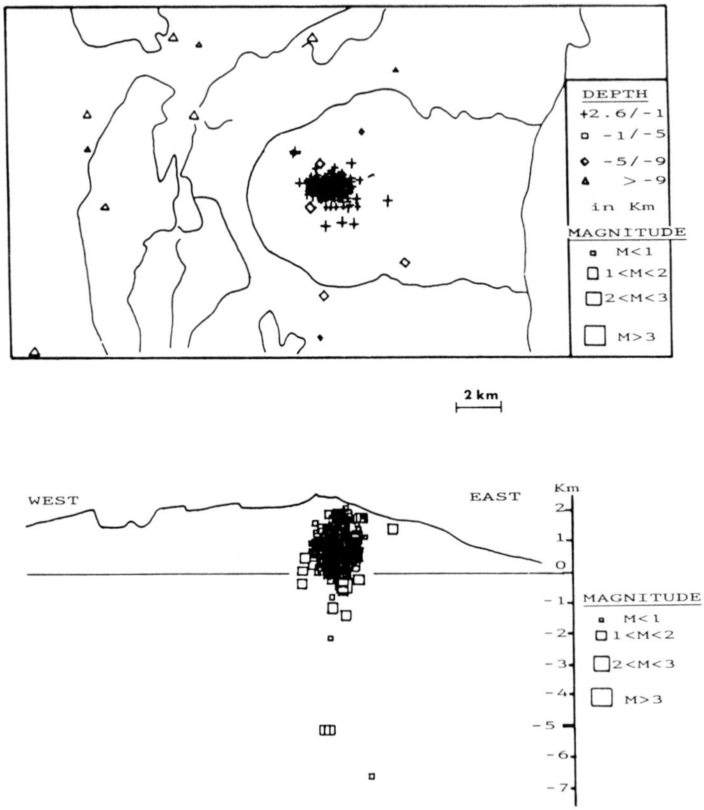

Fig. 16. Representation of all located earthquakes recorded between 1980 and late June 1985 (i.e. the July 1985 east flank crisis is excluded). Almost all the recorded seismicity was associated with the crises mentioned in this chapter, and occurs at shallow depths beneath the summit. The size and the shape of the hypocentral area beneath the summit is not uniformly accurate because the determinations improved as the seismic network was expanded during this period. However, the seismic activity was definitely confined to a restricted area of 1.5 km or so in diameter

flank. This model, inspired from the model derived for the south flank of Kilauea by Swanson et al. (1976) and Lipman et al. (1985), fits the surface structural observations (Bachèlery et al. 1983) and provides an explanation for some of the features of the crises described above; i.e. the lack of seismicity in the west part of the central zone (Figs. 9 and 14), the faulting mechanism of the 14 June 1985 intrusive crisis, and the 10 July 1985 east flank seismicity.

Most of the intrusive phase-related strains are accumulated in the central zone of the volcano (cf. Fig. 2). Figure 17 shows that successive intrusions add permanent deformation to the central cone. This accumulation of deformation must participate

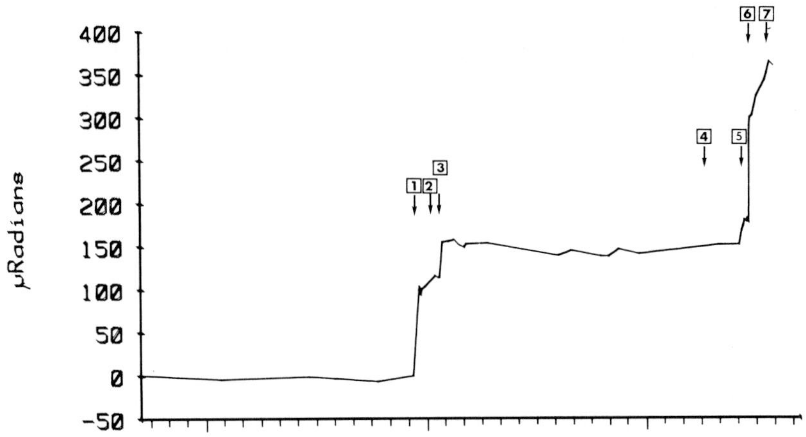

Fig. 17. Tilt record at BOR optical level tilt station from September 1982 to late July 1985. The main episodes of the activity are shown: **1** 4 December 1983 outbreak; **2** early January 1984 summit inflation: **3** 18 January 1984 outbreak; **4** early 1985 aseismic inflation; **5** May-June 1985 inflation associated with a pre-eruptive seismic crisis; **6** 14 June 1985 outbreak; **7** 9 July 1985 summit intrusion

in the growth of the central cone, where, owing to the frequency of eruptions (Fig. 3), large strains must also accumulate rapidly. There are two consequences of this dilatation of the central zone of the volcano. Firstly, the east flank tends to be pushed aside. This is confirmed by the structural pattern observed at the surface and by the presence of the Grand Brûlé slump.

We also favour the interpretation of the 10 July 1985 events as an episode of contractile strain release of the east flank through eastward displacement, although there is a lack of crucial deformation data in this area. Secondly, this seaward displacement of the Grand Brûlé area tends to also lower the strains in the direction perpendicular to the rift zones, which diverge north and south from the central cone. This would facilitate the development of intrusions down-rift.

The stresses within the edifice therefore result from a complex interplay between the gravitational stress field, the rapid accumulation of dilatation in the central area, the periodic release of contractile strains of the east flank and intrusions along the rift zones. The deformation monitoring effort thus far has been focussed on the central area: the network should now be expanded to the east flank, which is clearly a key area for the study of the tectonic behaviour of the volcano.

6 Conclusion

Piton de la Fournaise appears to be an interesting site to study oceanic intraplate basaltic volcanism. The data obtained during the recent years at the newly established volcanological observatory show that the volcano does not behave exactly as the volcanoes of Hawaii, the other well-studied site for this type of volcanism. However, 5 years of monitoring are obviously a period too short to provide a representative sample of the physical behaviour of Piton de la Fournaise. In particular, the monitoring of a large oceanite eruption should certainly provide crucial clues to the processes of large magma transfers at Piton de la Fournaise.

Acknowledgements. I wish to thank all my colleagues who have contributed to the collection of the data at the Piton de la Fournaise Volcanological Observatory since 1980. I am indebted to R.Y. Koyanagi and J. Latter who have significantly contributed to the improvement of the manuscript.

References

Bachèlery P (1981) Le Piton de la Fournaise (Ile de la Réunion): étude volcanologique, structural et pétrologique. Thesis, Univ Clermont II

Bachèlery P, Chevallier L, Gratier J-P (1983) Caractères structuraux des éruptions historiques du Piton de la Fournaise. CR Acad Sci Paris II 296:1345-1350

Boivin P, Bachèlery P, Benhamou G, Dupont P, Kornprobst J (1984) Etude pétrologique des laves de l'éruption de 1983-84 à l'île de la Réunion. Abstr Colloque PIRPSEV (CNRS ed) Dec 1984, p 19

Chevallier L, Bachèlery P (1981) Evolution structural du volcan actif du Piton de la Fournaise, Ile de la Réunion, Ocean Indien Occidental. Bull Volcanol, 44 4:723-741

Chevallier L, Vatin-Pérignon N (1982 Volcano-structural evolution of Piton des Neiges, Réunion Island, Indian Ocean. Bull Volcanol, 45 4:287-298

Delorme H, Delarue JF, Bachèlery P (1985) Reports from the Observatoire Volcanologique du Piton de la Fournaise. SEAN Bulletin, V10, n°8, 11, 12 (Vol. 10, No. 8, pp 2-3; Vol. 10, No. 11, p 13; Vol. 10, No. 12, pp 10-12)

Delorme H, Delarue JF, Bachèlery P (1986) Reports from the Observatoire Volcanologique du Piton de la Fournaise. SEAN Bulletin, V11, n°1, 2, 3, 6, 7, 11 et 12 (Vol. 11, No. 1, pp 8-10; Vol. 11, No. 2, pp 7-8; Vol. 11, No. 3, pp 13-16; Vol. 11, No. 6, p 11; Vol. 11, No. 7, p 16; Vol. 11, No. 11, pp 13-14; Vol. 11, No. 12, pp 3-5)

Duffield WA, Stieltjes L, Varet J (1982) Huge landslide blocks in the growth of Piton de la Fournaise, La Réunion and Kilauea volcano, Hawaii. J Volcanol Geotherm Res 12:147-0

Dzurisin D, Koyanagi RY, English TT (1984) Magma supply and storage at Kilauea volcano, Hawaii, 1956-1983. J Volcanol Geotherm Res 21:177-206

Einarsson P, Brandsdottir B (1980) Seismological evidence for lateral magma intrusion during the July 1978 deflation of Krafla volcano in NE-Iceland. J Geophys 47:160-165

Gillot P-Y, Nativel P (1982) K-Ar chronology of the ultimate activity of Piton des Neiges volcano, Réunion Island, Indian Ocean. J Volcanol Geotherm Res 13:131-146

Kieffer G, Tricot B, Vincent P (1977) Une éruption inhabituelle (Avril 1977) du Piton de la Fournaise (Ile de la Réunion): ses enseignements volcanologiques et stucturaux. CR Acad Sci Paris 285 D:957-960

Klein FW (1982) Patterns of historical eruptions at Hawaiian volcanoes. J Volcanol Geotherm Res 12:1-35

Klein FW, Koyanagi RY, Nakata JS, Tanigawa WI (1987) The seismicity of Kilauea's magma system. US Geol Surv Prof Paper 350:1019-1185

Kornprobst J, Boivin P, Lénat JF, Bachèlery P, Bonneville A, Dupont P, Lecointre J, Seidel JL, Thomas P, Vincent P (1984) Le Piton de la Fournaise, Ile de la Réunion. Abstr Colloque PIRPSEV (CNRS ed) Dec 1984, pp 75-82

Koyanagi RY, Swanson DA, Endo ET (1972) Distribution of earthquakes related to mobility of the south flank of Kilauea volcano, Hawaii. US Geol Surv Prof Paper 800-D:89-97

Lénat J-F (1987) Structure et Dynamique internes d'un volcan basaltique intraplaque océanique: le Piton de la Fournaise (Ile de la Réunion). Thesis, Univ Clermont II

Lénat J-F, Aubert M (1982) Structure of Piton de la Fournaise volcano (Réunion Island, Indian Ocean) from magnetic investigations. An illustration of the analysis of magnetic data in a volcanic area. J Volcanol Geotherm Res 12:361-392

Lénat J-F, Bachèlery P (1985) Dynamics of magma transfers at Piton de la Fournaise volcano (Réunion Island, Indian Ocean). Earth Evol Sci Special issue: Modeling of volcanic processes. Chi-Yu, Scarpa R (eds) Vieweg, Braunschweig Wiesbaden pp 57-72

Lénat J-F, Bachèlery P, Bonneville A, Tarrits P, Cheminee JL, Delorme H (in press -a-) The December 4, 1983 to February 18, 1984 éruption of Piton de la Fournaise (La Réunion, Indian Ocean): description and interpretation. J Volcanol Geotherm Res

Lénat J-F, Bachèlery P, Bonneville A, Hirn A (in press -b-) The beginning of the 1985-1987 eruptive cycle at Piton de la Fournaise; new insights in the magmatic and volcano-tectonic systems. J Volcanol Geotherm Res

Lipman PW, Lockwood JP, Okamura RT, Swanson DA, Yamashita MY (1985) Ground deformation associated with the 1975 magnitude-7.2 earthquake and resulting changes in activity of Kilauea volcano, Hawaii. US Geol Surv Prof Paper 1276, 45 pp

Ludden JN (1977) Eruptive patterns for the volcano Piton de la Fournaise, Réunion Island. J Volcanol Geotherm Res 2 4:385-395

MacDougall I (1971) The geochronology and evolution of the young volcanic island of Réunion, Indian Ocean. Geochem Acta 35:94-134

Simkin T, Siebert L, Mc Clelland L, Bridge D, Newhall C, Latter JH (1981) Volcanoes of the world. Hutchinson Ross, Stroundsburg

Swanson DA (1972) Magma supply rate at Kilauea volcano, 1952-1971. Science 175:169-170
Swanson DA, Duffield WA, Fiske RS (1976) Displacement of the south flank of Kilauea volcano: the result of forceful intrusion of magma into the rift zones. US Geol Surv Prof Paper 963, 39 pp
Tanguy J-C (1980) L'Etna, étude pétrologique et paléomagnétique. Implications volcanologiques. Thesis, Univ Paris VI
Tryggvason E (1980) Subsidence events in the Krafla area, north Iceland, 1975-1979. J Geophys 47:141-153
Vincent P, Kieffer G (1978) Hypothèse sur la strucutre et l'évolution du Piton de la Fournaise (Ile de la Réunion) aprés les éruptions de 1977. 6ème Réun Ann Sc Terre, Orsay (Soc Geol France ed) 407 pp
Wadge G, Guest JE (1981) Steady-state magma discharge at Etna, 1971-1981. Nature (Lond) 294:548-550

Volcano Monitoring in the Aeolian Islands (Southern Tyrrhenian Sea): The Lipari-Vulcano Eruptive Complex

S. Falsaperla[1], G. Frazzetta[1], G. Neri[1], G. Nunnari[1], R. Velardita[1] and L. Villari[2]

Abstract

Seismicity and ground deformation affecting the Lipari-Vulcano area (Aeolian Islands, Italy) have been continuously monitored for several years as part of a program of volcano surveillance in a densely populated area. Among the accumulated data, particular attention was paid to the evident correlation between ground deformation in the Lipari-Vulcano area and the seismic crisis which occurred in the Gulf of Patti, a few kilometres south of the island of Vulcano, culminating in a 5.5 magnitude shock which struck the region on 15 April, 1978. The seismicity was largely associated with the Tindari-Letoianni line (Ghisetti and Vezzani 1982), a large-scale fault system extending northward from northeastern Sicily to intersect the Aeolian Islands volcanic arc: along this tectonic trend are located the islands that form the Lipari-Vulcano eruptive complex. The occurrence of the seismic crisis was preceded by a marked deformation phase, which produced significant dilatation in the Lipari-Vulcano area. This phase had begun by the end of 1976 and culminated some time after the occurrence of the main shock on 15 April, 1978. At the same time, fumaroles on Vulcano Island underwent a temperature increase (more than 100°C) and unusual variations in the gas composition were detected (Martini et al. 1980; Carapezza et al. 1981). Even though the observed ground deformation and seismicity did not lead to an eruption, it is thought that the increasing tensional stress in the regional tectonic structures controlling the volcanism (as indicated by the incipient dilatation), together with more intense fracturing associated with the seismic crisis, represented a higher eruption probability.

1 Introduction

The Aeolian Islands volcanic arc in the southern Tyrrhenian Sea, Italy, is well known because of the active volcanoes, Stromboli and Vulcano, whose historical eruptions constitute a high level of risk in the region. This situation is becoming more serious as the population continues to increase close to the volcanoes.

---

[1] International Institute of Volcanology, CNR - Catania, Italy
[2] Earth Sciences Institute, University of Messina, Italy

IAVCEI Proceedings in Volcanology 1
J.H. Latter (Ed.), Volcanic Hazards

In particular, the islands of Vulcano and Lipari, which together make up a single volcanic complex, have recently experienced a great development of tourism, and many buildings have been constructed at the very foot of the eruptive cone of "la Fossa". As a result, research has been directed towards volcano surveillance, particularly geophysical monitoring methods.

The present paper describes the most significant data collected during the last decade and points to potential precursors of an impending eruption.

Possible relationships between regional tectonic activity and the internal dynamics of the volcanic system are investigated, following the hypothesis that the former could represent a potential trigger for an eruption.

2 Volcanological and Structural Framework

The Aeolian Archipelago, in the southern Tyrrhenian Sea, lies a few tens of kilometres north of the northern coast of Sicily. Five of the seven islands (Stromboli, Panarea, Salina, Filicudi and Alicudi), all entirely of volcanic origin, roughly describe an arcuate structure that extends for about 80 km, concave towards the NNW. The other two islands, which make up the Lipari-Vulcano eruptive complex, are sited along a transverse trend, intersecting the arc at the island of Salina. This trend coincides with the Tindari-Letoianni line (Ghisetti and Vezzani 1982), a well-defined lithospheric fault system extending southward into northeastern Sicily (Fig. 1).

The volcanic activity that formed the Aeolian Islands is mainly Pleistocene and Holocene in age, and the magmas show a marked orogenic affinity (from calc-alkaline to shoshonitic). The most recent eruptions occurred in historical times at Lipari (6th Century A.D.) and Vulcano (1888-1890), while eruptions are continous on the island of Stromboli.

The Aeolian Archipelago is also the site of shallow and deep seismicity, but very few earthquakes occur at depths between 50 and 200 km. The deep seismicity (h >200 km) lies along a steeply inclined plane, dipping to the NW (Caputo et al. 1970; Gasparini et al. 1982; Scarcella et al. 1985).

The petrogenetic affinity of the Aeolian magmas, the characteristics of the regional seismicity and the more general structural context which includes the volcanism of the Aeolian Islands, suggest that the volcanic activity could be the result of lithosphere convergence, i.e. the subduction of the African plate beneath the Eurasian one (Barberi et al. 1973, 1974). Alternative interpretations are proposed by Scandone (1979) and Wezel (1981).

Recent eruptions on the island of Vulcano took place at two well-defined volcanic edifices, Vulcanello and "la Fossa", in the northernmost part of the island. The volcanic products that form these structures mainly consist of pyroclastics, with

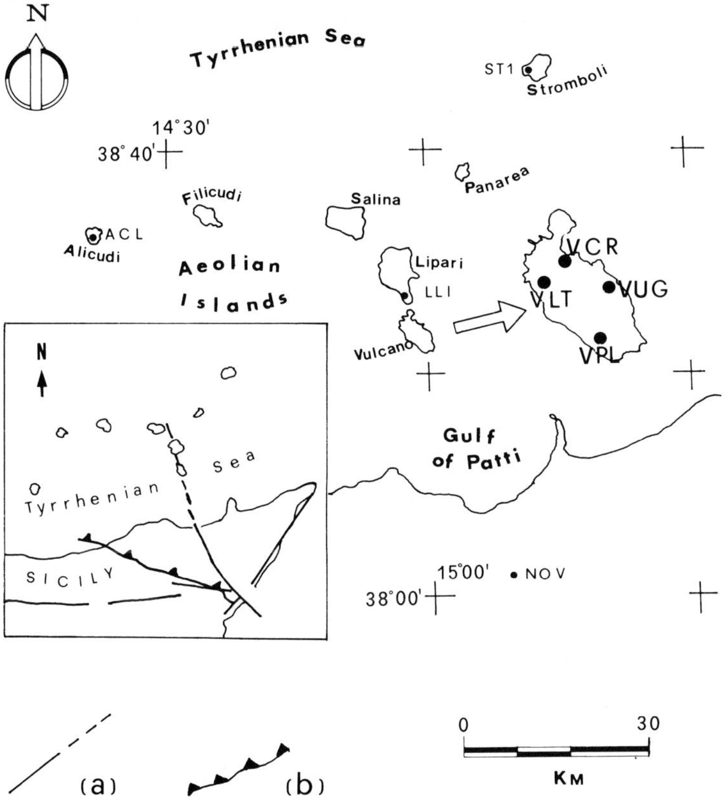

Fig. 1. Location map of the Aeolian Islands volcanic arc in the southern Tyrrhenian Sea. **Solid circles** indicate the stations of the local seismic network. The location of the main tectonic structures in the northeastern part of Sicily and the islands of Vulcano, Lipari and Salina is shown in the **inset**. **a** Large-scale fault systems; **b** thrust zone in the Appennine-Maghrebian chain

lesser volumes of lava flows. Among the outcropping pyroclastic deposits, numerous units have been observed that show clear textural evidence suggesting a shallow water-magma interaction (Frazzetta et al. 1983).

At present, volcanic activity is restricted to fumarolic degassing, mainly at the "la Fossa" volcanic structure, with maximum temperature normally ranging between 200° and 300°C (Martini et al. 1980).

Significant temperature increases have been episodically observed during historical times (e.g. 615°C in 1924; Sicardi 1940), in spite of the total lack of eruptions during the present century.

Recent drillings, carried out for geothermal research, have shown that the temperature at 2050-m depth exceeds 419°C and temperature gradients are sufficiently steep that magma could well be only a few hundred metres deeper (Faraone et al. 1986).

3 Geophysical Monitoring

Geophysical monitoring of the volcanoes of the Aeolian Islands is carried out by the International Institute of Volcanology, Catania, within the framework of the activities supported and coordinated by the National Group of Volcanology (G.N.V.) of the National Research Council of Italy (C.N.R.).

A seismic network (Fig. 1), consisting of eight stations and equipped with short period vertical seismometers (1 Hz), operates in the area. Data are telemetered to an observatory (Osservatorio Geofisico di Lipari) on the island of Lipari, and recorded on both paper and magnetic tape. The permanent local network is shown in Fig. 1; this has undergone various modifications since the date of its installation (Lo Giudice 1973). Only three seismic stations were continuously operating in the Lipari-Vulcano area between 1977 and 1982 (Del Pezzo and Martini 1981; Neri 1981), when the setting up of two other stations on the island of Vulcano (Falsaperla and Neri 1986b; U.O. "Sismologia Eolie" 1985) improved coverage around the main eruptive centres. The station presently located on Stromboli (Fig. 1) has been continuously operating since 1985, following a period of discontinuous recording.

Two categories of seismic events are usually distinguished and systematically investigated for volcano surveillance: the shallow tectonic earthquakes in the Archipelago and its vicinity, and the microseismicity more strictly related to the volcanic structures on Vulcano and Stromboli.

In the present paper, we are particularly concerned with the volcanic seismicity of Vulcano, analyzed by considering two different classes of events: (1) the shocks recorded at the VCR station, exceeding an empirically determined trace amplitude threshold of 2 cm; and (2) the earthquakes detectable at all the Lipari-Vulcano stations. The two above classes require different selection criteria, roughly based on the shock energy, in absence of an adequate magnitude scale for the smaller events ($M < 2.5$). Higher magnitude values are calculated using empirical magnitude-duration formulae (Di Prima et al., in press).

The geometry of the seismic network ensures adequate coverage of the area where the volcanic risk is highest (the northern part of Vulcano). Data from a neighbouring seismic network (Bottari et al. 1986a) in northeast Sicily helps to reduce the errors of location, at least for the events of higher magnitude ($M > 2.5$). Hypocentral computation is carried out with the EVM algorithm (Caccamo and Neri 1984) and two different models recently proposed for studying regional earthquakes (Bottari et al. 1979) and the shallow seismicity of Vulcano (Falsaperla et

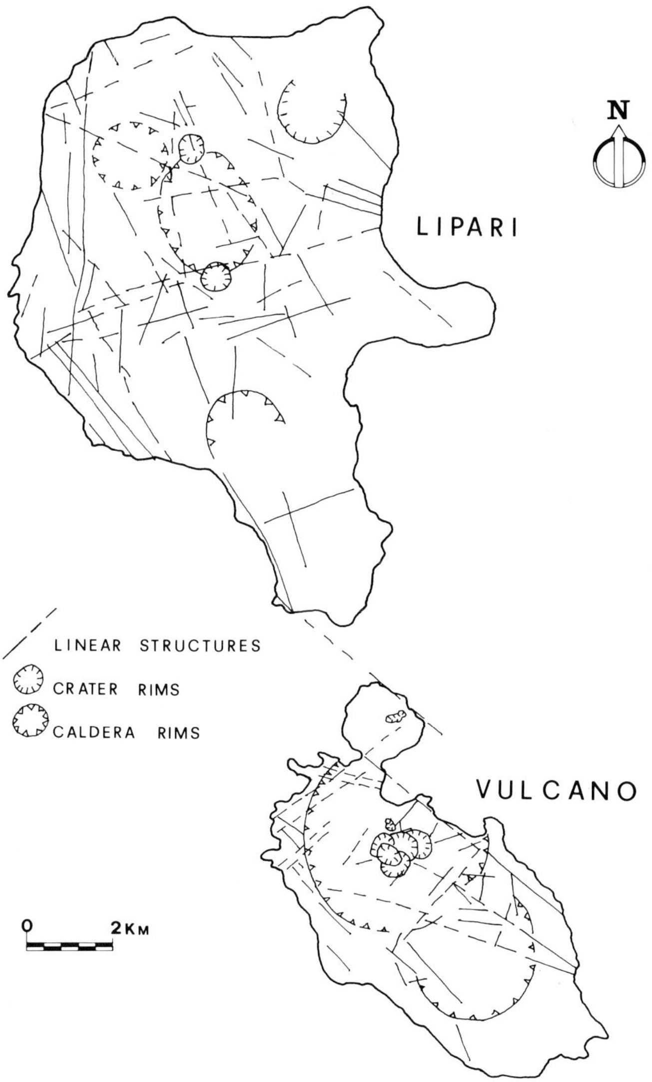

Fig. 2. Structural sketch map of the Lipari-Vulcano eruptive complex (Frazzetta et al. 1977)

al. 1985). Typical location errors are 2-3 km for the epicentre and 3-5 km for the focal depth for regional earthquakes. Considerably smaller uncertainties are obtained for the microearthquakes of Vulcano due to the shorter distances between nearby stations and to the low seismic wave velocity in the shallow layers beneath the island.

Geophysical monitoring in the Lipari-Vulcano area is also being carried out, by both discrete and continuous measurements of ground deformation (Falsaperla et al. 1985).

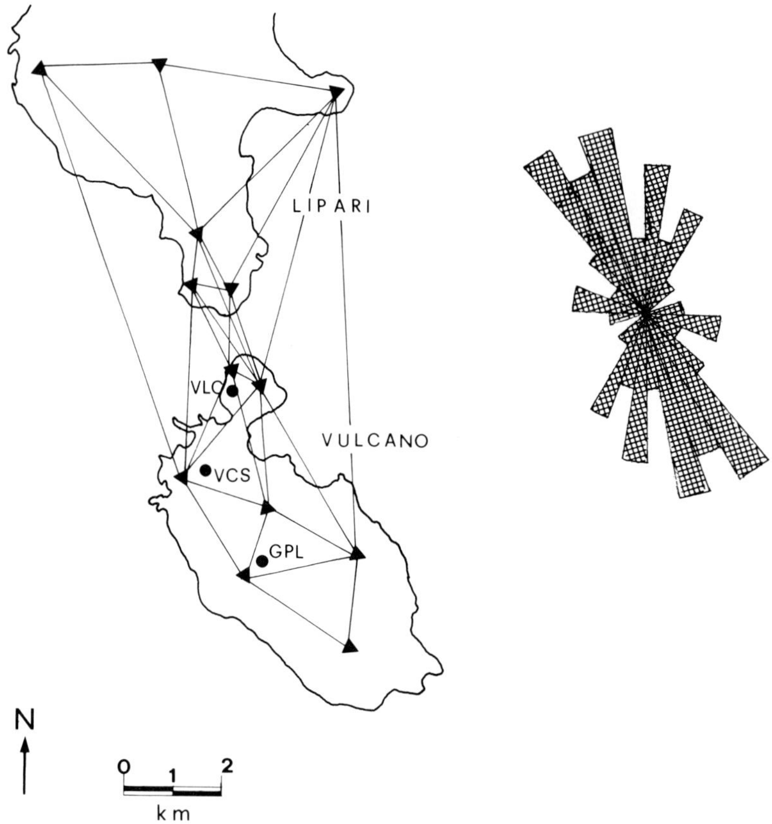

Fig. 3. The Geodimeter trilateration network set up in 1975 on the islands of Lipari and Vulcano. **Solid circles** indicate the three borehole tilt stations. The **rose diagram** refers to the azimuthal distribution of the measured lines

A Geodimeter trilateration network was designed and set up in 1975, covering the southern half of Lipari Island and the whole of Vulcano (Fig. 3). The network consists of 13 benchmarks and 37 lines, and it is normally measured twice a year, using an AGA 6BL laser Geodimeter. The number of the redundant lines is sufficient to confine the errors within acceptable limits, but the geometry of the network is constrained by the shape of the two islands, and affects the accuracy of the calculated strain parameters. Because of this, both displacement vectors and rotation are disregarded, and attention is focussed on the pattern analysis of the principal strain tensors and areal dilatation.

In recent years, continuously recording borehole tilt stations have been installed on Vulcano (Fig. 3): digital data are telemetered to the observatory on Lipari and recorded on magnetic tape. The stations have operated only intermittently, due both to instrumental problems, and to the poor stability of some of the selected sites (VLC in Fig. 3, for example). In the present study, the data have only been examined for the period April-August 1985, during which a local seismic swarm took place at Vulcano.

4 Local and Regional Seismicity

The Aeolian Islands area is subject to "moderate" (following Hagiwara's classification; Hagiwara 1964) seismicity distributed over a considerable range of depth, with recorded magnitudes up to 5.8. The largest energy release is associated with tectonic earthquakes. On the basis of the instrumental data from the past decade (Neri 1985; Bottari et al. 1986a) and information from the latest catalogues on historical seismicity (ENEL 1978; Neri 1985), the most active sectors appear to be the Gulf of Patti (south of Vulcano; Fig. 1) and the area west of the line joining Vulcano, Lipari and Salina.

The permanent local seismic networks, which were set up in the Aeolian Islands and neighbouring regions a decade ago, and the results from a series of short-term investigations on Vulcano and Lipari (Blot 1971; Latter 1971), make possible a more detailed picture than previously, above all for the microseismicity. There are substantial differences between sectors in the mechanism of seismic energy release (Neri 1985). The Gulf of Patti and the western sector show a tendency to a main shock-after shocks type of sequence, as was the case in 1978 and 1980 (Del Pezzo and Martini 1982; Bottari et al. 1986a), whereas the Lipari-Vulcano area is affected by smaller earth-

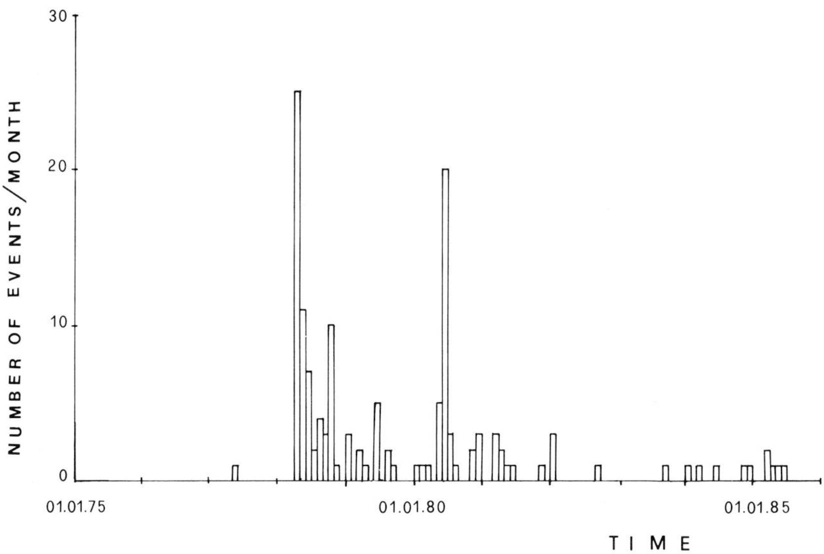

Fig. 4. Number of earthquakes (magnitude >3 and focal depth <50 km) vs time, in the area of the Aeolian Islands (latitude 38°00'-38°40'N and longitude 14°00-15°30'E), from January 1975 to June 1985

quakes often occurring as swarms (Del Pezzo and Martini 1981; Neri 1985; Falsaperla and Neri 1986b). The eastern sector appears to be rather inactive, although considerable explosion microshock activity has been recorded at Stromboli (Lo Bascio et al. 1973; Nappi 1975; Falsaperla and Neri 1986a), associated with its persistent eruptive state.

Different seismic processes coexist at Vulcano, such as fault-origin earthquakes (high frequency, clear onset and S phases) and shocks of purely volcanic origin (monochromatic waveforms) mostly connected with rising gases in the feeding system of the fumarolic field (Latter 1971; Badalamenti et al. 1986). Seismic swarms are almost exclusively confined (Falsaperla et al. in press 1988) to the northern part of the island at shallow depth (h <2 km) and are characterized by very low magnitudes (M <2.5).

The seismo-tectonic sequences of 1978 in the Gulf of Patti and of 1980 at Alicudi (Fig. 4) (in particular the former, close to

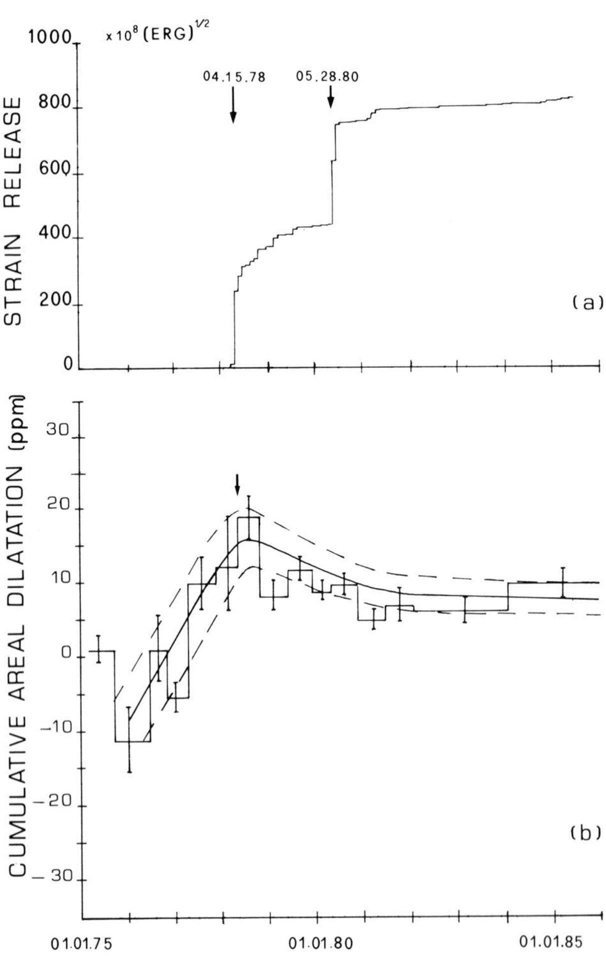

Fig. 5 a, b. a Cumulative strain release from earthquakes in the Aeolian Islands, and **b** cumulative areal dilatation at Lipari-Vulcano from the Geodimeter network data, beginning in 1975; **error bars** in the areal dilatation graph define an average confidence limit (**dashed lines**)

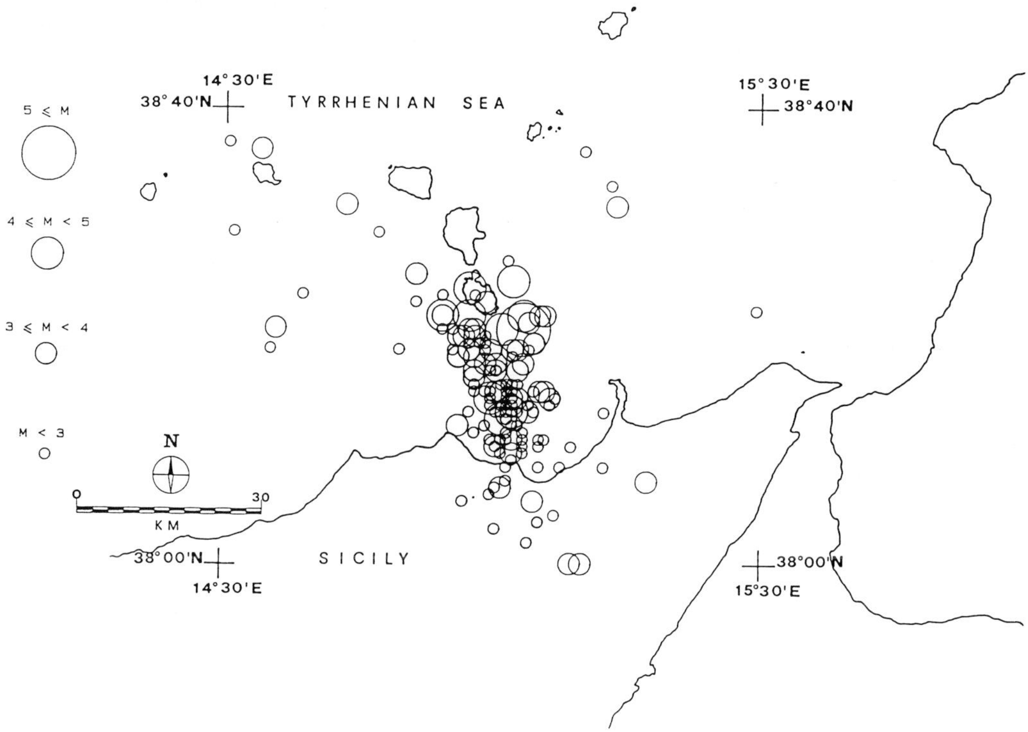

Fig. 6. Epicentral map of earthquakes (magnitude >2 and focal depth <50 km) of the Gulf of Patti tectonic sequence (1978-1979)

the active volcanic structures of Vulcano) and the Vulcano swarms in the past decade have been the subject of particular attention, because of the possibility of a renewal of volcanic activity. The 1978 sequence, which, unlike that of 1980, was accompanied by deformation in the Lipari-Vulcano area (Fig. 5) is apparently related to the Tindari-Letoianni line, a NNW-SSE regional structural system which affects northeastern Sicily and the islands of Vulcano, Lipari and Salina (Frazzetta et al. 1977, 1982). This relationship is suggested by the epicentre distribution (Fig. 6) and the main-shock focal mechanism (Gasparini et al. 1980). They proposed a right-lateral strike-slip fault plane solution for the main shock, which is consistent with the above NNW-SSE regional tectonic trend of the Tindari-Letoianni line. Furthermore, the depth distribution of the earthquake foci (Fig. 7) shows that the most intense shocks are localized in the medium to deep part of the crust (the Moho ranges from 20 to 30 km depth in this region; Morelli et al. 1975) and that nearly all the aftershocks occurred at a shallower depth than the main shock (M = 5.5; h = 28 km). Such a pattern is in agreement with the model of Mogi (1967) concerning the stress propagation following a strong crustal shock.

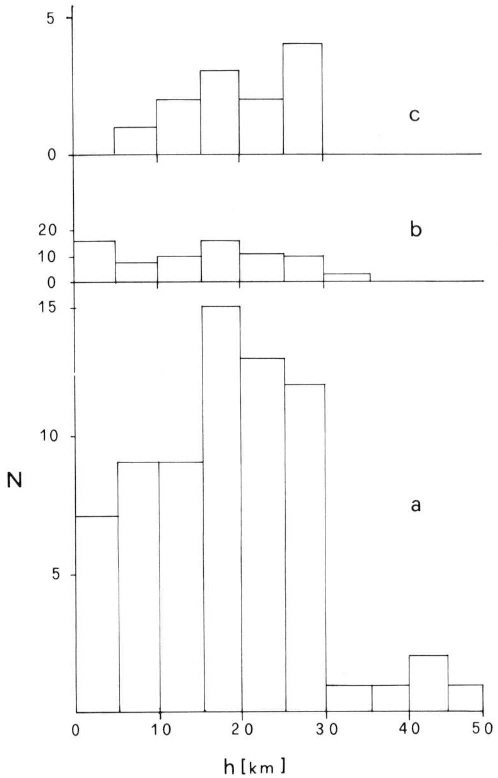

Fig. 7. Bar diagram of the depth (**h**) distribution of earthquake foci shown in Fig. 6: **a** = M >3; **b** = M <3; **c** = M >4

The Alicudi sequence began at the end of May 1980. Its main event was a magnitude 5.8 earthquake, almost the same size as the strongest shock in the 1978 sequence in the Gulf of Patti. The shock sequence showed an epicentral distribution trend approximately NW-SE, also related to the geostructural framework of the area (Bottari et al. 1986b; Finetti and Del Ben 1986), although it occurred on a different fault system which does not directly affect the Lipari-Vulcano area.

A further point of interest is the apparent relation between the tectonic seismicity in the immediate vicinity of Vulcano and the seismic swarms in the area of the main active crater. It has been observed on various occasions that the occurrence of events belonging to the former category is followed after a short period of time (1-2 weeks) by an increase in the second category of seismicity. Detailed analyses of the most evident episodes are given in Bottari et al. (1984) and Falsaperla and Neri (1986b), where the seismotectonic activity of magnitude greater than 2.5 occurring within a radius of 10 km from Gran Cratere of Vulcano is discussed and related to the occurrence of shallow volcanic microearthquake swarms. A statistical evaluation of this apparent relationship is under way, based on 1984-1986 data. The 1985 swarm (April-August; Fig. 8) was studied in greater detail, due to the higher number of stations operating at that time in the area. A map of the epicentres (computer locations) is given in Fig. 9 (Falsaperla and Neri

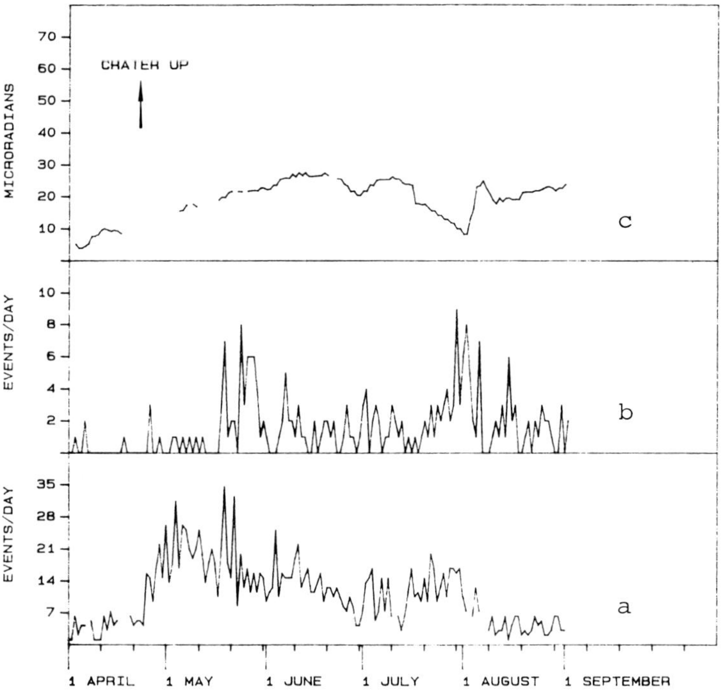

Fig. 8 a-c. a Daily frequency of local earthquakes recorded between April and August 1985 at the seismic station VCR (northern flank of the "La Fossa" cone); shocks are selected by using an assigned value of the trace amplitude threshold (20 mm). **b** Daily frequency of more energetic events recorded at all five seismic stations (LLI, VCR, VLT, VUG and VPL) in the Lipari-Vulcano area; **c** radial tilt recorded at a borehole tilt station (GPL) located at the foot of the southwestern slope of "La Fossa" cone

1986b). The hypocentres are almost entirely less than 2 km deep, and 70% are between 0 and 1 km deep. The waveform structure shows that within the swarms fault-origin earthquakes (high frequency, clear onsets and S phases) coexist with monochromatic shocks that, on the basis of the peculiar features of the sector (and in particular the presence of important fumarole fields), can be linked to rising gases in the fumarolic feeding system. This hypothesis has already been proposed by previous researchers (Blot 1971; Latter 1971), undertaking more detailed short-term investigations of this peculiar seismic phenomenon.

For various reasons (location, modest energy release and shock waveforms), the swarm in question seems comparable to previous ones recorded by both temporary field surveys (Blot 1971; Latter 1971) and the permanent network.

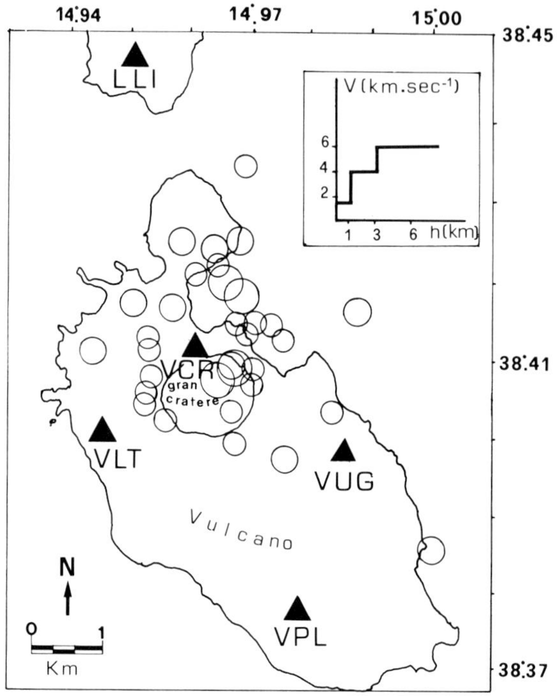

Fig. 9. Map of epicentres of microearthquakes recorded during the April-August 1985 seismic swarm at Vulcano; the velocity model used for computer locations (Falsaperla et al. 1985) is shown in the **inset.** A rough energy classification is adopted distinguishing the shocks in two classes (**different circle sizes**) respectively, above and below a given threshold of the mean value A of the trace maximum amplitude at the seismic stations

5 Ground Deformation

Geodimeter surveys carried out between 1975 and 1985 have shown that the Lipari-Vulcano area has undergone ground deformation, particularly horizontal movements.

These phenomena were clearly observed in relation to the April 1978 earthquake in the Gulf of Patti and the long series of aftershocks that lasted until 1979. The areal dilatation rate of the Lipari-Vulcano area, which at the end of 1976 underwent a noticeable and sharp increase, declined steadily during the following years (Fig. 10). From 1982 to December 1985, the whole area was quiet and showed no significant horizontal deformation. During the period 1975-1981 frequent oscillations between dilatation and contraction were observed. This effect was more conspicuous on Vulcano than over the rest of the trilateration network (Fig. 10).

A particularly well-defined sequence is illustrated by the cumulative areal dilatation trend (Fig. 5 b) which shows how, in spite of the presence of the dilatation and contraction phases, a clear overall areal dilatation preceded the occurrence of the Gulf of Patti seismic crisis of April 1978 by at least 2 years. It is also of interest to note that the tendency towards positive dilatation values clearly reversed only some time after the occurrence of the main shock on April 15 (between December 1978 and June 1979). The asymptotic trend towards a new equilibrium condition can be clearly seen in Fig. 5 b: this implies non-elastic behaviour of the region which definitely acquired a permanent deformation as a result of the earthquake.

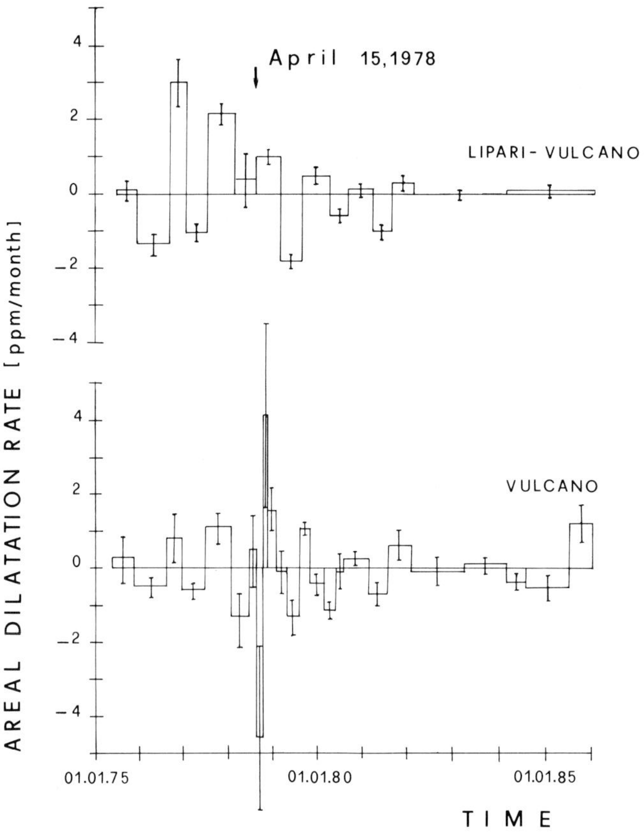

Fig. 10. Areal dilatation rate pattern for the entire surveyed area (Lipari-Vulcano) and for the island of Vulcano alone; error bars are also shown

Bearing this in mind, we may also note a further interesting characteristic of the main strain tensor pattern (Fig. 11). The prevailing orientations of the maximum (E1) and the minimum (E2) extension axes are almost exactly perpendicular to the Tindari-Giardini line (Fig. 1), suggesting a pronounced structural control over the deformation pattern by regional tectonic trends. The role of this structural control is also evident in the way the Gulf of Patti seismic crisis occurred, as previously pointed out. The prevailing orientation of the main strain axes does not appear to be affected by the azimuthal distribution of the measured lines in the trilateration network, in spite of the limit imposed by the physiographical constraints. The relative maxima of the azimuthal distribution of the measuring lines (Fig. 3) are quite far from the prevailing orientation of the main strain axes (Fig. 11); a closer coincidence would be expected if the line orientation in the trilateration network was seriously constraining the deformation analysis.

As mentioned above, the tilt data on Vulcano have been discontinuously collected and consequently do not permit any adequate analysis or comparison with the horizontal deformation.

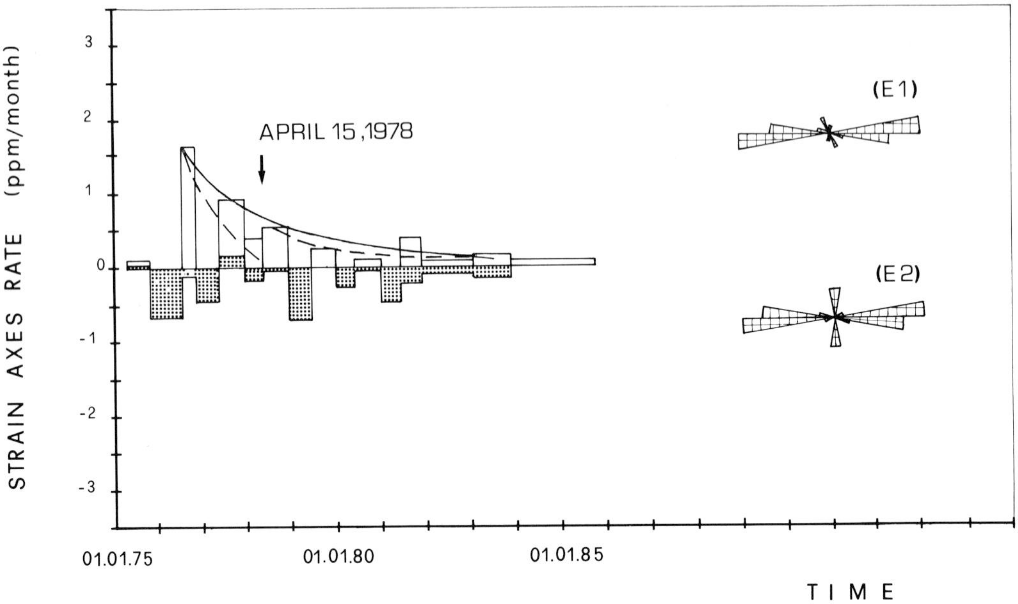

Fig. 11. Azimuthal distribution of the main strain axes (**right**) and temporal variation of strain rate (**left**). Note that minimum extension **E2** (**half tone** in the bar chart at **left**) mostly indicates compression. The **rose diagrams** (on the **right**) refer to the analysis of strain axis azimuths vs cumulative relative values of their modula

However, between April and August 1985, during the occurrence of a local seismic swarm on Vulcano, anomalous tilts were observed at one of the borehole tilt stations (GPL on Fig. 3). The tilt component radial to the main cone of "la Fossa" showed a continuous and progressive inflation of the structure during the microearthquake swarm. The overall inflation trend reversed as the seismic swarm progressively declined. This deflation phase was then perturbed by a sharp and abrupt inflation, coincident with the final part of the swarm, characterized by a higher occurrence of more energetic shocks recorded at all the Lipari-Vulcano stations (Fig. 8 b) and located in the whole northern sector of Vulcano (Fig. 9).

The observed time correlation, together with the characteristics of tilt and earthquakes (epicentres and hypocentres of the shocks, coexistence in the seismic swarm of fracturing and degassing mechanisms, small energy release, inflation and deflation of the volcanic structure), suggest a relevant role played by pressure variations within the shallow fluid system feeding the fumarolic fields.

6 Conclusions

The above observations have concentrated attention on an interesting working hypothesis regarding the relationship between regional tectonic activity and possible awakening of the dormant volcano on the island of Vulcano.

The April 1978 seismic crisis in the Gulf of Patti was notable for the following reasons:

1. The seismicity was largely confined to the Tindari-Letoianni line, running from northeastern Sicily northward to intersect the Aeolian Islands volcanic arc. Along this tectonic trend are located the islands that form the Lipari-Vulcano eruptive complex. The control of this fault system on the evolution of the volcanism at Lipari and Vulcano is clear not only from the location of the islands, but also from the distribution of calderas, craters and eruptive fractures (Frazzetta et al. 1977, 1982) along the same trend (Fig. 2).

2. The occurrence of the seismic crisis was preceded by a marked deformation phase, which produced a significant dilatation in the Lipari-Vulcano area. This phase had begun by the end of 1976 and culminated some time after the occurrence of the main shock on 15 April 1978. The subsequent return to a condition of dynamic equilibrium was characterized by an apparently permanent deformation (Fig. 5 b). The deformation rate pattern described by the maximum extension axis (El in Fig. 11) declined logarithmically (solid line in Fig. 11), and did not appear to be perturbed by the main shock on 15 April. Unfortunately, the presence of an expected perturbation could have been missed because of the infrequent Geodimeter measurements: it is possible that the real pattern of the extension rate could be qualitatively better described by two branches of the decrement curve (dashed line in Fig. 11). The prevailing orientation of the main strain axes, almost perpendicular to the regional tectonic trend, suggests that the latter is strongly controlling the deformation pattern. The fact that it is not exactly perpendicular emphasizes the role of the transcurrent component: this is also suggested by the focal mechanism solution of the earthquake on 15 April (Gasparini et al. 1980), and by the analysis of the linear structures surveyed on the two islands (Frazzetta et al. 1977, 1982). The alternating occurrence of deformation episodes of extension and contraction, similarly oriented with respect to the main strain axes, could be explained in terms of aseismic creep, which is highly probable in materials affected by a significant positive thermal anomaly (Faraone et al. 1986).

Increasing tensional stress in the Tindari-Letoianni line, the regional tectonic fault system controlling the volcanism, associated with a more intense fracturing due to a seismic crisis, could increase the probability of an eruption. The more direct effect on the fumarolic feeding system, which resulted in a temperature increase associated with detected variation of the gas composition (Martini et al. 1980; Carapezza et al. 1981), is tangible evidence of this interaction, in an unstable equilibrium connecting the active volcanic system to the regional tectonic structures.

The fact that increases in signs of volcanic activity (ground deformation, fumarole temperature increases and gas composition variation) only followed the 1978 Gulf of Patti earthquake swarm and not the 1980 Alicudi events, which were of a similar magnitude, demonstrates that as previously supposed, it is the Tindari-Letoianni line that is controlling volcanism at Vulcano and Lipari. Finally, there is an apparent relationship between local microseismic activity at Vulcano and the tilt pattern, as observed between April and August 1985. A more detailed investigation is presently under way to check the hypothesis that fluid pressure could be affected by even small stress changes, caused by tectonic earthquakes close to Vulcano.

References

Badalamenti B, Falsaperla S, Neri G, Nuccio PM, Valenza M, (1986) Confronto preliminare tra dati sismici e geochimici nell'area Lipari-Vulcano. Gruppo Nazionale per la Vulcanologia, CNR, Bollettino 1986, Roma

Barberi F, Gasparini M, Innocenti F, Villari L, (1973) Volcanism of the southern Tyrrhenian Sea and its geodynamic implications. J Geophys Res 78:5221-5232

Barberi F, Ferrara G, Keller J, Innocenti F, Villari L (1974) Evolution of the Aeolian arc volcanism. Earth and Planet. Sci Lett 21:269-276

Blot C (1971) Etudes sismologiques de Vulcano. Cahiers ORSTOM, Sér Geophys, 11 pp

Bottari A, Broccio F, Federico B, Lo Giudice E, (1979) Preliminary crustal model from seismological observations at the Messina Strait network. Ann Geofis 31:91-111

Bottari A, Caccamo D, Falsaperla S, Neri G (1984) Analisi preliminare dell'attività sismica con origine nell'area di Vulcano durante il primo semestre 1984. Open-File Rep 2/84, IIV, CNR, Catania

Bottari A, Caccamo D, Cefali F, Lo Giudice E, Neri G (1986a) Recent shallow seismicity in the southern Calabro-Peloritan arc region. Ann Geophys 4B 1:91-98

Bottari A, Caccamo D, Falsaperla S, Neri G (1986b) Confronto e correlazione tra dati di sismologia attiva e passiva nel Basso Tirreno. Atti V Convegno del Gruppo Nazionale di Geofisica della Terra Solida, 17-19 Nov, Roma 709-720

Caccamo D, Neri G (1984) A new analytical procedure to determine hypocentral parameters of local seismic events. Bull Seism Soc Am 74:665-677

Caputo M, Panza GL, Postpischil D (1970) Deep structure in the Mediterranean Basin. J Geophys Res 75 (26):4919-4923

Caputo M, Panza GL, Postpischil D (1972) New evidences about the deep structure of the Lipari arc. Tectonophysics 15:219-231

Carapezza M, Nuccio PM, Valenza M (1981) Genesis and evolution of the fumaroles of Vulcano (Aeolian Islands, Italy): a geochemical model. Bull Volcanol 44 3:547-564

Del Pezzo E, Martini M (1981) Seismic events under Vulcano, Aeolian Islands, Italy. Bull Volcanol 44 3:521-525

Del Pezzo E, Martini M (1982) The April 15, 1978 event of the Gulf of Patti, Sicily, and its aftershocks sequence. Boll Geof Teor Appl XXIV 94:109-119

Di Prima S, Falsaperla S, Neri G (in press) Determinazione della magnitudo nelle stazioni della rete sismica delle Isole Eolie mediante l'impiego della durata. Considerazioni critiche sull'applicazione di un siffatto criterio ai terremoti di origine vulcanica. Rendiconti Oss Geofis Reggino, Reggio Calabria

Ente Nazionale Energia Elettrica (ENEL) (1978) Catalogo dei terremoti italiani dal 1000 al 1975. Open-file Rep. Roma

Falsaperla S, Neri G (1986a) Seismic monitoring of volcanoes: Stromboli (southern Italy). Workshop on resumed activity in areas of quiescent volcanism: triggers and precursors. Firenze (Italy). 15-17 May, 1986. Per Mineralogia 55:153-163

Falsaperla S, Neri G (1986b) Seismic monitoring of volcanoes: Vulcano (Southern Italy). Workshop on resumed activity in areas of quiescent volcanism: triggers and precursors. Firenze (Italy). 15-17 May, 1986. Per Mineralogia 55:143-152

Falsaperla S, Nunnari G, Velardita R, Villari L (1985) Ground deformation and associated seismicity around Vulcano (Aeolian Islands) during the last ten years. (1985) IAVCEI Scientific Assembly, Giardini-Naxos, Italy, Sept 16-21, 1985. IIV Open File Report 4/87 CNR, Catania

Falsaperla S, Frazzetta G, Neri G (in press 1988) Problematiche sull'analisi di dati geofisici, geologici e vulcanologici a Vulcano. Presented at the workshop on "Aree sismogenetiche e rischio sismico in Italia", Erice, Italy, 26 Aug - 4 Sept

Faraone D, Silvano A, Verdiani G (1986) The monzogabbroic intrusion in the island of Vulcano, Aeolian Archipelago, Italy. Bull Volcanol 48:299-307

Finetti I, Del Ben A (1986) Geophysical study of the Tyrrhenian opening. Boll Geof Teor Appl

Frazzetta G, Villari A, Villari L (1977) Ground deformation and geological structures on Vulcano-Lipari volcanic complex (Aeolian Islands, Italy). IAVCEI-IASPEI Joint General Assemblies, Durham, Sept 1977, (Abstract)

Frazzetta G, Lanzafame G, Villari L (1982) Deformazioni e tettonica attiva a Lipari e Vulcano (Eolie). Mem Soc Geol It 24:293-297

Frazzetta G, La Volpe L, Sheridan MF (1983) Evolution of the Fossa Cone, Vulcano. J Volcanol Geotherm Res 17:329-360

Gasparini C, Iannaccone G, Scarpa R (1980) On the focal mechanism of the Italian earthquakes. Rock Mechanics Supl 9:85-91

Gasparini C, Iannaccone G, Scandone P, Scarpa R (1982) Seismotectonics of the Calabrian arc. Tectonophysics 84:267-286

Ghisetti F, Vezzani L (1982) Different styles of deformation in the Calabrian arc (southern Italy): implications for a seismotectonic zoning. Tectonophysics 85:149-165

Hagiwara T (1964) Brief description of the project proposed by the earthquake prediction research group of Japan. Proc US - Jpn Conf Res Relat Earthq Prediction Probl 1964:10-12

Latter JH (1971) Near surface seismicity of Vulcano, Aeolian Islands, Sicily. Bull Volcanol 35-1:117-126

Lo Bascio A, Luongo G, Nappi G (1973) Microtremors and volcanic explosions at Stromboli (Aeolian Islands). Bull Volcanol 37-4:596-606

Lo Giudice E (1973) L'Osservatorio Geofisico di Lipari e la rete sismica del Basso Tirreno. Stromboli 13:3-40

Martini M, Piccardi G, Cellini Legittimo P (1980) Geochemical surveillance of active volcanoes: data on the fumaroles of Vulcano (Aeolian Islands, Italy). Bull Volcanol 43 1:255-263

Mogi K (1967) Earthquakes and fractures. Tectonophysics 5-1:35-55

Morelli C, Giese P, Cassinis R, Colombi B, Guerra I, Luongo G, Scarascia S, Schutte KG (1975) Crustal structure of southern Italy. A seismic refraction profile between Puglia-Calabria-Sicily. Boll Geof Teor Appl 67:183-210

Nappi G (1975) Sull'attività recente dello Stromboli (Ottobre 1972 - Dicembre 1974) Boll Soc Geol It 94:465-478

Neri G (1981) Sismicità in ambiente vulcanico (Vulcano, Isole Eolie). Open-File Rep 2/81, IIV, CNR, Catania

Neri G (1985) Distribuzione spazio-temporale della sismicità delle Eolie. Atti IV Convegno del Gruppo Nazionale per la Geofisica della Terra Solida, Roma, 29-31 Oct, 1985, pp 825-837

Nuccio PM, Badalamenti B, Falsaperla S, Neri G, Valenza M, Velardita R, Villari L (1987) Geochemical and geophysical volcano monitoring at Vulcano island (Aeolian Archipelago, Italy) during 1985. Hawaii Symposium on How volcanoes work, Hilo, Hawaii, 19-25 Jan, 1987

Scandone P (1979) Origin of the Tyrrhenian Sea and the Calabrian Arc. Boll Soc Geol It 98:27-34

Scarcella G, Iannaccone G, Scarpa R (1985) Terremoti intermedi e profondi del Mar Tirreno. Rilocazione con il metodo JHD e meccanismi focali. Atti IV Convegno Gruppo Nazionale per la Geofisica della Terra Solida, Roma 29-31 Oct, 1985, pp 145-157

Sicardi L (1940) Il recente ciclo dell'attività fumarolica dell'Isola di Vulcano. Bull Volcanol VII:85-139

U.O. Sismologia Eolie (1985) Sismicità entro un raggio di 40 km intorno a Vulcano nel trimestre Ottobre-Dicembre 1984. Open-File Rep 1/85, IIV, CNR, Catania

Wezel CF (1981) The structure of the Calabro-Sicilian Arc: krikogenesis rather than subduction. In: Wezel CF (ed) Sedimentary basins of Mediterranean margins. CNR Italian Project of Oceanography, Tecnoprint, Bologna, pp 485-487

Thermal Infrared Surveys at Vulcano Island: An Experimental Approach to the Thermal Monitoring of Volcanoes

P.A. Brivio[1], E. Lo Giudice[2] and E. Zilioli[1]

Abstract

The last eruption on Vulcano Island dates back to the end of the 19th century. At present, its activity is confined to fumarole fields and sporadic limited phreatic eruptions; other types of surges have been recorded in the past. Two precise airborne infrared surveys were flown over the main cone of La Fossa volcano, where most of the fumarole fields occur, just before dawn and at noon, to give the most extreme temperature variations of the diurnal cycle. During the flights, meteorological data and ground radiance measurements were taken for calibration of the thermographs. Later on, the original aerial data were processed by analogue and digital techniques, in order to improve the heat sub-surface information. A similar method, recently used in the geothermal field of Travale (Italy) for energy exploitation (Zilioli et al. 1985), enables us to describe the surface of the volcano in terms of thermal inertia. The influence of the emissivity on various lithologic units has also been taken into account. The algorithm applied seems to offer a more reliable look at the actual residual thermal anomalies, which can be interpreted as surface manifestations of existing heat fluxes.

Introduction

The first experiments with infrared thermal surveillance at Vulcano Island, Italy, began on 5 May 1970. Since this time, a number of thermal observations has been made, both by aircraft and on the ground, so that an historical series of infrared measurements, taken under different seasonal conditions, is now available (see Table 1). The objective of these observations has been to identify locations of anomalous heat flow, and to explore the possibility of potential hazards prediction. A variety of instruments have been used including hand-held radiometers with a single field of view, video-recording thermal imaging systems and a calibrated airborne infrared scanner.

In 1980, a proper methodological approach was set up, and particular attention was paid to the calibration aspect of aerial

---

[1] Istituto per la Geofisica della Litosfera, CNR Via Ampere 56 - 20131 Milano, Italy

[2] Istituto Internazionale di Vulcanologia, CNR Viale Regina Margherita 6 - 95100 Catania, Italy

IAVCEI Proceedings in Volcanology 1
J.H. Latter (Ed.), Volcanic Hazards

Table 1. Log of thermal infrared surveys at Vulcano Island, sponsored by Consiglio Nazionale delle Ricerche. All data are placed on file at Istituto per la Geofisica della Litosfera, Milan

Date	Local time	Spectral bands (μ)	Altitude (ft.)	Platform	Instruments	Comments
5 May 1970	19:35	3-5	5,200	Dove aircraft	E.M.I. Airscan	Original data are on film
20 July 1973	Predawn	3-5/9-11				format 135. First tempe-
21 July 1973	Afternoon	3-5/9-11	6,000	Cessna aircraft	DS-1230	rature colour-coded map
						(Cassinis et al. 1970)
16 July 1974	19:40	3-5/9-11	10,000			Analysis on analogue
17 July 1974	05:45	3-5/9-11	5,000	Cessna aircraft	DS-1230	products for thermal in-
17 July 1974	05:50	3-5/9-11	6,600			terpretation and structu-
						ral assessment of La Fos-
17 July 1974	12:10	1 2/9-11	9,000			sa cone (Lechi et al.
						1971; Lo Giudice et al.
4 July 1976	22:00-24:00	2-5.6/6.5-20	---	Boat	Aga Thermovision 750	1975; Tonelli 1975)
5 July 1976	04:00-06:00	2-5.6/6.5-20	---	Boat	and Barnes PRT-5	
6 July 1976	21:00-23:00	2-5.6/6.5-20	---	Ground		
10 December 1976	07:40	3-5/9-11	11,000			
10 December 1976	07:50	3-5/9-11	6,400	Cessna aircraft	DS-1230	
10 December 1976	13:20	1-2/9-11	11,000			
10 December 1976	13:55	1-2/9-11	6,400			
2 April 1978 -						Sporadically out of order
21 September 1980	Hourly	6.5-20	---	Fixed		for some periods (Brivio
				automatic	Barnes Instatherm	et al. 1980)
22 June 1983 -	Hourly	6.5-20	---	station		
31 December 1983						
15 August 1980	06:50	3-5/9-11	6,000	Cessna aircraft	DS-1230	Survey with ground refe-
15 August 1980	06:00-09:00	6.5-20	---	Ground	Barnes Instatherm	rence data and digital
16 August 1980	12:00	1-2/9-11	6,000	Cessna aircraft	DS-1230	approach
16 August 1980	10:00-13:00	6.5-20	---	Ground	Barnes Instatherm	

surveys, so that quantitative measurements could be produced. Two precise airborne infrared surveys were carried out over the main cone of La Fossa volcano by means of a dual-channel analogue scanner DS-1230. While the second survey was near midday, when thermal radiance collected by sensors is greatly influenced by solar heating, the first one was just before dawn, when the diurnal cycle of ambient temperatures is at minimum, and the influence of local topography and albedo variations are moderate. During the overflights, ground reference data were recorded at different sites by an IR radiometer and by conventional thermometers, for calibration of the thermographs.

1 The Absolute Ground Temperature Map

The thermal surveillance of active volcanoes by means of aerial thermography implies the production of temperature maps. There are various, reasonably accurate, techniques for generating thermal maps from thermal infrared data. Final products range from the qualitative depiction of emitted thermal radiation to computer-generated, colour-coded maps of absolute temperatures. Let us see how an absolute temperature map can be obtained and how to evaluate its accuracy.

1.1 Image Pre-Processing and Calibration

Calibrated thermal images on transparencies, at a scale of 1 : 30 000, were obtained from the playback of the original analogue magnetic tape. These thermographs were converted into computer-compatible tapes, by means of an 8-bit digitization process, using a scanning microdensitometer Optronics C-4500, with a 200-μ resolution cell (see Fig. 1). This sampling interval was selected in order to match the ground spatial resolution. Because the scanner IFOV is 2.5 mrad, and the flight altitude was 1800 m, ground resolution is 4.5 m.

Internal blackbody sources, mounted in the scanner housing on either side of the angular field of view, were maintained at 293 K and 323 K respectively, for both surveys. For playback and calibration of the pre-dawn 9-11 μ thermography a six-level scale was used, while for the afternoon a two-level scale was used. These calibration samples were also digitized and the statistical analysis of the reference grey levels is given in Table 2.

At a specific spectral band, the relationship curve between radiant emittance, W, collected by the instrument, and the output digital numbers, DN, varies with the temperature range in question. Let us assume that, for a small temperature range, a linear fit of digital numbers to radiant emittance is appropriate; we then have:

$$DN = a + bW,$$

where a and b are parameters depending on the instrument specification.

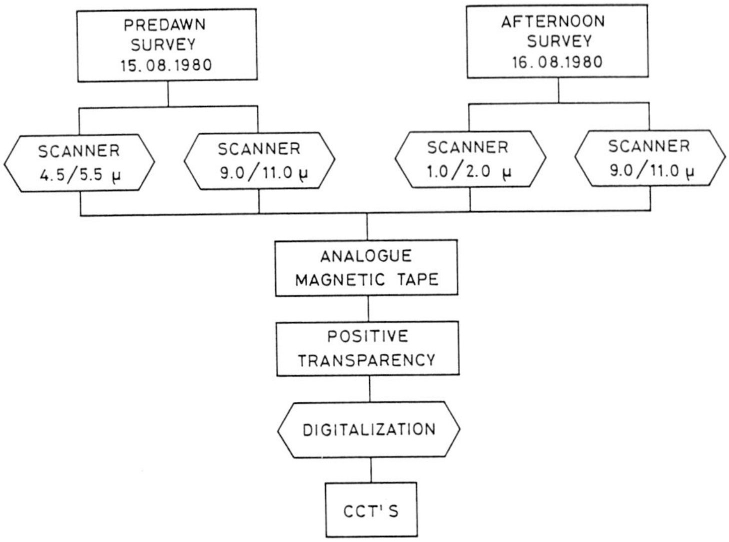

Fig. 1. Diagram of the data acquired and a pre-processing flow-chart

If we put $W = \epsilon \sigma T_k{}^4$, our relationship becomes a Stefan-Boltzman type equation:

$$DN = a + b\epsilon\sigma T_k{}^4,$$

where a is supposed to be negative, as it actually is in almost all real cases.

A more suitable form of the equation, as adopted by Lillesand and Kiefer (1979), is the following:

$$DN = a + \epsilon B\ T_k{}^4, \qquad (1)$$

where $B = b\sigma$.

The parameters a and B can be derived from internal blackbody calibration, when we consider the known reference temperatures T1 and T2, and the corresponding output digital numbers DN1 and DN2.

It is necessary to solve the two-equation system:

$$\begin{cases} DN1 = a + BT1; \\ DN2 = a + BT2; \end{cases}$$

where it is assumed that $\epsilon = 1$.

In the case of the afternoon survey, where T1 = 293 K, T2 = 323 K, DN1 = 56, and DN2 = 241, we obtain a = -332.83, and B = 0.53 x 10^{-7}.

This function appears to be almost linear for the temperature range considered here.

Table 2. Statistical values of the digitized blackbody temperature samples

			Predawn (9–11 μ)				Afternoon (9–11 μ)	
T (Kelvin)	283.24	284.24	285.24	286.24	287.24	288.24	293	323
N's of pixels	1600	1800	1800	2000	1500	1000	2800	2450
Arith. mean	13.93	37.59	63.46	94.00	127.22	150.45	56.03	241.50
SD	7.10	8.15	8.26	8.27	7.09	8.86	5.20	7.53
Median	13	37	63	94	127	151	56	241

For the pre-dawn survey, T1 = 283.24 K, T2 = 288.24 K, DN1 = 13, and DN2 = 151, so that a = -1890.13, and B = 2.95 x 10^{-7}.

Although relationship (1) is not linear, it is normally assumed to be so for small temperature ranges. Because six reference samples were available for the pre-dawn thermography (see Table 2), we have tested the regression curve by least square estimates for this particular temperature interval: the resulting correlation coefficient was 0.999, and the standard error of the estimate 0.091. In Fig. 2, the straight lines best fitting the equation are plotted for both surveys.

Thermal images collected during multi-temporal aerial surveys are naturally affected by geometric distortion. To make direct comparisons between the thermographs, it is necessary to spatially register these data to each other. Two separate algorithms are required: the first defines the spatial transformation, and the second defines the grey level value interpolation.

The 1-2 μ noon thermography was chosen as reference, and four control points were located (see Fig. 3). Spatial re-mapping of distorted images was carried out by bilinear interpolation to preserve continuity and connectivity (Castleman 1979); the "nearest neighbour" re-sampling technique was adopted for interpolation of the radiance values.

The images in Fig. 4 are the pre-dawn and afternoon brightness temperature maps, spatially registered, radiometrically calibrated and referenced to the same temperature-grey level scale.

1.2 Accuracy of Thermographic Data

Thermal sensitivity of the DS-1230 scanner used is nominally 0.5 K: the digitization of the analogue outputs has produced a

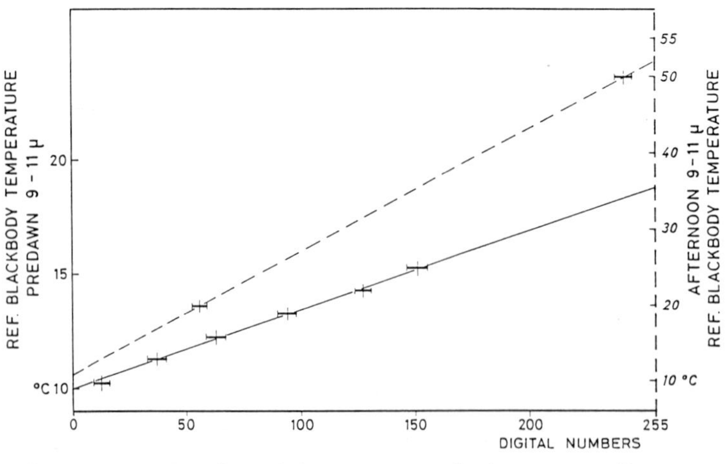

Fig. 2. Relationship curves between reference blackbody temperatures and the output digital numbers

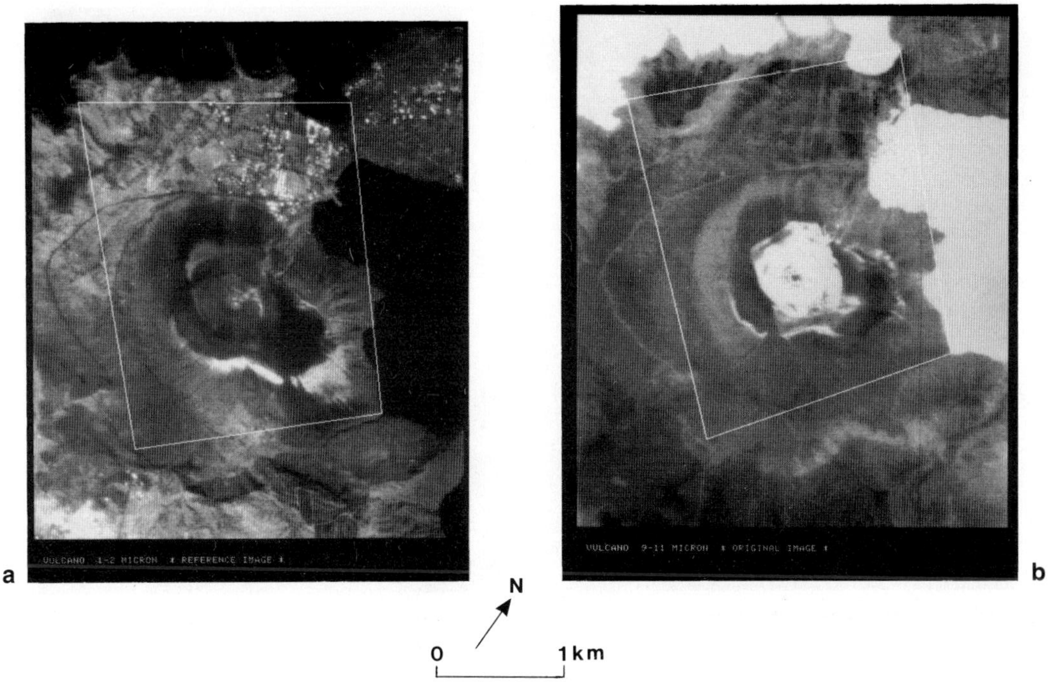

Fig. 3. Plate illustrating the geometric correction applied to the images: a reference image, b original image to be registered

DN range sensitive to 0.2 K. That does not mean that the radiometric resolution of the system has improved, but only that there is a finer interpolated description of it.

During the flight operations, a campaign of ground measurements for calibration was organized. Meteorological and radiometric data were collected at eight sites selected as: (1) easily identifiable on the thermography, (2) radiometrically homogeneous, (3) representative of main outcropping formations and (4) not directly influenced by strong thermal anomalies. Table 3 lists the measurements recorded and the data computed for each site considered, labelled according to the report by Lo Giudice and Longo (1980).

Kinetic and radiometric temperature ground measurements, T_k and T respectively, allowed us to compute emissivity coefficients ϵ, using the relationship:

$$\epsilon(\lambda,T) = \frac{\exp\,(c_2/\lambda T_k) - 1}{\exp\,(c_2/\lambda T\,) - 1},$$

which is derived from Planck's law, where c_2 = second radiation constant, 1.4388×10^4 µm K. From the brightness temperature values T_i on the images, assuming the specific emissivity previously evaluated, we derived the corresponding kinetic temperatures T_{ki} by means of the equation:

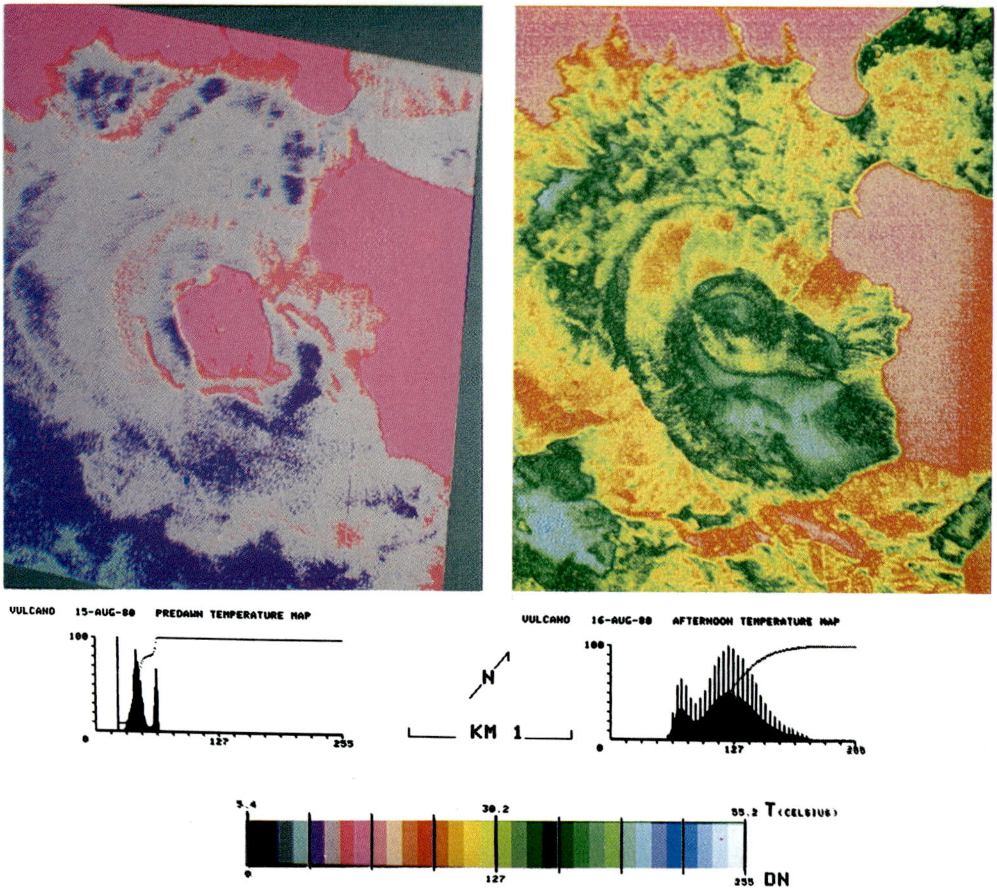

Fig. 4. Colour-coded representation of the thermographs calibrated and corrected; channel 9-11 μ, pre-dawn (left) and afternoon (right)

$$T_{ki} = c_2 \lambda^{-1} \left\{ \log\left[\exp(c_2/\lambda T_i) - \epsilon + 1\right] \right\}^{-1}.$$

Table 3 also lists discrepancies between the true ground and true aerial temperatures.

In order to estimate the temperature error due to the emissivity influence we recall the relation:

$$\frac{dW_\lambda}{W_\lambda} = \frac{d\epsilon}{\epsilon} + \frac{dT}{T} \cdot \frac{c_2}{\lambda T},$$

and we consider the case in which we have no variation of radiant emittance W_λ, so that we obtain the function:

$$dT = -\frac{d\epsilon}{\epsilon} \cdot \frac{\lambda T^2}{c_2},$$

Table 3. List of temperature and radiometric values recorded during the ground and aerial surveys at various selected test sites[a]

Sites	Latitude (N) Longitude (E)	Elevation a.s.l. (m)	T_{i1}	T_{ki1}	T_1	T_{k1}	$T_{k1,ki1}$	T_{i2}	T_{ki2}	T_2	T_{k2}	$T_{k2,ki2}$	10 μm
B	38°25'06'' 14°57'08''	0	290	290	-	298	8	295	295	301	301	6	0.999
C	38°24'38'' 14°57'02''	18	287	290	-	294	4	310	313	317	320	7	0.964
D[b]	38°29'33'' 14°57'22''	29	287	290	-	294	4	307	310	316	319	9	0.960
E	38°24'06'' 14°57'08''	50	286	288	-	293	5	306	308	323	323	15	0.979
F	38°23'53'' 14°57'163	60	287	290	-	293	3	307	310	321	324	14	0.965
G[b]	38°23'48'' 14°57'36''	85	287	291	-	294	3	308	313	317	322	9	0.944
H_a[b]	38°24'20'' 14°57'26''	225	286	287	294	295	8	-	-	-	-	-	0.980
H_b[b]	38°24'25'' 14°57'26''	150	287	289	294	295	6	-	-	-	-	-	0.980

[a] 1= Pre-dawn data; 2 = afternoon data. Temperatures are expressed in degrees Kelvin.

[b] Several measurements were available; average value is indicated: T_i = Brightness temperature on the thermography; T_{ki} = Kinetic temperature calculated by T_i and ϵ values; T = Brightness temperature obtained by radiometer Barnes-Instatherm at ground; T_k = Kinetic temperature measured directly on the test surfaces.

where temperature variations are shown to be dependent on emissivity, wavelength and temperature values. If we assume the average values λ = 10 μ, T_{k1} = 296 K (pre-dawn), T_{k2} = 311.25 K (afternoon), a variation $d\epsilon$ = 3.2% causes true temperature differences of 1.94 K and 2.15 K respectively. These values are much smaller than the discrepancies observed. Moreover, to overcome this error one should digitize a map of actual emissivity and use it, along with the derived temperature map, to produce another temperature map, corrected for emissivity influence. Indeed, this operation needs such a high accuracy in determining emissivity, that it is almost impossible, for practical purposes.

Two other components which strongly influence the remotely sensed data are the topographical elevation of the surfaces investigated and the atmospheric absorption. Correction factor for the conversion from aerial to ground temperature, due to a 1800-m air path, corresponds to T = 4.5 K approximately, as extrapolated by Schott and Tourin (1975).

Furthermore, a formula has been found by Bonneville et al. (1985) which relates the effects of atmospheric attenuation and elevation to the true temperature at the ground:

$$T_k = \beta^{-\frac{1}{4}} \cdot T + 0.006\ z,$$

where β is the attenuation coefficient due to the air mass and z is the elevation in metres of the points considered.

For T = 300 K, z = 150 m and ϵ = 0.95, we obtain ΔT = 4.6 K, which is comparable to the previous result.

Nevertheless, even with this correction, the temperature discrepancies listed in Table 3 are not sufficiently explained, as some of them are greater than 4.5 K and they are differently distributed above sea level.

If we sum the temperature differences, due to the emissivity influence (~2 K) and to the atmospheric path (4.5 K), we obtain a maximum total correction of about 6.5 K that does not yet sufficiently explain the discrepancies calculated.

That means that other local causes have yet to be identified. In particular one of these is the air humidity whose effects are very difficult to be considered because of the complexity of measuring and modelling.

In our case, there is also not a perfect identification of site locations on the thermographs: visual interpretation is difficult, because the ground spatial features are smoothed in the radiometric data representation. In effect, neighbouring values around one pixel often show differences in K as great as several degrees.

These arguments show that it is very difficult to obtain a true temperature map, although in many instances, patterns of relative temperatures maintain their own utility. This conclusion leads us to consider differential analysis as a more profitable tool for volcanic surveillance by remote sensing.

2 The Function of Apparent Thermal Inertia

Thermal distribution described on the thermographs depends on the response of surface materials to the heating produced by both exterior and interior sources, i.e. solar irradiance and geothermal flux.

Temperature patterns observed at different times can be interpreted as a function of a time-varying energy flux at the surface. The thermal property which describes this relationship and which can most easily be determined by remote sensing is the thermal inertia. This can be thought of as a measure that relates the magnitude of the temperature response of a material to an energy flux.

Thermal inertia is a property that is measured "not for the top micrometers or so of the earth but for the top few centimeters" (Gillespie 1980), since it depicts more properly the thermal history and the physical properties of the materials absorbing and re-emitting the solar radiation. On this basis, we suggest that thermal inertia can be a better candidate than the temperature data derived from the original thermographs for the characterization of solid surfaces, particularly in the context of remote sensing.

S being the solar flux absorbed by a surface, T1 and T2 the temperature responses of the surface to the incident radiant flux and t (expressed in seconds) the time elapsed between the two temperature measurements, the thermal inertia can be expressed as:

$$P = \frac{S}{t^{\frac{1}{2}} \cdot (T2 - T1)} \quad (\mathrm{cal\ cm^{-2}\ K^{-1}\ s^{-\frac{1}{2}}}).$$

For remote observation of thermal inertia, measurements of absorbed solar energy and extreme diurnal temperatures, are required; the albedo map is used to compute S according to the relation:

$$S = H\,(1 - \rho),$$

where H is the solar irradiance and ρ the surface reflectivity. Between the two surveys a 258.74 cal cm^{-2} solar irradiance H has been calculated, time elapsed is t = 18 600 s, and ρ is assumed to range between 0.05 and 0.5. We derived the albedo map on the basis of the noon survey in the spectral region of 1-2 μ, where the radiant energy is almost reflected. The resultant image displays relative reflectance surface behaviour: no absolute values are provided.

The algorithm actually implemented is:

$$FATI = 1.837\,\frac{(1 - \rho)}{\Delta T} \quad (\mathrm{cal\ cm^{-2}\ K^{-1}\ s^{-\frac{1}{2}}}),$$

which we call the Function of Apparent Thermal Inertia (FATI), since it slightly differs from the absolute thermal inertia P described above.

The immediate advantage of this differential approach is evident: for any given surface, the emissivity influence on the data remains constant for temperature changes (ΔT), and surface spectral characteristics are provided by the albedo map $(1 - \rho)$, which is more easily computable and more reliable than the emissivity map.

A proper linear transfer function allows us to correlate FATI outputs with the digital numbers DN, as shown in the reference scale of Fig. 5. A histogram of this image also shows that most of the FATI values are present in the range between 0.015 and 0.060 cal cm^{-2} K^{-1} $s^{-\frac{1}{2}}$, which corresponds well to the value range for soils and common rocks known in the literature, as shown in Table 4 (after Janza 1975).

Thermal inertia information helps to distinguish between materials of similar reflective properties (albedo), but different thermal properties: when supported by other remotely sensed data it can be of great value in resolving ambiguities, or in pointing out unsuspected anomalous zones.

2.1 Interpretation of FATI Map

The interpretation of a single thermal map is unable to give much good information. However, if we look at the FATI map of Fig. 5, interesting features can be deduced.

Black and bright-blue are off-scale values and they coincide with the fumarole fields and the sea-water respectively. If an analysis is made within the colour levels belonging to the 20-100 DN range (histogram in Fig. 5) some more profitable observations can be made.

The well-bedded ash-tuffs of "La Fossa Rossa" can be distinguished from the ancient and recent pyroclastics, but lava flows are not well discriminated: little detail is visible on the rhyolitic obsidian flow of Pietre Cotte. The FATI map so obtained is generally in good agreement with the lithologic map; in some case, it gives a finer description of the physical properties within the same lithologic unit. However, besides these lithologic correlations, three spots depicted in bluish and grey colours, within the red-coloured zones pointed out in the figure by numbers 1, 2 and 3, appear in the recent pyroclastic deposits on the main cone of La Fossa. Since they show high values of FATI, they can be interpreted, with a good chance of success, as sites of sub-surface anomalous geothermal flux.

The first site, on the western flank of La Fossa, may indicate an area where a lateral phreatic eruption is being prepared. In fact, under certain geostructural conditions, the interaction of the geothermal flux with phreatic water can be identified, when its sub-surface effect is sensible and detectable; in that case

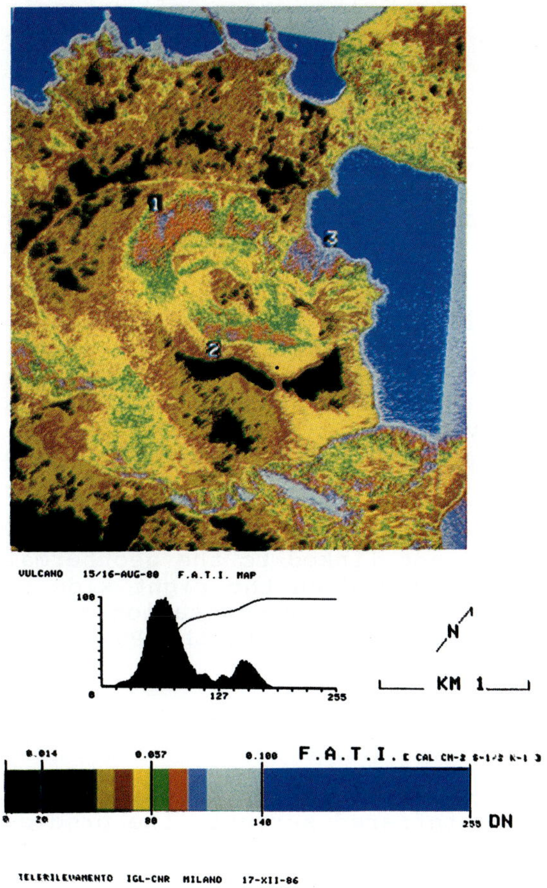

Fig. 5. Colour-coded FATI representation. Levels of greater interest range from red to grey corresponding to high values of relative thermal inertia

Table 4. Comparison between nominal values of thermal inertia P (Janza 1975) and FATI for some outcrops at La Fossa cone

Rock type	P	FATI
Sandy soil	0.024	0.028
Tuffs, welded	0.032	0.028
Obsidian	0.035	0.054
Gravel and Sandy gravel	0.041	0.057
Rhyolite	0.047	0.061

it becomes an important indicator of the stage of volcanic activity, in particular where the water table is not yet saturated with superheated steam.

The second zone, which borders the interior of the southern part of the main rim and fades along the top of the cone, may be due to sub-surface anomalous temperature in the underlying tuffs of La Fossa that belong to the Pietre Cotte and the modern eruptive cycle. In this area faint gas manifestations outcrop from fractures and reach the surface through the coating of altered ashes.

The last interesting anomaly, visible at site No. 3 to the north, corresponds to the area of the parasitic cone of Forgia Vecchia, densely fractured, which formed during the last eruption by wet-surge mechanisms, as described by Frazzetta et al. (1983). Nearby, hot springs not visible on these thermographs, occur in the sea to witness the present activity on that side of the volcano.

Besides the shoreline and the water mass of the sea, some other sites at very high values of FATI, not linked to the geothermal anomaly, appear to the south, distributed on the right side of the Rio Grande basin. That is due to the green vegetation coverage of the Mediterranean bush densely present in the Passo del Piano and Monte Rosso areas.

3 Conclusions

Any method for volcanic surveillance is based on precursor evidence of imminent eruption. Using infrared surveys, the predictive model stands on the assumption that some thermal anomalies are detectable and can be interpreted.

The data from the 1980 infrared surveys on Vulcano Island have shown that it is very hard to make a reliable temperature map from aerial thermographs. This is clearly demonstrated by the comparison of kinetic temperature values obtained by two kinds of measurements, on the ground and in the air. Discrepancies are too great to be acceptable. However, the map of FATI shows a good correlation with the rock types: furthermore, it has identified three thermal anomalies, which were masked by the pyroclastics of the last eruption.

In conclusion, we believe that a correct computation of all environmental influences on thermal remotely sensed data remains an arduous matter, far from a solution, for practical purposes. More calibration surveys on the ground need to be synchronized with the overflights.

The success of the differential approach leads us to recommend at least two flights, to be timed to coincide with the extremes of diurnal temperatures.

For a better understanding of internal geothermal flux by means of thermographs is the elaboration of a model based on the Poisson equation. Here, an internal heat source is assumed in order to simulate a map of the temperature gradient variation at the surface, both in the stationary case and when temperatures change over a given time interval.

Acknowledgements. This research has been supported and promoted by the Consiglio Nazionale delle Ricerche (CNR) in the framework of the activities for the volcanic surveillance.
The authors would like to thank Annalisa Fain and Massimo Righetti for their contribution in setting up suitable software for image geometric correction.
We are also grateful to Dr. John H. Latter for his valuable suggestions and revision of the manuscript.

References

Bonneville A, Vasseur G, Kerr Y (1985) Satellite thermal infrared observations of Mt. Etna after the 17th March 1981 eruption. J Volcanol Geotherm Res 24:293-313

Brivio PA, Tomasoni R, Zilioli E (1980) Vulcano Island: a new method for surficial heat control. Bull Bur Rech Géol Minières Sect IV 2:161-168

Cassinis R, Marino CM, Tonelli AM (1970) Ground and airborne thermal imagery on Italian volcanic areas. Geothermics Spec Iss 2:413-420

Castleman KR (1979) Digital image processing. Prentice-Hall, Englewood Cliffs, New Jersey, pp 110-135

Frazzetta G, La Volpe L, Sheridan MF (1983) Evolution of the Fossa cone, Vulcano. J Volcanol Geotherm Res 17:329-360

Gillespie AR (1980) Digital techniques of image enhancement. In: Siegal BS, Gillespie AR (eds) Remote sensing in geology. Wiley & Sons, New York, p 208

Janza FJ (1975) Interaction mechanisms. In: Reeves RG (ed) Manual of remote sensing. Falls Church, VA, Am Soc Photogrammetry, pp 83-85

Lechi GM, Marino CM, Tonelli AM (1971) Geological applications of remote sensing surveys in Italy with special reference to their utilization on the volcanic areas. Proc 24th Intern Geolog Congress, Ottawa

Lillesand MT, Kiefer RW (1979) Remote sensing and image interpretation. Wiley & Sons, New York, pp 425-440

Lo Giudice E, Longo E (1980) Vulcano, Agosto 1980. Rilievo aereo nell'I.R.termico. Rilevamento di misure al suolo. Rapporto N° 140 IIV CNR, Catania

Lo Guidice E, Marino CM, Tonelli AM (1975) Le tecniche di ripresa all'infrarosso termico applicate ai rilievi aerei dell'Isola di Vulcano. Analisi dei risultati in rapporto alla situazione vulcanologica locale. Boll Geofis Teor Appl Vol XVII 65, Trieste

Tonelli AM (1975) Application of the gradient/radiance characteristic function in the study of the thermal trend of volcanic areas. Proc 10th ERIM Symp, Ann Arbor, MI:1007-1015

Schott JR, Tourin RH (1975) A completely airborne calibration of aerial infrared water-temperature measurements. Proc 10th ERIM Symp, Ann Arbor, MI:477-484

Zilioli E, Brivio PA, Tomasoni R (1985) The infrared aerial surveys for heat flow anomalies mapping in the geothermal fields. Proc 4th ERIM Them Conf, San Francisco, 2:417-426

The Forecasting Significance of Chemical Indicators in Areas of Quiescent Volcanism: Examples from Vulcano and Phlegrean Fields (Italy)

M. Martini[1]

Abstract

Geochemical surveillance programs have been in progress for several years on Vulcano and the Phlegrean Fields in Italy. Specific changes simultaneously with the occurrence of geophysical phenomena have been observed. At Vulcano, an earthquake in 1978 and a seismic swarm in 1984 were both preceded by chemical variations of the fumaroles in the crater, and both were followed by consistent changes in the composition of phreatic waters. In the Phlegrean Fields, observations carried out before, during and after major ground uplift from 1982 to 1984, showed that changes in fumarole compositions have also preceded the decreasing trend and the stop of ground deformation.

An increased probability of new activity can be predicted on the basis of a statistical treatment of the results obtained at Vulcano.

1 Introduction

Erupting volcanoes have always represented an impressive phenomenon for populations living in the surrounding areas, and a fearful religious interpretation was placed in the past on any kind of volcanic activity. By the end of the 18th century, a French scientist (Dolomieu 1788) and an Italian researcher (Spallanzani 1792) provided the first attempts at a natural explanation of volcanism by recognizing the importance of gaseous species, and suggesting that the study of their chemical compositions would shed light on the evolution of activity. Two centuries later, the behaviour of volatiles still appears as a most promising method for the study of eruptive processes. The quantitative distribution of volatile species and gaseous compounds before, during and after eruptive events might represent a very effective parameter by which to monitor cycles of activity.

The available data for the chemical and isotopic composition of gases in different systems allow us to derive a few general rules. Carbon oxides can be considered as ubiquitous constituents; with few exceptions referring to seafloor spreading axes,

[1] Department of Earth Sciences, University of Florence, Italy

IAVCEI Proceedings in Volcanology 1
J. H. Latter (Ed.), Volcanic Hazards

where the mantle appears as the prevailing source, thus a strong contribution from thermal metamorphism of carbonaceous material in a crustal environment is very likely (Mizutani and Sugiura 1982; Allard 1983; Gerlach and Thomas 1986; Menyailov et al. 1986). The intrusion of a magmatic body into a crustal environment also gives rise to reactions between wallrocks and magma, which involve water as well. According to previous theoretical studies (Ellis 1957; Matsuo 1960; Heald et al. 1963; Gerlach and Nordlie 1975; Volkov and Ruzaykin 1976; Sabroux 1979) and available thermodynamic data, sulphur species and hydrogen appear to represent the main gaseous products of such processes; if salty water participates in the reactions, hydrogen chloride can also be produced. For any given system any change in the composition of crustal rocks reacting with the intruding magma is likely to require a long time, while the quantity and the salinity of water pertaining to aquifers participating in the reactions can vary at a much faster rate.

A general trend, which may be valid for any system, is illustrated by plotting C-S-(H_2 + HCl) as a ternary diagram (Fig. 1). Even if different origins can be envisaged for all components, the importance of the carbon species is always enhanced at lower temperatures. Accordingly, the carbon corner can be considered as a low temperature extreme, and sulphur species can be taken as the main indicator of magmatic conditions; hydrogen and hydrogen chloride should indicate the extent of the participation of water. The actual water content of fumarolic gases is not considered as related to deep processes, but mainly due to the influence of shallow and surface conditions. Several minor reactions can be triggered by the intruding magmas, but do not appear to affect the general picture here outlined.

The available information on volcanoes whose representative points are shown in Fig. 1 appears sufficiently consistent with these assumptions. However, the tentative evaluation of the stage of activity of a given system, on the basis of Fig. 1, can represent only a very first step, and a careful consideration of geochemical and geophysical changes recorded in the area is also needed for any attempt at prediction of new eruptive activity.

As an example, the results obtained through programs of geochemical surveillance at Vulcano and in the Phlegrean Fields are given here.

2 Vulcano

After the last eruption of 1888-1890, fumarolic activity of varying extent has been reported (Bergeat 1899; Washington and Day 1915; De Fiore 1924; Sicardi 1940; 1956; Le Guern 1972). A systematic study of the fumaroles at the crater and at the beach, and of the phreatic waters of the wells of Vulcano Porto (Fig. 2), has been in progress since 1977, and preliminary results have already been published (Martini 1980; Martini et al. 1980; Martini and Cellini Legittimo 1984). The data for the fumarolic gases at the crater, collected in October 1985, are given in Table 1.

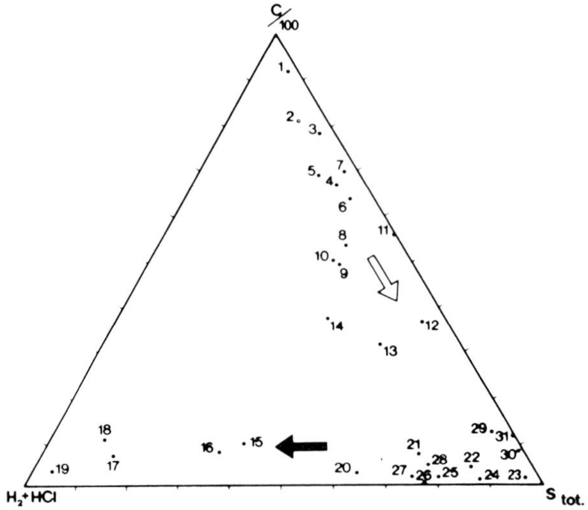

Fig. 1. Distribution of carbon, sulphur hydrogen and hydrogen chloride in fumarolic gases from different areas. **White arrow:** increasing equilibrium temperature; **black arrow:** increasing participation of water. 1 Caldeira Velha, S.Miguel, Azores; 2 Ribeira Grande, S.Miguel, Azores; 3 Lagoa das Furnas, S.Miguel, Azores; 4 Esguicho, S.Miguel, Azores; 5 Pisciarelli, Phlegrean Fields, Italy; 6 Solfatara, Phlegrean Fields, Italy; 7 Casa Diablo, California, USA; 8 Makaroyen, Sulawesi, Indonesia; 9 Lahendong, Sulawesi, Indonesia; 10 Tisate, Nicaragua; 11 S.Jacinto, Nicaragua; 12 Los Azufres, Mexico; 13 Lassen, California, USA; 14 Vulcano beach, Italy; 15 Vulcano crater, Italy; 16 Usu, Hokkaido, Japan; 17 Usu, Hokkaido, Japan; 18 Showashinzan, Hokkaido, Japan; 19 Tolbachik, Kamchatka, Soviet Union (Menyailov et al. 1980); 20 Momotombo, Nicaragua; 21 Noboribetsu, Hokkaido, Japan; 22 Nevado del Ruiz, Colombia; 23 San Cristobal, Nicaragua (Menyailov, pers. commun.); 24 Kilauea, Hawaii, USA (Greenland 1984); 25 Erta 'Ale, Ethiopia (Giggenbach and Le Guern 1976); 26 Arenal, Costa Rica (J.L. Cheminée, pers commun.); 27 Poas, Costa Rica (J.L. Cheminée, pers commun.); 28 White Island, New Zealand (Giggenbach 1975); 29 Ebeko, Kurile, Soviet Union (Menyailov et al. 1985); 30 Ontake-san, Honshu, Japan (Ossaka et al. 1980); 31 Popocatepetl, Mexico

A statistical cluster analysis (Davis 1973) of the chemical results of 94 samples of fumarolic gases enabled us to distinguish two main components, characterized by CO_2, SO_2, H_2, HF, higher values of temperature and HCl, B, Br, NH_4 respectively. Species pertaining to high temperature manifestations are represented in the first group, constituents for which a marine origin is very likely belong to the second group. A magmatic origin for the first component is thus strongly suggested, along with an important influence of the shallow environment on the second component. The actual data on the fluctuating activity of the fumaroles of Vulcano appear to confirm this assumption (Martini and Cellini Legittimo 1984). Accordingly, the

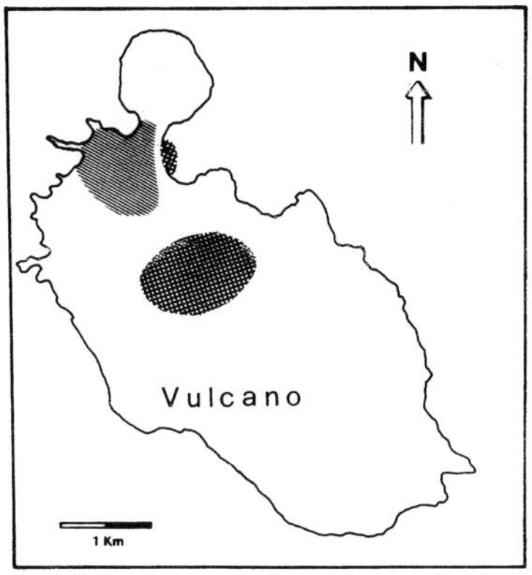

Fig. 2. The island of Vulcano. **Shaded area:** location of the wells studied. **Chequered area:** crater and beach fumaroles

comparison of any two variables from the different groups should indicate the variation in relative contributions of the two components. As can be seen in Fig. 3, two ratios showed consistent trends; other ratios give the same trends. An increased magmatic influence was observed from 1977 to 1979, followed by a declining trend and by a new increase starting in 1982.

The arrows in Fig. 3 indicate an earthquake (M = 5.3) which took place on 15 April 1978, and a seismic swarm (M = 3.9) occurred on 27 January 1984 (Bottari et al. 1984). Another important earthquake (M = 5.7) occurred 50 km to the west on 28 May 1980, but did not show any interference with the structure of Vulcano.

It is clear that the gas composition started to change before the seismic events. A possible causal link between chemical changes and seismic activity, due to a greater permeability at depth produced by strain accumulation, has already been suggested for the 1978 event (Martini 1983). The repetition of the phenomenon appears to strengthen this hypothesis.

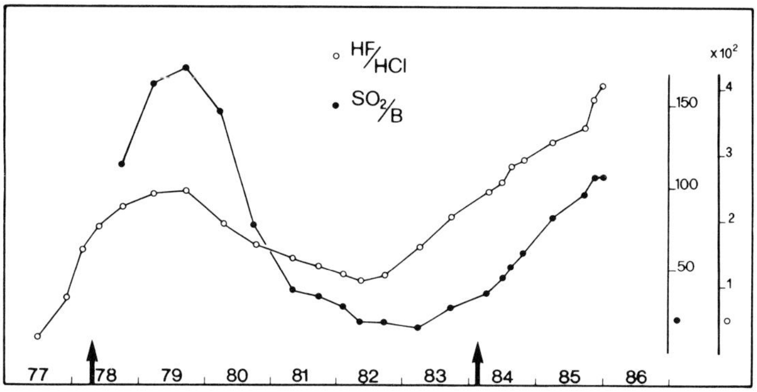

Fig. 3. Variation of the molar ratios HF/HCl and SO_2/B in the fumarolic gases collected at the crater of Vulcano. The **arrows** refer to seismic events (see text)

Table 1. Analytical results for the gases of different fumaroles in the crater of Vulcano, expressed in volume percent

Fumarole	Temp. (ºC)	H_2O	CO_2	H_2S	SO_2	HCl	HF	B	Br	NH_4	H_2	N_2	O_2
2	300	79.5	18.3	0.69	0.90	0.44	0.0211	0.0058	0.0006	0.0010	0.00063	0.084	0.00012
5	318	79.5	18.3	0.56	0.99	0.52	0.0192	0.0093	0.0007	0.0009	0.00057	0.086	0.00015
9	210	80.1	17.9	0.53	1.01	0.28	0.0033	0.0070	0.0002	0.0010	0.00048	0.093	0.00015
11	266	77.1	20.8	0.59	1.03	0.37	0.0055	0.0092	0.0004	0.0005	0.00038	0.088	0.00012
14	224	74.6	22.7	0.89	1.31	0.34	0.0016	0.0099	0.0004	0.0030	0.00036	0.090	0.00018

CO and CH_4 are below the detection limit.

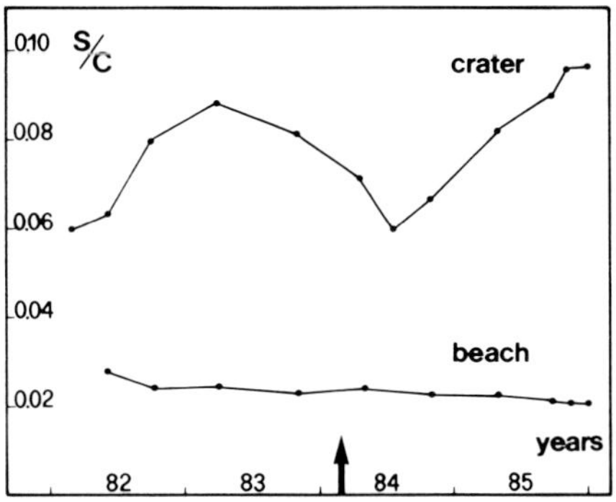

Fig. 4. Variations of the ratios S/C in fumarolic gases collected at the crater and at the beach of Vulcano. The **arrow** indicates a seismic swarm (see text). S = H_2S + SO_2; C = CO + CO_2

From Fig. 4 it is also possible to verify that, at least for the 1984 swarm, no important chemical variation was detected at the fumaroles near the beach. Even if the comparison is made using only total sulphur and total carbon species, because most of the other constituents are not found in the beach fumaroles, the strong influence of the surface environment (Martini et al. 1986) appears to overwhelm any possible fluctuation.

In order to analyze the response of the phreatic water body to the changes observed at the crater fumaroles, the variations of the mean values of HCO_3, Cl and SO_4 for 26 wells are shown in Fig. 5. Differing behaviour of the three constituents can easily be distinguished; chloride shows a minor degree of fluctuation; bicarbonate displays a progressive increase with time; and sulphate appears strictly related to the composition of the fumaroles (Fig. 3). The natural inertia of water bodies can explain the delay with which changes are observed in these phreatic waters. An almost continuous input of carbon dioxide into the shallow aquifers can be explained by the volcanic influ-

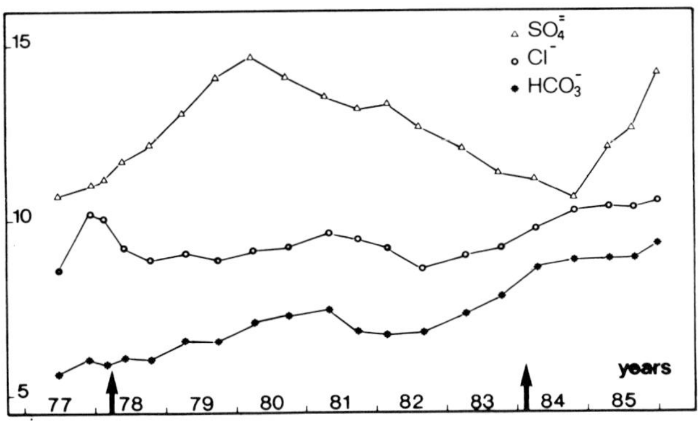

Fig. 5. Variations in time of the mean concentrations (milliequivalents per litre) of sulphate, chloride and bicarbonate ions in the waters of 26 wells from the area of Vulcano Porto. The **black arrows** refer to seismic events (see text)

ence, while sulphur species are mainly released when an increased magmatic influence produces a significant uprising of deep gases.

3 Phlegrean Fields

The comparison of changes of fumarolic gas composition with the rates of ground deformation and seismic energy release during the 1982-1984 crisis in the Phlegrean Fields (Barberi et al. 1984; see Fig. 6) indicates a different behaviour for the chemical species, compared with that at Vulcano, because the geological and hydrological settings are different.

As already argued elsewhere (Martini 1986), hydrogen sulphide and hydrogen are taken as an indication of an input into the shallower aquifers of gaseous species from the Phlegrean magma chamber. This magma chamber has been located at a depth of about 3 km (Armienti et al. 1983), and might undergo deformation because of tectonic movements; even if physical processes at the source cannot be inferred (Bonasia et al. 1984), opening fractures might allow an increased heat flow, producing volatile migration towards the surface.

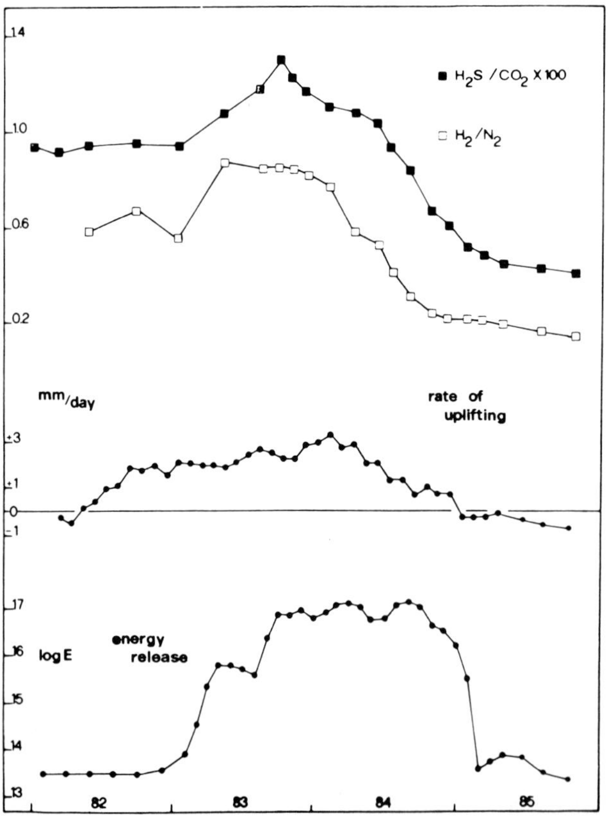

Fig. 6. Seismic energy release and rate of ground uplifting during the recent crisis in the Phlegrean Fields (monthly values from Osservatorio Vesuviano); correlation with the variations of chemical parameters used in geochemical surveillance (Martini 1986)

The analysis of ground uplift has been extensively carried out (Berrino et al. 1984; Del Pezzo et al. 1984; Gaudiosi and Iannaccone 1984) and is beyond the scope of this chapter. It is worth noting, however, that the changes in gas composition appear, as at Vulcano, to precede important seismic energy releases, and the exhaustion of the uplifting process.

4 Discussion

The systematic surveillance carried out at Vulcano and in the Phlegrean Fields has enabled the observation of significant changes in gas compositions at the fumaroles, which can be correlated with seismic activities or with deformation processes. On this basis, can we say that sufficient information is available to assess the degree of probability for new eruptive activities?

When comparing the areas considered, strong differences in gas composition are observed; evidently different degrees of variation have taken place, accompanied by different types of seismic activity. A unique interpretation is not possible. A similar conclusion arises if we extend our observation to other volcanic areas where some kind of surveillance has been carried out in recent times. That is, when considering the probability of new eruptions, variations in temperature or chemical composition of fumaroles, seismic activities or deformation processes, each can give only a partial answer, so that only an integrated examination of natural phenomena appears to promise a possible solution to the important problem of volcano prediction. For the areas considered here, more comprehensive chemical and geophysical data are required.

Starting from the results obtained from the cluster analysis, the bulk of chemical information collected at Vulcano over 9 years was used for a factor analysis procedure (Joreskog et al. 1976), which enabled an evaluation to be made of the extent to which a minimum number of new variables influences the behaviour of the system studied.

According to the results of Table 2, factor 1 appears to explain the observed variations to a greater extent than other factors. As shown in the simplified scheme of Fig. 7, factor 1 should correspond to the magmatic input, factor 2 to the effect of intermediate aquifers and factor 3 to the influence of shallow and surface environments.

The variation with time of the values of the scores of the different factors, reflecting the extent of their contributions to the system, is given in Fig. 8. An important magmatic input is reconfirmed in the periods before 1980 and after 1983, while a significant effect of aquifers at intermediate depths appears between 1979 and 1982. The influence of shallow and surface environments has progressively increased since 1979.

At present, the situation appears to be characterized by significant contributions of both factors 1 and 3. On this basis, a

Table 2. Main factors resulting from the statistical analysis; the values of the variance explained indicate the degree of the influence of a single factor on the observed chemical composition

Factor	Associated variables	Variance explained
1	CO_2, SO_2, HF, Temperature	0.510
2	HCl, Br	0.215
3	NH_4, H_2S, B	0.208

combined input of thermal energy and shallow water bodies could be derived, with a consequent possibility for a vapour pressure buildup.

Thus, according to the results obtained by the statistical procedure, conditions favourable to new eruptive activity are now prevailing at Vulcano. All this does not mean an eruption within a short time. However, since the resumption of activity in areas of quiescent volcanism is presumably a slow process, which evolves at a fast rate only in its terminal stage, the possibility of a timely detection of changes in progress within the system appears of great interest for forecasting purposes.

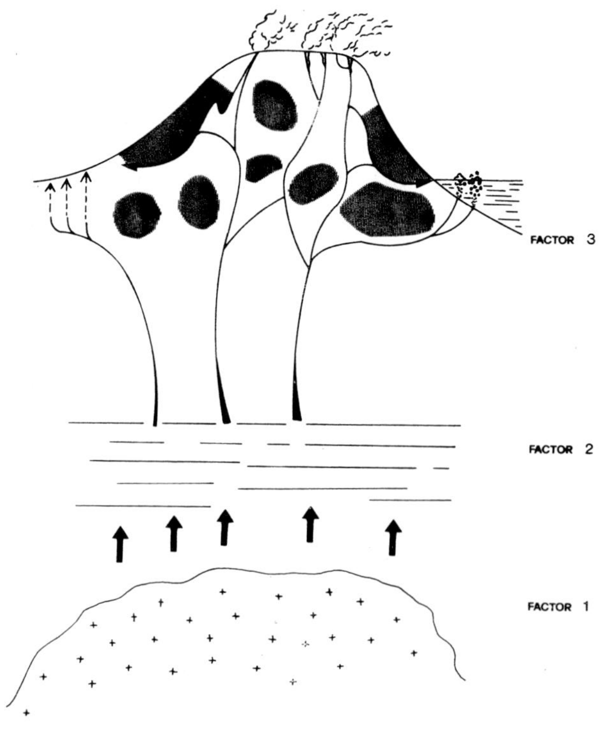

Fig. 7. Simplified scheme representing the different contributions to the fumaroles of Vulcano. A different set of components obtained by the statistical analysis has been attributed to each of these components (see text)

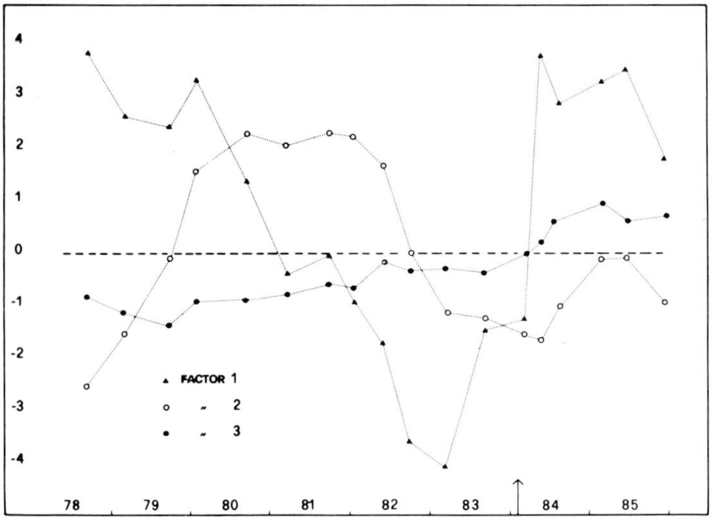

Fig. 8. Variations in time of the scores pertaining to the main factors obtained by the statistical analysis (see text). The arrow indicates a seismic swarm

In short, the probability of a new activity at Vulcano now appears greater than it was during the last few years, and any program of surveillance and mitigation of volcanic disasters should take this into account.

The statistical treatment ought also to include geophysical data, in order to provide a still more representative diagnosis. Communication between geophysicists and geochemists is, however, nowhere very easy. Because of this, an integrated systematic observation program, which would represent the best approach to the problem of prediction of renewed activity, is still missing. While we wait to fill this gap, new and unexpected eruptions may occur, thus producing new disasters.

Acknowledgement. This research was supported by C.N.R. - Centro di ricerca per la geochimica e mineralogia dei sedimenti.

References

Allard P (1983) The origin of hydrogen, carbon, sulphur, nitrogen and rare gases in volcanic exhalations: evidence from isotope geochemistry. In: Forecasting volcanic events. Tazieff H, Sabroux JC (eds) Elsevier, Amsterdam, pp 337-386

Armienti P, Barberi F, Bizouard H, Clocchiatti R, Innocenti F, Metrich N, Rosi M, Sbrana A (1983) The Phlegrean Fields: magma evolution within a shallow chamber. J Volcanol Geotherm Res 17:289-311

Barberi F, Corrado G, Innocenti F, Luongo G (1984) Phlegrean Fields 1982-1984: brief chronicle of a volcano emergency in a densely populated area. Bull Volcanol 47:175-186

Bergeat A (1899) Die Aeolischen Inseln. München, 274 pp
Berrino G, Corrado G, Luongo G, Toro B (1984) Ground deformation and gravity changes accompanying the 1982 Pozzuoli uplift. Bull Volcanol 47:187-200
Bonasia V, Pingue F, Scarpa R (1984) A fluid filled fracture as possible mechanism of ground deformation in the Phlegrean Fields volcanic area, Italy. Bull Volcanol 47:313-320
Bottari A, Caccamo D, Falsaperla S, Neri G (1984) Analisi preliminare dell'attività sismica con origine nell'area di Vulcano durante il primo semestre 1984. Open-File Rep, IIV, Catania
Davis JC (1973) Statistics and data analysis in geology. Wiley & Sons, New York, 543 pp
De Fiore O (1924) Brevi note sull'attività di Vulcano (Isole Eolie) dal 1890 al 1924. Bull Volcanol II:155-156
Del Pezzo E, De Natale G, Zollo A (1984) Space-time distribution of small earthquakes at Phlegrean Fields, south-central Italy. Bull Volcanol 47:201-208
Dolomieu D (1788) Mémoire sur les îles Ponces et catalogue raisonné des produits de l'Etna. Paris, 525 pp
Ellis AJ (1957) Chemical equilibrium in magmatic gases. Am J Sci 255:416-431
Gaudiosi G, Iannaccone G (1984) A preliminary study of stress pattern at Phlegrean Fields as inferred from focal mechanism. Bull Volcanol 47:225-232
Gerlach TM, Nordlie BE (1975) The C-H-O-S system: temperature, atomic composition and molecular equilibrium in volcanic gases. Am J Sci 275:377-394
Gerlach TM, Thomas DM (1986) Carbon and sulphur isotopic composition of Kilauea parental magma. Nature (Lond) 319:480-483
Giggenbach WF (1975) Variations in carbon sulfur and chlorine contents of volcanic gas discharge from White Island, New Zealand. Bull Volcanol 39:15-27
Giggenbach WF, Le Guern F (1976) The chemistry of magmatic gases from Erta 'Ale, Ethiopia. Geochim Cosmochim Acta 40:25-30
Greenland LP (1984) Composition of the January 1983 eruption at Kilauea volcano, Hawaii. Geochim Cosmochim Acta 48:193-195
Heald EF, Naughton JJ, Barnes IL (1963) The chemistry of volcanic gases. 2. Use of equilibrium calculations in the interpretation of volcanic gas samples. J Geophys Res 68:545-557
Joreskog KG, Klovan JE, Rayment RA (1976) Geologic factor analysis. Elsevier, Amsterdam, 178 pp
Le Guern F (1972) Etude dynamique sur la phase gazeuse éruptive. Thèse, CEA-CEN, Saclay, Gif-sur-Yvette
Martini M (1980) Geochemical survey on the phreatic waters of Vulcano (Aeolian Islands, Italy). Bull Volcanol 43:265-274
Martini M (1983) Variations in surface manifestations at Vulcano (Aeolian Islands, Italy) as a possible evidence of deep processes. Bull Volcanol 46:83-86
Martini M (1986) Thermal activity and ground deformation at Phlegrean Fields (Italy): precursors of eruptions or fluctuations of quiescent volcanism? A contribution of geochemical studies. J Geophys Res 91:12255-12260
Martini M, Cellini Legittimo P (1984) Relations between chemical species distribution and the fluctuating activity of Vulcano (Italy). J Volcanol Geotherm Res 22:123-132

Martini M, Piccardi G, Cellini Legittimo P (1980) Geochemical surveillance of active volcanoes: data on the fumaroles of Vulcano (Aeolian Islands, Italy). Bull Volcanol 43:255-263

Martini M, Cellini Legittimo P, Piccardi G, Giannini L (1986) The fumaroles of Vulcano: differences in chemical compositions produced by the surface environment. Geothermics 15:205-209

Matsuo S (1960) On the origin of volcanic gases. J Earth Sci Nagoya 8:222-245

Menyailov IA, Nikitina LP, Shapar Y VN (1980) In: Fedotov SA (ed.) Large Tolbachik fissure eruption, Kamchatka 1975-1976. Nauka, Moscow, 285-309 pp

Menyailov IA, Nikitina LP, Shapar VN (1985) Results of geochemical monitoring of the activity of Ebeko volcano (Kurile Islands) used for eruption prediction. J Geodyn 3:259-274

Menyailov IA, Nikitina LP, Shapar VN, Grinenko VA, Buachidze GI, Stoiber R, Williams S (1986) Khimicheski sostav, metallonosnoct' i isotopiya fumarol'nikh gasov vulkana Momotombo (Nicaragua) v 1982 g. Vulkanol Seismol 2:60-70

Mizutani Y, Sugiura T (1982) Variations in chemical and isotopic compositions of fumarolic gases from Showashinzan volcano, Hokkaido, Japan. Geochem J 16:63-71

Ossaka J, Hirabayashi J, Ozawa T, Kimishima K (1980) Geochemical study of the 1979 eruption of the On-take volcano (volcanic ashes and spring water). Rept Coord Comm Prediction Volcanic Eruption 18:12-17

Sabroux JC (1979) Equilibre thermodynamique en phase gazeuse volcanique. In Hautes températures et sciences de la terre. CNRS, Toulouse, pp 37-46

Sicardi L (1940) Il recente ciclo dell'attività fumarolica dell'isola di Vulcano. Bull Volcanol VII:85-139

Sicardi L (1956) Dell'esalazione solfidrico-solforosa nella attuale fase solfatarica dell'attività dell'isola di Vulcano e di alcune questioni ad essa inerenti. Bull Volcanol XVIII:159-168

Spallanzani L (1792) Viaggi alle Due Sicilie e in alcune parti dell'Appennino. Pavia, 272 pp

Volkov VP, Ruzaykin GI (1976) Equilibrium calculations in volcanic gases. Bull Volcanol 39:47-63

Washington HS, Day AL (1915) Present condition of the volcanoes of southern Italy. Bull Geol Soc Am 26:375-388

Geochemical Precursors at Solfatara Volcano, Pozzuoli (Italy)

R. Cioni[1], E. Corazza[1], M. Fratta[1], M. Guidi[2], G. Magro[1] and L. Marini[2]

Abstract

Solfatara is situated within the Phlegraean Fields volcanic area. Geological evidence and historical records show that volcanic activity is mainly explosive. The last eruption took place in 1538 (Monte Nuovo). In 1970-71 and in 1982-84 the area was affected by a shallow seismicity and by a ground uplift totalling 170 and 180 cm respectively. In both cases large sections of population had to be evacuated.

Complete chemical analyses of the Solfatara fumaroles have been carried out since 1978. Samplings and analyses were more frequent and regular after the ground had started inflating (August 1982), before the onset of the shallow seismicity (March 1983). During the period in which physical changes were observed, changes were also noticed in the chemical composition of the fumaroles: H_2O and H_2S increased before and during the first part of the crisis (until January-March 1984), then decreased together with H_2 and CH_4 and began declining a little earlier. N_2 exhibited a sudden increase at the same time as the decrease of H_2O and H_2S. Some geochemical variations, e.g. the gas/steam ratio, occurred in response to an increase of the heat flow from the magma chamber underlying the Solfatara volcano before the onset of both local seismicity and ground uplift. Hydrogen and methane returned to their "normal" conditions about 6 months before the seismicity and the ground uplift had given any sign of subsiding (January 1985). At the same time the gas composition allowed an interpretation to be made of the chemical equilibria in the geothermal system under Solfatara. The gas/steam variation was utilized in order to obtain a rough estimation of the P and T variations in the aquifer supplying the fumarolic fluids. A 10°C increase in temperature would be sufficient to explain the gas/steam decrease between the end of 1982 and the beginning of 1984. Yet it is not enough to explain the variations of some chemical parameters, e.g. H_2S, H_2 and CH_4. These changes occurred several months after the onset of both seismic activity and ground uplift. An input of high temperature fluids of deep origin should then be assumed. The delay was due to the buffer role played by the main aquifer pre-

---

1 Istituto di Geocronologia e Geochimica Isotopica, Pisa, Italy
2 Geotermica Italiana, Lungarno Mediceo, 16, Pisa, Italy

IAVCEI Proceedings in Volcanology 1
J.H. Latter (Ed.), Volcanic Hazards

sent underneath Solfatara, while other geochemical parameters, the Ar isotopic composition and the N_2/Ar ratio, accompanied discrete seismic swarms.

1 Introduction

The Phlegraean Fields represent a volcanically active fumarolic area near Naples (southern Italy). The recent volcanic history is characterized by mainly explosive events, the most remarkable of which was the emission of the Campanian ignimbrite (35,000 years ago); this eruption caused the formation of the Phlegraean caldera (with a diameter of about 12 km; Rosi et al. 1983). The last eruptive event occurred in 1538 when Monte Nuovo was formed. Volcanological (Rosi et al. 1983), petrological (Armienti et al. 1983) and geophysical evidence (Corrado et al. 1977) suggest that volcanic activity in the Phlegraean Fields is fed by a cooling magma body within a shallow chamber located near the town of Pozzuoli (Fig. 1). During the last 2000 years there has been a general subsidence of the area, with some short-lived reversals (Fig. 2). The only well-documented historical reversal is that preceding the eruption of Monte Nuovo (Parascandola 1947). The last but one reversal occurred in 1970-71 with a total uplift of 1.7 m, while the present one took place from August 1982 to December 1984, with 1.8 m uplift; the total uplift is in the range of 3.3 m (taking into account the slight deflations which followed inflations). The

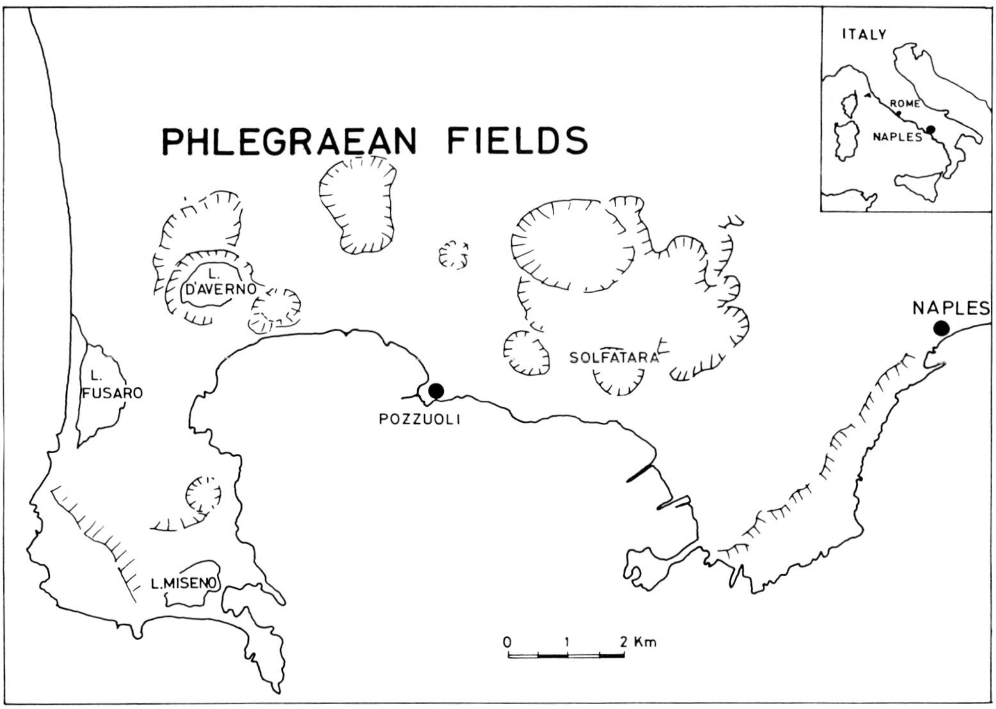

Fig. 1. Location map of Phlegraean Fields

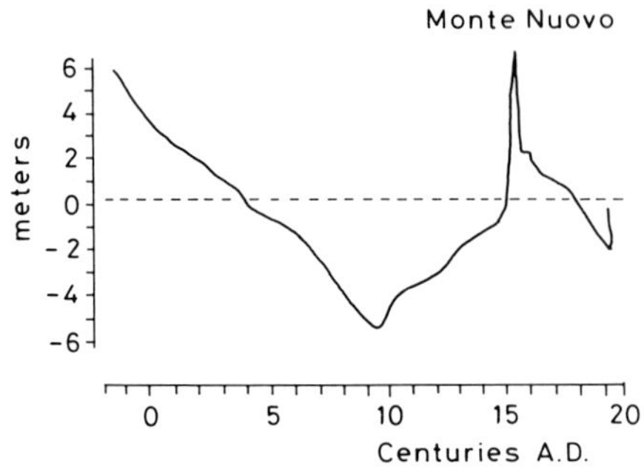

Fig. 2. Historical vertical ground movements at Serapeo, Pozzuoli, in the last two millennia (after Parascandola 1947 and Berrino et al. 1984)

area affected by the uplift is roughly circular, centered in the town of Pozzuoli and with a radius of approximately 6 km: such an area matches the Phlegraean caldera. In March 1983, several months after the beginning of the uplift, local shallow seismicity started in the area (the same pattern was observed also during the 1970-71 crisis (Barberi et al. 1984). Both phenomena ceased at the same time at the end of 1984.

The surveillance program was started by the authors in 1978, but frequent and more accurate data were collected only since 1983, when observations were intensified after the onset of the ground deformation. The general trend of the geochemical parameters shows that composition changes took place before the beginning of both ground uplift and local seismicity (Fig. 3), even though the few data do not allow one to pinpoint the moment of such changes. The intensive observations started in 1983 enabled the relationship between some short-lived details of chemical variations and the occurrence of seismic swarms to be examined, thanks to daily or hourly samplings.

2 Experimental Results

Geochemical surveillance was focussed on two fumaroles of the Solfatara volcano: Fumarola Circolare (FC) and Bocca Grande (BG) (Fig. 4). Only FC was sampled from 1978 to July 1983. Since then samples were collected also from BG as FC began to show signs of decline, leading to cessation of its activity in August 1984. Thus, later data refer to BG only. Even though there are some marked differences in the amplitude of chemical variations at the two fumaroles, their evolution with time is consistent (Fig. 3). Water (in total fluid) and hydrogen sulphide (in the uncondensable fraction) increased before and during the first part of the crisis until the maximum which was reached between January and March 1984. After a 10-month decline almost constant values have been observed since the spring of 1985. Hydrogen and methane began to decline a little earlier than water and hydrogen sulphide. At BG (Fig. 5) nitrogen exhibited a sudden increase at the same time as the decline in water and hydrogen sulphide. Carbon monoxide data are less accurate than major components, but CO also shows significant variations, although it is out of phase with other chemical species (Fig. 5). At BG the outlet temperature has remained constant at 157°C within the experimental errors (about ±1°C).

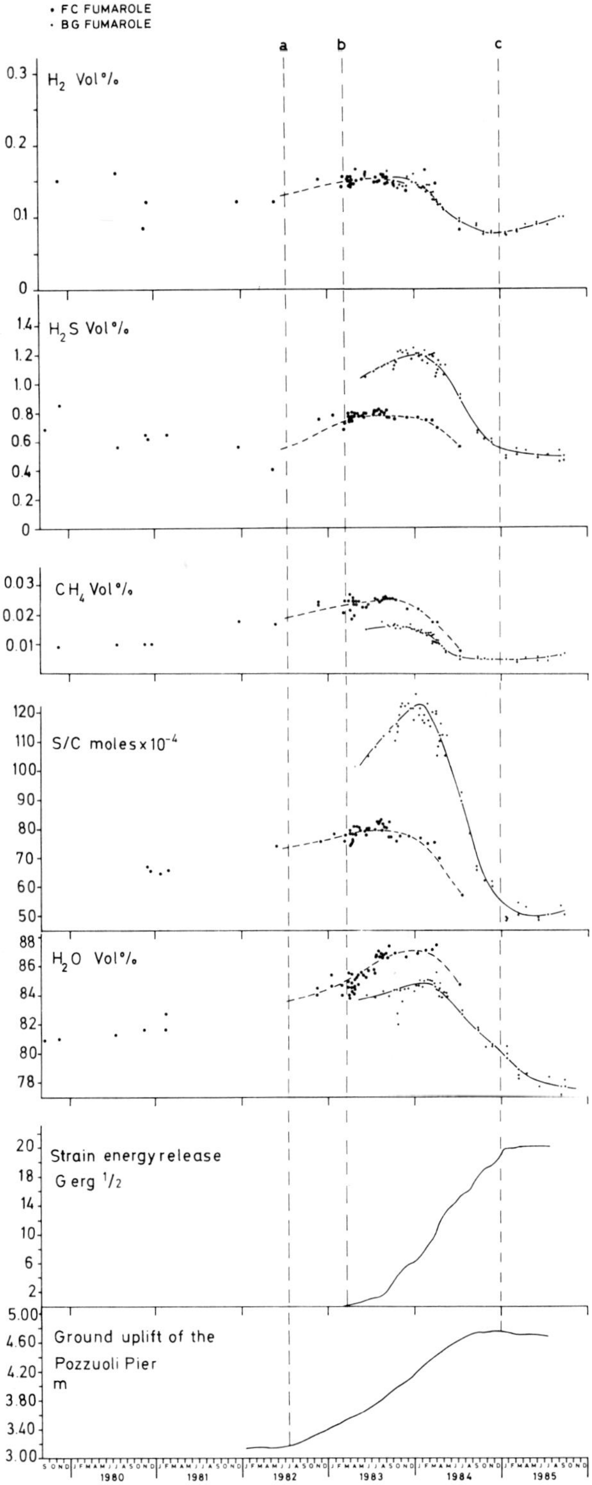

Fig. 3. Time variations of some components compared with physical phenomena. The **solid curve** refers to the BG fumarole; the **broken curve** to the FC fumarole. **Vertical broken lines** refer to the onset of ground uplift (**a**), to the onset of local seismicity (**b**) and to the end of both (**c**) respectively. Physical data were kindly supplied by Osservatorio Vesuviano, Napoli (Italy). H_2, H_2S and CH_4 refer to the uncondensable fraction; H_2O refers to the total fluid

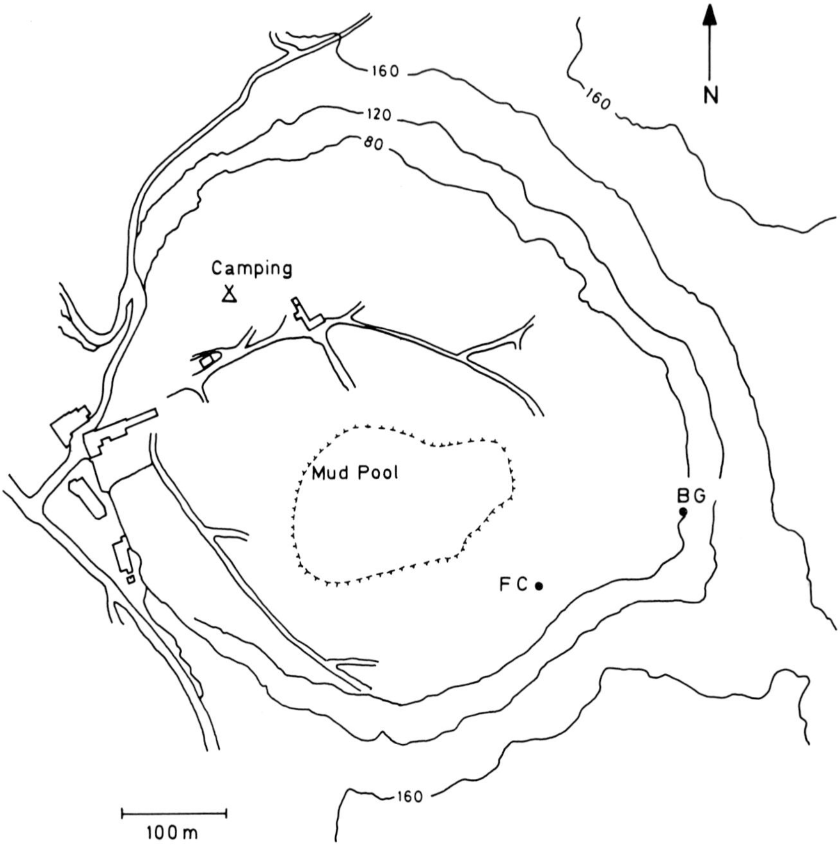

Fig. 4. Fumarole location within the Solfatara crater. **FC** denotes the Fumarola Circolare fumarole and **BG** denotes the Bocca Grande fumarole

The largest and best-documented variations of the chemical parameters took place between April and September 1984. A set of 18 chemical analyses, spanning this period, was therefore selected for the following discussion (Table 1 and Fig. 5).

3 Discussion

In order to interpret the changes in chemical composition of the vapour phase discharged by the Solfatara fumaroles (that is, to relate such changes with a variation somewhere in the whole hydrothermal-volcanic system), it is necessary to refer to a specific geochemical model. On the other hand, such a model has to be verified and refined on the basis of newly acquired data.

Let us first refer to the model proposed by Cioni et al. (1984) and try to insert the new pieces of information into it. According to this model the heat source of the Phlegraean hydro-

Table 1. Selected analyses (illustrated in Fig. 5) of BG fumarole. All components are expressed in volume; water and CO_2 are given in vol% of the total fluid; all other components are in ppm (volume) of the total fluid

Sample N°.	Date (d/m/y)	H_2O (%)	CO_2 (%)	H_2S (ppm)	N_2 (ppm)	H_2 (ppm)	CH_4 (ppm)	CO (ppm)
1	18/08/83	84.4	15.4	1730	343	240	28.2	0.55
2	11/09/83	84.0	15.8	1810	336	258	26.7	0.64
3	11/10/83	84.5	15.3	1780	326	226	26.2	0.34
4	13/10/83	82.8	16.9	1930	378	241	27.2	0.38
5	14/11/83	84.5	15.3	1880	310	219	25.0	0.39
6	30/11/83	84.6	15.2	1880	323	216	25.1	0.31
7	17/01/84	85.1	14.7	1790	313	212	21.8	0.22
8	21/02/84	85.1	14.7	1760	328	198	16.7	0.22
9	3/04/84	84.1	15.7	1700	318	192	18.8	0.27
10	10/05/84	83.9	15.9	1800	370	174	11.9	0.26
11	20/09/84	81.6	18.2	1210	699	158	10.7	0.20
12	24/10/84	81.1	18.7	1150	794	146	10.6	0.26
13	24/10/84	80.4	19.4	1200	843	149	10.6	0.29
14	27/11/84	80.5	19.3	1150	819	150	10.1	0.20
15	24/01/85	80.0	19.8	960	860	153	10.2	0.28
16	18/04/85	78.6	21.2	1160	877	186	12.8	0.47
17	2/06/85	77.8	22.0	1090	977	201	13.8	0.49
18	7/09/85	77.7	22.1	1160	1050	218	13.4	0.62

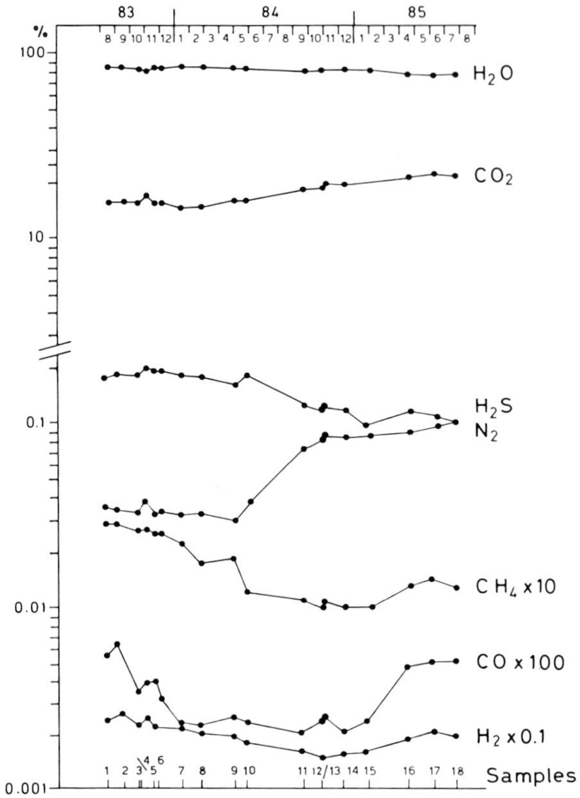

Fig. 5. Concentration of major and minor components at selected samplings during 1983-85. The sampling dates are those given in Table 1. All concentrations refer to the total fluid. Vertical scale is logarithmic

thermal system is a magma chamber located at a depth of approximately 3 km (Armienti et al. 1983). Fumarolic fluids are produced by the boiling of a main aquifer (box 2 in Fig. 6), the depth of its top being approximately 1450 m, according to a relatively old geothermal drilling (Minucci 1961). No relevant geochemical data are available; the temperature at the top of the aquifer before the production test was 225°C, while it was 290°C at 1728 m (Penta 1953). More recently, geothermal wells have been drilled in the Mofete and San Vito area (3 to 5 km to the west of Solfatara) by the Italian oil company (AGIP). Due to the remarkable differences observed in geological and geochemical features for such wells (Chelini 1984; Carella and Guglielminetti 1983) at distances of a few hundred meters, it is impossible to extrapolate the information provided by these recent drillings to the Solfatara subsurface. The presence of possible aquifers other than box 2, perhaps supplying it, cannot be ruled out.

The fumarolic fluids leaving box 2 enter box 1 (an intensively fractured area) where both complete separation of the vapour from the liquid phase and gas re-equilibration take place at about 236°C and 31 bar. This temperature is provided, on the one hand, by gas geothermometry in the system H_2O-CO_2-CH_4-H_2-CO: considering the data collected in the period February-July 1983 an average value of 234°C is obtained; the corresponding steam

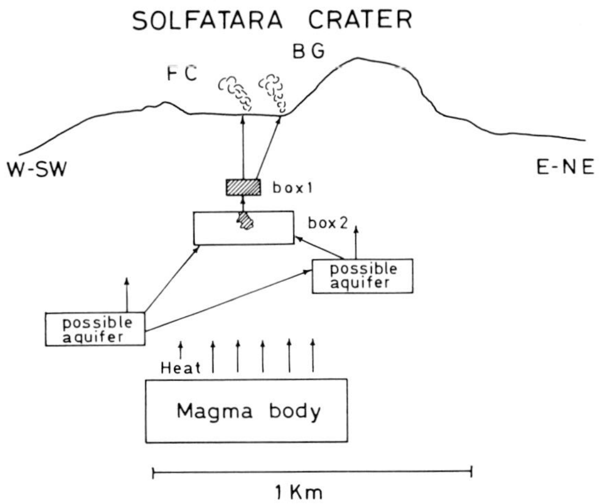

Fig. 6. A schematic geochemical model of the Solfatara system (Cioni et al. 1984). Shaded areas correspond to intensively fractured zones where the vapour is separated from the liquid phase. Vertical scale is arbitrary and reduced (exaggerated for topography)

fraction (y) is 1. On the other hand, the same temperature results from thermodynamic considerations based on the outlet temperature (157 ±1°C), assuming that the dry steam expands approximately isoenthalpically. In fact, a temperature slightly above 160°C at the atmospheric pressure is produced by the isoenthalpic expansion of dry steam separated from the liquid at 236°C and 31 bar (Keenan et al. 1978): these are the T and P conditions of saturated steam with the maximum enthalpy. The coincidence of temperature of final separation and gas equilibrium suggests that the gas equilibria are quenched upon separation of the steam from the liquid, which might act as a catalyzer for such equilibria (Giggenbach 1982).

It should be noted that even though the outlet temperature at BG remained constant at 157 ±1°C, the separation temperature in box 1 could not remain constant at 236°C. In fact, a decrease of the separation temperature of as much as 15°C would lead to a change in the outlet temperature within the experimental errors, assuming isoenthalpic expansion. Observed geochemical temperatures range from 217°C to 253°C (see Fig. 7, where the two logarithms of the products of molar fractions are plotted in order to estimate the gas equilibrium temperature and steam fraction in the gas equilibration zone; see also Bertrami et al. 1985 for further details). Such a departure of the gas geochemical temperature from the expected 236°C (far greater than the 5°C that the analytical errors would allow) could be simply explained by a non-equilibrium transient state of the system at that time. Hypothesizing that temperature in box 1 remained constant at 236°C from August 1983 to September 1985, the observed chemical changes must therefore be related (within the framework of the model by Cioni et al. 1984) to a variation of the temperature in box 2 (initial flashing temperature) and consequently to a change of the heat flow from the magma body, provided that the composition of the liquid phase in box 2 remained constant.

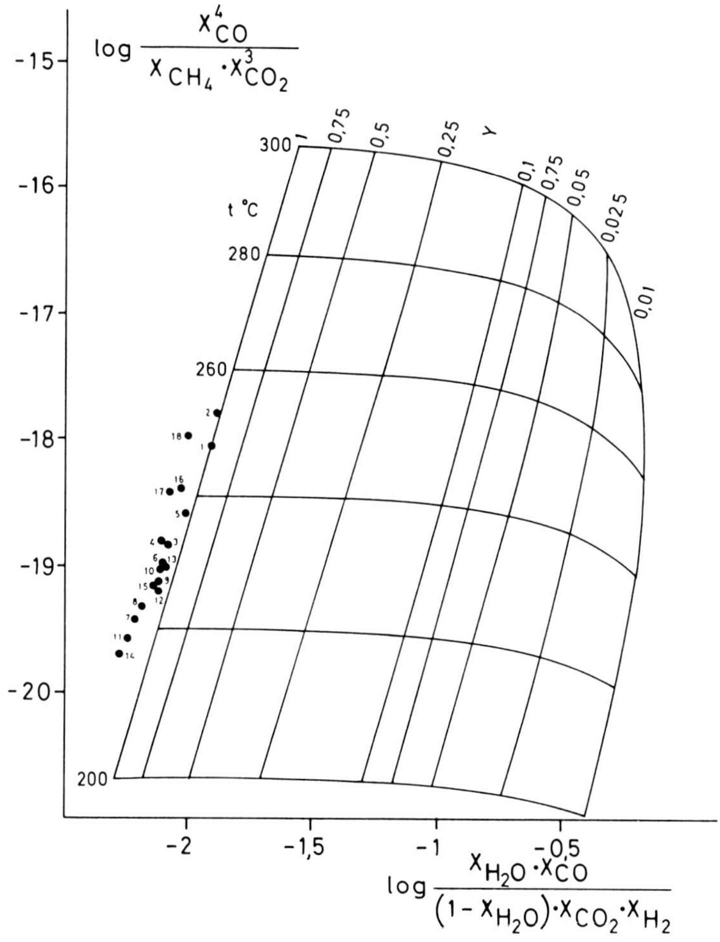

Fig. 7.
Plot of log $\frac{x^4_{CO}}{x_{CH_4} \cdot x^3_{CO_2}}$ versus log $\frac{x_{H_2O} \cdot x_{CO}}{(1 - x_{H_2O}) \cdot x_{CO_2} \cdot x_{H_2}}$ for the samples listed in Table 1. Isotherms and iso-(steam fraction) lines (y) are depicted

An increase in heat flow causes a temperature and pressure increase in box 2 (if the energy entering box 2 is greater than the energy leaving it). The steam phase released from box 2 will be richer in water and poorer in uncondensable gases; among these the low solubility components (H_2, CH_4, CO and N_2) decrease and the high solubility components (e.g. H_2S) increase; carbon dioxide exhibits an intermediate behaviour. Opposite variations of are expected when the heat flow decreases. The influence of pH variations in box 2 is not taken into consideration. This is equivalent to assuming that due to an effective buffer system the pH of the aqueous solution during flashing is maintained approximately constant.

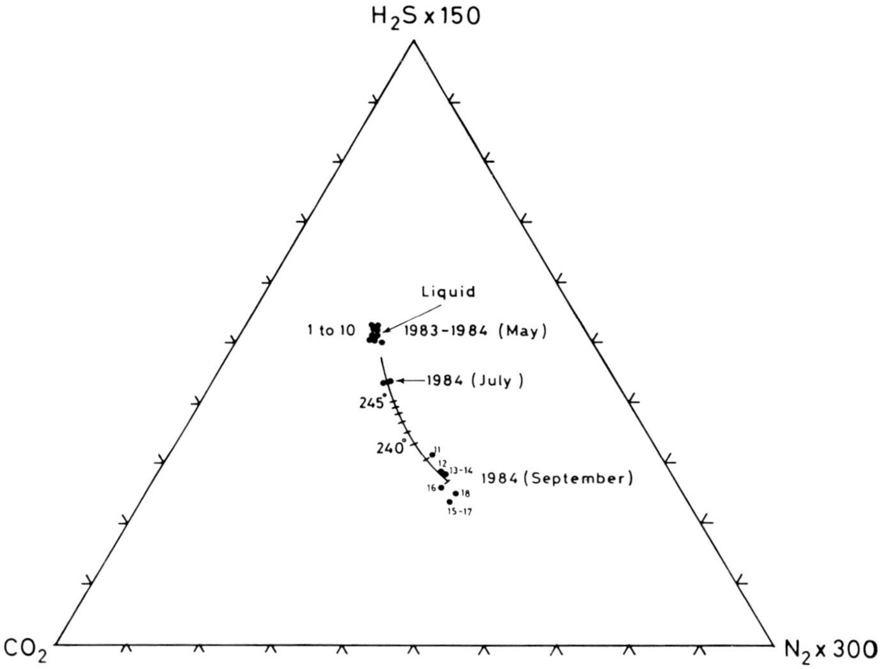

Fig. 8. The main components after water: H_2S, CO_2, N_2. Two families of experimental points are evident; the curve outlines the composition of the steam phase; initial flashing temperature is indicated; final separation temperature is 236°C. Two data points of July 1984 (not included in Table 1) are added. The liquid composition is hypothetical and the same as sample 1

In the CO_2-H_2S-N_2 triangular diagram (Fig. 8) experimental values are plotted along with a hypothetical original liquid phase present in box 2 (liquid in Fig. 8) and with the hypothetical vapour phases released at different temperatures from box 2 (238°, 239°, 240°, 245°C) and completely separated from the liquid in box 1 at 236°C. The distribution coefficients used in the calculations are from Giggenbach (1980) and Bertrami et al. (1985) for CO. Observed compositions match the hypothetical trend. Of course, the hypothetical original liquid could lie elsewhere in the diagram; yet the kind of trend would remain the same. On the other hand, it should be noted (Fig. 5) that only CO_2 and N_2 increased with respect to water, while H_2S decreased. Thus, H_2S changes do not agree with the model under discussion. Hence, one can conclude that observed variations only of H_2O, CO_2 and N_2 agree with variations of the flashing temperature in box 2.

In the CH_4-H_2-CO triangular diagram (Fig. 9) observed variations are much larger than the hypothetical ones. Moreover, the variations with time of H_2, CH_4 and CO do not agree with a model completely controlled by variations of flashing temperature in box 2 (Fig. 6).

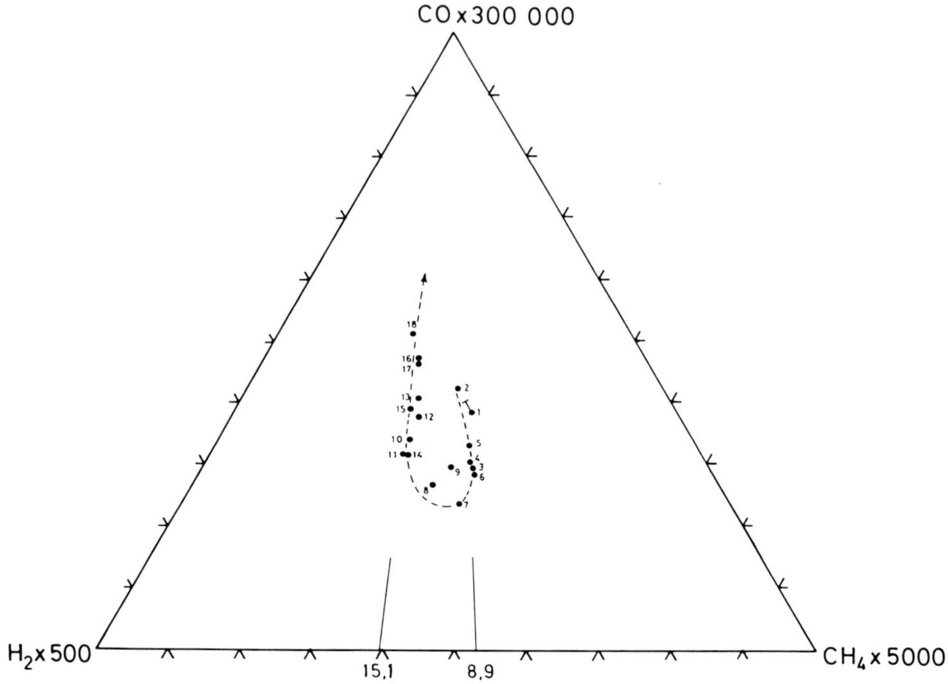

Fig. 9. Some minor components: CH_4, CO, H_2. In this diagram also experimental data fall into two different families. The H_2/CH_4 ratios of the two families are given on the **bottom line**. Starting from a hypothetical liquid (Fig. 8) with the same composition as sample 1, the inferred variation (for a temperature range in box 2 of Fig. 6 from 250° to 238°C) is outlined by the **small bar** pointing towards sample 2. The observed time trend is shown by the **broken line**

The important composition changes (Figs. 5 and 10) observed throughout 1984 (mainly the increase of CO_2/H_2, CO_2/CH_4, H_2/H_2S and H_2/CH_4 ratios) during the bradyseismic crisis can be explained by admitting an input of deep and hot fluids in box 2. The bradyseismic crisis itself could have been the cause of this evolution of the system. Such an evolution can be due either to a physical reason (increase of fracturing at depth determined by magma uplift) or to a chemical one (input of acidic gases, perhaps magmatic, into the overlying aquifers, decrease of their pH and release of CO_2). H_2S would not increase probably due to the buffering role played by silicates (Oskarsson 1978).

The doubling of N_2 content between April and September 1984 can be linked either to the deep and hot components entering box 2 and/or to air entering the shallow part of the system. The ground uplift determined, in fact, a remarkable deformation and fracturing near the surface, thus increasing the possibility of an air contamination of the fumarolic fluid. In the latter hypothesis the lack of oxygen in the emerging fluid would be caused by its consumption in the oxidation of reduced species,

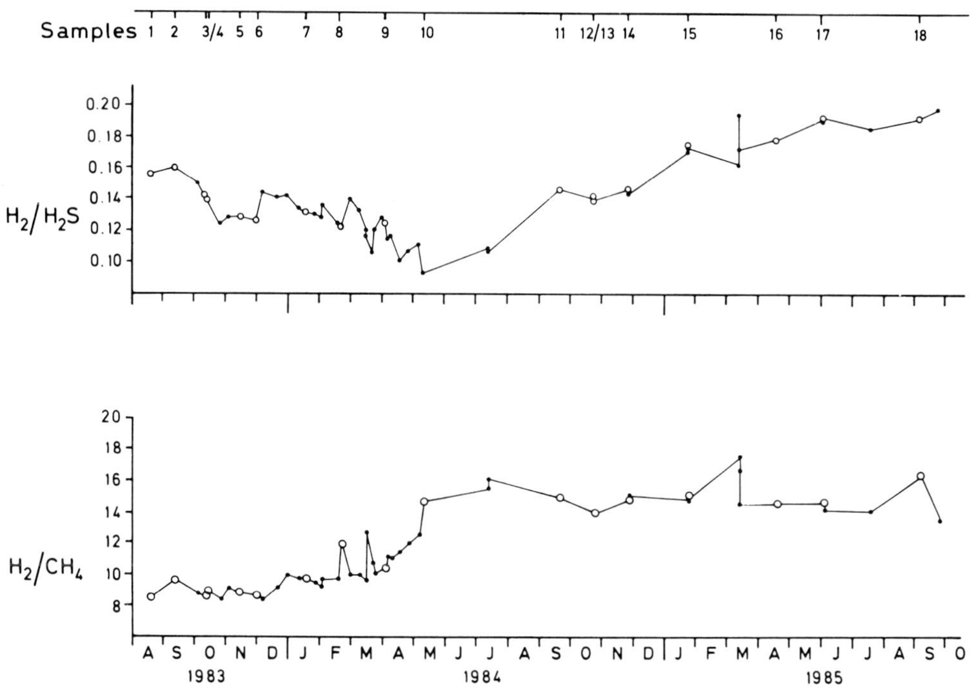

Fig. 10. Time plots of H_2/H_2S and H_2/CH_4 ratios for the BG fumarole. Samples from Table 1 are numbered and shown as **open circles**; additional samples are represented by **dots**

mainly H_2S. However, simple calculations show that the observed variation of H_2S is by far greater than the oxygen provided either by atmospheric air or by an oxygen-enriched gas stripped from shallow aquifers in which atmospheric air was dissolved.

The N_2/Ar and $^{40}Ar/^{36}Ar$ ratios are available until May 1984 and do not allow the meaning of the above mentioned N_2 increase to be clarified. These ratios are plotted versus time in Fig. 11; they are inter-related and higher than air values. A correlation between atmospheric components and the increase of tectonic disturbances was found for the regional earthquake of November 1980 (M = 6.8; epicenter distance about 100 km) and during the Phlegraean crisis of 1983-1984 (Fig. 11) (Magro et al. 1985). The first main variation in 1980 was followed by another important variation in 1983 when sampling was done every few days. The same pattern was found also during the seismic crisis of October 1983. At the same time water suddenly decreased. It must be pointed out that daily samplings would have been necessary to spot these changes. As the radiogenic fraction did not show any significant changes, the variations are to be attributed to the atmospheric component.

Since the aquifer(s) behaves like an atmospheric "buffer" when the system is in a steady state, the atmospheric component dominates over deep gases (low ratios); if earthquakes or ground

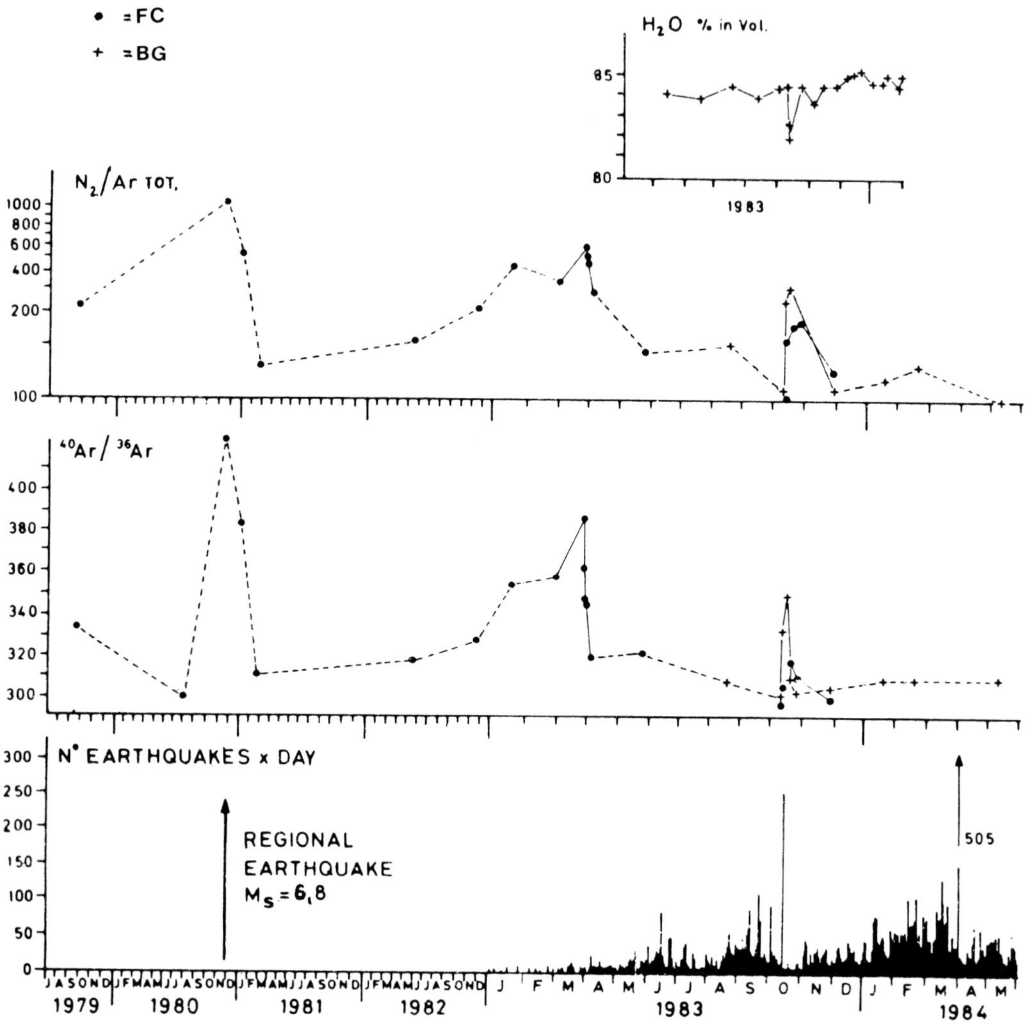

Fig. 11. Changes of $^{40}Ar/^{36}Ar$, N_2/Ar and H_2O in the Solfatara fumarolic uncondensable gases. **Dots** for FC fumarole; **crosses** for BG fumarole. Seismic activity is given as number of micro-shocks per day. Note variable time scale. N_2/Ar ratio in logarithmic scale

deformation change the permeability of the system, the deep inert components pass through more undisturbed (high ratios). The inverse correlation between atmospheric Ar and water content is in good agreement with the previous consideration.

4 Conclusions

Between April and September 1984 the discharged fumarolic fluids showed remarkable composition changes which are interpreted as variations in the state of activity of the Solfatara system at depth.

The concentration of the main components with the exception of H_2S (i.e. H_2O, CO_2 and N_2) could be controlled by a decrease of the boiling temperature of the main aquifer (box 2), even though alternative explanations for the changes of CO_2 and N_2 can be proposed. On the other hand, the remarkable increase of the H_2/CH_4 ratio (and of the H_2/H_2S ratio) almost certainly reflects a composition variation of the main aquifer, caused by an input of deep and high temperature fluids. The variation of the chemical composition in box 2, which can be classified as the most remarkable geochemical change of the bradyseismic crisis, occurred about 16 months after the beginning of the ground uplift and about 9 months after the onset of the seismic activity. This delay of the H_2/CH_4 ratio (and of the H_2/H_2S ratio) in showing wide variations most likely is due to the "buffer" role played by the main aquifer (box 2) present in the Solfatara subsurface.

It is worth pointing out that the chemical parameters that are expected to change in response to a variation of the heat flow from the magma body (such as the gas/steam ratio) behaved as true precursors: not of a real volcanic event, since no such event actually occurred, but of conspicuous physical phenomena such as ground uplift and local seismicity. In fact, a marked decrease of the gas/steam ratio has been actually detected since 1981 (Cioni et al. 1984), well before the onset of the physical phenomena.

It should, in any case, be pointed out that all these phenomena (both chemical and physical) are likely to be the common effect of a more general volcanic event (magma uplift) that affected the Solfatara system.

The non-reactive gases like Ar with its isotopes, and N_2 to a certain extent, were dependent instead on short-lived phenomena and changed at the same time as these physical phenomena.

Acknowledgements. The authors are greatly indebted to W. Giggenbach for his stimulating discussion during the preparation of the present work, as well as to P. Agostini and G. Giaconi for their skilled technical support. Many thanks are due also to R.B. Glover for his critical reviewing of the manuscript.

References

Armienti P, Barberi F, Bizouard H, Clocchiatti R, Innocenti F, Metrich N, Rosi M, Sbrana A (1983) The Phlegraean Fields: magma evolution within a shallow magma chamber. J Volcanol Geotherm Res 17:289-311

Barberi F, Corrado G, Innocenti F, Luongo G (1984) Phlegraean Fields 1982-1984: brief chronicle of the volcano emergency in a densely populated area. Bull Volcanol 47 2:175-185

Berrino G, Corrado G, Luongo G, Toro B (1984) Ground deformation and gravity changes accompanying the 1982 Pozzuoli uplift. Bull Volcanol 47 2:187-200

Bertrami R, Cioni R, Corazza E, D'Amore F, Marini L (1985) Carbon monoxide in geothermal gases. Reservoir temperature calculations at Larderello (Italy). Int Symposium on Geothermal Energy, Hawaii, 1985, Part 1: 299-303

Carella R, Guglielminetti M (1983) Multiple reservoirs in the Mofete field, Naples, Italy. Ninth workshop on geothermal reservoir engineering. Stanford, USA

Chelini W (1984) Alcuni aspetti geologico-petrografici sul sistema geotermico Flegreo. Rend Soc Ital Miner Petrol 39:387-391

Cioni R, Corazza E, Marini L (1984) The gas/steam ratio as indicator of heat transfer at the Solfatara fumaroles, Phlegraean Fields (Italy). Bull Volcanol 47 2:295-302

Corrado G, Guerra L, Lo Bascio A, Luongo G, Rampoldi R (1977) Inflation and microearthquake activity of Phlegraean Fields, Italy. Bull Volcanol 40 3:1-20

Giggenbach W (1980) Geothermal gas equilibria. Geochim Cosmochim Acta 44:2021-2032

Giggenbach W (1982) Carbon-13 exchange between CO_2 and CH_4 under geothermal conditions. Geochim Cosmochim Acta 46:159-165

Keenan JH, Keyes FG, Hill PG, Moore JG (1978) Steam tables. Thermodynamic properties of water including vapor, liquid and solid phases. Wiley & Sons, New York, pp 1-156

Magro G, Ferrara G, Giuliani O (1985) $^{40}Ar/^{36}Ar$ variations in the fumarolic gases of Vulcano (Aeolian Islands) and Solfatara (Phlegraean Fields). IAVCEI Scientific Assembly. Giardini-Naxos Italy, September 16-21, 1985

Minucci G (1961) La perforation "rotary" pour recherches d'énergie endogène. Conférence des Nations Unies sur les sources nouvelles d'énergie

Oskarsson N (1978) Effect of magmatic activity on fumaroles gas composition in the Namafjall-Krafla volcanic center, N-Iceland. Nordic Volcanol Inst 7803, University of Iceland

Parascandola A (1947) I fenomeni bradisismici del Serapeo di Pozzuoli. Genovese, Napoli

Penta F (1953) Ricerche S.A.F.E.N. negli anni 1951-1952-1953. Roma, Italy (unpublished)

Rosi M, Sbrana A, Principe C (1983) The Phlegraean Fields: structural evolution, volcanic history and eruptive mechanisms. J Volcanol Geotherm Res 17:273-288

Microgravity Changes and Ground Deformation at Rabaul Caldera, 1973–1985

C. McKee, J. Mori and B. Talai[1]

Abstract

A progressive gravity decrease of more than 410 μgal took place at Rabaul Caldera, Papua New Guinea, between August 1973 and July 1985. The gravity change correlates with ground deformation at the caldera, particularly with uplift of more than 1.8 m. The patterns of gravity, elevation, horizontal distance and tilt changes indicate a principal source of inflationary deformation beneath the east-central part of the caldera. These phenomena, together with increases in seismicity, are expressions of a new phase of caldera unrest which commenced in 1971 and intensified markedly during late 1983 and 1984.

The observed relationship between gravity and uplift is about -216 μgal m^{-1}. This relationship is consistent with an increase of sub-surface mass, presumed to be a magmatic intrusion. A simple model which accounts for the observed gravity and elevation changes is a "point" source at a depth of 1.8 km. The depth of the source remained constant suggesting that it is a well-established feature capable of accommodating the inferred physical change without becoming unstable, and/or that further ascent of magma was prevented by an overlying body of unyielding rock.

The scale of the monitored geophysical changes is generally indicative of a relatively small-scale intrusion. Calculations based on the gravity changes indicate a mass increase of 1.0×10^8 t. The magnitude of this intrusive event is well within the range of magnitudes of previous eruptions at the post-caldera vents.

1 Introduction

Rabaul Caldera, on the northeastern tip of New Britain Island, Papua New Guinea (Fig. 1), has been monitored continuously from 1950 to the present. No eruptions have occurred in this period but an intriguing episode of caldera unrest began in 1971 and intensified markedly in late 1983, lasting through to mid-1985: this latter period has been called the Rabaul Seismo-Deformational Crisis Period (Mori et al. Chap. 25, this Vol.)

---

[1] Rabaul Volcanological Observatory, Rabaul, Papua New Guinea

IAVCEI Proceedings in Volcanology 1
J.H. Latter (Ed.), Volcanic Hazards

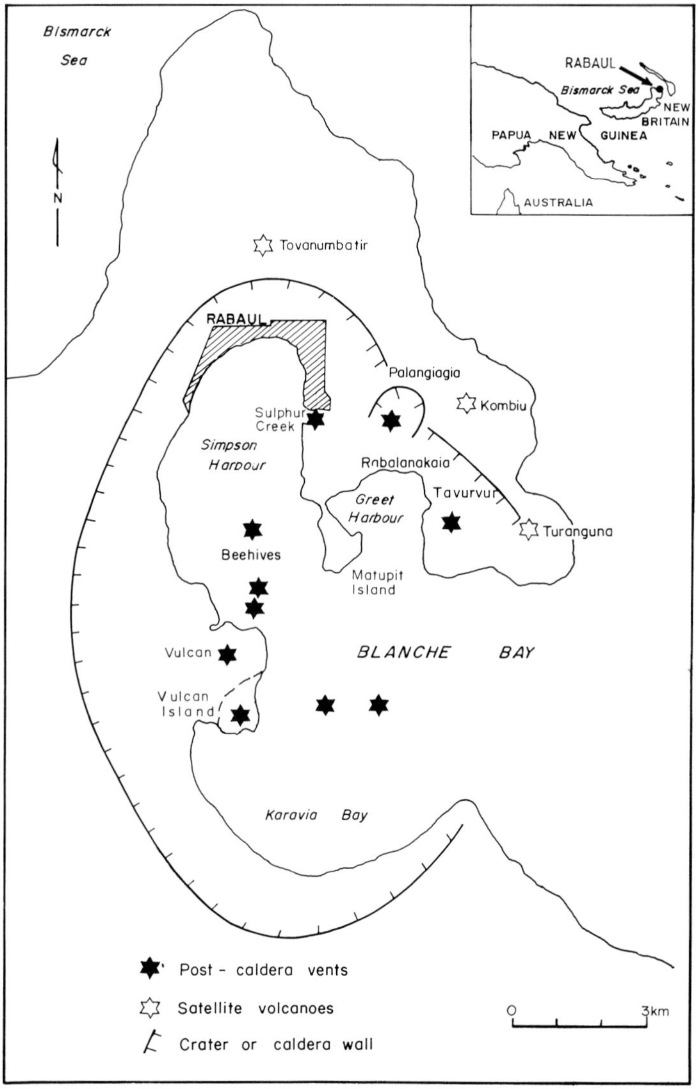

Fig. 1. Rabaul Caldera and locations of post-caldera and satellite volcanoes

Before November 1971, volcanic earthquakes within the caldera were rare, small events which could not be located precisely, but from 12 November 1971 onwards, larger events took place and became common (Cooke 1977). It was not until 1975 that reliable indications were found of structural deformation associated with the on-going caldera seismic activity. These indications came in part from gravity measurements.

A program of gravity observations, on a limited network in the caldera, was started in 1973 to check for possible temporal gravity changes. Levelling measurements on the gravity benchmarks commenced at the same time. Over the following 2 years, significant results were obtained indicating uplift of the ground and reduction in gravity values in a portion of the caldera. These results encouraged the establishment of the gravity study as part of the long-term volcano surveillance program at Rabaul.

This chapter summarizes the measured gravity changes and relates them to ground deformation and to the sub-surface physical processes taking place at Rabaul Caldera.

2 Geological Setting

Rabaul Caldera is a complex structure, measuring about 14 x 9 km, within a low relief ignimbrite shield volcano. A considerable portion of this volcano is practically inaccessible being submarine, but on-shore exposures indicate a diameter of about 60 km. Previous studies of the geology and volcanology of Rabaul (Fisher 1939; Heming 1974; Walker et al. 1981) have concentrated on its recent geological history which is dominated by a caldera-modifying event 1400 years B.P. Recent work (Nairn et al., in prep.) indicates a long eruptive history, perhaps several 100,000 years, that includes many ignimbrite-producing eruptions. These were the main determinants of the present aggradational morphology of Rabaul Volcano.

Immediately flanking Rabaul Caldera to the north is the deeply eroded extinct volcano Tovanumbatir (Fig. 1), and to the northeast and east are the more youthful volcanoes Palangiagia, Kombiu and Turanguna. Other nearby volcanoes are Watom Island, 13 km to the northwest and Varzin, 9 km to the south. These are all strato-volcanoes of basalt to andesite composition (Heming 1974). In contrast, most of the rocks produced by the vents within the caldera, including the ignimbrites which drape Rabaul Volcano, are typically more felsic, having andesite to rhyolite compositions.

The positions of the post-caldera vents shown in Fig. 1 define an incomplete ellipse which is slightly eccentric within the caldera. These vents are closely associated with the zone of caldera seismicity which in plan view forms an elliptical annulus (Fig. 2). The caldera seismicity is interpreted (Mori and McKee, 1987) to be an expression of stress release on a ring fault extending from the surface to a depth of about 4 km. This fault is believed to define the edge of a block which subsided during the latest major eruption, 1400 years B.P.

This subsidence event, and perhaps earlier ones also, has resulted in a weak gravity anomaly low, as seen in the Bouguer map produced by Brooks (1971) and reproduced here in Fig. 3. The center of the gravity anomaly, in the northern part of Blanche Bay, is near the middle of the area bounded by the caldera seismic zone. The Bouguer density used to prepare this map ($2.67\ t\ m^{-3}$) appears to be inappropriate as there is some correlation of gravity and topographic features. Nevertheless, the amplitude of the caldera gravity low may be of the order of

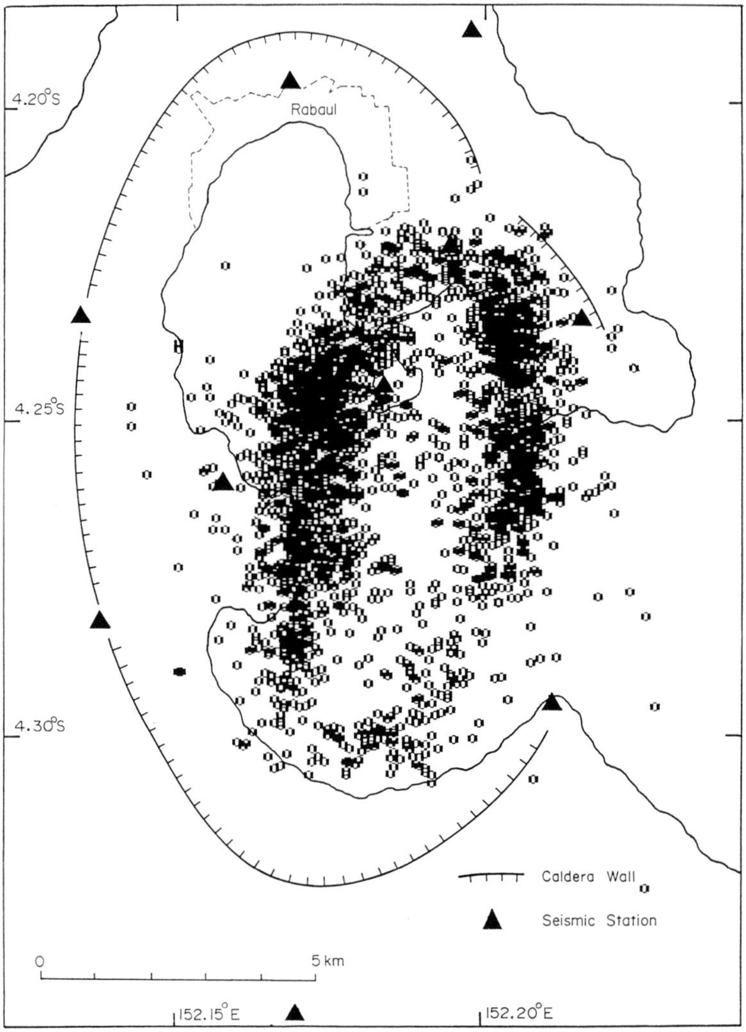

Fig. 2. Epicenters of over 2500 earthquakes for the Crisis Period from September 1983 to July 1985. All events plotted were recorded on seven or more stations and had horizontal errors of less than 1 km. A similar pattern of seismicity was evident for the period 1971 to August 1983 (Mori et al. Chap. 25, this Vol.)

5-10 mgal. This feature appears to cover an area similar to that of the caldera seismic zone. The gravity low may reflect partial filling of the topographic depression of the caldera with relatively low density tephra deposits. Such deposits are believed to have a thickness of up to about 3 km at many calderas where voluminous pyroclastic eruptions continued as caldera collapse took place (Walker 1980; Druitt and Sparks 1984; Lipman 1984). Seismic refraction studies (Cifali et al. 1969) indicate that the floor of Rabaul Caldera is underlain by up to about 3 km of low velocity material.

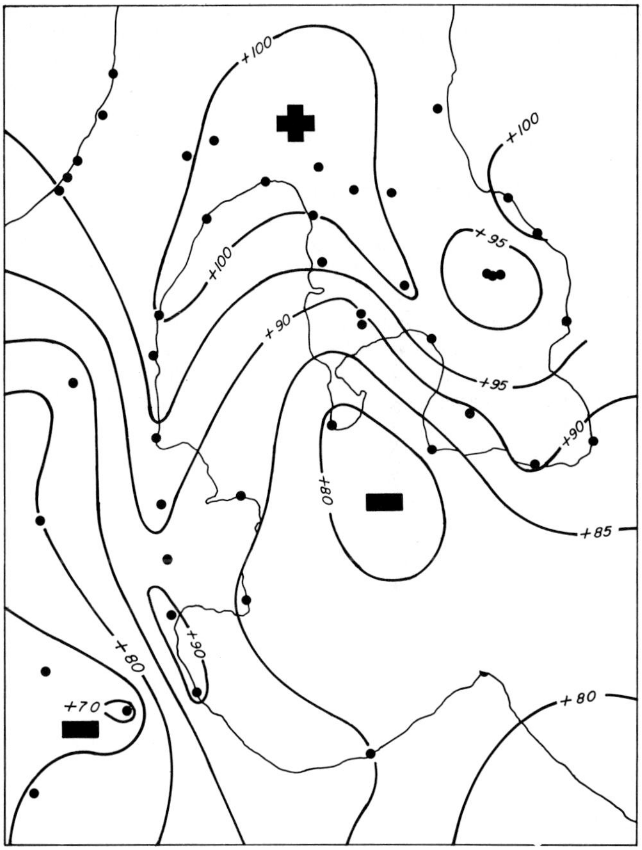

Fig. 3. Bouguer gravity map of Rabaul Caldera and environs. The contour interval is 5 mgal. Dots show locations of gravity measurement stations. Bouguer density = 2.67 t m^{-3} (after Plate 51 in Brooks 1971)

Apart from this gravity low, there is a gravity high in the northern part of the caldera. This feature is not correlated with topography and appears therefore to represent a lateral change in the sub-surface geology within the caldera. Higher seismic velocities are found at shallower depth in this part of the caldera (Cifali et al. 1969). This suggests that higher density rocks are closer to the surface here than elsewhere in the caldera. The northern part of the caldera appears to be an older and more stable area. This stability is indicated by a variety of geodetic measurements, as shown in the following section.

3 Ground Deformation

Ground deformation within Rabaul Caldera has been reported since the earliest European contacts (late 19th century) and has apparently occurred both at the times of eruptions and between them. Brown (1878) reported that at the time of the 1878 eruptions of Vulcan Island and Tavurvur, the Beehives subsided several feet, while small rocky islets near the site of the Vulcan Island eruption were raised several feet and new rocks were exposed. At about the same time, immediately southeast of Tavurvur, the coast was uplifted about 20 ft. and the coastline was shifted about 150 yards seawards. Several hours before the 1937 eruptions of Vulcan and Tavurvur, a number of small islets, up to 6 ft. above sea level, came into existence near the site of the Vulcan eruption, and a reef at the southern end of Matupit Island was exposed (Fisher 1939). This reef subsided during or soon after the eruption as did the whole Sulphur Creek-Matupit Island-Greet Harbour area. The greatest subsidence was over 3 ft. The Beehives also sank 3 ft. or more at the time of the eruption. The deformation at the Beehives in 1878 and 1937 and at the southeast coast of Tavurvur in 1878 has persisted to the present.

Instances of deformation between eruptions were less dramatic. Uplift in the Vulcan area some years before the 1878 eruption is indicated by reports of the pre-existence there of raised coral reef (Simpson 1873; Schleinitz 1889; Sapper 1910). In the 2-year period before the 1941-43 eruptions of Tavurvur, subsidence of about 75 mm took place in the Greet Harbour area (Fisher 1976).

Of particular interest is the Sulphur Creek-Matupit Island area which has a history of upheaval and subsidence between eruptions. Uplift was reported to have occurred in this area before an eruption at the head of Sulphur Creek in about 1850 (Brown 1878), although the time scale of this uplift is not known. Fisher (1939) cited a number of instances of elevation changes in the Sulphur Creek-Matupit Island area between the 1878 and 1937 eruptions of Tavurvur and Vulcan Island/Vulcan, and small-scale topographic changes have been observed in the Sulphur Creek area since the late 1940s (Crick 1975).

The fact that most of the caldera floor is covered by water has resulted in attention being focussed on the Sulphur Creek-Matupit Island area, while deformation in most other parts of the caldera would almost certainly have escaped notice in the past. Indeed, a major handicap in monitoring structural deformation of Rabaul Caldera is the inaccessibility of most of the caldera floor. However, sufficient land masses are present within the northern half of the caldera for measurements to allow a reasonable synthesis of the pattern of deformation there. The situation is less fortunate in the southern half of the caldera, resulting in a large deficiency in the coverage of our deformation monitoring program. Despite these problems, the latest episode of structural deformation at Rabaul Caldera appears to have been dominated by events near the mouth of Greet Harbour (Fig. 1), and this area has been adequately monitored since the early 1970s (McKee et al. 1984).

The most recent phase of deformation is an up-doming of the caldera floor. It appears that the up-doming began in late 1971 as shown by the results from tiltmeters near Sulphur Creek and

Tavurvur Volcanoes (P. de Saint Ours, pers. commun., Appendix by McKee in Scott 1982). The first indications of the scale of this up-doming came from levelling and gravity measurements in the mid-1970s. More recently, results from tilt and horizontal distance monitoring networks confirmed the trend of structural deformation and provided additional detail on the areal distribution of the deformation.

3.1 Uplift

The pattern of deformation in Rabaul Caldera is shown best by the record of elevation changes. These changes have been monitored mainly from measurements on a levelling line extending from the northern extremity of the caldera floor to the southern end of Matupit Island (Fig. 4). Measurements on this line commenced in August 1973, synchronously with the first gravity measurements. Surveys are usually run from a group of reasonably stable benchmarks near the northern end of this levelling line. The stability of these benchmarks is indicated by reference to sea level measurements from a tide gauge at the northern shore of Simpson Harbour which showed no significant elevation changes between 1969 and 1985 (Mori et al. 1987). Until the Crisis Period, re-occupations of this line were irregular, at intervals ranging from 3 to 27 months, but from November 1983, measurements were more frequent, at intervals of less than 1 month to about 3 months. All measurements except those between 1975 and 1979 were made by surveyors and are of second order. The quality of the 1975-79 data is somewhat doubtful. Although this does not affect interpretation of the long-term deformation, detailed analysis of both elevation and gravity changes between 1974 and 1981 is restricted.

The long-term pre-Crisis uplift rate at Matupit Island was about 100 mm per year, and the maximum measured uplift between August 1973 and June 1983 was 1028 mm at the southern end of the island. The greatest measured uplift for the 23-month Crisis Period was 767 mm, at a new benchmark at the southeastern end of the island. The total measured uplift at the southern end of the island between August 1973 and July 1985 was 1751 mm.

Less frequent elevation measurements on other levelling lines extending (1) from Rabaul Town along the western and southern margins of the caldera floor and (2) around the northern and eastern sides of Greet Harbour, have allowed the construction of a contour map of part of the deformation (Fig. 4). The main features of this deformation are a high degree of radial symmetry and a rapid decay of uplift with distance from the source. At a distance of about 4 km the uplift is reduced to 100 mm. Other features of the deformation surface are that it appears to be largely contained within the ring-shaped pattern of caldera seismicity (and inferred caldera ring fault, Fig. 2), and that the focal point of the deformation is near the middle of the region bounded by the ring fault.

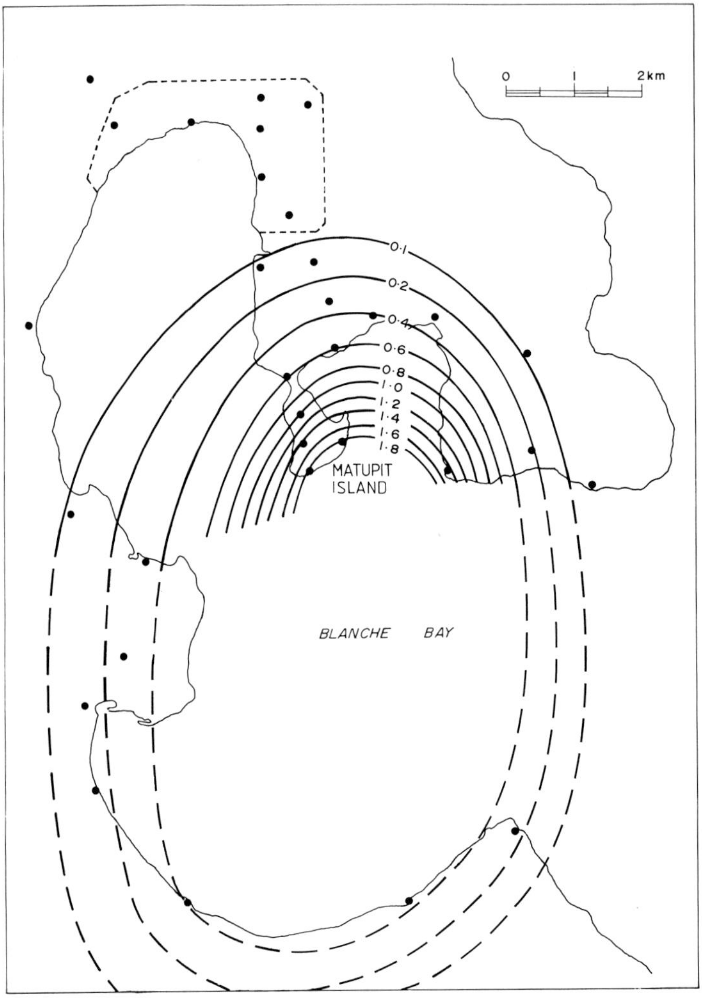

Fig. 4. Partial contours (0.2 m) of elevation changes for the period September 1973-July 1985. **Solid circles** denote benchmarks

3.2 Tilt

Surveillance of ground tilt in Rabaul Caldera before 1981 was limited to the measurements from four tiltmeter stations, of which only two were well situated to monitor volcanic tilt in the caldera. This network was expanded in late 1980 and early 1981 with the addition of five dry-tilt arrays.

Prior to the Crisis Period, tilt monitoring confirmed the trend of deformation which was registered by levelling measurements in the Sulphur Creek-Matupit Island area, and indicated that the source of deformation was near the mouth of Greet Harbour. Additionally, tilt measurements near Vulcan and at the southeastern shore of Karavia Bay hinted at the possibility of a second source of deformation, in the southwestern part of the caldera.

A major expansion of the network of dry-tilt arrays was carried out at the beginning of the Crisis Period. This allowed better resolution of the locations of the sources of deformation, particularly for the one at the mouth of Greet Harbour. The largest measured tilts in this area for the Crisis Period were almost 500 μrad, while tilts of up to about 150 μrad were recorded in the Vulcan area (Fig. 5).

Other significant results from the tilt monitoring program were the very low rates of tilt in the northern part of the caldera, in accord with the very small elevation changes there, and the negligibly small tilts from locations on the northern and western caldera walls. These results suggest that the deformation at Rabaul is strongly dominated by localized intracaldera sources.

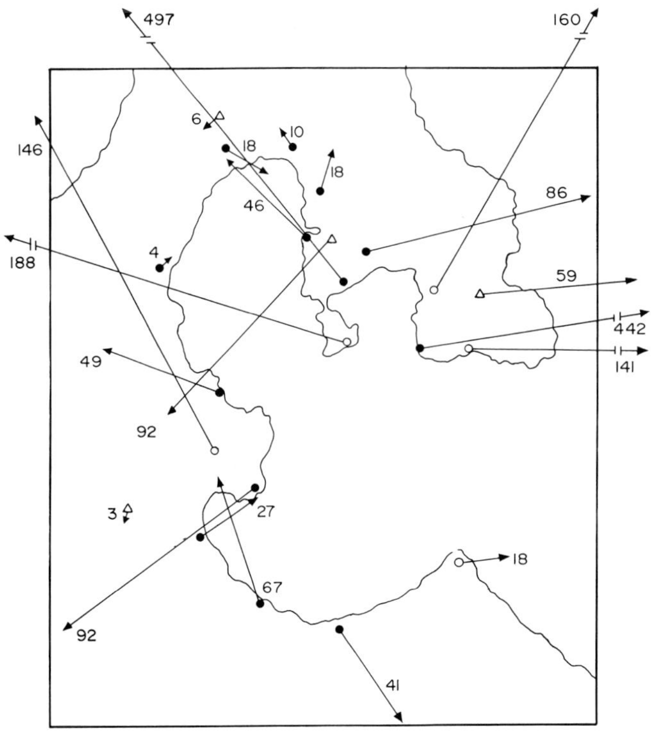

Fig. 5. Tilt vectors (μrad) at Rabaul Caldera for the period October 1983 to July 1985. Open triangles denote water-tube tiltmeters, open circles denote dry-tilt arrays installed 1980-81 and solid circles denote dry-tilt arrays installed September 1983

3.3 Horizontal Deformation

Horizontal deformation monitoring, in the form of electronic distance measurements (EDM), was introduced at Rabaul Caldera during the Crisis Period. The station network consists of two base stations, RVO and MUKA (Fig. 6), on the caldera rim from which measurements are made to reflectors located within the caldera and also to a number of stations outside the caldera. The pattern of horizontal deformation observed at Rabaul is divided into two time periods which exclude the events of the crisis of 3 March 1984 when the MUKA base station may have shifted.

A reasonably consistent pattern of horizontal deformation is observed for the northern half of the caldera, reflecting radially symmetric expansion centered at the mouth of Greet Harbour where the greatest individual station displacement was almost 400 mm (Fig. 6). The peak rates of expansion were about 25 microstrain per month across the mouth of Greet Harbour. EDM changes in the southern half of the caldera between December 1983 and March 1984 are reminiscent of the tilt changes in the same area, which suggest inflation about a source in the southwestern part of the caldera. For the period March 1984 to July 1985 the vectors for all of the stations in the caldera appear to radiate from a position near the mouth of Greet Harbour. Al-

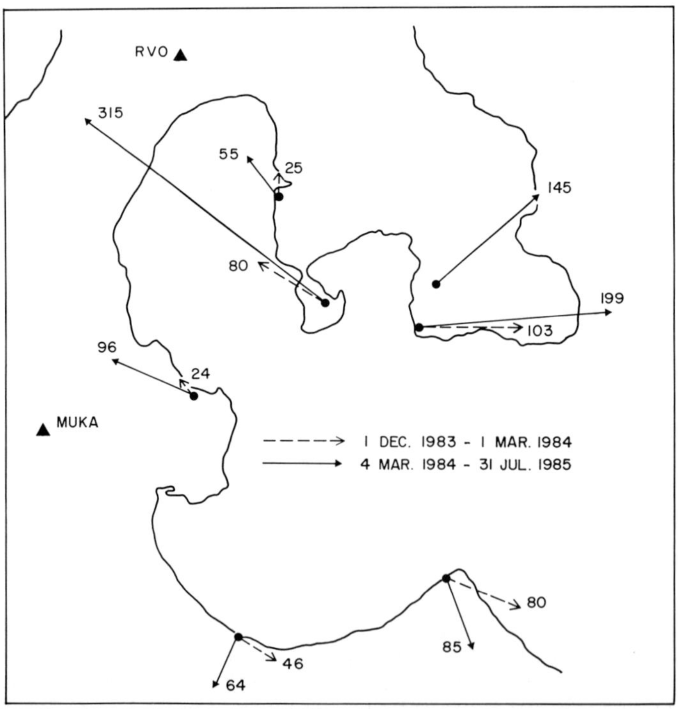

Fig. 6. Horizontal distance changes (mm) for two periods between 1 December 1983 and 31 July 1985. Solid triangles mark the EDM base stations

though this indicates the effect of a single deformation source, the vectors in the southern part of the caldera are larger than expected, and this, combined with a slight rotation of the vector for a station near Vulcan, may imply continuing inflation, albeit weak, from a shallow source in the southwestern part of the caldera, and/or deformation from the inferred large magma reservoir which underlies the caldera (Mori et al. Chap. 25, this Vol.).

4 Microgravity Data

4.1 Data Acquisition and Reduction

The gravity network at Rabaul initially consisted of about 20 stations in the northern and northeastern sectors of the caldera. Major expansion of the network took place in 1976, bringing the number of stations close to the present total of 43 (Fig. 7). The station at RVO is the gravity reference point.

Gravity surveys were initially carried out at 3- to 6-month intervals using LaCoste and Romberg gravity meter G252 on loan from the Australian Bureau of Mineral Resources. Seven rounds

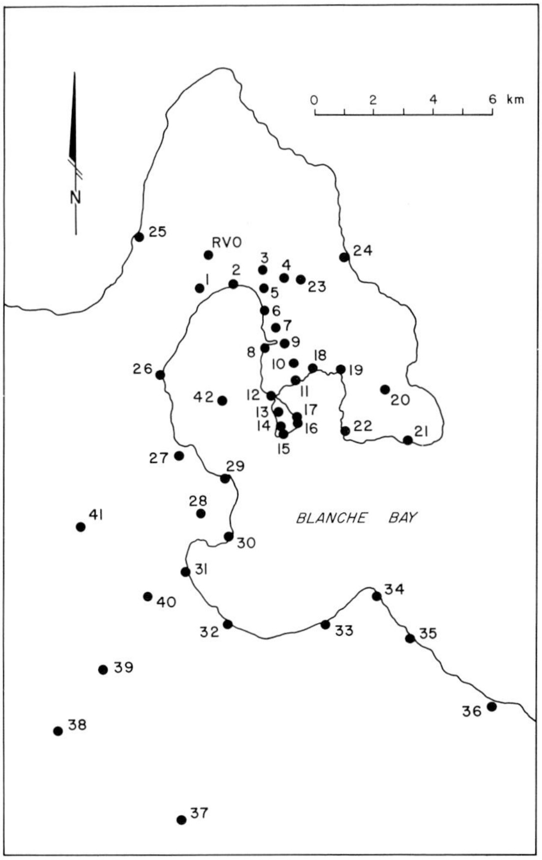

Fig. 7. Locations of stations in the Rabaul gravity network

of measurements were carried out with this instrument between 1973 and 1977. RVO obtained its own gravity meter (L&R G414) in December 1975. A looping technique of observations was employed beginning and ending at the base station, with loops having durations of 1-4 h. An average station gravity value for one survey was obtained from eight measurements (four from each of two observers). The survey procedure was altered in 1980 to single rounds of re-measurement on a monthly basis so that trends in gravity changes could be monitored more closely. This resulted in a reduction in the precision of the technique, but was offset by the improved surveillance capability.

Reduction of the gravity data is carried out one loop at a time, and firstly consists of conversion of meter readings to gravity units using the maker's tables and application of corrections for earth tides (based on the formulae of Longman 1959) and for meter drift. In the earth tide computations, a value of 1.2 is used for the tidal amplification factor and no phase-lag time is used.

Misclosure, the difference in reduced gravity at the base station between the measurements at the start and at the end of a loop, for G414 during the period 1980-85 has the value 21 μgal with a standard deviation of 18 μgal. This value for misclosure is believed to be representative of earlier measurements with this meter as well as the measurements with meter G252. Large misclosures were usually traced to tares and associated data were rejected.

Very large tares have occasionally affected G414, particularly in the first 7 years of its use at Rabaul. The largest of these may have been up to 5 mgal. The long-term drift of G414 is shown in Fig. 8. It is possible that some of these offsets are due to multiple tares. These offsets were usually produced when the meter was used for other surveys outside Rabaul, often involving helicopter transport. Notably, a number of these offsets were of opposite sign to the long-term drift. The general shape of the long-term drift curve for G414 indicates that this meter had settled in by 1982.

For the early period of multiple observations on stations in individual surveys, the typical standard deviation of a measured gravity interval between base and station was about 7 μgal (Cooke 1977). An estimate of the standard deviation for single station readings has been made following the method of Whitcomb et al. (1980). Two stations (40 and 23) were chosen whose gravity relative to the base did not change with time, and the standard deviations of these gravity intervals were calculated, viz., 21 μgal for station 40, and 17 μgal for station 23. Because two readings are required for each interval (the base station and the survey station), the single-reading standard deviations are: for station 40, 15 μgal; and for station 23, 12 μgal. This is in reasonable agreement with the estimates of Whitcomb et al. (1980) for surveys in southern California.

In order to compare the results from one meter to another, calibration factors must be determined for each station. This can

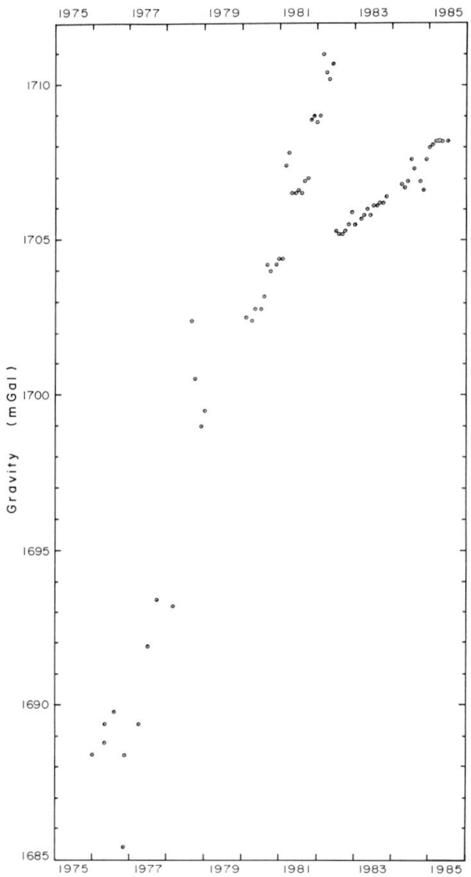

Fig. 8. Drift of LaCoste and Romberg gravity meter G414 at the RVO base station

be done from the data themselves where the two meters were used on simultaneous surveys. The data analyzed here are from December 1975-January 1976, May-June 1976 and June-July 1977. Figure 9 shows the results of this comparison which allow G252 data to be converted to G414 values.

4.2 Sources of Uncertainty

Possible sources of uncertainty include gravity meter scale slippage and scale non-linearities, ocean tides, approximations in the computation of corrections for solid earth tides, and variations in groundwater levels.

Scale slippage (or change in the scale correction factor) appears not to have occured with the two gravity meters used in this study. Measurements on parts of the Australian and Papua New Guinean gravity calibration ranges indicate long-term stability of the scales of the two meters (Table 1).

Scale non-linearities of a periodic nature are well documented for LaCoste and Romberg gravity meters (Groten 1983) and may

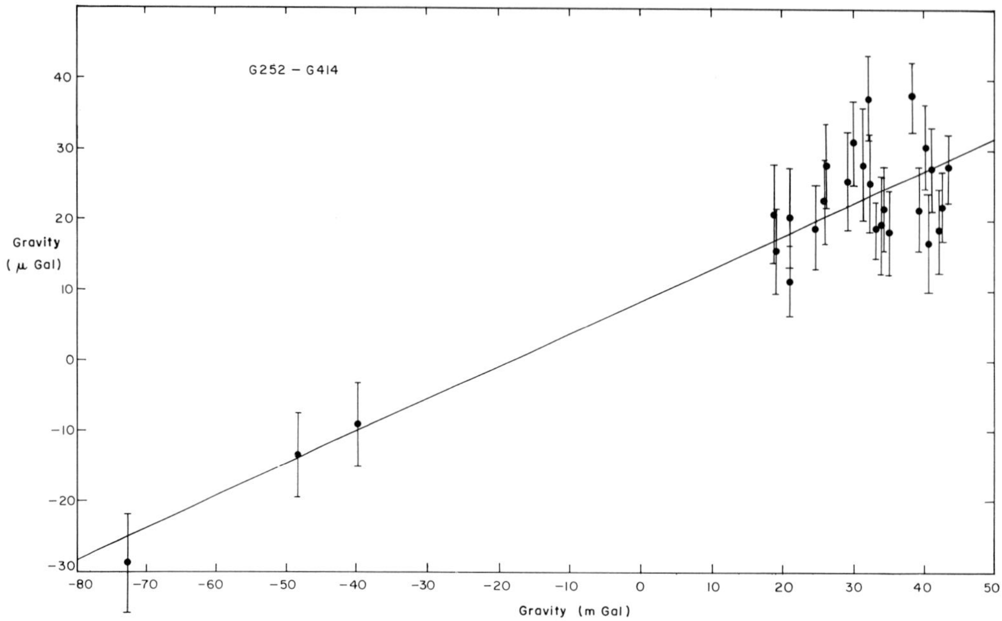

Fig. 9. Differences between G252 and G414 gravity readings as a function of the average gravity values at 27 stations. **Error bars** are + 1 SD. The **straight line** is a least squares fit to the data

have magnitudes of up to 20 μgal (for the 71-mgal period). The accumulated drift of meter G414 is about 20 mgal (Fig. 8) which indicates that some of the gravity intervals could contain errors of up to 20 μgal. For periods shorter than 71 mgal, the magnitudes of the periodic errors are less than 9 μgal. These

Table 1. Gravity meter scale correction factors (G252 data for 1973 and 1975 from P. Wellman, Australian Bureau of Mineral Resources)

Gravity meter	Time of calibration	Scale correction factor
G252	1973	1.000535
G252	1975	1.000517
G252	1976	1.000622
G414	1976	1.000541
G414	1980	1.000534
G414	1984	1.000550

errors are small compared with the gravity changes measured at Rabaul, and consequently no corrections have been made for them.

Owing to the geographical location of Rabaul, the gravitational effect of ocean tides could conceivably be significant. The maximum variation in tidal levels at Rabaul is about 1 m. However, repeated gravity measurements near the tide gauges at Rabaul through the full range of the ocean tides indicated a negligible effect of the ocean tides on the gravity readings.

As stated above, corrections for solid earth tides were calculated using the formula of Longman (1959), with an amplification factor of 1.2 and no phase lag (based on measurements at Port Moresby, about 800 km southwest of Rabaul). The solid earth tide at Rabaul has not been measured, and because of the complicated lithospheric structure in Papua New Guinea, the solid earth tide at Rabaul could be different to that at Port Moresby. Despite this uncertainty, errors in the applied solid earth tide corrections are probably less than 10 μgal as indicated by a set of repeated readings at the RVO base station.

Available information indicates that variations in the groundwater levels within the caldera at Rabaul are significant for periods of only a few days following heavy rains. Many of the gravity stations are near the coast so that the groundwater levels at these locations will be near sea level and will vary within a narrow range. However, there is a group of stations, including the base station RVO, at elevated locations where short- and long-term variations in groundwater levels are unknown. While the evidence is circumstantial, it is probable that variations in groundwater levels at these locations have affected the gravity readings. For the early period where observations from the two gravity meters can be compared, a significant pattern of systematic gravity differences of up to about 80 μgal between the coastal stations and the RVO base (and other elevated stations) is evident. Geodetic measurements to a benchmark at RVO indicate no significant movement of this location which could explain the gravity variations. Conversely, there is no indication of universal movement of the stations near sea level, and known secular variations in sea level result in gravity changes an order of magnitude smaller.

While data from water wells are not available, rainfall data may be a reasonable guide to groundwater levels. Annual rainfall totals for the period 1978-85 were plotted and compared with gravity data from stations such as station 2 whose gravity response was not complicated by ground deformation (Fig. 10). From this, we conclude that the gravity signal at the RVO station and most, if not all, of the elevated stations, was affected by variations in groundwater levels. To overcome this, the datum for determining station gravity values was transferred to a group of stations (including station 2) on the caldera floor at the head of Simpson Harbour. The stability of these stations is known from levelling measurements and as stated above, groundwater levels in this area vary within a narrow range determined by sea level.

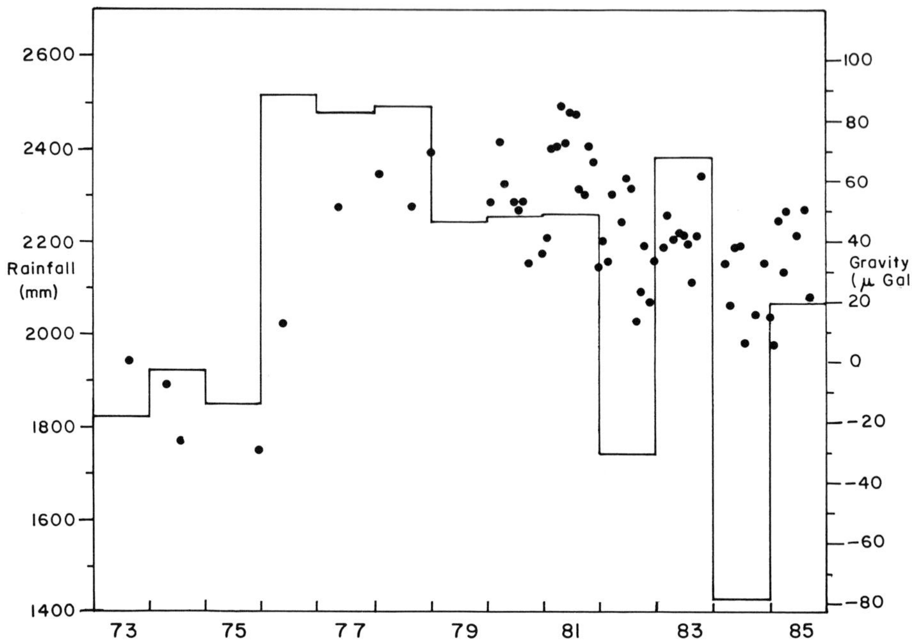

Fig. 10. Comparison of annual rainfall at Rabaul with gravity at station 2 at the head of Simpson Harbour

4.3 Distribution and Rate of Microgravity Changes

The gravity measurements in the period 1976 to 1985 indicate negative gravity changes (i.e. decreases in gravity relative to that at the datum) which were greatest near the mouth of Greet Harbour (Fig. 11). The distribution of changes is well defined only in the northern half of the caldera. It should be noted that the partial contours of -300 and -200 μgal near Tavurvur are inferred from the behavior of station 22 (Fig. 7) which was established during the Crisis Period. In the southern half of the caldera, the distribution of gravity changes is poorly known, but may be elongated towards the southwest. The approximate coincidence of the zero contour of Fig. 11 and the outer edge of the caldera seismic zone (Fig. 2) is noteworthy. Outside the caldera, the gravity changes were generally small, although the elevated stations all showed small positive changes which are interpreted to be due to a rise in groundwater levels.

Substantial variations in the rate of gravity changes appear to define two periods: 1976-83 and 1983-85.

1976-83. The area which experienced negative gravity changes in this period (Fig. 12) is essentially the same as for the whole period 1976-85, although large negative gravity changes were seen only on Matupit Island. The largest of these was 145 μgal, at the southern tip of the island. Moderately large positive gravity changes, suspected to be due to a rise in the water

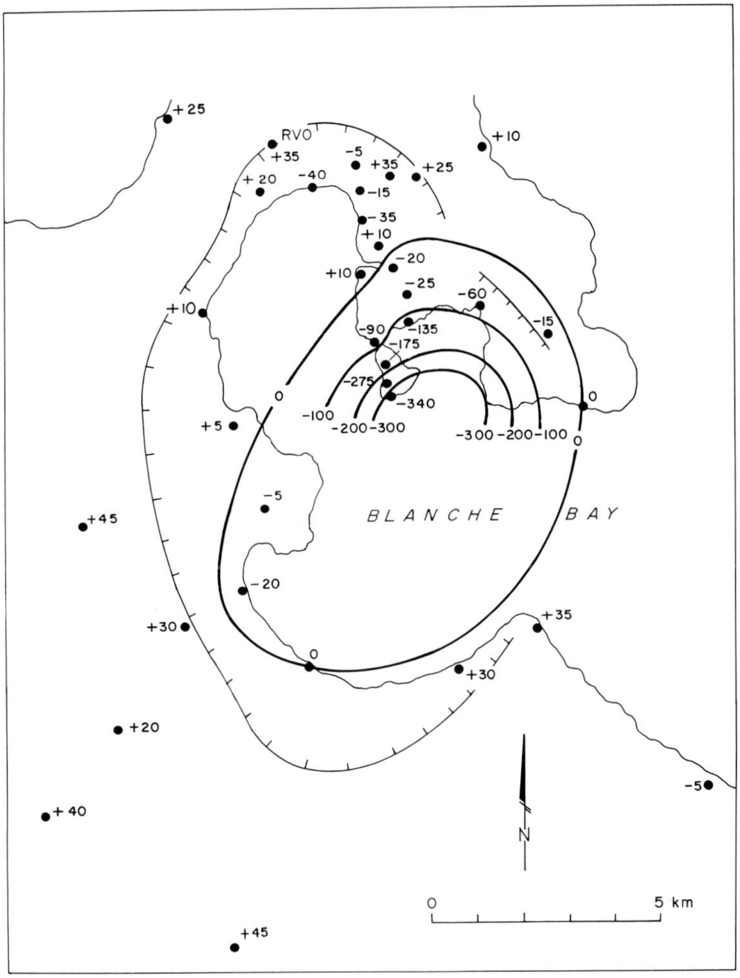

Fig. 11. Net gravity changes (μgal) for the period January 1976 to July 1985

table, affected the elevated stations on the northern rim of the caldera and those southwest of the caldera.

1983-85. This is the Crisis Period during which seismicity and rates of ground deformation greatly increased (McKee et al. 1984). Significant gravity changes in this period were confined to a relatively small part of the caldera, near its center. The greatest measured change was a decrease of 195 μgal at the southern end of Matupit Island (Fig. 13). Curiously, the gravity results showed little change in the Vulcan area and in the southern part of the caldera where tilt and horizontal distance measurements showed significant changes. However, measured elevation changes in these areas were small during this period. The negative gravity changes at the elevated stations are consistent with falling groundwater levels as indicated by reduced rainfall (Fig. 10).

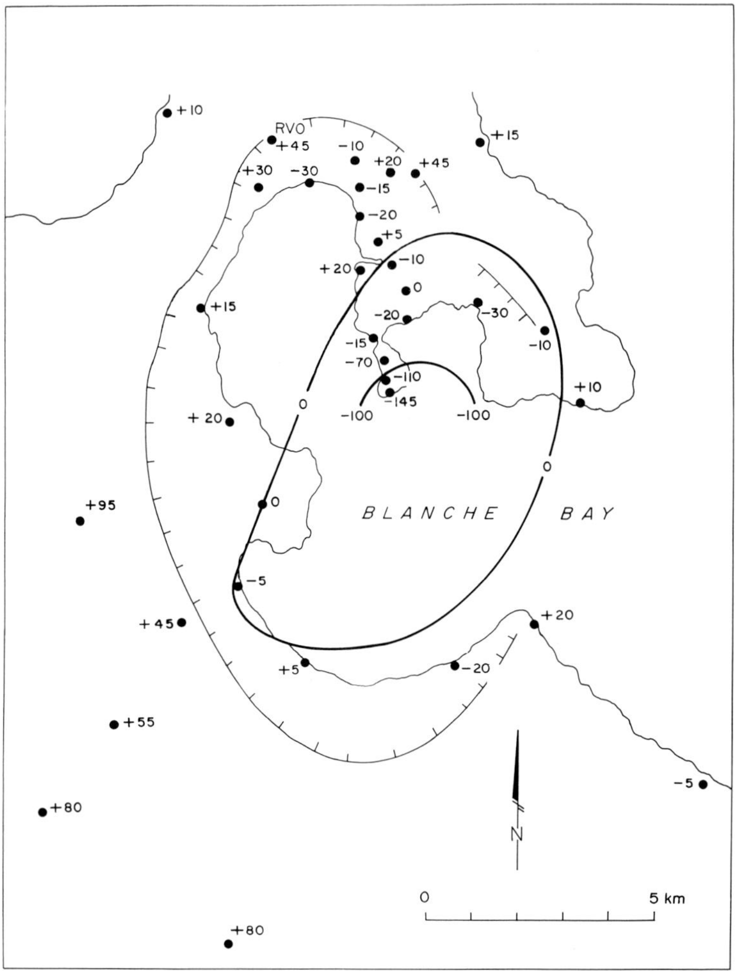

Fig. 12. Gravity changes (μgal) for the period January 1976 to June 1983

5 Interpretation

5.1 Gravity Versus Elevation Change

The largest total gravity change (about -410 μgal) was found at the southern end of Matupit Island (station 15) where uplift of about 1.8 m took place between September 1973 and July 1985. The temporal trends of gravity and elevation changes at the southern end of Matupit Island are clearly shown in Fig. 14. Between 1973 and 1983, almost linear rates of change of elevation and gravity were observed. However, at the onset of the Crisis Period, the rates of these changes increased exponentially. This pattern was sustained until about April 1984, after which time the changes did not increase as rapidly and appeared to follow a logarithmic trend.

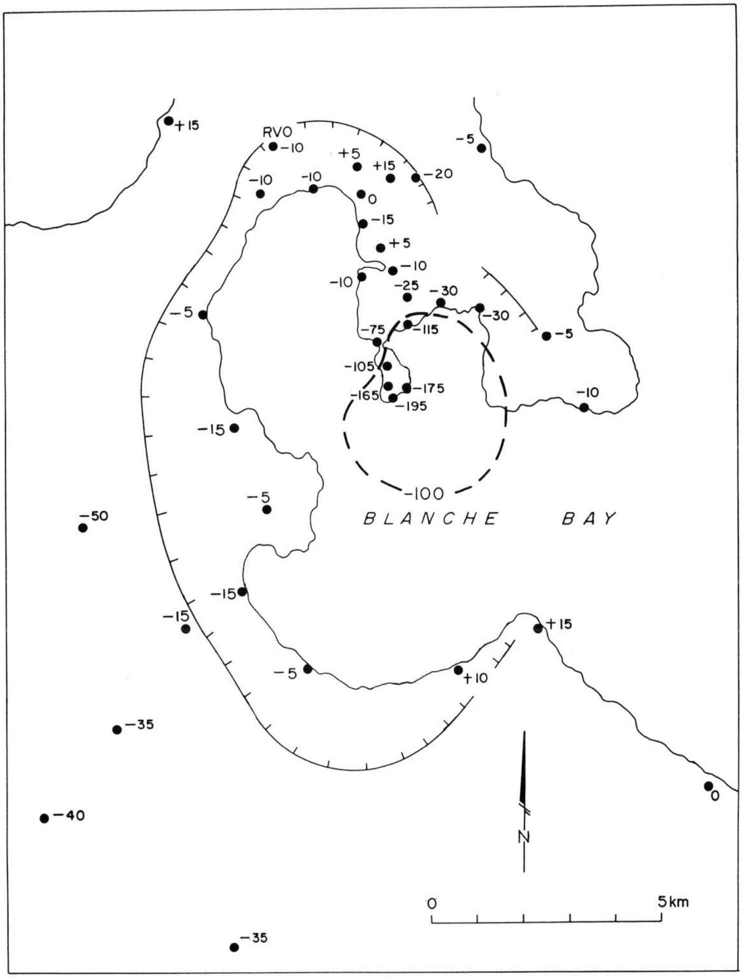

Fig. 13. Gravity changes (μgal) for the period June 1983 to July 1985

Gravity and elevation data along the Rabaul Town-Matupit Island line over different time periods since 1973 give a relationship between gravity and elevation changes ($\Delta g/\Delta h$) of -216 μgal m^{-1} with a standard deviation of 4 μgal m^{-1} (Fig. 15). If the relationship followed the free-air gradient (-309 μgal m^{-1}), the uplift could be interpreted as a result of increased volatile pressure in an underlying magma body or possibly thermal expansion of the caldera floor substrate. However, since $\Delta g/\Delta h$ does not follow the free-air gradient, sub-surface mass changes are indicated.

While the gravity and elevation data shown in Fig. 15 can be fitted quite well with a least squares straight line, it may be useful to consider separately the 1973-83 and 1983-85 data (Fig. 16). The slopes of these lines are: -199 ±7 μgal m^{-1}, for 1973-83, and -235 ±10 μgal m^{-1} for 1983-85, where the uncer-

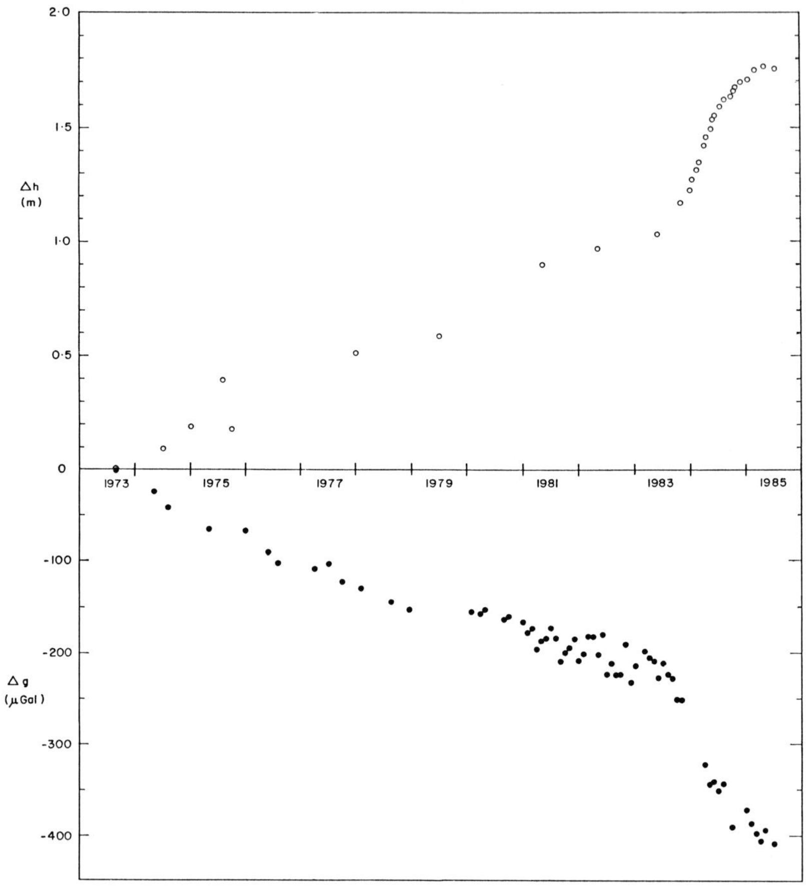

Fig. 14. Gravity vs elevation change at the southern tip of Matupit Island (station 15)

tainties are one standard deviation. Although these results should be treated with caution, they may suggest that the Crisis Period involved a density change in the caldera's magma reservoirs as well as an increase in the rate of mass input. If the change in $\Delta g/\Delta h$ is real and is due to a density change, it would signify a degree of vesiculation of the magma.

5.2 Modelling of Gravity and Elevation Changes

The strong negative correlation between the gravity and uplift data extends throughout the uplifted area (Fig. 17). The sharp decay in both quantities with distance from Matupit Island indicates a shallow source for these effects, which can be adequately modelled by a "point" source (Mogi 1958). The location of the source was determined from tilt data. As this location did not change appreciably with time, estimates of the depth of the source could be obtained from analysis of the levelling data.

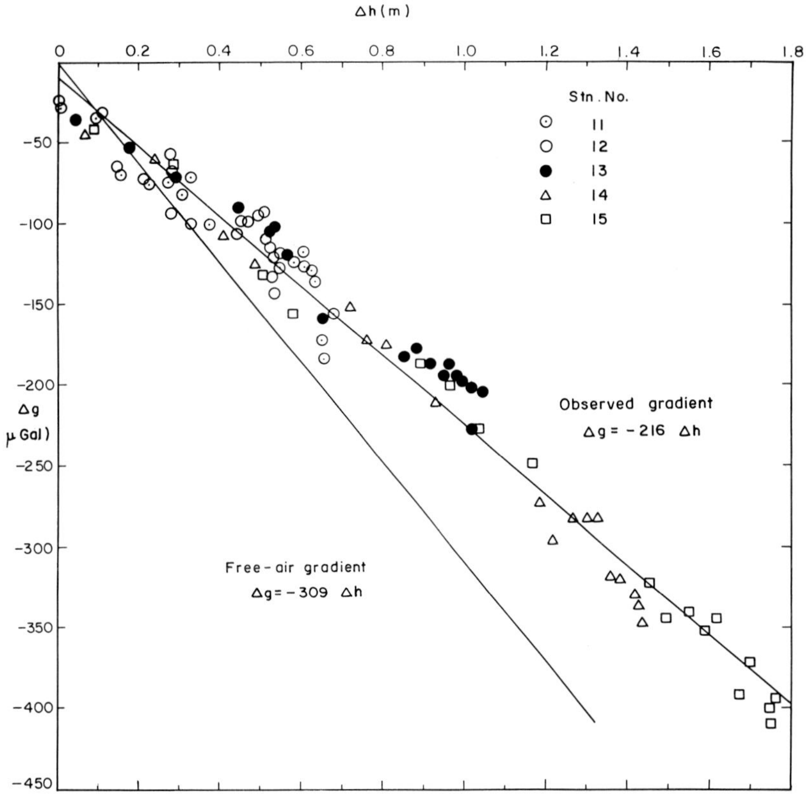

Fig. 15. Gravity vs elevation change for selected stations at intervals during the period 1973-1985. The free-air and observed relationships between gravity and elevation changes are shown

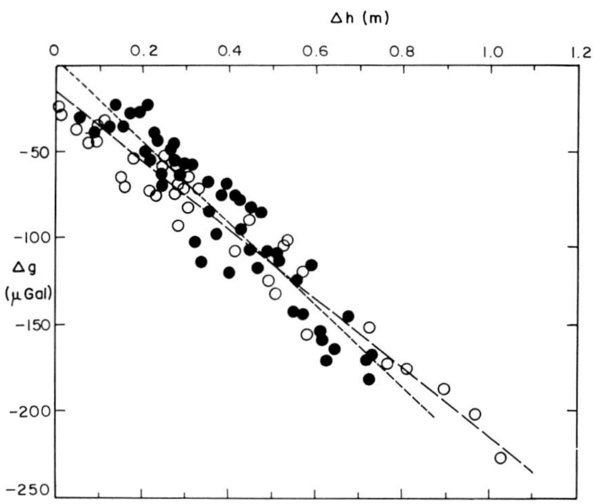

Fig. 16. Gravity vs elevation change for selected stations at intervals during the periods 1973-83 (open circles) and 1983-85 (solid circles). The observed relationships between gravity and elevation changes for the two time periods are shown

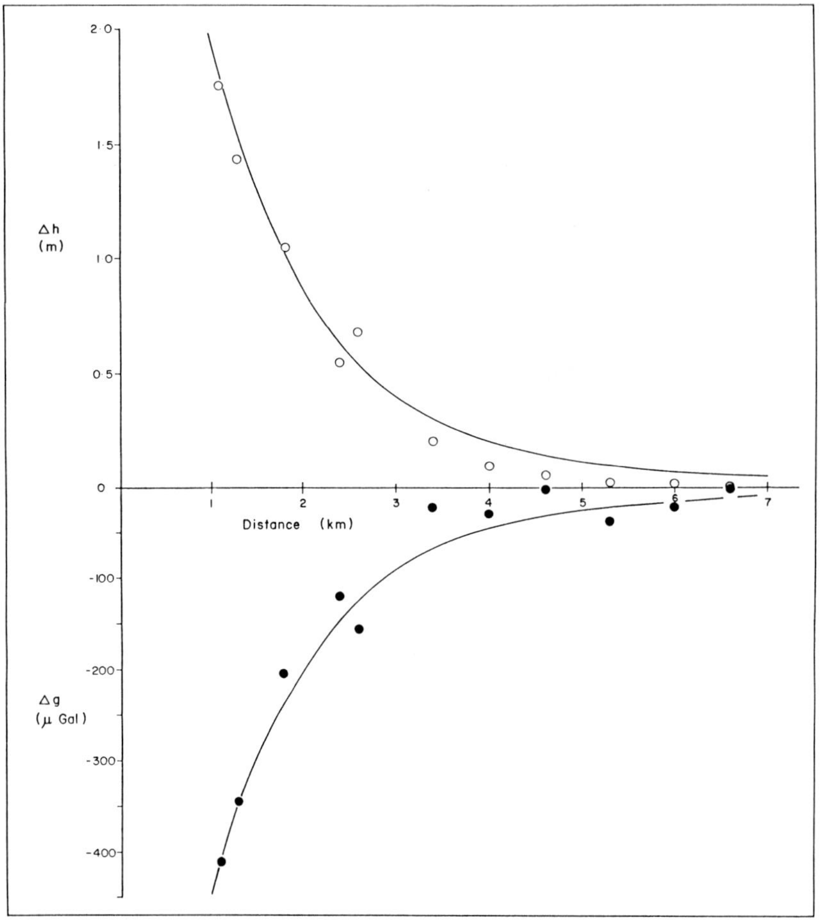

Fig. 17. Gravity and elevation changes vs radial distance from the assumed position of the source of these changes. The curved lines are the theoretical effects of a point source at a depth of 1.8 km

The depth of the source was estimated by comparing uplift data to point-source model curves in 0.1-km increments, and choosing the depth with the best fit using an RMS estimate of the error (Fig. 18). The depth estimates show a considerable degree of scatter for the period preceding 1981. This result is due to the variable quality of the early levelling data. The average of the depth estimates up to 1981 is 1.8 km which is the value consistently obtained for the Crisis Period. The scatter in the data gives a standard deviation of ±0.6 km for the depth estimate. Figure 17 shows the theoretical values of uplift and gravity change for a point source at a depth of 1.8 km located near the mouth of Greet Harbour.

Another plausible model, a vertical line source, was also considered. In this case, the fit to the measured uplift was not as good as for a point source, particularly at the farther

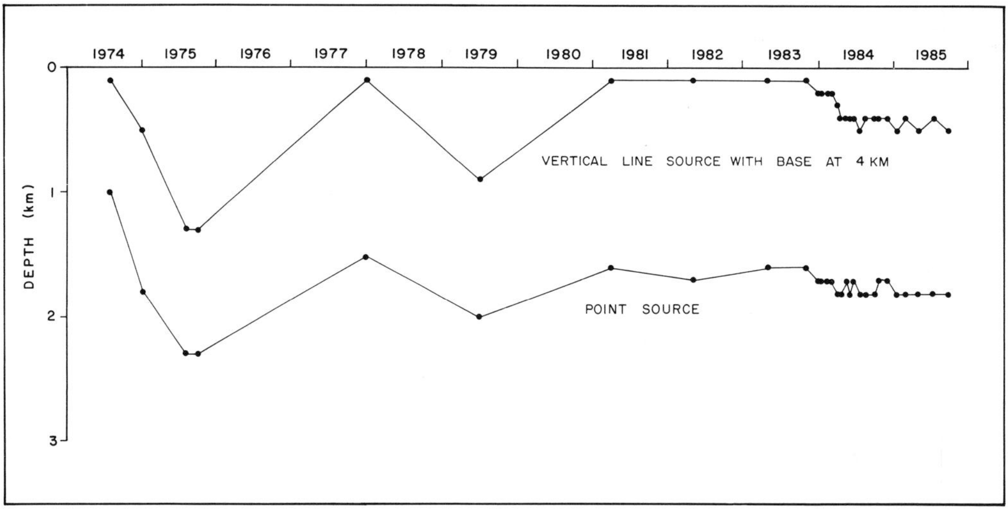

Fig. 18. Temporal variations in depth of a point source and the top of a vertical line source (whose base was held at a depth of 4 km) as determined from levelling data (see text for details)

distances. Depth estimates for the top of this source were also determined from the levelling data, using the same method as for a point source, while the base was held at a depth of 4 km (Fig. 18). These depth estimates up to 1981 are somewhat inconsistent but average out at about the value determined for the Crisis Period (0.5 km). The standard deviation of this value is ±1.0 km.

5.3 Excess Mass

The change in mass (excess mass) associated with the expanding source of the gravity changes and deformation can be estimated by application of Gauss Theorem (LaFehr 1965). The total change in mass (M) causing a gravity signal can be determined from the surface integral over the gravity field (g), regardless of the shape of the source, using:

$$M = 1/(2\pi\gamma)\int_S g \, ds, \tag{1}$$

where γ is the gravitational constant. Since we have shown that the gravity field is proportional to the uplift, and that the uplift can be modelled using a point source, we assume that the shape of the gravity field g(r) can be approximated by the expressions that describe the vertical deformation from a point source (Mogi 1958):

$$g(r) = g_o \, d \, (d^2 + r^2)^{3/2}, \tag{2}$$

where r is the radial distance, d is the source depth and g_o is the maximum gravity change directly above the source. Substi-

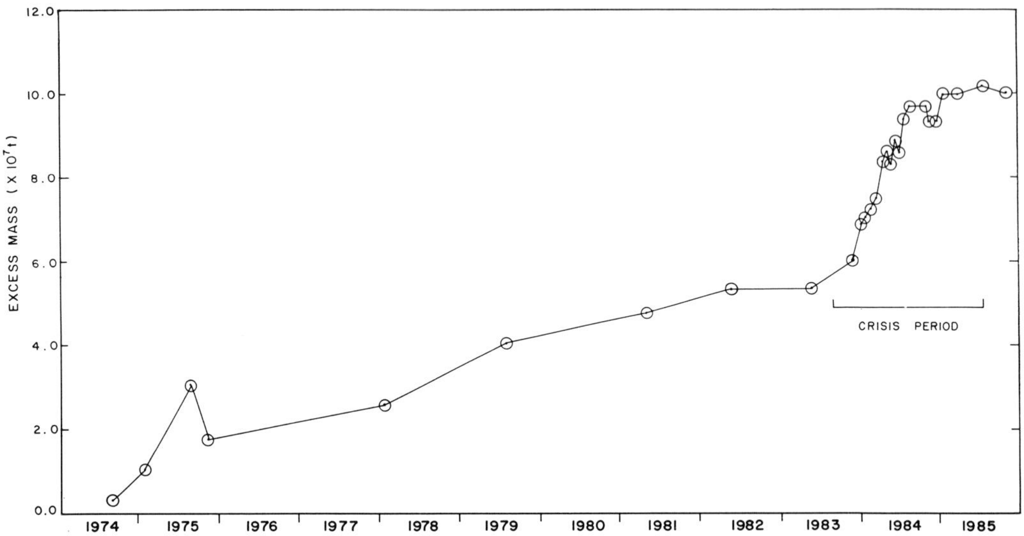

Fig. 19. Temporal variation of excess mass at the source of the deformation and gravity changes

tuting Eq. (2) into Eq. (1), one obtains a simple expression for the excess mass in terms of the observed maximum gravity change:

$$M = g_0/\gamma . \tag{3}$$

Assuming the depths of the Greet Harbour source are the values shown in Fig. 18, the maximum gravity change can be extrapolated from the data of each survey since 1973 using Eq. (2). From Eq. (3), this gravity change can be used to estimate the amount of excess mass accumulating at the Greet Harbour source (Fig. 19). The entire time period from 1973 to July 1985 yields a total of 1.0 x 10^8 t which is presumed to be the amount of magma which has accumulated during that 12-year period. Assuming a density of 2.5 t m^{-3}, which is appropriate for hydrous, intermediate composition magma at a depth of several kilometres (Kushiro 1978), the intruded magma for the period 1973-1985 would have a volume of 40 million m^3. This value is in reasonable agreement with previous estimates (50 million m^3 in McKee et al. 1984) derived directly from levelling data and using equations of Eaton (1962).

6 Discussion

6.1 Caldera Deformation and Structure

Analysis of the various geophysical data collected at Rabaul has yielded an interesting history of developments within the caldera since 1971. Over the long term, ground deformation data consistently showed that the area of greatest uplift was in the east-central part of the caldera, an area spatially separated from the zone of seismicity. When levelling and gravity data were compared a consistent negative correlation was found. The value of $\Delta g/\Delta h$ (-216 μgal m^{-1}) indicated that there was a mass increase associated with the ground deformation.

Modelling of ground uplift and gravity change in the central-northern part of the caldera has demonstrated that a Mogi point source produces a good fit to the observed changes. The fact that the observed changes at distances of several source depths were equal to or less than those due to a point source suggests the effect is due to a localized source and tends to obviate the need to try other, more complex, extended source models. Nevertheless, a vertical line source model was tried in order to gauge the effect of a geometrically extreme case. A reasonable fit was obtained using a finite line source with the base at a depth of 4 km and the top at a depth of 0.5 km. However, it is believed that an eruption would probably have taken place, especially during the Crisis Period, if the top of the magma body was actually at this shallow depth.

A significant result from the modelling was the consistency in time of the estimates of the depth of the source (about 1.8 km). This depth is slightly different to that obtained (less rigorously) in an earlier analysis of the ground uplift data (1.5 km in McKee et al. 1984). The consistency of the depth estimates suggests that the source is a well-established feature that can accommodate the input of small to moderate volumes of magma without causing an eruption. Past activity at this source could explain the previous reports of ground deformation in the Sulphur Creek-Matupit Island area between eruptions. Perhaps more importantly, pre-historical intrusion(s) in this area may be indicated by geological features exposed in a cliff at the southern end of Matupit Island, and detected offshore by marine seismic reflection surveys. Layered tephra deposits in the cliff have been arched upwards and the axial part of the resultant anticline has subsided along a set of stepped normal faults (Nairn et al., in prep.). Draping the faulted anticline is a thin layer of re-worked pumiceous sands and gravels, which contain coral and shell fragments. Up-bowing and north-south oriented faulting is also evident offshore, to the south of Matupit Island (Greene et al. 1986). This evidence of uplift and deformation is consistent with pre-historical caldera resurgence due to upward migration of magma within the caldera.

Consideration of the volume of magma intruded since 1971 and the observed surface deformation also indicates that the magma was added to a pre-existing system. If the new magma occupied a spherical volume, it would have a radius of about 220 m. The formation of a new magma body of this size would probably produce much greater surface deformation than that actually observed. Therefore, we conclude that the new magma was added to a pre-existing body whose expansion can be envisaged (for modelling purposes) as the addition of a relatively thin shell, which can add a sizeable volume of material while increasing the lateral dimensions only slightly.

Assuming that this magma source is a long-standing feature, the absence of a distinct vent directly above it would seem paradoxical. The known post-caldera vents are located near the margin of the caldera. Therefore, two possible magma feeding systems at shallow depth in the caldera can be envisaged. The post-caldera vents could be fed either from channel ways developed in the nearby caldera ring-fault zone, or from inclined feeding channels connected to the shallow centralized magma bodies. The first possibility may be ruled out because the post-caldera vents are in fact displaced from the caldera ring fault by 1 km or more (Mori and McKee 1987). Magma ascent from the shallow centralized magma bodies along inclined feeding channels could be feasible if cone-sheet fractures have developed in the roof rocks above the caldera magma storage system, in the manner predicted by Anderson (1936).

Another implication of stable depth of source estimates is that the overlying material has a strength greater than the stresses imposed so far by the increase in volume of the magma source. Although the existence of a gravity low over this part of the caldera suggests the presence of a body of relatively low density tephra, perhaps the basal section of this caldera fill is coherent, possibly due to welding, and acts as a barrier to ascending magma.

The scale of the recent sub-volcanic developments at Rabaul Caldera is small in comparison with the size of the caldera and the volume of magma erupted during the latest caldera collapse event. The caldera ring fault outlines an area of about 26 km^2 and indicates that the caldera magma reservoir is of similar lateral dimensions. The top of this large magma reservoir may be at a depth of about 4 km (Mori et al. Chap. 25, this Vol.). Walker et al. (1981) estimated that the volume of magma erupted during the 1400 years B.P. eruption was about 4 km^3 (DRE). This is about two orders of magnitude greater than the estimated volume increase of the shallow magma body beneath the northern part of Blanche Bay. Assuming applicability of the recharge rate of 1 km^3 per 1000 years of Smith (1979), about 30% of the volume of magma lost from the caldera magma reservoir during the 1400 years B.P. eruption may now have been replenished. The over-pressure within this system cannot be great since ground deformation outside the 1400 years B.P. caldera is minimal. Somewhat greater than expected horizontal distance changes were recorded at the southern margin of the caldera, but absolute uplift measured by a tide gauge at the northern extremity of Simpson Harbour (about 3 km distant from the active caldera ring fault) was extremely small (Mori et al. 1987). The horizontal expansion measured in the southern part of the caldera may contain a significant component due to the magma body which, as it has been suggested, exists about 3 km deep immediately offshore to the east of Vulcan (McKee et al. 1984), and/or a component due to the inferred deeper caldera magma reservoir.

6.2 Comparison with Other Volcanoes

The microgravity investigation at Rabaul is part of a growing application of the technique to volcano surveillance. Early studies were those at Mihara Volcano, Japan (Iida et al. 1952) and at Kilauea Volcano, Hawaii (Speith et al. 1975). The gravity changes recorded in these investigations were interpreted to be reflections of sub-volcanic magma movements. Another early study was at Izalco Volcano, El Salvador (Rose and Stoiber 1969), but no systematic gravity changes were observed associated with the eruption of about 9×10^5 m^3 of lava. Microgravity surveys at volcanoes became more common in the mid-1970s. At this time it became standard practice to carry out concurrent topographic surveys in order to separate the effects of elevation changes from those of changes in sub-surface distribution of density contrasts. A summary of rigorously documented studies of gravity changes and elevation changes (where available) over active volcanoes during the period 1975-1983 is given by Rymer and Brown (1986). Most investigations were carried out because the relevant volcano was undergoing eruption. Other recent volcano-gravity investigations have been carried out at non-erupting centers showing signs of unrest, for example, Long Valley, California (Jachens and Roberts 1985), Phlegraean Fields Caldera, Italy (Berrino et al. 1984), and Poas Volcano, Costa Rica (Rymer and Brown 1987).

Gravity studies have produced a wide range of $\Delta g/\Delta h$ relations, from about -1000 μgal m^{-1} at Pacaya Volcano, Guatemala, in 1979-80 (Eggers 1983), to positive $\Delta g/\Delta h$ values at Etna volcano, Sicily in 1979-81 (Sanderson et al. 1983). The extreme values of $\Delta g/\Delta h$ have been attributed to various combinations of magma density variations and/or sub-surface influx or drainage of magma accompanied by weaker than expected ground deformation. A few systems appear to have shown gravity and elevation changes consistent with simple Mogi-type inflation/deflation cycles in which $\Delta g/\Delta h$ values were in the range of about -200 to -230 μgal m^{-1}. These are: Krafla Caldera Iceland in the period 1975-79 (Torge 1981), Long Valley Caldera in 1981-84 (Jachens and Roberts 1985) and Phlegraean Fields Caldera in 1981-84 (Berrino et al. 1984). Rabaul Caldera can now be added to this list.

7 Conclusions

This study has confirmed that repeated high-precision gravity measurements can be a useful adjunct to the traditional geodetic methods employed in volcano surveillance. The close correlation of gravity and elevation changes at Rabaul suggest that gravity measurements alone could be used as indicators of structural deformation. While the maximum precision in measuring deformation by gravity is not as great as that obtainable from optical levelling, it is not dependent upon distance from the datum point as levelling is. A gravity program is simpler to operate, and potentially faster than levelling, being based upon a series of spot measurements by one person, rather than on a linear traverse by a party of several. The capacity for measuring direct from datum to a critical point also allows individual station checks to be made between complete re-surveys of a network

The advantage of maintaining concurrent programs of gravity and elevation surveillance on volcanoes is that comparison of the changes detected by the two methods places constraints on the nature of deformation sources. The results of these programs in Rabaul have shown that the deformation is due to sub-surface mass increase. Other possible causes of the deformation, such as increased volatile pressure in a magma body and thermal expansion of the caldera substrate, are ruled out because in these cases the gravity change should follow the free-air gradient since there is no additional mass. We conclude that the increase in sub-surface mass at Rabaul is magmatic and amounts to about 1.0×10^8 t. It appears that the magma was added to a pre-existing system from which material may be tapped for a future eruption.

Acknowledgements. The late Rob Cooke planned and established the Rabaul gravity monitoring project and carried out a share of the gravity observations until his untimely death in 1979. Richard Almond also contributed to this work in making some of the gravity observations between 1977 and 1980, and writing the computer program for data reduction. Brad Scott assisted in 1980 with some of the gravity observations. Jonathan Kuduon helped with some of the data processing. Dan Dzurisin provided useful comments on the data in 1984 and presented preliminary results on our behalf at the AGU Fall Meeting in 1985. Geoff Brown, Trevor Hunt, John Latter, Peter Lowenstein, Hazel Rymer, Patrice de Saint Ours and Peter Wellman provided detailed reviews of the manuscript. The authors publish with the permission of the Secretary of the Papua New Guinea Department of Minerals and Energy.

References

Anderson EM (1936) The dynamics of the formation of cone sheets, ring-dikes and cauldron subsidences. R Soc Edinburgh Proc 56:128-163

Berrino G, Corrado G, Luongo G, Toro B (1984) Ground deformations and gravity changes accompanying the 1982 Pozzuoli uplift. Bull Volcanol 47:187-200

Brooks JA (1971) Investigations of crustal structure in the New Britain/New Ireland Region, 1969. Part 1: Geophysical and geological data. Aust Bur Min Res, Record 71/131 (unpublished)

Brown G (1878) Journal of the Rev. G. Brown, 1860-1902, (11 volumes) (unpublished). Mitchell Library, Sydney Aust

Cifali G, d'Addario GW, Polak EJ, Wiebenga WA (1969) Rabaul preliminary crustal seismic test, New Britain 1966. Aust Bur Min Res, Record 69/125 (unpublished)

Cooke RJS (1977) Rabaul Volcanological Observatory and geophysical surveillance of the Rabaul Volcano. Aust Phys 1977:27-30

Crick IH (1975) Investigation into a change in topography at Rabaul Golf Course. Geol Surv Papua New Guinea Report 75/6 (unpublished)

Druitt TH, Sparks RSJ (1984) On the formation of calderas during ignimbrite eruptions. Nature (Lond) 310:679-681

Eaton JP (1962) Crustal structure and volcanism in Hawaii. Am Geophys Union Monogr 6:13-29

Eggers AA (1983) Temporal gravity and elevation changes at Pacaya Volcano, Guatemala. J Volcanol Geotherm Res 19:223-237

Fisher NH (1939) Geology and vulcanology of Blanche Bay, and the surrounding area, New Britain. Territory of New Guinea Geol Bull 1:1-68

Fisher NH (1976) 1941-42 Eruption of Tavurvur Volcano, Rabaul, Papua New Guinea. In: Johnson RW (ed) Volcanism in Australasia. Elsevier, Amsterdam, pp 201-210

Greene HG, Tiffin DL, McKee CO (1986) Structural deformation and sedimentation in an active caldera, Rabaul, Papua New Guinea. J Volcanol Geotherm Res 30:327-356

Groten E (1983) Report on high precision gravimetry Volume II. International Association of Goedesy Special Study Group 3.37. Verlag des Institutes für Angewandte Geodasie. Frankfurt A.M. 103 p

Heming RF (1974) Geology and petrology of Rabaul Caldera, Papua New Guinea. Geol Soc Am Bull 85:1253-1264

Iida K, Hayakawa M, Katayose K (1952) Gravity survey of Mihara Volcano, Oosima Island, and changes in gravity caused by eruption. Geol Surv Jpn Rep 152:1-28

Jachens RC, Roberts CW (1985) Temporal and areal gravity investigations at Long Valley Caldera, California. J Geophys Res 90:11210-11218

Kushiro I (1978) Density and viscosity of hydrous calc-alkalic andesite magma at high pressures. Carnegie Inst, Washington, Year Book, 77:675-677

LaFehr TR (1965) The estimation of the total amount of anomalous mass by Gauss's theorem. J Geophys Res 70:1911-1919

Lipman PW (1984) The roots of ash flow calderas in western North America: windows into the tops of granitic batholiths. J Geophys Res 89:8801-8841

Longman IM (1959) Formula for computing the tidal acceleration due to the moon and the sun. J Geophys Res 64:2351-2355

McKee CO, Lowenstein PL, de Saint Ours P, Talai B, Itikarai I, Mori JJ (1984) Seismic and ground deformation crises at Rabaul Caldera: prelude to an eruption? Bull Volcanol 47:397-411

Mogi K (1958) Relations between the eruptions of various volcanoes and the deformations of the ground surface around them. Bull Earthq Res Inst 36:94-134

Mori J, McKee C (1987) Outward dipping ring-fault structure at Rabaul Caldera as shown from microearthquake locations. Science 235:193-195

Mori J, McKee CO, de Saint Ours P, Itikarai I (1987) Sea level measurements for inferring ground deformations in Rabaul Caldera. Geo Mar Lett 6:241-246

Nairn IA, Talai B, Wood CP, McKee CO (1988) Geology of Rabaul Caldera (in preparation)

Rose Jr. WI, Stoiber RE (1969) The 1966 eruption of Izalco volcano, El Salvador. J Geophys Res 74:3119-3130

Rymer H, Brown GC (1986) Gravity fields and the interpretation of volcanic structures: geological discrimination and temporal evolution. J Volcanol Geotherm Res 27:229-254
Rymer H, Brown GC (1987) Causes of microgravity change at Poas Volcano, Costa Rica: an active but non-erupting system. Bull Volcanol 49:389-398
Sanderson TJO, Berrino G, Corrado G, Grimaldi M (1983) Ground deformation and gravity changes accompanying the March 1981 eruption of Mt. Etna. J Volcanol Geotherm Res 16:299-315
Sapper K (1910) Beiträge zur Kenntnis Neupommerns und des Kaiser-Wilhelmsland. Petermanns Mitt 56:189-193, 255-256 (in German)
Schleinitz GEG (1889) Die Forschungsreise S.M.S. "Gazelle" in den Jahren 1874 bis 1876 unter Kommando des Kapitän zur See Freiherrn von Schleinitz. Herausgegeben von dem Hydrographischen Amt des Reichs-Marine-Amts. I. Teil des Reiseberichts. Mittler, Berlin (in German)
Scott BJ (1982) Tiltmeter recordings at Rabaul Caldera Papua New Guinea: 1963-1979. Geol Surv Papua New Guinea Report 80/13:1-12 (unpublished)
Simpson CH (1873) Hydrographical extract from a six months cruise among the South Sea Islands. Report to the Admiralty, 11 December 1872, Great Britain, Parliamentary Accounts and Papers, Session 6 February - 5 August 1873, Colonies and British Possessions, pp 1-50
Smith RL (1979) Ash-flow magmatism. In: Chapin CE, Elston WE (eds) Ash-flow Tuffs. Geol Soc Am Spec Paper 180:5-27
Speith MA, Stuart WD, Kinoshita WT (1975) Time-dependent gravity and elevation changes on Kilauea Volcano. EOS Trans Am Geophys Union 56:1071
Torge W (1981) Gravity and height variations connected with the current rifting episode in northern Iceland. Tectonophysics 71:227-240
Walker GPL (1980) The Taupo pumice: product of the most powerful known (Ultraplinian) eruption? J Volcanol Geotherm Res 8:69-94
Walker GPL, Heming RF, Sprod TJ, Walker HR (1981) Latest major eruptions of Rabaul Volcano. In: Johnson RW (ed) Cooke-Ravian Volume of Volcanological Papers, Geol Surv Papua New Guinea Mem 10:181-193
Whitcomb JR, Franzen WO, Given JW, Pechmann JC, Ruff LJ (1980) Time-dependent gravity in southern California, May 1974 to April 1979. J Geophys Res 85:4363-4373

Earthquakes of the Rabaul Seismo-Deformational Crisis September 1983 to July 1985: Seismicity on a Caldera Ring Fault

J. MORI, C. MCKEE, I. ITIKARAI, P. LOWENSTEIN, P. de SAINT OURS and B. TALAI[1]

Abstract

During the Rabaul Seismo-Deformational Crisis from September 1983 through July 1985, the seismicity in the caldera intensified to levels that were much higher than at any other time since the seismic network began recording in the late 1960's. Earthquakes increased steadily for the first 8 months, peaking in April 1984, and then gradually declining to pre-Crisis levels by the end of July 1985. Over 100,000 events were recorded, although most were of small magnitude. All of the earthquakes were high-frequency events resembling tectonic earthquakes: the largest was an M_L 5.1 event. For the Crisis period, the total seismic energy release was 2.5 x 10^{19} ergs and the total moment release was 1.1 x 10^{24} dyne-cm. The seismicity concentrated on an elliptical ring fault system 10 x 5 km at depths of 0 to 4 km. Cross-sections through the seismicity pattern show a trend that slopes away from the center of the caldera, thus the earthquakes outline a central caldera block with outward-dipping sides. This central block is thought to have subsided during the last major caldera eruption 1400 years B.P. Focal mechanisms of some of the larger events indicate that the central block is now rising, and this is also confirmed by ground deformation measurements. The active magma which is causing stresses to reactivate the ring fault is thought to be located in shallow intrusions within the central block and also in a large caldera reservoir underlying this block. There is no indication of magma intrusion into the ring fault, and the zone of active seismicity is therefore separate from the present region of magma accumulation. The pattern of ring fault seismicity, itself, is probably not a short term precursor to an eruption. Instead, seismicity which reflects actual magma movements from the shallow magma bodies towards the surface would be a stronger indication of an imminent eruption.

1 Introduction

Rabaul caldera, located on the northeastern end of New Britain Island in Papua New Guinea, is a volcanic complex associated with the subduction of the Solomon Sea plate beneath the South Bismarck Sea plate. The present caldera is the dominant feature

---

[1] Rabaul Volcanological Observatory, Rabaul, Papua New Guinea

IAVCEI Proceedings in Volcanology 1
J.H. Latter (Ed.), Volcanic Hazards

of a Quaternary shield volcano that was partly destroyed in a series of large ignimbrite eruptions of which the latest has been dated at 1400 years B.P. (Fisher 1939; Heming 1974). The volcanic cones (Mother, North Daughter, South Daughter and Watom Island) on or outside the caldera rim are probably older satellite volcanoes to the main volcanic structure. Since the last major eruption 1400 years B.P. the volcanic vents inside the caldera (such as Sulphur Creek, Rabalanakaia, Tavurvur, Beehives, Vulcan, Vulcan Island and several submarine vents) were formed and modified in small to moderate eruptions. These events include the historical eruptions of 1767 (exact location unknown), Tavurvur 1791, Sulphur Creek ~1850, Tavurvur and Vulcan Island (formed Vulcan Island) 1878, Tavurvur and Vulcan (formed Vulcan cone) 1937 and Tavurvur 1941-43. The locations of the volcanic vents are shown in Fig. 1. Eruptive products from the caldera include both mafic rocks (basaltic lava flows and scorias) and felsic material (dacitic and rhyolitic ignimbrites). Compositions of recent eruptive products from the caldera suggest that the vents on the western side of the caldera may erupt more acidic magma (3-5% higher silica contents) than vents on the eastern side (Heming, 1974).

The first sign of volcanic unrest since the 1941-43 eruptions of Tavurvur was the increase of seismicity within the caldera in 1971. Seismicity continued to increase gradually through the 1970's and early 1980's. Measurements of ground uplift along a 15-km level line were initiated in 1973 and from that time to 1982 showed uplifts of 7 to 13 cm yr^{-1}.

In September 1983 there was a dramatic increase in earthquake activity. The high level of seismicity and accelerated rates of ground deformation continued to intensify throughout 1983 and into 1984, suggesting the possibility of an imminent eruption. In April 1984 the seismicity and ground deformation reached their highest rates (over 1000 earthquakes per day and 8-cm uplift for the month) and plans for the evacuation of Rabaul township were waiting to be put into action. However, there were no signs of an eruption during this peak of activity and subsequently the seismicity and ground deformation rates began to gradually decline. The earthquake activity returned to the pre-September 1983 level by the end of July 1985.

The period from September 1983 through July 1985 is termed the Rabaul Seismo-Deformational Crisis of 1983-85 and this chapter presents the seismological data collected during that time. The earthquake data are used to determine details of the subsurface structure of the caldera and have enabled us to present interpretations of the events that occurred during this period of volcanic unrest.

2 Seismicity from 1967 to 1983

Early studies of the seismicity within the caldera from 1940 were reported by Latter (1966) and seismic activity associated with the 1937 and 1941-42 eruptions has been described by Fisher (1939; 1976) and Kizawa (1951); however, the constant monitoring

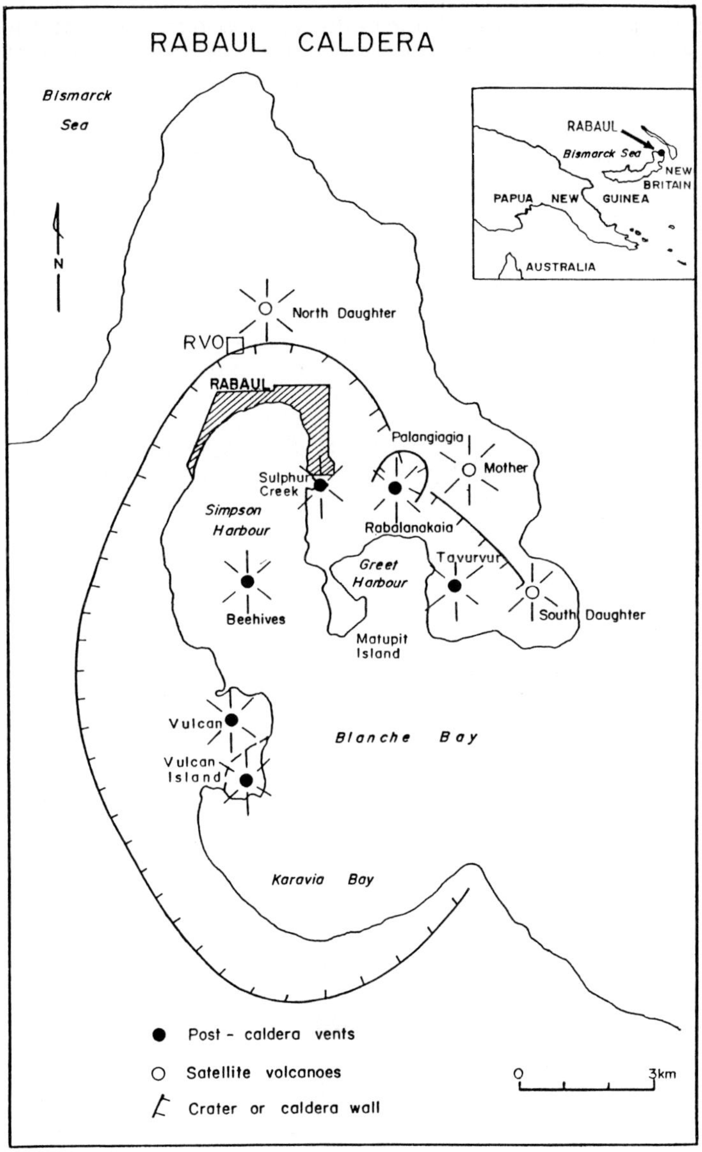

Fig. 1. Rabaul caldera with locations of the various volcanoes. Not shown are Watom and Varzin which are located 22 km NW and 15 km SW respectively, from the center of the caldera. RVO indicates the location of Rabaul Volcanological Observatory

of microearthquake activity using a permanent short-period seismic network did not begin until 1967. From 1967 to late 1971 the activity was fairly constant with 20 to 100 small (generally, M_L <0.5) events per month. Starting in November 1971 the activity increased (Cooke 1977) with the appearance of earthquake swarms that have since occurred irregularly at intervals of a few months. These swarms typically lasted from 10 min to several hours and included hundreds of events, although the records usually became saturated so that accurate event counts were difficult. All the earthquakes recorded were high-frequency events resembling tectonic earthquakes. Through the 1970's the swarms tended to have increasing numbers of events (McKee et al. 1984) as evidenced by the gradual rise of the monthly totals shown in Fig. 2. The largest earthquakes prior to the Crisis period were an M_L 5.2 event in October 1980 and an M_L 5.1 event in March 1982.

The locations of over 1000 events from 1971 to 1983 are plotted in Fig. 3. These locations were determined using the FASTHYPO program by Herrmann (1975) and the velocity structure (after Almond and McKee 1982) given in Table 1. This velocity structure and station corrections were determined from an experiment using known explosion sources within the caldera. Using other velocity structures (Cifali et al. 1969; Finlayson and Cull 1973), the locations may change by up to 2 km (J.H. Latter, pers. commun.). The epicenters shown in Fig. 3 are for events that were recorded on six or more stations and which had horizontal uncertainties of less than 2 km. The events plotted in Fig. 3 represent about 3% of the total number of earthquakes counted during this time period.

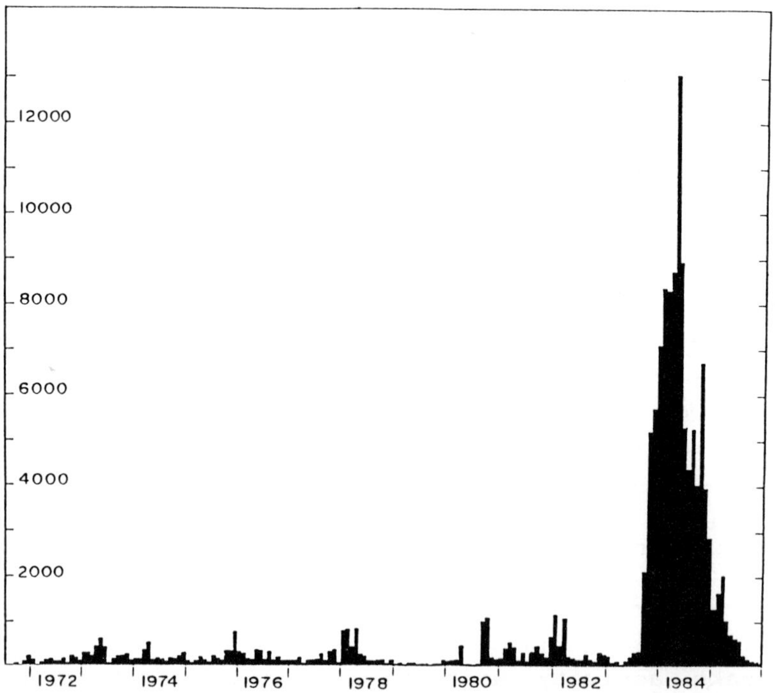

Fig. 2. Monthly number of caldera earthquakes recorded on two or more stations of the Rabaul Harbour network. The smallest events are about M_L 0.5

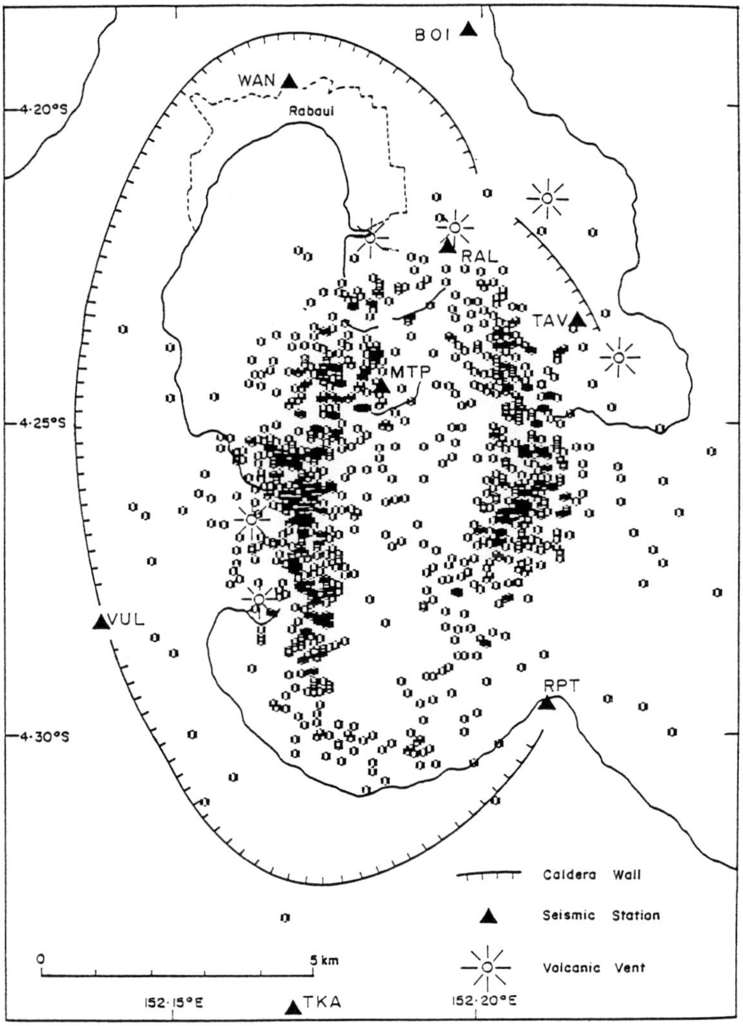

Fig. 3. Epicenters of over 1000 earthquakes from the pre-Crisis period, 1971 through August 1983. All events plotted were recorded on six or more stations and have horizontal errors of less than 2 km. The **triangles** show the eight-station configuration of the Rabaul Harbour network during that time

The striking annular pattern of seismicity seen in Fig. 3 was first noted by Cooke (1977), although the shape and extent of his pattern was slightly different to the present one. McKee et al. (1984) also showed a ring pattern similar to Fig. 3 based on the data from 1977 to 1982. Figure 4 shows an east-west cross-section through the center of the ring of seismicity from 4.24°S to 4.28°S. The events plotted in the cross section had depth uncertainties of less than 2 km. This figure shows that the earthquakes were generally located at depths less than 4 km. The details of the seismicity pattern are discussed below in conjunction with data from the Crisis period.

Table 1. Rabaul velocity structure (after Almond and McKee 1982)

Layer	Depth (km)	P wave velocity (km s^{-1})
1	0.0	1.70
2	0.25	2.06
3	0.50	2.43
4	0.75	2.79
5	1.0C	3.15
6	1.25	3.51
7	1.50	3.88
8	1.75	4.24
9	2.00	4.60
10	4.00	6.11

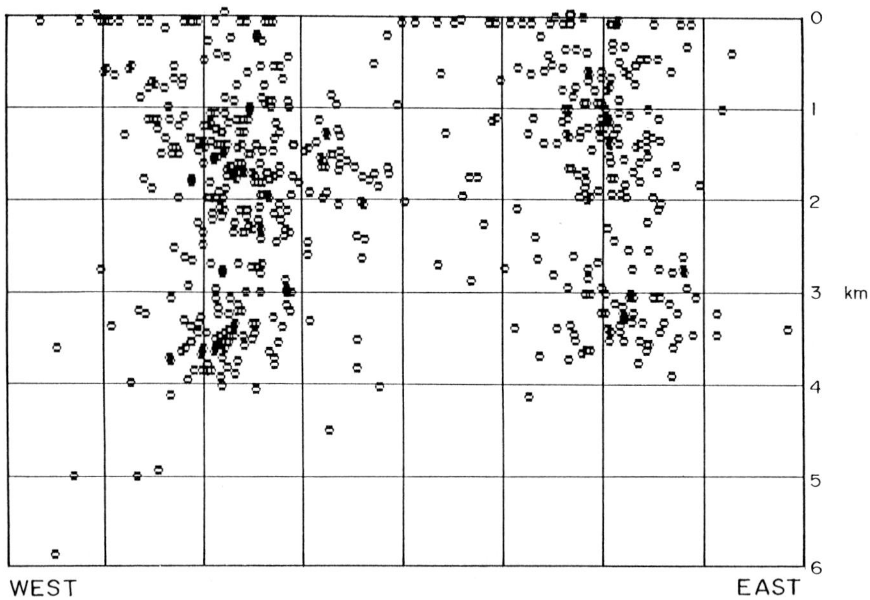

Fig. 4. East-west cross-section of the pre-Crisis seismicity from 4.24°S to 4.28°S. All events plotted have vertical and horizontal errors less than 2 km. The grid lines are spaced at 1 km with no vertical exaggeration

3 Events of the Crisis Period

To facilitate the description of the earthquake sequence, the region of active seismicity has been divided into five areas and named as shown in Fig. 5. The locations of the two sources of uplift that were identified from the ground deformation data (McKee et al. 1984) are also shown. Figure 6 shows the daily

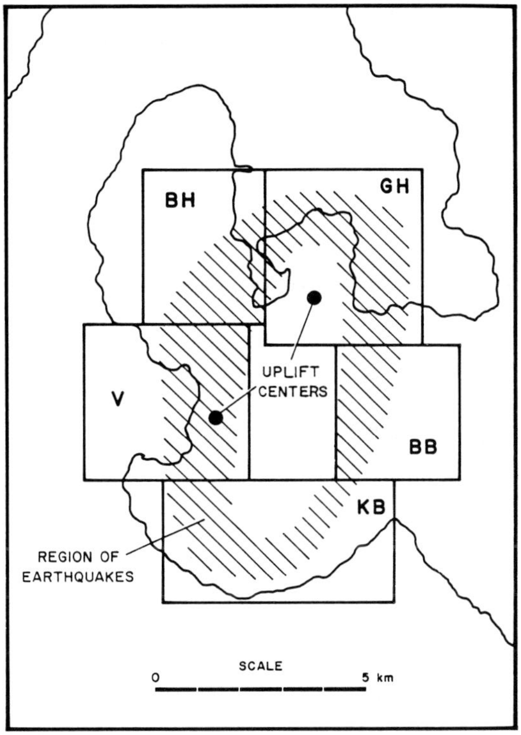

Fig. 5. The five regions into which the area of active seismicity has been divided. **BH** Beehives; **GH** Greet Harbour; **V** Vulcan; **BB** Blanche Bay; and **KB** Karavia Bay. The **solid circles** show the two centers of uplift. The northern one is the Greet Harbour source and the western one is the Vulcan source

number of earthquakes in each region. Because of the variation in the distribution and sensitivity of the stations, there is a bias towards the Greet Harbour area in the event counts. For the Greet Harbour area the smallest magnitude event counted is about 0.5. The threshold for the Beehives and Vulcan areas is about 1.0 and for the Blanche Bay and Karavia Bay areas about 1.3. The seismicity shifted to various portions of the ring pattern during the Crisis period, but no systematic patterns are apparent. Sometimes only one region would be seismically active. For example, in April 1984, when the most intense seismicity was occurring in the Greet Harbour area, all of the other areas were relatively quiet (Fig. 6). At other times more than one region was active simultaneously, as in October 1984, when all five regions had increased levels of seismicity (Fig. 6).

Much of the earthquake activity occurred in swarms when tens to hundreds of events were recorded within periods of a few minutes to an hour. The most intense swarms, when 150 or more events were recorded on four or more stations within 24 h, have been designated "seismic crises". Usually near the beginning of a seismic crisis the earthquakes tended to be concentrated in a

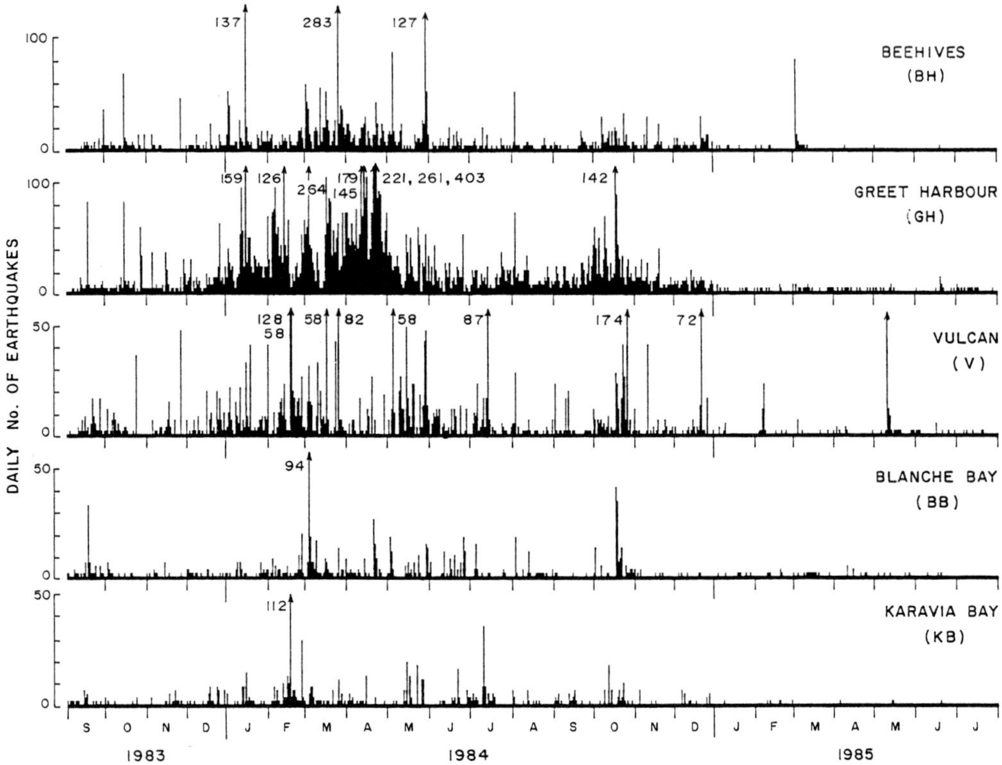

Fig. 6. Daily counts of earthquakes for each of the five regions within the caldera during the Crisis period

fairly local area, and included one or more events of magnitude greater than 3.0. During the following 24 h the seismic activity usually remained at a relatively high level, during which earthquakes would occur in many different areas. Table 2 lists information on the 25 seismic crises that occurred during the Crisis period. Almost all the earthquakes recorded were high-frequency tectonic-like events. The lower frequency content seen in a few events can probably be attributed to propagation effects from shallow sources.

Summarized below are the significant events of the Crisis period. More detailed information is given in the monthly summaries and the monthly seismicity plots in Mori et al. (1986).

The increase of seismicity, in hindsight, appeared to start in the last week of August 1983 when earthquake counts rose from about 5 events/day to 10-20 events/day. The seismicity continued to increase throughout September, reaching 50-80 events/day. The first seismic crisis occurred on the 19th and included an M_L 4.2 event. From October 1983 through January 1984 the seismicity continued to increase with daily counts fluctuating between 50 and 400 events/day, excluding the 4 days of seismic crises when 400-900 events were recorded. During these first

Table 2. Seismic crises. The event counts are for a 24-h period following the start of the crises

Date	Location	No. events (4 or more stations)	Largest event (M_L)	Seismic energy release (ergs)
19 Sep.83	Blanche Bay	210	4.2	3.8×10^{17}
15 Oct.83	Greet Harbour/ Beehives	334	3.6	0.4
28 Oct.83	Greet Harbour	195	4.0	1.8
26 Nov.83	Beehives/Vulcan	212	2.9	0.07
15 Jan.84	Greet Harbour/ Beehives	342	4.6	16.7
13 Feb.84	Greet Harbour	164	2.6	---
18 Feb.84	Karavia Bay/Vulcan	212	4.3	5.1
27 Feb.84	Blanche Bay	159	3.0	0.03
3 Mar.84	Blanche Bay	484	5.1	125.5
17 Mar.84	Vulcan	219	2.1	---
25 Mar.84	Beehives/Vulcan	398	3.5	0.4
11 Apr.84	Greet Harbour	217	2.0	---
13 Apr.84	Greet Harbour	212	2.2	---
20 Apr.84	Greet Harbour	302	3.0	0.03
21 Apr.84	Greet Harbour	287[a]	3.7	1.0
22 Apr.84	Greet Harbour	485	4.8	36.3
4 May 84	Beehives/Vulcan	168	3.5	0.2
29 May 84	Beehives/Vulcan	323	3.6	0.7
13 Jul.84	Vulcan	151	3.5	0.2
2 Aug.84	Beehives	242	3.4	0.2
17 Oct.84	Greet Harbour	233[a]	4.9	54.2
18 Oct.84	Blanche Bay	158	3.8	0.7
26 Oct.84	Vulcan	209	3.2	0.1
3 Mar.85	Beehives	333	3.6	0.7
10 May 85	Vulcan	**[b]	3.7	0.6

[a] Event count for less than 24 h because another seismic crisis occurred.
[b] Event count not possible because of regional earthquakes.

4 months, tilts of 10-40 microradians/month were recorded in the Greet Harbour area and were associated with the uplift source, which had been active since 1973 or earlier.

In February and March 1984 the total numbers of earthquakes remained about the same as in January, although there was a shift in the concentrations of the seismicity. Compared to the previous months, there were relatively more events in the Blanche Bay, Vulcan and Karavia Bay areas, and fewer events in the Greet Harbour and Beehives areas. The seismic crisis in Blanche Bay on March 3rd, one of the strongest of the entire Crisis period, was initiated by an M_L 5.1 event, which was the largest event of the Crisis period. There were tilts of 10-60 microradians/month in the Greet Harbour area and several re-occupations of the Matupit Island level line indicated that the southern tip of the island was rising at about 3.5 cm/month. In the Vulcan area there were tilts of 10-35 microradians. These significant deformations for the first time identified an uplift source in the Vulcan area.

In April the activity again centered in the Greet Harbour area and reached its most intense level. There were five seismic crises during the month, all in the Greet Harbour area. Daily earthquake counts were 250-400 events/day and reached 1700 events/day during the seismic crisis of the 22nd, which included an M_L 4.8 event. Ground deformation rates were also very high for the month in the Greet Harbour region, with up to 8-cm uplift on Matupit Island.

Following the strong activity in April, the seismicity and the ground deformation rates declined gradually during the period from May to September 1984. During these 5 months, the earthquake counts varied between 50 and 300 events/day, and there were four seismic crises. The ground deformation rates also slowed, with 10-30 microradians of tilt per month in the Greet Harbour area, and less in the Vulcan area. The measured uplift of Matupit Island was 6 cm/month in May, and declined to 1.5 cm/month by September.

There was a slight resurgence of activity in October which had three seismic crises, including the second largest earthquake of the Crisis Period, an M_L 4.9 event in the southern Greet Harbour area. Ground deformation rates also increased again, with significant tilts of 30-127 microradians in the Greet Harbour area, and 6-26 microradians in the Vulcan area. There was an uplift of 3.9 cm for the month at the southern tip of Matupit Island.

After October the seismicity and ground deformation smoothly declined with only two seismic crises, in March and May 1985. The May seismic crisis was interesting because it was apparently triggered by a regional earthquake 180 km away which was strongly felt in Rabaul (MM V). The seismicity returned to pre-Crisis levels by the end of July 1985. In July, another very strong regional earthquake, 60 km away (MM VI-VII in Rabaul) had apparently no effect on the caldera seismicity.

4 Earthquake Locations

Earthquake locations for the Crisis period were determined using the same procedure described in the section on the pre-Crisis seismicity. The addition of two stations on the western rim of the caldera has, however, improved the location accuracy, particularly in determining depths for earthquakes on the western side. Figure 7 shows the epicentral locations for over 2500 events that were recorded on seven or more stations and had horizontal uncertainties of 1 km or less. Since these locations represent only larger events, the earthquake distribution

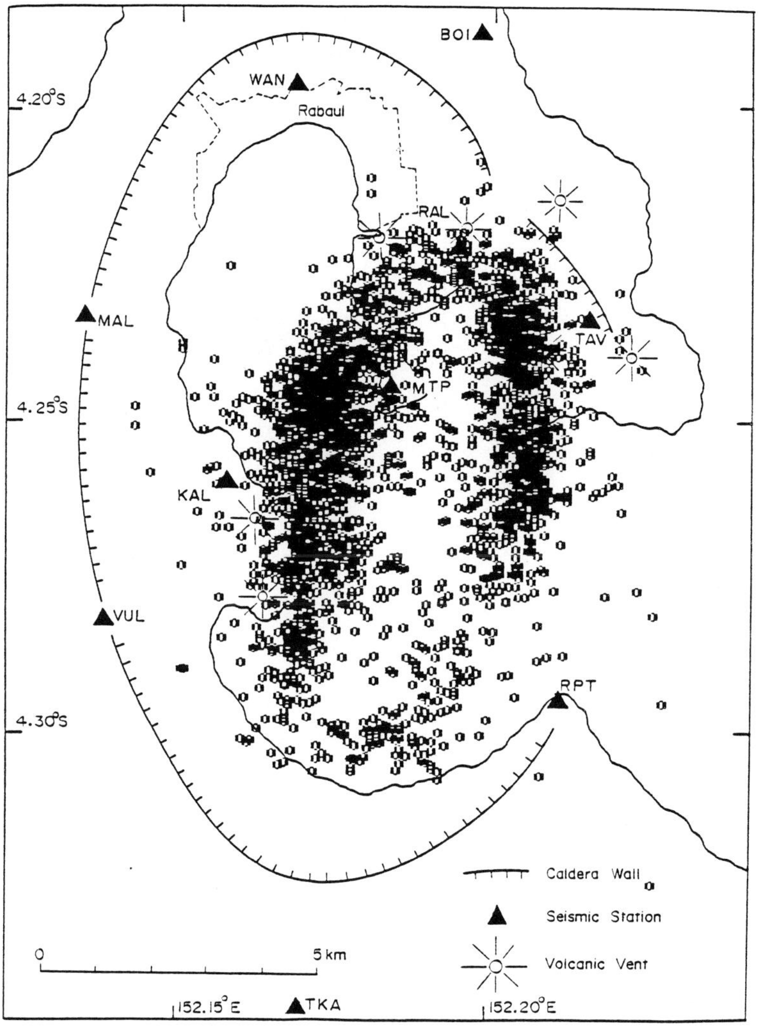

Fig. 7. Epicenters of over 2500 earthquakes for the Crisis period, September 1983 through July 1985. All events plotted were recorded on seven or more stations and had horizontal errors of less than 1 km

should not be affected by differences in magnitude threshold levels in various parts of the caldera. The striking ring pattern that was seen in the pre-Crisis seismicity (Fig. 3) is still present, although the areas of concentrated seismicity shifted. Figure 8 shows the seismicity of the two time periods contoured by the number of events per 0.25 km^2. The pre-Crisis data showed concentrations in the Vulcan and Blanche Bay areas, while the seismicity of the Crisis period shifted to the north so that the strongest concentrations were in the Beehives and Greet Harbour areas. The northward shift of seismicity may be associated with the large uplift centered in Greet Harbour, which became particularly active in early 1984.

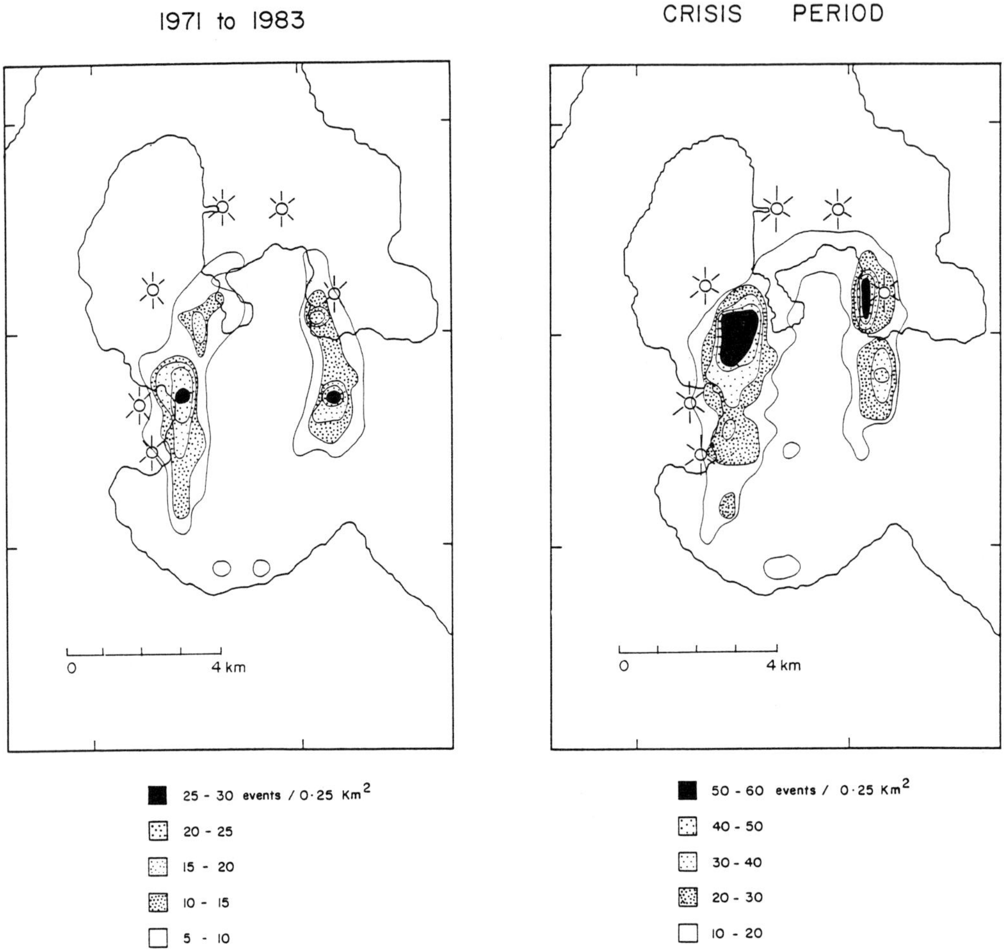

Fig. 8. Contours of the number of events located per 0.25 km^2 ($\frac{1}{2}$ x $\frac{1}{2}$ km squares) in the caldera. Left-hand map is for the pre-Crisis period from 1971 through August 1983. Right-hand map is for the Crisis period from September 1983 through July 1985

Figure 9 shows three east-west cross-sections through the middle of the ring pattern and three radial sections through the northern end. Events which have vertical uncertainties of 1.5 km or less have been plotted in these sections. Although the elliptical pattern shown in Fig. 7 trends slightly east of north, the strong concentrations of earthquakes on the eastern and western sides appear to trend almost due north-south: accordingly, east-west cross-sections were plotted through this central section. Note that in most of the cross-sections the trend in the locations dips steeply outward from the middle of the caldera, with the only exception being section B. Figure 10 is a composite of the three east-west cross-sections and clearly shows this outward-sloping shape on both sides of the ring structure. The cross-section from the pre-Crisis seismicity

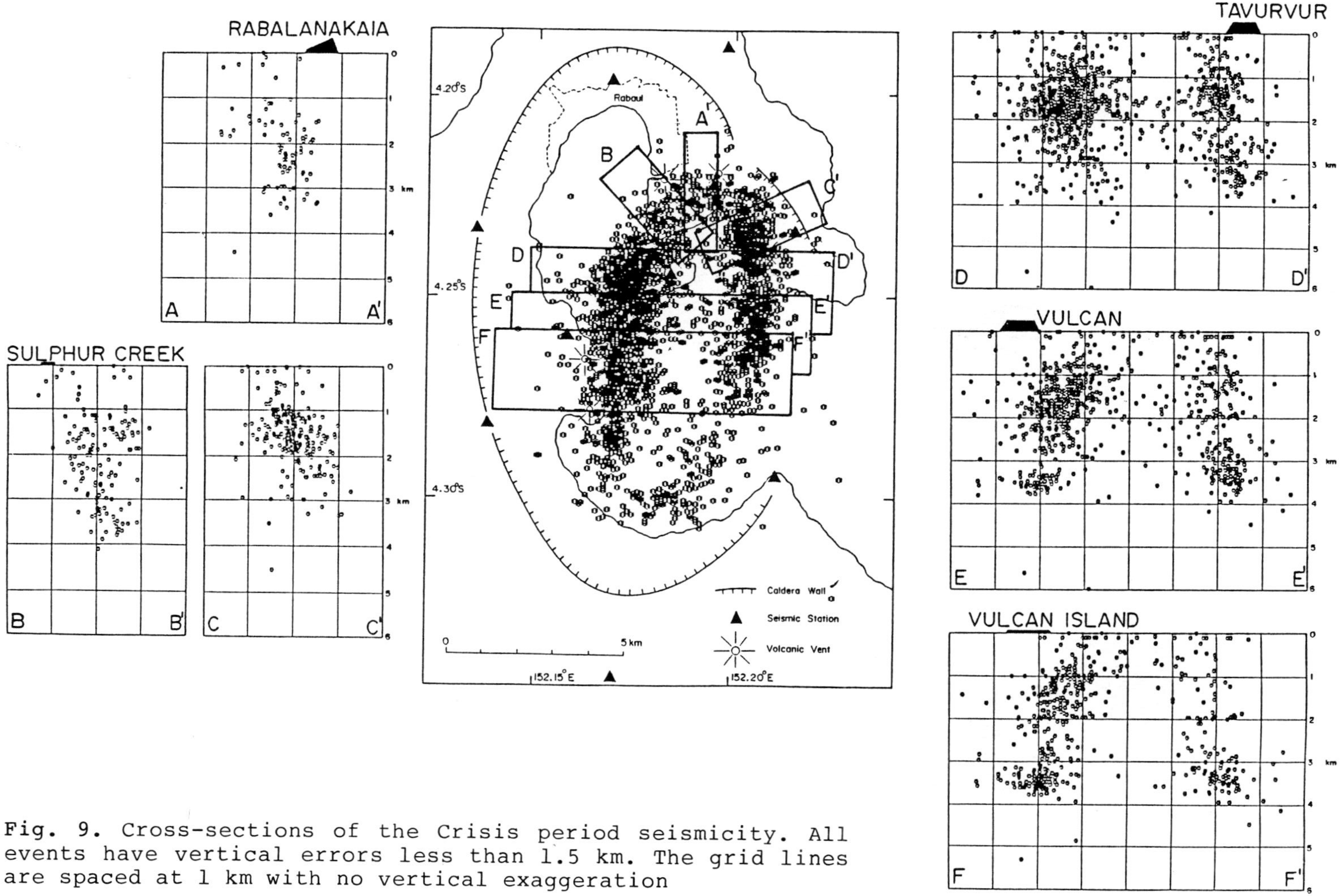

Fig. 9. Cross-sections of the Crisis period seismicity. All events have vertical errors less than 1.5 km. The grid lines are spaced at 1 km with no vertical exaggeration

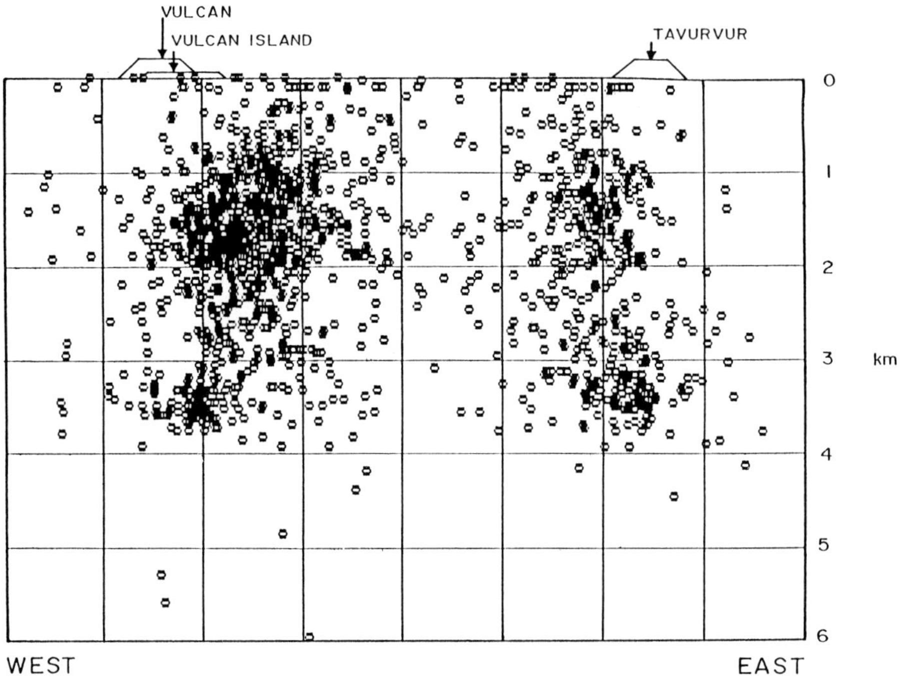

Fig. 10. Composite of the three east-west cross-section of Fig. 9. This includes all the well-located events between 4.24°S and 4.28°S

(Fig. 4) also shows a similar pattern, although not as clearly as the improved data from the Crisis period. The similarity of Figs. 4 and 10 indicates that there was no significant change in the depth distribution of the earthquakes between the two time periods.

It is possible that the outward-dipping structure may be due to systematic location errors, if there is a low-velocity region within the central part of the caldera. This is not unlikely since ground deformation data indicate a shallow magma body located near the mouth of Greet Harbour (McKee et al. 1984). A more detailed study of the arrival time residuals is necessary to resolve this problem.

There were also no systematic changes in the depths of the earthquakes during the Crisis period, as can be seen in Fig. 11, which shows the average depth of the seismicity for each month. The average depth remained constant, usually within a few tenths of a kilometer, especially through the period of intense seismicity in early 1984.

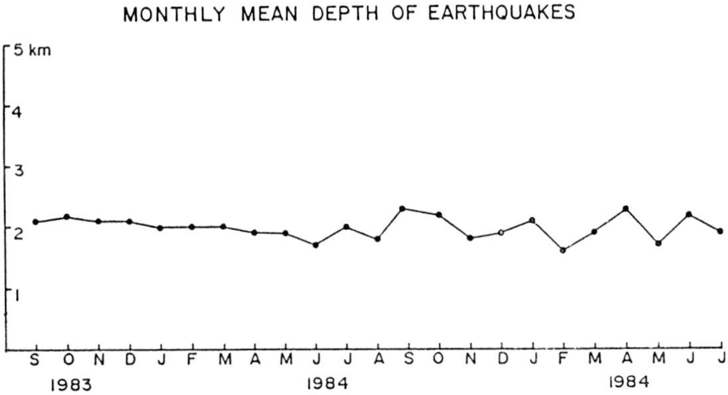

Fig. 11. Average depths of earthquakes for each month of the Crisis period

5 Magnitude Distribution and Seismic Energy Release

The magnitudes of the larger events (M_L >2.0) were determined from a single-component, electronically simulated Wood-Anderson seismograph 25 km east of the caldera. The magnitudes greater than M_L 4.5 were determined from a standard Wood-Anderson instrument 750 km away in Port Moresby. The magnitudes of the smaller events were determined from estimates of the coda duration on the station RAL or TAV of the Harbour Network. There are no sensitive seismic stations outside the caldera, and accordingly the duration measurements had to be made using a station within the caldera. Unfortunately, this means that local structures and the depth of an earthquake may significantly affect its coda magnitude. Also, the coda magnitude is difficult to use during swarms, when earthquakes are occurring in rapid succession, and the coda length cannot be accurately measured.

Generally, the magnitudes of the tens of thousands of events recorded during the Crisis period were quite small. There was one M_L 5.1 earthquake, seven earthquakes with magnitudes from 4.0 to 5.0 and 56 earthquakes with magnitudes from 3.0 to 4.0. The seismic energy release from these earthquakes was calculated using the relationship of Richter (1958):

$$\log E = 9.9 + 1.9\, M_L - 0.02\, M_L{}^2. \tag{1}$$

The cumulative sum and the daily sums of the square root of seismic energy release are plotted in Fig. 12. The rate of seismic energy release increased from the beginning of the Crisis period until early March 1984, when the M_L 5.1 event occurred. The rate then progressively declined till the end of the Crisis period, except for a small burst in October 1984. The total seismic energy release for the entire Crisis period amounted to 2.5 x 10^{19} ergs.

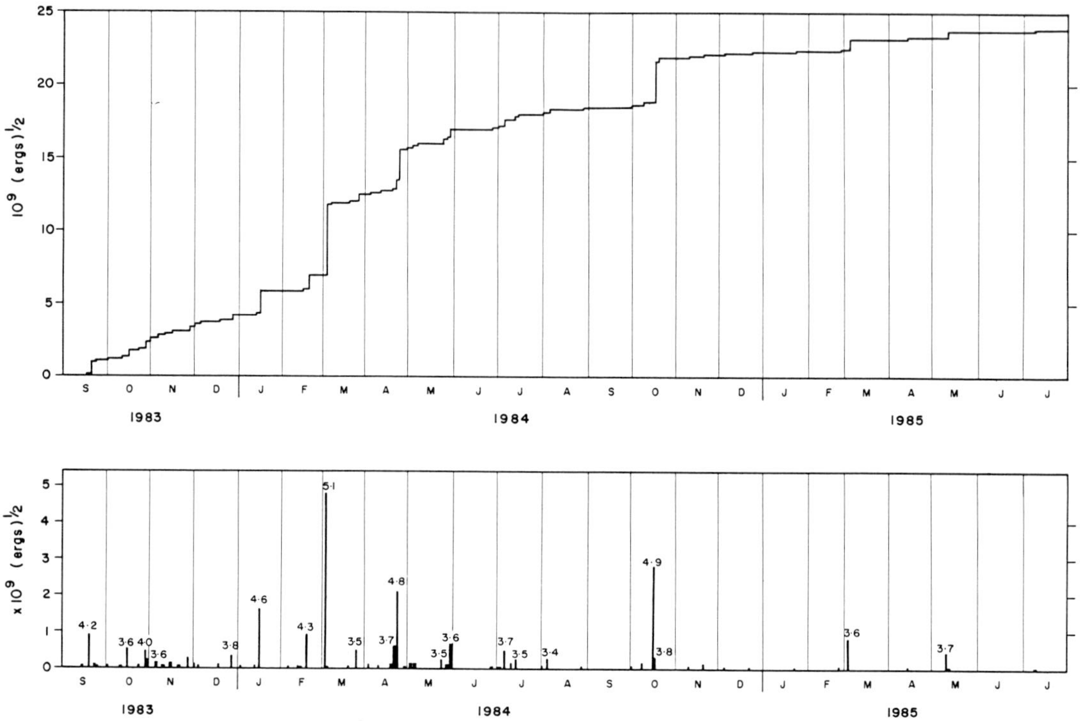

Fig. 12. Plots of cumulative and daily total of the square root of the seismic energy release for the Crisis period

The peak in the total number of earthquakes in late April 1984 (Fig. 6) was not associated with the large energy release at the beginning of March. This is indicative of the trend of increasing earthquake numbers but decreasing size. The average magnitudes of the earthquakes were calculated for the various regions of the caldera (Fig. 13), except for the Karavia Bay area which had too few events. The average magnitude was determined for samples of about 300 events for the Beehives, Greet Harbour and Vulcan areas. For the Blanche Bay area samples of 100 events were used because of the smaller number of earthquakes recorded. The number in each sample was not strictly 300 or 100 because all the events of a particular swarm or seismic crisis were grouped together into one sample. Note that there is a definite decrease in the average size of the earthquakes for all regions during the first 8 months of the Crisis period. When the numbers of earthquakes were at a peak in March and April, as indicated by the closer spacing of the points, the average magnitude had already diminished considerably from the beginning of the Crisis.

These data are presented in terms of average earthquake's size instead of b-values, because the magnitude range was not very great (usually less than two magnitude units), and with large numbers of small events the b-value determination was rather

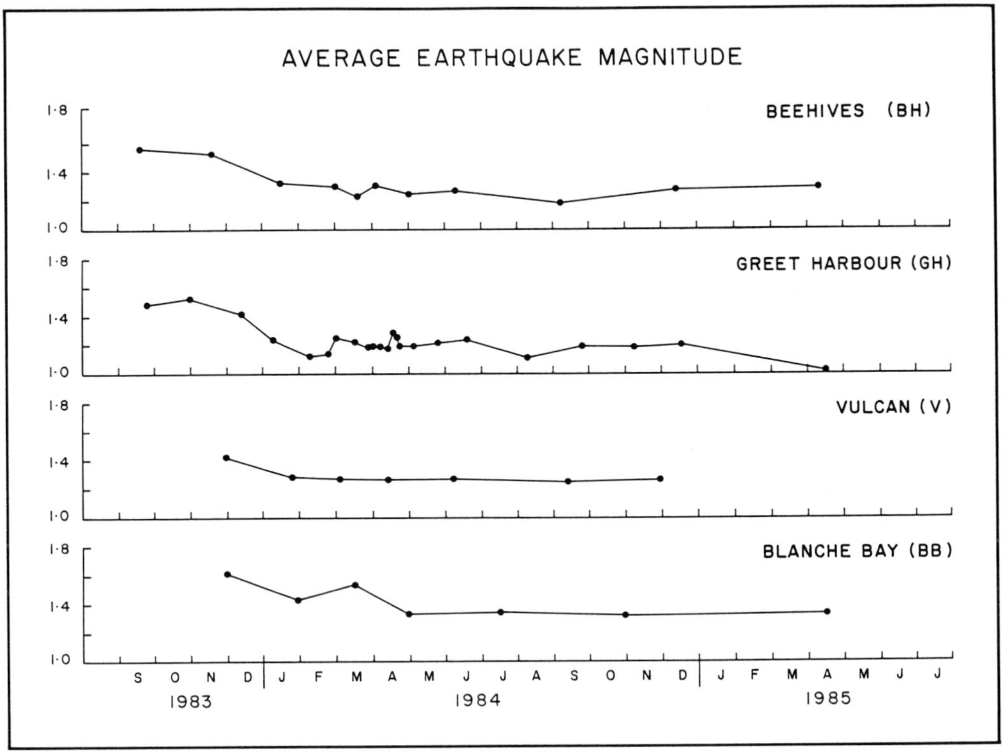

Fig. 13. Average magnitudes of earthquakes for each region during the Crisis period. The averages were calculated for sets of about 300 events for the Beehives, Greet Harbour and Vulcan areas. Sets of about 100 events were used for the Blanche Bay area

unstable, especially for these small samples. The average magnitudes varied from 1.2 to 1.5, which would correspond to b-values of 2.2 to 0.8. If the data for the entire Crisis period are compiled, there is enough information to calculate a reasonable average b-value for the whole Crisis period, for each region. The values for all the regions, taken together, are found to be between 0.8 and 1.1. Such values fall in the range for tectonic earthquakes, and are relatively low compared to values often reported for volcanic earthquakes (Shimozuru, 1971).

The spatial distribution of the seismic energy release is shown in Fig. 14. The strongest concentration is in the Blanche Bay area on the east side of the ring structure, reflecting the location there of the M_L 5.1 event of March 1984 and two magnitude 4 events. The other concentrations of seismic energy release are associated with the locations of the other larger magnitude events (M_L 4.0-4.9) of the Crisis period.

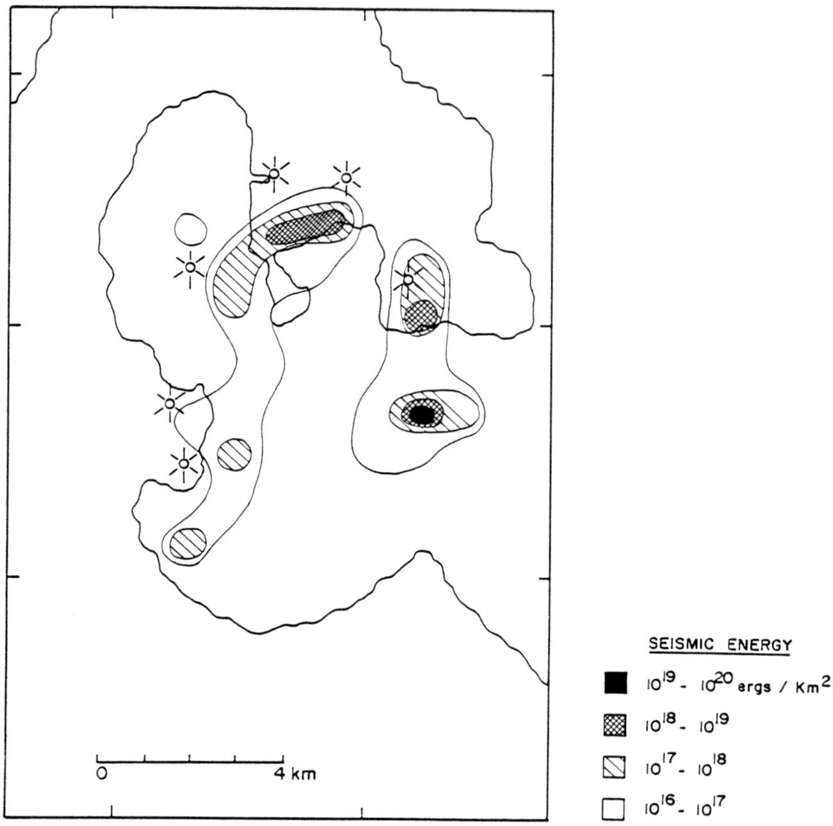

Fig. 14. Distribution of seismic energy release per square kilometer during the Crisis period

6 Coda Durations

The daily total of the coda durations measured for all the events is plotted in Fig. 15 along with the daily total number of earthquakes. The plot of durations follows the plot of total earthquake numbers fairly consistently, although the duration estimate gives more emphasis to larger events. The bottom plot of Fig. 15 shows the average duration for each day. The large peaks during September 1983 and March 1984 are probably due to the combination of larger average size of the events (Fig. 13) and the location of the events. The events from Blanche Bay have relatively longer coda durations as recorded on the TAV and RAL stations.

7 Focal Mechanisms and Slip on the Ring Fault

Focal mechanisms (Fig. 16) for 13 of the larger events were determined using first motion data from the Rabaul Harbour Network. Take-off angles were calculated using the velocity structure of Table 1. Unfortunately, the station coverage for

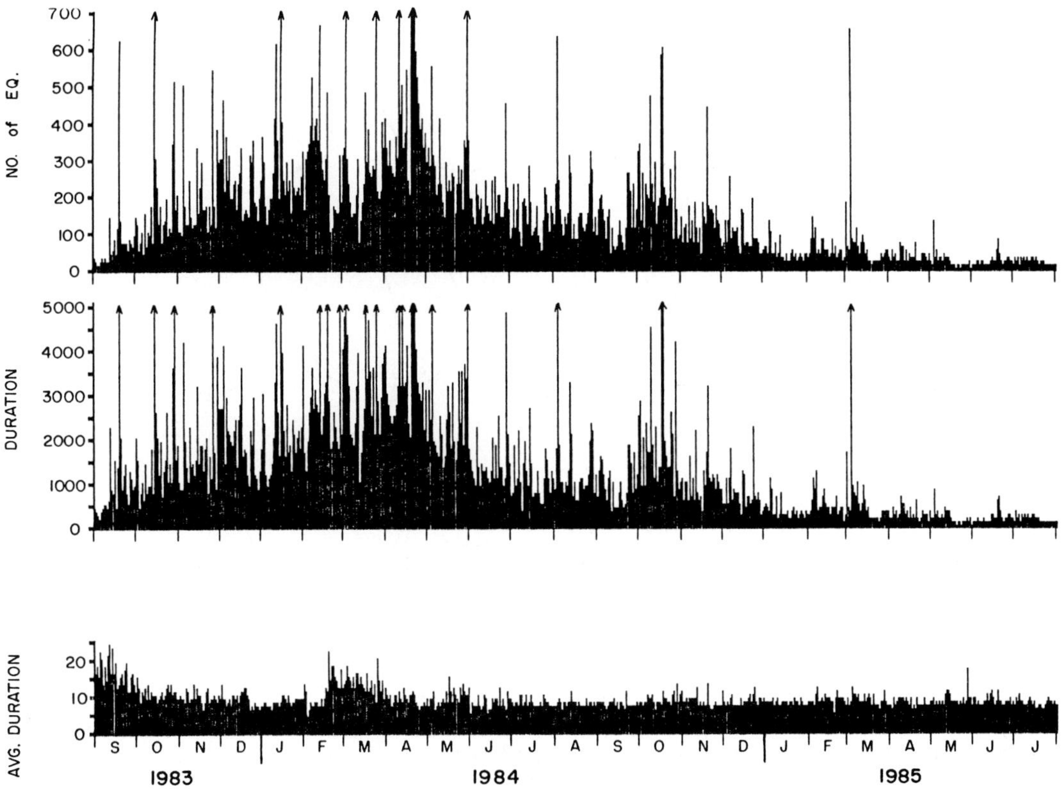

Fig. 15. Plots of the total number of events counted, the total coda duration and the average duration for each day of the Crisis period. The smallest events counted are about M_L 0.5

some of these events is not adequate to produce well-constrained mechanisms, as indicated by the solutions with the dotted nodal planes. However, even for the poorly constrained solutions there is enough information to infer the type of faulting (i.e. normal, strike-slip or thrust) and the general orientation. Solutions 1 to 9 are for normal faults and 10 to 13 are for thrust faults. The information about the fault plane solutions is summarized in Table 3. The focal mechanisms are shown with their respective locations within the seismicity pattern in Fig. 17. Since many of these solutions are not well constrained, they are drawn as pure thrust or pure normal faults. If more information were available in determining the mechanisms, the true pattern might not be as simple as indicated by the figure.

The general pattern consists of mainly normal faulting events with orientations in agreement with the trends on the ring of seismicity. The outward-dipping nodal planes of the focal mechanisms are chosen as the fault planes, to be consistent with the cross-sections of Fig. 9, which show faults steeply dipping away from the center of the caldera. This choice of

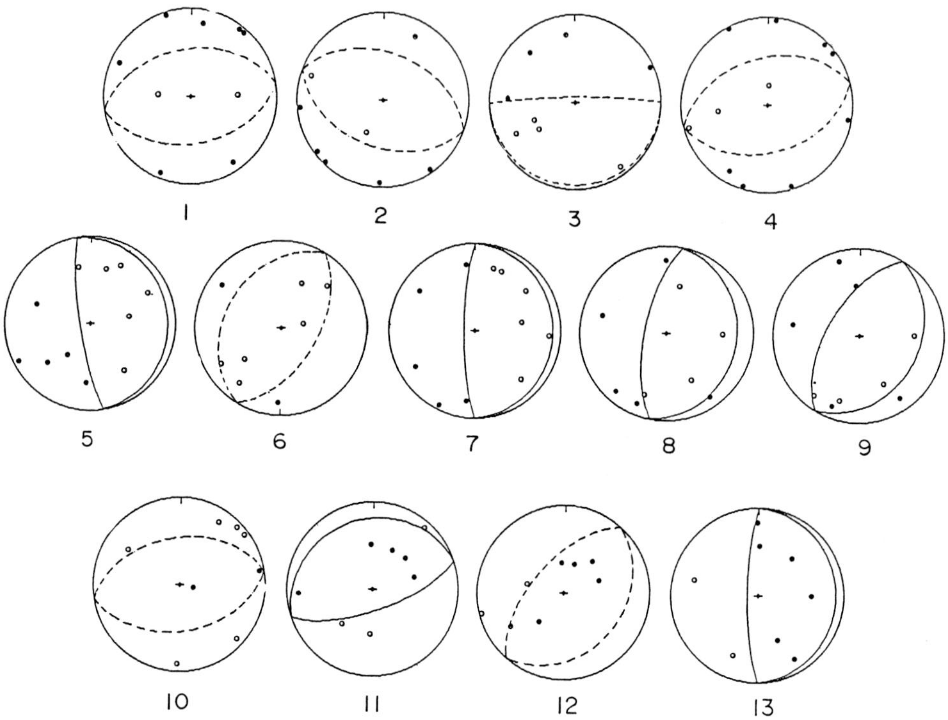

Fig. 16. Lower hemisphere plots of the first motion data for some of the larger events. Dotted nodal planes indicate focal mechanisms with poorly constrained solutions. The locations of the earthquakes are shown in Fig. 17

fault planes implies that the region bounded by the ring-fault is moving upward relative to its surroundings. The presence of the thrust-fault solutions, however, indicates that there is a complexity to the movement and slip may be taking place on faults with various orientations. Three of the thrust-fault solutions are located on the northern portion of the ring structure, where cross-section B of Fig. 9 does not outline a clear pattern in the seismicity. This may be an indication of complicated fault structures in the area.

We can estimate the amount of slip (S) that has occurred on the ring faults from an estimate of the seismic moment Mo (Aki, 1966):

$$Mo = \mu A S, \quad (2)$$

where μ is the rigidity of the material and A is the area of the fault. The moment was estimated from the local magnitudes (M_L) using the relationship of Thatcher and Hanks (1973):

$$\log Mo = 1.5 M_L + 16.0. \quad (3)$$

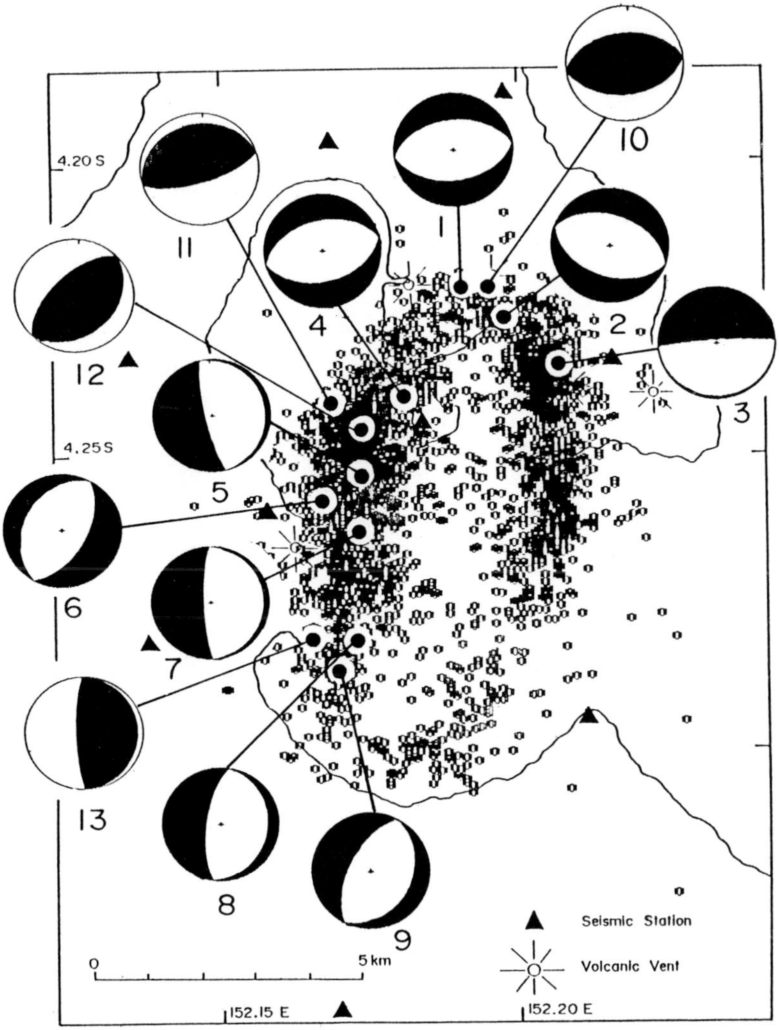

Fig. 17. Focal mechanisms and locations of the same earthquakes as those in Fig. 16

Since the occurrence of earthquakes is not constant around the ring-fault, the moment release has been determined for each km^2 of the region. A map of the distribution of the moment release during the Crisis period is shown in Fig. 18.

Using the values for moment release, we can estimate the slip with Eq. (2). The fault is assumed to extend from the surface to a depth of 4 km at an angle of 75° as indicated by the cross-sections of Fig. 9. This gives a fault area of 4.1 km^2 for each km^2 region. There is no direct information on S-wave velocities to determine the rigidity. Using the P-wave velocity of 4.6 km s^{-1} and assuming a Poisson ratio of 0.25 (which may be reasonable if magma has not intruded the fault), the S-wave

Table 3. Focal mechanisms

No.	Date	Time (GMT)	Lat.	Long.	Depth	Mag.	Strike	Dip.	Slip
1.	19 Nov. 1984	1731	-4.2217	152.1878	2.2	2.9	80	-45	90
2.	6 Nov. 1984	1257	-4.2282	152.1958	2.2	2.2	110	-45	90
3.	27 Jun. 1984	1440	-4.2359	152.2055	1.8	1.5	90	-85	90
4.	25 Oct. 1984	552	-4.2411	152.1835	2.1	1.5	75	-45	90
5.	10 Nov. 1984	1635	-4.2509	152.1788	1.6	1.9	-10	-80	90
6.	2 Dec. 1984	1540	-4.2399	152.1667	3.3	1.5	30	-55	90
7.	26 Oct. 1984	151	-4.2623	152.1756	1.8	1.5	0	-80	90
8.	10 Jul. 1984	1207	-4.2852	152.1704	3.7	2.3	10	-70	90
9.	4 Jul. 1984	901	-4.2861	152.1685	3.3	1.5	30	-60	90
10.	19 Nov. 1984	1737	-4.2224	152.1982	2.2	1.5	80	-45	-90
11.	30 Oct. 1984	2212	-4.2457	152.1696	1.0	1.5	70	-20	-90
12.	10 Nov. 1984	1643	-4.2478	152.1694	1.3	1.5	40	-60	-90
13.	10 Jul. 1984	1205	-4.2833	152.1694	3.3	2.9	0	-80	-90

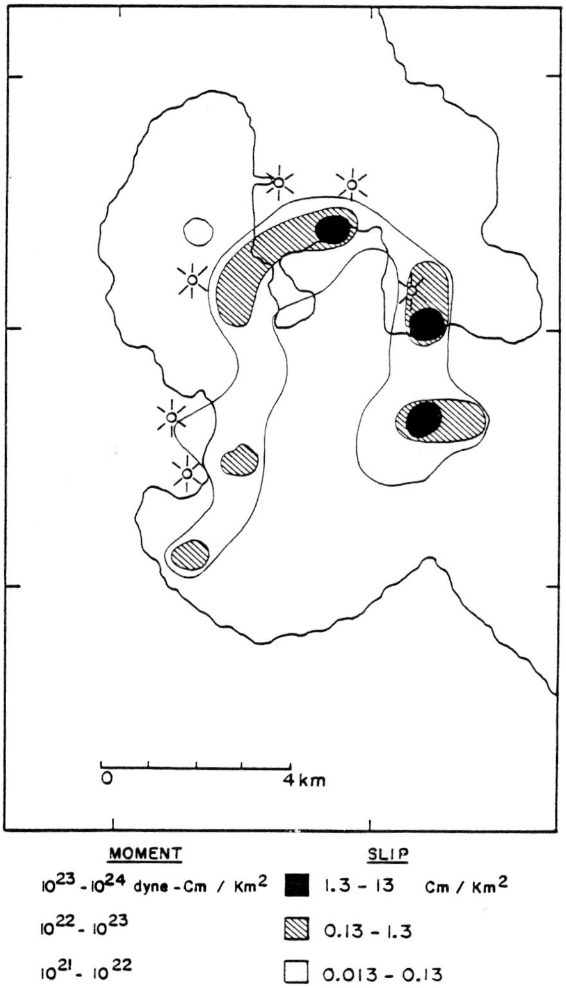

Fig. 18. Seismic moment release and fault slip per square kilometer, calculated from the earthquakes of the Crisis period

velocity (V_S) would be 2.66 km s^{-1}. With a density (ρ) 2.5 g cm^{-3} the rigidity (μ) is estimated to be 1.8 x 10^{10} Pa using the relation:

$$V_S = (\mu/\rho)^{\frac{1}{2}}. \tag{4}$$

With the above values for fault area and rigidity, the slip determined for each km^2 region is directly proportional to the moment and has the values shown in Fig. 18. These estimates are probably maximum values since they assume that all the earthquakes had the same slip directions on the same fault plane. In actuality there is probably a complexity of faults that slip in various directions, as indicated by the mixture of fault plane solutions (Fig. 17) and the complicated seismicity patterns (Fig. 9), so that the net slip would be smaller than our value, which is calculated directly from the total moment. Also, the rupture lengths of some of the larger events may be somewhat larger than the 1-km grid spacing, in which case the moment should be distributed over a larger area, giving a smaller value of slip.

Just as in the map of seismic energy release (Fig. 14), the largest amounts of slip are concentrated on the east and north-east sides of the ring fault. The largest amount of slip was estimated to have been 6.1 cm in the Blanche Bay area. The two concentrations of slip in Greet Harbour were 3.1 and 2.1 cm. Other portions of the ring fault had much less slip.

8 Correlation of Seismic Activity with Ground Deformation

From the distribution of ground tilts there appeared to be two shallow sources of uplift within the caldera (Fig. 5): one at the mouth of Greet Harbour and another off the east coast of Vulcan (McKee et al. 1984). Comparisons of the deformation produced by these sources with the monthly number of earth-quakes in the surrounding areas are shown in Figs. 19 and 20. Figure 19 is the monthly number of earthquakes, in the Vulcan region, with the monthly values of ground tilt measured at the Baluan dry-tilt station, on the southeast flank of the Vulcan

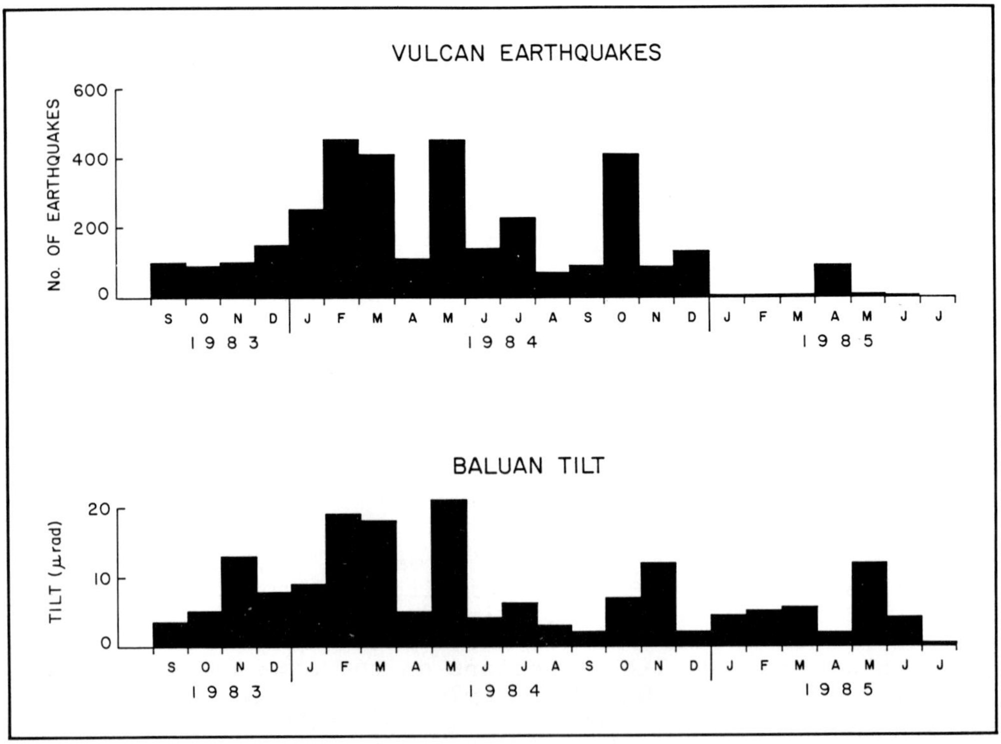

Fig. 19. Comparison of the monthly number of earthquakes in the Vulcan region (**top**) with the ground tilt at the Baluan dry-tilt array (**bottom**), which is located on the southern part of the Vulcan headland

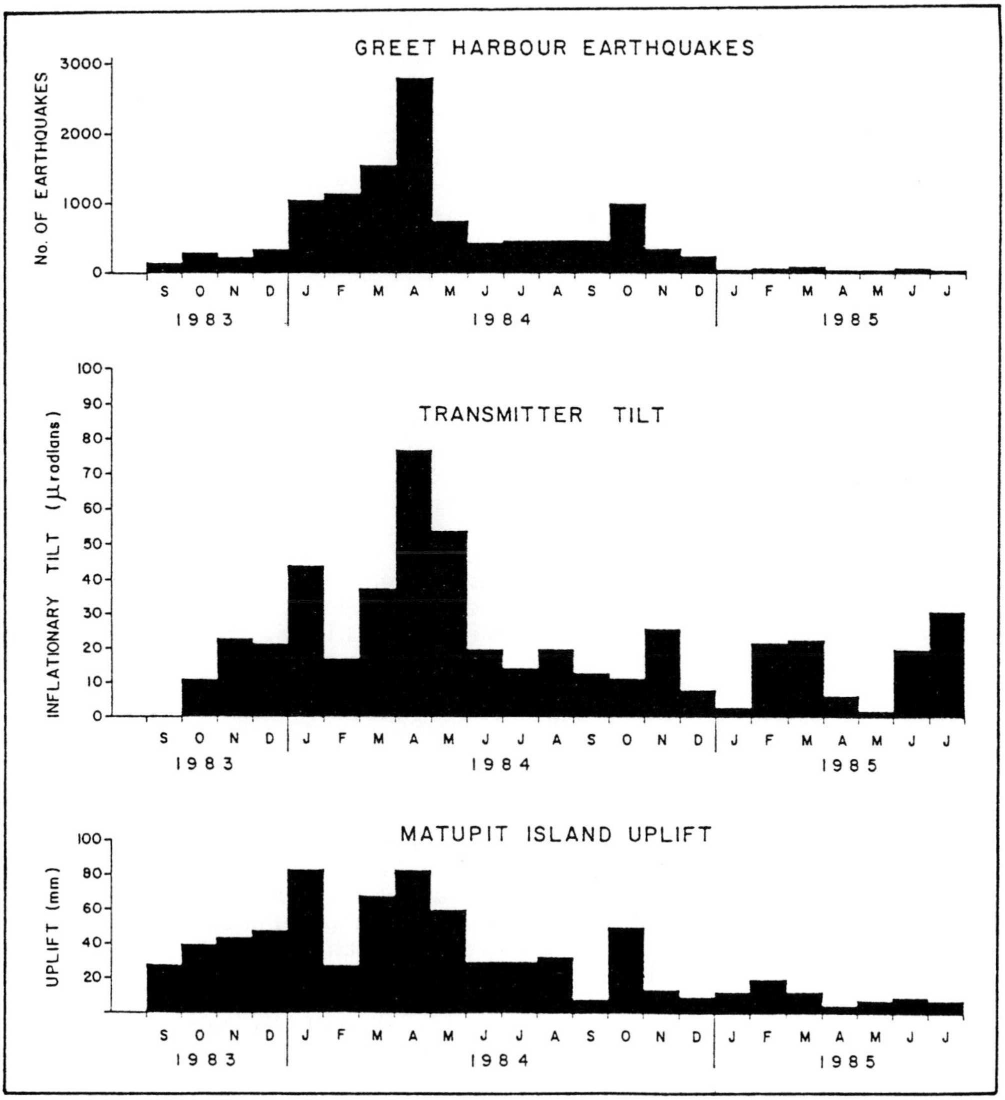

Fig. 20. Comparison of the monthly number of earthquakes in the Greet Harbour region (**top**) with the ground tilt recorded at the head of Greet Harbour (**middle**) and the ground uplift measured at the southern tip of Matupit Island (**bottom**)

cone. All the tilts except for those of May 1984 were inflationary, with the ground tilting downwards away from the source of uplift, indicating intrusion of material into the source. The deflationary tilt and associated seismicity of May 1984 may indicate a time when magma was moving away from the Vulcan source. Figure 20 shows the monthly number of earthquakes in the Greet Harbour region, together with the monthly values of tilt at the head of Greet Harbour and the uplift at the southern tip of Matupit Island. For both the Greet Harbour and Vulcan uplift sources, the seismic activity and the rates of ground deformation track fairly consistently. Also note that the two areas are not always active concurrently. For example, in April 1984 when the Greet Harbour seismicity and the Matupit Island uplift were strongest, the Vulcan seismicity and ground deformation were at quite low levels. This observation indicates that the ground deformation is due to two separate sources of uplift that sometimes act independently of each other, and is not due to a single extended source within the caldera.

9 Correlation of Seismicity with Ocean Tides

Lowenstein (1986) stated that there was a good correlation between the occurrences of earthquakes and the times of low ocean tide. The difference in the vertical normal stress between low and high tide, due to the weight of the water, is about 0.1 bar, which resolves to 0.025 bar of shear stress on a fault dipping at 75°. Since the triggering ocean tide produces only this small amount of stress change, many points on the ring fault must already be very close to failure. However, the occurrences of the seismic crises, which are thought to be directly associated with the stresses due to rising magma, did not appear to correlate with the ocean tides. The ocean tides do not, therefore, appear to affect the magma body directly, and the observed correlation is probably due to the response of highly stressed faults to the ocean load.

10 Discussion

Caldera Structure and Location of Magma

The pattern outlined by the earthquake locations in Figs. 7, 9 and 10 shows quite clearly a coherent ring structure with sides that slope outward from the center of the caldera and extend from the surface to about 4-km depth. As discussed by Mori and McKee (1987), the seismicity pattern is interpreted as the outline of a caldera block which subsided in the last major eruption, 1400 years B.P. The sharp cutoff in the seismicity below 4 km indicates the depth to which faulting occurred during the subsidence, and thus probably represents the top of the 1400 years B.P. magma reservoir. Assuming that the size of the magma reservoir has not changed appreciably since that time, the top of the magma reservoir may still be at this depth.

The ring fault is a zone of weakness which is currently being reactivated by the stresses due to increased magma pressure. At present there is no evidence for intrusion of the ring fault by magma. There have been no low-frequency volcanic earthquakes and no harmonic tremor has been recorded. The tectonic-like appearance of the earthquakes and the b-value of close to 1.0 suggest that the seismicity is due simply to slip on preexisting faults. The amount of slip was up to several centimeters, distributed around the ring fault as shown in Fig. 18.

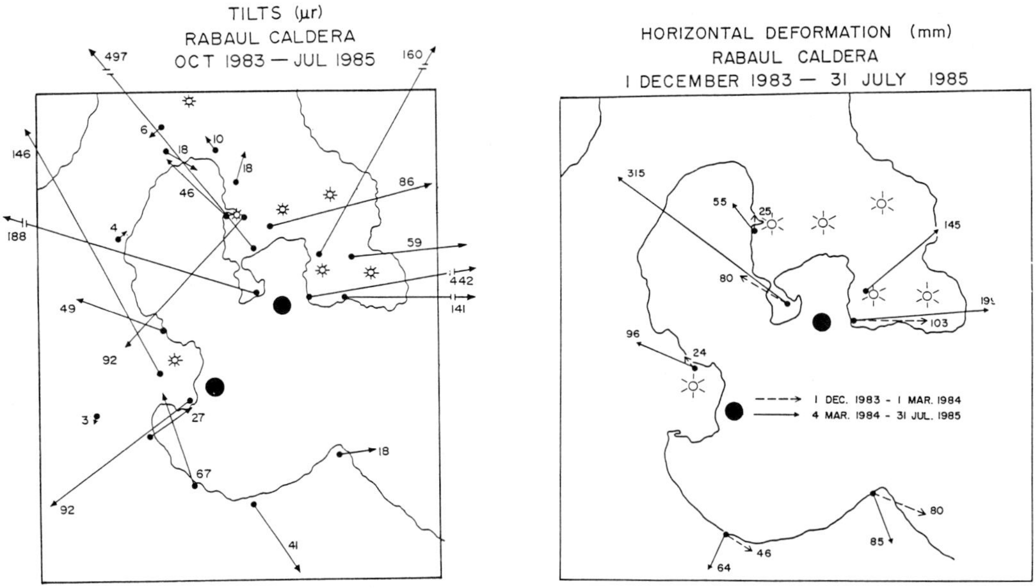

Fig. 21. Maps showing the distribution of tilt and horizontal distance changes for the Crisis period. The **solid circles** are the two centers of uplift identified from the tilt data

The ground deformation data give a different picture of the subsurface structure (Fig. 21). The measured tilts and ground uplifts are generally much larger than would be associated with the inferred fault slips and, instead, indicate two localized areas of ground uplift that are not on the ring fault. As a first approximation, the measured deformations were adequately modelled as the elastic response to shallow intrusions of magma, at depth of 1 to 3 km, without including slip on the ring fault and without possible deformations from a large underlying magma reservoir (McKee et al. 1984). The interpretation of small localized magma intrusions is further supported by gravity data which show a localized increase of subsurface mass (5×10^{10} kg) in the Greet Harbour area during the Crisis period (McKee et al. 1988).

Assuming the validity of both the seismic and ground deformation results, the conclusion is drawn that the earthquake locations are quite separate from the locations of the magma. The ring fault seismicity represents a zone of weakness which focuses the release of stresses generated by magmatic activity elsewhere in the caldera. In order to explain the nearly complete ring pattern, we suggest two alternative interpretations.

1. The seismicity is completely due to stresses generated by the two shallow magma bodies. If the central block has high rigidity, the stresses can be transmitted to the entire ring fault from these localized sources of uplift. The higher concentrations of earthquakes which occur close to the uplift sources near the northern and western parts of the ring fault may support this idea.

2. The seismicity represents the combined effects of stresses produced by the shallow magma bodies and the large active magma reservoir beneath the caldera. The caldera reservoir may be exerting pressure on the entire central block, producing a low level of seismicity on the entire ring fault. Superimposed on this pattern are the higher concentrations of seismicity caused by the shallow magma bodies.

Of these two possibilities, present evidence favors the second interpretation. Closer study of the horizontal distance measurements (Fig. 21) indicates that deformations in the southern part of the caldera are larger than would be expected from the two shallow magma bodies. These changes may be part of a broad deformation of low amplitude caused by the caldera reservoir. In the northern part of the caldera, this effect is masked by the large deformations from the shallow magma bodies. Thus, both the seismicity pattern and the ground deformations can be explained in terms of the combined stresses from shallow magma bodies and a large caldera reservoir.

Another feature of the caldera structure is that the post-caldera vents are not situated on the ring fault. In the cross-sections of Figs. 9 and 10, it can be seen that they are displaced from the fault at distances of 1.0 to 1.5 km. This suggests that the zone of weakness has not been used as the main channel of magma for post-caldera eruptions. It is proposed that magma moved from the shallow magma bodies within the central block to the post-caldera vents along cone sheets or radial fractures, as depicted in Fig. 22. There is, however, no strong evidence for the existence of cone sheets or radial fractures, although there is a suggestion of a cone-sheet trend in the seismicity of cross-section D in Fig. 9.

Figure 22 summarizes our interpretation of the caldera structure. There is a central block with outward sloping sides that is being pushed upward from below and from a diapir within the block. The upward pressure causes slip on the ring fault as indicated by the arrows. The slip direction is consistent with the slip observed in normal faulting solutions of some of the larger earthquakes (Fig. 17). The horizontal arrows above the ring fault represent the caldera-wide expansion that has been observed in the recent horizontal distance measurements (Fig. 21). This deformation is thought to be caused by the rising of the central block, which, due to its shape, pushes outwards on the surrounding material. The question marks indicate the possible paths along cone-sheet structures or radial fractures that the magma might travel in order to reach the known post-caldera vents.

10.1 Seismic Activity

The significant increase in seismicity during the early part of the Crisis, from September 1983 to April 1984, appears to have been related to the increase in ground deformation, which in turn has been interpreted as due to localized magma intrusion. Judging from both the seismic and ground deformation data, the

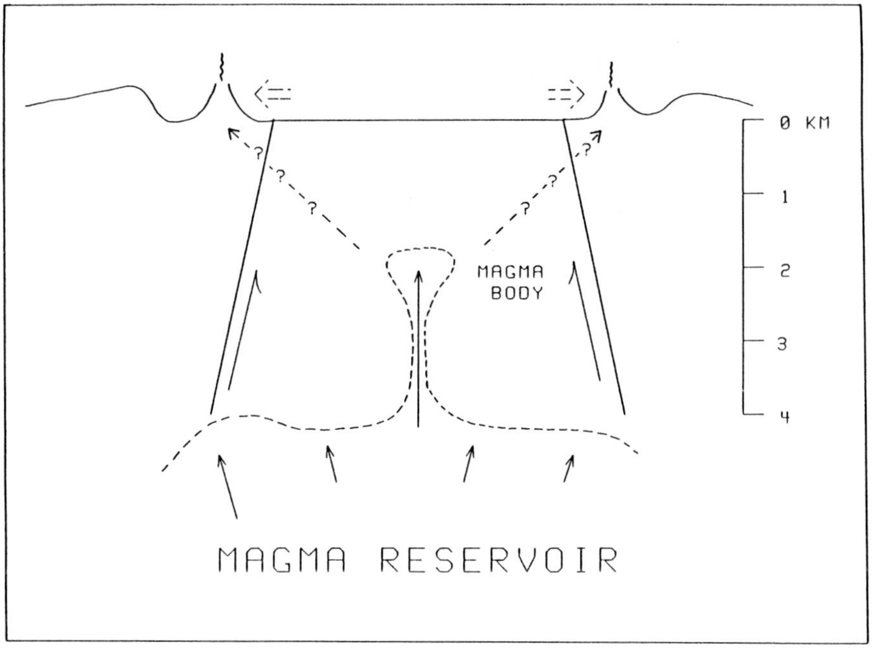

Fig. 22. Diagram showing the interpreted structure of the caldera. See text for full explanation

times of greatest magma intrusion were in April 1984 for the Greet Harbour area and in February-March 1984 for the Vulcan area. Since the location of the ring fault seismicity is displaced from the magma intrusion, it is not surprising that there were no systematic changes in earthquake depths (Fig. 11) or in epicentral locations (Fig. 6), such as are often associated with magma transport. Although the earthquake locations cannot be used to pinpoint the location of the magma, they do identify the parts of the caldera being stressed. The shifts in earthquake concentrations around the ring fault are presumed to indicate where changes in magma pressures are taking place. The distribution of earthquake concentrations over a large part of the ring pattern during the Crisis period is interpreted to mean that the stress was not localized at any one spot, but was distributed over a large portion of the ring fault.

Although there were very large numbers of earthquakes, most of the events were quite small. In comparison with other volcanoes, the seismic energy release was one-sixth of that recorded before the 1980 Mount St. Helens eruption (Endo et al. 1981), 1/25 of the seismic energy release at Fernandina caldera, Galapagos Islands, during the caldera subsidence event of 1968 (Filson et al. 1973). However, the seismic energy release was 18 times that of the 1982-84 bradyseismic crisis at Campi Flegrei, Italy (Scandone, pers. commun., 1986), which was caused by a caldera unrest very similar to the Rabaul Seismo-Deformational Crisis.

Figure 18 shows that the largest fault movements occurred in the northern and eastern portions of the ring fault near the Greet Harbour uplift source. One would expect that the north-western portion of the ring fault would also have had a significant amount of slip, since it is also relatively close to the Greet Harbour source. As can be seen from the distribution of earthquakes during the Crisis period (Fig. 8), the strongest concentration of earthquake numbers was in this area, showing that this region had many small earthquakes, but none of the larger magnitude events which produced relatively large amounts of slip. One possibility for the lack of observed slip is that the movement was occurring aseismically, because of different rock properties on this portion of the ring fault. Another possibility is that the strain energy is still building up and, therefore, this region is a likely site for a larger magnitude earthquake in the future.

There was only a small amount of slip on the ring fault in the Vulcan region (Fig. 18). However, throughout the Crisis period the surface ground deformations around Vulcan were considerably smaller than those around Greet Harbour, indicating a deeper and/or weaker intrusion source, which may not have put as much stress on the ring fault as in the Greet Harbour area.

One change in the seismicity that was noted during the Crisis was the trend of decreasing magnitudes (Fig. 13). This might be explained in terms of the strength on the ring fault. When the stresses from the rising magma were first transmitted to the fault, the earthquakes occurred on old fault surfaces that had not moved for some time. These faults would have been relatively strong, thus producing larger (higher stress drop) events. As the seismicity increased through the Crisis period, the fault surfaces were weakened and allowed smaller earthquakes, with lower stress drop, to occur. This change in earthquake size distribution is opposite to that which has sometimes been observed at other volcanoes. At Mount St. Helens (Endo et al. 1981), Usu (Okada 1983) and Bezymianny (Tokarev, 1971), there were trends towards larger earthquakes as stresses built up towards an eruption. The difference in the present situation might be due to the well-developed fault system which controls the seismicity at Rabaul, which is not observed at some of these other volcanoes. Another point, which is discussed in the next section, is that the activity of the 1983-85 Crisis period did not culminate in an eruption. Seismicity before an actual eruption may have different characteristics from that observed during the Crisis period.

10.2 Future Eruptive Activity

It is uncertain how close the situation came to an eruption during the peak of activity from February to April 1984. One notable point is that during the Crisis period, although a considerable amount of material was moving up into the shallow magma bodies, there were no indications of continued magma movement from the shallow magma bodies towards the surface or towards any of the post-caldera vents. In order for an eruption to occur it appears that even higher magma pressures (due to more heat or higher volatile content) would be needed. In this case, one might expect even more dramatic precursory activity. This evidently was the situation before the 1937 Vulcan eruption when many earthquakes began to be felt and reefs were raised about 2 m above sea level during the day before the eruption (Fisher 1939). There are local reports that the felt earthquakes and observed ground uplift began several weeks before the eruption.

The inferred paths from the shallow magma bodies to the post-caldera vents do not follow the ring fault, so that, for an eruption to begin, channels to the surface, which are not presently seismically active, will have to be created or re-opened. In such a high stress situation, one might expect to see precursory seismicity. Therefore, the first important sign of an imminent eruption may be concentrations of seismicity off the ring fault. Secondly, since this seismicity would not be occurring on an already weakened fault system, there may be a trend towards increasing size of the earthquakes. A further sign may be the appearance of low-frequency volcanic earthquakes and harmonic tremor, which might indicate magma flow towards the surface. If these changes in seismicity did appear, however, there would still be the difficulty of estimating how long it would take for magma to reach the surface and initiate an eruption.

Historical volcanism at Rabaul has consisted of small to moderate eruptions at the post-caldera vents. Material for these eruptions probably came from shallow magma bodies such as the ones which have been recently identified from the ground deformation data. The volumes of these shallow magma bodies are most likely not larger than one or two cubic kilometers, which is probably too small to produce a major caldera eruption. However, interaction with the large magma reservoir below could cause a moderate eruption to evolve into a catastrophic one. The nearly complete ring pattern of seismicity, and the overall ground deformation pattern, are thought to indicate caldera-wide stresses acting on the central block. The eruptions in 1878 and 1937, during which vents on opposite sides of the caldera were erupting simultaneously, suggest that there already is a degree of coordination between the shallow magma bodies and the larger reservoir. If the central block lost support because of eruption or withdrawal of magma from the reservoir below, the subsidence of the block could open the ring fault, allowing large volumes of material to be channelled to the surface producing another major caldera eruption.

11 Conclusions

The characteristics of the seismicity during the Rabaul Seismo-Deformational Crisis are summarized as follows:

1. The earthquakes were high-frequency, tectonic-like events with b-values close to 1.0

2. The earthquakes occurred predominantly on an elliptical ring fault system measuring about 10 x 5 km. The active faults extend from the surface to a depth of 4 km and slope outwards from the central block of the caldera.
3. The earthquakes often occurred in swarms and seismic crises, with hundreds of events within an hour.
4. The largest earthquake was an M_L 5.1 event, and there were seven events with magnitudes between 4.0 and 5.0. The total seismic energy release for the Crisis period was 2.5 x 10^{19} ergs. The total seismic moment release was 1.1 x 10^{24} dyne-cm.
5. The numbers of earthquakes steadily increased from the beginning of the Crisis in September 1983 and reached a peak for the Vulcan area in February-March 1984 and for the Greet Harbour area in April 1984. After April 1984 the seismicity gradually decreased to pre-Crisis levels by the end of July 1985. The gradual decrease was interrupted by occasional bursts of activity in October 1984, March 1985 and May 1985.
6. For the whole Crisis period, most of the events took place in the Beehives and Greet Harbour areas. The strongest concentrations of seismic energy release, however, were in the Blanche Bay and Greet Harbour areas.
7. There were no systematic changes in earthquake depths or epicentral locations during the Crisis period.
8. There was a trend of decreasing average magnitude during the first 8 months of the Crisis.
9. Focal mechanisms showed predominantly normal faulting, which was inferred to occur on planes dipping steeply outwards from the caldera, indicating the rise of the central block relative to its surroundings.
10. Slip on the ring faults was estimated to range from less than a millimeter to several centimeters. The largest amounts of slip occurred on the northern and eastern portions of the ring fault.
11. There was a good correlation between local ground deformation and the seismicity rates on nearby portions of the ring fault.
12. There was a good correlation between the level of background seismicity and the times of low ocean tides. The occurrences of seismic crises, however, did not appear to correlate with the ocean tides.

The following interpretations about the caldera structure and the events during the Crisis period have been made:

1. The earthquakes outline a ring-fault system on which caldera subsidence occurred during the last major eruption, 1400 years B.P. The sharp cutoff in seismicity below 4 km is believed to indicate the top of the 1400 years B.P. magma chamber, and may still represent the top of an active magma reservoir.
2. The earthquakes are occurring on the ring fault in response to stresses created by magma intrusions into the central block and increased magma pressures on the bottom of the block.

3. The large amount of slip that occurred on the northern and eastern portions of the ring fault is associated with the intrusions at the Greet Harbour uplift source. The Vulcan source produced only small amounts of slip.
4. The ring fault is not being intruded by magma, and therefore the locations of the earthquakes are separated from the locations of the active magma.
5. The trend in decreasing earthquake size may be explained by a progressive reduction in the strength of the ring fault.
6. The times of greatest magma intrusion during the Crisis period were during April 1984 for the Greet Harbour area and February-March 1984 for the Vulcan area.
7. The magma did not move from the shallow magma bodies towards the surface, nor towards any of the post-caldera vents, during the Crisis period.
8. The paths to the surface or the post-caldera vents from the shallow magma bodies do not follow the ring fault. Therefore, for an eruption to occur, magma must travel along channels that are presently not defined by the seismicity.
9. If magma begins to move towards the surface, the locations of the seismicity may change to areas off the ring fault. Low frequency volcanic events and tremor may also appear.
10. The nearly complete ring pattern of seismicity may indicate that caldera-wide stresses are acting on the central block. This raises the possibility of another major caldera eruption in the future.

Acknowledgements. C. Matupit, J. Kuduon, L. Topue, P. Daimbari, R. Pakat, A. Pidik, and D. Bual helped with the collection and analysis of the data. J. Latter provided many helpful comments. The authors publish with permission of the Secretary of the Dept. of Minerals and Energy.

References

Aki K (1966) Generation and propagation of G waves from the Niigata earthquake of June 16, 1964. 2. Estimation of earthquake movement, released energy, and stress-strain drop from G wave spectrum. Bull Earthq Res Inst 44:23-88

Almond RA, McKee CO (1982) Location of volcano-tectonic earthquakes within the Rabaul caldera. Geol Surv Papua New Guinea Report 82/19

Cifali G, d'Addario GW, Polak ES, Wiebanga WA (1969) Rabaul preliminary crustal seismic test, New Britain 1966. Aust Bur Min Res, Record 69/125

Cooke RJS (1977) Rabaul Volcanological Observatory and geophysical surveillance of the Rabaul volcano. Aust Phys 1977:27-30

Endo ET, Malone SD, Noson LL, Weaver CS (1981) Location, magnitudes, and statistics of the March 20-May 18 earthquake sequence, in The 1980 Eruption of Mount St. Helens, Washington. US Geol Surv Prof Paper 1250:93-107

Filson J, Simkin T, Lei-kuang L (1973) Seismicity of a caldera collapse: Galapagos Islands 1968. J Geophys Res 78:8591-8623

Finalyson DM, Cull JP (1973) Time-term analysis of New Britain - New Ireland island arc structures. Geophys JR Astr Soc 33:265-280
Fisher NH (1939) Geology and vulcanology of Blanche Bay, and the surrounding area, New Britain. Territory of New Guinea, Geol Bull 1:1-68
Fisher NH (1976) 1941-42 Eruption of Tavurvur volcano, Rabaul, Papua New Guinea. In: Johnson RW (ed) Volcanism in Australasia. Elsevier, Amsterdam, pp 201-210
Heming RF (1974) Geology and petrology of Rabaul caldera, Papua New Guinea. Geol Soc Am Bull 85:1253-1264
Herrmann RB (1975) FASTHYPO - A hypocenter location program. Earthq Notes 50:25-35
Kizawa T (1951) Volcanic tremor and tilting of the ground (first paper). Kenshin Jiho 15:18-34 (in Japanese)
Latter JH (1966) Notes on near earthquakes, interference, and the problem of volcanic tremors at Rabaul, New Britain. Aust Bur Min Res, Record 1966/19 (unpublished)
Lowenstein P (1986) Tidal triggering of volcanic earthquakes during the Rabaul seismic crisis of 1983-84. In: Abstracts. Int Volcanol Congr, New Zealand, 1-9 Feb 1986:255
McKee CO, Lowenstein PL, de Saint Ours P, Talai B, Itikarai I, Mori JJ (1984) Seismic and ground deformation crisis at Rabaul caldera: prelude to an eruption? Bull Volcanol 47:397-411
McKee C, Mori J, Talai B (1988) Temporal microgravity changes accompanying structural deformation at Rabaul caldera, Papua New Guinea. Bull Volcanol (this volume)
Mori J, McKee C (1987) Outward dipping ring-fault structure at Rabaul Caldera as shown by earthquake locations. Science 235:193-195
Mori J, McKee C, Itikarai I, Lowenstein P, de Saint Ours P, Talai B, (1986) Account and interpretation of the seismicity during the Rabaul seismo-deformational crisis, September 1983 to July 1985. Geol Surv Papua New Guinea Rep 86/26
Okada H (1983) Comparative study of earthquake swarms associated with major volcanic activities. In: Shimozuru D, Yokoyama I (eds) Arc volcanism: physics and tectonics. Terra Sci Publ, Tokyo, pp 43-62
Richter CF (1958) Elementary seismology. Freeman, San Francisco, 768 pp
Shimozuru D (1971) Seismological approach to the prediction of volcanic eruptions. In: Surveillance and prediction of volcanic activity. UNESCO, Paris, pp 19-45
Thatcher W, Hanks TC (1973) Source parameters of southern California earthquakes. J Geophys Res 78:8547-8576
Tokarev PI (1971) Forecasting volcanic eruptions from seismic data. Bull Volcanol 35:243-250

Some Seismic Precursors to Eruptions at Pavlof Volcano, Alaska, October 1973 – April 1986

S. R. McNutt[1]

Abstract

This chapter examines 13 years of seismicity data (1973-1986) at Pavlof Volcano for eruption precursors. Changes in the rate of occurrence of earthquakes give the most consistent information. Tests for changes in rate of occurrence were performed using different values for short-term (STA) and long-term (LTA) average seismicity, and STA/LTA thresholds. The best parameters of LTA = 3 days, STA = 2 days, threshold = 3x, and absolute number of events >150 per day yielded 46% precursors, 33% false alarms, and the remainder missed eruptions. Seismic activity at Pavlof is monitored by an array of five seismometers which are located at distances of 4-11 km from the summit vent. The numbers of earthquakes increase from a background level averaging 33 low-frequency or B-type events per day to an average of 453 events per day during phreatomagmatic eruptions, and to 1919 events per day during magmatic eruptions (volcanic tremor, which occurs during magmatic eruptions, is converted to an equivalent number of B-type earthquakes, and included in this total). Data were also systematically searched for eight other general types of seismic precursors: (1) changes in event locations; (2) changes in event sizes; (3) changes in frequency content; (4) changes in b-value; (5) seismic quiescence; (6) changes in focal mechanisms; (7) changes of event types; and (8) onset of tidal (or other) periodicities. None of these gave results as good as changes in rate of occurrence. Seismic monitoring alone has provided some useful data on eruption precursors, but, based on comparisons with other volcanoes, geodetic monitoring of the volcano is considered likely to improve the short-term eruption forecasting capability.

1 Introduction

The purpose of this chapter is to examine 13 years of seismicity data (1973-1986) at Pavlof Volcano for eruption precursors. Previous works have discussed Pavlof's eruptions (McNutt 1985, 1987a), seismicity (McNutt 1986), and velocity structure (McNutt and Jacob 1986), but no work has been devoted to systematic analysis of eruption precursors.

---

[1] California Division of Mines and Geology, 630 Bercut Drive, Sacramento, CA 95814, USA

IAVCEI Proceedings in Volcanology 1
J. H. Latter (Ed.), Volcanic Hazards

Eruptions at Pavlof Volcano (lat. 55° 25'N, Long. 161° 54'W) are of two types: magmatic and phreatomagmatic. Most magmatic eruptions are of 1-2 days duration and consist of Strombolian lava fountaining. Nine of ten magmatic eruptions during the time period October 1973-April 1986 occurred in the fall, between 9 September and 20 November. Phreatomagmatic eruptions are of days to months duration and consist of series of small Strombolian explosions. The 18 phreatomagmatic eruptions during 1973-1986 occurred throughout the year, and also in association with magmatic eruptions.

Volcanic risk at Pavlof is relatively low, because the volcano is located in a sparsely populated area near the west end of the Alaska peninsula. While few inhabitants live near the volcano itself, Pavlof Bay, 15 km to the east (usually downwind), is an active commercial fishing area. Sand Point, a village of about 1,000 people, 89 km to the east, and Cold Bay, a village of 200 people 58 km SW, have been subjected to ash falls of as much as several centimeters. Further, Cold Bay is the site of an international airport, used chiefly as a refueling stop for commercial aircraft flying polar routes. Ash clouds from Pavlof have caused some local rerouting of aircraft in the last 13 years. Lastly, the oil potential of the Bering Sea and Gulf of Alaska may require the siting of refineries or storage facilities on the Alaska peninsula. Seven wells have already been drilled within 150 km NE of Pavlof (J. Taber, pers. commun., 1987). Such facilities could be subjected to contamination or damage by volcanic ash from Pavlof.

In this chapter, the seismic precursors to magmatic and phreatomagmatic eruptions, defined in terms of numbers of earthquakes, are systematically examined and statistics are compiled as a guide for evaluating future activity of the volcano. For those precursors which are suitable, the use of computer algorithms to predict eruptions is discussed. Finally, data from Pavlof are compared with data from other volcanoes to determine constraints on the use of combined data to forecast activity of volcanoes with no historic eruptions.

2 Data

Pavlof has been seismically monitored since October 1973 by Lamont-Doherty Geological Observatory of Columbia University, in cooperation with the University of Alaska. Figure 1 shows locations of microearthquake stations operating near the volcano during the study period (October 1973-April 1986). A single station (PVV) was used to monitor the volcano from October 1973-August 1976; thereafter, a network of 5-12 stations has been operating (see Fig. 1 caption). PVV remains the long-term monitoring station. When it malfunctions, other stations are substituted after correcting for differences in distance, magnification, and site response. The standard seismic system response is shown in McNutt (1986, pp. 156, Fig. 2).

Three types of seismic events are recorded at Pavlof: B-type earthquakes (Minakami 1960), explosion earthquakes, and volca-

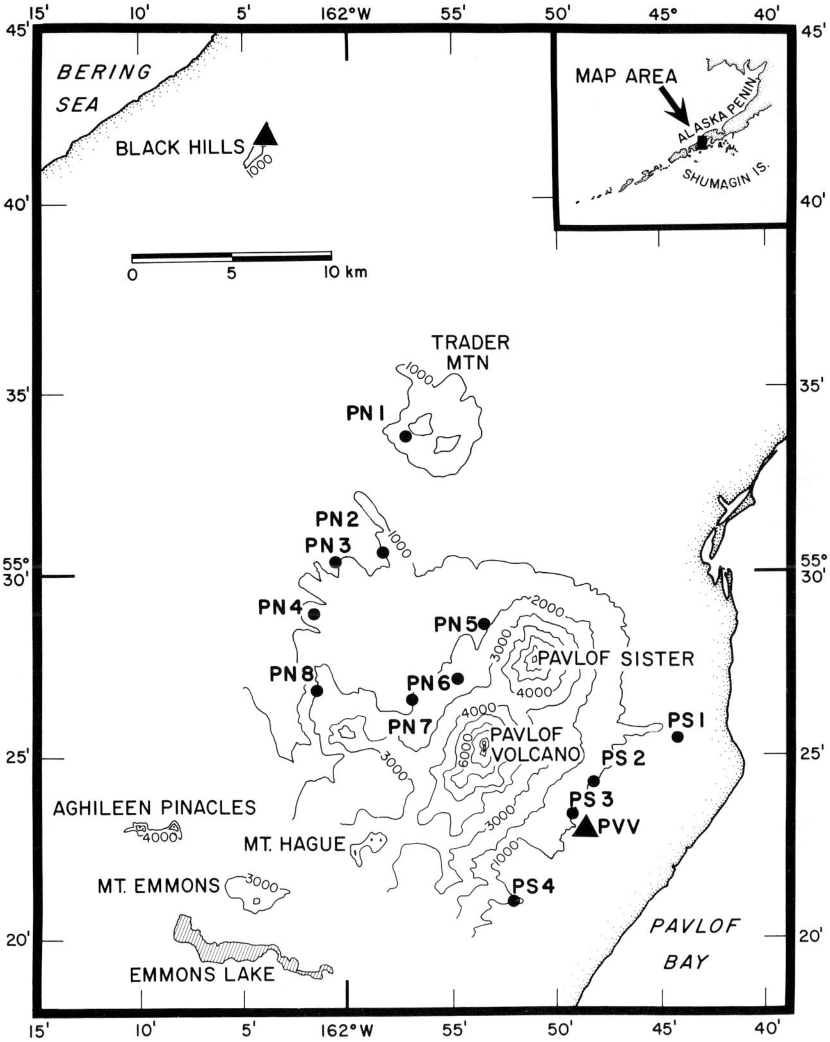

Fig. 1. Index map of seismic stations around Pavlof Volcano. **PVV** and **Black Hills** were installed in 1973; all others in 1976. **PN1** and **PN5** were removed in 1979; **PN3** in 1980; and **PN4**, **PN7**, **PN2**, **PS2**, and **PS3** in 1981. **PVV** and **Black Hills** (**triangles**) were upgraded to three-component stations in 1982

nic tremor. Locatable high-frequency or A-type events have not been recorded at Pavlof. Examples of the three types of events are shown in Fig. 2.

Data reduction consisted of counting the daily number of B-type events and explosion earthquakes over a threshold of M = 1.0 (based on amplitude, hence equivalent to M_L; coda durations are about 5-10 s for the smallest events), and converting volcanic tremor episodes into an equivalent number of events based on typical B-type event coda durations of 20-25 s. (Thus, 1 h of

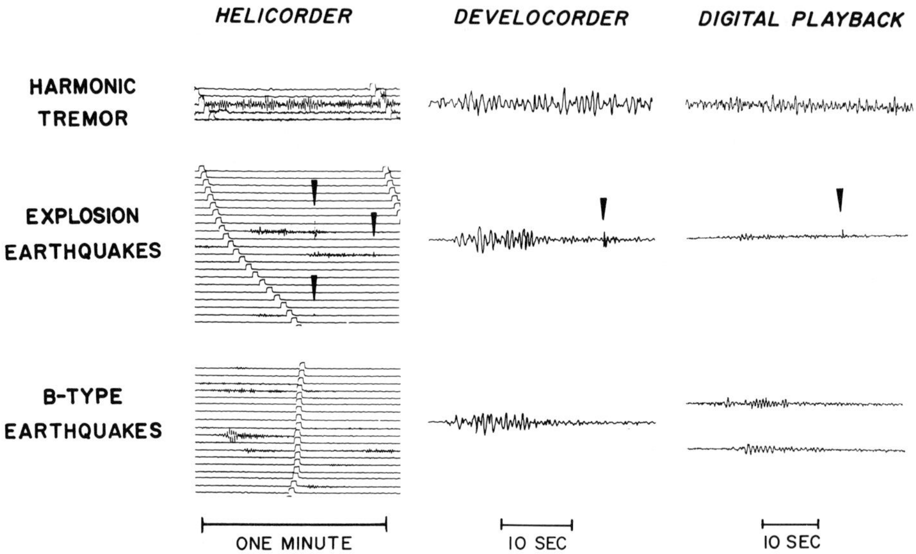

Fig. 2. Examples of tremor, explosion earthquakes, and B-type events from station PVV vertical component, as recorded on helicorder, develocorder, and digital playback units. **Arrowheads** point to the air-shock phase of explosion earthquakes. Note that the dominant frequency in all cases is between 1.0 and 1.5 Hz. The tremor episode, as displayed on the helicorder, is less than one drum revolution (= 15 min) in duration

continuous tremor is equivalent to 144-180 events; the mean value of 162 events per hour of tremor was used for consistency. Because the uncertainty is high, and also to keep the plots at standard scale, maximum values of 2,200 events per day are shown in Fig. 3.)

There are several reasons for converting tremor to the equivalent number of B-type events. First, the B-type events, explosion earthquakes, and tremor have common attributes such as spectra, trace amplitude ratios at different station pairs, and overall signal characteristics; all occur in higher numbers and generally at higher amplitudes during eruptions, and all are believed to be related to degassing (McNutt 1986). Second, during tremor episodes, individual events cannot be seen, so during times of highest seismic energy release, the number of discrete events would be zero. It is desired to display on a single plot (e.g., Fig. 3) a value which quantifies the seismicity. A single tremor episode of 30-min duration has approximately the same seismic energy as 81 discrete events of the same amplitude, thus, the conversion achieves the stated purpose. Alternatively, separate counts can be made of B-type events, explosion earthquakes, and either number or cumulative duration of tremor episodes. Such separate counts were made for subsets of the total data set, then all data were systematically reduced.

Plots of the daily number of events, reduced according to this scheme, are shown in Fig. 3. Each magmatic eruption accompanied by a tremor episode is marked with a vertical arrow above the data, and each phreatomagmatic eruption consisting of a series of explosion earthquakes is marked with a horizontal bar above the data. All magmatic eruptions except October 18-22, 1976, were witnessed, usually by pilots, whereas the start and end dates of phreatomagmatic eruptions correspond to the first and last explosion earthquakes recorded on seismograms. There were no confirmed magmatic eruptions that occurred without accompanying tremor. Detailed information on the eruptions is given in McNutt (1985, 1987a, b), as well as SEAN Bulletins (1975-1986).

3 Analysis and Results

The present work focuses on an aspect of Pavlof seismicity that has previously been neglected: a systematic study of the changes in seismicity which preceded eruptions. Seismic precursors that might occur include: (1) systematic changes in event locations; (2) systematic changes in event sizes; (3) systematic changes in frequency content of individual events; (4) systematic changes in the frequency of occurrence vs magnitude relation (b-value); (5) seismic quiescence; (6) systematic changes in focal mechanisms; (7) changes of event types; (8) tidal (or other) periodicities; and (9) systematic changes in the rate of occurrence of earthquakes. The Pavlof data were examined for each of these possibilities, and each is treated separately below:

1. Event Locations. Changes in event locations could not be evaluated from 1973-1975 because only a single station (PVV) was in existence. Thereafter, examination of seismograms and locations showed no significant changes over time. All the B-type events, explosion earthquakes, and tremor episodes appear to originate at shallow depths in the upper 0.3 to 1.6 km of the volcanic edifice (McNutt 1986). Volcanic tremor could not be located using standard techniques, but tremor displayed the same relative amplitudes at different stations as did explosion earthquakes and B-type events, and hence is assumed to have a similar hypocentral location. Prior to April 1986, no high-frequency or A-type events were recorded; however, several may have occurred in April 1986 (J. Taber, written commun., 1986), but they failed to trigger the event-triggered automatic recording system, and hence could not be located. (Only one Pavlof station, PVV, is recorded continuously; all other stations are connected to the event-triggered system).

2. Event Sizes. Pavlof B-type events span the restricted magnitude range (M_L equivalent) from 0.5 to 2.3 (McNutt 1986), but the dynamic range of the continuous recording system (helicorder) at any given time is only about 26 dB, or 1.3 magnitude units. (The dynamic range of the event-triggered digital system is 60 db, but it does not record continuous data.) Within the narrow range of 1.3 magnitude units, no systematic changes have been observed, except for the September 1981 eruption, when average event size increased about 2 weeks before the eruption,

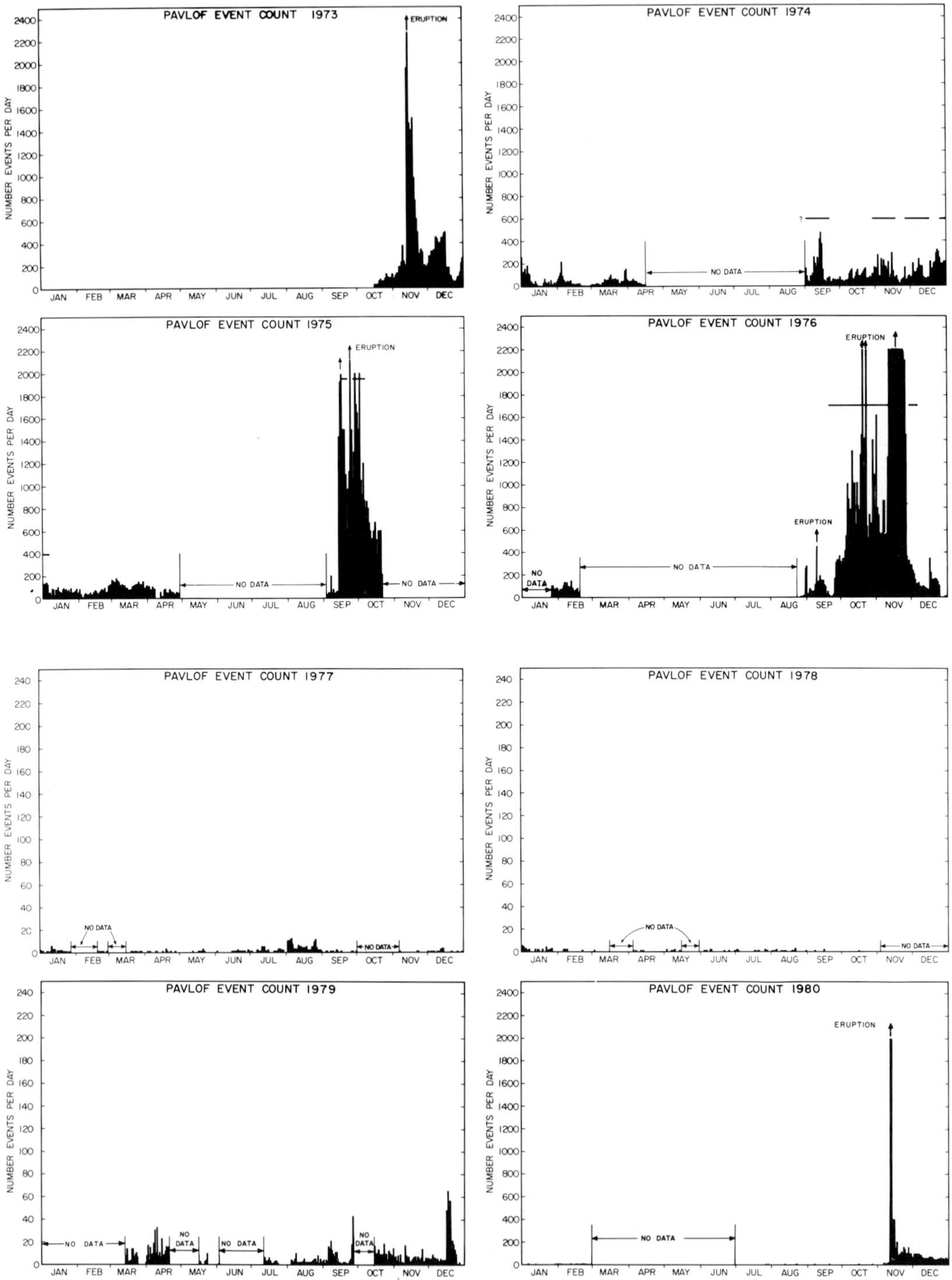
PAVLOF EVENT COUNT 1973
ERUPTION
PAVLOF EVENT COUNT 1974
NO DATA
PAVLOF EVENT COUNT 1975
ERUPTION
NO DATA
NO DATA
PAVLOF EVENT COUNT 1976
ERUPTION
ERUPTION
NO DATA
NO DATA
PAVLOF EVENT COUNT 1977
NO DATA
NO DATA
PAVLOF EVENT COUNT 1978
NO DATA
NO DATA
PAVLOF EVENT COUNT 1979
NO DATA
NO DATA
NO DATA
NO DATA
PAVLOF EVENT COUNT 1980
ERUPTION
NO DATA
NUMBER EVENTS PER DAY
JAN FEB MAR APR MAY JUN JUL AUG SEP OCT NOV DEC

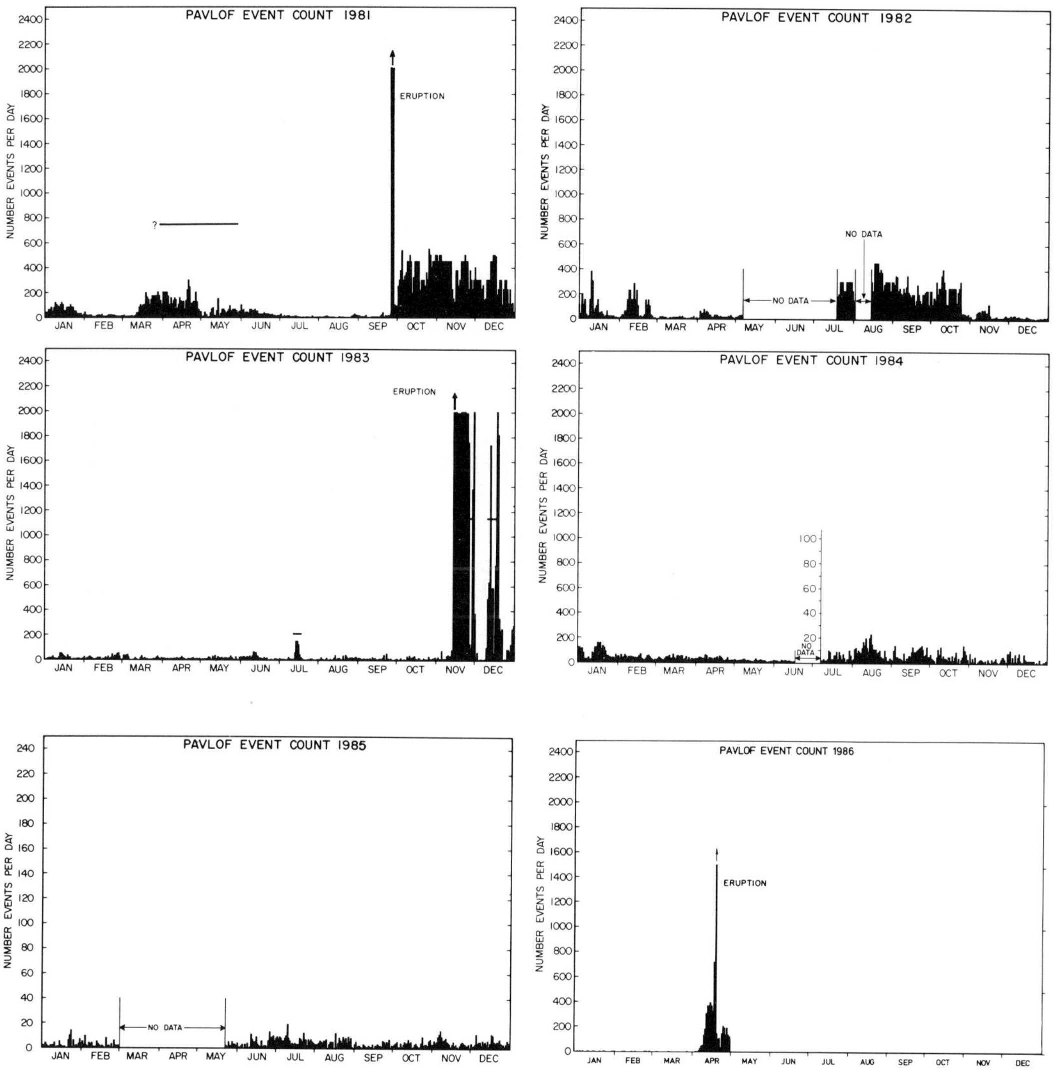

Fig. 3. Yearly plots of the numbers of earthquakes at Pavlof Volcano for the period October 1973 to April 1986. Plots show the number of B-type events and explosion earthquakes per day. Periods of volcanic tremor are normalized to the equivalent number of discrete events as explained in the text. **Vertical arrows** indicate eruptions accompanied by lava fountaining; **horizontal bars** above the data indicate time periods during which explosion earthquakes were occurring at a rate of more than one per day. Note that the years 1977, 1978, 1979, July-December 1984, and 1985 are plotted at a tenth of the vertical scale used for the other years. Data gaps are indicated by **no data**

then decreased again a few days before the eruption. No statistical tests were performed, however, to assess the significance of this observation. In general, no correlation between the magnitude of events, and either volume or violence of the subsequent eruptions, has been observed. The April 1986 eruption was preceded by 8 days of events with long durations (about 3-6 min) but normal amplitudes.

3. Frequency Content. All significant spectral peaks for Pavlof B-type events, explosions, and tremor lie in the narrow range from 0.8-3.0 Hz (McNutt 1986, 1987b). While fluctuations do occur in the positions and relative amplitudes of some spectral peaks, the fluctuations over a time period of only 15 min can be nearly as great as those between events a year or more apart (McNutt 1987b). No systematic and statistically reliable changes in frequency content have been observed prior to eruptions. Rather, the very occurrence of volcanic tremor, regardless of its exact frequency content, is itself a precursor to eruptions with reliability comparable to rates of occurrence of discrete events (see below).

4. Frequency of Occurrence vs Magnitude (b-Value). The b-values, the coefficients in the frequency-magnitude relation log N = a - bM, were determined for several different time periods using the maximum likelihood method of Utsu (1965). The range of only 1.8 magnitude units at Pavlof is on the low side for Utsu's method, but is sufficient for comparing similar data sets to each other, as is done here. Separate determinations of the b-value were made for background, preeruptive, eruptive, and posteruptive periods (Table 1). For these determinations, preeruptive and posteruptive periods were chosen to be 4 days in length, which is the time period during which seismicity changed from background to eruptive levels (or vice versa) for

Table 1. Pavlof Volcano 1974 eruption[a]

	Maximum likelihood b-values			
Background	2.59 ±0.10 Jan	2.02 ±0.07 Feb	1.91 ±0.05 Mar	2.10 ±0.08 Apr
Preeruptive	2.28 ±0.09			
Eruptive	2.45 ±0.10 Oct	2.40 ±0.08 Nov	2.18 ±0.09 Dec	
Posteruptive	2.28 ±0.09			

[a] A sample of background seismicity = 1 month of data; samples of preeruptive, and posteruptive seismicity = 4 days of data, and samples of eruptive seismicity = actual duration of eruption (details are given in the text, Table 3, and McNutt and Beavan 1981).

the 1974 eruptions. Unfortunately, b-values observed during background periods alone range from 1.9 ± 0.1 to 2.6 ± 0.1, while those preceding and during eruptions range from 2.2 ± 0.1 to 2.5 ± 0.1. Thus, values observed during these background periods span all the observed values, including values measured throughout the time period 1973-1986, regardless of eruptive status. The b-value is thus of limited use as an eruption precursor at Pavlof.

5. **Seismic Quiescence.** Sudden seismic quiescence either during a background period or during a period of increased activity has preceded some eruptions (Shimozuru 1986; Newhall and Endo 1987). Long-term relative quiescence preceding eruptions has occurred at Pavlof (compare the quiet periods of July-September 1981 or August-October 1983 with the high background levels of January-April 1975 or July-October 1982 in Fig. 3.) No short-term abrupt periods of quiescence, lasting for hours or days, have however, preceded any of the eruptions at Pavlof.

6. **Focal Mechanisms.** Since Pavlof has not had any locatable high-frequency (A-type) events, it was not possible to evaluate this parameter by conventional means (i.e., plotting first arrivals on a stereo net). The B-type events and explosion earthquakes have similar mechanisms as evidenced by similar (but emergent) first arrivals, codas (except for the air wave), and amplitude ratios at different stations. Alternatively, these parameters may indicate that the two types of events occur more or less in the same place, rather than that their mechanisms are the same. No systematic changes have been observed in these parameters prior to eruptions. Selected three-component seismograms from station PVV (after July 1982) were examined to test for possible changes in amplitude between vertical and horizontal components. No such changes were observed.

7. **Changes of Event Types.** Strictly speaking, B-type events, explosion earthquakes, and volcanic tremor are different types of events, although all share common seismogram characteristics and all are believed to be related to degassing (McNutt 1986). If only B-type events are being recorded, and then an explosion earthquake occurs, this is a change of event type, and also signals the start of a phreatomagmatic eruption (it <u>is</u> the seismic expression of one pulse of the eruption). Six of the ten magmatic eruptions were preceded by explosion earthquakes, while the other four began without any explosion earthquakes (see Table 3). There were 18 phreatomagmatic eruptions in all. Thus, the occurrence of explosion earthquakes, considered here as a change of event type, was a magmatic eruption precursor in 6 of 18 cases (33%).

Changes of event type as described here would probably be noted by seismologists or other trained personnel, but would be difficult to discern using a computer or event detector (also see below). Use of a short-period microbarograph would help in identifying explosion earthquakes instrumentally, by being able to associate the direct air wave with the seismograms.

8. Tidal (or Other) Periodicities. Tidal or other short-period periodicities (<24 h 50 min) are sometimes observed in seismicity data from volcanic regions before eruptions (e.g., Mauk and Kienle 1973; McNutt and Beavan 1981, 1984). To test for such periodicities requires that the data be sampled more frequently than once per day, preferably every 1-3 h. Only the data for the time period September 1974-April 1975 were systematically sampled every 2 h. Tidal periodicities in seismicity were observed for subsets of data of length 3-4 days before and 3-4 days after periods of explosion earthquakes in fall 1974; results are described in detail in McNutt and Beavan (1981). The remaining data for the time period 1973-1986 were sampled once per day (for example, Fig. 3 shows the number of events per day), and were not suitable for tidal analyses. The helicorder records still exist, so finer sampling could be carried out, but this has not been done.

Two other considerations, however, rule out tidal periodicities for the other eruptions: (1) the seismicity rate increased over a time period less than a tidal period for many of the magmatic eruptions; and, (2) low seismicity rates produce many hours with zero or only a few events. Statistical significance tends to be low for such data sets.

The numbers of earthquakes increased abruptly for seven eruptions, and seismicity rates were too low to test for six eruptions. Seismograms were examined for the remainder, but no regular patterns were observed, except as described in McNutt and Beavan (1981).

In addition to testing for tidal periodicities, subsets of seismicity time series of 2 to 4 days duration were compared with atmospheric pressure time series, recorded at Sand Point (80 km E) and Cold Bay (35 km SW). No systematic trends were observed.

9. Rate of Occurrence. Data to evaluate this parameter (Fig. 3) are plentiful, relatively easy to evaluate, and yield the most consistent and reliable results. The basic idea is that background seismicity is low, but numbers of earthquakes increase before and during eruptions. The task is then to quantify the seismicity and find the best ways to test for precursory activity. (Data were analyzed for the time period October 1973-June 1984 and April 1986. Data for July 1984-March 1986 are shown in Fig. 3, but were not available at the time of analysis.) First, basic seismic data on the eruptions are tabulated. The average daily numbers of earthquakes during background periods, phreatomagmatic eruptions, and magmatic eruptions are shown in Table 2. The average background numbers increase by a factor of 14 during phreatomagmatic eruptions, and by a factor of 58 during magmatic eruptions. The background periods, however, consist of three distinct ranges which differ by a factor of 21. Background seismicity is computed from months when no explosions or magmatic eruptions occurred. Average background seismicity was much higher during October 1973-December 1976 and November 1980-June 1984, time periods including eruptions, than it was during January 1977-October 1980, a time period of no eruptions.

Table 2. Pavlof Volcano average daily numbers of earthquakes

Eruptive period	Number of samples[a]	Mean daily numbers of earthquakes	Standard deviation[b]
Background:			
All	57	33	41
Oct. 1973-Dec. 1976	14	84	48
Jan. 1977-Oct. 1980	15	4	4
Nov. 1980-June 1984	28	23	13[c]
Phreatomagmatic eruptions:			
All	18	498	592
Phreatomagmatic only	10	185	230
Mixed only	8	890	685
Magmatic eruptions:	10	1919	535

[a] A sample of background seismicity = 1 month of data; whereas a sample of eruptive seismicity = actual duration of eruption, regardless of length.

[b] Three of the standard deviations are quite large, indicating that a normal distribution does not adequately model the data. In these cases the standard deviations are not meaningful.

[c] 68% of values (= 19 of 28) are within 13 of mean. Calculator gave error condition. Sample distribution appears bimodal.

Next, detailed data on individual eruptions are examined. Table 3 lists the dates of each eruption, the average number of events per day during the eruption, and the average number of events per day during the 5 days preceding the eruption. (Precursor time ranged from 1-10 days, with an average of about 5 days.) The average number of earthquakes per day for the 5 days preceding eruptions in 389, which is a factor of 11 greater than the average for background periods. Some of the eruptions were preceded by seismicity that was less than the average background level. For all but five eruptions, seismicity was higher during the eruption than it was before the eruption. The five exceptions are those eruptions that closely followed another eruption.

During all the eruptions but two (12-14 March 1974 and 13 January 1975), the numbers of earthquakes exceeded 150 events per day for at least 2 consecutive days during the eruption, or 300 events during a single day (= average of 150 per day for 2 days). The average values during six eruptions, however, were less (Table 3). Thus, the data shown in Fig. 3 were searched to find all the dates on which the numbers of earthquakes exceeded 150 events per day for 2 consecutive days, or 300 events in 1

Table 3. List of Pavlof Volcano eruptions, October 1973-April 1986, and Accompanying Seismicity

Date	Type of eruption	Average No. of events per day during eruption	Average No. of events per day during previous 5 days	No. of explosions (E) or No. hours volcanic tremor (T)
Nov. 13-14, 1973	Magmatic	2084	245	35 T
Mar. 12-14, 1974	Phreatomagmatic	50	22	12 E[d]
Before Sept. 1-21, 1974	Phreatomagmatic	143	19[a]	54 E
Oct. 29-Nov. 17, 1974	Phreatomagmatic	137	77	208 E
Nov. 25-Dec. 16, 1974	Phreatomagmatic	97	32	310 E
Dec. 25-Jan. 6, 1975	Phreatomagmatic	143	215	393 E
Jan. 13, 1975	Phreatomagmatic	109	77	3 E
Sept. 13-15, 1975	Magmatic	1784	56	23 T
Before Sept. 18-21, 1975	Phreatomagmatic	869	1395	15 E[d]
Sept. 23-24, 1975	Phreatomagmatic	1800	820	19 T
Sept. 25-Oct. 6, 1975	Phreatomagmatic	782	1020	30 E[d]
Sept. 8, 1976	Phreatomagmatic	117	67	5 E
Sept. 9, 1976	Magmatic (?)	452	79	2 T
Sept. 19-Nov. 20, 1976	Phreatomagmatic	601[b]	117	568 E
Oct. 18-22, 1976	Magmatic (?)	1916	904	96 T
Nov. 10-22, 1976	Magmatic	2557	679	240 T
Nov. 28-Dec. 4, 1976	Phreatomagmatic	245	1770	20 E[d]
Nov. 11, 1980	Phreatomagmatic	2200	10	1 E

Table 3 (cont.)

Date	Type of eruption	Average No. of events per day during eruption	Average No. of events per day during previous 5 days	No. of explosions (E) or No. hours volcanic tremor (T)
Nov. 11-12, 1980	Magmatic	2200	10	40 T
Before? Mar. 30-May 28, 1981	Phreatomagmatic	47	5[c]	277 E
Sept. 26-27, 1981	Magmatic	2000	12	49 T
July 12-18, 1983	Phreatomagmatic	64	9	119 E
Nov. 11-13, 1983	Phreatomagmatic	539	20	15 E
Nov. 14-16, 1983	Magmatic	2200	28	73 T
Nov. 21-30, 1983	Phreatomagmatic	1608	2100	49 E
Dec. 11-18, 1983	Phreatomagmatic	818	22	47 E
April 16, 1986	Phreatomagmatic (?)	400	329	?
April 18-19, 1986	Magmatic	1115	370	8 T

[a] This value is for the last 5 days of April 1974; the Sept. 1974 eruption was already in progress when seismic recording resumed on Sept. 1.
[b] Excluding magmatic eruptions Oct. 18-22 and Nov. 10-20.
[c] First 10 days of March only.
[d] Estimated value.

day. There were 40 such episodes; these are listed in Table 4. Only 12 (30%) were precursors, varying from 2 h to 10 days before eruptions. For ten cases (25%), the seismicity increased to 150 events per day after the onset of the eruption. False alarms accounted for 17 cases (43%) since no eruption occurred, and one was a special case (March 24, 1981), since a previous case had occurred only 5 days before, and there is uncertainty concerning the start date of the eruption. Three eruptions were missed, the 12-14 March 1974 and 13 January 1975 phreatomagmatic eruptions cited above, and the 10 November 1976 magmatic eruption, which began during a period of high seismicity accompanying a phreatomagmatic eruption which was already in progress.

Table 4. Pavlof seismic precursors to eruptions[a]

Date that seismicity exceeded 150 events/day for 2 consecutive days	Date of eruption start (M = magmatic, P = phreatomagmatic)	Comment
November 8, 1973	November 13, 1973 M	Precursor, 5 days
December 29, 1973	None	False alarm
----	March 12, 1974 P	No precursor
September 1, 1974 (?)	September 1, 1974 P	Precursor, <1 day (?)
September 7, 1974	September 1, 1974 P	Seismicity increased after eruption started
October 31, 1974	October 28, 1974 P	Seismicity increased after eruption started
November 6, 1974	October 28, 1974 P	Seismicity increased after eruption started
December 7, 1974	November 26, 1974 P	Seismicity increased after eruption started
December 20, 1974	December 26, 1974 P	Precursor, 6 days
March 1, 1975	None	False alarm
March 6, 1975	None	False alarm
September 13, 1975	September 13, 1975 M	Precursor, <1 day
August 30, 1976	September 8-9, 1976 P,M	Precursor, 9 days
September 12, 1976	September 19, 1976 P	Precursor, 7 days
September 25, 1976	September 19, 1976; Oct. 18, 1976 M (?)	Seismicity increased after eruption started
----	November 10, 1976 M	No precursor
December 16, 1976	None	False alarm
November 11, 1980	November 11, 1980 M	Precursor, <1 day

a Underlined dates in column one are those for which seismicity was also two times greater than the average of the previous 3 days.

Table 4 (cont.)

Date that seismicity exceeded 150 events/day for 2 consecutive days	Date of eruption start (M = magmatic, P = phreatomagmatic)	Comment
March 19, 1981	March 27, 1981 ? P	Precursor, 8 days (?)
March 24, 1981	March 27, 1981 ? P	Special case (see text)
April 1, 1981	March 27, 1981 ? P	Seismicity increased after eruption started
April 20, 1981	March 27, 1981 ? P	Seismicity increased after eruption started
September 26, 1981	September 26, 1981 M	Precursor, <1 day
October 2, 1981	None	False alarm
November 16, 1981	None	False alarm
December 9, 1981	None	False alarm
December 23, 1981	None	False alarm
January 9, 1982	None	False alarm
February 7, 1982	None	False alarm
July 19, 1982 (?)	None	False alarm
September 19, 1982	None	False alarm
September 24, 1982	None	False alarm
September 30, 1982	None	False alarm
October 3, 1982	None	False alarm
October 22, 1982	None	False alarm
July 14, 1983	July 12-18, 1983 P (?)	Seismicity increased after eruption started
November 14, 1983	November 14, 1983 M	Precursor, <1 day
November 29, 1983	November 21-30, 1983 P	Seismicity increased after eruption started
December 11, 1983	December 11, 1983 P	Precursor, <1 day

Table 4 (cont.)

Date that seismicity exceeded 150 events/day for 2 consecutive days	Date of eruption start (M = magmatic, P = phreatomagmatic)	Comment
December 17, 1983	December 11-18, 1983 P	Seismicity increased after eruption started
December 30, 1983	None	False alarm
April 12, 1986	April 16, 1986 M (?)	Precursor, 4 days

The tabulation described above does not consider the number of earthquakes per day on preceding days. If we add to the entries shown in Table 4 the further restriction that the numbers of earthquakes should also be twice the average of the previous 3 days, then the number of cases reduces to 24. These are underlined in Table 4. For these 24 cases, eleven (46%) were precursors, five (21%) were increases in seismicity after the eruption onset, eight (33%) were false alarms, and four eruptions were missed. For the restricted data set relative to the original, the precursor percentage increased, the false alarm percentage decreased, and the percentage of cases where numbers of earthquakes increased after the beginning of an eruption was approximately the same. However, the restricted data set missed more eruptions, indicating that there is a trade-off between successes, failures, and false alarms.

These searches were based on absolute numbers of events above a given threshold per day. A second set of searches was also made, based on relative numbers of events. For these searches, the number of events each day was compared to the average of the 3 previous days, and tabulated if the numbers on the fourth day exceeded the average by a factor of 3 (Table 5). Thirty-four such cases occurred; 16 (47%) preceded 12 eruptions by 2 h to 10 days (4 eruptions had a pair of precursors each), 14 (41%) were false alarms, 3 (9%) were increases in seismicity after the onset of eruption, 1 (3%) was a special case (a small steam emission may have occurred on 31 December, 1979, but no explosion earthquakes were recorded), and 7 eruptions were missed. If the threshold is increased so that seismicity on the fourth day must be a factor of 5 higher than the average of the previous 3 days, then 22 cases remain (underlined values in Table 5). Of these, 11 (50%) preceded 10 eruptions, 7 (32%) were false alarms, 2 (9%) were increases in seismicity after the eruption onset, and 1 (5%) was the special case already mentioned. Nine eruptions were missed. For the higher threshold, therefore, the percentage of precursors increased slightly, the percentage of false alarms decreased, the percentage of increases in seismicity after eruption onset was about the same, and the number of missed eruptions increased.

Table 5. Pavlof seismic precursors to eruptions

Date seismicity increased 3 x above previous 3 days	Date of eruption start (M = magmatic; P = phreatomagmatic)	Comment
November 13, 1973	November 13, 1973 M	Precursor, <1 day
September 1, 1974 ?	September 1, 1974 P	Precursor, <1 day (?)
December 18, 1974	December 26, 1974 P	Precursor, 8 days
September 7, 1975	September 13, 1975 M	Precursor, 6 days
September 13, 1975	September 13, 1975 M	Precursor, <1 day
August 30, 1976	September 8, 1976 P	Precursor, 9 days
September 9, 1976	September 9, 1976 M	Precursor, <1 day
September 25, 1976	September 19, 1976 P	Seismicity increased after eruption started
November 11, 1976	November 10, 1976 M	Seismicity increased after eruption started
December 16, 1976	None	False alarm, may be artifact
August 2, 1977	None	False alarm
September 5, 1979	None	False alarm
December 16, 1979	December 31, 1979 (?)	Possible eruption (?), but no explosions recorded
November 6, 1980	November 11, 1980 M	Precursor, 5 days
November 11, 1980	November 11, 1980 M	Precursor, <1 day
March 11, 1981	March 27, (?) 1981 P	Precursor, 16 days
August 29, 1981	None	Marginal, within measurement error
September 26, 1981	September 26, 1981 M	Precursor, <1 day
January 3, 1982	None	False alarm
January 9, 1982	None	False alarm
February 7, 1982	None	False alarm
February 21, 1982	None	False alarm

Table 5 (cont.)

Date seismicity increased 3 x above previous 3 days	Date of eruption start (M = magmatic; P = phreatomagmatic)	Comment
April 3, 1982	None	False alarm
July 19, 1982 (?)	None	False alarm
November 6, 1982	None	False alarm
January 13, 1983	January 13, 1983 (?)	Explosions (?), possible eruption - not confirmed
July 8, 1983	July 12, 1983 P	Precursor, 4 days
July 13, 1983	July 12, 1983 P	Seismicity increased after eruption started
November 5, 1983	November 14, 1983 M	Precursor, 9 days
November 14, 1983	November 14, 1983 M	Precursor, 1 day
December 10, 1983	December 11, 1983 P	Precursor, 1 day
December 26, 1983	None	False alarm
January 12, 1984	None	Marginal; within measurement error
April 12, 1986	April 16, 1986 P-M?	Precursor, 4 days

Other parameter values were tested as well, only one of which is discussed here. The data were searched for times when the numbers of earthquakes for a 2-day period increased by a factor of 2 above the average of the previous 5 days. For this set of parameters, 70 episodes were found (Table not shown); 19 (27%) were precursors to 13 eruptions, 38 (54%) were false alarms, 12 (17%) were increases that took place after eruption onset, and 4 eruptions were missed. Compared with the previous test, three eruptions had earlier precursors using the latter parameters. The percentage of precursors was lower, however, and the false alarm rate was higher, indicating that the same general tradeoff exists between successes, failures, and false alarms.

Limitations. The approach used in this section was retrospective; parameters obtained from the data (e.g., number of events per day = 150) were used to test the same data. Retrospective data fitting is quite common in geophysics, but always introduces the problem of self-deception. Gabrielov et al. (in press, 1988) show an example in which one set of random numbers was used to "predict" a second set of random numbers. With proper "adjustment" of the parameters, surprisingly good results were obtained.

It is desirable, then, to derive the model parameters from one suite of data and test them on a separate data set. Fortunately, Pavlof provided such an opportunity because eruptions continued in 1986 after the cutoff date (April 30, 1986) for data used in this work. Using the best parameters above LTA = 3 days, STA = 2 days, threshold = 3, and absolute number of events greater than 150 per day, the May-December 1986 data yielded five triggers. Three were precursors, one a false alarm, and one eruption was missed. While the number of cases is too small to yield valid statistics, the results are in general agreement with those obtained above, and hence lend support to the choice of parameters. (Data on the 1986 eruption are being prepared for publication separately.)

More rigorous methods of testing for seismicity rate fluctuations exist, including Haberman's (1983) z-statistic and the cumulative sum method. These can and will be applied to the Pavlof data, but have been deferred here because extensive reforming of the data will first be required.

4 Discussion and Conclusions

4.1 Pavlof in Relation to Other Volcanoes

When a volcano which has had no previous historic activity begins to have earthquake swarms, what is the probability that an eruption will occur? Attempts to answer this question, a fundamental one in volcano seismology, generally involve study and review of seismicity at other, similar volcanoes that eventually erupted. Decker (1986) reviews seismic and other techniques used to forecast eruptions. A few well-studied volcanoes with frequent eruptions such as Kilauea (Klein 1984; SEAN Bulletins), Mt. Etna (Cosentino et al. 1982), Ruapehu (Latter 1981; SEAN Bulletins), Aso (SEAN Bulletins), Asama (Minakami et al. 1970), Sakurajima (SEAN Bulletins), and recently Mount St. Helens (Malone et al. 1981; Endo et al. 1981; Swanson et al. 1983, 1985) have contributed greatly to the available data set. Data for Pavlof should be included in the worldwide sample.

A brief statement of caution, however, is necessary. Pavlof is different from the other volcanoes cited above in that it has virtually no A-type or high-frequency earthquakes, but instead has thousands of B-type or low-frequency events. In this regard, Pavlof is atypical, and its seismic activity has physical implications as well, (e.g., Pavlof may be open-vent), which may preclude direct comparison of precursors between Pavlof and other volcanoes.

For most of the others, A-type earthquakes occur at various depths presumably because magma is being injected into cracks or dikes at or near those depths. Earthquakes are thought to occur either at crack tips (e.g., Chouet 1981; Aki 1984) or on faults in the adjacent region which experiences local changes in ambient stresses (Hill 1977). The physical basis for prediction is thus that magma moves from depth towards the vent and, in so doing, generates stresses along the way which produce earthquakes. This line of thought illustrates why earthquakes should precede eruptions, and also why migration of earthquake foci from deep to shallow depths is sometimes observed (e.g., Eaton and Murata 1960).

Pavlof events, on the other hand, all occur in the upper few hundred meters of the volcanic edifice, and appear to be related to degassing (McNutt 1986). Thus, there is no detectable depth migration, and since degassing is a fundamental eruptive process, the events occur mostly during the times of eruptions, rather than before. The data shown in Tables 2-5 support these observations. Pavlof's eruptive behavior and seismicity, then, for the time period October 1973-April 1986, have been of the open-vent type, and may best be used to address a separate, but related, question: Once a volcano erupts and becomes open-vent, what is the probability of subsequent eruptions within a given (short) time period?

4.2 Algorithms for Detecting Seismicity Rate Changes

The strategies used above to identify seismicity rate changes can be easily converted to computer algorithms for use on automatic seismic data acquisition and analysis systems. The first, simply whether or not some chosen number of events above a given threshold occurred on a given day, is the easiest to test for. In fact, as an alternative, technicians could be trained to reduce the data, or event counters could also be used (e.g., Ward et al. 1974). The effects of changes in magnification, instrument type or location, and varying background noise, however, are nonlinear and would be different for each type of counter.

The other tests, described above, involved computation of a short-term average number of events (STA), a long-term average number of events (LTA), and a comparison to test whether or not the ratio of STA/LTA had exceeded a given threshold value. The tests were: LTA = 3 days, STA = 1 day, threshold = 3; LTA = 3 days, STA = 1 day, threshold = 5; and LTA = 5 days, STA = 2 days, threshold = 2. Each produced triggers, some of which were precursors, some false alarms, and some increases in seismicity after the onset of eruption. The three examples shown all gave triggers during the time period January 1977-October 1980, when no eruptions occurred and seismicity was very low. This time period illustrates the differences between the two general schemes used to search the data. The absolute test required the number of events to exceed 150 per day, a condition not met between January 1977-October 1980. The relative test (STA/LTA) tests only for changes in seismicity rate, regardless of the absolute level. Also, as discussed above, the various parameters produced trade-offs in the number of precursors, number of false alarms, and number of missed eruptions: strictly speaking, the category of "seismicity increased after eruption started" can be treated as a missed eruption.

Of the values given above, those that gave best performance were a combination of the absolute and relative schemes: LTA = 3 days, STA = 2 days, threshold = 3, and absolute number of events greater than 150 per day. These parameters yielded 46% precursors of 2 h to 10 days duration, 33% false alarms, and the remainder missed eruptions (in five of which the trigger occurred after eruption began, and four which did not trigger at all). Of the four eruptions whose seismicity did not trigger the algorithm, one was the smallest of all (12-14 March 1974) and posed little hazard, and the other three (26 November 1974; 19 September 1976; and 11 November 1976) occurred in close association in time with other eruptions.

4.3 Volcanic Tremor

Volcanic tremor alone, as opposed to the equivalent number of B-type or low-frequency events, can be treated as a separate precursor. Details of the occurrence and characteristics of Pavlof tremor are given in McNutt (1987b). Briefly, 23% of the 488 tremor episodes occurred during eruptions, 42% occurred during 10 days preceding eruptions (although 60% of these preceded a single eruption), 16% occurred during 10 days following eruptions, and 26% occurred during background periods. Percentages do not add up to 100 because of overlapping time periods preceding and following some eruptions. Tremor occurrence is thus comparable to numbers of earthquakes, as discussed above, in terms of its usefulness as a precursor.

Trained personnel can easily convert tremor to the equivalent number of B-type events, but computer algorithms are less successful. Typical event-detected seismic data acquisition systems with STA = 1 s, LTA = 1 min, will trigger only once at the onset of a tremor episode, assuming the tremor onset is impulsive, but will then return to normal because the LTA will reset at the higher tremor value. If the onset of tremor is gradual or emergent, event-detected systems may not trigger at all. For such cases, an analog device which simply measures absolute amplitude within the narrow frequency range of tremor would suffice as a trigger (see McNutt (1987b), for a more complete discussion of this topic). Event counters (e.g., Ward et al. 1974) do not work as well, with either volcanic tremor or low-frequency (B-type) events, as they do with impulsive, high-frequency (A-type) events.

4.4 Other Data

The results presented above demonstrate that with data on numbers of earthquakes alone it was possible to forecast eruptions 2 h to 10 days in advance with probabilities of 27-50%. The number of missed eruptions varied from two to nine, or 7-32% of the 28 eruptions. The reliability may be difficult to improve upon at Pavlof if only seismicity data are recorded in the future. At other volcanoes, such as Kilauea (Klein, 1984) and Mount St. Helens (Swanson et al. 1983, 1985), tiltmeters, strainmeters, gas sampling, and other techniques, used in conjunction with seismic data, greatly improve both the forecasting capability and the physical understanding of eruptions and related phenomena. It is thus strongly recommended that strainmeters and tiltmeters, or at least, dry-tilt figures, be installed at Pavlof as soon as possible.

Acknowledgments. I thank M. Reichle and C. Cramer for critical reviews, J.H. Latter and an anonymous referee for helpful comments, and C. Pingree for word-processing the manuscript. P. Reasanburg kindly provided a preprint of his related work. Seismic data was recorded by Lamont-Doherty Geological Observatory of Columbia University in cooperation with the University of Alaska. I thank K. Jacob and J. Taber for continued cooperation in using new data (1984-1987).

References

Aki K (1984) Evidence for magma intrusion during the Mammoth Lakes earthquakes of May 1980 and implications of the absence of volcanic (harmonic) tremor. J Geophys Res 89:7689-7696

Chouet B (1981) Ground motion in the near field of a fluid-driven crack and its interpretation in the study of shallow volcanic tremor. J Geophys Res 86:5985-6016

Cosentino M, Lombardo G, Patane G, Schick R, Sharp ADL (1982) Seismological researches on Mount Etna: state of art and recent trends, Mem Soc Geol It 23:159-202

Decker RW (1986) Forecasting volcanic eruptions. Ann Rev Earth Planet Sci 14:267-291

Eaton JP, Murata KJ (1960) How volcanoes grow. Science 132: 925-938

Endo ET, Malone SD, Noson LL, Weaver CS (1981) Locations, magnitudes, and statistics of the March 20-May 18 earthquake sequence. In: Lipman PW, Mullineaux DR (eds) The 1980 eruptions of Mount St. Helens, Washington. USGS Prof Paper 1250:93-107

Gabrielov AM, Dmitrieva OE, Keilis-Borok VI, Kosobokov VG, Kuznetsov IV, Levshina TA, Mirzoev KM, Molchan GM, Negmatullaev SK, Pisarenlo VF, Prozoroff AG, Rinehart W, Rotvain IM, Shebalin PN, Shnirman MG, Schreider SY (1988) Algorithms of long-term earthquakes' prediction. Rev Geofis (in press)

Haberman RE (1983) Teleseismic detection in the Aleutian Island Arc. J Geophys Res 88:5056-5064

Hill DP (1977) A model for earthquake swarms. J Geophys Res 82:1347-1352

Klein FW (1984) Eruption forecasting at Kilauea Volcano, Hawaii. J Geophys Res 89:3059-3073

Latter JH (1981) Volcanic earthquakes, and their relationship to eruptions at Ruapehu and Ngauruhoe volcanoes. J Volcanol Geotherm Res 9:293-309

Malone SD, Endo ET, Weaver CS, Ramey JW (1981) Seismic Monitoring for Eruption Prediction. In: Lipman PW, Mullineaux DR (eds) The 1980 eruptions of Mount St. Helens, Washington. USGS Prof Paper 1250:803-813

Mauk FJ, Kienle J (1973) Microearthquakes at St. Augustine Volcano, Alaska, triggered by earth tides. Science 182:386-389

McNutt SR (1985) The eruptive activity, seismicity, and velocity structure of Pavlof Volcano, eastern Aleutians. Ph.D. Thesis, Columbia University, NY, 202 pp

McNutt SR (1986) Observations and analysis of B-type earthquakes,explosions, and volcanic tremor at Pavlof Volcano, Alaska. BSSA 76:153-175

McNutt SR (1987a) Eruption characteristics and cycles at Pavlof Volcano, Alaska, and their relation to regional earthquake activity. J Volcanol Geotherm Res 31:239-267

McNutt SR (1987b) Volcanic tremor at Pavlof Volcano, Alaska, October 1973-April 1986. PAGEOPH 125, 6:1051-1077

McNutt SR, Beavan RJ (1981) Volcanic earthquakes at Pavlof Volcano correlated with the solid earth tide. Nature (Lond) 294:615-618

McNutt SR, Beavan RJ (1984) Patterns of earthquakes and the effect of solid earth and ocean load tides at Mt. St. Helens prior to the May 18, 1980 eruption. J Geophys Res 89:3075-3086

McNutt SR, Jacob KH (1986) Determination of large-scale velocity structure of the crust and upper mantle in the vicinity of Pavlof Volcano, Alaska. J Geophys Res 91:5013-5022

Minakami T (1960) Fundamental research for predicting volcanic eruptions (I) Earthquakes and crustal deformations originating from volcanic activities. Bull Earthquake Res Inst 38:497-544

Minakami T, Utibori S, Miyazaki T, Hiraga S, Gyoda N, Utsunomiya T (1970) Seismometrical studies of Volcano Asama (Part I). Seismic and volcanic activities of Asama during 1934-1969. Bull Earthq Res Inst 48:235-301

Newhall C, Endo ET (1987) Sudden seismic calm before eruptions: illusory or real? Abstract Volume, Hawaii Symposium on How Volcanoes Work, Hilo Hawaii, p 190

SEAN Bulletins (1975-86) Vol 1-11, Smithsonian Inst, Washington, DC

Shimozuru D (1986) Some significant features of pre-eruption volcanic earthquakes. Abstracts Volume, Int Volcanol Congr New Zealand, 1-9 February 1986, p 280

Swanson DA, Casadevall TJ, Dzurisin D, Malone SD, Newhall CG, Weaver CS (1983) Predicting eruptions at Mount St. Helens, June 1980 through December 1982. Science 221:1369-1376

Swanson DA, Casadevall TJ, Dzurisin D, Holcomb RT, Newhall CG, Malone SD, Weaver CS (1985) Forecasts and predictions of eruptive activity at Mount St. Helens, USA: 1975-1984. J Geodyn 3:397-423

Utsu T (1965) A method for determining the value of b in a formula log n = a-bM showing the magnitude-frequency relation for earthquakes. Geophys Bull Hokkaido Univ 13:99-103

Ward PL, Endo ET, Harlow DH, Allen R, Marquez D, Eaton JP (1974) Development and evaluation of a prototype global volcano surveillance system utilizing the ERTS-1 satellite data collection system. USGS Open-File Rep 74/124

Volcanic Tremor of Kliuchevskoi Volcano (1984 Eruption)

E. I. GORDEEV, Y. Y. MELNIKOV, V. I. SINITSYN and V. N. CHEBROV[1]

Abstract

Results of volcanic tremor studies carried out with portable seismic stations at Kliuchevskoi Volcano in July and August 1984 are presented. Analysis of spectra over a long period of observations has shown the stability of the principal spectral characteristics of the wave field through time, which implies the stability of the parameters of the volcanic tremor source. Significant spatial variations in the wave field of tremor are related to the complex structure of the upper layers of the medium within which the surface waves are propagated. Based on the results of numerous observations we have obtained the absorption coefficients for frequency in the range from 0.5 to 3 Hz, for both the horizontal and vertical components. It is shown that the absorption for compressional waves exceeds the absorption for shear waves. Phase velocities of horizontal and vertical components were determined. The values of absorption coefficients and phase velocities of volcanic tremor agree well with the previous determinations at other volcanoes. The detailed analysis of the temporal variation in volcanic tremor intensity helped to reveal the eruption regime with characteristic periods of 5-10 h. The most probable mechanism of such a regime is the excitation of overheated, gas-saturated magma which is in a metastable state. Results of volcanic tremor studies show that tremor may be one of the most reliable, quantitative ways of estimating the energy of eruption.

1 Introduction

A summit eruption of Kliuchevskoi, the largest Eurasian volcano and the most active volcano in the Kurile-Kamchatka volcanic belt, commenced in March 1984. A previous strong flank eruption lasted from March to June 1983 (Khrenov et al. 1985). Successful prediction of this flank eruption (Tokarev 1983) allowed the organization of a wide range of investigations. Unfortunately, however, the high altitude of the eruption (2875 m), and bad winter weather, made it impossible to carry out special seismological observations to study in detail the seismic phenomena accompanying the eruption.

---

[1] Institute of Volcanology, Petropavlovsk-Kamchatsky, 683006, USSR

IAVCEI Proceedings in Volcanology 1
J.H. Latter (Ed.), Volcanic Hazards

Routine seismological observations on the Kliuchevskoi volcanic group have been carried out since 1946 when the first seismic station was installed in the settlement of Kliuchi (Tokarev 1966). At present, the seismic network in this region consists of five permanent stations of analogue recording (regional type). The aim of this network is to monitor the seismicity. From the onset of seismic observations, volcanic tremor was recorded during intensified periods of activity at Kliuchevskoi (Gorshkov 1954). A detailed study of volcanic tremor at the regional stations was rather prohibitive for technical reasons, and accordingly the first results on the wave structure of volcanic tremor were obtained in 1961 after the organization of a special seismic experiment (Tokarev 1964).

The eruption of 1984 was one of the most suitable for carrying out special seismological studies on volcanic tremor. The long duration (mainly in summer months), and the power of the eruption, permitted us to establish different observation systems, both in the near and distant zones. Previous studies on volcanic tremor (Gordeev et al. 1978; Tokarev 1982; Cosentino et al. 1982) at different volcanoes made it possible to specify the problems and design an appropriate experiment for the investigations.

The main purposes of the field observations of volcanic tremor in July-August 1984 in the Kliuchevskoi region were:

1. Studies of spectral and wave composition as well as spatial and temporal variations in spectra of volcanic tremor;
2. Determination of the site of the volcanic tremor source;
3. Determination of phase velocities and the parameters of seismic wave attenuation of the volcanic tremor.

2 Experimental Methods and the Processing of the Results

In July and August 1984, observations of seismic events in the region of Kliuchevskoi volcano were conducted using five three-component seismic stations. To solve the different problems, several experiments were organized. The observation points and the region under study are shown in Fig. 1. The first siting of seismic stations was made along a profile oriented radially to the summit crater of Kliuchevskoi Volcano. Observations were conducted at points 1-5 from 17 to 19 July, and at points 5-9 from 19 to 21 July (see Fig. 1). On 25 July a long profile was made with recording taken in turns at points 10-23. Simultaneously, a continuous recording was made at points 9, 10 and 16. The last and the longest experiment was made by siting five stations at points 9, 10, 24, 25 and 26 (see Fig. 1). In total, the recording during this experiment was carried out from 24 July to 4 August. As a result of all experiments, we have obtained a continuous three-component analogue recording of volcanic tremor on a magnetic tape. In further processing, an analogue spectrum analyzer was used and computer calculations of encoded analogue recordings of volcanic tremor were made.

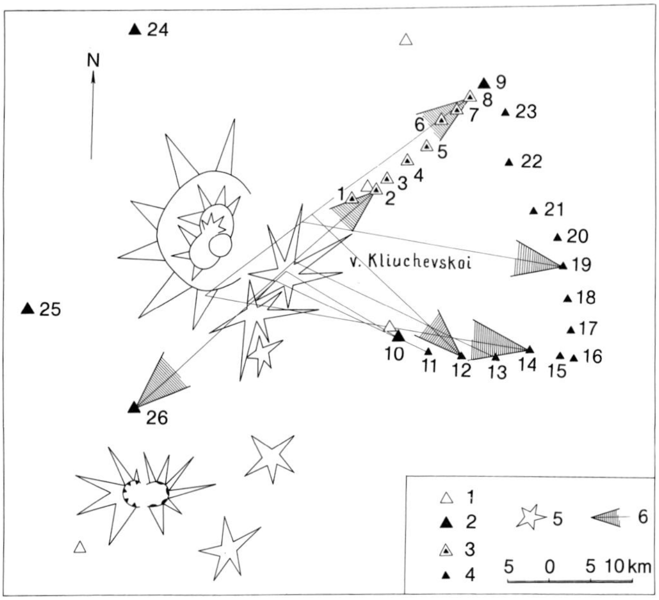

Fig. 1. Scheme showing the position of seismic stations in the region of Kliuchevskoi volcano in July and August 1984.
1 Seismic stations of the Kamchatka regional network; 2 points of observations with portable seismic stations from 24 July to 4 August 1984; 3 points of observations with portable seismic stations in profile from 17 July to 21 July 1984; **4** points of observations with one portable seismic station on 25 July 1984; 5 volcanoes; **6** direction towards the tremor source with mean-square error

The calculation of spectra with the analyzer was carried out on-line. An instantaneous spectrum was determined from a sample of recording with a duration of about 40 s and with a digitization step of 0.067 s. It is possible, using a spectrum analyzer, to obtain the average linear, square-law and logarithmic spectra by averaging the instantaneous selective estimates. The number of the averaged instantaneous spectra may vary from 1 to 1024. In order to analyze volcanic tremor, which is a random process, it is more reasonable to consider the averaged spectral estimates after determining the duration of the averaged interval yielding a stable spectral estimate.

The average linear spectra of volcanic tremor are shown in Fig. 2. These spectra are the spectra of the velocity of the ground motion for the range of 0.5-5 Hz.

The maximum duration of the averaged interval was 340 min. The average spectral estimates at intervals with duration of more than 10 min differ insignificantly and these estimates may be considered representative of the long-term spectra. Below, in studying the properties of volcanic tremor by spectra, we shall use the average spectra at intervals with a duration of not less than 40 min (see Fig. 2, spectra 4).

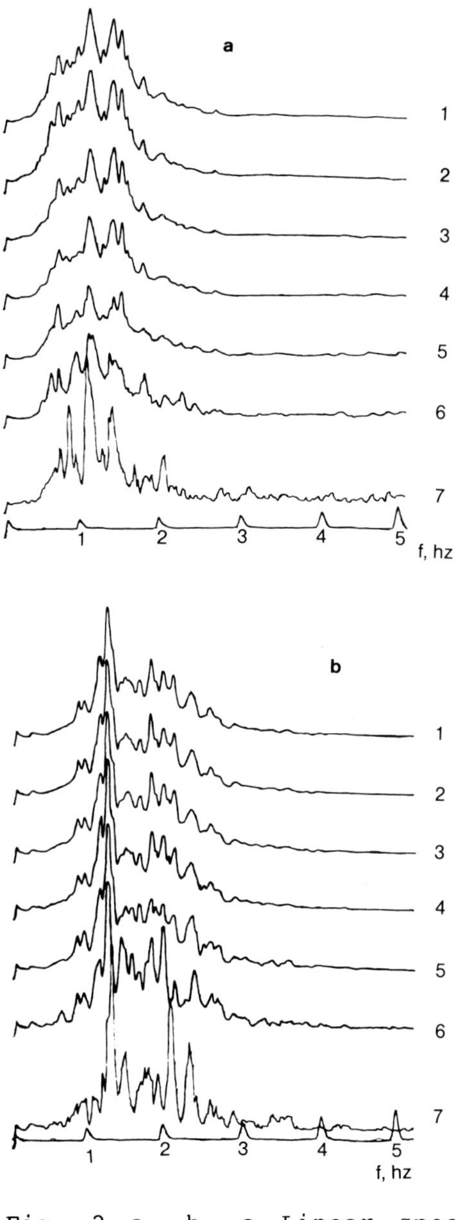

Fig. 2 a, b. a Linear spectra of the vertical component of volcanic tremor at point 5 (see Fig. 1) on 19 July 1985 with different durations of averaging: **1** spectrum obtained by averaging of 512 successively selected estimates, each being 40 s in duration. The total duration of averaging is 340 min; **2** 256 selected estimates were averaged, the total duration is 170 min; **3** 128 averagings, the total duration is 85 min; **4** 64 averagings, the total duration is 42.5 min; **5** 16 averagings, the total duration is 10.6 min; **6** 4 averagings, the total duration is 2.66 min; **7** instantaneous spectrum, the duration is 40 s. **b** The same for volcanic tremor recorded at point 10 (see Fig. 1) on 25 July 1985

We have shown earlier that a random process of volcanic tremor is quasi-stationary (Gordeev et al. 1978). A constancy of spectra through time is one of the main criteria for a stationary process. Comparison of spectra for successive intervals of recording with a duration of 40 min were carried out every 3 h. In total, a comparison was made for 27 h. The average linear spectra of the vertical component of tremor are presented in Fig. 3. It should be noted that the spectral estimates remain entirely similar. An insignificant change in the value of spectral peaks is caused by a variation in the level of volcanic tremor, which is closely related to the eruption regime. The stability of spectra through time confirms that the process of volcanic tremor remains almost stationary.

3 Relation of Volcanic Tremor to the Eruption Regime

Continuous observations of volcanic tremor were conducted at points 5, 9, 10, 25, 24 and 26 (see Fig. 1). In order to analyze temporal variations in tremor we selected points 5, 9 and 26. The intensity of volcanic tremor was determined using the spectrum analyzer by determining the area beneath the average intensity spectrum at intervals of 42.5 min within the range of 0.3-5.0 Hz. This value corresponds to the value of tremor energy. The values of the area consecutively determined for the observation period represent variations in intensity of volcanic tremor through the averaged intervals. Variations in the vertical component of tremor at point 5 from July 17 to 20 and at points 9 and 26 from July 27 to August 3 1984 are shown in Fig. 4 (dates are GMT throughout). Observations of the state of the volcano during that period were irregular, largely because of unstable weather. Nevertheless, Ya.D. Muraviev and A.Yu.

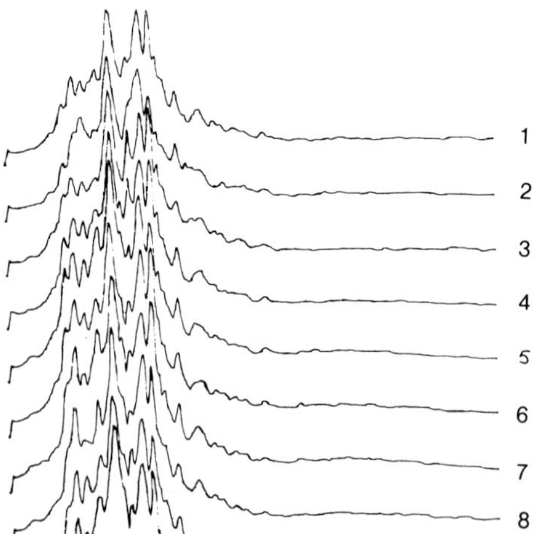

Fig. 3. Linear spectra of the vertical component of volcanic tremor recorded at point 5 (see Fig. 1) at 3-hour intervals from 22.00 h 19 July to 01.00 h 21 July (**curves 1-10**). Times and dates are GMT throughout. Spectra are obtained by averaging 64 successively selected estimates, each being 40 s in duration

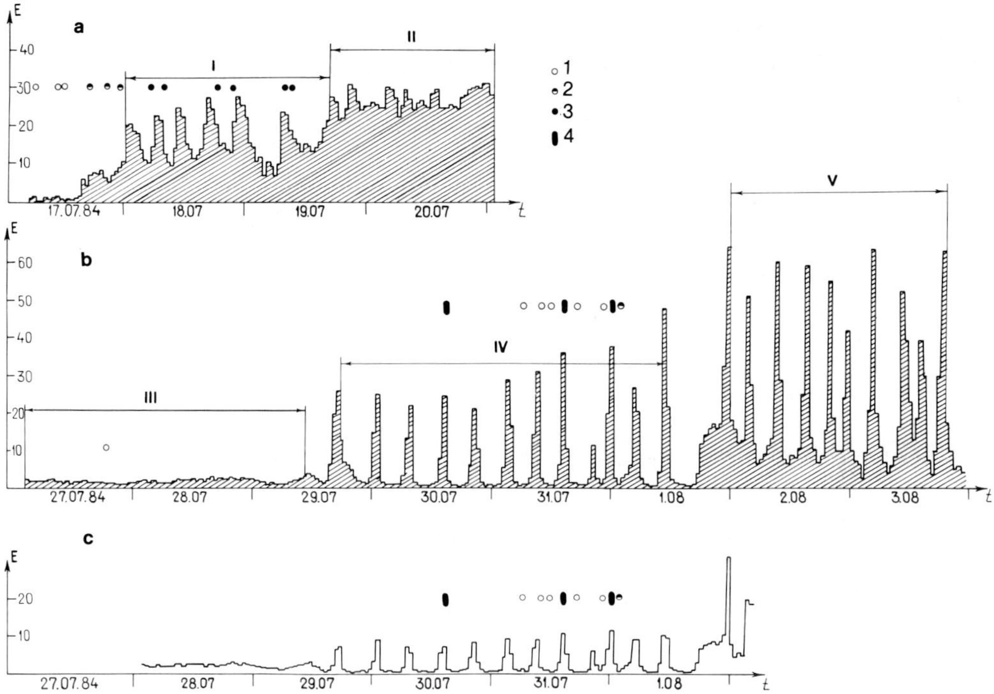

Fig. 4 a-c. Variations in energy of the vertical component of volcanic tremor through time from observations at points 5 **(a)** 26 **(b)** and 9 **(c)** The state of the active crater is denoted as follows: **1** steam-gas emissions; **2** steam-gas emissions saturated with ash; **3** ejections of ash and volcanic bombs; **4** fountaining of melted material

Ozerov made some accurately timed observations of the state of the summit crater (see Fig. 4). Of most interest in Fig. 4 are the periodicities of volcanic tremor intensities corresponding to the periodicities of the eruption activity. This is most pronounced from July 29 to August 3. Isolated observations at two points demonstrate the complete identity of the evolution of volcanic tremor (Fig. 4 b, c). Figure 5 a and b gives the compressed tremor recordings at point 26 on August 1 and 2. Detailed recordings show that during the active period a continuous intense tremor is recorded; during pauses, against a background of slight tremor, seismic vibrations occur (see Fig. 5 c, d). They are related to isolated explosions in the crater (from visual observations). By comparing tremor with visual observations of the activity of the volcano, we may state that a low level of tremor exists during steam-gas emissions. The increase in the activity of the eruption is associated with the increase in the level of tremor, and the maximum level corresponds to continuous fountaining of incandescent material in the crater (see Fig. 4). Comparison of the eruption regime and tremor intensity points to a direct connection between the level of tremor and the activity of eruption. Earlier, the confirmation of this connection was obtained from the analysis of tremor and the evolution of the Alaid eruption (Gordeev 1984).

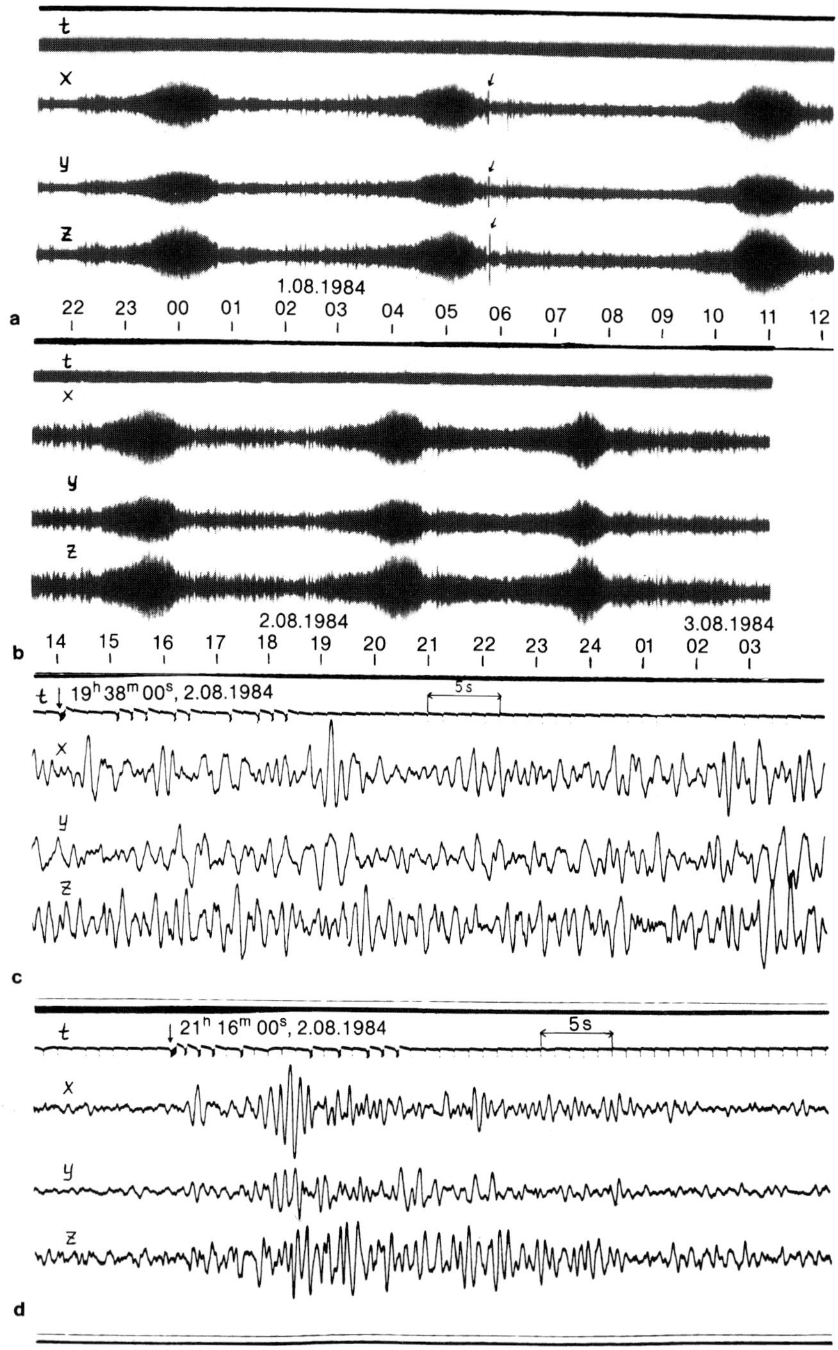

Fig. 5 a-d. Recordings of volcanic tremor at point 26 (see Fig. 1). a b Compressed recordings made on 1 and 2 August 1984; c tremor recording during the active period of eruption; d tremor recording during the pause. Arrow denotes calibration signals

In order to verify the possible reasons for the origin of the eruption periodicity a detailed analysis of volcanic tremor periodicities in terms of energy was made. Since a direct relationship between the eruption regime and the level of volcanic tremor was observed, we may conclude that all variations of volcanic tremor correspond to variations in eruption.

Figure 6 shows changes in time of periods, total energy and power of each cycle for intervals having obvious periodicities (see regimes I, IV and V in Fig. 4).

Regime I is characterized by the gradual increase in energy and periods of a cycle, the power remaining almost constant. During regimes IV and V the energy, periods and power do not vary significantly.

Figure 6 also demonstrates the boundary levels of the average intensity of tremor for regimes II and III (see Fig. 4). At that time constant tremor occurred, corresponding to slight steam-gas emissions (regime III) and to continuous ash-gas flow (regime II).

Thus, there are two stable eruption regimes differing more than 20 times in intensity of volcanic tremor. The other three types of the activity of the volcano (regimes I, IV and V, see Fig. 4) occupy an intermediate place between the upper and lower levels, in terms of the average intensity, although for regime V the tremor intensity during the active stage of the cycle is higher than the maximum intensity for regime II.

Now we shall consider the probable reasons for the existence of the summit eruption periodicity. We speculate that when the supply of the material from the depths is constant, a periodi-

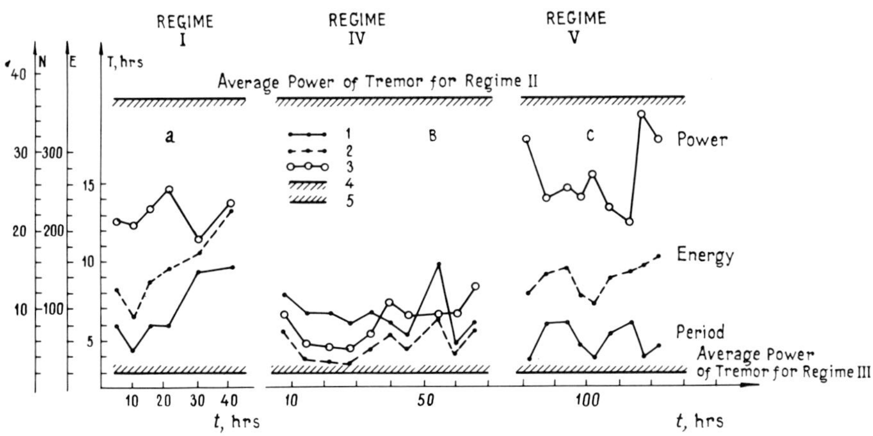

Fig. 6 a-c. The dependence of period (1), energy (2) and power (3) of the cycles of the periodic regime of volcanic tremor on time for regimes I (**a**), IV (**b**) and V (**c**) from Fig. 4; **4** the average power of tremor for regime II, **5** the average power for regime III

cal capping and breakthrough of the formed plug occurs when a definite energy threshold is achieved. In this case the properties of the capping should remain the same and the energy threshold should be stable, and, as a result, the periods, energy and power for cycles within different regimes will be constant. In practice, the periods and energy within separate regimes and for different regimes may vary more than twofold, and the power varies almost fivefold (see Fig. 4, regimes IV and V).

Regime I (see Fig. 4) is characterized by the absence of a period of repose, and therefore the conduit is open and the periodicity must be due to other reasons. It is possible that such a periodicity is caused by the properties of a two-phase flow. However, for regimes IV and V the periodicity cannot be explained by the properties of two-phase flow because of the existence of repose periods when the conduit is closed and the two-phase flow is absent. Moreover, the characteristic time of flow fluctuations (from 5 to 10 h) is too great for the real velocities of magma issuing from the crater (of the order of a few meters per second).

It is possible that the eruption periodicity may be caused by the unstable state of the gas-saturated material in the upper part of the conduit. The overheated two-phase medium may be in the metastable state and the characteristic periods of its excitation may depend upon many factors (volume, temperature, gas composition, etc.). Therefore, the variations in periods and energy may be due to the change in the parameters of the medium. This may be caused by the regime of the material supplied from the depths, by variations in geometry of the conduit, or by the content of the gas component, etc.

The ambiguous nature of interpretation of data on the regime of volcanic tremor is due to the lack of comprehensive data on the activity of the summit crater, the lack of quantitative estimates of magma discharge during the eruption and the lack of information on the regime of lava flows during volcanic tremor monitoring.

4 Temporal and Space Properties of Spectra of Volcanic Tremor

Variations in spectra of tremor in the quasi-stationary regime are given in Fig. 3. The stability and similarity of the spectra in this regime (regime II in Fig. 4) are unquestionable. Of interest is a comparison of spectral estimates for the regimes having significant variations in volcanic tremor intensity (e.g. regimes IV and V, see Fig. 4). In the work on the Alaid volcano (Gordeev 1984) it is shown that no noticeable variations in spectra are produced by the 20-fold changes in level of tremor. In our case, comparison of spectra was made for regime IV (see Fig. 4). The tremor intensity varied more than 30-fold during the active period and pause. Linear spectra of the vertical component of tremor recorded at point 10 on 25 July 1984 are shown at identical scales in Fig. 7 a. The maximum and minimum levels differ more than sixfold. Note the pre-

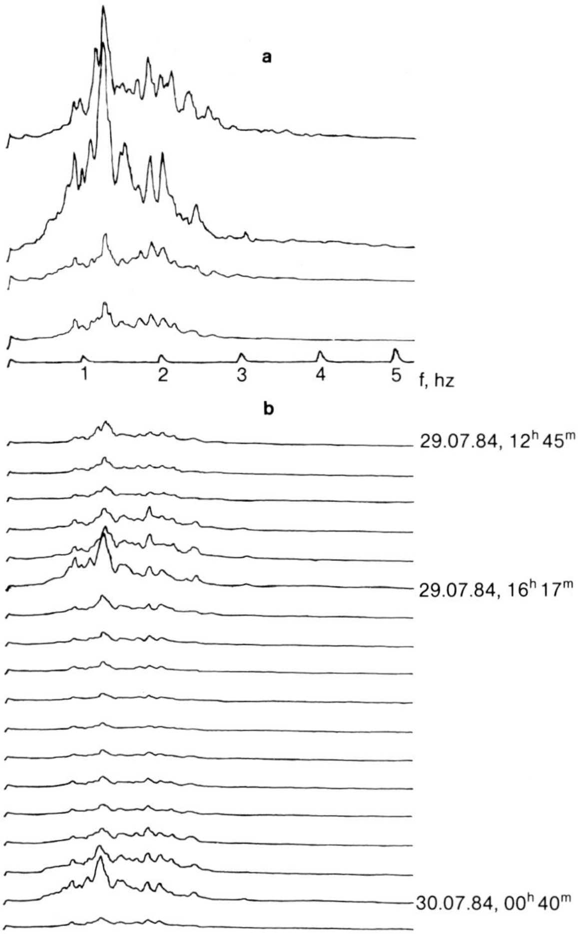

Fig. 7 a, b. Linear spectra of the vertical component of volcanic tremor at point 10 (see Fig. 1). a For different tremor intensities on 25 July 1984. b The average spectra for successive intervals during the periodic regime. Everywhere the averaging was made at intervals of 42.5 min. Scales are identical

servation of the main spectral properties. The correlation between the different values of spectral peaks remains constant. Consequently, the source of volcanic tremor in this dynamic range has linear properties. A similar conclusion was drawn at the Alaid volcano (Gordeev 1984). The current variation in spectra between the active periods in regime IV is shown in Fig. 7 b. If the average spectra during the pauses consist of instantaneous spectra of separate seismic signals, triggered by explosions in the crater, comparison of spectra in Fig. 7 a, b implies a similarity of spectra of continuous tremor and signals triggered by separate explosions.

Analysis of temporal variations in spectra of tremor for different regimes of the eruption shows that the structure of the wave field of volcanic tremor is stable through time. Of particular interest are spatial characteristics of the wave field. Figure 8 shows the linear spectra for the observation sites most remote and nearest to the active crater (Fig. 8 a and b respectively). Spectra in both cases represent a narrow-band process. The spectra of the horizontal and vertical components are almost identical, even when the character of spectra has a more complicated pattern. As an example, the current linear spectra of three components at point 5 are given in Fig. 8 c. The more complicated character of the spectra than that shown,

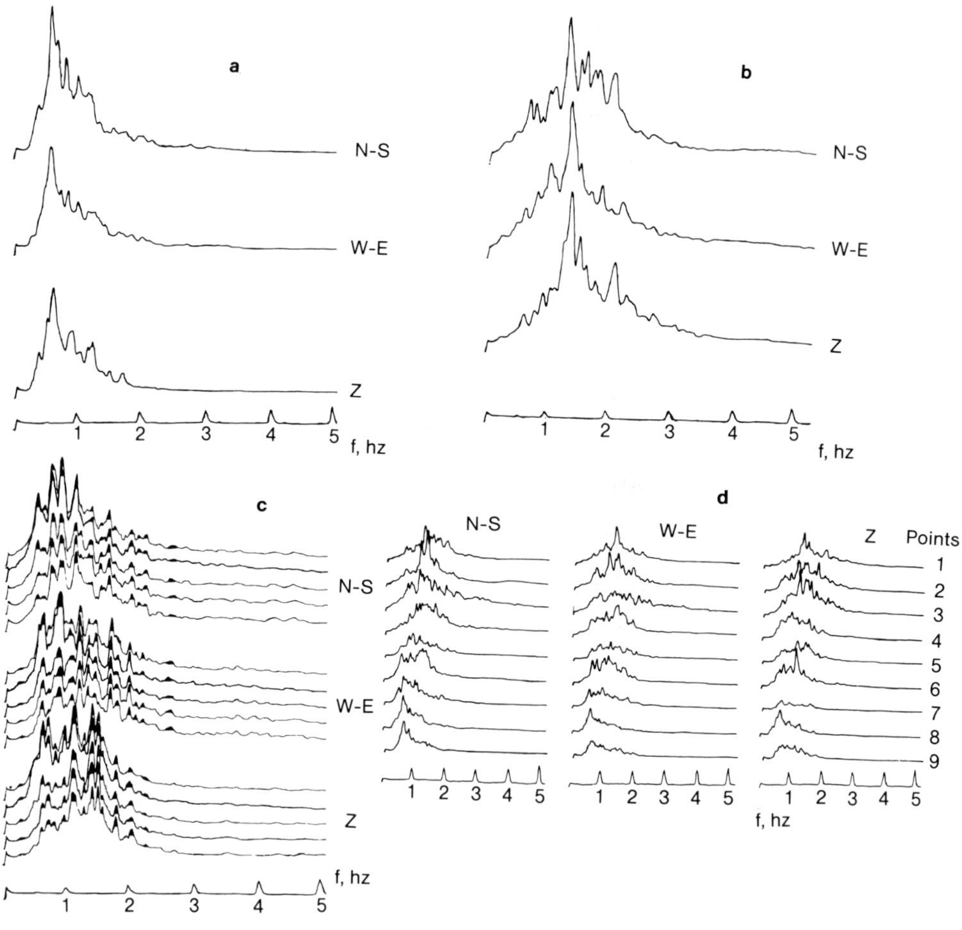

Fig. 8 a-d. Linear spectra of volcanic tremor: **a** at point 8 (see Fig. 1), 08.00 h, 20 July 1984; the averaging was made at intervals of 170 min; **b** at point 1 (see Fig. 1), 07.00 h, 18 July 1984; the averaging was made at intervals of 170 min;
c spectra averaged at successive intervals of 42.5 min, 23.00 h, 18 July 1984 at point 5 (see Fig. 1); **d** spectra averaged at intervals of 170 min along the profile at points 1-9 (from top to the bottom)

for example in Fig. 8 a, is characteristic of all the components.

Figure 8 d demonstrates the variations in spectra for three components along the profile at points 1-9 (see Fig. 1). It is difficult to correlate definite spectral peaks along the whole profile, although at some sites the peaks may be distinguished unambiguously at one and the same frequencies. As is shown in Fig. 8 d, the high-frequency part of the tremor propagating along the profile is more strongly attenuated.

5 Absorption and Phase Velocities of Volcanic Tremor

The absorption coefficients may be determined only if there is information on the siting of the source. Data on comparison of the eruption regime and tremor intensity, as well as the decrease of tremor amplitude with increasing distance from the volcano, indicate that the source is directly connected with the active crater. This is in accord with the results obtained earlier on other volcanoes (Gordeev et al. 1978; Cosentino et al. 1982; Gordeev 1984). To confirm this conclusion, we have determined the direction towards the source by the method of distinguishing the polarization planes of seismic waves (Gordeev et al. 1978). In Fig. 1 the indicated directions towards the tremor source are shown at several points with their mean-root-square errors. Taking into account all the directions, we may confirm that the source is located in the area of the active crater. The large mean-root-square error points to the complicated nature of the wave composition, which is likely to be due to the complicated structure of the medium.

The absorption coefficients were determined for three tremor components from observations in profile 1-9, as well as at points 10, 24 and 25. In calculating the absorption coefficients, all observational data were normalized in accordance with the tremor regime in Fig. 4.

Figure 9 shows the dependence of absorption on frequency calculated from the relation of spectral components at points 24, 25 and 10 to point 1, as well as in profile 1-9, with the tremor source assumed to be at the center of the summit crater. The absolute values of absorption coefficients agree fairly well with earlier determinations (Tanaka 1968; Guerra et al. 1976; Gordeev et al. 1978). A significant increase in the absorption coefficients with frequency is evident. For the average dependence, which reflects the average properties of the medium, this increase has a linear dependence on frequency (see Fig. 9 e, f). The individual properties of each propagation path yield different absolute coefficient values (see Fig. 9 c, d), as well as different frequencies at which absorption minima caused by the layered structure of the medium are observed. Figure 10 shows, for comparison, accurate absorption curves for the path towards point 10 (see Fig. 10 b) and the average for the paths towards points 24 and 25 (see Fig. 10 a).

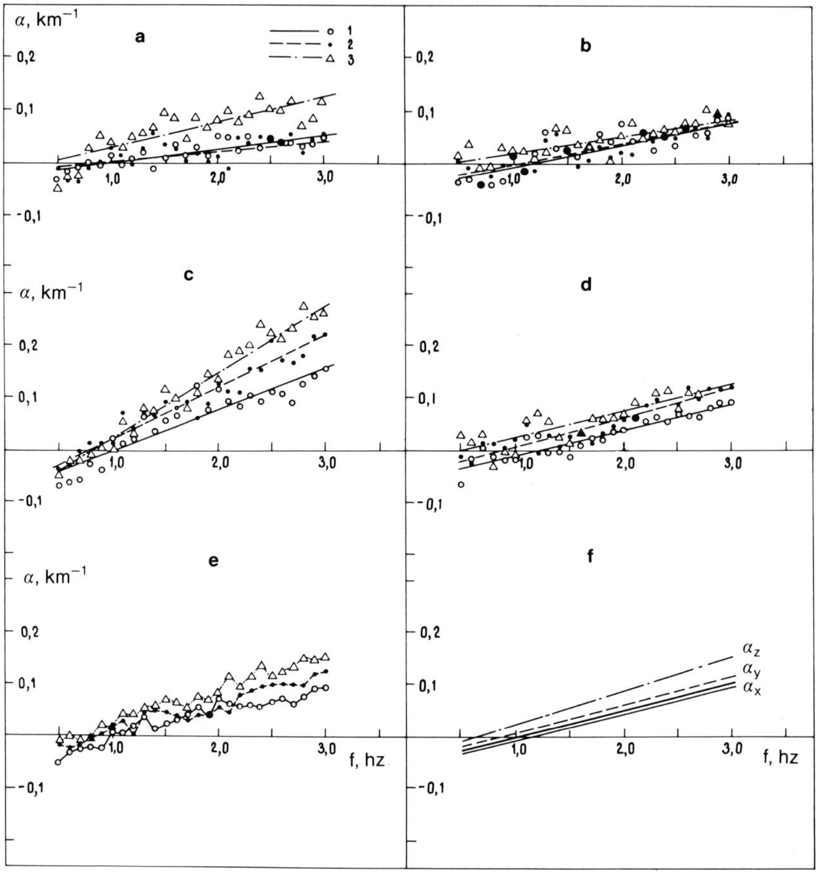

Fig. 9 a-f. The dependence of the absorption coefficient of volcanic tremor on frequency, from observations at points 24 **a**, 25 **b**, in profile 1-9 **c** and at point 10 **d**. The average dependence for all observation points **e** and a linear function approximation of the average dependence using a method of least squares **f** for components N-S (1), W-E (2) and for the vertical component (3)

For comparison, the absorption curve for volcanic tremor of the great Tolbachik fissure eruption is given in Fig. 10 a (Gordeev et al. 1978). On all curves there are minima, corresponding most likely to the fundamental layer in which the seismic waves of volcanic tremor are propagated. Assuming that the waves are surface waves, thicknesses derived for the main layers along the paths, using the average data on propagation of the Love waves, are within the ranges of 300 to 500 m. One of the most interesting results in estimating absorption is a stable difference in absorption for vertical and horizontal components. In almost all cases, the absorption of the vertical components exceeds the absorption of the horizontal components. On the whole, the absorption for compressional waves exceeds the ab-

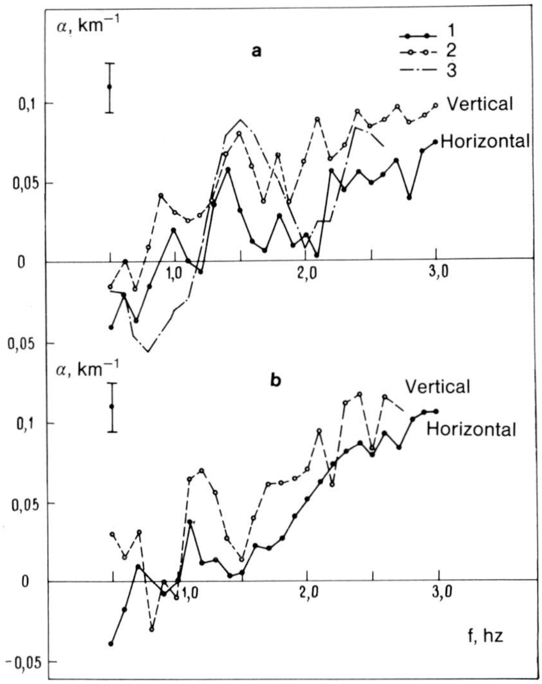

Fig. 10 a, b. The dependence of the absorption coefficient of volcanic tremor on frequency, from observations at points 24, 25 (**a**) and 10 (**b**). The average (1) for the horizontal components, (2) for the vertical component and (3) for volcanic tremor during the great Tolbachik fissure eruption in 1975 (horizontal components). Vertical boundaries denote the mean-square deviation

sorption for shear waves because in the vertical component the greater proportion is of compressional waves and in the horizontal component the greater proportion is of shear waves. The same result was obtained for the absorption of volcanic tremor of the great Tolbachik fissure eruption (Gordeev et al. 1978). Q-factor of the medium averages 70 for horizontal components and 45 for the vertical component. Phase velocities were determined in profile 1-9 between points 2 and 3, and 8 and 9 (see Fig. 1). A cross-correlation of separated volcanic tremor sequences was calculated for each component. On the basis of a shift of the maximum correlation, the phase velocities at the beginning (points 2, 3) and at the end of the profile (points 8, 9) respectively, are 0.98 km s^{-1} and 1.16 km s^{-1} for the vertical component, and 0.85 km s^{-1} and 0.99 km s^{-1} for the horizontal component. In a similar volcanic area of the great Tolbachik fissure eruption the seismic wave velocities determined experimentally in the 500-m-thick upper layer were 2.0-2.2 km s^{-1} and 1.2-1.3 km s^{-1} for P- and S-waves respectively (Balesta et al. 1978). The velocity values in our case are in good agreement with the estimates obtained by other authors (Tokarev 1964; Gordeev et al. 1978) and conform to surface waves. It is noteworthy that at distances of the order of two wavelengths the wave field of volcanic tremor is only slightly correlated. This is due to a complex structure of the upper horizons of the volcanic edifice. The wave field instability in space may be noted from the change in spectra of the tremor along profile 1-9 (see Fig. 8 d).

6 Discussion

Observations of volcanic tremor of the summit eruption of Kliuchevskoi Volcano were carried out at distances of 12-36 km from the source, and therefore the results and conclusions may be considered pertinent to observations of the wave field distorted by the effects of the propagating medium. Analysis of spectra of tremor at different observation points provides evidence that the medium strongly affects the dynamics of the wave field. Evidence from phase velocities suggests that tremor consists of surface waves. The source is located in the area of the active crater. The great number of spectral peaks which are stable through time at each observation point indicates that the medium is multilayered, and that its properties change so that even over short baselines many spectral properties are uncorrelated. At almost all observation points the upper layers of the medium are composed of intercalated volcanogenic deposits, consisting of pyroclastics and lava flows, the disposition and alternation of which show no orderliness. Along with wedged-out and newly formed layers the lens-shaped heterogeneities of any size may occur along the path of seismic wave propagation. Only in the case of very distant observation points we may expect the existence of a layer of sedimentary rocks, stable in space, which, apparently, may determine the main spectrum of a signal.

Figure 11 shows the variation in some properties of the wave field of volcanic tremor and the medium, along the profile 1-9 (see Fig. 1). Based on geological data, the boundaries of three zones were recognized in the profile. The first zone is located in the region of the main cone. It consists of cinder and ash material and glaciers. The boundary of the first zone lies at a distance of about 10 km from the active crater. The second is the zone of the flank cones. This zone has numerous lava flows intercalated with cinder and ash. The third is the zone in which a large sedimentary layer exists. These rocks overlie the consolidated old edifice of the more ancient Dalnaya and Blizhnyaya Ploskaya volcanoes. The boundary between the second and third zones lies between 20 and 25 km from the active crater (see Fig. 11 d). Apparently, all the parameters of the wave field of tremor change their properties sharply in the region of the boundaries between the second and third zones. This is connected with the reconstruction of the wave field when it passes into the medium with a new layered structure.

Unfortunately, we are unable to estimate the spectral composition of volcanic tremor near the source, but from the analysis of variations in the spectra envelope shape we suggest that near the source the spectrum is caused by a layered structure of the medium in the vicinity of the active crater, and that the frequency range is much wider there than that observed by us. In all cases during volcanic tremor propagation the initial signal is filtered, and the more homogeneous the structure of the medium along the propagation path, the more stable are the properties of the wave field of volcanic tremor. Medium studies using tremor require, in most cases, a great number of detailed

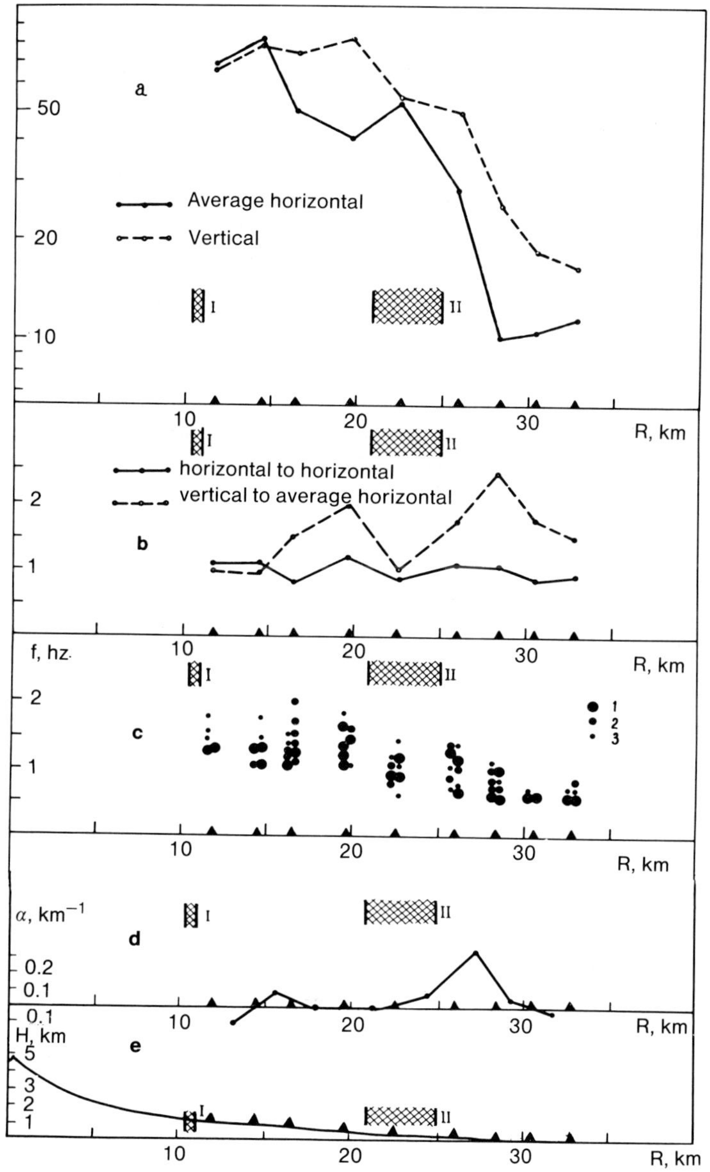

Fig. 11 a-e. Variation in a volcanic tremor amplitude for average horizontal and vertical components; b amplitude ratio of horizontal components to each other and of vertical to the average horizontal component with increasing distance from the active crater along profile 1-9; c variation of frequency of the spectral peaks of the vertical and horizontal components with respect to the maximum value of spectral peaks 0.9-1.0 (1), 0.8-0.9 (2) and 0.7-0.8 (3); d variation in the average absorption coefficient with increasing distance; e profile of Kliuchevskoi volcano. I and II are the boundaries between the zones with different geological structure of the upper layers. **Triangles** denote the position of seismic stations in profile 1-9 (see Fig. 1)

observations, because of the significant heterogeneity of the medium.

The relationship between volcanic tremor and the eruption regime arouses more interest. As was shown, the tremor intensity is closely related to the eruption regime, and, as a result, study of tremor may become one of the more reliable quantitative ways of estimating the energy of eruption. However, in order to use tremor for routine volcanological observations, we need a normalizing scale for estimating the eruption energy. In general, fairly reliable characteristics of the eruption regime may be obtained from records of a wide-band seismic station.

7 Conclusions

Study of the properties of volcanic tremor of Kliuchevskoi Volcano has shown the complexity of the wave composition and the spatial instability of the wave fields of tremor, which are, most likely, due to the existence of significant spatial heterogeneities in the upper volcanogenic-sedimentary deposits of the volcanic edifice.

Study of the detailed temporal variations in the level of volcanic tremor, by comparing it with the eruption regime, gives evidence that volcanic tremor may become one of the most reliable quantitative factors characterizing the regime (energy) of the eruption. In order to estimate the energy of eruption it is necessary to conduct a detailed comparison of variations in tremor with the eruption regime for each type of eruption.

A stability of the characteristic spectral properties of volcanic tremor through time suggests that the main parameters of volcanic tremor are also stable.

Acknowledgements. The authors are grateful to L.K. Ivanenko and S.A. Shevchenko who have aided them with computer calculations, as well as to O.P. Nechiporenko and T.P. Khubunaya for the preliminary processing of the material.

References

Balesta ST, Kargapoltsev AA, Grigoryan GB (1978) The structure of the New Tolbachik Volcanoes from seismic data. In: Fedotov SA, Markhinin EK (eds) The Great Tolbachik fissure eruption, geological and geophysical data 1975-1976. Nauka, Moscow, pp 225-233 (in Russian)

Cosentino M, Lombardo G, Patane G, Schick R, Sharp DL (1982) Seismological researches on Mt. Etna: state of art and recent trends. Mem Soc Geol It 23:159-202

Gordeev EI (1984) Volcanic tremor of Alaid volcano (eruption of 1981). Volcanol Seismol 1:98-102 (in Russian)

Gordeev EI, Pheophilaktov VD, Chebrov VN (1978) A study of volcanic tremor during the Tolbachik eruption. In: The Great Tolbachik fissure eruption, geological and geophysical data 1975-1976. Nauka, Moscow, pp 151-163 (in Russian)

Gorshkov GS (1954) Seismic observations in 1949. Bull Volcanol 21:19-39 (in Russian)
Guerra I, Lo Bascio A, Luongo G, Scarpa R (1976) Seismic activity accompanying the 1974 eruption of Mt. Etna. J.Volcanol Geotherm Res 1 4:347-362
Khrenov AP, Ozerov AYu, Litasov NE, Slezin YuB, Muraviev YaD, Zharinov NA (1985) The flank eruption of Kliuchevskoi volcano (Predskazanny breakthrough, 1983). Volcanol Seismol 1:2-20 (in Russian)
Tanaka Y (1968) A seismometrical study of Izu-Oshima (I). Pap Met Geophys 19 4:627-650
Tokarev PI (1964) On the types of seismic waves during volcanic tremor and on the phase velocities of their propagation. In: Rudich KN (ed) Recent volcanism of north-eastern Siberia. Nauka, Moscow, pp 103-109 (in Russian)
Tokarev PI (1966) Eruptions and seismic regime of the volcanoes of the Kliuchevskoi group (1949-1963). Nauka, Moscow, pp 1-118 (in Russian)
Tokarev PI (1982) Low frequency volcanic tremor. Volcanol Seismol 6:56-78 (in Russian)
Tokarev PI (1983) Prediction of the Kliuchevskoi flank eruption in March 1983. Volcanol Seismol 5:3-9 (in Russian)

Some Significant Features of Pre-Eruption Volcanic Earthquakes

D. SHIMOZURU[1] and T. KAGIYAMA[2]

Abstract

Precursory changes in the nature of volcanic earthquakes prior to explosive eruptions of Asama volcano have been studied. The main results are that before eruptions:

1. High-frequency earthquakes become dominant;
2. The b-value markedly decreases;
3. Seismic quiescence is sometimes observed;
4. Relative amplitude ratio of the horizontal to the vertical component tends to decrease; and
5. Peculiar tremor of monochromatic frequency with logarithmic amplitude decay sometimes appears. For these empirical observations, possible interpretations are given.

1 Introduction

At Asama volcano, Minakami (1960, 1974) made pioneering studies of the relationship between the number of shallow volcanic earthquakes and the occurrence of explosive eruptions. Based on the statistical analysis of the daily number of earthquakes and the date of the eruptions, he succeeded in making a vulnerability curve for the expected eruptions. This curve remained valid up to the period of the 1961 eruptive activity. After an 11-year quiescent period, Asama became active on 1 February 1973. We tried to apply his vulnerability curve, but found that vulnerability decreased just before the first eruption, since seismic quiescence appeared prior to the eruption. Such an aseismic period preceding an individual eruption was not taken into consideration in Minakami's statistical study. After the 1973 eruptions, there were three further occasions when explosions took place at Asama. Seismic equipment at the University of Tokyo's Asama Volcano Observatory has been much improved since Minakami's study, and we have been able to obtain high quality seismic data which have enabled us to study the nature of seismic waves in detail.

---

1 Tokyo University of Agriculture, Physics Department, Sakuraga-Oka 1-1-1, Setagaya-ku, Tokyo 156 Japan
2 Asama Volcano Observatory, University of Tokyo, Japan

IAVCEI Proceedings in Volcanology 1
J. H. Latter (Ed.), Volcanic Hazards

The state of the art knowledge is that the change of frequency of earthquakes or of released seismic energy are by themselves insufficient for the quantitative prediction of volcanic eruptions. Furthermore, for mitigating disasters,immediate prediction is much more in demand than generalized probablistic forecasting. To meet this social demand, we have carefully examined the seismic data, and have identified some significant features of pre-eruption volcanic earthquakes observed at the Asama Volcano Observatory in recent years.

2 Seismic Evidence and Its Interpretation

2.1 Evidence of Time Variation in Seismic Wave Frequency

After a 9-year quiescent period lasting from 1973, explosive eruptions took place at Asama volcano on 26 April and 2 October 1982, and on 8 April 1983. On these occasions, however, the daily number of precursory earthquakes was not greatly increased. In the case of the 1983 eruption (Fig. 1), seismicity increased in March, ending with a dramatic decrease at the end of the month. With the beginning of April, high-frequency earthquakes began to appear in increasing numbers, and a week later the eruption occurred. A sample seismogram (Fig. 2) shows that the earthquakes in March were low-frequency, whereas those in April were high-frequency. Before the 2 October 1982 eruption also, the dominant frequency increased, and after the eruption decreased, as shown in Fig. 3.

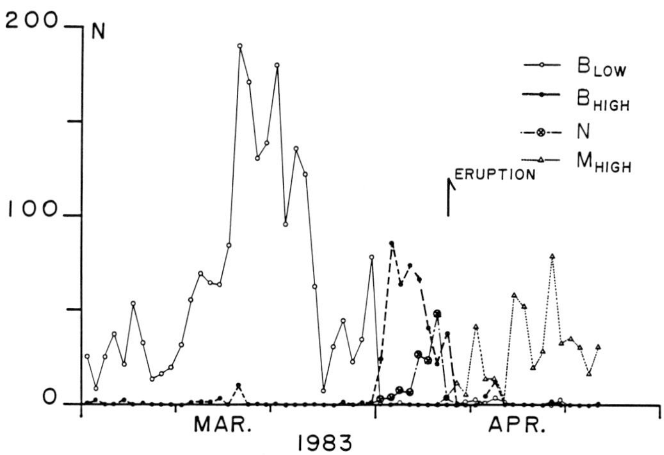

Fig. 1. Daily number of volcanic earthquakes observed at Asama Volcano Observatory. B_{LOW} and B_{HIGH} are shallow low-frequency and high-frequency events. **N** indicates tremor like earthquakes with long coda of monochromatic frequency. M_{HIGH} is high-frequency seismic noise

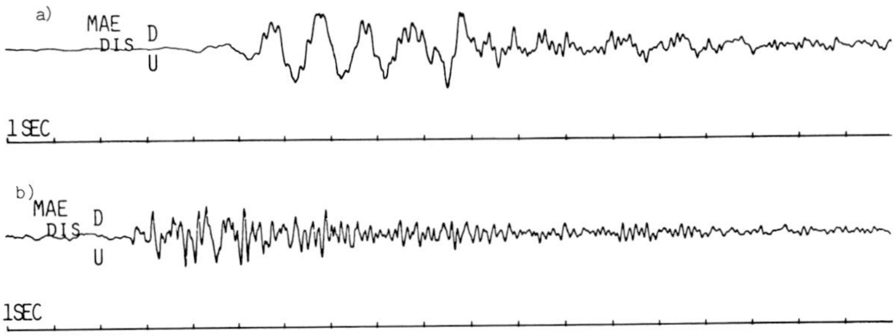

Fig 2. a, b. Typical seismograms (displacement) of low-frequency earthquakes and high-frequency earthquakes observed at a site 1.2 km east of the summit crater of Asama volcano

Interpretation. According to Ohnaka and Mogi (1982), low-frequency earthquakes are generated by closing or shearing of pre-existing cracks with a low aspect ratio which surround the plumbing system. After the cracks are closed, intensive formation of new cracks may occur, generating high-frequency seismic waves. These seismic episodes are generally associated with magma intrusion and consequent strong degassing.

Contrary to the above, low-frequency earthquakes became dominant immediately before the 1977 eruption of Usu volcano (Suzuki and Kasahara 1979), and also before the 1980 eruption of Mount St. Helens (Malone et al. 1981). The low-frequency

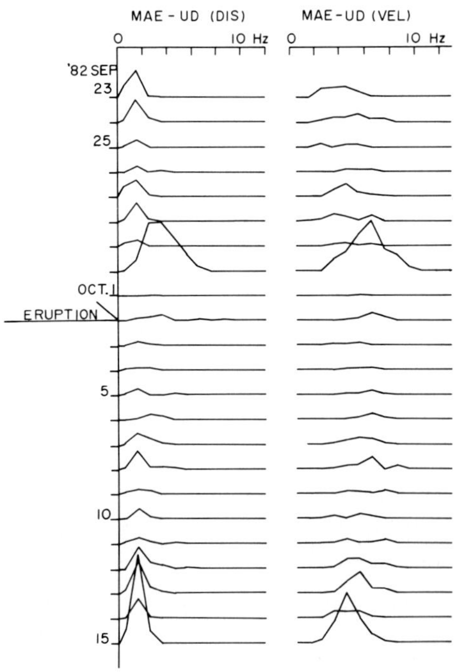

Fig. 3. Typical variation of the spectral amplitude of earthquakes before and after the 2 October 1982 eruption of Asama volcano. Data from displacement and velocity-type seismograms are shown separately. The vertical scale is arbitrary

earthquakes at Usu were interpreted as having a dip-slip faulting mechanism (Okada et al. 1981), with low stress drop, associated with the forceful intrusion of very viscous dacitic magma.

2.2 Evidence of Change of b-Value with Time

The four recent explosive eruptions of Asama volcano showed a sharp decrease with time of the b-value (slope of the magnitude-frequency logarithmic plot) for volcanic earthquakes prior to each eruption. Instead of the b-value, we have plotted the trace amplitude frequency on a logarithmic scale. The slope gives Ishimoto-Iida's (1939) m-value, $m = 1 + b$.

The four cases of the past 16 years are shown in Fig. 4. We find in every case that the m-value begins to increase (corresponding to low-frequency events), and that this is followed by a decreasing value (corresponding to high-frequency events) prior to the eruptions.

Interpretation. The results of a rock fracture experiment under hydrostatic pressure (Scholz 1968) suggested that the b-value of microfracturing decreases with increasing differential stress. Especially, it drastically decreases immediately before the ultimate fracture. His result is shown in Fig. 5. Taking his experimental result into consideration, our evidence can be interpreted as follows. During the early stage of increasing stress in the shallow part of the volcanic edifice, the stress level is low, and consequently the effective cross-section of closing or shearing of pre-existing cracks is small. This process generates small earthquakes of low frequency, which means an increasing m-value. However, immediately before an eruption, the stress level becomes high, and consequently larger cracks are produced. As a result, the ratio of the number of large earthquakes to that of small ones increases, and this is reflected by a decrease in the m-value.

2.3 Evidence of Seismic Quiescence

On many occasions, increase of precursory seismicity is followed by a marked seismic quiescence just before eruptions. An example is shown in Fig. 6 for the case of the 1973 eruption of Asama volcano.

Interpretation. The increasing pressure of magma in the reservoir and the volcanic conduits produces an increasing level of stress. Initially , existing open cracks are crushed or sheared, causing the increasing seismicity. The seismic quiescence comes when most of the pre-existing cracks have been closed: this corresponds to the disappearance of low-frequency events, as described in Section 2.1. This phenomenon is also demonstrated by the rock fracture experiment shown in Fig. 5. According to this experiment, during a certain regime of differential

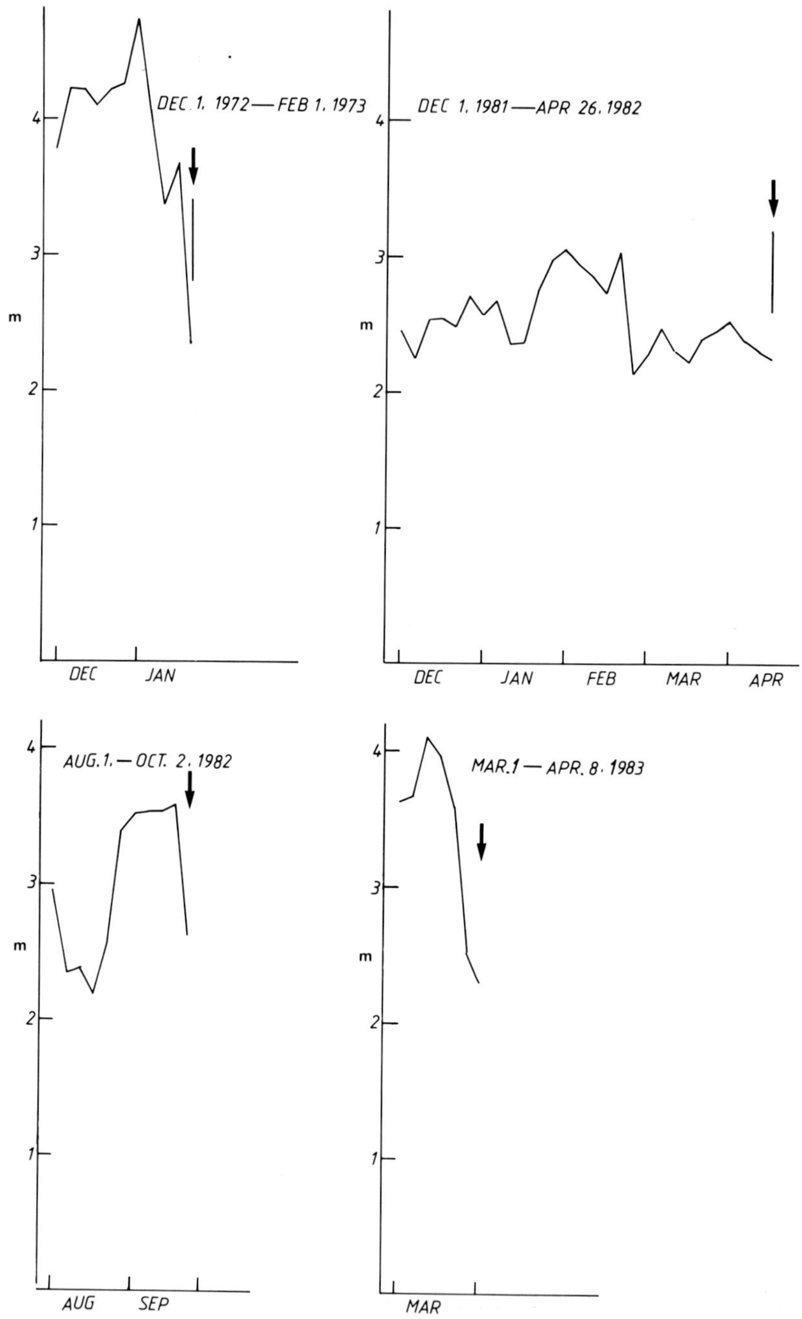

Fig. 4. Change of m-value (m = 1 + b) before eruptions of Asama volcano

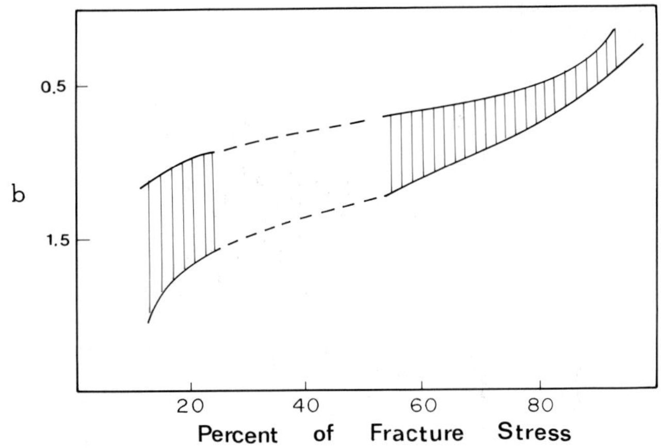

Fig. 5. The b-values as a function of normalized stress for five rocks in uniaxial compression. The **dashed parts** of the curves are in the region where few events were detected (simplified from Scholz 1968)

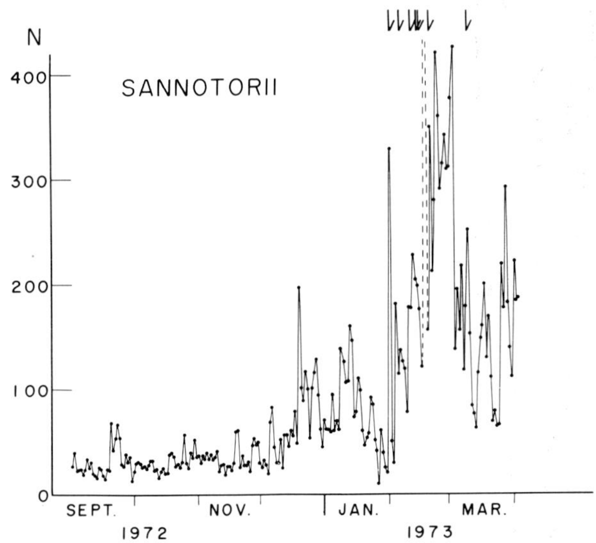

Fig. 6. Variation of the daily number of volcanic earthquakes associated with the February-March 1973 eruptions of Asama volcano **(arrows)**, observed at a site 2.5 km east of the summit crater

stress, no acoustic emission is observed. This fact could be a possible interpretation of the seismic quiescence immediately before the eruptions.

2.4 Evidence of Change of Seismic Amplitude Ratio H/V

The ratio of the maximum trace amplitude of the horizontal component to the vertical component in each earthquake tends to become smaller in pre-eruption earthquakes. Examples are shown in Figs. 7 and 8. All these seismic episodes are of the swarm type. However, the upper illustrations in Figs. 7 and 8 are of episodes without eruptions. Those illustrated in the lower figures show smaller values of H/V, and were accompanied by eruptions in each case.

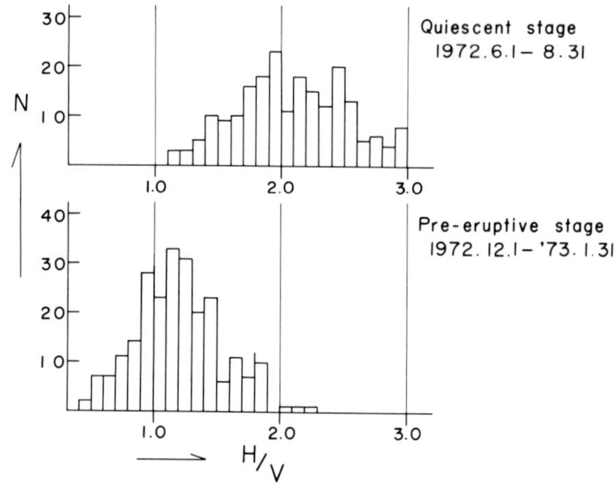

Fig. 7. Histogram of the ratio of the maximum trace amplitude of the horizontal to the vertical component, for earthquakes in a quiescent period, and at a pre-eruptive stage

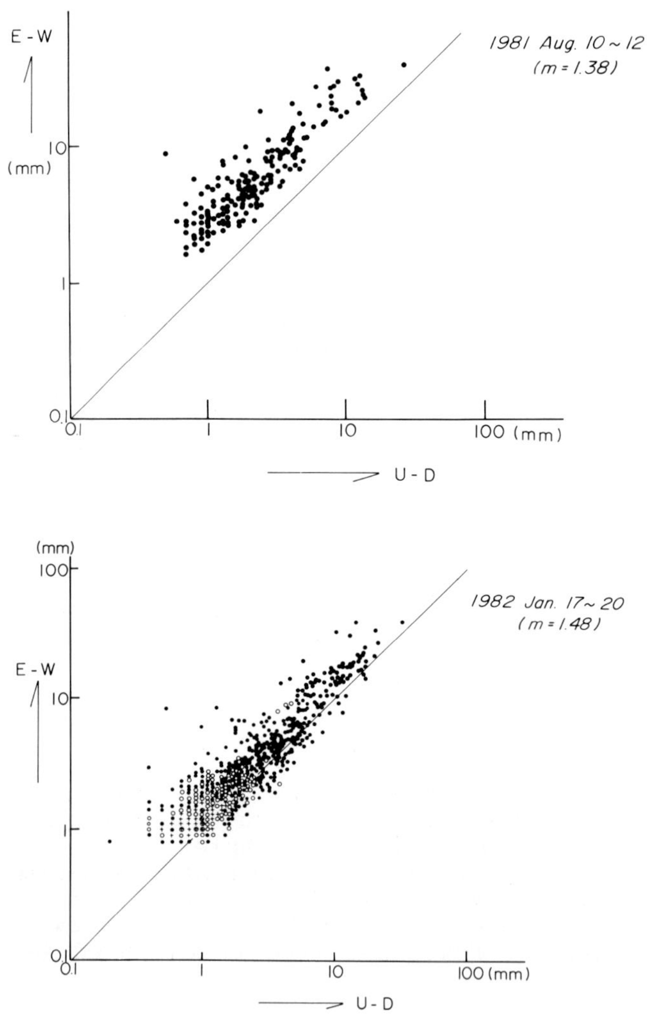

Fig. 8. An alternative way of illustrating the relationship shown in Fig. 7, for a different set of swarm earthquakes at Asama volcano

Interpretation. This evidence indicates the relatively large amplitude of SV waves for precursory earthquakes. At present, the reason is obscure, but may be related to a change of the focal mechanism or to a change of focal depth.

2.5 Evidence of Appearance of Tremor with Monochromatic Frequency

Sometimes, earthquakes of shallow origin occur which are characterized by very long coda waves of monochromatic frequency and smooth logarithmic amplitude decay, as shown in Fig. 9. The frequency histogram shows a pseudo-Gaussian distribution ranging between 1 and 5 Hz with a median of 2 Hz. The initial motion sometimes resembles that of high-frequency earthquakes.

Interpretation. The slope of the logarithmic plot of the successive amplitudes of the coda differs according to the locality of the seismic stations, and also between individual events. The occurrence of this peculiar tremor could be interpreted as a forced oscillation of a part of the volcanic conduit, or as an excitation of oscillation of layered elastic media in the volcanic edifice. This assumes that earthquakes occur at a place suitable for exciting free oscillation of an elastic plate.

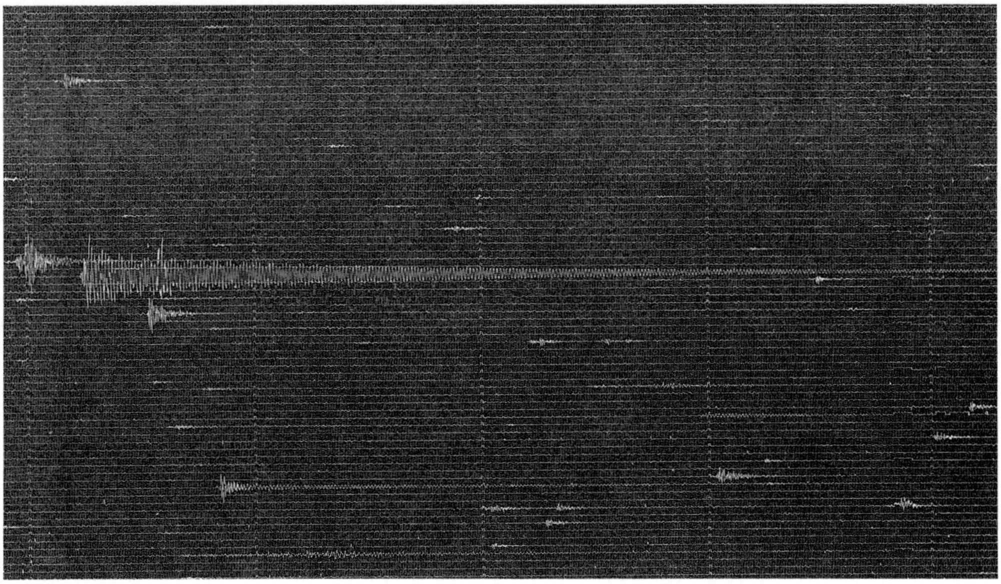

Fig. 9. Sample seismograms of tremor-like volcanic earthquakes with codas of long duration and monochromatic frequency

3 Concluding Remarks

For protection of life against volcanic hazards, prediction of volcanic eruptions based on instrumental monitoring is, needless to say, indispensable. Especially necessary are data which can be interpreted as immediate precursors of a state of imminent danger at an individual volcano, essential for issuing evacuation orders in the course of emergency management. Foremost among such data are those of seismicity.

We have described five significant features of pre-eruption volcanic earthquakes observed at Asama volcano. However, the conclusion cannot be drawn that these results are useful for other volcanoes of a similar explosive nature. We hope that every volcanological institution will endeavour to find the useful nature of precursory earthquakes, especially those that occur before a state of imminent danger, and compare them with those of other volcanoes. Thus we can determine what evidence is common, and what is specific to individual volcanoes.

References

Ishimoto M, Iida K (1939) Observations sur les séismes enregistrés par le microsismographe construit dernièrement. Bull Earthq Res Inst, Tokyo Univ 17:443-478 (in Japanese with French abstract)

Malone SD, Endo ET, Weaver CS, Ramey JW (1981) Seismic monitoring for eruption prediction. In: The 1980 eruptions of Mount St. Helens, Washington. US Geol Surv Prof Paper 1250:803-813

Minakami T (1960) Fundamental research for predicting volcanic eruptions. Bull Earthq Res Inst, Tokyo Univ 38:497-544

Minakami T (1974) Prediction of volcanic eruptions. In: Civetta L et al. (eds) Physical volcanology. Elsevier, Amsterdam, pp 314-333

Ohnaka M, Mogi K (1982) Frequency characteristics of acoustic emission in rocks under uniaxial compression and its relation to the fracturing process to failure. J Geophys Res 87:3873-3884

Okada Hm, Watanabe H, Yamashita H, Yokoyama I (1981) Seismologial significance of the 1977-1978 eruptions and the magma intrusion process of Usu Volcano, Hokkaido. J Volcanol Geotherm Res 9:311-334

Scholz CH (1968) The frequency-magnitude relation of microfracturing in rock and its relation to earthquakes. Bull Seism Soc Am 58:399-415

Suzuki S, Kasahara M (1979) Seismic activity immediately before and in the early stage of the 1977 eruption of Usu Volcano, Hokkaido, Japan. J Fac Sci, Hokkaido Univ, Ser VII, 6:239-254

The Taupo Seismic System

A.W. HURST[1], S. SHERBURN[2] and V.M. STAGPOOLE[2]

Abstract

The Taupo Volcanic Zone is an area of recent volcanism, with considerable shallow seismicity, including frequent earthquake swarms. The New Zealand Seismograph Network had only a few stations in this area, and only one station of low gain was near Lake Taupo, the source of a major eruption approximately 1800 years ago.

An improved seismic network was needed firstly to monitor the normal tectonic seismicity of the region, and secondly to enable accurate earthquake locations to be calculated quickly. In an area of known volcanic risk, the rapid determination of locations of felt earthquakes improves public confidence. Seismic signals from 13 vertical seismometers and one horizontal seismometer in the Taupo Volcanic Zone are transmitted to the DSIR Wairakei office. Three of these seismometers monitor seismicity in the Ohaaki area, where a geothermal power station is currently being built.

At the Wairakei recording centre, up to 16 seismic signals are sampled by a microprocessor-based digitizing system. Eight frequency-domain detectors monitor selected channels by continuously calculating spectra to see if the incoming signal has an earthquakelike spectrum. If any detector is triggered, the readings from all channels for a period starting 20 s before the trigger are recorded on magnetic tape, and the first arrival time for èach channel is printed and sent to the analysis computer. These arrival times come from a time-domain phase picker ('P-picker') monitoring all channels.

A DEC MicroPDP-11 computer is used for earthquake analysis. Immediate locations can be calculated from the P-picker times, or an interactive graphics program can be used for more complete analysis later. Two visual recorders are used for setting up and for checking the system performance. The total capital cost of the system is less than half that of on-site visual recording, the analysis costs are much less, and locations can be calculated rapidly, rather than weeks after an earthquake.

---

1 Geophysics Division, D.S.I.R., Wellington, New Zealand
2 Geophysics Division, D.S.I.R., Wairakei, New Zealand

IAVCEI Proceedings in Volcanology 1
J.H. Latter (Ed.), Volcanic Hazards

1 Introduction

The purpose of this seismic system is to monitor the seismic activity in the Taupo Volcanic Zone as part of the volcanic monitoring program of the Geophysics Division, D.S.I.R. This area has been the site of many very large volcanic eruptions and Lake Taupo was the site of the Taupo Pumice eruption about 1800 years ago. A number of smaller eruptions have occurred in the area since then, including the Tarawera eruption of 1886.

The system is designed to give rapid locations of local earthquakes. This is particularly necessary for the earthquake swarms which are common in this area. Major swarms occurred in 1922 (largest event M_L = 5.5-6) and 1964-65 (largest event M_L = 4.6), with two smaller swarms in 1983 (Smith and Webb 1986) and one in 1984. Although there has been no major volcanic activity in the area covered by the network this century, any swarm in this area in which the activity is very shallow could be a possible precursor of an eruption.

The system is also used to monitor any earthquake activity associated with the Ohaaki Power Station (Fig. 1). The extraction of geothermal fluid and the reinjection of the waste fluid into the ground are expected to produce seismic activity when electricity generation commences in 1988.

The seismic system consists of a network of seismometers, which send their signals by radio links to the Wairakei office of the Geophysics Division, where seismic events are identified and recorded by computer (see network diagram, Fig. 1). An in-

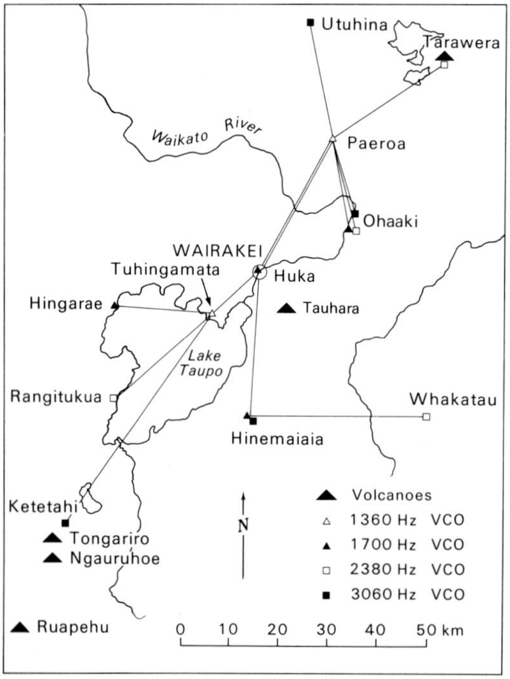

Fig. 1. Location diagram for the Taupo Seismic System, North Island, New Zealand, showing station sites and VCO frequencies. (The 3060-Hz VCO at Hinemaiaia is for a horizontal seismometer.) All radio links are one-way towards Wairakei

teractive analysis program is used to correct the computed arrival times if this is necessary, and to calculate the earthquake locations.

2 Seismic Outstations

These are conventional telemetering seismic stations and use normal analogue techniques. Each station has a vertical seismometer, a preamplifier and a voltage controlled oscillator to modulate an audio tone with the seismic signal. The preamplifiers have a variable gain stage, which reduces the gain by a factor of 10 if the input signal exceeds a certain value. This increases the dynamic range by a factor of 5. The modulated audio tones are transmitted to Wairakei by UHF radio links, in some cases using repeaters. Up to four audio tones can share one radio channel.

The network has a station spacing of about 20 km, which should enable accurate locations to be calculated for earthquakes within the area. A temporary station, which can quickly be installed in an area of earthquake activity, is being assembled. Its signal will be telemetered to Wairakei through the repeater at Paeroa, which can receive signals from a wide area of the Taupo Volcanic Zone.

3 Recording

Analogue recording of the seismic signals was considered but rejected. Conventional drum seismographs are now very expensive, and there is a large cost for reading and analysis of the records. Analogue recording on magnetic tape was excluded because of the excessive time needed to read the tapes, while the photographic material and chemicals for photographic film systems made them too costly.

Therefore it was decided to use a digital recording system, which only records into permanent storage when earthquake events are detected, thus minimizing data storage and handling problems. The system currently used, known as SNARE (Seismic Network Automatic Recording Equipment), was developed by the Geophysics Division for recording seismic network recording data (Gledhill and Randall 1986). It uses three 6809 microprocessors which control the digitizing, detection and storage of the seismic records.

The main microprocessor oversees the conversion of the analogue signals into digital form, and controls the storage of the digital readings in two 64 kbyte ring buffers of solid state memory. Each channel is sampled 50 times per second using a 13-bit analogue to digital conversion system. The digital reading for each channel requires two bytes of storage, so the ring buffers will each hold 41 s of data in a 16-channel system. The storage microprocessor transfers the data in the temporary ring buffers to permanent storage on magnetic tape on command from the main microprocessor.

The detection microprocessor calculates frequency spectra for eight of the input channels. Every second, spectra are calculated for 1 s-long records from all eight channels. For each channel, the total power in each of five frequency bands is compared with trigger levels. The trigger levels are dependent on the previous noise level in that band. For each band that exceeds its trigger level, a preset weight is added to the total spectral weight. The different frequency bands have different weights, reflecting the importance of each band for detecting an earthquake. If the total spectral weight exceeds its preset trigger level long enough, then an event is declared. The weights of the different bands can be altered depending on whether local (high-frequency) or regional (mid-frequency) earthquakes are wanted (Gledhill 1985). Only when this processor has declared an event will the readings in the ring buffers be written out to the magnetic tape cartridge. When an event is declared, the contents of one or more buffers are written to magnetic tape, starting at a fixed time (usually 20 s) before the time at which the event was confirmed, and finishing at the end of the current buffer once 'end of coda' has been signalled on all channels.

At the same time the control microprocessor is operating the phase-picking routines, which are based on an algorithm by Allen (1978). This algorithm uses a short-term average/long-term average criterion to detect and time seismic arrivals. This method can give more accurate arrival times than the spectral detector, but it is more affected by noise. If more than a set number of channels detect arrivals, or if the spectral detector detects an event, information on the phase arrivals is printed and sent to the DEC MicroPDP-11 computer. The information for each arrival is the channel number and station code, phase (P or S), arrival time to 1/100 s, quality of pick (0 very good, to 4 poor), direction of first motion, amplitude and duration. Vertical seismometers all detect P-phases, while horizontal seismometers can be used to detect an S-phase by only activating picking on that channel when a P-pick has been declared from an adjacent vertical seismometer.

Normally, information is recorded on tape only when the detection microprocessor detects an earthquake, although the system can be preset to record at certain times, such as when explosions are to be set off.

4 Analysis System

An interactive analysis and location program is operated on a DEC microPDP-11 computer. In the initial version of the recording and analysis system, when only eight channels were installed, the microPDP-11 also ran the seismic picking and recording, using an analogue/digital converter board and the Allen (1978) phase-picker program.

The microPDP-11 is now used for location calculations and interactive graphic analysis of earthquakes. If the picks from SNARE for an earthquake are of good quality, a location routine is called which automatically calculates the earthquake location and time using an appropriate velocity model. Phases from stations outside the network can be added to improve locations.

The interactive graphics program uses the digitized readings from the SNARE recording system. At present this is done by taking the magnetic tape from the SNARE system and reading it into the microPDP-11 using another drive, but it would be possible to send the readings directly to the microPDP-11 as well as to magnetic tape if instant analysis is needed. The seismic traces from any three channels can be displayed on a graphics screen at one time. The times at which each channel was picked by the phase picker are lined up vertically, and the maximum amplitudes and first motions are shown at the start of each trace. This information comes to the microPDP-11 from the phase picker, and is not on the magnetic tape with the readings. The trace from any channel can be displayed in an enlarged form at the bottom of the screen. A manual picking of arrival time can be made by the adjustment of cross-hairs in relation to the enlarged trace, if a seismic arrival was not picked by SNARE, or was picked at the wrong time. A typical display on the graphics screen is shown in Fig. 2.

Two visual recorders are operated to assist in setting up the system and to monitor the quality of the seismic records from the stations. One recorder is normally used with the Ketetahi seismometer, to watch for volcanic tremor from the Tongariro National Park volcanoes.

5 Operation

There are two main factors which affect the ability of this system to record small local earthquakes. The first is the ground conditions at each outstation, which affect the usable gain for that seismometer. The use of radio telemetry has meant that usually sites away from cultural noise could be used, and accordingly fairly high gains (40 000-70 000 at 10 Hz) have been achieved at most sites.

In the volcanic area, sites are often found to have anomalously high attenuation for earthquake waves, because of hard rock on the surface overlying unconsolidated pumiceous material. All sites were tested and those which had high levels of ground noise or anomalous attenuation were rejected. In some areas where there was no solid rock at the surface, seismometers were put down boreholes to depths of about 50 m in order to improve the signal to noise ratio. Current indications are that shallow earthquakes with a magnitude (M_L) greater than 2, within the area of the network, will be detected and recorded.

The second factor is the ability of the earthquake detection system to distinguish between earthquakes and noise. At present the detection processor has not been in use long enough to give reasonable statistics, but initial tests show that the frequency detector is giving better discrimination between earthquakes and noise than the Allen (1978) picker previously used.

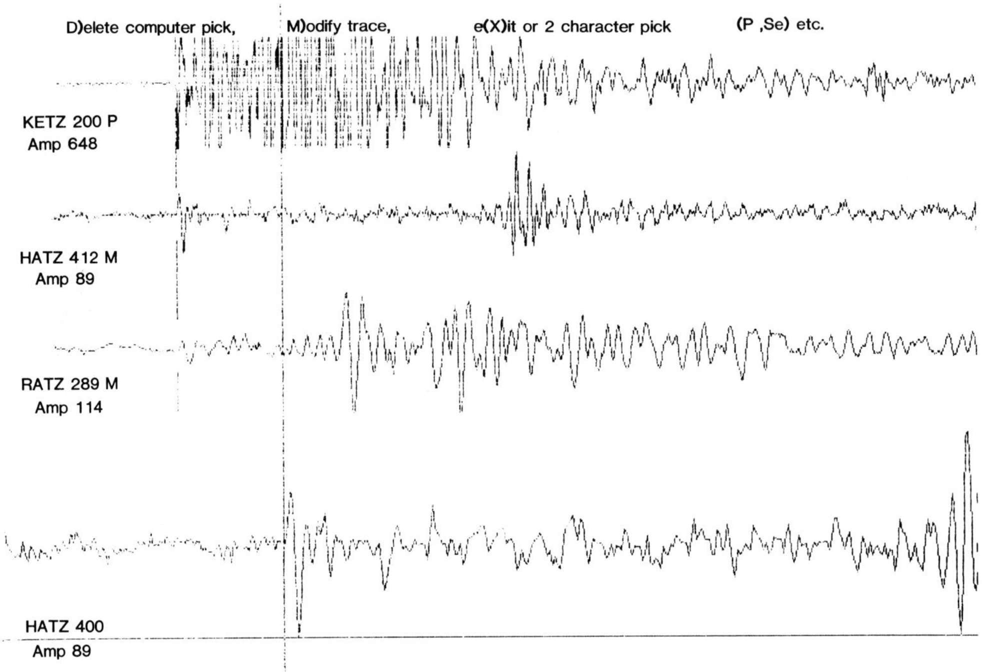

Fig. 2. Printer copy of screen display while processing earthquake. This was a magnitude 2.2 earthquake, origin time 1985 November 5, 08 h 04 min 10.4 s. Location about 10 km northeast of Ketetahi. Depth restricted to 5 km. The line of text at top of screen shows what commands are valid at any time. The three traces shown are labelled by station codes (**KETZ** Ketetahi, **HATZ** Hinemaiaia, **RATZ** Rangitukua), the reading at which the first pick for that channel is currently recorded (**P if computer, M** if manual), and the peak trace amplitude. The enlarged trace of **HATZ (below)** shows the vertical line of the screen cross-hairs aligned with the P-arrival, ready for manual picking

No volcanic tremor has ever been recorded in the Taupo area, although it is common at Ruapehu and Ngauruhoe and occasionally occurs at other locations near these volcanoes. If volcanic tremor is commonly recorded at any seismometer, it will be necessary to process it separately. It would be possible to have another spectral microprocessor, running algorithms similar to those used currently at the Chateau Observatory (Hurst 1985), calculating tremor power spectra and recording average at convenient intervals. Any volcanic tremor at a new site would be taken as a sign of a possible impending eruption.

The total cost of the basic system of eight channels, including outstations radio links and recording system, is about $US 28 000, and a complete 16- or 32-channel system costs about $US 57 000 or $US 93 000 respectively. Installation costs would be additional. The system is eminently suitable for low cost surveillance and monitoring of volcanic areas where eruptions are possible at a large number of places.

Acknowledgements. The seismic outstations around Ohaaki, and their radio links, as well as a contribution towards increasing the capacity of the recording system, were funded by Design Power, a subsidiary of the Electricity Corporation of New Zealand.

W.M. Kohler, of the U.S. Geological Survey, supplied the picking program used in the initial eight-channel system.

References

Allen RV (1978) Automatic earthquake recognition and timing from a single trace. Bull Seism Soc Am 68:1521-1532

Gledhill KR (1985) An earthquake detector employing frequency domain techniques. Bull Seism Soc Am 75:1827-1835

Gledhill KR, Randall MJ (1986) SNARE: an earthquake detecting and recording system for small seismograph networks. Bull Seism Soc Am 76:1485-1489

Hurst AW (1985) A volcanic tremor monitoring system. J Volcanol Geotherm Res 26:181-187

Smith EGC, Webb TH (1986) The seismicity and related deformation of the Central Volcanic Region, North Island, New Zealand. In: Smith IEM (ed) Late Cenozoic Volcanism in New Zealand. Roy Soc New Zealand, Bulletin 23, Wellington 112-133

Classification of Volcanic Earthquakes at Mount Erebus Antarctica

K. SHIBUYA[1], R.R. DIBBLE[2], M. BABA[3] and P.R. KYLE[4]

Abstract

On the basis of infrasonic and seismic waveforms, we classify volcanic earthquakes at Erebus volcano into Group-I events accompanied by infrasonic signals of eruptions, and Group-II without eruption signals.

Group I is subdivided as follows:

α-Types have a sharp infrasonic pressure pulse of 0.05-0.35 s, and a rarefaction tail of 1.5-3.5 s duration. The seismic power spectrum extends up to 15 Hz. Reported focal depths range from 3 to 5 km.

β-Types have a large onset of both seismic and infrasonic waves with a sinusoidal waveform at the summit station. The onset portion of the seismic power spectrum, defined as having the duration of the infrasonic signal, has a dominant frequency lower than 3-4 Hz, whereas the coda has a dominant frequency of 5-10 Hz. Reported focal depths range from 1.5-4 km below the summit.

γ-Types have a large infrasonic but very small seismic onset at the summit station. They are considered to be generated by the bursting of 200-300 m diameter low pressure bubbles at shallow depth in the lava lake.

Group II is subdivided as follows:

Tremor types have an emergent onset, a narrow spectral peak of 5.0-5.4 Hz, and a duration of 20-30 s. Although they have no infrasonic onset, they occur during medium to high infrasonic noise levels.

[1] National Institute of Polar Research, Kaga 1-9-10, Itabashi-ku, Tokyo 173, Japan
[2] Department of Geology, Victoria University of Wellington, Wellington, New Zealand
[3] Department of Earth Science, Faculty of Science, Chiba University, 33, Yayoicho 1-chome, Chiba 260, Japan
[4] Department of Geoscience, New Mexico Institute of Mining and Technology, Soccoro, New Mexico 87801, USA

IAVCEI Proceedings in Volcanology 1
J.H. Latter (Ed.), Volcanic Hazards

High-frequency types have ordinary earthquake onsets, and a dominant frequency of 12 Hz. Reported focal depths range from 4 km below the summit (sea level) to 15 km.

Groups I and II are connected by a hybrid type, which has a seismic signal of tremor type, and an infrasonic signal of β-type. Hence, there must be a transition in the generating mechanism from β- to hybrid to tremor type. Considering all types together, the dominant seismic frequency increases in a general way with reported focal depth, and a similar increase in the spectrum of the source time function is postulated.

1 Introduction

Erebus is a 3794-m-high anorthoclase-phonolite volcano on Ross Island, West Antarctica, with a small active lava lake (2000 m^2) and an explosive vent. Erebus erupted an average of 3.6 times a day during observations between 1974 and 1982 (Dibble et al. 1984). Its volume of 1670 km^3 puts it among the 50 largest volcanoes of the world (Katsui 1971). The lava has a higher silica content (56%, Kyle et al. 1982) and a higher calculated viscosity (10^3 Pas at 1200°C, Dibble et al. 1984) than any other known persistent lava lake.

The steady-state convection of magma observed for nearly 12 years is thought to have slowed or stopped just before the 1984-85 Antarctic field season, allowing the surface of the lake to crust over. Cooling of the system is believed to have triggered the strong explosive activity which commenced abruptly on 13 September 1984, and was especially violent for 7 days.

Why the volcanic activity of Mt. Erebus is so high in a seismically inactive Antarctica (so believed) is an open question, and several attractive working hypotheses have been presented. For example, Hamaguchi and Zana (1985) regarded Mt. Erebus as one of four main hot spots on the earth which are distributed 90° apart on a great circle passing through the present geomagnetic poles. The other three are Mt. Nyriagongo, Iceland which is coincident with a ridge, and Hawaii. Hamaguchi and Zana (1985) pointed out that the effect of the main geomagnetic field originating in the earth's core could be playing an important role in the existence of volcanism with persistent lava lake activity.

Interpretation of the volcanic activity of Mt. Erebus in a regional tectonic context requires geophysical data covering adequate ranges of time and area with sufficient resolution. This was the motivation for the International Mount Erebus Seismic Study (IMESS) which started in 1980. In the 1981-82 field season there were five telemetering seismic stations, four temporary seismographs, and one telemetry infrasonic station (Fig. 1). The overall response of the telemetry geophones to ground particle velocity is almost constant from 1 to 20 Hz (Shibuya et al. 1983). In the 1981-82 field season a new infrasonic sensor was installed on the main crater floor (Dibble et al. 1984) and telemetered to Scott Base. This consisted of two Philips

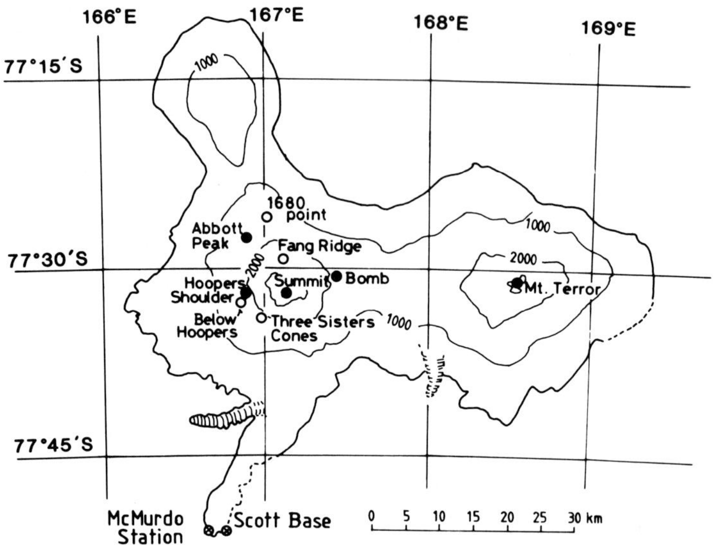

Fig. 1. Seismic network on Ross Island in the 1981-82 field season. **Solid circles** denote radio-telemetering stations; **open circles** show temporary stations. The temporary stations were operated during November-December in 1981. (After Shibuya et al. 1983)

Table 1. List of the events for display and/or spectrum analysis

Event No.	Date and time of occurrence (GMT)	Event type
350	352d 02h 42m 1981	ß
453	355d 05h 50m 1981	α
7	13d 00h 29m 1982	Hybrid
14	14d 16h 20m 1982	Hybrid
50	27d 18h 10m 1982	ß
83	35d 22h 05m 1982	ß
108	47d 13h 54m 1982	High-frequency
137	57d 08h 30m 1982	ß
158	62d 16h 53m 1982	Tremor
164	63d 22h 08m 1982	Tremor
168	65d 07h 31m 1982	Tremor
199	73d 13h 13m 1982	Hybrid
211	77d 17h 47m 1982	Tremor
215	78d 13h 48m 1982	α
221	79d 19h 38m 1982	γ
280	97d 14h 13m 1982	Tremor
394	137d 21h 17m 1982	Tremor

Hi-Z speakers mounted on opposite sides of a 50-liter sealed enclosure so as to cancel seismic response.

This chapter is a study of the correlation of seismic and infrasonic waveforms and spectra at the summit station. It attempts to classify volcanic earthquakes into several types and to correlate these with specific magmatic processes. The spectral analyses were made by using IMESS analog magnetic tapes recorded in 1981-82, which were edited and digitized at the National Institute of Polar Research in Tokyo. The event numbers cited in the following sections are summarized in Table 1 together with the type name.

2 Volcanic Earthquakes

Various types of earthquakes have been recorded by the IMESS network. Frequency of occurrence versus seismic power of Erebus earthquakes has been summarized by Dibble et al. (1984) and is shown in Fig. 2. The events appear to be a mixture of two populations. The b-value is 1.5-1.7 for earthquakes smaller than 100-W radiated seismic power; this is within the range considered normal for volcanic earthquakes. For earthquakes larger than 100 W, the b-value is around 0.6; this is more typical of tectonic earthquakes. Volcanic earthquakes recorded on the uncalibrated IMESS net occur at a fairly constant rate of 15.7 ± 5.7 per day. This rate is higher than the 3.6 ± 2.7 Strombolian eruptions per day, but not as high as previously recorded by Dibble et al. (1984) with temporarily higher gain seismographs.

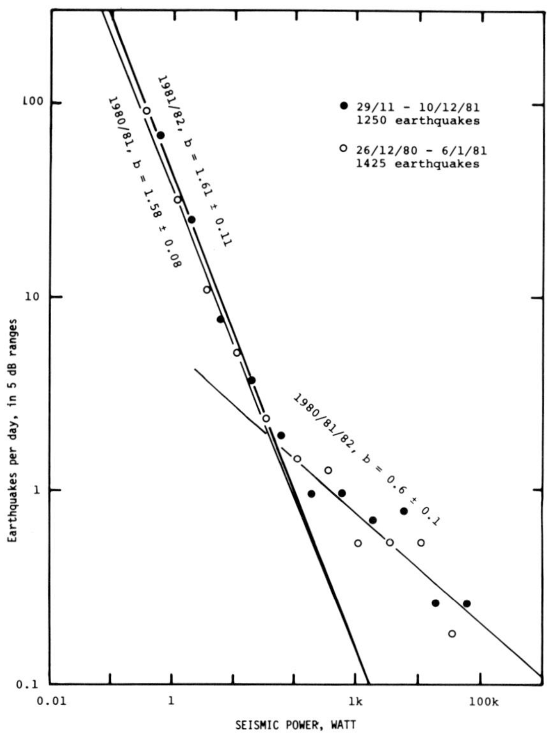

Fig. 2. Frequency of occurrence versus seismic power of Erebus earthquakes. (After Dibble et al. 1984)

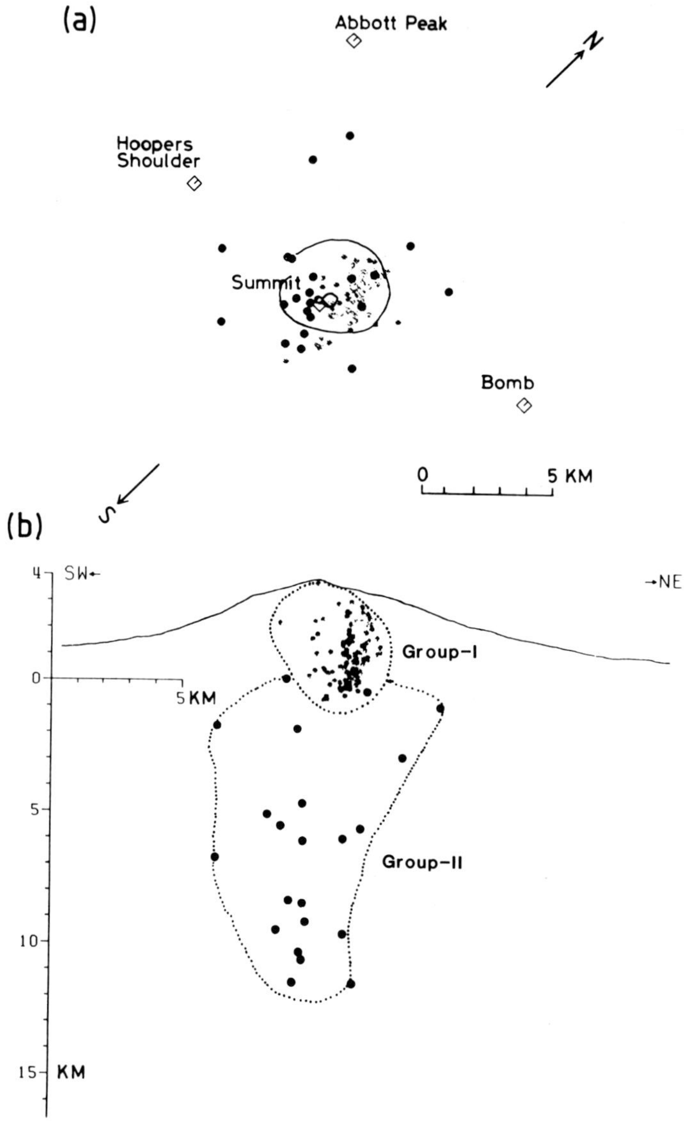

Fig. 3 a, b. Hypocentral distribution of seismic events during the 1981-82 field season, **a** epicenters, **b** hypocenters, projected onto an NE-SW section across Mt. Erebus. (After Shibuya et al. 1983)

Shibuya et al. (1983) showed that the hypocenter distribution of Erebus earthquakes can be grouped into two subsets depending on whether or not they are accompanied by infrasonic signals. Figure 3 taken from Shibuya et al. (1983) illustrates epicenter and hypocenter distributions obtained from data recorded in the 1981-82 season. Group-I events, which accompany infrasonic waves, scatter in a pipelike vertical distribution beneath the summit extending to sea level (4 km deep). The hypocentral depths of Group-II events, which are not accompanied by infrasonic waves, extend to 15 km below the summit. The number of Group-II events is smaller by an order of magnitude than that of Group-I events, and their hypocenters are more scattered.

The earthquake distribution shown in Fig. 3 is based on the oversimplified assumption of a uniform and homogeneous semi-infinite medium with a constant P-wave velocity of 2.1 km s^{-1}. The examination of arrival time differences between seismic and infrasonic onsets at the summit station indicates that the P-wave velocity in the shallowest part of the structure at the summit area may be as low as 900-1600 m s^{-1} (Dibble et al. 1984).

In the 1984-85 field season, a seismic explosion experiment was carried out (Kienle et al. 1985; Rowe and Kienle 1986) which provided a refined average velocity model of the volcano consisting of a linear velocity gradient above a constant velocity half space.

Earthquakes located with this model show similar features in the earthquake distribution (Takanami et al. 1983; Ueki et al. 1984; Kienle et al. 1985; Rowe and Kienle 1986; Baba et al. 1985).

3 Classification of Group-I Events

Group-I events are B-type volcanic earthquakes (Minakami 1963), and are believed to be of the same general type as explosion earthquakes, since eruptions observed visually (for example, Table 1 in Shibuya et al. 1983; Appendix 4 in Dibble et al. 1984) were always accompanied by earthquakes of this type at the summit station. Group-I events can further be classified into three types according to the waveform of the infrasonogram; α-type, ß-type and γ-type (bubble explosion type) (Shibuya et al. 1983; Kaminuma et al. 1985).

Figure 4 shows examples of seismograms and infrasonograms of each type. α-Type events (Fig. 4 a) are characterized by impulsive infrasonic pressure increases lasting 0.05-0.35 s, followed by a recovery tail lasting 1.5-3.5 s. ß-Type events (Fig. 4 b) have an infrasonic signal that is characterized by sinusoidal oscillations of frequencies below 3 Hz. The predominant frequency in the seismic power spectrum for ß-type events (Fig. 5 b) is usually lower than that for α-type events (Fig. 5 a) at all of the flank stations, and for both these types, the frequencies at the flank station Hoopers Shoulder (dotted curves in Fig. 5 a and b) are lower than at the summit station (solid curves), due to the effect of dispersion on transient waveforms.

γ-Type events (Fig. 4 c) are characterized by a large amplitude infrasonic signal and a comparatively small amplitude seismic signal at the summit station. The seismic onset can

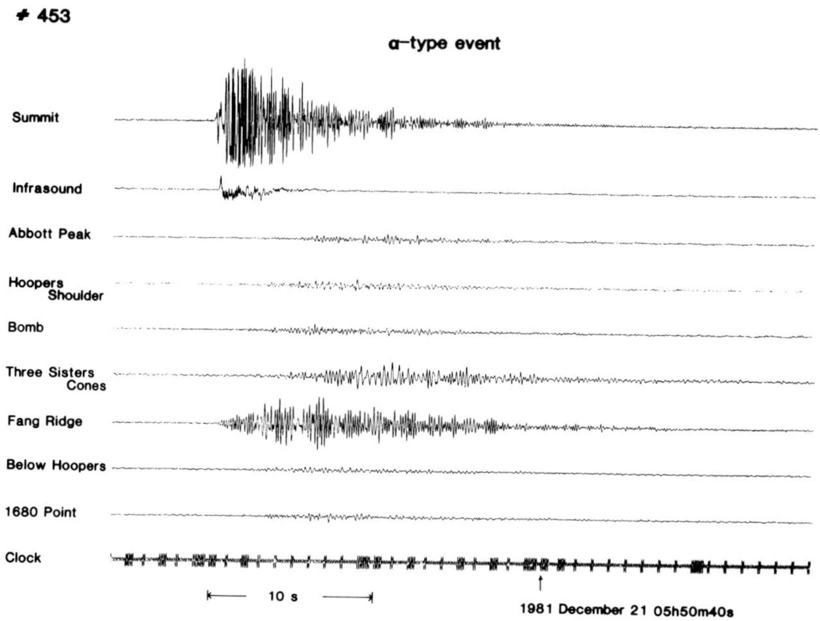

Fig. 4 a

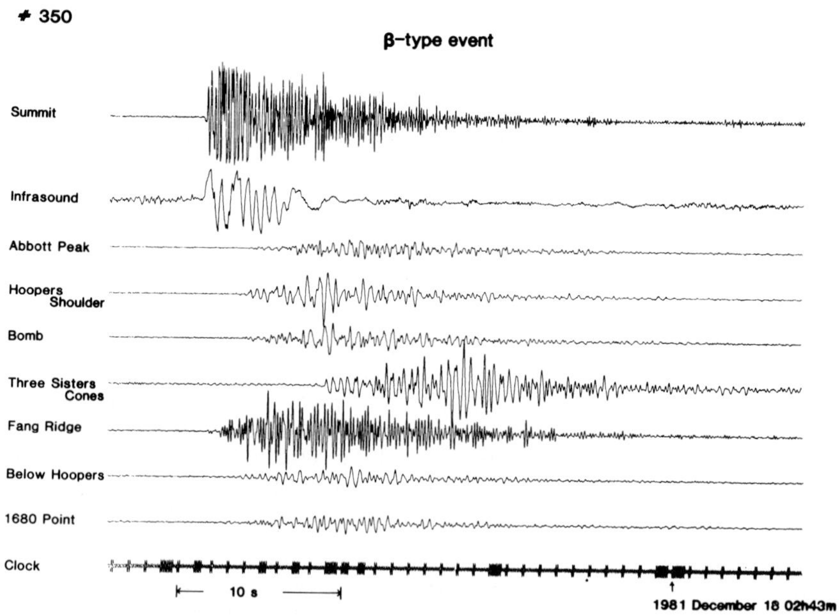

Fig. 4 b

Legend see p. 528

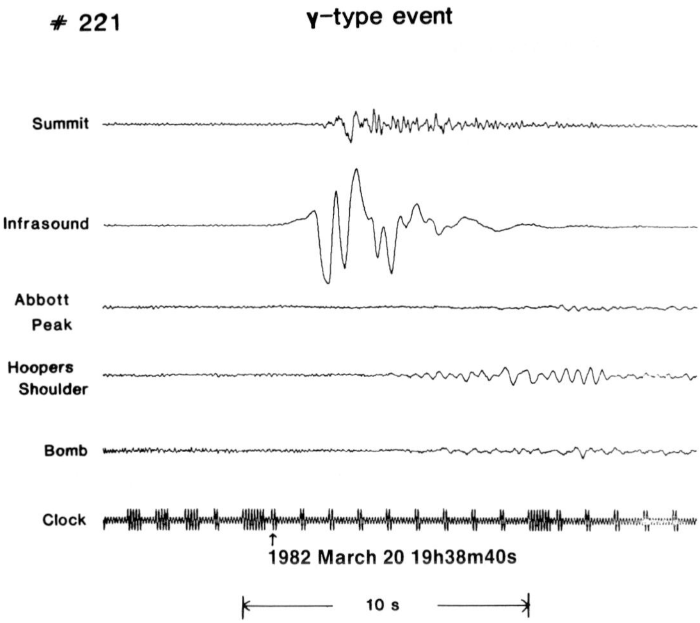

Fig. 4 c

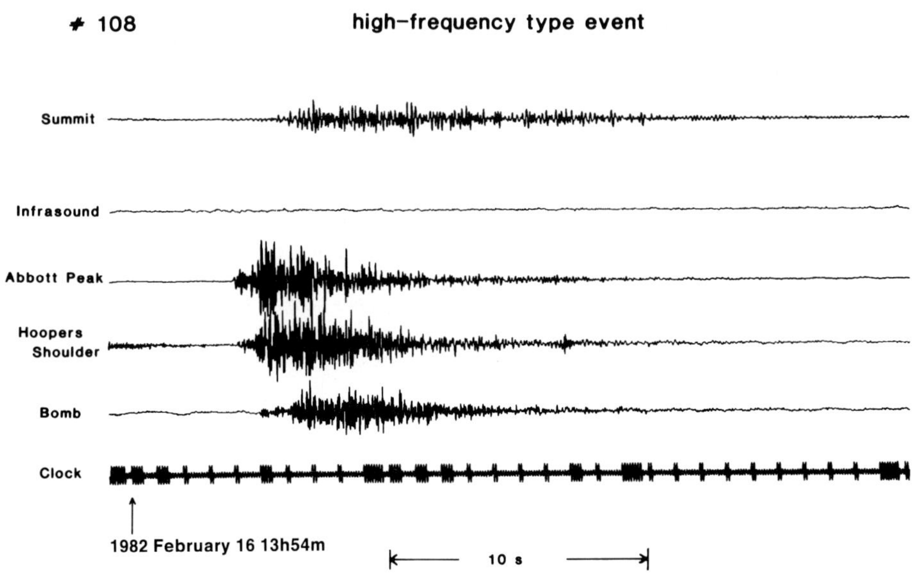

Fig. 4 d

Legend see p. 528

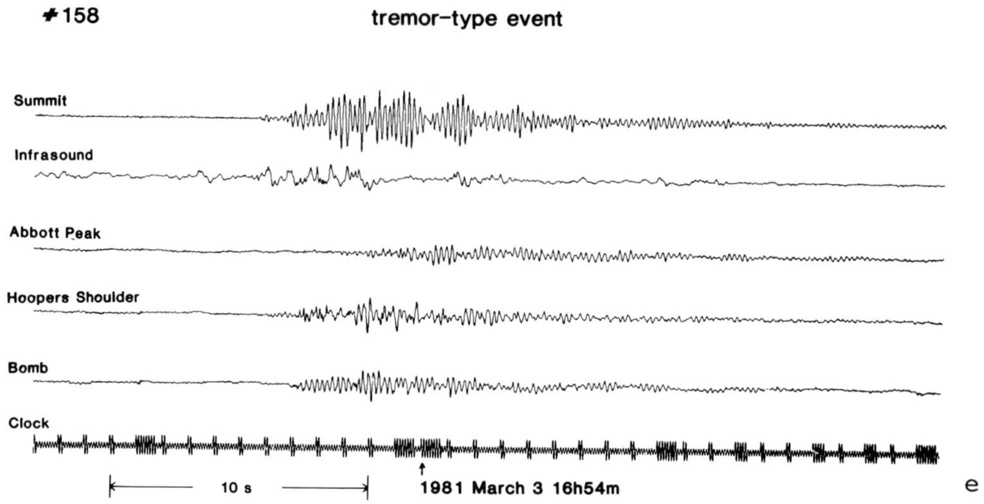

Fig. 4 a-e. Examples of each event type. **a** #453 for α-type; **b** #350 for ß-type; **c** #221 for γ-type (bubble explosion type); **d** #108 for high-frequency type; **e** #158 for tremor type

hardly be detected at the flank stations. The polarity of the infrasonic onset is either a sharp pressure increase (explosive) or a very gradual pressure increase followed by a large implosive signal. The seismic onset at the summit station occurs 0.5-1.0 s after the arrival time of the infrasonic implosion.

On the whole, α-type events occur twice as often as ß-type events. γ-Type events are least common, occurring only one-fifth as often as ß-type events. Kaminuma et al. (1985) showed that the focal depths of ß-type events range from 1.5-4 km below the crater level and are on average shallower than those of the α-type which range from 2.5 to 5 km. Though focal depths of γ-type events cannot be determined because the onsets are so weak, the infrasonic and seismic waveforms are similar to those features of bubble explosions in the lava lake described by Dibble et al. (1984), and therefore these events may have the shallowest focal depths.

4 Volcanic Tremor and Tremor-Type Group-II Events

Volcanic tremor is observed on many active volcanoes. Individual volcanoes have different predominant frequencies which vary from 0.1 to 10 Hz (Kubotera 1974). Kyle et al. (1982) noted that volcanic tremor at Mt. Erebus was not observed during annual short-term visits to the summit between 1974 and 1980. This lack was noted for data recorded both inside and outside the summit crater.

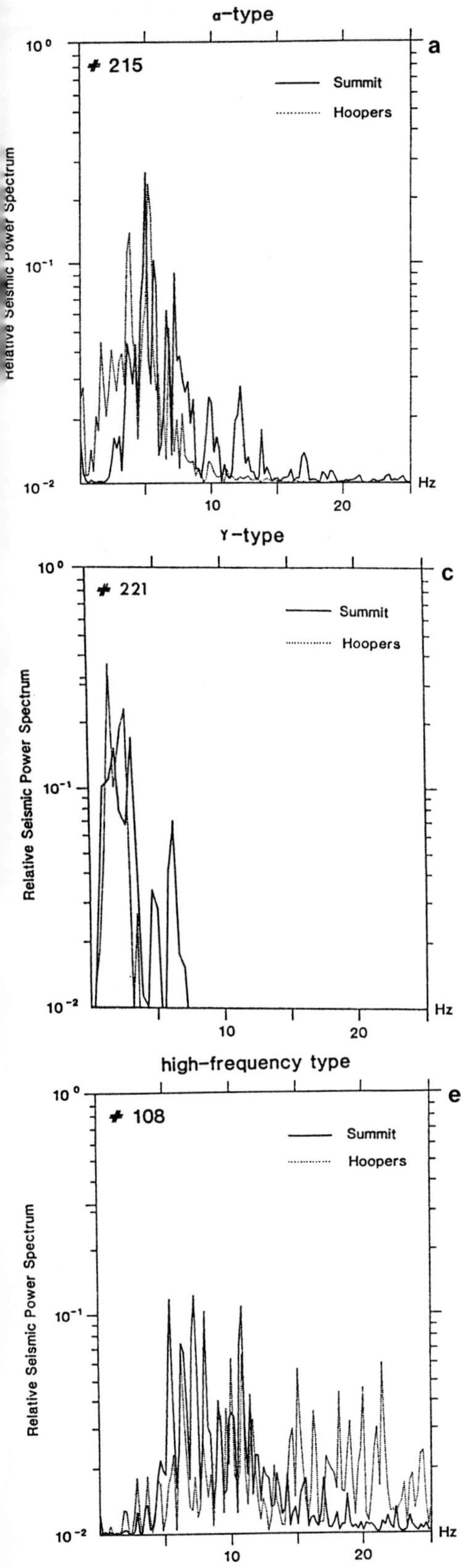

Fig. 5 a-e. Seismic power spectra for examples of volcanic earthquakes of each event type. The **solid curve** shows that observed at the summit station and the **dotted curve** shows that observed at the flank station Hoopers Shoulder. **a** α-type, #215; **b** ß-type, #50; **c** γ-type, #221; **d** tremor type, #211; **e** high-frequency type, #108

One of the results of the long-term continuous monitoring of the IMESS network has been the detection of volcanic tremor and tremor-type events. Ueki et al. (1984) observed a 1-h-long period of monochromatic tremor with a characteristic period of 0.9 s which occurred after 8 October 1982. Tremor-type events (Fig. 4 e) occur during medium to high infrasonic noise levels and have 20-30 s seismic durations. They have no clear infrasonic onset corresponding to the seismic onset. The events have earlier and larger first arrivals at the summit station than at the flank stations.

Figure 6 a shows six tremor-type events observed at the summit station. They have narrow-band seismic power spectra around a predominant frequency of 5.0-5.4 Hz (Fig. 6 b), and the flank and summit stations record similar spectra (Fig. 5 d). Hence, the seismic power spectrum of the tremor-type event must be strongly dependent on the shape of the source time function itself, rather than the highly attenuative/dispersive properties of the volcano. The seismic power spectrum is notably different from that of α- and ß-type events in Fig. 5 a and b.

5 High-Frequency Type Group-II Events

These are not accompanied by infrasonic waves and are usually characterized by higher frequency seismic waves at all stations (Figs. 4 d and 5 e). We therefore refer to these as high-frequency types. Scattered distribution and greater depths may indicate that these events are of volcano-tectonic or tectonic origin (Rowe and Kienle 1986).

6 Discussion

Figure 5 compared seismic power spectra of α-, ß-, γ-, tremor-type, and high-frequency type events at the summit (solid curve), and at the flank station Hoopers Shoulder (dotted curve). The spectrum was calculated for each type of seismogram shown in Fig. 4. It was noted that the predominant frequency increases to higher values as the event type shifts from ß- to α- to high-frequency type, at both summit and flank stations. In general, the shift corresponds to an increase in reported focal depths. In spite of the increase in path length through the highly dispersive/attenuative volcanic medium, the power spectrum for deeper events is enriched in high-frequency spectral components, indicating a different source mechanism, or an entirely different path avoiding the dispersive material.

The generating mechanism for tremor-type events must be similar to that of ß-type events, since there exists a hybrid type between these two. Examples of such hybrid types are shown in Fig. 7 a (Nos. 7, 14, and 199) and compared with typical ß-types in Fig. 7 b (Nos. 50, 83, and 137). These events are all displayed at the same arbitrary magnification.

Figure 8 illustrates a comparison of the infrasonic power spectra of a typical γ-type event (No. 221 by the solid curve),

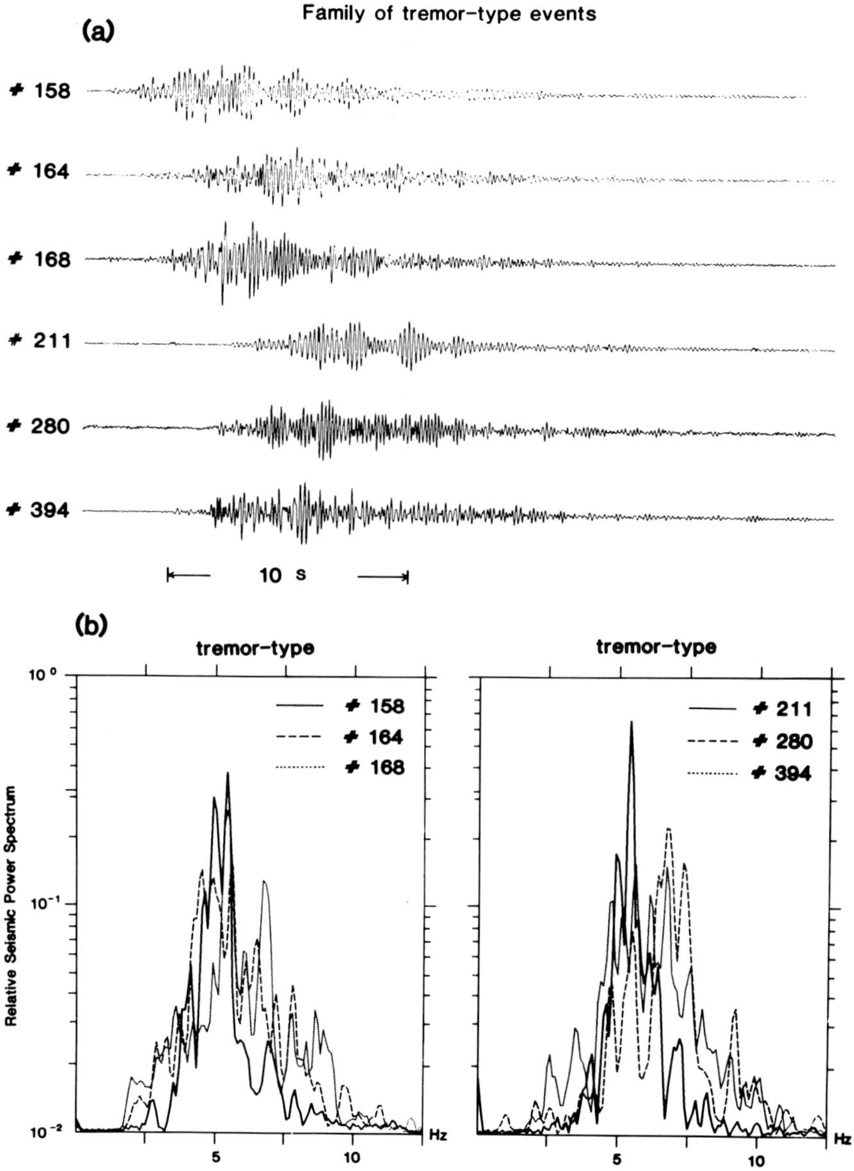

Fig. 6 a, b. a A family of tremor-type events and b their seismic power spectra at the summit station

a ß-type event (No. 50 by the broken curve), and a hybrid-type event (No. 14 by the dotted curve). They all have the fundamental frequency of 0.5-0.6 Hz and exhibit second and third harmonic overtones. For frequencies below 2.5 Hz, there is little difference in the infrasonic power spectra of the three types of events.

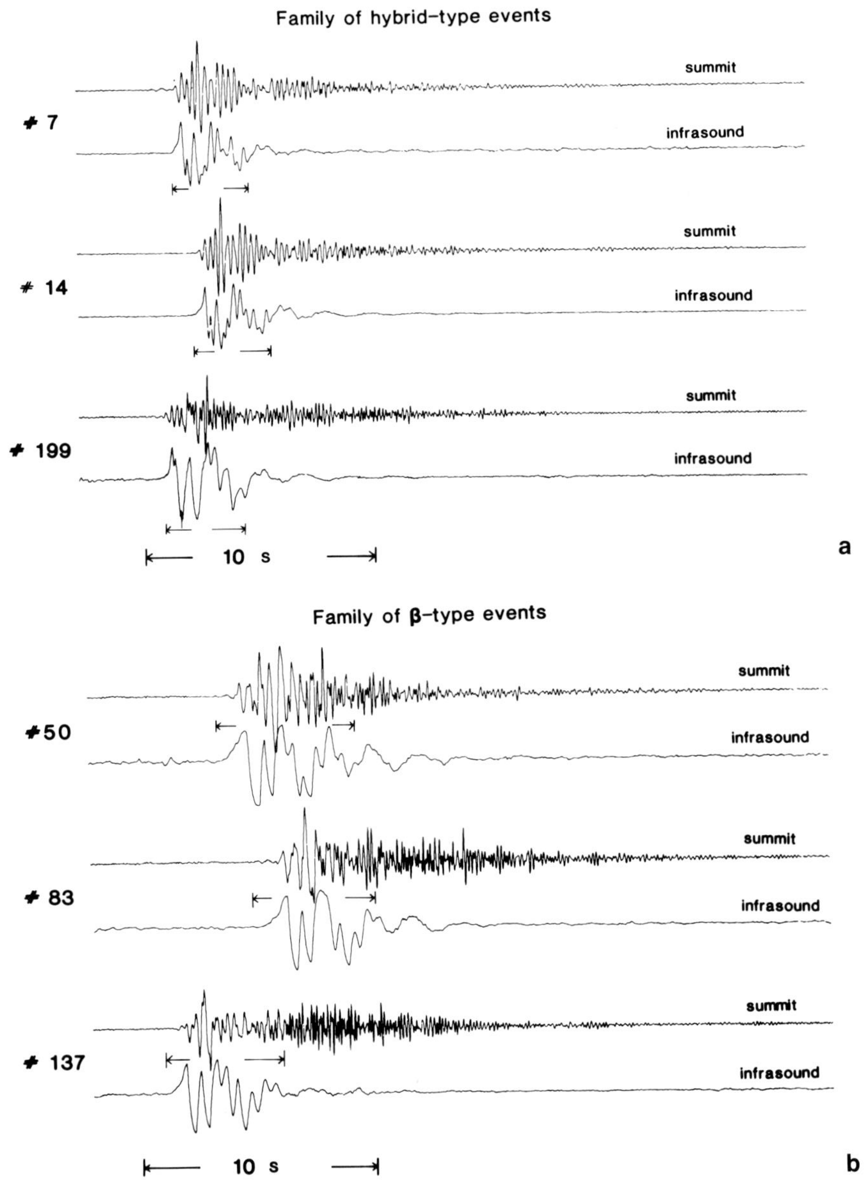

Fig. 7 a, b. Examples of seismograms and infrasonograms for a hybrid-type events (#7, #14, and #199) and b ß-type events (#50, #83, and #137). For each pair, the upper trace shows seismic, and the lower trace shows infrasonic waves

Figure 7 b shows qualitatively that seismic signals for ß-type events can be divided into two portions of differing spectral composition: (1) the onset portion, which contains lower frequencies, and (2) the coda portion which contains higher frequencies. The adopted division between the onset and the coda por-

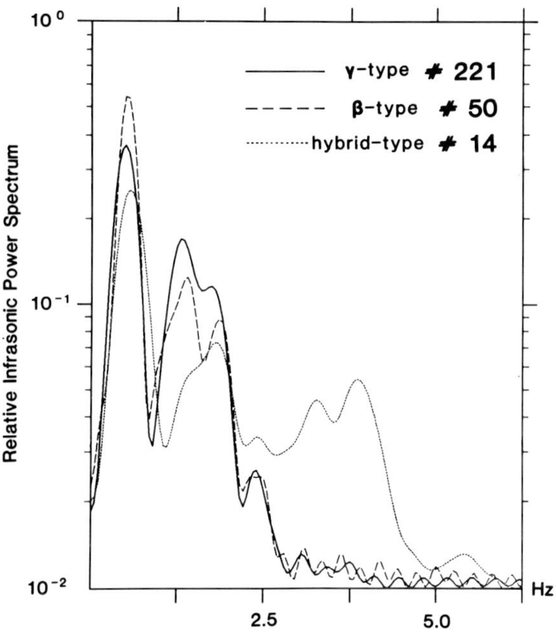

Fig. 8. Infrasonic power spectrum for γ-type event (#221 with solid curve), ß-type (#50 with broken curve) and hybrid type (#14 with dotted curve)

tion is the fifth to sixth oscillation of the infrasonic signal (about 5 s). In Figure 9 the power spectra of the onset and coda portions of the seismic signals in ß- and hybrid-type events are compared. The power spectrum of the onset portion for ß-type events (solid curves) has its peak at 2.2 Hz: the corresponding spectrum for the coda portion (dotted curves) is rather broad in the frequency range of 3-10 Hz (Fig. 9 a). For hybrid-type events, however, the seismic power spectrum is shown to have a similar pattern for both the onset and coda portions (Figs. 7a and 9 b). The 3-dB band width of the normalized maximum in the spectrum of hybrid-type events is 1.3 Hz, and is comparable to that of the onset portion of the ß-type events (1-1.5 Hz); it is wider than the 0.3-0.5 Hz band width of tremor-type events. There must be a transition in the generating mechanism from ß- to hybrid- to tremor-type events, and as the type shifts, the predominant frequency in the seismic power spectrum, which reflects the source time function, shifts from the lower 2.2 Hz of ß-type, to the higher 5.4 Hz of tremor-type events.

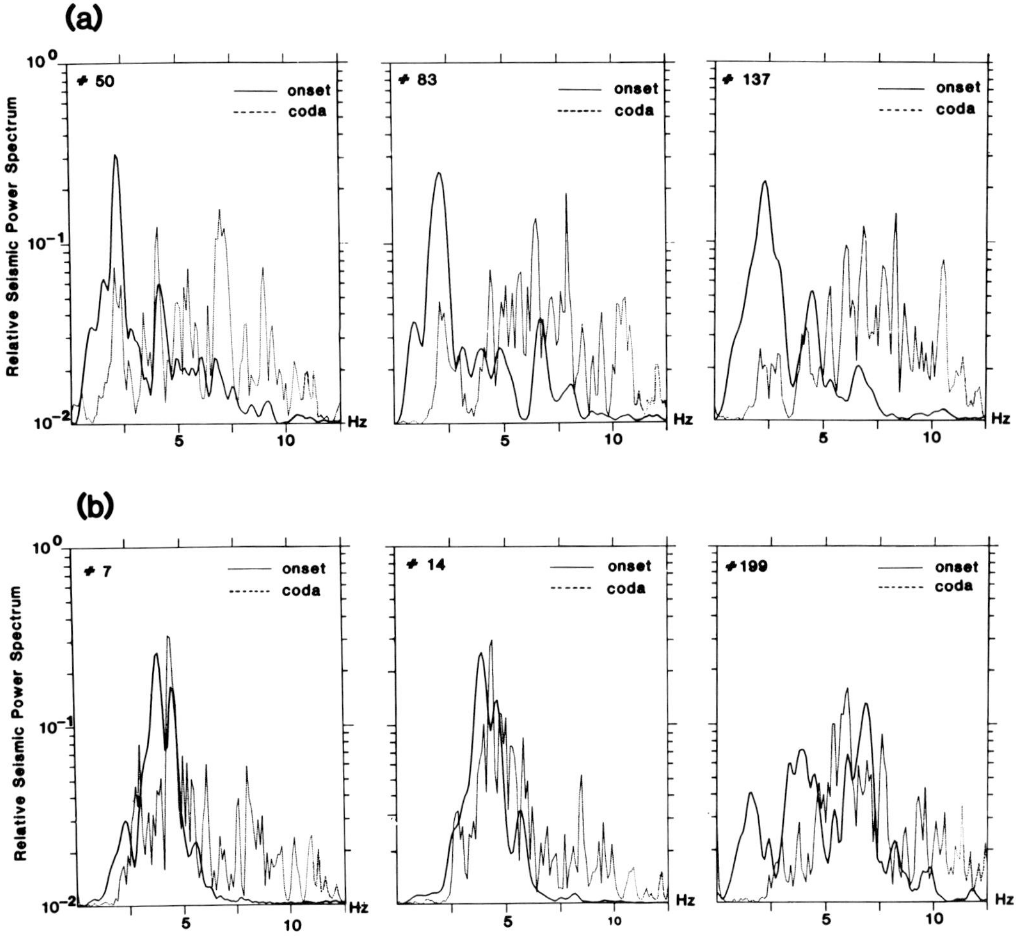

Fig. 9 a, b. a. The seismic power spectrum of the onset portion for ß-type events shown by a **solid curve** and that of the coda portion by a **dotted curve**. Examples are for events #50, #83, and #137. b Similar spectrum for hybrid-type events. Examples are for events #7, #14, and #199

Acknowledgements. The authors express their sincere thanks to the people at Scott Base for their devotion to the maintenance of the IMESS recording system. Many scientists and technicians from USA, New Zealand and Japan support this international co-operation, though individual field participant names are not mentioned here. K Shibuya thanks Prof K Aki of the Massachusetts Institute of Technology for discussions on volcanic tremor. IMESS is financially and logistically supported by the Ministry of Education, Science and Culture through the National Institute of Polar Research, Japan, by Victoria University of Wellington and the Department of Scientific and Industrial Research, New Zealand, and by the National Science Foundation, Division of Polar Programs, USA (Grants DPP-81-17968 and 83-00185 to J. Kienle; grants DPP-79-20316 and 82-18.Y93 to P.R. Kyle).

References

Baba M, Kaminuma K, Koyama E, Ueki S (1985) Seismic activity of Mount Erebus, Antarctica in 1983-1984. Mem Natl Inst Polar Res, Spec Iss 37:29-39

Dibble RR, Kienle J, Kyle PR, Shibuya K (1984) Geophysical studies of Erebus volcano, Antarctica, from 1974 December to 1982 January. NZ J Geol Geophys 27:425-455

Hamaguchi H, Zana N (1985) Distribution of hotspot volcanoes: "HINEP" lines as an evidence for deep mantle plumes. Presented at the 23rd general assembly of IASPEI meeting, F10-20, Tokyo

Kaminuma K, Baba M, Shibuya K, Dibble RR (1985) Explosion earthquakes of Mount Erebus, Antarctica. Mem Natl Inst Polar Res Spec Iss 37:40-47

Katsui Y (ed) (1971) List of the world active volcanoes. Special issue of bulletin of volcanic eruptions. Volcanol Soc Jpn, Tokyo, 160 pp

Kienle J, Rowe CA, Kaminuma K, Shibuya K, Kyle PR, McIntosh WE, Dibble RR (1985) Eruption of Mount Erebus, and Ross Island seismicity (1984-1985). 1985 Review, Antarct J US 20 (5):25-28

Kubotera A (1974) Volcanic tremors at Aso volcano. In: Civetta L, Gasparini P, Luongo G, Rapolla A (eds) Physical Volcanology. Developments in solid earth geophysics 6, Elsevier, Amsterdam, pp 27-47

Kyle PR, Dibble RR, Giggenbach WF, Keys JR (1982) Volcanic activity associated with the anorthoclase phonolite lava lake, Mount Erebus, Antarctica. In: Craddock C (ed) Antarctic Geoscience. Madison, Univ Wisconsin pp 735-745

Minakami T (1963) A note on earthquakes originating from volcanoes. In: Geophysical papers dedicated to Professor Kenzo Sassa. Kyoto Univ Geophys Inst pp 277-283

Rowe CA, Kienle J (1986) Seismicity in the vicinity of Ross Island, Antarctica. J Geodyn 6:375-385

Shibuya K, Baba M, Kienle J, Dibble RR, Kyle PR (1983) A study of the seismic and volcanic activity of Mount Erebus, Antarctica, 1981-1982. Mem Natl Inst Polar Res, Spec Iss 28:54-66

Takanami T, Kaminuma K, Terai K, Osada N (1983) Seismological observations on Mount Erebus, Ross Island, Antarctica, 1980-1981. Mem Natl Inst Polar Res, Spec Iss 28:46-53

Ueki S, Kaminuma K, Baba M, Koyama E, Kienle J (1984) Seismic activity of Mount Erebus, Antarctica in 1982-1983. Mem Natl Inst Polar Res, Spec Iss 33:29-40

Infrasonic Recordings of Strombolian Eruptions of Erebus, Antarctica, March – December 1984, Covering the Jump in Activity on 13 September 1984

R. R. Dibble[1]

Abstract

Continuous infrasonic recordings made at Windless Bight, 26 km from the summit crater of Erebus, between March and December 1984, show an abrupt change both in the average interval between explosions and in the maximum infrasonic amplitudes on 13 September. Before the change the average interval was 2.8 days, and the maximum infrasonic amplitude was 6.3 microbar peak, and afterwards the average interval was about 1.5 h, and the maximum amplitude on 17 September was 32 microbar peak. From 13 September to 8 December, when activity ceased for 13 days, about 1000 explosions occurred.

The b-value graphs of infrasonic magnitude in 5-day groups show flat peaks instead of the negative slope normal for earthquakes, and the peak occurs at higher magnitude during higher activity. Grouping all data gives an apparent b-value near 0.6, similar to that previously found for the larger earthquakes at Erebus.

The most frequent interval between the first 1000 explosions was 15-20 min (4.6%), but Poisson distributions of average interval 1.85 h for intervals less than 70 min, and 1.09 h for intervals exceeding 70 min, fit the data well. This indicates a recovery time of 70 min. Before 13 September, 1984, there were gaps in activity from 4 to 25 May and 4 July to 3 August. The penultimate explosion was on 8 September.

Most infrasonic signatures had a duration of 6-20 s, and consisted of 1 to 1.5 cycles with the wave period increasing with time from initial values of 4-10 s, independent of amplitude. A longer period may indicate a longer eruption. The infrasonic energy of the largest explosion (1984 September 17d 16h 57m UT) was 1-2.7 x 10^9 J and the total to 8 December exceeded 1.3 x 10^{11} J. Assuming infrasonic energy is that of volcanic gas expansion, the largest explosion released 1-5 x 10^4 m^3 and the total to 8 December was at least 2 x 10^6 m^3 of gas after adiabatic expansion to atmospheric pressure.

---

1 Victoria University of Wellington, New Zealand

IAVCEI Proceedings in Volcanology 1
J.H. Latter (Ed.), Volcanic Hazards

The study demonstrates that independent monitoring of explosive eruptions by infrasonic arrays can determine explosion energy and minimum estimates of erupted gas volumes at nearby volcanoes. The other unique advantage is a superior far-field detection ability which offers the greatest potential for systematically reporting volcanic explosion energy worldwide. Such arrays monitored automatically as events take place, could be used to warn commercial airliners against flying into ash clouds from otherwise unmonitored volcanoes.

1 Introduction

Erebus is an anorthoclase-phonolite stratovolcano of 3794 m elevation, situated at 77.53°S, 167.15°E on Ross Island (Fig. 1), near the western shore of the Ross Sea. The main central crater of 550-m diameter and 120-m depth contains an inner crater of 220-m diameter and 95-m depth which normally has an active lava lake of about 2000 m^2 in its northwestern half, and

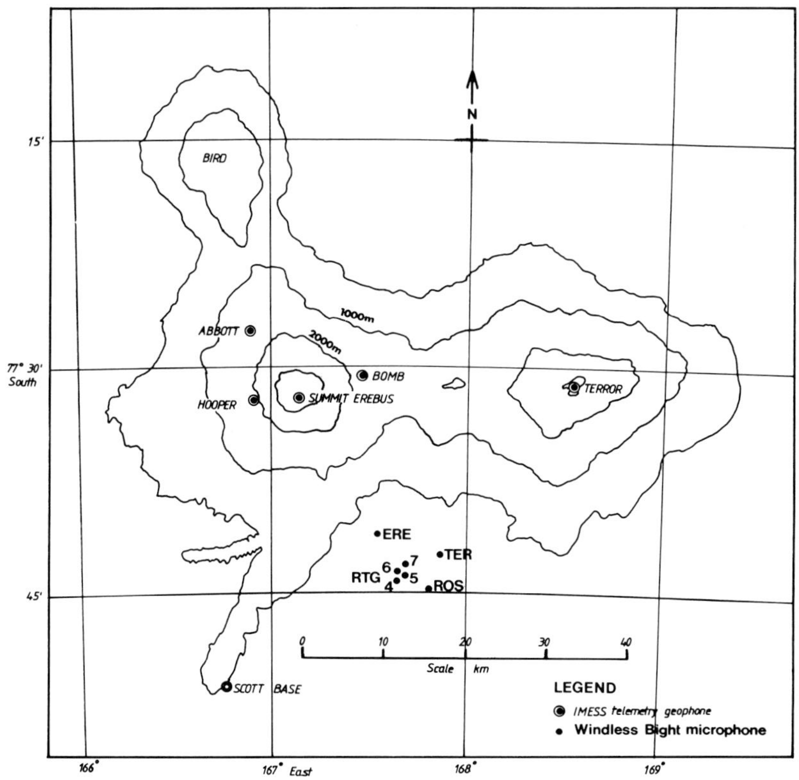

Fig. 1. Map of Ross Island showing the microphone locations in the Windless Bight Infrasonic Array. **RTG**, **ERE**, **TER** and **ROS** form the large, long-period (7-70 s) array. **RTG4**, **5**, **6** and **7** form the smaller short-period (1-10 s) array. **RTG** and **RTG4** represent the same microphone feeding two channels with different pass bands. Some of the telemetry geophone sites are also shown

an explosive vent, known as the active vent, on a bench in the southeastern half. Strombolian eruptions from the active vent or the lava lake occur several times a day, at least during the summer field season, but they are not normally evident at any distance from the mountain.

From 13 September 1984 at 05h 06m UT, powerful explosions from Erebus volcano were seen and heard at Scott Base and McMurdo Base, 37 km away. No comparable explosions had been reported since the first observed activity in 1841 (Kyle et al. 1982).

The normal explosions had been recorded and studied since 1974, initially with geophones and infrasonic microphones recording near the crater during summer expeditions (Kyle et al. 1982; Dibble et al. 1984), and since 1980 with radiotelemetry geophones and infrasonic microphones recording year round at Scott Base (Kienle et al. 1981, 1982, 1983, 1984; Shibuya et al. 1983; Nagata 1982; Ueki et al. 1984; Dibble et al. 1984; Dibble 1985). In practice the battery/solar panel power supplies usually die for about a month at the end of the winter darkness and the infrasonic microphones ice up in autumn, so our knowledge of eruptive as distinct from seismic activity has been restricted to about 4 months of the year.

However, the Windless Bight Infrasonic Array operated by the University of Alaska for the study of micro-baroms, mountain-eddy waves and auroral waves, operates reliably throughout the year at a distance of about 26 km from the crater. It had recorded waves from major eruptions of volcanoes worldwide, at great distances (McKibben 1983), but was not regarded as sensitive to the small explosions of Erebus.

Nevertheless, the author had already arranged to search the recordings held at the University of Alaska, Fairbanks, when the increase in activity at Erebus became obvious on the Windless Bight Array recordings. Thanks to the interest and dedication of B. Lawson and B. McKibben, technicians at Scott Base and McMurdo Base respectively, it was immediately evident that the eruptions had been recorded on the Erebus seismic net, the Scott Base seismograph and the South Pole tidal gravity meter: accounts have already been published by Kaminuma et al. (1985), Dibble (1985) and Kienle et al. (1984).

The present study takes advantage of the unusually high quality and quantity of the infrasonic data to study the increased eruptions and forerunning activity independently of other data, and thus entirely on the basis of the release of gas and energy into the atmosphere.

2 Instrumentation and Processing of Data

Windless Bight (Fig. 1) is an area of the Ross Ice Shelf just south of Ross Island, and was named by Scott's expedition for its still air. This appears due partly to the shelter provided by Ross Island and partly to the smooth, flat surface of soft, porous snow which promotes laminar air flow. The infrasonic array, designed by Spell, was installed at Windless Bight (Wilson and Spell 1981). It consisted of seven condenser-type microphones arranged in two quadrilateral arrays (Fig. 1), one large (RTG, ERE, TER, ROS) and one small (RTG4, 5, 6, 7), with one microphone common to each. Each microphone had an acoustic filter consisting of a 91-m-long pipe with holes spaced along it, for the purpose of reducing wind noise. The pipes became buried in snow during the winter, increasing their effectiveness in this respect.

Pre-amplifiers, VCO modulators and telemetry transmitters were powered by a Radioisotope-Thermoelectric-Generator (RTG) in the hut at the common microphone, ensuring continuous operation. Calibration of the whole system, except for the buried filter pipes, was by motor driven pumps on the microphones. The signals were transmitted to the recording hut, which was located between McMurdo Base and Scott Base. The signals were demodulated in analogue form, and filtered to give a band pass of 1 to 10 s for the small array and 7 to 70 s for the large array. To provide a telemetry quality check, all channels of the filtered analogue signals from the large array were recorded on strip charts, and three of the four channels from the small array were recorded on a drum recorder. These visual records proved invaluable during the eruptions, but the primary data collection was digital, using a PDP11/03 computer. Each channel of the small array was sampled four times a second, and once a second for the large array, and amplitudes were recorded as relative numbers of digital counts. Data were collected in 128-s blocks, with on-line computer processing to determine cross-correlation coefficients between pairs of microphones in each array, for a range of lags corresponding to signals coming from all points of the compass, at apparent velocities between 200 and 600 m s^{-1}. A data-adaptive filter designed by Olson (Wilson et al. 1981) then filtered out the least correlated parts of the waveforms, and the cross-correlations were repeated to find the azimuth and apparent velocity of the best correlated waves in each 128-s block. The original data and the results of the processing were recorded on 5-day magnetic tapes.

A variety of PDP/11 programs were available to read the tapes. The fastest programs listed only the block number, time, azimuth and apparent velocity of arrival of the best correlated signal over the block, and the correlation coefficients. The printout could be restricted to list only waves within a chosen range of azimuths and apparent velocities, with a correlation coefficient above a chosen minimum, so that a list of apparent Erebus events could be quickly obtained. Maximum positive and negative amplitudes on each channel, correlation coefficients between each pair before and after filtering and the dominant spectral peak in each block could be listed nearly as quickly, but less clearly.

The results were excellent for signals not much shorter than the block length, but brief explosions from Erebus were discriminated against unless they were much stronger than continuous signals from other sources. Thus the quick printout was reliable and complete during the vigorous stage of the eruptions, but before that stage few of the apparent eruptions on the chart records were picked by the computer. On the other hand, wind-generated eddies from the summit of Erebus were sometimes picked by the computer. They could be recognized by a low ratio of peak to mean amplitudes in the block. In this chapter, peak amplitude means half the peak to peak amplitude.

Reprocessing the digitized waveforms with programs designed to select signals from Erebus was slower - complete reprocessing took nearly as long as the original recording - but B. McKibben kindly modified a program at the author's request to perform the cross-correlations over a small range of azimuths and apparent velocities, and to compute the ratio of maximum to mean RMS amplitude in each block, as a means of recognizing blocks containing brief signals. This took 9 s per block to run, giving a nominal computing time of 462 h for the 9 months' records that I wished to reprocess. I had almost completed this by the end of my 4-week stay at the University of Alaska.

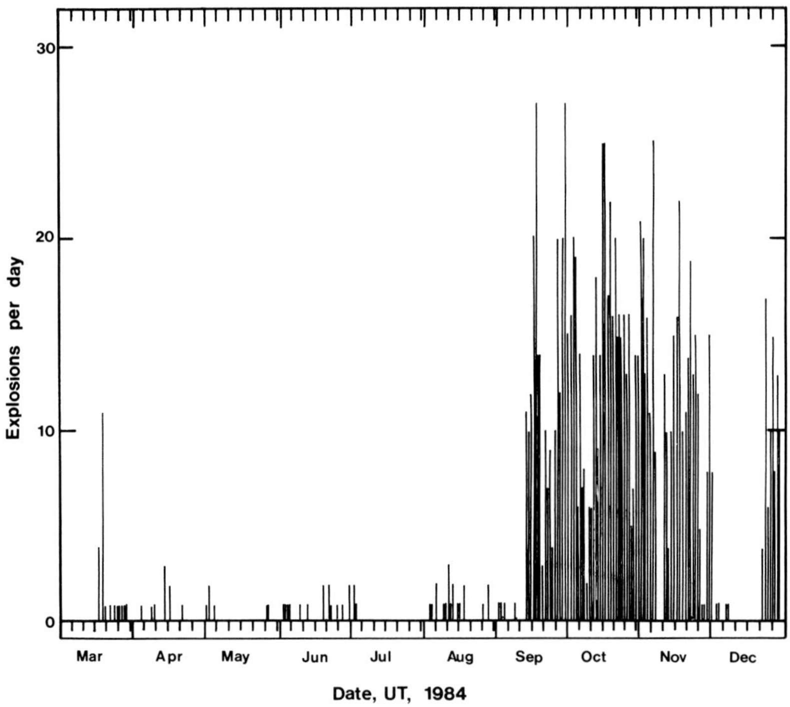

Fig. 2. Number of Erebus explosions per day detected by the Windless Bight Infrasonic Array in 1984. Most of the small events prior to 13 September were found by scanning the analogue records, but from 13 September, the explosions were reliably flagged by the digital recording computer

3 Variation in the Amplitude and Number of Explosions per Day

The numbers of explosions per day (Fig. 2), recorded by the small Windless Bight Array, were counted from the drum recordings prior to the step increase in activity on 13 September, and from the computer listing after the increase. In the 6 months prior to the increase the maximum number per day was three, and the average was 0.36 ± 0.64 per day, which is well below and less regular than the 3.6 ± 2.7 per day observed during summer expeditions (Dibble et al. 1984). The longest period without detectable eruptions was 30 days in July 1984, and the last explosion before the increase was on 8 September 1984, and was quite ordinary. Pressure amplitudes at RTG4 (Fig. 3) averaged 2.1 ± 0.85 microbar peak before the eruption, and the maximum was 6.3 microbar. After the step increase at 05h 07m UT on 13 September 1984, there were 13.2 ± 6.4 explosions per day in the 88 days to 1 December 1984, when activity suddenly decreased again. There was no strong trend in the number of explosions in the first 88 days, but the amplitudes of the strongest explosions at RTG4 decreased rapidly from a maximum of 32-microbar peak on 17 September 1984, levelling at the pre-increase level after 1 November 1984.

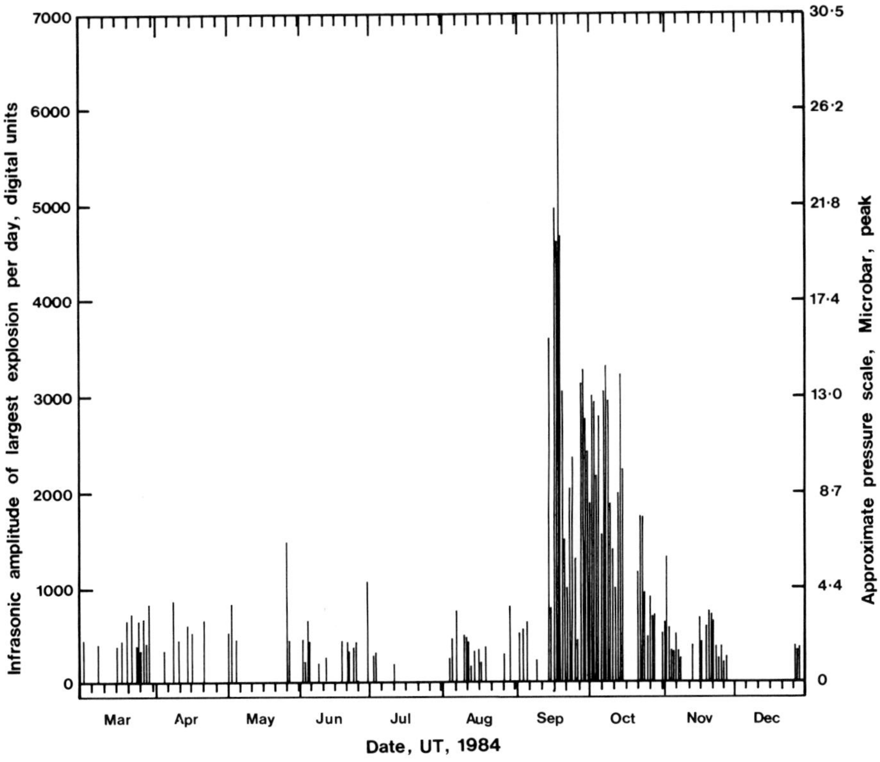

Fig. 3. Pressure amplitude, peak, of the largest Erebus explosion each day, recorded by the Windless Bight Infrasonic Array in 1984

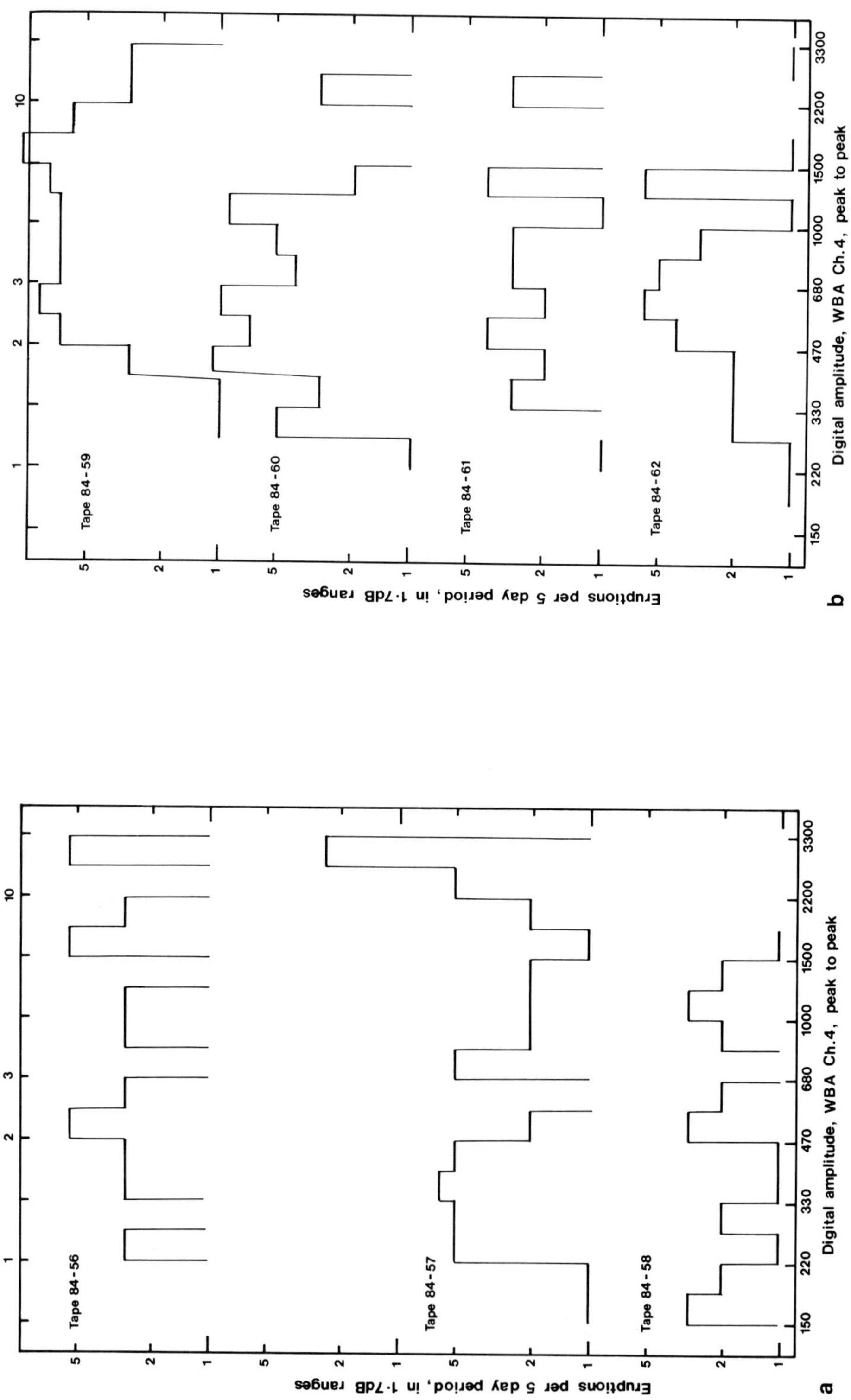
Tape 84-59
Tape 84-60
Tape 84-61
Tape 84-62
Eruptions per 5 day period, in 1·7dB ranges
Digital amplitude, WBA Ch.4, peak to peak
b
Tape 84-56
Tape 84-57
Tape 84-58
Eruptions per 5 day period, in 1·7dB ranges
Digital amplitude, WBA Ch.4, peak to peak
a

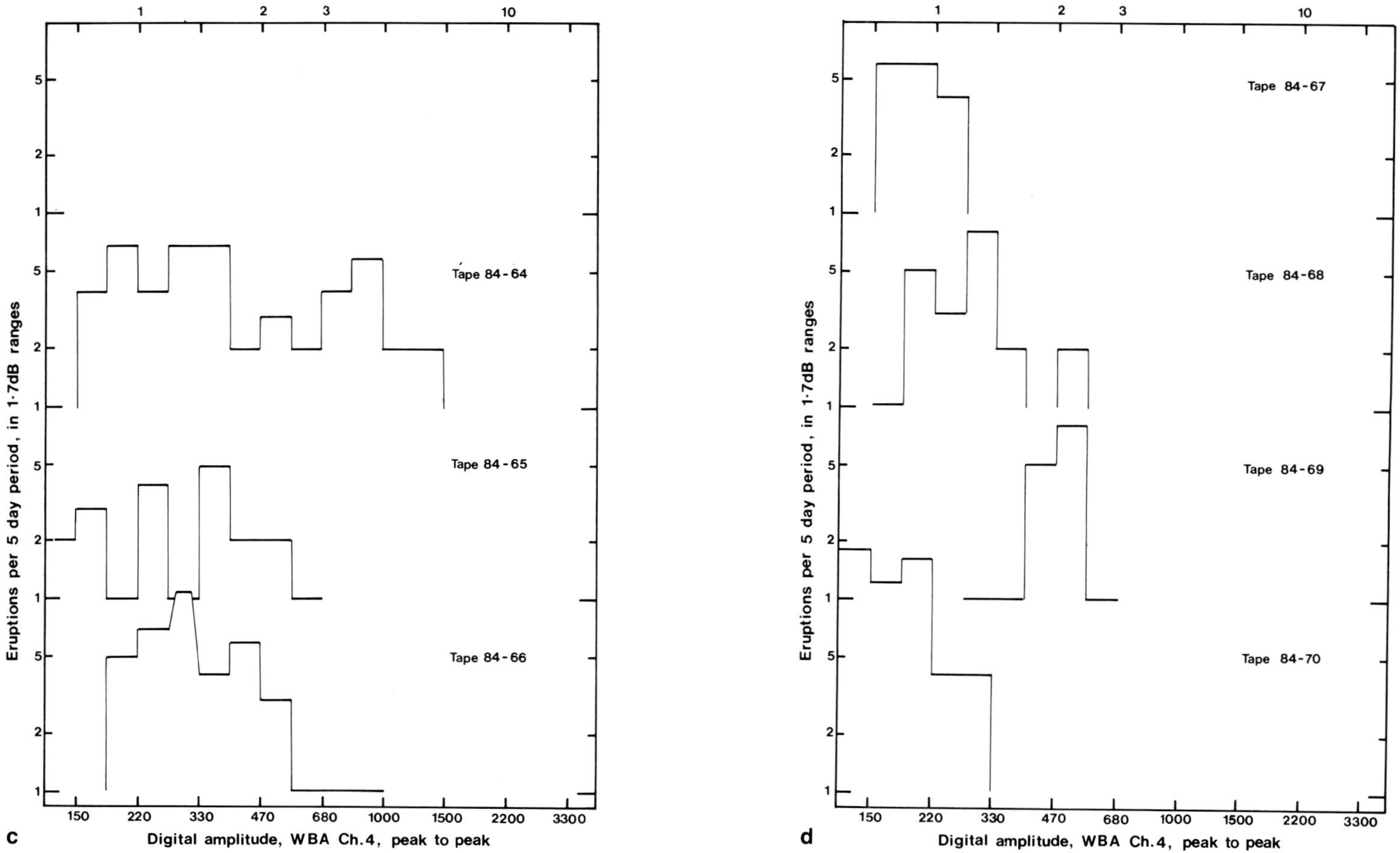

Fig. 4 a-d. Number of explosions versus infrasonic amplitude, in microbar peak at the **top**, and digital counts peak to peak at the **bottom**, for each 5 day tape record. The log-log plots correspond to seismic b-value plots where a 10x change in amplitude is a unit of magnitude. The graphs do not show the linear negative slope usual with earthquakes. **a** 10 to 25 September 1984; **b** 25 September to 14 October 1984; **c** 20 October to 6 November 1984; **d** 6 to 27 November 1984

The determination of pressure amplitudes was not straightforward because many of the explosions recorded by the small array were clipped to some degree, with channel 7 the most affected, and channel RTG4 the least affected. Considering that air turbulence can modify the recorded amplitudes considerably, the mean amplitude of all four channels would normally be preferable, but channel RTG4 only was used below its clipping level. For events clipped on channel RTG4, the mean amplitude on the chart records of the large array were converted to equivalent small array amplitudes, using the ratio of amplitudes determined from small well-recorded events. From individual comparisons, this ratio was observed to vary by a factor up to 2, which was ascribed to turbulence effects.

4 Frequency of Occurrence Versus Explosions Magnitude

In order to determine whether volcanic explosions have a frequency/magnitude relationship (b-value) similar to that for earthquakes, graphs of the number of Erebus explosions, in each 1.7 dB of infrasonic amplitude range, were constructed for each 5-day recording tape between 13 September 1984 (tape 84-56) and 15 January 1985 (85-03), except for tapes 84-63, 84-71 to 84-75, and 85-01. Adopting a factor of 10 change in amplitude (i.e. 20 dB) for the magnitude unit, as for earthquakes, then 1.7 dB is one-twelfth of a magnitude unit.

The graphs (Fig. 4) do not have a consistent slope. They are roughly flat (b = 0) over amplitude ranges which are restricted, and change from tape to tape. For instance, tape 84-59 (Fig. 4 b) has a flat graph between 2 and 10 microbar, with lesser numbers of smaller and larger explosions, while tape 84-66 (Fig. 4 c) is fairly flat between 0.8 and 2.4 microbar. The b-values above and below the flat tops are large and of opposite sign, quite different to the normal b-value graph for earthquakes.

When all the data are included in a single graph (Fig. 5), they appear to have a b-value of about 0.6, as found by Dibble et al. (1984) for the explosion earthquakes at Erebus. However, the projection of this line above 14-microbar amplitude would imply that more than 100 explosions of between 14 and 400 microbar peak should have occurred, whereas only one (of amplitude 32 microbar) actually occurred. Thus the b-value must be large above 14 microbar, and may be large and of opposite sign below 0.7 microbar, but even with over 600 data points the values are indeterminate.

5 Repose Period Analysis

The random appearance of the graph of explosions per day (Fig. 2) invites a repose period analysis of the whole series of 1000 events, following the methods of Wickman (1966a, b). A histogram (Fig. 6) of repose periods between successive explosions shows that the most frequently occurring interval was 15-20 min. However, the number in this interval range (4.6% of

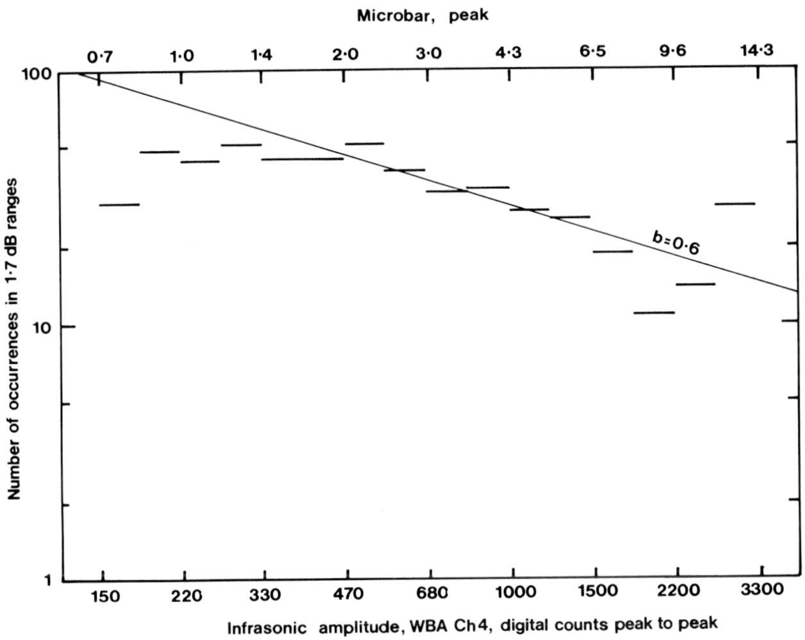

Fig. 5. Number of explosions versus infrasonic amplitude in microbar peak at the top, and digital counts peak to peak at the bottom, for all events recorded on the Windless Bight array between 13 September and 1 December 1984. A b-value line of 0.6 is superimposed

total) is only 0.7% above that expected for a Poisson distribution with an average interval of 1.85 h, which is a good fit to the data. On a log-cumulative-number versus repose-period graph (Fig. 7), a Poisson distribution would plot as a straight line of slope:

$$s = -2.303\ \tilde{t},$$

where $\tilde{t}$ is the average interval between events.

In plotting this graph, ten intervals exceeding 500 min were ascribed to recording/processing problems, and were omitted.

The data between 13 September and 1 December 1984 appear to plot on two straight lines, representing Poisson processes with average intervals of 1.85 h for intervals less than 70 min, and 1.09 h for intervals exceeding 70 min. Although this is a useful model for calculating probabilities of the next eruption occurring within a certain time, truly Poisson processes have no such memory effects, and the distribution is therefore of a hybrid type, best explained by a 70-min recovery period from exhaustion after each explosion.

From the standard expression for the probability of a random occurrence (Wickman 1966a), the risk of an explosion in any

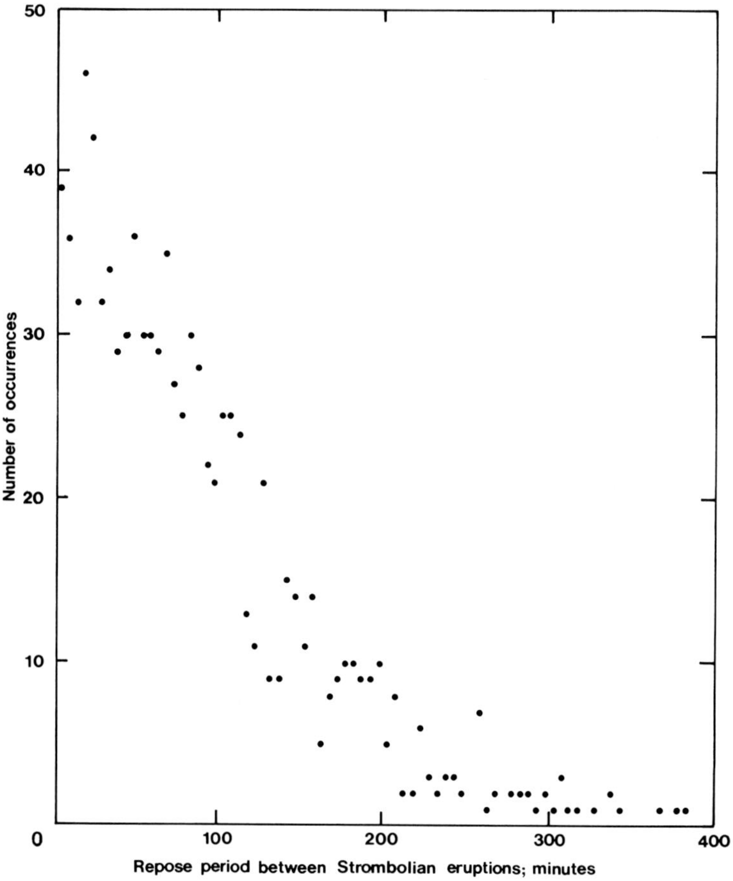

Fig. 6. Histogram of intervals between Erebus explosions from 13 September and 1 December 1984. The most frequent repose was 15-20 min (4.6%), but the data are erratic

10-min period within 70 min of the previous event was 8.6%, but after 70 min it increased to 14.2%.

6 Infrasonic Waveform and Energy

The waveforms with best resolution are obtained from the digital data from the small array. An example is shown in Fig. 8 with all channels corrected for the theoretical travel time differences of waves arriving at sonic velocity from the azimuth of Erebus, and displayed superimposed. Almost all events show three strong half-cycles of total duration 6-7 s, and period 3.7 s, with a small coda of 20-s duration. Three of the channels have similar amplitudes and are assumed to be correctly calibrated. The fourth (ch 7) shows amplitudes at least double the rest, and appears to contain higher frequencies. The cause of the difference may have been a broken filter pipe on

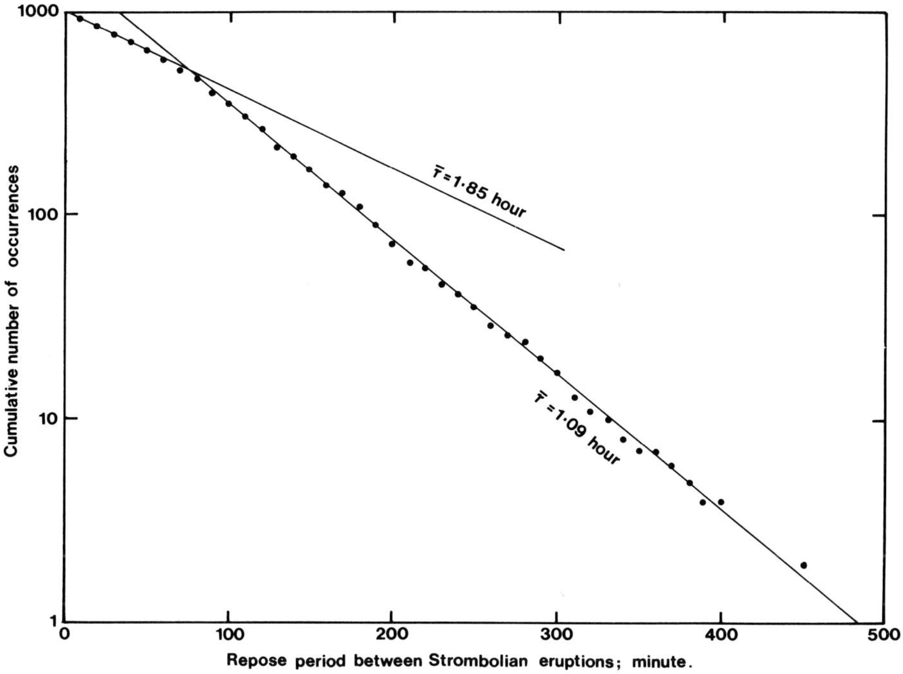

Fig. 7. Cumulative number versus time intervals between explosions of Erebus from 13 September to 1 December 1984. The **straight lines** indicate a random Poisson process, with average interval (Tau) changing from 1.85 to 1.09 h for intervals exceeding 70 min

one microphone, but the log book entries to establish this have not been supplied.

A very few of the explosions show longer period waves on the small array (Fig. 9). They are not of larger amplitude than those of normal period, and their recorded waveforms are consistent with the longer period being due to longer eruption time rather than to a more intense explosion.

For the strongest signals, which are heavily clipped on the small array recordings, it is necessary to refer to the strip chart recordings of the large array (Fig. 10). These show explosions as a single cycle of initial period 4 to 10 s (from the rise time) and duration 14 s. Some doubt exists as to the waveform which would be recorded on a single wide-band system. Amplitudes on the large array are clearly attenuated by its restricted high-frequency response. The third half-cycle on the small array may be partly the result of the restricted low-frequency response of this array. The waveform undoubtedly includes many cycles of low audio frequency, because the larger explosions were heard at Scott and McMurdo Bases. However, the recordings show very little energy at the 1-Hz band edge, and although absorption is proportional to frequency squared, it is

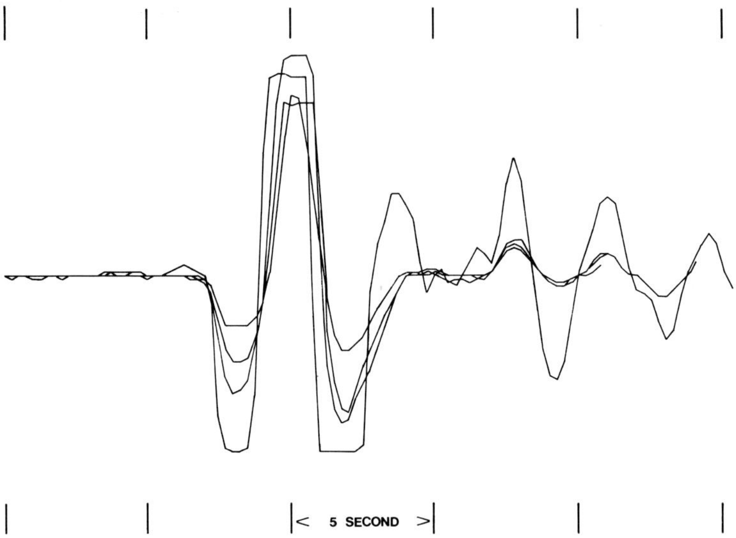

Fig. 8. Digital record from the 1-10 s period array of the explosion at 06h 38m UT on 13 September 1984. The spectral peak was at the 3.8-s period. The channels were superimposed after time shifts for an arrival from Erebus. Channel RTG4 shows the least amplitude and clipping, and corresponds to 9.5 microbar (0.95 Pa) peak. Channel 7 shows large amplitudes, possibly due to a broken filter pipe

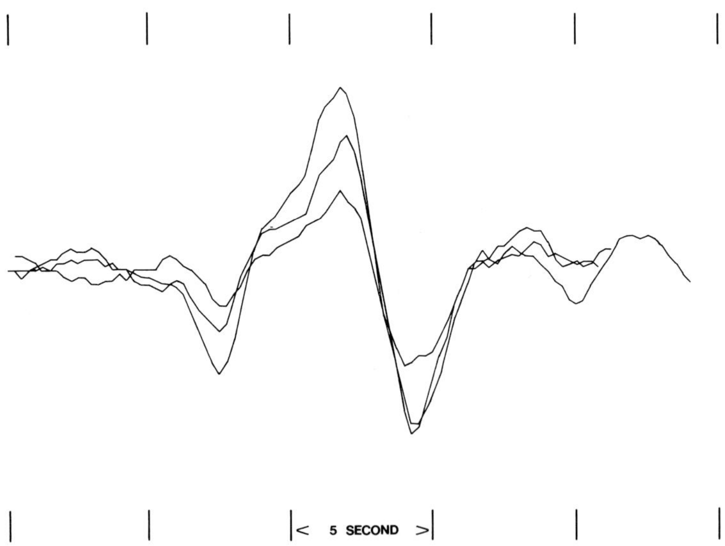

Fig. 9. Digital record from the 1-10 s period array of the long-period explosion on 14 September 1984 at 11h 50m UT. The spectral peak was at the 9.1 s period. The digital amplitude for channel RTG4 corresponds to 2.2 microbar (0.22 Pa) peak. The mean for channels RTG4, 5 and 6 was 2.9 ± 0.7 microbar using 1983 calibrations for channels 5 and 6

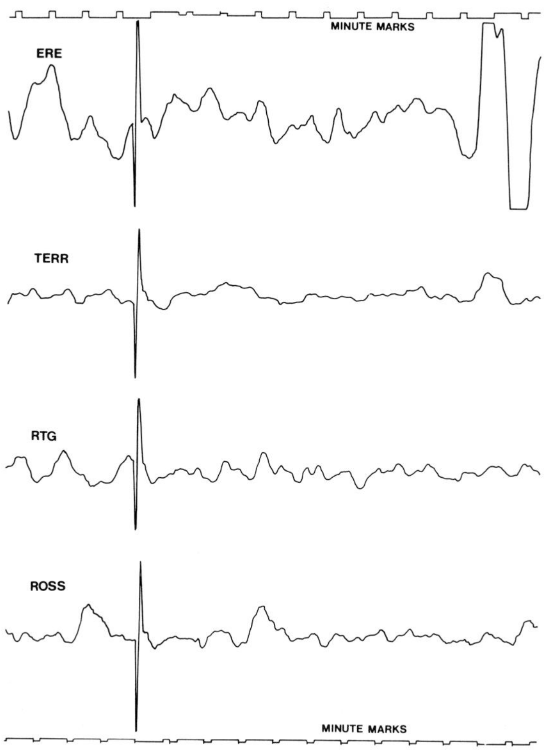

Fig. 10. Infrasonograms of the initial explosion of the increased Erebus activity on 13 September 1984 at 05h 07m UT., recorded on analogue charts from the large 7-70 s period array. Nominal amplitude was 8 ± 1.4 microbar (0.8 Pa) peak, but the maximum amplitudes were outside the pass band. Scaling from smaller explosions recording without clipping on the short-period array, the amplitude on channel RTG4 would have been 15.5 microbar peak

small over distances to 26 km. Hence it is assumed that energy above the spectral peak may be neglected.

The infrasonic energy of the explosions was calculated assuming hemispherical propagation through a standard atmosphere, after Shimozuru et al. (1975) and Dibble (1985). Erebus was well within the 75-km distance to which Hunt et al. (1960) found that propagation was simple, and the following simple equation for sinusoidal waves can be used:

$$E = \frac{\pi \Delta^2 P^2 T}{p c}, \qquad \text{S.I. units}$$

where Δ is distance (26,600 m to Windless Bight);
p is air density (1.394 kg m^{-3} at -20°C and 101.3 kPa);
c is sound velocity (319 m s^{-1} at -20°C);
P is pressure amplitude (peak), Pa;
T is duration of signal, s;
E is infrasonic energy, J.

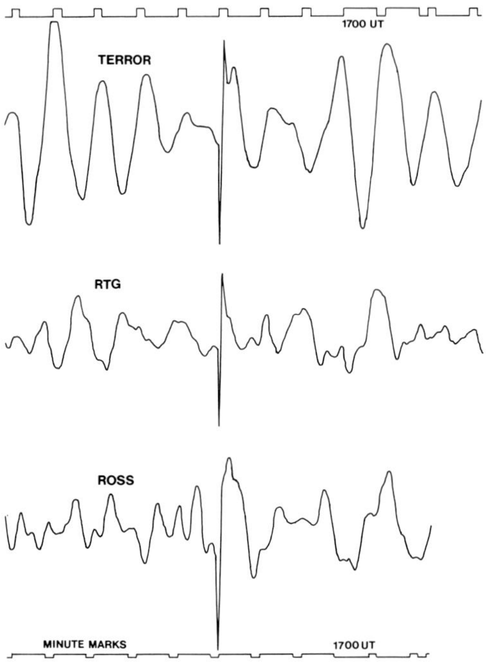

Fig. 11. Analogue infrasonograms of the largest explosion, on 17 September 1984 at 16h 57m UT recorded on the 7-70 s period array. The duration of the low-frequency tail could be as long as 52 s, but was obscured by wind noise. Nominal amplitude was 8.6 ± 0.7 microbar (0.86 Pa) peak. Scaling from small explosions, the equivalent amplitude on channel RTG4 of the short-period array was 32 ± 5 microbar (3.2 Pa) peak

Infrasonograms of the largest explosion (Fig. 11) show a single cycle of 32 microbar peak amplitude, apparently with a slow decay of the negative half-cycle over about 31-52 s. However, it is equally probable that this is partly due to the background wind noise, and that the true duration is about 14 s, as for most of the other explosions in the sequence. None of the records show a short positive half-cycle followed by the long negative half-cycle typical of a shock front, although this had previously been recorded by Dibble et al. (1984) at the crater rim from much smaller explosions. Presumably the shock front had dissipated during propagation, and the calculated energy range for the largest explosion, 0.8 to 2.7 x 10^9 J, is a minimum estimate. The total infrasonic energy between 13 September and 1 December was obtained by summing the squares of the amplitudes, and assuming that the durations were all 14 s. The sum exceeded 1.3 x 10^{11} J.

A simplistic model for the generation of waves of these energies is the injection of hemispherical volumes of gas into the atmosphere against atmospheric pressure. Scorer (1950), Hunt et al. (1960) and Gossard and Hooke (1975) have derived expressions for the waves from given injection rates, and found that they are insensitive to details of the injection, provided it is brief enough. Thus, while the erupted volumes might be determined reliably from the infrasonic signal, the rates and durations of eruption cannot be reliably estimated. The model of a point source of intense heat as in a nuclear bomb, the yield of which can be estimated from the dominant period, is clearly inappropriate. Instead, we can equate the infrasonic energy with the work done against the atmosphere by the eruption and expansion of volcanic gas to atmospheric pressure: $E = pV$, where p is atmospheric pressure (ca. 63 000 Pa at the summit of Erebus), and V is the volume after adiabatic expansion. Over the duration of the infrasonic signal, no appreciable transfer of heat to the atmosphere will have taken place, and even when it does, it will not affect the new total volume of the atmosphere unless changes of state are involved. Thus, the volume of gas erupted in the largest explosion was 12 000 to at least 43 000 m^3, and the total volume between 13 September and 1 December was at least 2×10^6 m^3.

The temperature, and hence the mass of gas, is unknown.

7 Conclusion

It is found that an infrasonic array at a safe distance from a volcano is suitable for monitoring explosive eruptions independently of other monitoring systems, provided that it has sufficient resolution of azimuth and apparent velocity of wave arrivals. It has the unique advantage of measuring gas eruptions directly, without regard to visibility, and putting them in order of magnitude. From the air wave energy, it is simple to estimate step increases in erupted gas volumes, but not the rates of continued eruption.

Although infrasonic amplitudes are modified by wind and precipitation, whereas seismic amplitudes are not, the transmission system response of the atmosphere over short distances is more easily and reliably determined than that of the earth. Corrections for wind effects can be made from the perturbations in azimuth and apparent velocity determined by the array.

In these regards, therefore, an infrasonic array can be considered as a valuable addition to a seismic network at a volcano, but it has another unique advantage. Its ability to detect eruptions worldwide is limited only by the quietness of the site. The rapid but quiet eruption of large clouds of ash is difficult to detect at unmonitored volcanoes except during good visibility. Commercial aircraft need to fly near many such volcanoes in all visibilities, and many hundreds of passengers have had narrow escapes after flying into ash clouds. A global reporting system of infrasonic arrays at established volcano observatories would be much quicker and cheaper than creating many additional observatories.

As Mauk (1983) has stated, the independence from near-field data is of fundamental importance, and it offers the greatest potential for systematically reporting volcanic explosive energy.

Acknowledgements. To Prof. C.R. Wilson, project leader of the infrasonic program, Mr. B. McKibben, Prof. B.D. Spell, Prof. J. Kienle and Miss K. Driscoll, all of the Geophysical Institute, University of Alaska, Fairbanks, for (respectively) inviting me to Fairbanks to search the Windless Bight recordings for Erebus eruptions; for operating the array, telexing data to me on the new activity and later assisting in the computer search; for providing a computer and programs for the search; for including me in the U.S. Antarctic Research Program to Erebus during the new activity; and for familiarizing me with the array. Also to the Geophysical Institute, Fairbanks, and the National Institute of Polar Research, Tokyo, for facilities and hospitality during my working visits, and to Victoria University of Wellington for granting me Research Leave. The author is grateful to the last two institutions for financing his leave.

References

Dibble RR (1985) New eruption parameters and spectral relationships between seismic and infrasonic signals from Erebus volcano, Antarctica. Proc 5th Symp on Antarctic Geosciences, 1984. Mem Nat Inst Polar Res 37:22-28

Dibble RR, Kienle J, Kyle PR, Shibuya K (1984) Geophysical studies of Erebus volcano, Antarctica, from 1974 December to 1982 January. NZ J Geol Geophys 27:425-455

Gossard EE, Hooke WH (1975) Waves in the atmosphere. Developments in atmospheric science, 2. Elsevier, Amsterdam, 456 pp

Hunt JN, Palmer R, Penney W (1960) Atmospheric waves caused by large explosions. Phil Trans R Soc Lond, Ser A 252:275-315

Kaminuma K, Baba M, Shibuya K, Dibble RR (1985) Explosion earthquakes of Mount Erebus, Antarctica. Proc 5th Symp on Antarctic Geosciences, 1984. Mem Nat Inst Polar Res 37:40-47

Kienle J, Kyle PR, Estes SA, Takanami T, Dibble RR (1981) Seismicity of Mount Erebus, 1980-1981. Antarct J US 16(5):35-36

Kienle J, Marshal DL, Estes SA, Dibble RR, Shibuya K, Kyle PR (1982) Seismicity of Mount Erebus, 1981-1982. Antarct J US 17(5):29-31

Kienle J, Marshal DL, Kyle PR, Kaminuma K, Shibuya K, Dibble RR (1983) Volcanic activity and seismicity of Mount Erebus, 1982-1983. Antarct J US 18(5):41-44

Kienle J, Kaminuma K, Dibble RR (1984) Seismicity of Mount Erebus and vicinity, 1983-1984. Antarct J US 19(5):25-27

Kyle PR, Dibble RR, Giggenbach WF, Keys JR (1982) Volcanic activity associated with the anorthoclase phonolite lava lake, Mount Erebus, Antarctica. In: Craddock C (ed.) Antarctic geoscience. IUGS B 4:735-745

Mauk JK (1983) Utilisation of seismically recorded infrasonic-acoustic signals to monitor volcanic explosions: The El Chichon sequence 1982 - a case study. J Geophys Res 88 (B12):385-410

McKibben B (1983) Atmospheric infrasound from volcanic eruptions in 1982. Antarct J US 18(5):257-259

Nagata T (1982) Japanese earth science in the McMurdo Sound region, 1981-1982. Antarct J US 17(5):26-28

Scorer RS (1950) The dispersion of a pressure pulse in the atmosphere. Proc R Soc Lond A 201:137-157

Shibuya K, Baba M, Kienle J, Dibble RR, Kyle PR (1983) A study of the seismic and volcanic activity of Mount Erebus, Antarctica, 1981-1982. Proc 3rd Symp on Antarctic Geosciences, 1982. Mem Nat Inst Polar Res 28:54-66

Shimozuru D, Utibori S, Gyoda N, Koyama E, Miyazaki T, Matsumoto T, Osada N, Terao H (1975) The explosive activity of Asama volcano; general description of explosive and seismic events. Bull Earthq Res Inst 50(1):115-151

Ueki S, Kaminuma K, Baba M, Koyama E, Kienle J (1984) Seismic activity of Mount Erebus, Antarctica in 1982-1983. Proc 4th Symp on Antarctic Geosciences, 1983. Mem Nat Inst Polar Res 33:29-39

Wickman FE (1966a) Repose period patterns on volcanoes. 1. Volcanic eruptions regarded as random phenomena. Ark Mineral Geol 4(7):291-301

Wickman FE (1966b) Repose period patterns on volcanoes. 5. General discussion and a tentative stochastic model. Ark Mineral Geol 4(11):351-367

Wilson CR, Spell BD (1981) Digital infrasonic system installation in Antarctica. GIR 81-2 under AFOSR 80-0125, Geophys Inst, Univ of Alaska, Fairbanks

Wilson CR, Olson JV, Spell BD (1981) Infrasonic signal enhancement by a pure state filter. GIR 81-1 under AFOSR 80-0125, Geophys Inst, Univ of Alaska, Fairbanks

Deformation Monitoring of Mt. Erebus, Antarctica, 1980–1985

G. H. BLICK, P. M. OTWAY and B. J. SCOTT[1]

Abstract

Deformation studies of the summit area of Mt. Erebus, Antarctica, commenced in December 1980. Triangulation and tilt-levelling networks have been surveyed annually, in conjunction with the International Mount Erebus Seismic Studies (IMESS) project.

Only minor deformation has been detected. Measurements from the triangulation network on the summit indicate cumulative expansion of approximately 50 mm across the crater. Most of this appears to have occurred in 1982 and 1984, the latter period coinciding with increased volcanic activity. Tilts of more than 15 microradians/yr have been recorded on several of the tilt-levelling patterns, but their directions appear random and do not correlate with volcanic activity. Tilts recorded during 1985 are larger than those recorded previously. The significance of this is not yet understood. They are thought to be due to local ground instability rather than to volcanic deformation.

1 Introduction

Mt. Erebus, located on Ross Island in McMurdo Sound, is one of several active volcanoes in Antarctica. Volcanic deformation studies commenced in December 1980 when triangulation and tilt-levelling networks were installed around the summit of the volcano. The deformation studies were designed to complement the International Mount Erebus Seismic Studies, a joint USA, Japanese and New Zealand project designed to monitor the seismic activity of Mt. Erebus. The initial survey was made during December 1980-January 1981 with resurveys during December 1981, December 1982, November-December 1983, November-December 1984 and December 1985.

[1] N.Z. Geological Survey, Department of Scientific and Industrial Research, P.O. Box 30368, Lower Hutt, New Zealand

IAVCEI Proceedings in Volcanology 1
J. H. Latter (Ed.), Volcanic Hazards

2 Geological Structure and Eruptive Activity

Mt. Erebus is an anorthoclase-phonolitic stratovolcano whose summit rises to an elevation of 3794 m. The summit Main Crater is 550 m in diameter with a mainly flat floor, 120 m below the crater rim. An Inner Crater, 250 m across and 100 m deep, is situated at its northern end. An anorthoclase-phonolite lava lake has been present in the Inner Crater since before December 1972 and Strombolian-type eruptions have occurred from the lava lake and surrounding active vents (Giggenbach et al. 1973; Kyle et al. 1982). Strombolian eruptions have averaged over three per day since 1974 (Dibble et al. 1984). Eruptive activity in September-October 1984 was significantly stronger than any previously observed, many eruptions throwing lava bombs up to 500 m from the crater. It resulted in emptying of the lava lake and modification of the topography of the Inner Crater. By December 1985 a small lava lake again occupied the floor of the Inner Crater.

The alignment of four craters, Inner, Main, Side and Western (Fig. 1), and Tower Ridge, a 500 m-long linear fumarolic feature, suggests the presence of a rift system trending at 60° across the summit of the volcano.

3 Survey Network

Figure 1 shows the location of the network in relation to the summit craters. The initial survey network installed in 1980 consisted of the stations A-H, U-Z and the tilt-levelling patterns. Due to increased volcanic activity during September-October 1984, the network was modified in November by the addition of marks N-S to monitor possibly more extensive deformation, and to shorten exposure time on the crater rim during periods of more violent activity. The network is designed to monitor horizontal deformation of the summit area of the volcano. Indications of tilting are provided by three tilt-levelling (dry tilt) patterns. The alignment of marks U-Z monitors local ground instability near the Mt. Erebus Hut. All marks consist of 12 mm steel rods driven to refusal and held in place by frozen water (Antarctic cement).

4 Surveying Methods

Surveys of Mt. Erebus have been made annually since 1980 during the months of November-December, although the initial survey extended into January.

Angle observations have been made with a theodolite observing to steel tube targets in daylight (Blick and Scott 1986). The theodolite was optically plumbed over each station. Three sets of vertical angles and six sets of horizontal angles were recorded. From adjustments, the standard error (SE) of a direction is estimated to be 2.5".

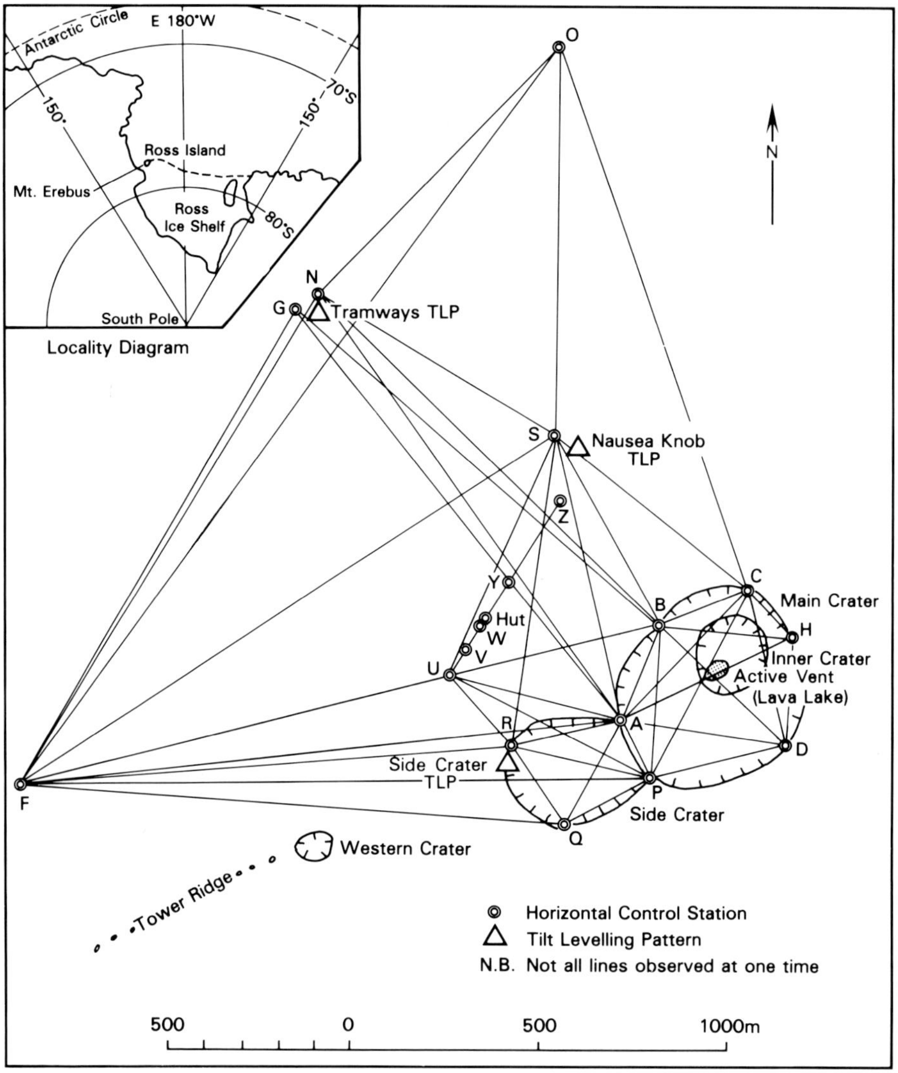

Fig. 1. Mt. Erebus deformation monitoring network and volcanic craters

Distance measurements were made in 1981, 1984 and 1985 using infrared distance meters measuring to glass reflectors. Meteorological parameters were measured at the end of lines to correct for ambient air temperature and pressure. From adjustments, the SE of distance measurements is estimated to be ±5 mm ±3 ppm.

Tilt-levelling patterns have been surveyed using an automatic level with a parallel plate micrometer observing to a 3-m invar staff. The error on the magnitude of tilt is estimated to be 10 microradians. This is attributed to a combination of survey error and site noise, the latter due to ground instability and refraction.

5 Surveying Results

The displacement of stations around the Main Crater has been derived. Scale in 1981, 1984 and 1985 was provided by distance measurements. The measured distance of F-G showed no significant difference when measured in 1981, 1984 and 1985, and therefore, for the 1982 and 1983 surveys it was assumed to be the same as in 1981. As angular observations were not completed from F and G during 1980 and distances were not measured, the 1981 value of A-B has been used for scale in 1980. Figure 2 shows the "intrinsic" displacement solutions in which the

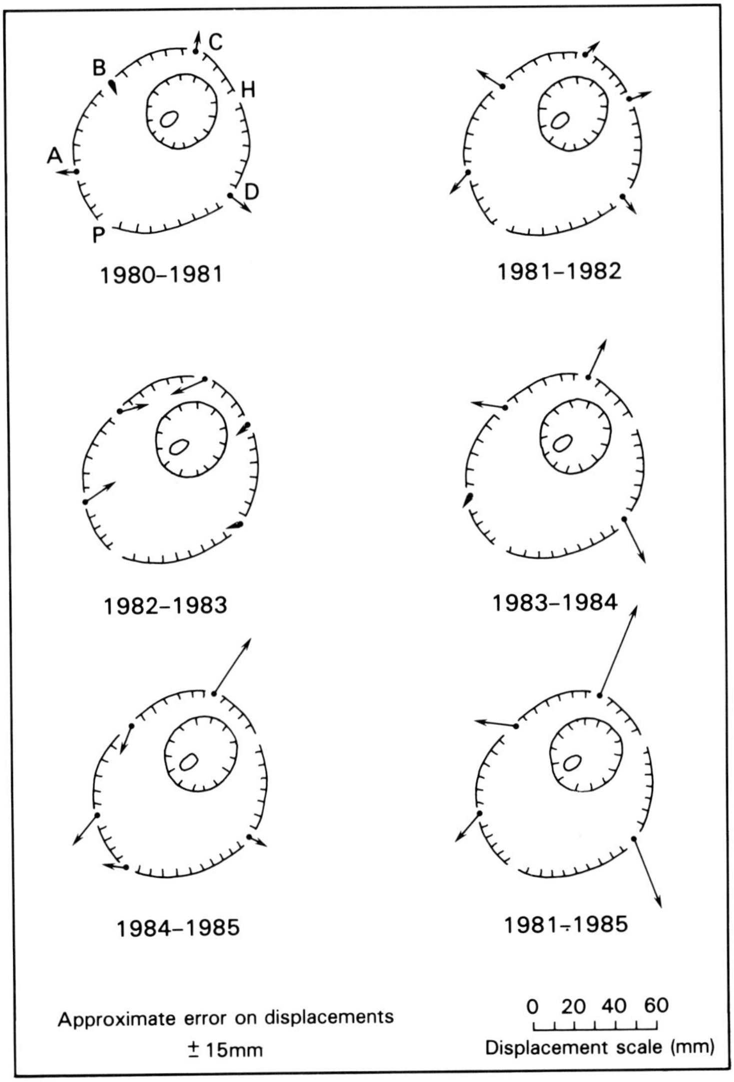

Fig. 2. Displacement of stations around the Main Crater for the intervals indicated

translation and rotation of the network between surveys is calculated to minimize the sum of squared lengths of the displacement vectors of the stations around the Main Crater. The estimated SE of the displacements is approximately ±15 mm. The results indicate relatively small displacements of up to 40 mm between surveys resulting from a possible expansion across the Main Crater in 1982 and 1984 (the latter being the period of increased volcanic activity). A large outward movement of station C is apparent for the last interval of 1984-1985. Displacements for the total period 1981-1985 are shown in the final diagram (1981 has been selected as a base, as the line F-G was first included in that survey). This indicates that during the total period there has been an overall expansion across the Main Crater.

The displacement of stations has been resolved into the line-length changes (with SE) of the lines A-B, B-D and C-D (Fig. 3). Line-length changes have been computed assuming a fixed baseline A-B for 1980, and F-G for 1982 and 1983 at which time no distances were measured. The results indicate a possible expansion during 1982 and a more pronounced expansion during 1984. The line C-D has continued to expand during 1985. The line-length changes are consistent with an expansion across the Main Crater.

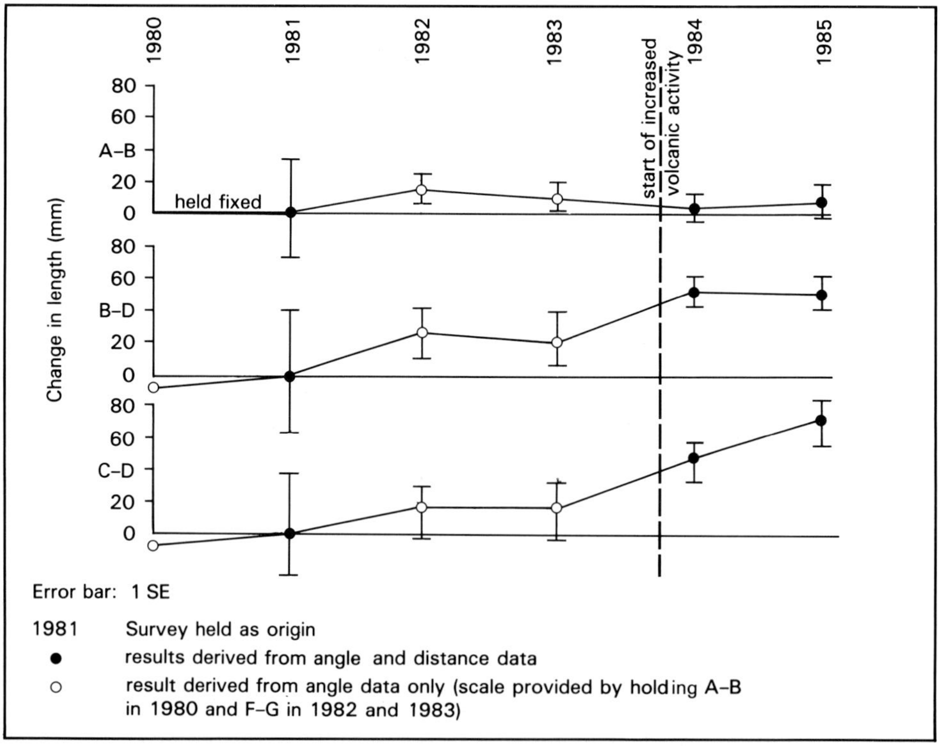

Fig. 3. Line-length changes across the Main Crater

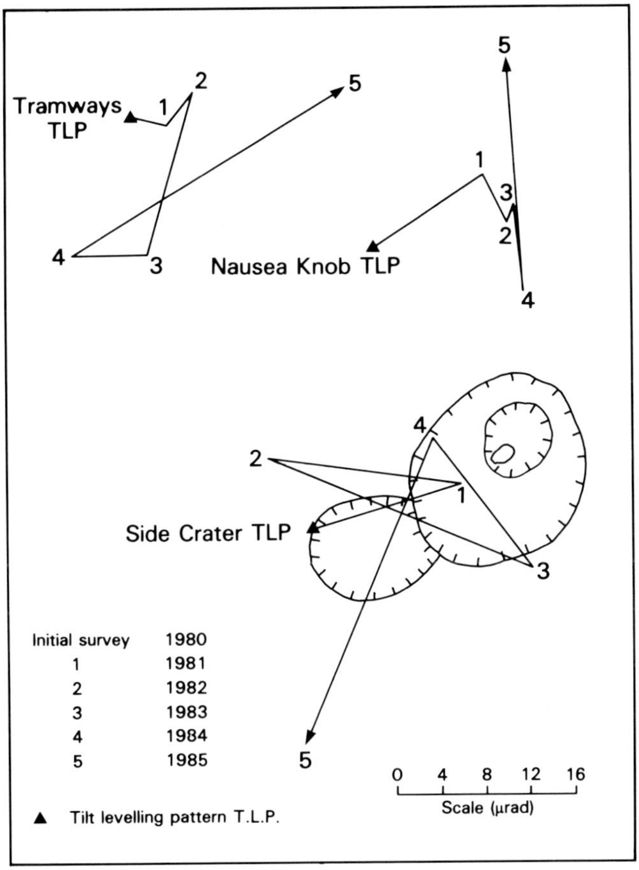

Fig. 4. Tilt-levelling wander plots showing magnitude and direction of downward tilts between surveys

Figure 4 shows the magnitude and direction of downward tilts. The tilts prior to 1985 appear to be random, with little relationship to the Main Crater and Active Vent or to the inferred rift. Tilts recorded during the 1985 survey are significantly larger than earlier results. While the tilts from the Side Crater and Nausea Knob are away from the Main Crater, that from Tramways is tangential to the Main Crater. The significance of the 1985 results is unclear. Tilts do not appear to decrease with distance from the Main Crater. These results do not show a regional pattern of tilt.

The alignment of marks U-Z shows small movements (<20 mm). The movements are consistent with local ground instabilities.

6 Summary

Only minor deformation has been detected on Mt. Erebus from geodetic surveys. The network at the summit indicates cumulative expansion of approximately 50 mm across the Main Crater. Most of this appears to have occurred in the years 1982 and 1984, the latter period coinciding with increased volcanic activity. The results in 1985 indicate possible outward movements of station C, but otherwise little change. Tilts have generally been less than 10 microradians except during 1985 when they were significantly greater. The directions of tilting appear random and uncorrelated with volcanic activity. They are possibly due to local ground instability rather than volcanic deformation.

Acknowledgements. The authors would like to thank the many people who made this project possible: in particular Antarctic Division DSIR; R. Williams, N.Z. Geological Survey and S. Currie, Ministry of Works and Development, who assisted with field work.

References

Blick GH, Scott BJ (1986) Volcanic activity and deformation of the summit of Mt. Erebus, Antarctica, the 1985 Survey. NZ Geol Surv Rep EDS 101:39

Dibble RR, Kienle J, Kyle PR, Shibuya K (1984) Geophysical studies of Erebus Volcano, Antarctica, from 1974 December to 1982 January. NZ J Geol Geophys 27:425-455

Giggenbach WF, Kyle PR, Lyon GL (1973) Present volcanic activity on Mount Erebus, Ross Island, Antarctica. Geology 1:135-136

Kyle PR, Dibble RR, Giggenbach WF, Keys J (1982) Volcanic activity associated with the anorthoclase-phonolite lava lake, Mount Erebus, Antarctica. In: Craddock (ed) Madison, Antarctica Geoscience. Univ Wisconsin, pp 735-745

Vertical Deformation Monitoring by Periodic Water Level Observations, Lake Taupo, New Zealand

P. M. OTWAY[1]

Abstract

A system of monitoring lake levels to detect earth deformation has been developed at Lake Taupo, New Zealand. By using a low cost, portable gauge and taking basic precautions, relative height changes of comparable accuracy with those obtained by precise levelling can be quickly determined. This system may have worldwide applications for similar lakes in volcanically or tectonically active areas.

Since 1979 five periods of deformation affecting the shoreline of Lake Taupo have been identified. These include three episodes of differential uplift and subsidence of up to 50 mm associated with earthquake swarms as well as a period of secular subsidence averaging 13 mm yr^{-1} without any abnormal seismicity and a 2.5-year period of both low deformation and seismicity. No major shallow seismicity at Lake Taupo has occurred since 1979 without associated deformation. The deformation is in general agreement with the concept of regional extension and rifting of the Taupo Volcanic Zone.

1 Introduction

Lake Taupo, with its history of pumice eruptions, earthquake swarms and faulting, has proved an ideal location to develop a system for monitoring vertical deformation by systematic lake level observations. Lake-level surveys were introduced experimentally in April 1979 (Otway 1980) to complement tilt-levelling (dry tilt) surveys at 11 nearby sites, mainly north and east of the lake (Otway 1978, 1979). A portable water level gauge was used initially at five stations around the lake shore, and on a specially constructed platform at Horomatangi Reef (Fig. 2), to determine their heights relative to lake level and to the origin station (Rainbow Point). The number of stations was increased to 15 by mid-1984 (Fig. 7). The tilt-levelling surveys meanwhile were reduced to annual checks. Little success was achieved before 1983 in correlating data from the tilt and lake-level surveys, as the apparently random variations in localized tilt masked the much lower level of regional deformation indicated by the lake-level surveys.

[1] N.Z. Geological Survey, D.S.I.R., Geothermal Research Centre, Private Bag, Taupo, New Zealand

IAVCEI Proceedings in Volcanology 1
J.H. Latter (Ed.), Volcanic Hazards

An episode of deformation in 1983, first reported by the author (P.M. Otway, unpubl., five reports 7 April 1983 to 28 October 1983), involved widespread uplift, detected by lake-level surveys at the northern end of Lake Taupo preceding a strong earthquake swarm in June 1983 centred in the same area and was followed by complex deformation. These reports and the associated surface faulting on the Kaiapo Fault are discussed by Otway et al. (1984), Grindley and Hull (1986) and Otway (1986).

Four additional periods of deformation identified since the scheme commenced in 1979 are:

1. A period of low deformation rate (<4 mm yr^{-1}), from May 1979 to October 1981, accompanied by little seismicity;
2. Minor subsidence, totalling ca. 20 mm, of the southern end of the lake, over a 10-month period preceding an earthquake swarm from February to April 1984. The swarm was accompanied and followed by a rebound of the same magnitude over a further 10-month period;
3. Minor uplift, totalling 10-15 mm, of a section of the northeastern shoreline over a 3-4 month period, followed by a reversal coincident with several small felt earthquakes near the centre of the area (February/March 1985);
4. Persistent secular subsidence of the northern end of the lake averaging 13 mm yr^{-1} from late 1981 to late 1985 without any abnormal seismicity.

Conversely, major shallow seismicity at Lake Taupo has occurred since 1979 without associated deformation.

2 Method

The water level gauge in use, a 0.6-m-long, 50-mm-wide, galvanized iron cylinder (Fig. 1), hangs vertically by a chain from a hook and fills with water through a 1.5-mm orifice, effectively acting as a stilling well that damps out normal wave action (<0.5-m amplitude). When the tap is closed the gauge can be lifted and the inside water level read at eye level to within 1 mm by means of a clear viewing tube set against a scale on the outside of the gauge. A modification was made in early 1985 by mounting a pressure transducer on top of the gauge with one port extending below water level and the other open to the atmosphere, to provide an electrical readout capable of being recorded automatically. If this proves dependable, it will be used for future observations.

The gauge is hung from a hook mounted on a wharf piling, sheer rock face or tripod set up temporarily in shallow water. The height difference between the horizontal surface on top of the gauge and the reference mark on the piling, rock face or ground above it is obtained either by direct taping or by precise levelling. The gauge is positioned by choosing the appropriate link in its suspension chain so as to be approximately half submerged. The tap is opened, and the gauge allowed to stabilize for several minutes. Readings are then made optically every 5 min or electronically every 2.5 min for a 30-min period

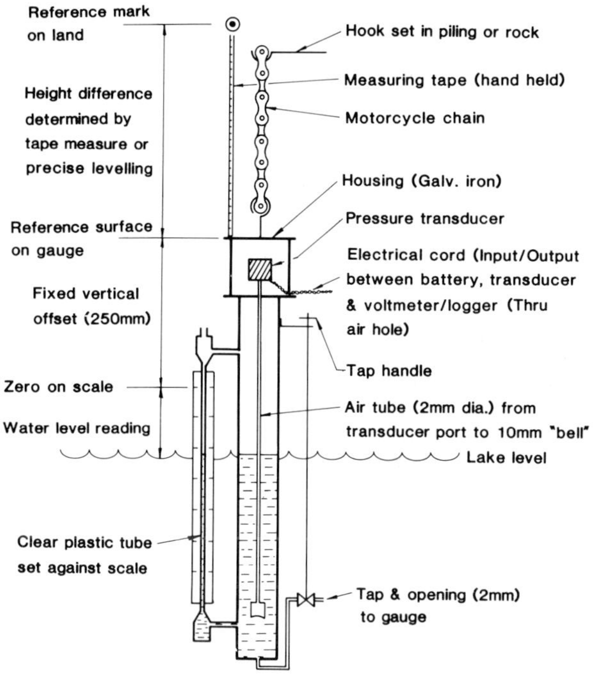

Fig. 1. Portable water level gauge used in Taupo lake level surveys. Pressure transducer added in early 1985

(one cycle of the predominant seiche on Lake Taupo). The readings are scanned before moving on to the next site. The reference marks are connected, where possible, to nearby benchmarks by precise levelling and checked periodically.

To carry out a survey, readings commence at the origin station, Rainbow Point. A set of readings is made at Rainbow Point to determine the lake level at that time relative to a nearby benchmark. The gauge is then moved to various stations around the lake (Fig. 2) taking a 30-min set of readings at each one. Up to eight stations can conveniently be observed by one observer in a day before returning to the origin to obtain a second set. This enables the lake level to be interpolated (assuming a constant rate of change) for each set of observations at stations around the lake. The height of each reference mark relative to the origin is thus obtained simply from the interpolated lake level at that time. The effects of the barometric gradient and strong wind on the lake surface are generally avoided by operating only during times of uniform atmospheric pressure and low wind velocities. Spurious variations in the rate of lake-level change can be avoided by not making observations during or immediately following heavy rain, and by ensuring that the control gates at the outlet of the lake have not been adjusted during the day.

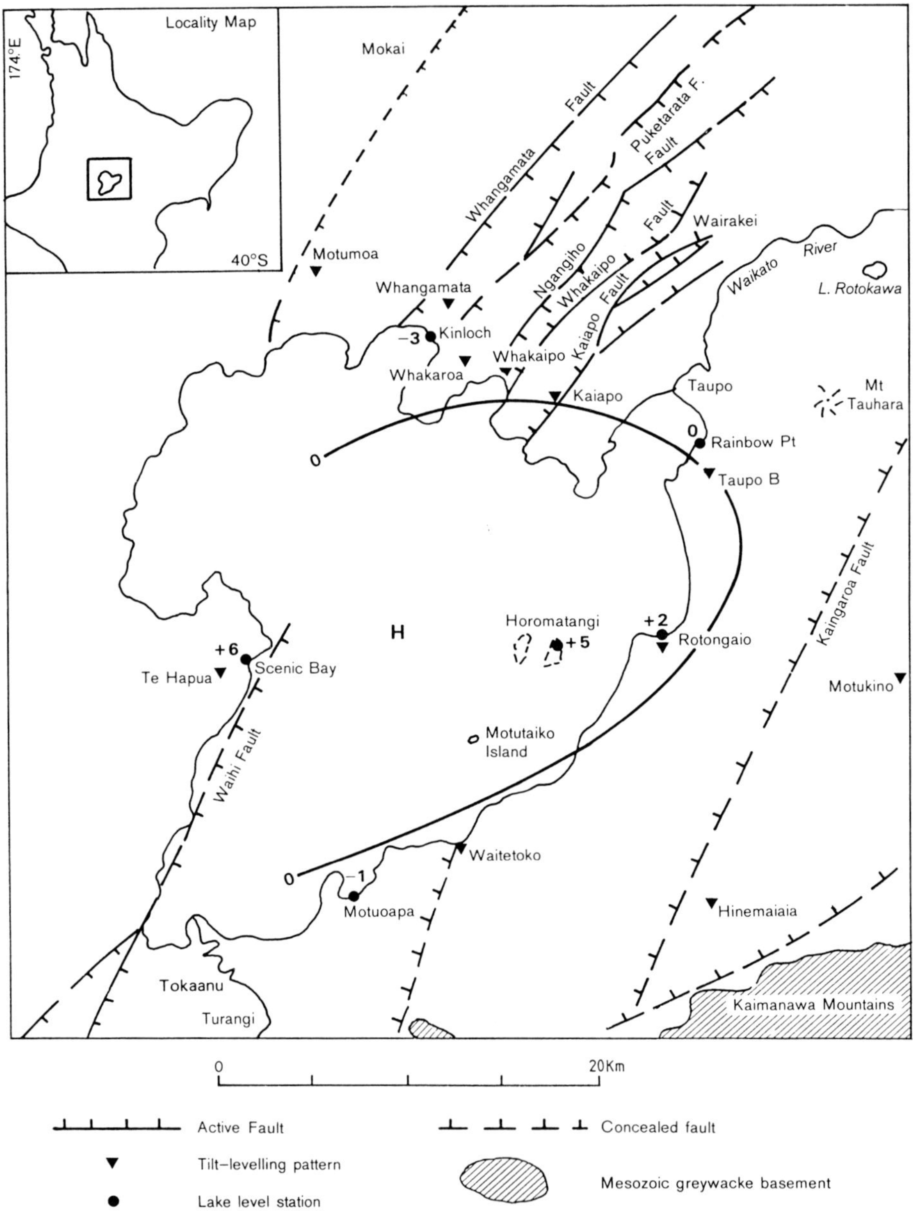

Fig. 2. Vertical displacement in millimetres relative to Rainbow Point, 23 May 1979-30 Oct. 1981

Examination of repeated observations made during the first few years of operation indicates that the apparent errors vary from about 2 mm for stations near Rainbow Point to 7 mm for distant ones (39 km away) and could be approximated by the formula SD in mm = $1.1 \sqrt{D}$, where D is the distance from the origin in km. This is equivalent to 0.2 to 0.4 microradians over the distances involved and is comparable with errors related to precise levelling over similar distances. It is an order of magnitude more accurate than that obtained at the 11 tilt-levelling patterns around the lake.

3 Discussion of Observed Deformation Over Selected Periods

Lake-level measurements over successive periods from 1979 until 1985 are shown in Figs. 2-8.

1979-1982. From establishment of the lake-level scheme in April 1979 until the October 1981 survey, none of the stations exhibited significant change (Fig. 2). Over the following year, October 1981-October 1982 (Fig. 3), significant deformation commenced, with a trough of relative subsidence (>10 mm) developing between Kinloch and Horomatangi.

1983. In early February 1983 an earthquake swarm took place at the northern end of Lake Taupo centred approximately 7 km WNW of Kinloch. No significant tilt was disclosed by tilt-levelling surveys made soon afterwards but a routine lake-level survey in mid-March showed Kinloch station to have risen 33 (±4) mm since October 1982, and Rangatira 13 mm since late January 1983, relative to Rainbow Point. Further surveys showed continuing uplift, and six new stations were added around the lake, mostly at its northern end. By 4 June 1983, Kinloch stood 43 mm above its October 1982 level and Rangatira 18 mm, while Motuoapa, at the southern end of the lake, had fallen 14 mm. Lesser changes were recorded at Scenic Bay, Horomatangi and Rotongaio (Fig. 4). Meanwhile, further checks of the tilt-levelling patterns, less frequent than the lake-levelling surveys, indicated only slight tilt (3 microradians), consistent with uplift SE of Kinloch.

On 16 June 1983 a second, much more strongly felt earthquake swarm started. Early reports from residents located the earthquakes between Kinloch and Acacia Bay, confirmed in a study by Latter and Balsillie (1985) which also computed the magnitudes at $\geq M_L$ 3.2. Epicentres computed from an array of portable seismographs set out on 29 June were, however, generally 2-6 km farther southeast, along the extension of the Kaiapo Fault in Lake Taupo (Hull and Grindley 1984). Lake-level surveys made on 20 and 21 June, after strong winds on the lake had died down, showed that Kinloch had subsided 37 mm since the previous survey a week earlier, while Whakaipo, a new station, was down 17 mm in 2 weeks. Rangatira, on the other hand, had risen a further 8 mm in 2 weeks (Fig. 9).

Field inspection early on 24 June 1983 confirmed reports from residents that displacement had occurred on the Kaiapo Fault within the past 24 h. The southeast side of the fault was upthrown by up to 50 mm, with ca. 30-mm extension (Hull and Grindley 1984). Precise levelling on 28 June of most of the Kaiapo tilt-levelling pattern revealed a 47-mm vertical displacement across the fault scarp 250 m NE of the fault break; the levelling also disclosed an associated 33-microradians SSE tilt

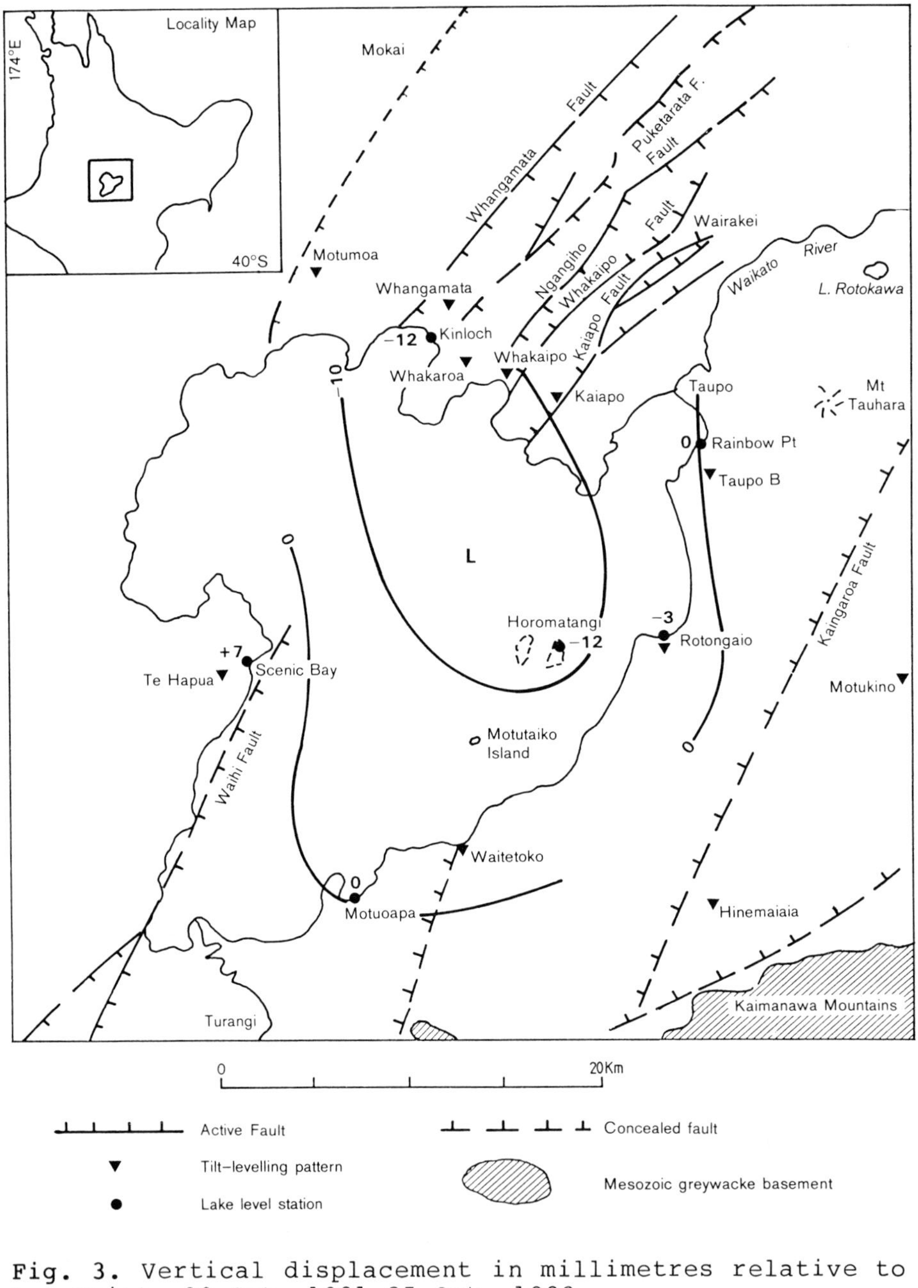

Fig. 3. Vertical displacement in millimetres relative to Rainbow Point, 30 Oct. 1981-25 Oct. 1982

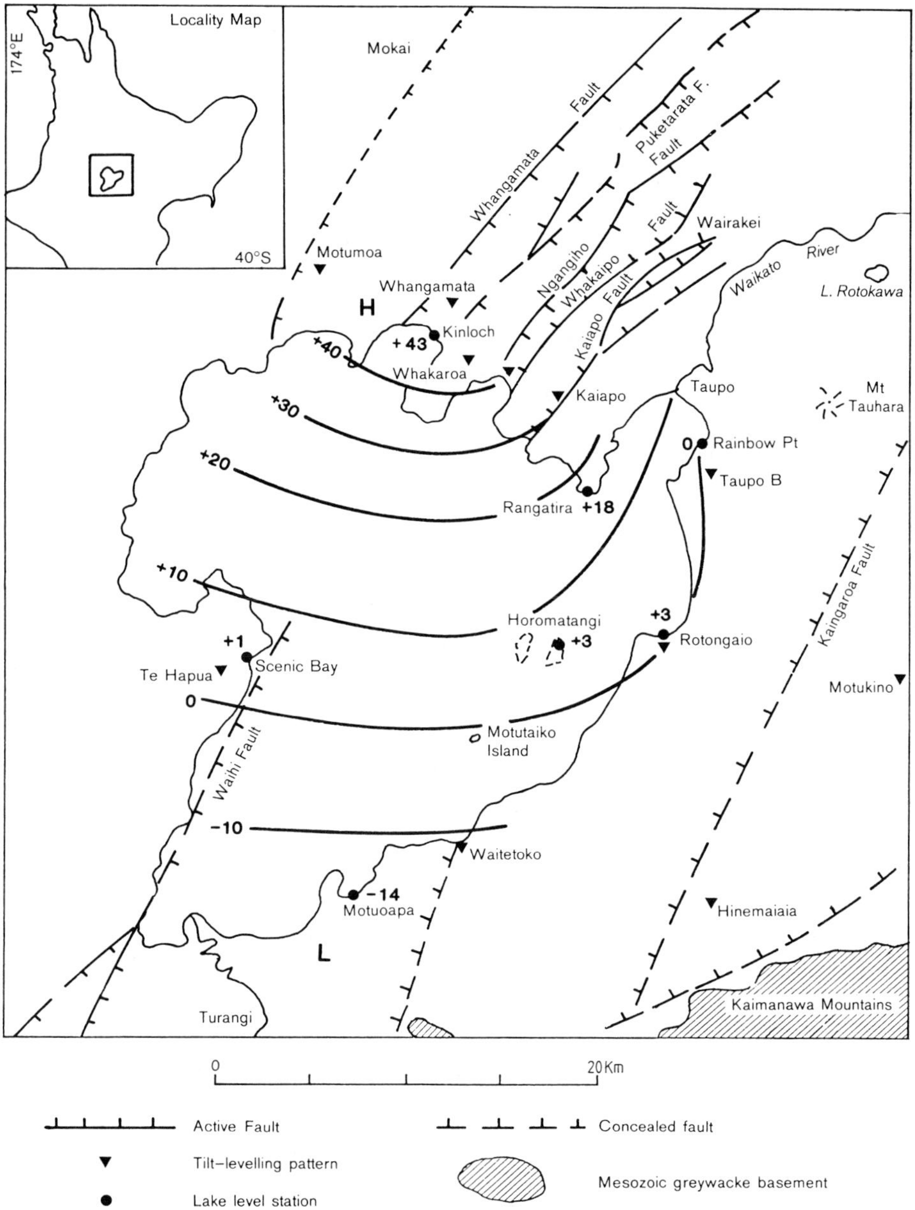

Fig. 4. Vertical displacement in millimetres relative to Rainbow Point, 25 Oct. 1982-4 June 1983

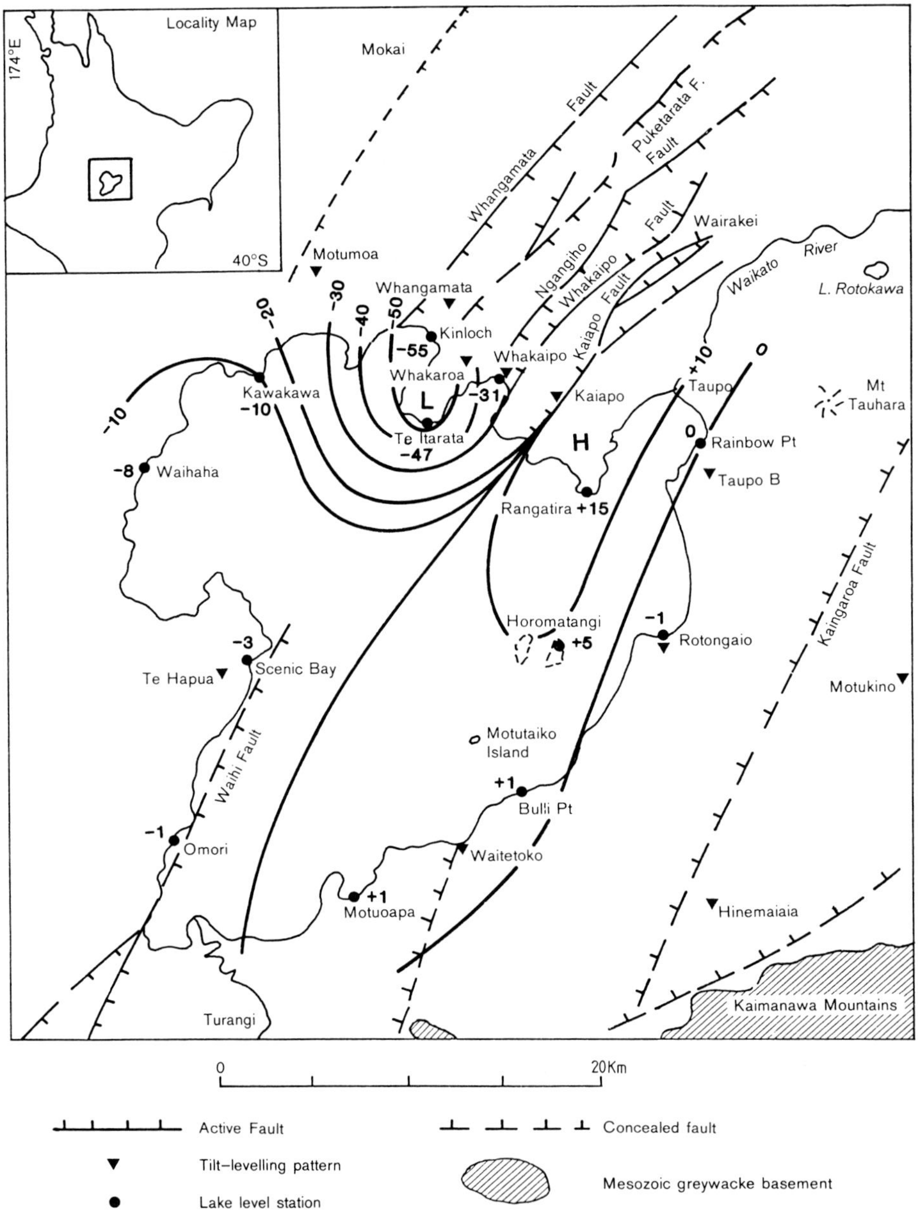

Fig. 5. Vertical displacement in millimetres relative to Rainbow Point, 4 June 1983-19 July 1983

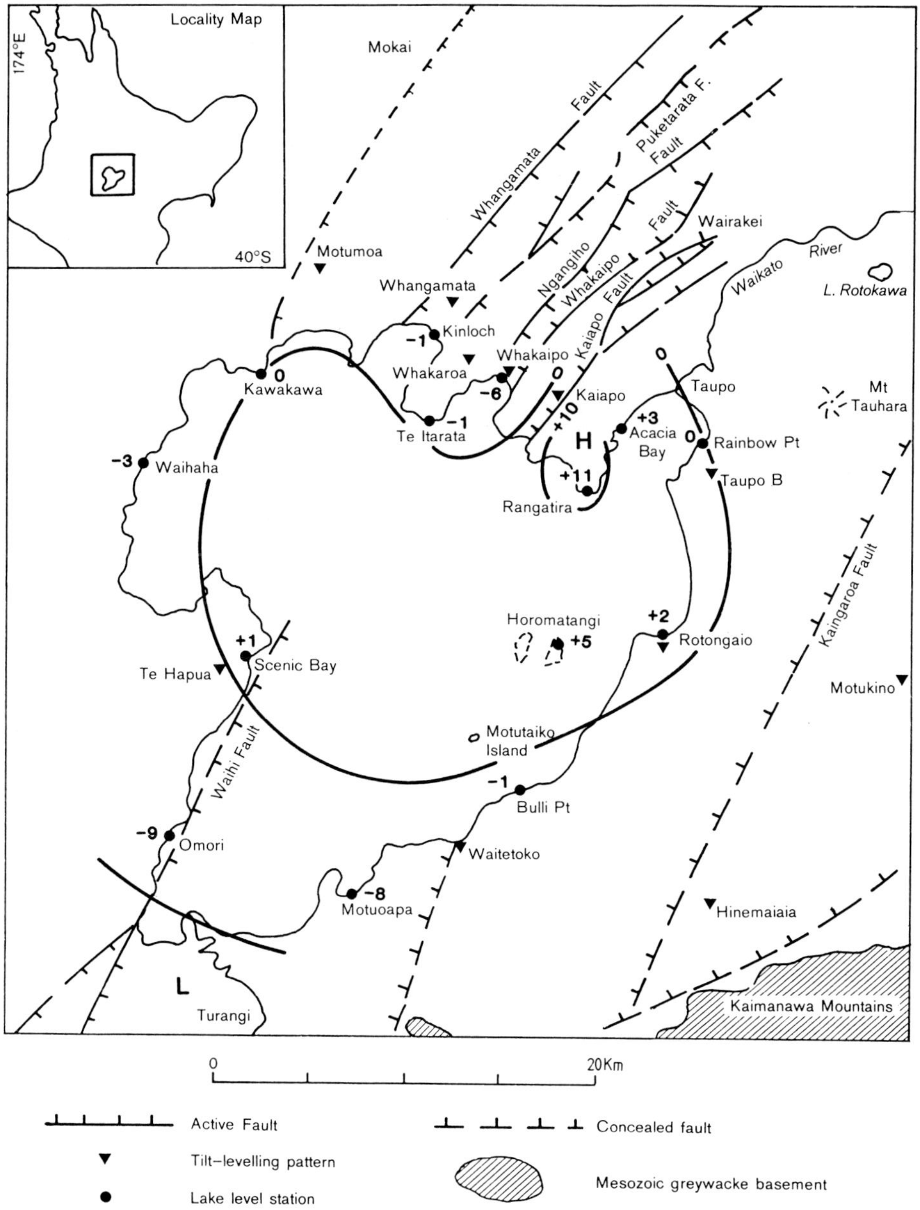

Fig. 6. Vertical displacement in millimetres relative to Rainbow Point, 19 July 1983–1 Jan. 1984

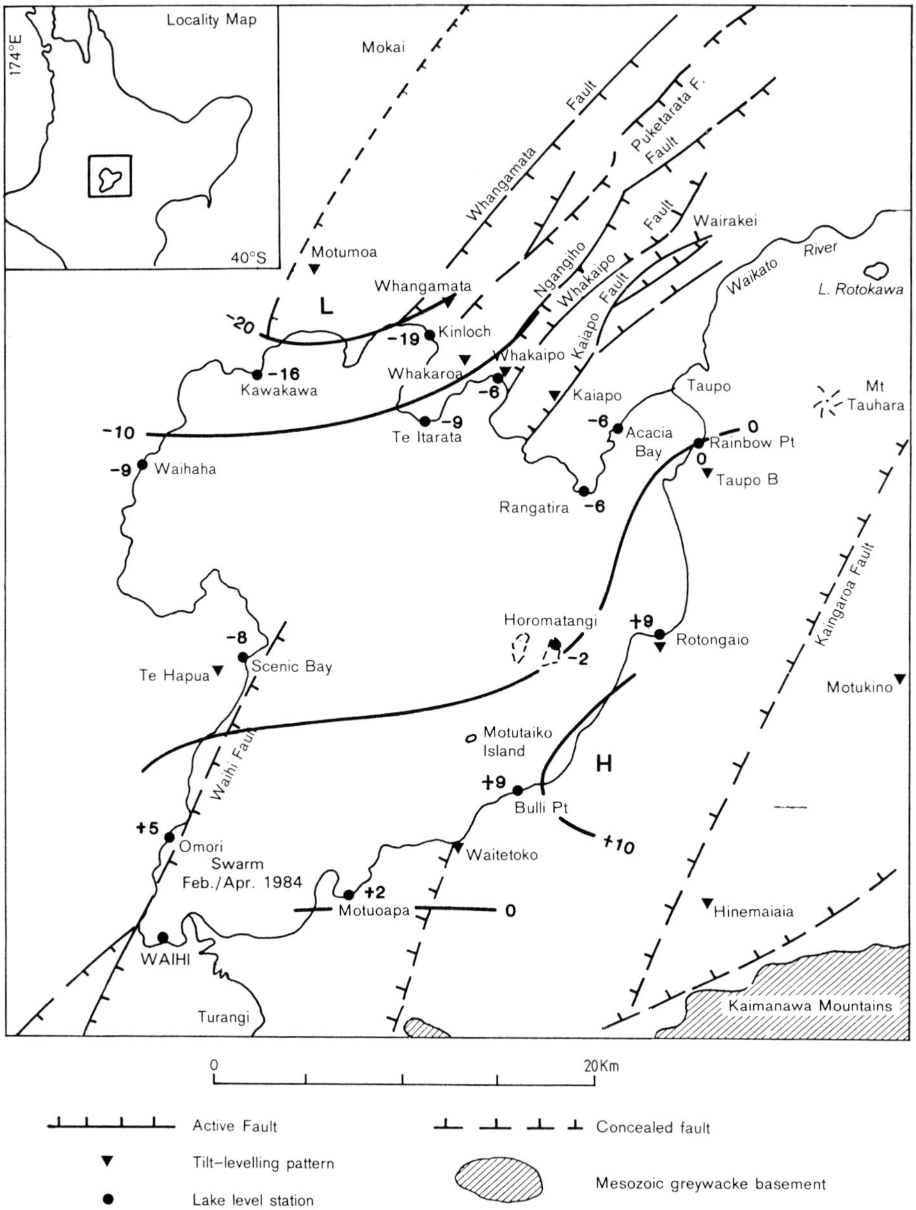

Fig. 7. Vertical displacement in millimetres relative to Rainbow Point, 1 Jan. 1984-1 Feb. 1985

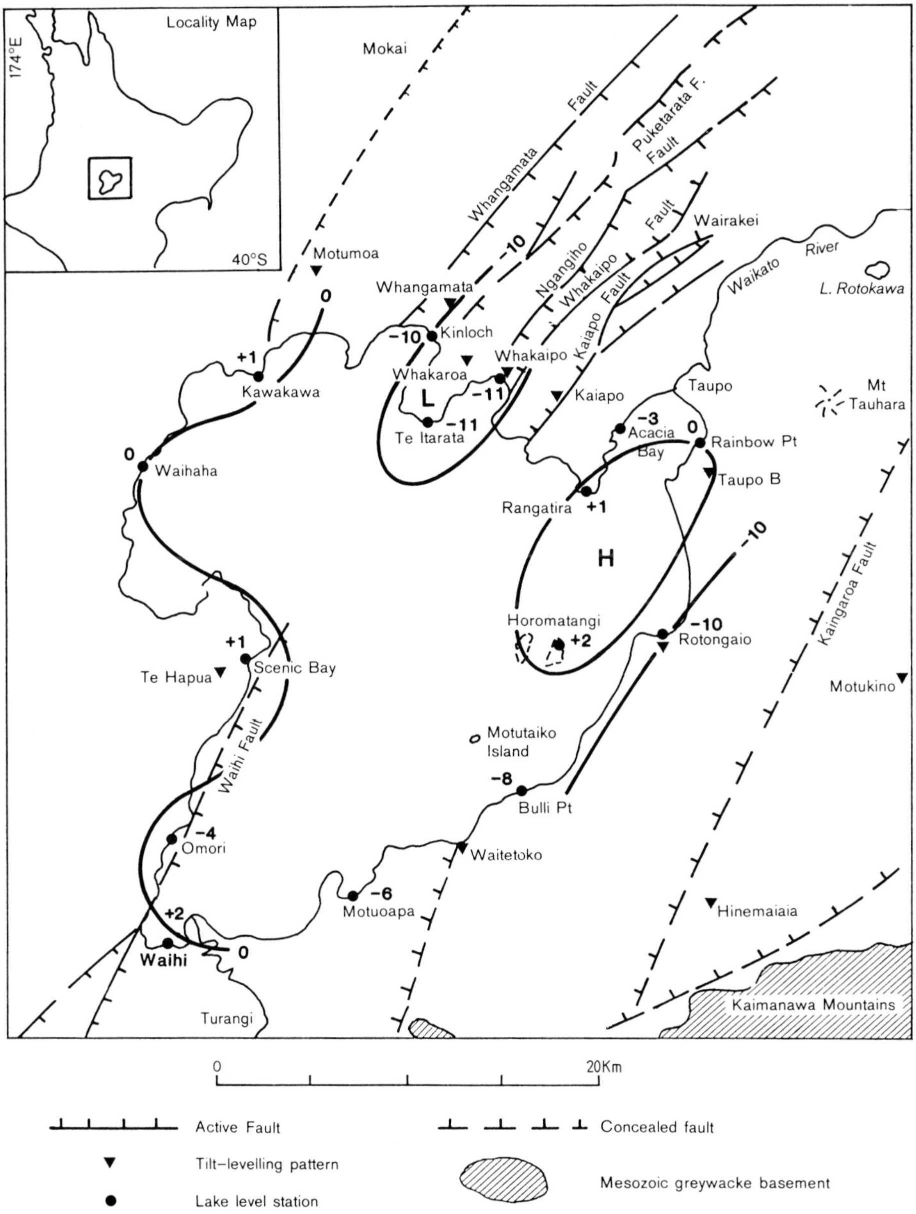

Fig. 8. Vertical displacement in millimetres relative to Rainbow Point, 1 Feb. 1985-10 Dec 1985

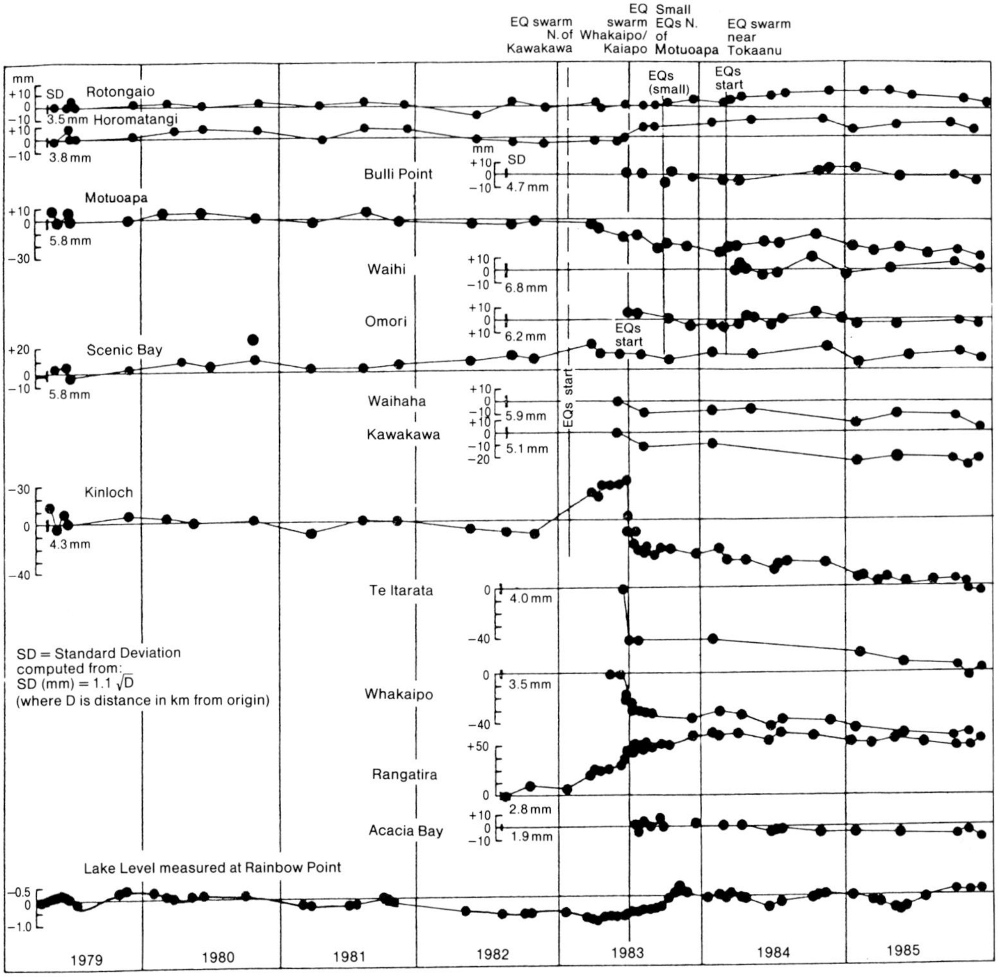

Fig. 9. Apparent changes in benchmark heights relative to Rainbow Point, Lake Taupo, from 1979 to 1985

of the Kaiapo Graben. Checks in late June of the other tilt-levelling patterns revealed clear reversals of tilt (5-10 microradians) at most patterns during the 2-month period spanning the June swarm, but little further change occurred during the following 2 months apart from a 10-microradian reversal at Kaiapo. Continuing subsidence near Kinloch and uplift at Rangatira were observed during lake-level monitoring in early July.

When a comprehensive lake-level survey was made on 19 July total deformation over the previous 6.5-week period was found to consist of a 17-km-wide depression centred just south of Kinloch, 70 mm lower than the SW-trending ridge over Rangatira and separated by the Kaiapo Fault (Fig. 5). Full surveys of all tilt-levelling patterns in September, although less precise, were generally consisten with this broad picture but also clearly showed tilting of the Kaiapo Graben (Otway 1986). Relative stability appeared to have been achieved at Kinloch by the end of August and, at Rangatira, by December. Relevelling in September of sections of an existing first-order traverse north of the lake near major faults revealed little change, as did more extensive levelling in 1983/84 by the Lands and Survey Department.

During the remainder of 1983 (Fig. 6), Rangatira continued to rise slowly while the southern shore of the lake continued to subside at a similar rate. A possible 5-10-mm upward step occurred at Motuoapa and Bulli Point in September/October coinciding with two short earthquake swarms, which were not accurately located, but were thought, from felt reports, to be centred just north of Motuoapa.

1984-Late 1985. The three southernmost stations appeared to stop subsiding and start a slow recovery in late February 1984, at about the time when a strong earthquake swarm began, felt most strongly in the Tokaanu area, which persisted into May. The frequency of surveys was again increased and, a little later, a new station, Waihi, was installed to assist in monitoring possible differential movement of the nearby Waihi Fault. All four southern stations exhibited minor oscillations during 1984, reaching a high in about October, but no significant movement was noted across the Waihi Fault (Fig. 9). At the northern end of the lake, the long-term subsidence, which appeared to have started in late 1981 (Fig. 9), continued at Kinloch and the other northern stations. During 1984 it totalled 19 mm at Kinloch with lesser amounts at neighbouring stations (Fig. 7) but, by the end of 1985 (Fig. 8), had apparently become localized between Kinloch and Whakaipo, totalling about 10 mm in this area during the year. Cumulative subsidence at Kinloch since 1981 reached 54 mm, a mean rate of 13.5 mm yr^{-1}. Except for the swarm activity in 1983, seismicity in this area has remained generally low throughout. Another, possibly significant, development during 1984 was minor uplift south of Rotongaio of about 15 mm relative to Rangatira in late 1984/early 1985 followed by a recovery between February and May 1985 (Figs. 7-9). Localized small earthquakes were felt at Rotongaio, on the northern edge of a sparsely inhabited area, in February/March 1985 but were not recorded on the closest operating seismograph, 28 km WNW, and hence were inferred to have been shallow and of low magnitude.

4 Conclusion

The current pattern of long-term deformation, taking the form of subsidence at the northern and southern ends of the lake relative to the eastern and western sides, is in general agreement with the widely held concept of regional extension and rifting of the Taupo Volcanic Zone as a whole. Superimposed on this pattern, however, are short-term variations which have been suspected in volcanic areas but seldom before monitored in detail (Grindley 1975). Their relationship to the more pronounced deformation, both reversible and non-reversible, which accompanies major seismic episodes, is not yet known, nor is their significance in relation to the episodic volcanism of the region yet apparent (Grindley and Hull 1986).

Acknowledgements. Thanks are due to members of the Chemistry and Geophysics Divisions for assistance in developing and testing the equipment, N.Z. Geological Survey and U.S. Geological Survey personnel for field assistance, the Department of Internal Affairs and the Division of Marine and Freshwater Science personnel for logistic support on the lake and to G.W. Grindley, P.R. Wood, J.G. Gregory, I.G. Speden, J.H. Latter and D.A. Swanson for reviewing the original manuscript.

References

Grindley GW (1975) Relationship of volcanism to earth movement, Bay of Naples, Italy. In: Gonzalez-Ferrano (ed) Symposium on Andean and Antarctic volcanology problems. Bull Volcanol, pp 598-612

Grindley GW, Hull AG (1986) Historic Taupo earthquakes and earth deformation 1922, 1964, 1983. R Soc New Zealand, Wellington, pp 173-186

Hull AG, Grindley GW (1984) Active faulting near Taupo. EOS, Feb 14, 1984

Latter JH, Balsillie FH (1985) Shallow seismicity of the Taupo volcanic zone, Mt Egmont area, and the Auckland Volcanic Field. NZ Volcanol Record 13:8-28

Otway PM (1978) Tilt-levelling monitoring in the Taupo volcanic zone. NZ Volcanol Record 7:20

Otway PM (1979) Tilt-levelling monitoring in the Taupo volcanic zone. NZ Volcanol Record 8:27-28

Otway PM (1980) Taupo lake level survey. NZ Volcanol Record 9:21-22

Otway PM (1986) Vertical deformation associated with the Taupo earthquake swarm, June 1983. Bull R Soc NZ 24:187-200

Otway PM, Grindley GW, Hull AG (1984) Earthquakes, active fault displacement and associated vertical deformation near Lake Taupo, Taupo Volcanic Zone. NZ Geol Surv Rep 110:73

Geodetic and Geophysical Monitoring of the 1886 Tarawera Rift

B.J. Scott[1]

Abstract

The 1886 Tarawera Rift, a series of subparallel left-stepping, en echelon eruptive fissures 17 km long formed on 10 June 1886, occupies the axial position of the Tarawera Vent Lineation. The 1886 Tarawera Rift fracture pattern is consistent with dextral/normal movement on an underlying fault striking at 57° east of north.

Geodetic monitoring is conducted by means of two horizontal deformation patterns which straddle the rift, three tilt-levelling patterns and three lake-level recorders. Lake-levelling studies commenced in 1976, tilt-levelling in 1978 and horizontal monitoring in 1983. Significant engineering shear strain rates have been detected by the Lake Rotomahana pattern (1.66 ± 1.16) ppm/year (95% confidence) which are high (probably a manifestation of the short time interval, 2.5 years). The significance of the deformation is not fully understood. The Mt. Tarawera pattern and two of the tilt-levelling patterns have detected no significant deformations, but the lake-levelling study and the Tarawera South tilt-levelling pattern indicate about 1.2 microradians per year of tilt down to the ENE.

A single component, direct-write seismograph has been operated at several locations along the 1886 Tarawera Rift since October 1980. Repeated earthquake swarms located on the Paeroa Fault have been recorded, but only minor activity associated with the 1886 Tarawera Rift.

Selected hydrothermal features within the Waimangu Hydrothermal Field at the southern end of the 1886 Tarawera Rift have been monitored since 1970. Five natural disturbances, characterized by short-lived, intense heat flow to the surface, have occurred since monitoring commenced. Two of these disturbances have resulted in small localized hydrothermal eruptions.

None of the monitoring programs associated with the 1886 Tarawera Rift have recorded any activity suggesting future volcanism.

[1] N.Z. Geological Survey, Department of Scientific and Industrial Research (DSIR), P.O. Box 499, Rotorua, New Zealand

IAVCEI Proceedings in Volcanology 1
J.H. Latter (Ed.), Volcanic Hazards

1 Geological Setting

Okataina Volcanic Centre (OVC) is the most recently active of the four major rhyolite eruptive centres within the linear volcano-tectonic depression of the Taupo Volcanic Zone. The Centre is defined (Nairn 1981) to include all the recently active vents which lie to the east of Rotorua Caldera (Fig. 1). It includes the large volcanic complexes of Haroharo and Tarawera, with other eruptive centres at Edgecumbe, Okareka and Rotoma. All of these eruptive complexes have grown to their present forms within the past 20 000 years, during which more than 80 km^3 of magma has been erupted as pyroclastics and lavas.

The Tarawera Volcanic Complex has been constructed within the southern part of the Haroharo Caldera during four major rhyolite eruptive episodes between ca. 17 000 and 700 years ago (Cole 1979; Nairn 1981). Basalt is known to be associated with two of these rhyolite eruptions.

Although the multiple rhyolitic vents at Tarawera are scattered over a wide area of the complex, all vents (acid and basic) lie within a 38 km long, 5 km wide zone, striking at 57° which probably overlies a fundamental deep-seated crustal fracture (Malahoff 1968). This vent zone extends beyond the Haroharo Caldera boundaries, from the basaltic/phreatic explosion craters at Waimangu through Lake Rotomahana and Mt. Tarawera, to the andesite/dacite cones of Puhipuhi and Edgecumbe (Nairn and Cole 1981).

The fracture zone has been termed the "Tarawera Vent Lineation" (Nairn 1981) and is not a direct continuation of any of the regional faults or structures of the Taupo Volcanic Zone southwest of Tarawera, most of which strike at about 40°. The Tara-

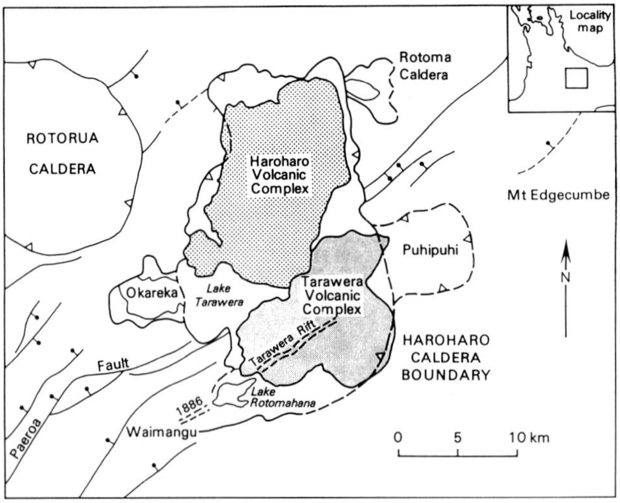

Fig. 1. Simplified regional tectonic map of the Okataina Volcanic Centre (after Nairn 1981) showing the caldera boundary, major faults, volcanic complexes and 1886 Tarawera Rift

wera Rift, a series of subparallel, left-stepping, en echelon eruptive fissures, occupies the axial position of the Tarawera Vent Lineation. The structural interpretation (Nairn and Cole 1981) suggests that this is a fracture pattern which would be produced by dextral/normal movement on an underlying fault striking at 57°. Geodetic and geophysical studies by the New Zealand Geological Survey (NZGS) and Geophysics Division of DSIR have concentrated on the Tarawera Volcanic Complex.

2 Geodetic Monitoring

2.1 Horizontal Deformation Patterns

In 1983 two horizontal deformation patterns were established across the Tarawera Rift; one centred on Lake Rotomahana, and the other on Ruawahia Crater (Fig. 2).

The Lake Rotomahana pattern consists of a series of interconnected quadrilaterals and a 5.7 km long alignment of stations trending about 130° (Fig. 2). A detailed description of the pattern including monument construction techniques and site access is contained in Scott and Chadwick (1983). The second pattern is located across the northern portion of the Tarawera Rift, centred on Ruawahia Crater. The pattern consists of two braced quadrilaterials (Fig. 2). A detailed description of the pattern, including monument construction techniques, is contained in Scott (1984a).

The patterns have been occupied approximately annually, usually with triangulation and trilateration surveys being carried out

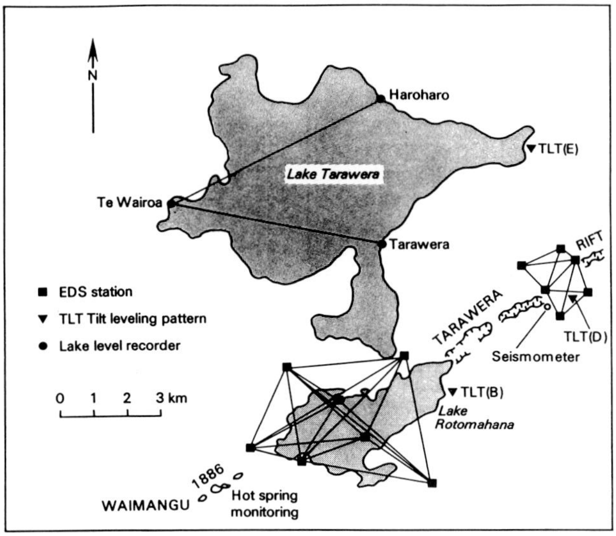

Fig. 2. Sketch plan showing the geodetic patterns about and across the 1886 Tarawera Rift. Also shown is the location of the seismometer, and Waimangu

at the same time. Horizontal angle observations are made in daylight using a theodolite, and all stations are occupied. Distance measurements are made with an infrared distance meter to glass prisms. Atmospheric wet and dry bulb temperatures and air pressures are measured at each end of a line during distance measurements. Each station is occupied by the instrument, hence all lines are measured twice. Following each survey, a survey report is issued in which observational data are tabulated, along with station coordinates computed using a least squares variation of the coordinates method (Scott 1984b).

All observational data are held on the NZGS Earth Deformation Section (EDS) geodetic data base, which has been developed in order to store and update data safely and easily. With this rapid manipulation and computation can be carried out.

Up to the present time (Jan 1986), five surveys have been conducted on the Lake Rotomahana pattern. These were in March and May 1983, January 1984, January 1985 and December 1985. All observational data (triangulation and trilateration) have been combined, and analyzed for the existence of engineering shear strain using the method of simultaneous reduction, to give the following results:

Dilatation	-4.33 ± 0.90 ppm yr^{-1}
Maximum extension	-1.33 ± 0.70 ppm yr^{-1}
Minimum extension	-3.00 ± 0.76 ppm yr^{-1}
Maximum shear strain	1.66 ± 1.16 ppm yr^{-1}
Principle axis of relative contraction	$54 \pm 21°$

Note that all errors are at the 95% confidence limit.

These results indicate that significant shear strain and a high rate of dilatation have been detected by the Lake Rotomahana pattern. The negative dilatation prevents an interpretation of the strain as just an extension orthogonal to the rift axis. The significance is not yet understood. Errors associated with the survey data can produce apparent deformations when the surveys are closely spaced (G.H. Blick, pers. commun.). Hence, the high strain rates may be a manifestation of the short time interval over which monitoring has been conducted.

Only two surveys have been conducted on the Mt. Tarawera pattern. These were in July 1984 and August 1985. An analysis for shear strain between these two surveys indicates no significant deformation, but has shown that there are problems with the design of the network. Further stations are to be added before the next survey.

2.2 Tilt Levelling

There are three tilt-levelling patterns being currently monitored near the 1886 Tarawera Rift (Fig. 2). These are Tarawera South (B), installed in June 1978; Tarawera Airstrip (D), installed in May 1982; and Tarawera Outlet (E), installed in January 1982. Surveys are normally made at three to four monthly intervals (Otway 1979; Scott and Otway 1985).

The Airstrip and Outlet patterns, both of which are 30 m triangles, have shown no consistent trends since installation. The Tarawera South pattern, a 150-m triangle, which has been in use for 5 years, has recorded an apparent tilt since installation, averaging 1.2 microradians downward per year at 68° which is in agreement with the lake-levelling result discussed below.

2.3 Lake Tarawera Lake Levelling

Three water level recorders on Lake Tarawera (two analog and one digital) have been used since January 1976 to monitor relative vertical displacements of the Tarawera and Haroharo Volcanic Complexes, within the Okataina Volcanic Centre, (Fig. 2).

Throughout this period, mean daily water levels in millimetres for each station have been determined. Differences in water level between the Te Wairoa recorder and the Haroharo or Tarawera recorders are calculated and expressed as a tilt in microradians. The data have been presented annually in the New Zealand Volcanological Records, 6 to 13. In November 1984 the two analog recorders were replaced by digital recorders.

Apparent monthly mean tilts between Te Wairoa and Haroharo have ranged from +2.1 to -7.8 microradians during the recording period, while the tilts between Te Wairoa and Tarawera have ranged from +1.6 to -7.9 microradians (Fig. 3). The Te Wairoa-Haroharo leg is 7220 m long and trends at 39°, while the Te Wairoa-Tarawera leg is 6450 m long and trends at 100°. Short-term apparent tilts, up to 2 microradians in amplitude, are often induced by adverse meteorological conditions.

A total of about 8 microradians of down tilt relative to Te Wairoa has now accumulated on the Haroharo leg, mostly since 1981, while about 7 microradians of down tilt has accumulated on the Tarawera leg since 1979 (Fig. 3). The resultant cumulative tilt is 8.4 ± 1.2 microradians at an azimuth of 72°, an average rate of 0.9 microradians per year. This agrees well with tilts measured at Tarawera South, which have amounted to 1.2 microradians per year at 68°. These increments may indicate a regional tilt, the significance of which is not understood.

3 Seismic Monitoring

Since October 1980 NZGS and the University of Auckland (Physics Department) have operated a single component direct-write seismograph at several places along the 1886 Tarawera Rift. In July 1984 NZGS and the Geophysics Division (DSIR) established a telemetry link between Mt. Tarawera and Rotorua, and recordings commenced at the NZGS office in Rotorua on 15 July 1984.

During the operation of the seismograph the area has been subjected to repeated earthquake swarms, the epicentres of which

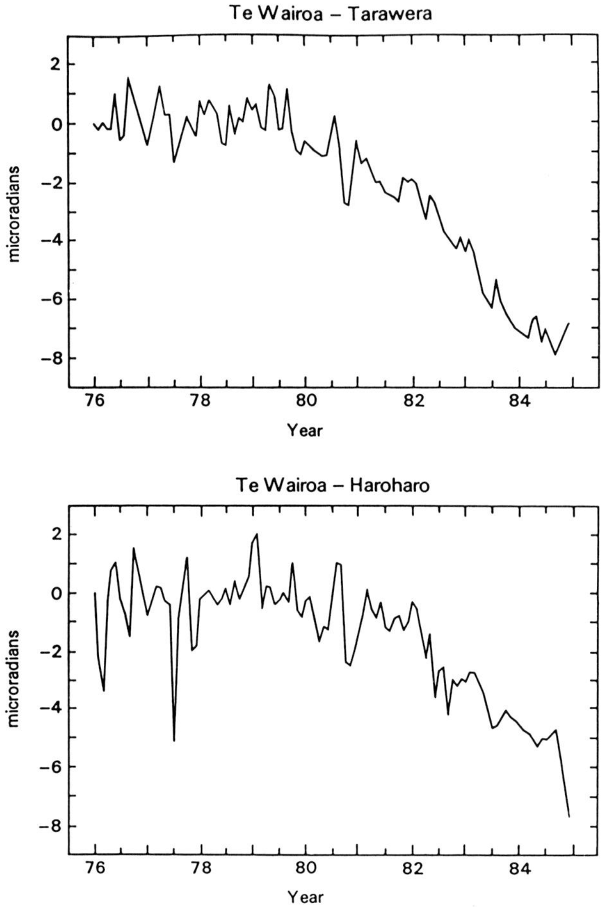

Fig. 3. Plots of monthly mean tilts across Lake Tarawera. The **upper plot** displays tilt between Te Wairoa and Tarawera, while the **lower plot** displays tilt between Te Wairoa and Haroharo. See Fig. 2 for location of recorders. All tilts are expressed in microradians

have been located on the Paeroa Fault (8 km west of Lake Rotomahana), within the Okataina Volcanic Centre, and in the Rotorua Caldera to the west. Earthquake activity on the 1886 Tarawera Rift has been minor, and is difficult to quantify with only one recorder. Microearthquake activity on the Paeroa Fault and within the Okataina Volcanic Centre was detected during a microearthquake survey in January-February 1981 (Rowe 1984).

Earthquakes located by the NZ Seismological Network show that an increase in activity occurred on the Paeroa fault, commencing with a ML 4.1 event on 30 April 1982, and apparently ending with a ML 5.1 event on 14 December 1983 (Smith et al. 1984).

4 Hydrological Heat Flow Studies (Waimangu)

Instruments were installed in the Waimangu Hydrothermal Field in 1970 to monitor the discharge rate and discharge temperature of Frying Pan Lake, and the temperature, water level and intermittent discharge rate of Inferno Crater Lake (Fig. 4). The instruments were funded by the Tourist and Publicity Department and by NZGS, and the continuing research has been jointly conducted by NZGS and the University of Auckland (Physics Department).

Short-term objectives of the program were to verify suspected hydrological connections between the two features, and to provide data to establish a qualitative model of the system (Stanton 1978). Although these short-term objectives have been

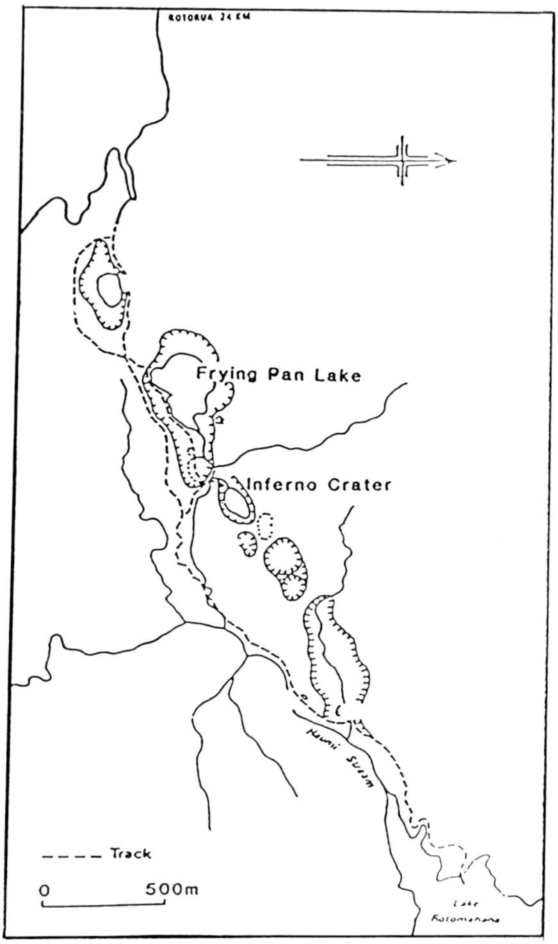

Fig. 4. Sketch plan of Waimangu Valley showing the location of the craters formed along the 1886 Tarawera Rift in that area. Hydrological studies are concentrated on Frying Pan Lake and Inferno Crater

achieved, measurements are still being conducted because they have been found to provide valuable data for surveillance of the hydrothermal system and will prove useful in the long term for volcanic surveillance.

The water in Inferno Crater Lake exhibits remarkable quasi-cyclic level fluctuations oscillating between overflow and as much as 12 m below overflow (Fig. 5), the period between overflows is about 6 weeks. The water temperature varies directly with water level, ranging from about 35°C when the lake is low, to over 70°C when overflowing.

The discharge of Frying Pan Lake exhibits quite large fluctuations (Fig. 5), some are of hydrothermal origin, while others are meteorologically induced. A regular long-term low amplitude

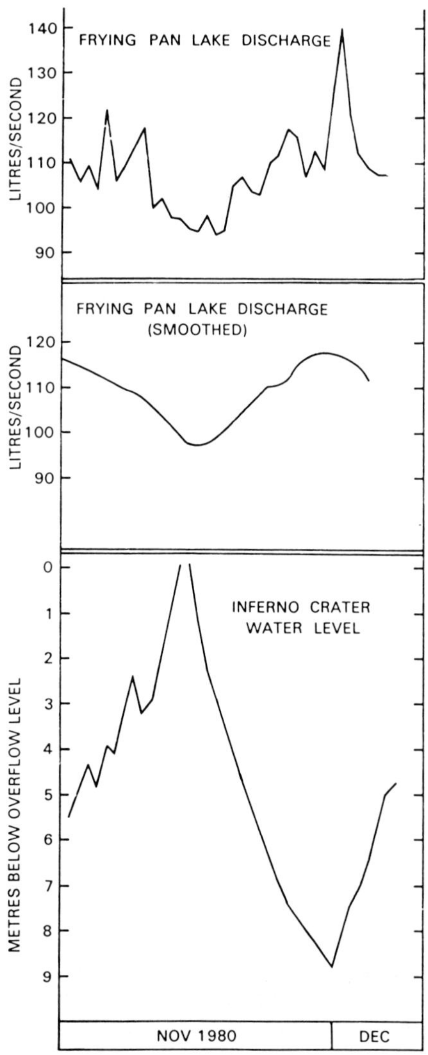

Fig. 5. A plot of the monitored discharge of Frying Pan Lake and the water level of the Inferno Crater Lake. The two upper portions show a 10-20 litre s^{-1} fluctuation in the discharge of Frying Pan Lake which shows an inverse correlation with the water level of the Inferno Crater Lake. The short-term high amplitude fluctuations in the uppermost plot are attributed to rainfall-induced runoff

fluctuation, with flows ranging from about 100 litres s^{-1} to over 125 litres s^{-1} shows a remarkable inverse correlation with the water level of Inferno Crater Lake. A short-term high amplitude fluctuation is induced by rainfall. The temperature of Frying Pan Lake outflow displays a strong diurnal fluctuation (up to 6°C) due to meteorological effects. The average daily temperatures show an annual range of about 9°C (46-55°C), which is also attributed to seasonal meteorological variations. No significant temperature changes of Frying Pan Lake outflow attributable to hydrothermal/volcanic effects have been noted during the monitoring program.

Since monitoring commenced in 1970, Inferno Crater Lake has responded to four natural disturbances in the Waimangu hydrothermal system by missing flows (Scott and Lloyd 1982). These disturbances were the 1973 Trinity Terrace eruption (Lloyd and Keam 1974), the 1975 Spring N disturbance, the 1978-79 Mud Rift activity and the 1981 Mud Rift eruption. All four events were characterized by short-lived, intense heat flow to the surface, almost certainly resulting in temporary reductions in aquifer pressure in the affected volume, which were transmitted to other parts of the hydrothermal system. On a fifth occasion (the 1978 Raupo Pond-Inferno Crater disturbance), the heat flow to both Mud Rift and Inferno Crater intensified. A marked shortening of the Inferno Crater Lake cycle ensued, lasting for several months.

Several detailed short-term investigations of seismic noise, solar and atmospheric contributions to the heat flow and lake bathymetry have been conducted by NZGS and the University of Auckland (Physics Department) and are described by Scott (1986).

5 Summary

Geodetic monitoring, in particular of the horizontal deformation pattern at Lake Rotomahana, has shown that significant deformation is occurring about the 1886 Tarawera Rift. The Lake Tarawera lake-levelling study and the Tarawera South tilt-levelling pattern appear to be registering a regional tilt, which commenced in 1980-81.

A significant increase in seismic activity, centred on the Paeroa Fault (8 km west of Rotomahana) commenced in early 1982. No significant seismicity associated with the rift has been recorded.

The hydrothermal-heat flow studies at Waimangu have recorded five disturbances, characterized by short-lived intense heat flow to the surface. These have been of geothermal rather than volcanic origin.

None of the monitoring programs associated with the 1886 Tarawera Rift have recorded activity which suggests impending future volcanism.

Acknowledgements. The author would like to thank P.M. Otway and the staff of EDS for field assistance and technical comments on the geodetic studies and Prof. R.F. Keam and students for assistance with hydrological studies at Waimangu.

References

Cole JW (1979) Structure, petrology and genesis of Cenozoic volcanism, Taupo Volcanic Zone, New Zealand - a review. NZ J Geol Geophys 22:631-657

Lloyd EF, Keam RF (1974) Trinity Terrace hydrothermal eruption, Waimangu, New Zealand. NZ J Sci 17:511-528

Malahoff A (1968) Origin of magnetic anomalies over the central volcanic region of New Zealand. In: Knopoff L, Hast PG. (eds) The crust and upper mantle of the Pacific area. Geophysical Monograph 12. Am Geophys Union, pp 218-240

Nairn IA (1981) Some studies of the geology, volcanic history and geothermal resources of the Okataina Volcanic Centre, Taupo Volcanic Zone, New Zealand. Ph.D.thesis, lodged in the library, Victoria Univ Wellington, New Zealand (unpublished)

Nairn IA, Cole JW (1981) Basalt dikes in the 1886 Tarawera Rift. NZ J Geol Geophys 24:585-592

Otway PM (1979) Tilt-level monitoring in the Taupo Volcanic Zone. NZ Volcanol Record 8:27-28

Rowe C (1984) Activity at Mt Tarawera, North Island, New Zealand. Results of 1981 microearthquake survey. BSc Hons. paper, Univ of Alaska, Fairbanks (unpublished)

Scott BJ (1977) Okataina Volcanic Centre, tilt recordings. NZ Volcanol Record 6:19-20

Scott BJ (1984a) Mt Tarawera strain monitoring pattern site report (Tarawera Volcanic Complex). NZ Geol Surv Rep EDS 93

Scott BJ (1984b) Survey report, Lake Rotomahana strain monitoring pattern (Tarawera Volcanic Complex) March 1983, May 1983. NZ Geol Surv Rep EDS 95

Scott BJ (1986) Monitoring at Okataina Volcanic Centre. In: Gregory JG Watters WA (eds) Volcanic hazards assessment in New Zealand. NZ Geol Surv Record 10:49-54

Scott BJ, Chadwick WW (1983) Lake Rotomahana triangulation and strain monitoring pattern site report (Tarawera Volcanic Complex). NZ Geol Surv Rep EDS 88

Scott BJ, Lloyd EF (1982) Hydrological measurements, Frying Pan Lake - Inferno Crater Lake 1970-1981. Waimangu Hydrothermal Field. NZ Geol Surv Rep EDS 67

Scott BJ, Otway PM (1985) Okataina Volcanic Centre, deformation surveys. NZ Volcanol Record 13:36-39

Smith EGC, Scott BJ, Latter JH (1984) The Waiotapu earthquake of 1983, December 14. Bull NZ Soc Earthquake Engineering 17(4):272-279

Stanton TP (1978) Seismic noise characteristics of Inferno Crater Waimangu. MSc thesis, lodged in the library, Univ of Auckland (unpublished)

A Preliminary Experiment on Automated Judgement of the Stages of Eruptive Activity Using Tiltmeter Records at Sakurajima, Japan

K. Kamo and K. Ishihara[1]

Abstract

A project was initiated to collect and evaluate high-resolution tilt data to forecast eruptive activity at Sakurajima Volcano in southern Japan. To implement the experiment, an underground tunnel was constructed in the lava dome 2.8 km northwest of the summit crater. The tunnel was designed to house a two-component, water-tube tiltmeter and a three-component extensometer with a span of 28 m on each side. Characteristic inflation prior to an eruption and deflation after the activity was observed by the tiltmeter. The inflationary tilt in the radial direction to the active crater ranged from 0.01 to 0.2 μrad and lasted from 10 min to 7 h.

The ground tilt data in the radial component were continuously monitored and processed in real time by a computer in order to determine the eruption potential from the rate and the amount of the ground tilt. The state of activity was categorized into five stages: Pre-eruption stage (inflation), eruption stage (deflation), non-eruption stage (minimal tilt change), warning stage (sustained inflation) and critical stage (critical level of sustained inflation). The statistical results of a 26-day experiment indicate that the volcano was declared to be in a critical stage before 27 of 39 actual explosive eruptions, 4 of 5 short eruptions and 7 of 22 continuous eruptions. There were 21 critical stages reported when no significant eruptions occurred. Most of these exceptions were caused by temporary relaxation in tilt due to minor events, for example, gas emission.

1 Introduction

Inflation of the ground is expected to be caused by storage of magma or gas prior to eruption of a volcano. Geophysicists have made efforts to detect this phenomenon by ground tilt measurements and to utilize the data to forecast volcanic eruptions (Jaggar and Finch 1927; Sassa 1936; Minakami 1942). At the Hawaiian Volcano Observatory, tilt measurements by short-span,

---

1 Sakurajima Volcanological Observatory, Disaster Prevention Research Institute, Kyoto University, Sakurajima-cho, Kagoshima-ken, 891-14, Japan

IAVCEI Proceedings in Volcanology 1
J.H. Latter (Ed.), Volcanic Hazards

water-tube tiltmeters (Eaton 1959) have been continuously carried out since 1956, and Dzurisin et al. (1984) clarified magma supply and magma transport-storage systems at Kilauea Volcano using reliable tilt data. The recent improvements in instrumental sensitivity and accuracy have made it possible to use tilt data for forecasting the eruptive activity at Mount St. Helens (Swanson et al. 1985).

Sakurajima Volcano is an andesitic stratovolcano in southern Japan. The eruptive activity has been at a high level since October 1955. The continuing activity at the summit crater has included more than 5400 explosive eruptions as of the end of 1985. Most of the explosive eruptions are of vulcanian type, which is accompanied not only with the emission of volcanic bombs and ash, but also with the occurrence of shock waves and an explosion quake (Ishihara 1985). The eruptive activity has also included non-explosive eruptions and continuous emission of volcanic ash, which are accompanied with volcanic earthquakes or tremors.

Summit eruptions are classified into three types by Sakurajima Volcanological Observatory (SVO) from seismic observation as follows;

1. Explosion (explosive eruption): The maximum amplitude of an explosion quake is equal to or more than 10 μ and the intensity of air shocks is equal to or more than 0.1 mb at the Harutayama Branch (HAR) which is located 2.8 km northwest of the active crater. Non-explosive eruptions which do not satisfy the conditions of explosion are classified into the next two types.
2. Eruption (short eruption): The duration of an explosion quake or volcanic tremor does not exceed 100 s.
3. Continuous eruption: The duration of a volcanic tremor is more than 100 s.

Gas emission, which contains little volcanic ash, is not counted as an eruption at SVO.

A typical forerunner of eruptive activity is a swarm of volcanic earthquakes, which originates at a shallow part beneath the crater (Kamo 1977; Nishi 1984; Iguchi 1985). Explosions have successively taken place for a week or more after the occurrence of an earthquake swarm. Kagoshima Meteorological Observatory has conducted seismic observations at Sakurajima and has issued the official statements on the activity of the volcano when volcanic earthquake swarms occurred.

The purpose of the present chapter is to demonstrate the possibility of the practical use of tilt data for automated judgement of the stages of eruptive activity and for forecasting individual explosive eruptions.

Tilt measurements at the volcano were started in 1957 by Yoshikawa (1962), and were later taken over by Eto (1966). They observed precursory ground tilt associated with strong eruptions or swarms of explosions by using horizontal pendulum tiltmeters.

The characteristics of the tilt observed by them were abnormal. Namely, the direction of the ground tilt did not indicate the center of tilt to be at the summit crater. This peculiarity was puzzling, because levelling surveys showed that the center of inflation or deflation was located at the center of the volcano (Eto 1965). Yoshikawa (1965) considered that the direction of the ground tilt might be affected by the existence of faults or cracks near the tilt stations.

Before a remarkable eruptive event in May 1976, inflation of the summit was clearly indicated by a biaxial tiltmeter installed in a shallow vault at the Harutayama Branch of SVO. The precursory inflation was 10 arc-seconds in tilt for a 2-week interval of time. Explosive eruptions successively occurred in the deflationary stage (Kamo 1978).

These previous studies illustrate the relationship of ground tilt and eruptions, and that high-resolution tilt data could be used for prediction of volcanic eruptions. The early researchers pointed out that tilt observation by water-tube tiltmeters should be improved and continued, to collect more reliable tilt data. Hence, the construction of an underground tunnel was planned to obtain tilt data with a higher resolution and sufficient sensitivity to detect ground tilt associated with an individual summit eruption.

The location of the tunnel was chosen with the following considerations. First, the distance of the site from the crater must be in the range of 2 to 3 km to observe the maximum amount of tilt, based on the results of levelling surveys that reveal the center of deformation at 3 to 5 km beneath the active crater (Ishihara and Eto 1978). Secondly, to eliminate the influ-

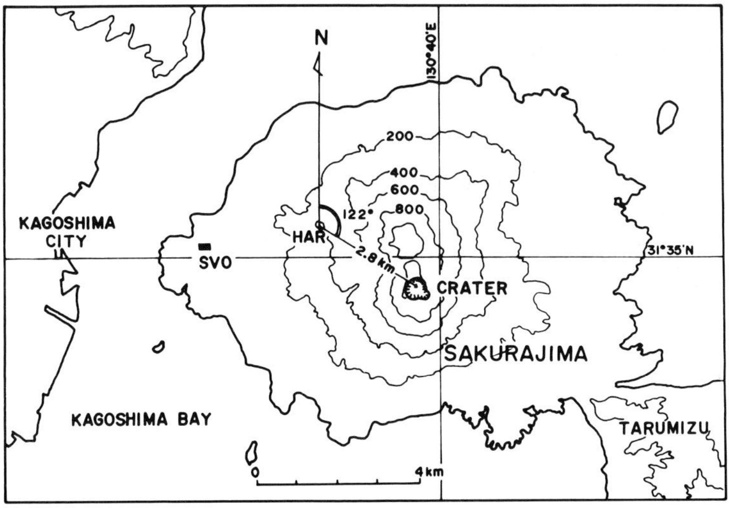

Fig. 1. Location of Sakurajima Volcanological Observatory (SVO) and Harutayama Branch (HAR)

ence of the outdoor environment on the instruments, the topography of the site should be suitable for the construction of an underground tunnel with a depth of more than 50 m. As one of the best sites for the construction of the tunnel, we selected Harutayama lava dome, which is located 2.8 km northwest of the summit crater.

After 1.5 years spent constructing the tunnel, a two-component, water-tube tiltmeter and a three-component extensometer were installed in August 1985. We immediately obtained records of the ground deformation associated with individual summit eruptions.

Fig. 2. Aerial photograph of Harutayama lava dome. The house on the base of the hill (1) is the entrance room of the underground tunnel, and the buildings at the top of the hill (2) are Harutayama Branch. The active crater is located at the summit of Minami-dake (3) of Sakurajima

2 Observations

Harutayama lava dome is located 2.8 km northwestward from the summit crater of Sakurajima Volcano, as illustrated in Fig. 1. At the top of the dome, the altitude of which is 408 m, the office building of Harutayama Branch (HAR) is situated. The underground tunnel was constructed in the dome from the northwestern base (1) to the center of the dome (2) to align the long axis of the tunnel in the direction to the summit crater (3), as shown in Fig. 2.

The geological structure of the dome and the arrangement of the instruments are illustrated in Fig. 3. The lava dome underlies pumice and volcanic ash with a thickness of 25 m and the bottom of the dome overlies tuff breccia. The water-tube tiltmeter (WTT) and the extensometer (EX) are installed at a depth of 65 m beneath Harutayama Branch. At the side of the entrance room of the tunnel, a borehole with a depth of 290 m was drilled to install seismometers, a two-component tiltmeter and a thermometer at the bottom.

The water-tube tiltmeter and the extensometer, denoted by WTT-R and EX-1, are installed in the radial direction to the crater and WTT-T and EX-1 are in the tangential direction. The amplifiers of these instruments, the recorders and the telemetry are installed in the room attached to the entrance of the tunnel. The data received at the main observatory (SVO) are monitored by analog recorders, digitized and processed by a computer.

Basements for the installation of instruments are attached directly to the ground of the tunnel independently of each other to detect the real deformation of the ground.

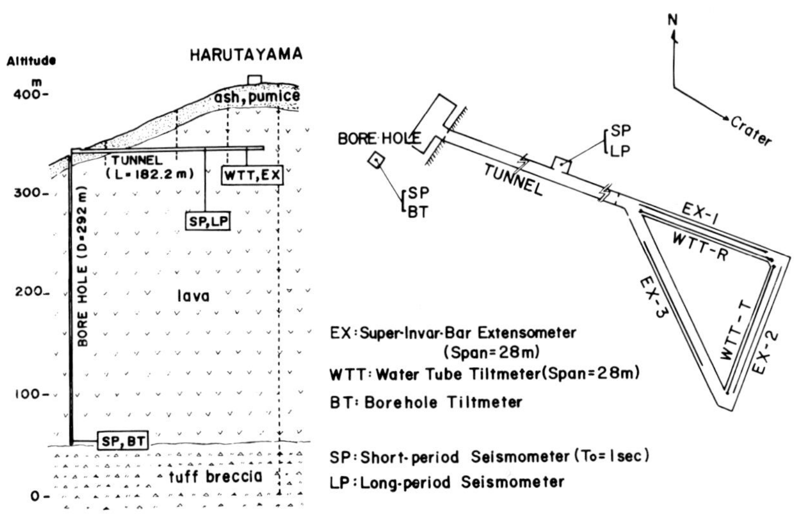

Fig. 3. Geological structure of Harutayama lava dome shown as a vertical cross-section and the arrangement of the instruments

The basic specification of the water-tube tiltmeter and the extensometer are similar to those commonly used in Japan (for example, Kato 1977). The water-tube tiltmeter is composed of a silica glass tube (28 m in length and 15 mm in diameter), and two water tanks on both sides of the tube. The water level in each tank is detected by a float and transformed into voltage by a magnetic sensor.

The standard scale of the extensometer is made of a super-invar bar 28 m in length and 10 mm in diameter. One edge of the bar is fixed on a basement and the other is free from the ground. The bar is hung by a stainless steel wire at 1-m interval.

The horizontal displacement of the ground is measured by a differential transformer attached at the free edge.

The sensitivities of the instruments are listed in Table 1.

Table 1. Constants of the instruments

Instrument	Component and azimuth	Span (m)	Sensitivity	Transducer
Water-tube tiltmeter	WTT-R N 111°E	28	0.56 V/μrad	Magnetic sensor
	WTT-T N 159°W	28	0.56 V/μrad	
Super-invar bar extensometer	EX-1(R) N 111°E	28	2.80 V/10-6	Differential transformer
	EX-2(T) N 159°W	28	2.80 V/10-6	
	EX-3 N 156°E	28	2.80 V/10-6	

3 Characteristics of Ground Tilt

A typical example of the records is illustrated in Fig. 4. Note the radial component of the tiltmeter (WTT-R) indicating inflation of the summit by an upward tilt that began about 1.5 h before the explosion at 06:43 JST, 14 October 1985. The amount of the tilt was 0.12 μrad. At the onset of the explosion, the

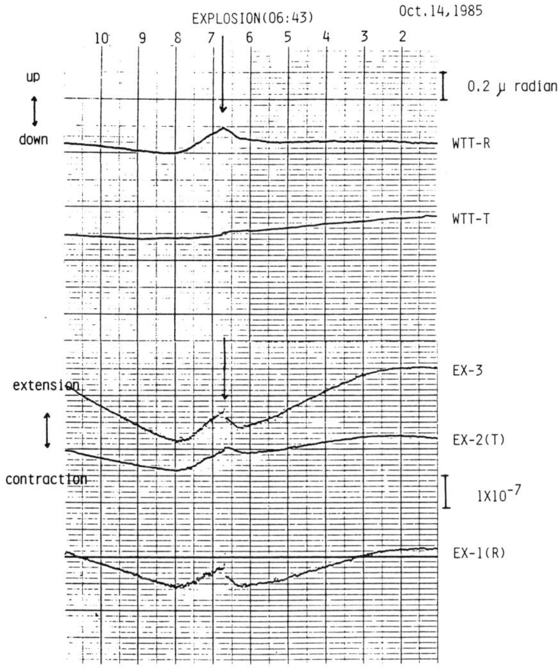

Fig. 4. Typical records, associated with an explosive eruption, of water-tube tiltmeters and extensometers

inflationary tilt reversed to deflation. The gradual deflation indicated by the downward tilt continued during the emission of volcanic ash.

On the extensometer records, the background amplitude of earth tides is larger than that recorded by the tiltmeter. However, extension before eruptions, and contraction after eruptions, was clearly recognized, corresponding to inflation and to deflation respectively, as indicated by the tilt data.

In Fig. 5, the vector diagrams of the ground tilt are illustrated for two cases. The direction of the tilt which should be towards the crater is slightly deflected due to the influence of the earth tide. Compensating for the effects of the earth tide, we infer that the inflation and deflation indicated by the tilt data are caused by pressure changes beneath the summit crater. This inference is strongly supported by seismic observation and by levelling surveys. The hypocenters of explosion quakes are distributed vertically at a depth of 1 to 3 km beneath the active crater (Kamo 1978). Reduction of levelling data indicates a pressure source located 3 to 5 km beneath the summit crater (Ishihara and Eto 1978).

Comparable records were obtained for 110 explosions, amounting to about 60% of the total events during the period of observation from August to December of 1985. Furthermore, similarities in the changes of ground tilt and strain were clearly observed

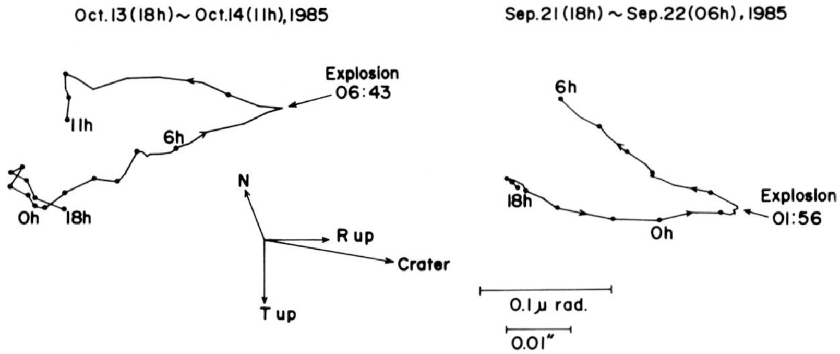

Fig. 5. Vector diagrams of ground tilt

for the non-explosive eruptions, which emitted large amounts of volcanic ash.

The relationship between the amount of precursory tilt and its duration is illustrated in Fig. 6. The precursory tilt ranges from 0.01 to 0.2 μrad in amount and from 10 min to 7 h in duration. The amount of tilt is not necessarily proportional to the intensity of eruptions, as indicated by the intensity of shock waves and the ejection velocities of volcanic rocks at the onset of an explosion. On the other hand, the amount of precur-

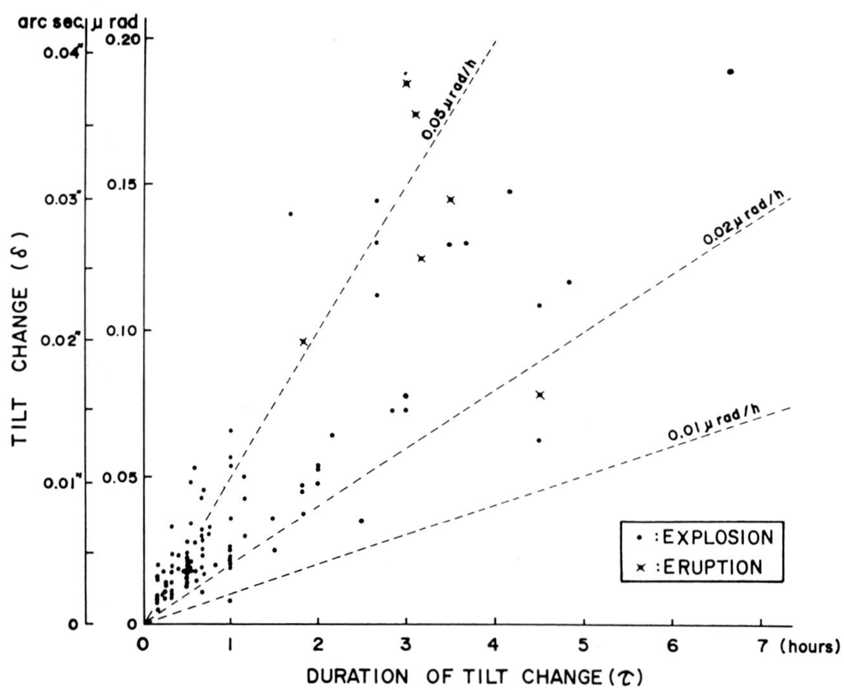

Fig. 6. The amount of precursory tilt versus duration

sory ground tilt appears to be related to the total energy of eruptive activity stored beneath the crater, because a larger precursory tilt is often accompanied by a non-explosive eruption which ejects a great amount of volcanic ash.

The amount of tilt is generally proportional to the duration. In most cases, the rate of change ranges from 0.01 to 0.2 μrad h^{-1}, which is greater than the maximum change in rate due to earth tides. Therefore, the precursory eruption tilt may be easily distinguished from tidal tilt on the radial component without first eliminating the effects of tides.

4 Data Processing

An experiment on automating the judgement of precursory tilt was carried out in real time using a personal computer. The tilt data of WTT-R (radial component) were continuously monitored and processed by a computer with a 12-bit analog-to-digital converter. The tilt data are sampled every 10 s and digitized with a unit of 0.5 mv per digit, which corresponds to about 0.0009 μrad/digit. The digitized data were smoothed by a running mean method at intervals of 5 min to eliminate short-period variations of tilt data. After this preliminary data processing, the stage of the eruptive activity is judged from the rate and the amount of the ground tilt.

A flowchart of the operational program is outlined in Fig. 7. The stages of activity defined by the tilt data are summarized in Table 2. Pre-eruption stage and eruption stage correspond to the inflationary and deflationary state respectively. The non-eruption stage is indicated by a negligible amount of tilt change.

Table 2. Stages of eruptive activity defined by tilt data

Stage	Condition[a]
Pre-eruption	$0 < a(= 0.02$ μrad $h^{-1}) <$ (Rate of tilt change) : Inflation
Eruption	(Rate of tilt change) $< b(= -0.02$ μrad $h^{-1}) < 0$: Deflation
Non-eruption	$b \leq$ (Rate of tilt change) $\leq a$
Warning	$L_w(= 0.01$ μrad$) \leq$ (Amount of tilt since the onset of pre-eruption stage) $< L_{cr}$
Critical	$L_{cr}(= 0.02$ μrad$) \leq$ (Amount of tilt since the onset of pre-eruption stage)

[a] The values in parentheses were used in the practice of the preliminary experiment.

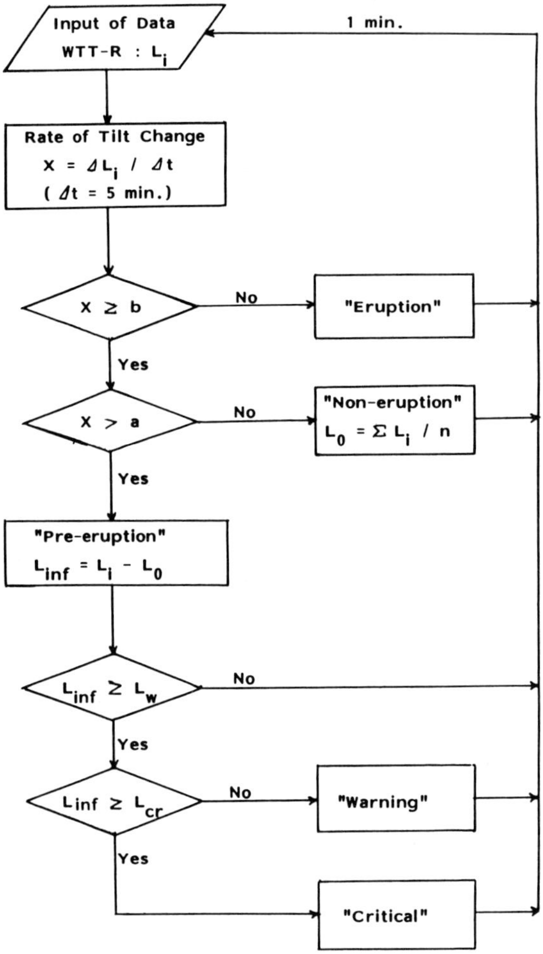

Fig. 7. Flowchart of the preliminary program for forecasting an explosive eruption

The program is composed of three parts:

1. The rate of tilt during the past 5 min is calculated.
2. The present stage is judged from the tilt rate. If the present stage is the non-eruption stage, the calculation of a mean level of the tilt is carried out continuously.
3. In the pre-eruption stage, the amount of the tilt from the onset of the stage is continuously compared with two other levels, warning and critical. If the amount of tilt is equal to or more than the warning level, the comment "Warning" is printed out. Furthermore, the comment "Critical" is printed out, when the amount of tilt reaches the critical level.

The parameters for the detailed condition of each stage used in the program were as follows:

1. Pre-eruption stage: Inflationary tilt at a rate of more than 0.02 μrad h^{-1} is sustained for 5 min.
2. Eruption stage: Deflationary tilt at a rate of more than 0.02 μrad h^{-1} is sustained for 5 min.
3. Non-eruption stage: The tilt changes less than 0.02 μrad h^{-1} continues for 5 min.
4. Warning stage: The accumulated inflation since the onset of the pre-eruption stage reaches a warning level of 0.01 μrad.
5. Critical stage: The accumulated inflation since the onset of the pre-eruption stage reaches a higher level (i.e. critical level) of 0.02 μrad.

5 Results

The program was put into practice for the period 3 to 29 December 1985. An example of the record is illustrated in Fig. 8, in which the stages of the eruptive activity judged from tilt data are shown in comparison with the actual activity and the tilt data of the radial component. In this case, three explosions and a short eruption occurred successively within 12 h after a continuous eruption. The tilt data indicated gradual inflation for 10 h and a sudden deflation after the last explosion. During the process of gradual inflation one clear inflation-deflation episode and other minor deflations occurred. Critical stages were reported six times. Although three of the six times corresponded to explosions, the other three critical stages were released by temporary relaxation of tilt (i.e. deflation) which continued for 5 min or more. Most of these minor and temporary deflationary tilts corresponded to minor seismic events

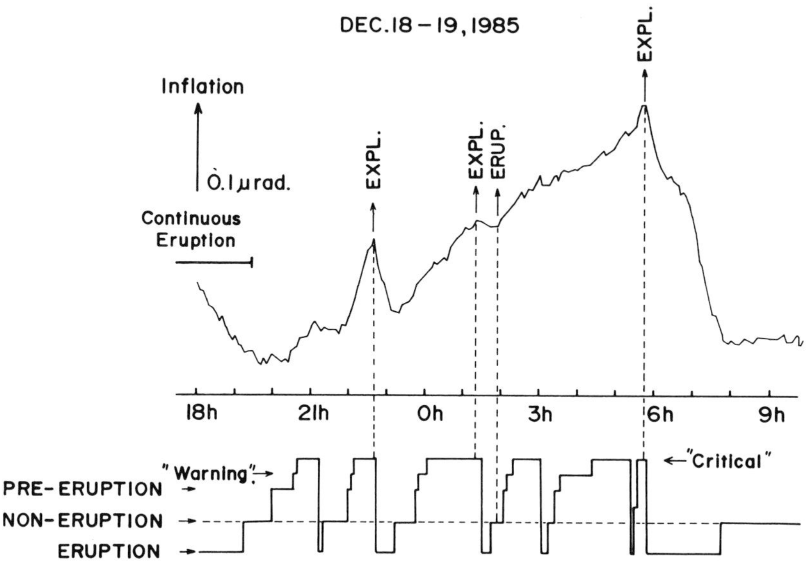

Fig. 8. An example of the test program. The stages judged by the computer are compared with the actual eruptive activity and the radial component of tilt data

Table 3. Summary on testing of the preliminary program

Actual eruptive activity		Number of events which occurred in the critical stage		Comments and remarks
Type	Number	Total; 59 times		[a] The number of the explosions which occurred in each stage; 27 : critical stage 8 : warning stage 3 : pre-eruption stage 1 : non-eruption stage
Explosion	39[a]	27		
Eruption	5	4	38	
Continuous eruption	22	7		[b] "No event" means that no remarkable events could be recognized on the seismograms
No event[b]	-	21	21	

The program was put into practice from December 3rd to 29th in 1985.

and gas emissions. In other words, "false eruptions" had cancelled the statement "critical stage" before the actual explosive eruptions occurred. To prevent the cancellation of the critical stage due to these temporary deflations of tilt, it is necessary to introduce new parameters into the program of judgement.

Statistical results of the experiment are summarized in Table 3. During 25 days of the experiment, 39 explosions occurred. Except for one explosion, the explosions occurred in the pre-eruption stage or further advanced stages.

The number of the critical stages reported by the computer was 59 times, 38 of which corresponded to eruptive events. However, there were 21 critical stages issued when no significant events occurred. Most of these exceptions were caused by temporary deflation in tilt due to minor eruptions or gas emission, as mentioned in the previous paragraph.

6 Conclusions

The underground tunnel constructed into a lava dome at Sakurajima Volcano has made it possible to observe inflationary ground deformation prior to individual summit eruptions, using water-tube tiltmeters and extensometers. A preliminary experiment on automated judgement of the stages of eruptive activity was carried out in real time using the tilt data for a month. The results of the experiment have demonstrated the feasibility of an automated forecast of summit eruptions with the aid of a computer. In a continuing effort to refine the eruption forecasting program at Sakurajima, our preliminary experiment using tilt data will be expanded to include horizontal deformation data presently collected from the extensometers.

Acknowledgements. We acknowledge Dr. R.Y. Koyanagi of Hawaiian Volcano Observatory, U.S.G.S. for critically reading this manuscript. The program of data processing was made in cooperation with M. Iguchi of SVO and the instruments were installed with the aid of the staffs of Sakurajima Volcanological Observatory. This study was supported by the Volcanic Eruption Prediction Plan of Japan.

References

Dzurisin D, Koyanagi RY, English TT (1984) Magma supply and storage at Kilauea Volcano, Hawaii, 1956-1983. J Volcanol Geotherm Res 21:177-206

Eaton JP (1959) A portable water-tube tiltmeter. Bull Seism Soc Am 49:301-316

Eto T (1965) Volcanic crustal deformations (1). Annuals Disast Prev Res Inst Kyoto Univ 8:71-82 (in Japanese)

Eto T (1966) Volcanic crustal deformations (2). Annuals Disast Prev Res Inst Kyoto Univ 9:83-98 (in Japanese)

Iguchi M (1985) Analysis of volcanic "shallow earthquakes" observed near the active crater. J Volcanol Soc Jpn 30:1-10 (in Japanese)

Ishihara K (1985) Dynamical analysis of volcanic explosion. J Geodyn 3:327-349
Ishihara K, Eto T (1978) Surface deformation associated with eruptive activity at the summit crater of Sakurajima Volcano. Annuals Disast Prev Res Kyoto Univ 21B-1:153-162 (in Japanese)
Jaggar TA, Finch RH (1927) Tilt records for thirteen years at the Hawaiian volcano. Bull Seism Soc Am 19:38-51
Kamo K (1978) Some phenomena before the summit eruptions at Sakurajima Volcano. J Volcanol Soc Jpn 23:53-64 (in Japanese)
Kato M (1977) Observation of crustal movements by newly-designed horizontal pendulum and water-tube tiltmeters with electromagnetic transducers (1). Bull Disast Prev Res Inst Kyoto Univ 27:155-171
Minakami T (1942) On volcanic activities and tilting of the earth's surface. Bull Earthq Res Inst 20:431-504
Nishi K (1984) Volcanic B-type earthquake swarm preceding volcanic explosion. Annuals Disast Prev Res Inst Kyoto Univ 27B-1:29-34 (in Japanese)
Sassa K (1936) Micro-seismometric study on eruptions of the Volcano Aso. Mem Coll Sci Kyoto Univ A19:11-56
Swanson DA, Casadevall TJ, Dzurisin D, Holcomb RT, Newhall CG, Malone SD, Weaver CS (1985) Forecasts and predictions of eruptive activity at Mount St. Helens, USA: 1975-1984. J Geodyn 3:397-423
Yoshikawa K (1962) On the crustal movement accompanying with the recent activity of the Volcano Sakurajima (Part 2). Bull Disast Prev Res Inst Kyoto Univ 50:1-26
Yoshikawa K (1965) The volcanic deformation. J Volcanol Soc Jpn 10:110-118 (in Japanese)

Subject Index